INTERNATIONAL UNION OF PURE AND APPLIED CHEMISTRY

ANALYTICAL CHEMISTRY DIVISION
COMMISSION ON SOLUBILITY DATA

SOLUBILITY DATA SERIES

Volume 27/28

METHANE

SOLUBILITY DATA SERIES

Selected Volumes in Preparation

NOTICE TO READERS

Dear Reader

If your library is not already a standing-order customer or subscriber to the Solubility Data Series, may we recommend that you place a standing order or subscription order to receive immediately upon publication all new volumes published in this valuable series. Should you find that these volumes no longer serve your needs, your order can be cancelled at any time without notice.

Robert Maxwell
Publisher at Pergamon Press

SOLUBILITY DATA SERIES

Editor-in-Chief
A. S. KERTES

Volume 27/28

METHANE

Volume Editors

H. LAWRENCE CLEVER
Emory University
Atlanta, Georgia, USA

COLIN L. YOUNG
University of Melbourne
Parkville, Victoria, Australia

Contributors

RUBIN BATTINO
Wright State University
Dayton, Ohio, USA

WALTER HAYDUK
University of Ottawa
Ottawa, Ontario, Canada

DENIS A. WIESENBURG
Naval Ocean Research & Development Activity
NSTL, Mississippi, USA

PERGAMON PRESS

OXFORD · NEW YORK · BEIJING · FRANKFURT
SÃO PAULO · SYDNEY · TOKYO · TORONTO

U.K.	Pergamon Press, Headington Hill Hall, Oxford OX3 0BW, England
U.S.A.	Pergamon Press, Maxwell House, Fairview Park, Elmsford, New York 10523, U.S.A.
PEOPLE'S REPUBLIC OF CHINA	Pergamon Press, Room 4037, Qianmen Hotel, Beijing, People's Republic of China
FEDERAL REPUBLIC OF GERMANY	Pergamon Press, Hammerweg 6, D-6242 Kronberg, Federal Republic of Germany
BRAZIL	Pergamon Editora, Rua Eça de Queiros, 346, CEP 04011, Paraiso, São Paulo, Brazil
AUSTRALIA	Pergamon Press Australia, P.O. Box 544, Potts Point, N.S.W. 2011, Australia
JAPAN	Pergamon Press, 8th Floor, Matsuoka Central Building, 1-7-1 Nishishinjuku, Shinjuku-ku, Tokyo 160, Japan
CANADA	Pergamon Press Canada, Suite No. 271, 253 College Street, Toronto, Ontario, Canada M5T 1R5

First edition 1987

Library of Congress Cataloging in Publication Data
The Library of Congress has cataloged this serial title as follows:
Solubility data series. - Vol. 1 — Oxford; New York: Pergamon, c 1979–
v.; 28 cm.
Separately cataloged and classified in LC before no. 18.
ISSN 0191–5622 - Solubility data series.
1. Solubility-Tables-Collected works.
QD543.S6629   541.3'42'05–dc19   85–641351
AACR 2   MARC-S

British Library Cataloguing in Publication Data
Methane.—(Solubility data series;
v.27/28).
1. Methane—Solubility
I. Clever, H. Lawrence   II. Young, Colin L.
III. Battino, Rubin   IV. Hayduk, Walter
V. Wiesenburg, Denis A.   VI. Series
547'.41104542   QD305.H6
ISBN 0–08–029200–3

Printed in Great Britain by A. Wheaton & Co. Ltd., Exeter

CONTENTS

FOREWORD

If the knowledge is
undigested or simply wrong,
more is not better

How to communicate and disseminate numerical data effectively in chemical science and technology has been a problem of serious and growing concern to IUPAC, the International Union of Pure and Applied Chemistry, for the last two decades. The steadily expanding volume of numerical information, the formulation of new interdisciplinary areas in which chemistry is a partner, and the links between these and existing traditional subdisciplines in chemistry, along with an increasing number of users, have been considered as urgent aspects of the information problem in general, and of the numerical data problem in particular.

Among the several numerical data projects initiated and operated by various IUPAC commissions, the *Solubility Data Project* is probably one of the most ambitious ones. It is concerned with preparing a comprehensive critical compilation of data on solubilities in all physical systems, of gases, liquids and solids. Both the basic and applied branches of almost all scientific disciplines require a knowledge of solubilities as a function of solvent, temperature and pressure. Solubility data are basic to the fundamental understanding of processes relevant to agronomy, biology, chemistry, geology and oceanography, medicine and pharmacology, and metallurgy and materials science. Knowledge of solubility is very frequently of great importance to such diverse practical applications as drug dosage and drug solubility in biological fluids, anesthesiology, corrosion by dissolution of metals, properties of glasses, ceramics, concretes and coatings, phase relations in the formation of minerals and alloys, the deposits of minerals and radioactive fission products from ocean waters, the composition of ground waters, and the requirements of oxygen and other gases in life support systems.

The widespread relevance of solubility data to many branches and disciplines of science, medicine, technology and engineering, and the difficulty of recovering solubility data from the literature, lead to the proliferation of published data in an ever increasing number of scientific and technical primary sources. The sheer volume of data has overcome the capacity of the classical secondary and tertiary services to respond effectively.

While the proportion of secondary services of the review article type is generally increasing due to the rapid growth of all forms of primary literature, the review articles become more limited in scope, more specialized. The disturbing phenomenon is that in some disciplines, certainly in chemistry, authors are reluctant to treat even those limited-in-scope reviews exhaustively. There is a trend to preselect the literature, sometimes under the pretext of reducing it to manageable size. The crucial problem with such preselection - as far as numerical data are concerned - is that there is no indication as to whether the material was excluded by design or by a less than thorough literature search. We are equally concerned that most current secondary sources, critical in character as they may be, give scant attention to numerical data.

On the other hand, tertiary sources - handbooks, reference books and other tabulated and graphical compilations - as they exist today are comprehensive but, as a rule, uncritical. They usually attempt to cover whole disciplines, and thus obviously are superficial in treatment. Since they command a wide market, we believe that their service to the advancement of science is at least questionable. Additionally, the change which is taking place in the generation of new and diversified numerical data, and the rate at which this is done, is not reflected in an increased third-level service. The emergence of new tertiary literature sources does not parallel the shift that has occurred in the primary literature.

With the status of current secondary and tertiary services being as briefly stated above, the innovative approach of the *Solubility Data Project* is that its compilation and critical evaluation work involve consolidation and reprocessing services when both activities are based on intellectual and scholarly reworking of information from primary sources. It comprises compact compilation, rationalization and simplification, and the fitting of isolated numerical data into a critically evaluated general framework.

The *Solubility Data Project* has developed a mechanism which involves a number of innovations in exploiting the literature fully, and which contains new elements of a more imaginative approach for transfer of reliable information from primary to secondary/tertiary sources. *The fundamental trend of the Solubility Data Project is toward integration of secondary and tertiary services with the objective of producing in-depth critical analysis and evaluation which are characteristic to secondary services, in a scope as broad as conventional tertiary services.*

Fundamental to the philosophy of the project is the recognition that the basic element of strength is the active participation of career scientists in it. Consolidating primary data, producing a truly critically-evaluated set of numerical data, and synthesizing data in a meaningful relationship are demands considered worthy of the efforts of top scientists. Career scientists, who themselves contribute to science by their involvement in active scientific research, are the backbone of the project. The scholarly work is commissioned to recognized authorities, involving a process of careful selection in the best tradition of IUPAC. This selection in turn is the key to the quality of the output. These top experts are expected to view their specific topics dispassionately, paying equal attention to their own contributions and to those of their peers. They digest literature data into a coherent story by weeding out what is wrong from what is believed to be right. To fulfill this task, the evaluator must cover *all* relevant open literature. No reference is excluded by design and every effort is made to detect every bit of relevant primary source. Poor quality or wrong data are mentioned and explicitly disqualified as such. In fact, it is only when the reliable data are presented alongside the unreliable data that proper justice can be done. The user is bound to have incomparably more confidence in a succinct evaluative commentary and a comprehensive review with a complete bibliography to both good and poor data.

It is the standard practice that the treatment of any given solute-solvent system consists of two essential parts: I. Critical Evaluation and Recommended Values, and II. Compiled Data Sheets.

The Critical Evaluation part gives the following information:

(i)   a verbal text of evaluation which discusses the numerical solubility information appearing in the primary sources located in the literature. The evaluation text concerns primarily the quality of data after consideration of the purity of the materials and their characterization, the experimental method employed and the uncertainties in control of physical parameters, the reproducibility of the data, the agreement of the worker's results on accepted test systems with standard values, and finally, the fitting of data, with suitable statistical tests, to mathematical functions;

(ii) a set of recommended numerical data. Whenever possible, the set of recommended data includes weighted average and standard deviations, and a set of smoothing equations derived from the experimental data endorsed by the evaluator;

(iii) a graphical plot of recommended data.

The Compilation part consists of data sheets of the best experimental data in the primary literature. Generally speaking, such independent data sheets are given only to the best and endorsed data covering the known range of experimental parameters. Data sheets based on primary sources where the data are of a lower precision are given only when no better data are available. Experimental data with a precision poorer than considered acceptable are reproduced in the form of data sheets when they are the only known data for a particular system. Such data are considered to be still suitable for some applications, and their presence in the compilation should alert researchers to areas that need more work.

The typical data sheet carries the following information:

(i) components - definition of the system - their names, formulas and Chemical Abstracts registry numbers;

(ii) reference to the primary source where the numerical information is reported. In cases when the primary source is a less common periodical or a report document, published though of limited availability, abstract references are also given;

(iii) experimental variables;

(iv) identification of the compiler;

(v) experimental values as they appear in the primary source. Whenever available, the data may be given both in tabular and graphical form. If auxiliary information is available, the experimental data are converted also to SI units by the compiler.

Under the general heading of Auxiliary Information, the essential experimental details are summarized:

(vi) experimental method used for the generation of data;

(vii) type of apparatus and procedure employed;

(viii) source and purity of materials;

(ix) estimated error;

(x) references relevant to the generation of experimental data as cited in the primary source.

This new approach to numerical data presentation, formulated at the initiation of the project and perfected as experience has accumulated, has been strongly influenced by the diversity of background of those whom we are supposed to serve. We thus deemed it right to preface the evaluation/compilation sheets in each volume with a detailed discussion of the principles of the accurate determination of relevant solubility data and related thermodynamic information.

Finally, the role of education is more than corollary to the efforts we are seeking. The scientific standards advocated here are necessary to strengthen science and technology, and should be regarded as a major effort in the training and formation of the next generation of scientists and engineers. Specifically, we believe that there is going to be an impact of our project on scientific-communication practices. The quality of consolidation adopted by this program offers down-to-earth guidelines, concrete examples which are bound to make primary publication services more responsive than ever before to the needs of users. The self-regulatory message to scientists of the early 1970s to refrain from unnecessary publication has not achieved much. A good fraction of the literature is still cluttered with poor-quality articles. The Weinberg report (in 'Reader in Science Information', ed. J. Sherrod and A. Hodina, Microcard Editions Books, Indian Head, Inc., 1973, p. 292) states that 'admonition to authors to restrain themselves from premature, unnecessary publication can have little effect unless the climate of the entire technical and scholarly community encourages restraint...' We think that projects of this kind translate the climate into operational terms by exerting pressure on authors to avoid submitting low-grade material. The type of our output, we hope, will encourage attention to quality as authors will increasingly realize that their work will not be suited for permanent retrievability unless it meets the standards adopted in this project. It should help to dispel confusion in the minds of many authors of what represents a permanently useful bit of information of an archival value, and what does not.

If we succeed in that aim, even partially, we have then done our share in protecting the scientific community from unwanted and irrelevant, wrong numerical information.

A. S. Kertes

PREFACE

This volume of the Solubility Series presents original data and evaluations of the solubility of methane gas in liquids. Collected here are the solubility data of methane in water, seawater, aqueous electrolyte solutions, mixed solvents, hydrocarbons, alcohols, ketones, carboxylic acids, esters, halocarbons, sulfur and nitrogen containing organic substances and other liquids at all temperature and pressures from papers published in the scientific literature through mid 1985. The publication of the methane volume completes the compilation and evaluation of the solubility data of the first five members of the alkane hydrocarbon series, C_nH_{2n+2}. The ethane solubility data are in Solubility Series Volume 9, and the propane, butane, and 2-methylpropane data are in Volume 24, both edited by W. Hayduk.

Methane melts at 90.68 K, boils at 111.42 K at 0.101325 MPa, and has critical temperature, pressure, and density of 190.55 K, 4.595 MPa and 162 kg m^{-1}, respectively, Some selected values of the second virial coefficient and methane molar volumes at 0.101325 MPa are:

T/K	B/cm^{3}mol^{-1}	V/cm^{3}mol^{-1}	T/K	B/cm^{3}mol^{-1}	V/cm^{3}mol^{-1}
115	-300.5	9,126	300	- 40.7	24,576
150	-176.2	12,130	350	- 25.5	28,694
200	-100.9	16,310	400	- 14.7	32,808
250	- 63.4	20,450	450	- 6.6	36,919
273.15	- 51.7	22,362	500	- 0.4	41,028
298.15	- 41.4	24,424	550	+ 4.4	-

The second virial constants and some of the other physical data above are from the International Thermodynamic Tables of the Fluid State, Volume 5, Methane, edited by S. Angus, B. Armstrong, and K.M. De Reuck, Pergamon Press, 1978. The methane molar volumes at 273.15 and 298.15 K are 0.23 and 0.17% less than the ideal gas values. The corrections for non-ideal gas behaviour have been applied in only a few cases of highly accurate gas solubility measurements.

Definitions of the many gas solubility units in use are covered in the introductory section on the Solubility of Gases in Liquids. Also discussed in the section is a thermodynamically consistent equation to use to fit the solubility as a function of temperature, and the thermodynamic changes that can be derived from the equation. The accuracy of most gas solubility measurements justifies use of only a two, or occasionally a three constant equation. These equations often require six to eight digits to properly reproduce a gas solubility value to three significant digits. When the equation constants are used to calculate thermodynamic changes the thermodynamic changes should be scaled back to no more than three significant digits.

In addition to the introductory section on the Solubility of Gases in Liquids the user will find helpful sections on Thermodynamic Considerations of Gas Solubility by E. Wilhelm, and on the Sechenov Salt Effect Parameter by H.L. Clever in Solubility Series Volume 10 on Nitrogen and Air. The original data is usually recalculated in several other units to better compare data from different sources. These unit conversions require density data which is not directly referenced except in special cases. Our most common source of denisty data is Organic Solvents, J.A. Riddick and W.B. Bunger, (Technique of Chemistry, Volume 11, A. Weissberger, Editor), Wiley-Interscience, New York, 1970, 3rd. Ed. The preface of Solubility Series, volume 1, Helium and Neon lists additional references for density data.

The editors renew their plea that authors publishing gas solubility data should always report the primary experimental observations of temperature, pressure, volume, etc., and that they should indicate the precise method used to calculate the solubility value. Much of the value of an Ostwald coefficient is lost if the pressure at which the measurement is made is not reported. Henry's constants have been defined and calculated in a number of ways and the precise significance of a particular value is often lost if pressure measurements are not given.

The editors are grateful for advice and help given by fellow members of the IUPAC Commission on Solubility Data. The editors wish to express their appreciation to Marian Iwamoto for technical assistance and to Carolyn Dowie, Cherryl Parrish, and Lesley Flanagan for typing the final manuscript.

H. Lawrence Clever   Colin L. Young

Atlanta   Melbourne

1986, March

THE SOLUBILITY OF GASES IN LIQUIDS

Introductory Information

C. L. Young, R. Battino, and H. L. Clever

INTRODUCTION

The Solubility Data Project aims to make a comprehensive search of the
literature for data on the solubility of gases, liquids and solids in
liquids. Data of suitable accuracy are compiled into data sheets set out
in a uniform format. The data for each system are evaluated and where
data of sufficient accuracy are available values are recommended and in
some cases a smoothing equation is given to represent the variation of
solubility with pressure and/or temperature. A text giving an evaluation
and recommended values and the compiled data sheets are published on
consecutive pages. The following paper by E. Wilhelm gives a rigorous
thermodynamic treatment on the solubility of gases in liquids.

DEFINITION OF GAS SOLUBILITY

The distinction between vapor-liquid equilibria and the solubility of gases
in liquids is arbitrary. It is generally accepted that the equilibrium
set up at 300K between a typical gas such as argon and a liquid such as
water is gas-liquid solubility whereas the equilibrium set up between
hexane and cyclohexane at 350K is an example of vapor-liquid equilibrium.
However, the distinction between gas-liquid solubility and vapor-liquid
equilibrium is often not so clear. The equilibria set up between methane
and propane above the critical temperature of methane and below the critical temperature of propane may be classed as vapor-liquid equilibrium or
as gas-liquid solubility depending on the particular range of pressure
considered and the particular worker concerned.

The difficulty partly stems from our inability to rigorously distinguish
between a gas, a vapor, and a liquid; a subject which has been discussed
in numerous textbooks. We have taken a fairly liberal view in these
volumes and have included systems which may be regarded, by some workers,
as vapor-liquid equilibria.

UNITS AND QUANTITIES

The solubility of gases in liquids is of interest to a wide range of scientific and technological disciplines and not solely to chemistry. Therefore
a variety of ways for reporting gas solubility have been used in the primary literature. Sometimes, because of insufficient available information,
it has been necessary to use several quantities in the compiled tables.
Where possible, the gas solubility has been quoted as a mole fraction
of the gaseous component in the liquid phase. The units of pressure used
are bar, pascal, millimeters of mercury, and atmosphere. Temperatures are
reported in Kelvins.

EVALUATION AND COMPILATION

The solubility of comparatively few systems is known with sufficient accuracy to enable a set of recommended values to be presented. This is true
both of the measurements near atmospheric pressure and at high pressures.
Although a considerable number of systems have been studied by at least
two workers, the range of pressures and/or temperatures is often suffi-
ciently different to make meaningful comparison impossible.

Occasionally, it is not clear why two groups of workers obtained very
different sets of results at the same temperature and pressure, although
both sets of results were obtained by reliable methods and are internally
consistent. In such cases, sometimes an incorrect assessment has been given.
There are several examples where two or more sets of data have been classified as tentative although the sets are mutually inconsistent.

Many high pressure solubility data have been published in a smoothed form.
Such data are particularly difficult to evaluate, and unless specifically
discussed by the authors, the estimated error on such values can only be
regarded as an "informed guess".

Many of the high pressure solubility data have been obtained in a more
general study of high pressure vapor-liquid equilibrium. In such cases a
note is included to indicate that additional vapor-liquid equilibrium data
are given in the source. Since the evaluation is for the compiled data,
it is possible that the solubility data are given a classification which is
better than that which would be given for the complete vapor-liquid data
(or vice versa). For example, it is difficult to determine coexisting
liquid and vapor compositions near the critical point of a mixture using
some widely used experimental techniques which yield accurate high pressure
solubility data. As another example, conventional methods of analysis
may give results with an expected error which would be regarded as suffi-
ciently small for vapor-liquid equilibrium data but an order of magnitude
too large for acceptable high pressure gas-liquid solubility.

It is occasionally possible to evaluate data on mixtures of a given substance with a member of a homologous series by considering all the
available data for the given substance with other members of the homologous
series. In this study the use of such a technique has been limited.

The estimated error is often omitted in the original article and sometimes
the errors quoted do not cover all the variables. In order to increase the
usefulness of the compiled tables *estimated* errors have been included even
when absent from the original article. If the error on *any* variable has
been inserted by the compiler, this has been noted.

PURITY OF MATERIALS

The purity of materials has been quoted in the compiled tables where given
in the original publication. The solubility is usually more sensitive to
impurities in the gaseous component than to liquid impurities in the liquid
component. However, the most important impurities are traces of a gas dissolved in the liquid. Inadequate degassing of the absorbing liquid is
probably the most often overlooked serious source of error in gas solubility measurements.

APPARATUS AND PROCEDURES

In the compiled tables brief mention is made of the apparatus and procedure.
There are several reviews on experimental methods of determining gas
solubilities and these are given in References 1-7.

METHODS OF EXPRESSING GAS SOLUBILITIES

Because gas solubilities are important for many different scientific and
engineering problems, they have been expressed in a great many ways:

The Mole Fraction, x(g)

The mole fraction solubility for a binary system is given by:

$$x(g) = \frac{n(g)}{n(g) + n(l)}$$

$$= \frac{W(g)/M(g)}{\{W(g)/M(g)\} + \{W(l)/M(l)\}}$$

here n is the number of moles of a substance (an *amount* of substance),
W is the mass of a substance, and M is the molecular mass. To be unambigu-
ous, the partial pressure of the gas (or the total pressure) and the temperature of measurement must be specified.

The Weight Per Cent Solubility, wt%

For a binary system this is given by

$$wt\% = 100 \; W(g)/\{W(g) + W(l)\}$$

where W is the weight of substance. As in the case of mole fraction, the
pressure (partial or total) and the temperature must be specified. The
weight per cent solubility is related to the mole fraction solubility by

$$x(g) = \frac{\{wt\%/M(g)\}}{\{wt\%/M(g)\} + \{(100 - wt\%)/M(1)\}}$$

The Weight Solubility, C_w

The weight solubility is the number of moles of dissolved gas per gram of
solvent when the partial pressure of gas is 1 atmosphere. The weight
solubility is related to the mole fraction solubility at one atmosphere
partial pressure by

$$x(g) \text{ (partial pressure 1 atm)} = \frac{C_w M(1)}{1 + C_w M(1)}$$

where M(1) is the molecular weight of the solvent.

The Moles Per Unit Volume Solubility, n

Often for multicomponent systems the density of the liquid mixture is not
known and the solubility is quoted as moles of gas per unit volume of
liquid mixture. This is related to the mole fraction solubility by

$$x(g) = \frac{n \, v^o(1)}{1 + n \, v^o(1)}$$

where $v^o(1)$ is the molar volume of the liquid component.

The Bunsen Coefficient, α

The Bunsen coefficient is defined as the volume of gas reduced to 273.15K
and 1 atmosphere pressure which is absorbed by unit volume of solvent (at
the temperature of measurement) under a partial pressure of 1 atmosphere.
If ideal gas behavior and Henry's law are assumed to be obeyed, then

$$\alpha = \frac{V(g)}{V(1)} \, \frac{273.15}{T}$$

where V(g) is the volume of gas absorbed and V(1) is the original (starting)
volume of absorbing solvent. The mole fraction solubility x is related to
the Bunsen coefficient by

$$x(g, 1 \text{ atm}) = \frac{\alpha}{\alpha + \dfrac{273.15}{T} \, \dfrac{v^o(g)}{v^o(1)}}$$

where $v^o(g)$ and $v^o(1)$ are the molar volumes of gas and solvent at a pressure
of one atmosphere. If the gas is ideal,

$$x(g) = \frac{\alpha}{\alpha + \dfrac{273.15R}{v^o(1)}}$$

Real gases do not follow the ideal gas law and it is important to establish
the real gas law used for calculating α in the original publication and to
make the necessary adjustments when calculating the mole fraction
solubility.

The Kuenen Coefficient, S

This is the volume of gas, reduced to 273.15K and 1 atmosphere pressure,
dissolved at a partial pressure of gas of 1 atmosphere by 1 gram of solvent.

The Ostwald Coefficient, L

The Ostwald coefficient, L, is defined as the ratio of the volume of gas absorbed to the volume of the absorbing liquid, all measured at the same temperature:

$$L = \frac{V(g)}{V(l)}$$

If the gas is ideal and Henry's Law is applicable, the Ostwald coefficient is independent of the partial pressure of the gas. It is necessary, in practice, to state the temperature and total pressure for which the Ostwald coefficient is measured. The mole fraction solubility, x(g), is related to the Ostwald coefficient by

$$x(g) = \left[\frac{RT}{P(g)\ L\ v^{o}(l)} + 1 \right]^{-1}$$

where P is the partial pressure of gas. The mole fraction solubility will be at a partial pressure of P(g). (See the following paper by E. Wilhelm for a more igorous definition of the Ostwald coefficient.)

The Absorption Coefficient, β

There are several "absorption coefficients", the most commonly used one being defined as the volume of gas, reduced to 273.15K and 1 atmosphere, absorbed per unit volume of liquid when the total pressure is 1 atmosphere. β is related to the Bunsen coefficient by

$$\beta = \alpha(1 - P(l))$$

where P(l) is the partial pressure of the liquid in atmospheres.

The Henry's Law Constant

A generally used formulation of Henry's Law may be expressed as

$$P(g) = K_{H}x(g)$$

where K_{H} is the Henry's Law constant and x(g) the mole fraction solubility. Other formulations are

$$P(g) = K_{2}C(l) \qquad \text{or} \qquad C(g) = K_{c}C(l)$$

where K_{2} and K_{c} are constants, C the concentration, and (l) and (g) refer to the liquid and gas phases. Unfortunately, K_{H}, K_{2} and K_{c} are all sometimes referred to as Henry's Law constants. Henry's Law is a *limiting law* but can sometimes be used for converting solubility data from the experimental pressure to a partial gas pressure of 1 atmosphere, provided the mole fraction of the gas in the liquid is small, and that the difference in pressures is small. Great caution must be exercised in using Henry's Law.

The Mole Ratio, N

The mole ratio, N, is defined by

$$N = n(g)/n(l)$$

Table 1 contains a presentation of the most commonly used inter-conversions not already discussed.

For gas solubilities greater than about 0.01 mole fraciton at a partial pressure of 1 atmosphere there are several additional factors which must be taken into account to unambiguously report gas solubilities. Solution densities or the partial molar volume of gases must be known. Corrections should be made for the possible non-ideality of the gas or the non-applicability of Henry's Law.

TABLE 1   Interconversion of parameters used for reporting solubility

$$L = \alpha(T/273.15)$$

$$C_w = \alpha/v_0 \rho$$

$$K_H = \frac{17.033 \times 10^6 \rho(\text{soln})}{\alpha\, M(1)} + 760$$

$$L = C_w\, v_{t,\text{gas}}\, \rho$$

where v_0 is the molal volume of the gas in $cm^3 mol^{-1}$ at 0°C, ρ the density of the solvent at the temperature of the measurement, ρ_{soln} the density of the solution at the temperature of the measurement, and $v_{t,\text{gas}}$ the molal volume of the gas ($cm^3 mol^{-1}$) at the temperature of the measurement.

SALT EFFECTS

Salt effect studies have been carried out for many years. The results are often reported as Sechenov (Setchenow) salt effect parameters. There appears to be no common agreement on the units of either the gas solubility, or the electrolyte concentration.

Many of the older papers report the salt effect parameter in a form equivalent to

$$k_{scc}/\text{mol dm}^{-3} = (1/(c_2/\text{mol dm}^{-3}))\, \log\, ((c_1^o/\text{mol dm}^{-3})/(c_1/\text{mol dm}^{-3}))$$

where the molar gas solubility ratio, c_1^o/c_1, is identical to the Bunsen coefficient ratio, α^o/α, or the Ostwald coefficient ratio, L^o/L. One can designate the salt effect parameters calculated from the three gas solubility ratios as k_{scc}, $k_{sc\alpha}$, k_{scL}, respectively, but they are identical, and $k_{scc}/\text{dm}^3\ \text{mol}^{-1}$ describes all of them. The superzero refers to the solubility in the pure solvent.

Recent statistical mechanical theories favor a molal measure of the electrolyte and gas solubility. Some of the more recent salt effects are reported in the form

$$k_{smm}/\text{kg mol}^{-1} = (1/(m_2/\text{mol kg}^{-1})\, \log\, ((m_1^o/\text{mol kg}^{-1})/(m_1/\text{mol kg}^{-1}))$$

In this equation the m_1^o/m_1 ratio is identical to the Kuenen coefficient ratio, s_1^o/s_1, or the solvomolality ratio referenced to water, A_{sm}^o/A_{sm}. Thus the salt effect parameters k_{smm}, k_{sms}, and $k_{smA_{sm}}$ are well represented by the $k_{smm}/\text{kg mol}^{-1}$.

Some experimentalists and theoreticians prefer the gas solubility ratio as a mole fraction ratio, x_1^o/x_1. It appears that most calculate the mole fraction on the basis of the total number of ions. The salt effect parameters

$$k_{scx}/\text{dm}^3\ \text{mol}^{-1} = (1/(c_2/\text{mol dm}^{-3}))\, \log\, (x_1^o/x_1)$$

and

$$k_{smx}/\text{kg mol}^{-1} = (1/(m_2/\text{mol kg}^{-1}))\, \log\, (x_1^o/x_1)$$

are both in the literature, but k_{scx} appears to be the more common.

The following conversions were worked out among the various forms of the salt effect parameter from standard definitions of molarity, molality, and mole fraction assuming the gas solubilities are small.

$$k_{smc} = (c_2/m_2) \, k_{scc} = (c_2/m_2) \, k_{scm} + F_{1m}$$

$$k_{scm} = k_{scc} - F_{1c} = (m_2/c_2) \, k_{smc} - F_{1c} = (m_2/c_2) \, k_{smm}$$

$$k_{scx} = (m_2/c_2) \, k_{smx} = (m_2/c_2) \, ksmm + F_{2c}$$

$$k_{smm} = k_{smx} - F_{2m} = (c_2/m_2) \, k_{scx} - F_{2m}$$

$$k_{smx} = (c_2/m_2) \, k_{scx} = (c_2/m_2) \, k_{scc} + F_{3m}$$

$$k_{scc} = k_{scx} - F_{3c} = (m_2/c_2) \, k_{smx} - F_{3c}$$

where

$$F_{1m} = (1/m_2) \, \log \, [(\rho°/\rho) \, (1000 + m_2 M_2)/1000]$$

$$F_{1c} = (m_2/c_2) \, F_{1m}$$

$$F_{2m} = (1/m_2) \, \log \, [(1000 + \nu m_3 M_3)/1000]$$

$$F_{2c} = (m_2/c_2) \, F_{2m}$$

$$F_{3m} = (1/m_2) \, \log \, [(1000\rho + (\nu M_3 - M_2) \, c_2)/1000\rho°)]$$

$$F_{3c} = (m_2/c_2) \, F_{3m}$$

The factors F_{1m}, F_{1c}, F_{2m}, F_{2c}, F_{3m}, and F_{3c} can easily be calculated from aqueous electrolyte data such as weight per cent and density as found in Volume III of the International Critical Tables. The values are small and change nearly linearly with both temperature and molality. The factors normally amount to no more than 10 to 20 per cent of the value of the salt effect parameter.

The symbols in the equations above are defined below:

Component	Molar Concentration c/mol dm^{-3}	Molal Concentration m/mol kg^{-1}	Mole Fraction x	Molecular Weight M/g mol^{-1}
Nonelectrolyte	$c_1^°$, c_1	$m_1^°$, m_1	$x_1^°$, x_1	M_1
Electrolyte	c_2	m_2	x_2	M_2
Solvent	c_3	m_3	x_3	M_3

The superscript "°" refers to the nonelectrolyte solubility in the pure solvent. The pure solvent and solution densities are $\rho°$/g cm^{-3} and ρ/g cm^{-3}, respectively. They should be the densities of gas saturated solvent (water) and salt solution, but the gas free densities will differ negligibly in the $\rho°/\rho$ ratio. The number of ions per formula of electrolyte is symbolized by ν.

The following table gives estimated errors in k_{scc} for various salt concentrations and a range of random errors in the gas solubility measurement

Error in k_{scc}/dm^3 mol^{-1} [a]

c_2/ mol dm^{-3}	Random Error in gas solubility Measurement				
	±2%	±1%	±0.5%	±0.1%	±0.05%
1	±18%	±9%	±5%	±1.5%	±1%
0.1	±175%	±87%	±43%	±9%	±4%
0.05	±350%	±174%	±87%	±17%	±9%
0.01	±1750%	±870%	±435%	±87%	±43%

[a] Based on a k_{scc} value of 0.100.

AQUAMOLAL OR SOLVOMOLAL, $A_{\delta m}$ or $m_i^{(\delta)}$

The term aquamolal was suggested by R. E. Kerwin (9). The unit was first used in connection with D_2O and $H_2O + D_2O$ mixtures. It has since been extended in use to other solvents. The unit represents the numbers of moles of solute per 55.51 moles of solvent. It is represented by

$m_i^{(\delta)}$/mol kg^{-1} = $(n_1 M_2/w_2)(w_2/M_0)$ = $m_i(M_2/M_0)$ where an amount of n_i of solute i is dissolved in a mass w_2 of solvent of molar mass M_2; M_0 is the molar mass of a reference solvent and m_i/mol kg^{-1} is the conventional molality in the reference solvent. The reference solvent is normally water.

TEMPERATURE DEPENDENCE OF GAS SOLUBILITY

In a few cases it has been found possible to fit the mole fraction solubility at various temperatures using an equation of the form

$$\ln x = A + B / (T/100K) + C \ln (T/100K) + DT/100K$$

It is then possible to write the thermodynamic functions $\Delta \bar{G}_1^\circ$, $\Delta \bar{H}_1^\circ$, $\Delta \bar{S}_1^\circ$ and $\Delta \bar{C}_{P_1}$ for the transfer of the gas from the vapor phase at 101,325 Pa partial pressure to the (hypothetical) solution phase of unit mole fraction as:

$$\Delta \bar{G}_1^\circ = -RAT - 100\,RB - RCT \ln (T/100) - RDT^2/100$$

$$\Delta \bar{S}_1^\circ = RA + RC \ln (T/100) + RC + 2\,RDT/100$$

$$\Delta \bar{H}_1^\circ = -100\,RB + RCT + RDT^2/100$$

$$\Delta \bar{C}_{P_1}^\circ = RC + 2\,RDT/100$$

In cases where there are solubilities at only a few temperatures it is convenient to use the simpler equations

$$\Delta \bar{G}_1^\circ = -RT \ln x = A + BT$$

in which case A = $\Delta \bar{H}_1^\circ$ and -B = $\Delta \bar{S}_1^\circ$

REFERENCES

1. Battino, R.; Clever, H. L. *Chem. Rev.* 1966, 66, 395.

2. Clever, H. L.; Battino, R. in *Solutions and Solubilities*, Ed. M. R. J. Dack, J. Wiley & Sons, New York, 1975, Chapter 7.

3. Hildebrand, J. H.; Prausnitz, J. M.; Scott, R. L. *Regular and related Solutions*, Van Nostrand Reinhold, New York, 1970, Chapter 8.

4. Markham, A. E.; Kobe, K. A. *Chem. Rev.* 1941, 63, 449.

5. Wilhelm, E.; Battino, R. *Chem. Rev.* 1973, 73, 1.

6. Wilhelm, E.; Battino, R.; Wilcock, R. J. *Chem. Rev.* 1977, 77, 219.

7. Kertes, A. S.; Levy, O.; Markovits, G. Y. in *Experimental Thermochemistry* Vol. II, Ed. B. Vodar and B. LeNaindre, Butterworth, London, 1974, Chapter 15.

8. Long, F. A.; McDevit, W. F. *Chem. Rev.* 1952, 51, 119.

9. Kerwin, K. E., Ph.D. Thesis, University of Pittsburgh 1964.

Revised: April 1982 (R.B., H.L.C.)

COMPONENTS:	EVALUATOR:
(1) Methane; CH_4; [74-82-8] (2) Water; H_2O; [7732-18-5]	Rubin Battino Department of Chemistry Wright State University Dayton, OH 45435 USA 1985, January

CRITICAL EVALUATION:

AN EVALUATION OF THE SOLUBILITY OF METHANE IN WATER BETWEEN TEMPERATURES OF 273.15 and 523.15 K AT A METHANE PARTIAL PRESSURE OF 0.101325 MPa.

Normally only one equation and table of smoothed data are given in an evaluation. This evaluation gives three equations and two tables of smoothed data. Equation (1) represents the best evaluation of data prior to 1981. Equation (2) and Table 1 represent the recommended evaluation between temperatures of 273.15 and 328.15 K based on the precise measurements of Rettich *et al.* (22). Equation (3) and Table 2 represent tentative values between temperatures of 273.15 and 523.15 K based on data from Rettich *et al.* (22) and Crovetto *et al.* (9). For applications around room temperature the data represented by Equation (2) and Table 1 are recommended.

The solubility of methane in water is reported in many papers (1-33). Most of these measurements are at *ca.* atmospheric pressure, but there are many measurements reported at larger pressures which will be evaluated separately. Many of the papers contain additional data on the solubility of methane in aqueous electrolyte solutions and in mixed solvents.

The thirty-three papers that were examined for the evaluation are listed alphabetically in the reference list. The recent results reported by Rettich *et al.* (22) are at least an order of magnitude more precise than the data reported in any other papers. Seven papers published prior to 1981 were judged to contain more reliable data (3, 4, 7, 20, 28, 31, 32) than the other papers. A linear regression of the 36 data points in the 274 to 313 K temperature interval from the seven papers gave the equation

$$\ln x_1 = -78.1584 + 104.4791/\tau + 29.7802 \ln \tau \qquad (1)$$

with a standard deviation of 0.0050 in $\ln x_1$ or about 0.50 percent in x_1 at the middle of the temperature range. The sixteen data points from Rettich *et al.* (22) give the equation for the 275 to 328 K temperature range of

$$\ln x_1 = -115.647716 + 155.575631/\tau + 65.2552591 \ln \tau - 6.616975729 \tau \quad (2)$$

with a standard deviation in x_1 of 0.056 percent. In both equations $\tau = (T/100 \text{ K})$. In Equation (1) the mole fraction solubility, x_1, is for a partial pressure of 0.101325 MPa methane. In Equation (2) the mole fraction is for a fugacity of 0.101325 MPa methane. Rettich *et al.* (22) used a rigorous thermodynamic approach described in their paper to convert their

COMPONENTS:	EVALUATOR:
(1) Methane; CH_4; [74-82-8] (2) Water; H_2O; [7732-18-5]	Rubin Battino Department of Chemistry Wright State University Dayton, OH 45435 USA 1985, January

CRITICAL EVALUATION:

experimental measurements to mole fraction solubility values. The mole fraction solubilities calculated from Equation (2) are 1 - 2 percent larger at 273.15 K and 0.20 percent larger at 313 K. Both equations have been ex-trapolated to 373 K. At the larger temperatures Equation (2) gives smaller solubility values than Equation (1). At 373 K Equation (2) gives a solubil-ity value that is eleven percent smaller than the value calculated from Equation (1). Table 1 contains the smoothed data calculated from Equation (2). Equation (2) and the data smoothed at 5 K intervals from 273.15 to 328.15 K are *recommended* values of the mole fraction solubility of methane in water at a fugacity of 0.101325 MPa. The extrapolated values between 333.15 and 373.15 K are tentative. Also given in Table 1 are ideal gas Ostwald coefficients and enthalpy, entropy and constant pressure heat capa-city changes for the transfer of one mole of methane from the gas phase to the infinitely dilute solution in water. The mole fraction solubility at 0.101325 MPa fugacity methane shows a minimum at 363 K.

Recently Crovetto *et al.* (9) measured the solubility of methane in water from 297 to 518 K over a fugacity range of 1.3 to 6.5 MPa with an es-timated precision in the Henry's constant of 1 - 2 percent. The evaluator computed mole fraction methane solubility values in water at a partial pres-sure of 0.101325 MPa assuming Henry's law is obeyed. The seven values from the work of Crovetto *et al.* (9) were combined with the 16 values of Rettich *et al.* (22) in a linear regression to yield the equation for the mole frac-tion solubility of methane at a partial pressure of 0.101325 MPa (1 atm) over the 273.15 to 523.15 temperature range of:

$$\ln x_1 = -99.14188 + 132.821/\tau + 51.91445 \ln \tau - 4.25831 \tau \qquad (3)$$

where again $\tau = (T/100 \text{ K})$. The standard deviation in $\ln x_1$ is 0.015, and the percent error in x_1 at the middle of the temperature range is 1.5 per-cent. Smoothed tentative values of the mole fraction solubility and the thermodynamic changes on solution at 10 K intervals from 273.15 to 523.15 K are given in Table 2. Equation (3) and the smoothed data in Table 2 are for the use of workers who need values in the 373 to 523 K temperature range at a methane partial pressure of 0.101325 MPa. For values at larger partial pressures see the evaluation of high pressure methane in water solubility data.

COMPONENTS:	EVALUATOR:
(1) Methane; CH_4; [74-82-8] (2) Water; H_2O; [7732-18-5]	Rubin Battino Department of Chemistry Wright State University Dayton, OH 45435 USA 1985, January

CRITICAL EVALUATION:

TABLE 1.   Solubility of methane in water at a methane partial pressure of 0.101325 MPa.   Recommended [a] values of the mole fraction and Ostwald solubility, and of the partial molar thermodynamic changes on solution as a function of temperature.

T/K	Mol Fraction $10^5 x_1$	Ostwald[b] Coefficient L/cm^3 cm^{-3}	$\Delta H_1^0 /$ kJ mol^{-1}	$\Delta S_1^0 /$ J K^{-1} mol^{-1}	$\Delta C_{p_1}^0 /$ J K^{-1} mol^{-1}
Recommended					
273.15	4.6666	0.058055	-19.426	-154.0	262.3
278.15	4.0221	0.050959	-18.127	-149.3	257.2
283.15	3.5192	0.045377	-16.854	-144.8	252.1
288.15	3.1224	0.040946	-15.606	-140.4	246.9
293.15	2.8062	0.037405	-14.384	-136.2	241.8
298.15	2.5523	0.034559	-13.188	-132.2	236.7
303.15	2.3469	0.032267	-12.018	-128.3	231.5
308.15	2.1802	0.030420	-10.873	-124.5	226.4
313.15	2.0445	0.028936	- 9.754	-120.9	221.3
318.15	1.9340	0.027752	- 8.660	-117.5	216.1
323.15	1.8442	0.026821	- 7.592	-114.1	211.0
328.15	1.7717	0.026103	- 6.550	-110.9	205.9
Tentative					
333.15	1.7138	0.025569	- 5.533	-107.9	200.8
338.15	1.6683	0.025196	- 4.542	-104.9	195.6
343.15	1.6336	0.024966	- 3.577	-102.1	190.5
348.15	1.6082	0.024862	- 2.637	- 99.3	185.4
353.15	1.5911	0.024873	- 1.723	- 96.7	180.2
358.15	1.5815	0.024990	- 0.835	- 94.2	175.1
363.15	1.5785	0.025206	+ 0.028	- 91.9	170.0
368.15	1.5817	0.025513	0.865	- 89.6	164.9
373.15	1.5905	0.025909	1.676	- 87.4	159.7

[a] The data are classed as recommended over the 273.15 to 328.15 K temperature interval of the experimental data.   The extrapolated data over the 333.15 to 373.15 K temperature interval are classed as tentative.

[b] The Ostwald coefficients were not corrected for non-ideal behavior.

COMPONENTS:	EVALUATOR:
(1) Methane; CH_4; [74-82-8] (2) Water; H_2O; [7732-18-5]	Rubin Battino Department of Chemistry Wright State University Dayton, OH 45435 USA 1985, January

CRITICAL EVALUATION:

TABLE 2.   The solubility of methane in water at unit fugacity. Tentative values of the mole fraction solubility, and thermodynamic changes on solution as a function of temperature between 273.15 and 523.15 K. Values based on the experimental data of Rettich *et al*. (22) and Crovetto *et al*. (9). The Crovetto *et al*. data were extrapolated to a pressure of 0.101325 MPa assuming Henry's law is obeyed.

T/K	Mol Fraction $10^5 x_1$	$\Delta G_1^0/$ kJ mol^{-1}	$\Delta H_1^0/$ kJ mol^{-1}	$\Delta S_1^0/$ J K^{-1} mol^{-1}	$\Delta C_{p_1}^0/$ J K^{-1} mol^{-1}
273.15	4.625	22.669	-18.947	-152.4	238.2
283.15	3.507	24.149	-16.600	-143.9	231.1
293.15	2.803	25.549	-14.324	-136.0	224.1
298.15	2.550	26.219	-13.213	-132.3	220.5
303.15	2.344	26.871	-12.119	-128.6	217.0
313.15	2.037	28.132	- 9.985	-121.7	209.9
323.15	1.832	29.307	- 7.922	-115.2	202.8
333.15	1.695	30.428	- 5.929	-109.1	195.7
343.15	1.609	31.491	- 4.007	-103.4	188.6
353.15	1.560	32.498	- 2.156	- 98.1	181.6
363.15	1.542	33.454	- 0.375	- 93.2	174.5
373.15	1.548	34.362	+ 1.334	- 88.5	167.4
383.15	1.577	35.226	2.973	- 84.2	160.3
393.15	1.625	36.047	4.541	- 80.1	153.2
403.15	1.691	36.829	6.038	- 76.4	146.2
413.15	1.776	37.575	7.464	- 72.9	139.1
423.15	1.878	38.288	8.819	- 69.6	132.0
433.15	1.998	38.969	10.104	- 66.6	124.9
443.15	2.137	39.621	11.318	- 63.9	117.8
453.15	2.295	40.247	12.461	- 61.3	110.8
463.15	2.472	40.848	13.533	- 59.0	103.7
473.15	2.670	41.427	14.534	- 56.8	96.6
483.15	2.889	41.986	15.465	- 54.9	89.5
493.15	3.131	42.526	16.325	- 53.1	82.4
503.15	3.395	43.049	17.113	- 51.5	75.4
513.15	3.683	43.557	17.832	- 50.1	68.3
523.15	3.995	44.052	18.479	- 48.9	61.2

COMPONENTS:	EVALUATOR:
(1) Methane; CH_4; [74-82-8] (2) Water; H_2O; [7732-18-5]	Rubin Battino Department of Chemistry Wright State University Dayton, OH 45435 USA 1985, January

CRITICAL EVALUATION:

References

1. Amirijafari, B.; Campbell, J. M. *Society Petroleum Engrs. J.* 1972, 21.

2. Barone, G.; Castrunuovo, G.; Volpe, D.; Elia, V.; Grassi, L. *J. Phys. Chem.* 1979, 83, 2703-14.

3. Ben-Naim, A.; Wilf, J.; Yaacobi, M. *J. Phys. Chem.* 1973, 77, 95-102.

4. Ben-Naim, A.; Yaacobi, M. *J. Phys. Chem.* 1974, 78, 170-5.

5. Bunsen, R. W. *Ann. Chem. Pharm.* 1855, 93, 1-50.

6. Christoff, A. *Z. Phys. Chem.* 1906, 55, 622-34.

7. Claussen, W. F.; Polglase, M. F. *J. Am. Chem. Soc.* 1952, 74, 4817-9.

8. Cosgrove, B. A.; Walkley, J. *J. Chromatogr.* 1981, 216, 161-7.

9. Crovetto, R.; Fernandez-Prini, R.; Japas, M. L. *J. Chem. Phys.* 1982, 76, 1077-86.

10. Eucken, A.; Hertzberg, G. *Z. Phys. Chem.* 1950, 195, 1-23.

11. Feillolay, A.; Lucas, M. *J. Phys. Chem.* 1972, 76, 3068-72.

12. Lannung, A.; Gjaldbaek, J. C. *Acta Chem. Scand.* 1960, 14, 1124-8.

13. Matheson, I. B. C.; King, A. D. *J. Coll. Interface Sci.* 1978, 66, 464-9.

14. McAuliffe, C. *Nature* 1963, 200, 1092.

15. McAuliffe, C. *J. Phys. Chem.* 1966, 70, 1267.

16. Morrison, T. J.; Billett, F. *J. Chem. Soc.* 1948, 2033.

17. Morrison, T. J.; Billett, F. *J. Chem. Soc.* 1952, 3819-22.

18. Moudgil, B. M.; Somasundaran, P.; Lin, I. J. *Rev. Sci. Instrum.* 1974, 45, 406-9.

19. Mishnina, T. A.; Avdeeva, O. I.; Bozhovskaya, T. K. *Inf. Sb., Vses. Nauchn-Issled. Geol. Inst.* 1962, No. 56, 137-45. *Chem. Abstr.* 1964, 60, 8705g.

20. Muccitelli, J. A.; Wen, W.-Y. *J. Solution Chem.* 1980, 9, 141-61.

21. Namiot, A. Yu. *Zh. Strukt. Khim.* 1961, 2, 408-17. *J. Struct. Chem. (Engl. Transl.)* 1961, 2, 381-9.

22. Rettich, T. R.; Handa, Y. P.; Battino, R.; Wilhelm, E. *J. Phys. Chem.* 1981, 85, 3230-7.

23. Rudakov, E. S.; Lutsyk, A. I. *Zh. Fiz. Khim.* 1979, 53, 1298-1300. *Russ. J. Phys. Chem.* 1979, 53, 731-3.

24. Schröder, W. *Z. Naturforsch.* 1969, 24b, 500-8. *Chem.-Ing.-Tech.* 1973, 45, 603-8.

25. Shoor, S. K. *Ph.D. thesis*, University of Florida 1968.

COMPONENTS:	EVALUATOR:
(1) Methane; CH_4; [74-82-8] (2) Water; H_2O; [7732-18-5]	Rubin Battino Department of Chemistry Wright State University Dayton, OH 45435 USA 1985, January

CRITICAL EVALUATION:

26.   Shoor, S. K.; Walker, R. D., Jr.; Gubbins, K. E.   *J. Phys. Chem.*
      <u>1969</u>, *73*, 312-7.

27.   Tokunaga, J.; Kawai, M.   *J. Chem. Eng. Japan* <u>1975</u>, *8*, 326-7.

28.   Wen, W.-Y.; Hung, J. H.   *J. Phys. Chem.* <u>1970</u>, *74*, 170-80.

29.   Wetlaufer, D. B.; Malik, S. K.; Stoller, L.; Coffin, R. L.   *J. Am.
      Chem. Soc.* <u>1964</u>, *86*, 508-14.

30.   Winkler, L. W.   *Chem. Ber.* <u>1901</u>, *34*, 1408-22.

31.   Yaacobi, M.; Ben-Naim, A.   *J. Solution Chem.* <u>1973</u>, *2*, 425.

32.   Yamamoto, S.; Alcauskas, J. B.; Crozier, T. E.   *J. Chem. Eng. Data*
      <u>1976</u>, *21*, 78-80.

33.   Yano, T.; Suetaka, T.; Umehara, T.; Horiuchi, A.   *Kagaku Kogaku* <u>1974</u>,
      38, 320-3.

COMPONENTS:	ORIGINAL MEASUREMENTS:
(1) Methane; CH_4; [74-82-8] (2) Water; H_2O; [7732-18-5]	Bunsen, R. W. *Ann. Chem. Pharm.* 1855, *93*, 1-50. [The Journal's title later changed to *J. Liebigs Ann. Chem.*]
VARIABLES: $T/K = 279.35 - 298.75$	PREPARED BY: H. L. Clever

EXPERIMENTAL VALUES:

Temperature		Mol Fraction	Bunsen Coefficient
$t/^0C$	T/K	$10^5 x_1$	$10^2 \alpha / cm^3 (STP) cm^{-3} atm^{-1}$
6.2	279.35	3.821	4.742
9.4	282.55	3.587	4.451
12.5	285.65	3.326	4.126
18.7	291.85	2.894	3.586
25.6	298.75	2.522	3.121

The compiler calculated the Kelvin temperature and the mole fraction solubility values.

The complete paper was translated into English by a Mr. Roscoe and published by Bunsen (1). Two long abstracts of the paper were published (2, 3). One was translated into French by M. Verdet (2). The two abstracts presented only the interpolation formula and a table of smoothed values at one degree intervals from 0 to 20 ^0C. The complete set of experimental values, the interpolation equation and smoothed values are also given in Bunsen's book on gasometric methods (4).

Bunsen's interpolation equation for the data is

$$\alpha/cm^3 (STP) cm^{-3} atm^{-1} = 0.05449 - 0.0011807 \, (t/^0C) + 0.000012078 \, (t/^0C)^2.$$

The paper reports solubility measurements of nitrogen, hydrogen, methane, ethane, ethene, carbon monoxide, oxygen and air in water made chiefly by a Dr. Pauli. Only the methane data agree well with modern values.

AUXILIARY INFORMATION

METHOD/APPARATUS/PROCEDURE:	SOURCE AND PURITY OF MATERIALS:
Bunsen's original apparatus and procedure were used. They are described in detail in the paper. Many of the data reported in this paper are of only historical interest. The methane solubility values have stood the test of time fairly well. They are within 2 to 5 percent of modern values.	(1) Methane. Natural gas sample from the mud-volcanoes of Bulganack in the Crimea. Treated with K to remove CO_2. (2) Water. Boiled briskly under vacuum to remove dissolved air. Transferred to the apparatus without contact with air.

REFERENCES:

1. Bunsen, R. W. *Phil. Mag.* 1855, *9*, 116-30, 181-201, plus plate.

2. Bunsen, R. W. *Ann. Chim. Phys.* [*3*] 1855, *43*, 496-508.

3. Bunsen, R. W. *Arch. Sci. Phys. Nat.* [*1*] 1855, *28*, 235- .

4. Bunsen, R. W. *GASOMETRISCHE METHODEN*, II Ausgabe, Braunschweig, 1858, p. 214.

COMPONENTS:	ORIGINAL MEASUREMENTS:
(1) Methane; CH_4; [74-82-8] (2) Water; H_2O; [7732-18-5]	Winkler, L. W. *Chem. Ber.* <u>1901</u>, *34*, 1408-22.

EXPERIMENTAL VALUES:

Temperature		Pressure	Water Volume	Methane Volume (STP)	Bunsen Coefficient
$t/^0C$	T/K	p_1/mmHg	v_2/cm^3	v_1/cm^3	$10^2\alpha$/cm^3(STP)cm^{-1}atm^{-1}
0.25	273.40	431.81	2066.43	64.90	5.528
0.30	273.45	431.90	2066.43	64.86	5.523
0.23	273.38	444.58	2098.12	67.69	5.515
0.27	273.42	444.55	2098.12	67.69	5.516
0.28	273.43	444.62	2098.12	67.65	5.511
9.98	283.13	469.59	2066.71	53.35	4.178
10.00	283.15	469.42	2066.71	53.43	4.186
10.00	283.15	469.71	2066.71	53.28	4.171
20.08	293.23	505.46	2069.78	45.52	3.307
20.00	293.15	505.26	2069.75	45.58	3.312
20.00	293.15	505.09	2069.75	45.68	3.321
19.98	293.13	504.99	2069.75	45.54	3.311
20.05	293.20	505.19	2069.77	45.49	3.305
20.02	293.17	505.34	2069.76	45.39	3.298
20.00	293.15	523.86	2101.49	47.83	3.302
20.00	293.15	523.66	2101.49	47.88	3.307
20.00	293.15	523.86	2101.49	47.79	3.299
29.95	303.10	538.90	2074.92	40.67	2.764
30.10	303.25	539.25	2075.00	40.65	2.761
30.00	202.15	539.10	2074.95	40.61	2.759
40.02	313.17	573.54	2082.09	37.31	2.375
40.03	313.18	573.95	2082.10	37.15	2.363
40.00	313.15	573.77	2082.08	37.21	2.367
50.08	323.23	608.54	2090.93	35.70	2.132
50.00	323,15	608.24	2090.86	35.77	2.137
50.00	323.15	608.54	2090.86	35.65	2.129
60.03	333.18	644.69	2101.14	35.09	1.969
59.93	333.08	645.77	2101.04	34.61	1.939
60.00	333.15	644.45	2101.11	35.20	1.977
59.95	333.10	676.49	2133.29	36.93	1.945
60.02	333.17	676.23	2133.47	37.07	1.953
59.95	333.10	676.65	2133.29	36.87	1.941
70.00	343.15	682.54	2112.78	35.03	1.846
70.00	343.15	683.94	2112.78	34.45	1.816
70.05	343.20	684.12	2112.84	34.43	1.811
80.00	353.15	725.54	2125.82	35.78	1.763
80.02	353.17	722.20	2125.84	35.69	1.767
79.97	353.12	721.44	2125.79	35.91	1.779

COMPONENTS:	ORIGINAL MEASUREMENTS:
(1) Methane; CH_4; [74-82-8] (2) Water; H_2O; [7732-18-5]	Winkler, L. W. *Chem. Ber.* <u>1901</u>, *34*, 1408-22.
VARIABLES: T/K = 273.38 - 353.17 p_1/kPa = 57.570 - 96.731	PREPARED BY: H. L. Clever

EXPERIMENTAL VALUES:

The temperatures and Bunsen coefficients below are the average values given by Winkler from the experimental data on the preceeding page.

Temperature		Mol Fraction	Bunsen Coefficient	Ostwald Coefficient
$t/^0$C	T/K	$10^5 x_1$	$10^2 \alpha/cm^3$ (STP) $cm^{-3} atm^{-1}$	$10^2 L/cm^3$ cm^{-3}
0.27	273.42	4.451	5.519	5.524
9.99	283.14	3.490	4.178	4.331
20.01	293.16	2.864	3.307	3.549
30.02	303.17	2.479	2.761	3.064
40.02	313.17	2.205	2.368	2.715
50.03	323.18	2.058	2.133	2.524
59.98	333.13	1.953	1.954	2.383
70.02	343.17	1.889	1.825	2.293
80.00	353.15	1.897	1.770	2.288

The mole fraction solubility at 101.325 kPa methane partial pressure was calculated by the compiler using a methane molar volume of 22,360.4 cm^3 (STP) mol^{-1}.

The Ostwald coefficients were calculated by the compiler. The Kelvin temperatures were added by the compiler.

AUXILIARY INFORMATION

METHOD/APPARATUS/PROCEDURE:	SOURCE AND PURITY OF MATERIALS:
The original Bunsen absorption method (ref 1) was used. The method and apparatus are described in earlier papers (ref 2).	(1)Methane. Prepared by the decomposition of dimethyl zinc by air free water. (2) Water. Distilled.

ESTIMATED ERROR:

δT/K = ± 0.01
$\delta \alpha/\alpha$ ± 0.01 (compiler)

REFERENCES:

1. Bunsen, R. W.
   Gasometrische Methoden, 2nd. ed.,
   Braunschweig, <u>1858</u>.

2. Winkler, L. W.
   Chem. Ber. <u>1893</u>, *24*, , 3602.

COMPONENTS:	ORIGINAL MEASUREMENTS:
1. Methane; CH_4; [74-82-8] 2. Water; H_2O; [7732-18-5]	Wetlaufer, D. B.; Malik, S. K.; Stoller, L.; Coffin, R. L. *J. Am. Chem. Soc.* <u>1964</u>, *86*, 508-514.

VARIABLES:	PREPARED BY:
Temperature	C. L. Young

EXPERIMENTAL VALUES:

T/K	10^3 Conc. of methane[†] in soln./mol dm^{-3}	Mole fraction[*] of methane x_{CH_4}
278.2	0.00219	0.0000396
298.2	0.00141	0.0000255
318.2	0.00107	0.0000193

[†] at a partial pressure of 101.3 kPa.

[*] calculated by compiler.

AUXILIARY INFORMATION

METHOD/APPARATUS/PROCEDURE:	SOURCE AND PURITY OF MATERIALS:
Modified Van Slyke-Neill apparatus fitted with a magnetic stirrer. Solution was saturated with gas and then sample transferred to the Van Slyke extraction chamber.	1. Matheson c.p. grade, purity 99 mole per cent or better. 2. Distilled.

ESTIMATED ERROR:

$\delta T/K = \pm 0.05$; $\delta x_{CH_4} = \pm 2\%$.

REFERENCES:

COMPONENTS:	ORIGINAL MEASUREMENTS:
(1) Methane; CH_4; [74-82-8] (2) Water; H_2O; [7732-18-5]	Morrison, T. J.; Billett, F. *J. Chem. Soc.* <u>1952</u>, 3819 - 22.

| VARIABLES:
$T/K = 285.1 - 348.4$
$p_1/kPa =]0].325$ | PREPARED BY:

H. L. Clever |

EXPERIMENTAL VALUES:

Temperature		Mol Fraction	Solubility
$t/^0C$	T/K	$10^5 x_1$	$S/cm^3 (STP) kg^{-1}$
11.9	285.1	3.215	39.90
14.7	287.9	3.014	37.41
20.5	293.7	2.655	32.96
25.2	298.4	2.422	30.06
35.0	308.2	2.042	25.35
41.0	314.2	1.919	23.82
46.8	320.0	1.808	22.44
51.8	325.0	1.722	21.38
62.3	335.5	1.600	19.86
71.5	344.7	1.539	19.10
75.2	348.4	1.518	18.84

[a] The compiler calculated the mole fraction solubility at 101.325 kPa (1 atm) using the real gas molar volume of 22,360.4 cm^3 at standard conditions of 273.15 K and 101.325 kPa.

[b] The solubility reported by the authors is the same as 10^3 times the Kunen coefficient, or $10^3 s/cm^3 (STP) g^{-1}$.

The authors gave the smoothing equation (ref 1).

$$\log_{10}(S/cm^3(STP)kg^{-1}) = -77.067 + 4090/(T/K) + 26.20 \log_{10}(T/K)$$

AUXILIARY INFORMATION

METHOD/APPARATUS/PROCEDURE:	SOURCE AND PURITY OF MATERIALS:
The original apparatus was described in references (1, 2). Degassed solvent flows in a thin film through the gas down an absorption helix. The gas absorbed and the solvent volume used are read on burets. NOTE: In the authors earlier paper (ref 2) they report five methane solubility values in water at 298.2 ± 0.1 K of 27.9, 27.8, 27.9, 27.8 and 27.9 $cm^3 dm^{-3}$ (gas volume at STP). The results are about 7 per cent smaller than the results reported in this paper. They are considered preliminary results and no data sheet was prepared for them.	(1) Methane. Prepared by the authors from Grignard reagent. (2) Water. Distilled.

| ESTIMATED ERROR:
$\delta T/K = \pm 0.01$
$\delta S/S = \pm 0.01$ (compiler) |

| REFERENCES:
1. Morrison, T. J.
   J. Chem. Soc. <u>1952</u>, 3814.
2. Morrison, T. J.; Billett, F.
   J. Chem. Soc. <u>1948</u>, 2033. |

COMPONENTS:	ORIGINAL MEASUREMENTS:
(1) Methane; CH_4; [74-82-8] (2) Water; H_2O; [7732-18-5]	Claussen, W. F.; Polglase, M. F. *J. Am. Chem. Soc.* <u>1952</u>, *74*, 4817-9.

VARIABLES: $T/K = 274.8 - 312.8$ $p_1/kPa = 101.325$	PREPARED BY: H. L. Clever

EXPERIMENTAL VALUES:

Temperature		Mol Fraction	Bunsen Coefficient	Ostwald Coefficient
$t/^0C$	T/K	$10^5 x_1$	$10^2\alpha$ cm^3(STP)cm^{-3} atm^{-1}	$10^2 L/cm^3$ cm^{-3}
1.6	274.8	4.41	5.52, 5.42 Av. 5.47	5.50
2.0	275.2	4.33	5.42, 5.38, 5.32 Av. 5.38	5.42
10.5	283.7	3.45	4.25, 4.29, 4.28 4.316, 4.251 Av. 4.28	4.44
19.8	293.0	2.83	3.488, 3.251, 3.509 Av. 3.51	3.76
30.4	303.6	2.34	2.886, 2.897 Av. 2.89	3.21
39.6	312.8	2.07	2.542, 2.563 Av. 2.55	2.92

The authors reported the Bunsen coefficients and their average. The compiler calculated the mole fraction and the Ostwald coefficient values. A methane volume of 22,360.4 cm^3STP) mol^{-1} was used to calculate the methane mole fraction at 101.325 kPa.

AUXILIARY INFORMATION

METHOD/APPARATUS/PROCEDURE:

   The solubility was determined by a micro combustion technique. Methane was bubbled through the water via a sintered glass disc to saturate the water.

   The methane in the saturated solution was removed by bubbling oxygen. The train for analysis was composed of an oxygen tank to sweep out the dissolved gas, pressure regulators, mercury manometer, preheater, absorption U-tube containing ascarite and anhydrone, aerator, combustion tube containing copper oxide at 973 K, weighing tubes containing ascarite and anhydrone, and a Marriotte flask.

SOURCE AND PURITY OF MATERIALS:

(1) Methane. Phillips Petroleum Co. 99.7 % with N_2 the greatest impurity by mass spectrometry.

(2) Water. Doubly distilled.

ESTIMATED ERROR:
   $\delta T/K = \pm 0.1$
   $\delta\alpha/\alpha = \pm 0.01$

REFERENCES:

COMPONENTS:	ORIGINAL MEASUREMENTS:
(1) Methane; CH_4; [74-82-8] (2) Water; H_2O; [7732-18-5]	Lannung, A.; Gjaldbaek, J. C. *Acta Chem. Scand.* <u>1960</u>, *14*, 1124-8.

VARIABLES:	PREPARED BY:
$T/K = 291.15 - 310.15$ $p_1/kPa = 101.325$	J. Chr. Gjaldbaek

EXPERIMENTAL VALUES:

Temperature		Mol Fraction	Bunsen Coefficient
$t/^0C$	T/K	$10^3 x_1$	$\alpha/cm^3(STP)cm^{-3}atm^{-1}$
18	291.15	2.82 2.85	0.0350 0.0354
25	298.15	2.55 2.50	0.0315 0.0310
37	310.15	2.10 2.11	0.0259 0.0261

AUXILIARY INFORMATION

METHOD/APPARATUS/PROCEDURE:

The apparatus is a calibrated all-glass combined manometer and bulb enclosed in an air thermostat (ref 1). The entire apparatus is shaken until equilibrium is reached.

The absorbed volume of gas is calculated from the initial and final amounts, both saturated with solvent vapor. The amount of solvent is determined by the weight of displaced mercury. Details in the reference.

SOURCE AND PURITY OF MATERIALS:

(1) Methane. Generated from magnesium methyliodide. Purified by fractional distillation. Specific gravity corresponing with the molecular weight 16.08.

(2) Water. Redistilled. Specific conductivity 2×10^{-7} $(\Omega\ cm)^{-1}$.

ESTIMATED ERROR:

$$\delta T/K = \pm\ 0.05$$
$$\delta x_1/x_1 = \pm\ 0.015$$

REFERENCES:

1. Lannung, A.
   J. Am. Chem. Soc. <u>1930</u>, *52*, 68.

COMPONENTS:	ORIGINAL MEASUREMENTS:
(1) Methane; CH_4; [74-82-8] (2) Water; H_2O; [7732-18-5]	Namiot, A. Yu. *Zh. Strukt. Khim.* <u>1961</u>, *2*, 408-17. *J. Struct. Chem. (Engl. Transl.) <u>1961</u>, *2*, 381-9.

VARIABLES:	PREPARED BY:
T/K = 273, 283 p_1/kPa = 101.3	H. L. Clever

EXPERIMENTAL VALUES:

Temperature		Henry's Constant	Mol Fraction at One Atm (compiler)
$t/^0C$	T/K	K/atm	$10^5 x_1$
0	273.15	22500	4.44
10	283.15	29000	3.45

Henry's constant, K/atm = $(p_1/atm)/x_1$.

AUXILIARY INFORMATION

METHOD/APPARATUS/PROCEDURE:	SOURCE AND PURITY OF MATERIALS:
No experimental details are given. The paper does not make clear whether these are new experimental values or literature values. The paper does contain literature values of the partial molar volume of the gas in water and other thermodynamic information.	No details.
	ESTIMATED ERROR:
	REFERENCES:

COMPONENTS:	ORIGINAL MEASUREMENTS:
(1) Methane; CH_4; [74-82-8] (2) Water; H_2O; [7732-18-5]	Wen, W.-Y.; Hung, J. H. *J. Phys. Chem.* 1970, *74*, 170 - 80.
VARIABLES: T/K = 278.15 - 308.15 p_1/kPa = 101.325	PREPARED BY: H. L. Clever

EXPERIMENTAL VALUES:

T/K	Mol Fraction $10^5 x_1$	Bunsen Coefficient $10^3 \alpha /$ $cm^3(STP) cm^{-3} atm^{-1}$	Ostwald Coefficient $10^3 L /$ $cm^3 cm^{-3}$	Kunen Coefficient $10^3 S /$ $cm^3(STP) g^{-1} atm^{-1}$
278.15	3.987	49.48	50.39	49.48 ± 0.08
288.15	3.100	38.46	40.57	38.49 ± 0.11
298.15	2.526	31.26	34.12	31.35 ± 0.10
308.15	2.136	26.35	29.73	26.51 ± 0.08

The authors reported the solubility of methane as $cm^3(STP) kg^{-1}$. This is the same as 10^3 time the Kunen coefficient reported above.

The compiler calculated the mole fraction, Bunsen coefficient, and Ostwald coefficients using the real gas molar volume of 22,360.4 $cm^3 mol^{-1}$ for methane at standard conditions of 273.15 K and 101.325 kPa (1 atm).

AUXILIARY INFORMATION

METHOD/APPARATUS/PROCEDURE:	SOURCE AND PURITY OF MATERIALS:
The apparatus was similar to that described by Ben-Naim and Baer (ref 1). Teflon needle valves were used in place of stopcocks. The apparatus consists of three parts, a dissolution cell of 300 to 600 cm^3 capacity, a gas volume measuring column, and a manometer. The solvent is degassed in the dissolution cell, the gas is introduced and dissolved as the gas is stirred by a magnetic stirrer. Dissolution of the gas results in the change in the height of a column of mercury which is measured by a cathetometer.	(1) Methane. Matheson Co., Inc. Stated to be better than 99.9 percent pure. (2) Water. Distilled from an all glass apparatus. Specific conductivity 1.5 x 10^{-6} (ohm cm)$^{-1}$
	ESTIMATED ERROR: $\delta S/S$ = ± 0.003 $\delta T/K$ = ± 0.005
	REFERENCES: 1. Ben-Naim, A.; Baer, S. *Trans. Faraday Soc.* 1963, *59*, 2735.

M—B

COMPONENTS:	ORIGINAL MEASUREMENTS:
(1) Methane; CH_4; [74-82-8] (2) Water; H_2O; [7732-18-5]	Ben-Naim, A.; Wilf, J.; Jaacobi, M. *J. Phys. Chem.* <u>1973</u>, *77*, 95 - 102.

VARIABLES:	PREPARED BY:
$T/K = 278.15 - 298.15$ $p_1/kPa = 101.325$	H. L. Clever

EXPERIMENTAL VALUES:

Temperature		Mol Fraction	Ostwald Coefficient
$t/^0C$	T/K	$10^5 x_1$	$L/cm^3\ cm^{-3}$
5	278.15	3.97	0.0502
10	283.15	3.48	0.0448
15	288.15	3.10	0.0405
20	293.15	2.78	0.0370
25	298.15	2.53	0.0342

The compiler added the Kelvin temperatures.

The compiler calculated the mole fraction solubilities for 1 atm (101.325 kPa) using real methane molar volumes. The real molar volumes increased the mole fraction solubility by about 0.22 percent.

AUXILIARY INFORMATION

METHOD/APPARATUS/PROCEDURE:	SOURCE AND PURITY OF MATERIALS:
The method of Ben-Naim and Baer (ref 1) was used. The apparatus was modified by the addition of Teflon stopcocks. The degassed solvent in a volumetric container is forced by a stirrer created vortex up side arms and through tubes containing solvent vapor saturated gas. The gas uptake is deterimned on a buret at constant pressure.	(1) Methane. Matheson Co., Inc. Purity 99.97 percent. (2) Water. Doubly distilled.

	ESTIMATED ERROR:
	$\delta L/L = \pm 0.005$ (compiler)

	REFERENCES:
	1. Ben-Naim, A.; Baer, S. *Trans. Faraday Soc.* <u>1963</u>, *59*, 2735.

COMPONENTS:	ORIGINAL MEASUREMENTS:
(1) Methane; CH_4; [74-82-8] (2) Water; H_2O; [7732-18-5]	Ben-Naim, A.; Yaacobi, M. *J. Phys. Chem.* 1974, *78*, 170 - 5.

VARIABLES:	PREPARED BY:
$T/K = 283.15 - 303.15$ $p_1/kPa = 101.325$	H. L. Clever

EXPERIMENTAL VALUES:

Temperature		Mol Fraction	Ostwald Coefficient
$t/^0C$	T/K	$10^5 x_1$	$L/cm^3\ cm^{-3}$
10	283.15	3.482	0.04480
15	288.15	3.096	0.04051
20	293.15	2.781	0.03700
25	298.15	2.530	0.03420
30	303.15	2.326	0.03192

[a] Mole fraction solubility at 101.325 kPa (1 atm) partial pressure methane calculated by the compiler using the second viral coefficient of methane. The values are about one per cent larger than the values obtained when ideal gas behavior is assumed.

[b] The same Ostwald coefficients appear in the author's paper Yaacobi, M.; Ben-Naim, A. *J. Soln. Chem.* 1973, *2*, 425. They appear to be the same as published by Ben-Naim, Wilf and Yaacobi *J. Phys. Chem.* 1973, *77*, 95, and by Yaacobi and Ben-Naim *J. Phys. Chem.* 1974, *78*,175, but the values are rounded to three digits in the two papers.

AUXILIARY INFORMATION

METHOD/APPARATUS/PROCEDURE:	SOURCE AND PURITY OF MATERIALS:
Used the method of Ben-Naim and Baer (ref 1) as modified by Wen and Hung (ref 2). Degassed liquid in a volumetric container is forced by a stirrer created vortex up side-arms and through tubes containing gas saturated with liquid. Gas absorption is measured on a buret at a constant gas pressure.	(1) Methane. Matheson Co., Inc. Stated to be 99.97 per cent methane. (2) Water. Passed through an ion exchanger and double distilled.

ESTIMATED ERROR:
$\delta L/L = \pm 0.005$ (compiler)

REFERENCES:
1. Ben-Naim, A.; Baer, S. *Trans. Faraday Soc.* 1963,*59*,2735.

2. Wen, W.-Y.; Hung, J. H. *J. Phys. Chem.* 1970, *74*, 170.

COMPONENTS:	ORIGINAL MEASUREMENTS:
(1) Methane; CH_4; [74-82-8] (2) Water; H_2O; [7732-18-5]	Moudgil, B. M.; Somasundaran, P.; Lin, L. J. *Rev. Sci. Instrum.* 1974, *45*, 406-9.

VARIABLES:
    T/K = 298.15
    p_1/kPa = 101.325
	PREPARED BY:
	
	    H. L. Clever

EXPERIMENTAL VALUES:

Temperature		Solubility/	Mol Fraction	Ostwald Coefficient
$t/^0C$	T/K	cm^3(STP) kg^{-1}	$10^5 x_1$	L/cm^3 cm^{-3}
25.0	298.15	31.67	2.546	0.03447

The mole fraction and Ostwald coefficient values were calculated by the compiler assuming ideal gas behavior.

AUXILIARY INFORMATION

METHOD/APPARATUS/PROCEDURE:	SOURCE AND PURITY OF MATERIALS:
The apparatus is based on the design of Ben Naim and Baer (1). The apparatus consists of an absorption cell, a gas measuring column and the pressure control system. The pressure control system is automated.	(1) Methane. Matheson Co., Inc. Stated to be 99.9 percent. (2) Water. Triple distilled, specific conductivity 1.5×10^{-6} Ω^{-1} cm^{-1}

	ESTIMATED ERROR: $\delta T/K$ = ± 0.1
	        $\delta p/kg$ cm^{-2} = ± 0.1
	    Maximum error ± 0.4 percent
	    (authors)

	REFERENCES: 1. Ben Naim, A.; Baer, S. *Trans. Faraday Soc.* 1963, *59*, 2735.

COMPONENTS:	ORIGINAL MEASUREMENTS:
(1) Methane; CH_4; [74-82-8] (2) Water; H_2O; [7732-18-5]	Yamamoto, S.; Alcauskas, J. B.; Crozier, T. E. *J. Chem. Eng. Data* <u>1976</u>, *21*, 78 - 80.

EXPERIMENTAL VALUES:

Temperature		Mol Fraction[1]	Bunsen Coefficient
t ^{0}C	T/K	$10^5 x_1$	$10^2 \alpha / cm^3$ (STP) cm^{-3} atm^{-1}
0.76	273.91	4.506	5.592
0.78	273.93	4.522	5.612
0.79	273.94	4.523	5.613
0.80	273.95	4.512	5.600
0.80	273.95	4.505	5.591
0.81	273.96	4.515	5.604
4.92	278.07	3.994	4.957
4.94	278.09	3.999	4.963
4.95	278.10	4.068	4.973
4.96	278.11	3.998	4.963
4.97	278.12	4.003	4.968
10.90	284.05	3.413	4.235
10.93	284.08	3.398	4.216
10.93	284.08	3.419	4.242
10.94	284.09	3.404	4.224
10.94	284.09	3.401	4.220
10.94	284.09	3.416	4.238
10.95	284.10	3.404	4.224
10.95	284.10	3.406	4.226
10.96	284.11	3.427	4.252
17.99	291.14	2.909	3.606
18.00	291.15	2.907	3.603
18.01	291.16	2.920	3.619
18.02	291.17	2.910	3.607
18.02	291.17	2.909	3.606
24.10	297.25	2.582	3.196
24.11	297.26	2.579	3.192
24.15	297.30	2.576	3.189
24.16	297.31	2.594	3.211
24.17	297.32	2.585	3.200
29.52	302.67	2.358	2.915
29.54	302.69	2.357	2.913
29.54	302.69	2.358	2.215
29.54	302.69	2.355	2.911
29.55	302.70	2.347	2.901

[1] The mole fraction values were calulated by the compiler using a methane molar volume of v_1/cm^3 mol^{-1} = 22,360.4 and water density values from the *SMOW* tables.

The Bunsen ceofficients are repeated on the sea water data sheet from the paper.

COMPONENTS:	ORIGINAL MEASUREMENTS:
(1) Methane; CH_4; [74-82-8] (2) Water; H_2O; [7732-18-5]	Yamamoto, S.; Alcauskas, J. B.; Crozier, T. E. *J. Chem. Eng. Data* <u>1976</u>, *21*, 78-80.

VARIABLES:	PREPARED BY:
T/K = 273.91 - 302.70 p_1/kPa = 101.325	H. L. Clever

EXPERIMENTAL VALUES:

The author's equation

$$\ln\ (\alpha/cm^3(STP)cm^{-3}atm^{-1}) = -67.1962 + 99.1624/(T/100\ K)$$

$$+\ 27.9015\ \ln\ (T/100\ K)$$

was obtained by the method of least squares from the solubility data
in water and in sea water at the various temperatures. Only the pure
water part of the equation is given above.

AUXILIARY INFORMATION

METHOD/APPARATUS/PROCEDURE:	SOURCE AND PURITY OF MATERIALS:
Solubility measurements were made by the Scholander microgasometeric method (ref 1) as modified by Douglas (ref 2). The author's procedure was described in an earlier paper (ref 3). The solubilities were corrected for the effect of dissolved gas on the volume of the aqueous phase by using a value of 37 cm^3 for the partial molal volume of methane in water. The correction increased the Bunsen coefficients by about 0.16 percent. The standard deviations of a single measurement at a constant temperature and pressure ranged from 0.09 to 0.53 percent.	(1) Methane. Linde Specialty Gas. Research grade, 99.99 percent purity. The gas was passed through Ascarite to remove CO_2 prior to use. (2) Water. Distilled.

	ESTIMATED ERROR:
$\delta\alpha/\alpha$ = ± 0.003 (authors)

REFERENCES:
1. Scholander, P. F.
   J. Biol. Chem. <u>1947</u>, *167*, 235.
2. Douglas, E.
   J. Phys. Chem. <u>1964</u>, *68*, 169.
3. Crozier, T. E.; Yamamoto, S.
   J. Chem. Eng. Data <u>1974</u>, *19*, 242.

COMPONENTS:	ORIGINAL MEASUREMENTS:
(1) Methane; CH_4; [74-82-8] (2) Water; H_2O; [7732-18-5]	Muccitelli, J. A.; Wen, W.-Y. *J. Solution Chem.* 1980, *9*, 141 - 61.

VARIABLES: $T/K = 278.15 - 298.15$ $p_1/kPa = 101.325$	PREPARED BY: H. L. Clever

EXPERIMENTAL VALUES:

T/K	Mol Fraction $10^5 x_1$	Bunsen Coefficient $10^3 \alpha/cm^3$(STP)$cm^{-3}atm^{-1}$	Ostwald Coefficient $10^3 L/cm^3\ cm^{-3}$
278.15	4.044	50.19	51.12 ± 0.35
283.15	3.501	43.44	45.04 ± 0.17
288.15	3.086	38.27	40.39 ± 0.18
293.15	2.813	34.85	37.42 ± 0.10
298.15	2.535	31.37	34.26 ± 0.15

The compiler calculated the mole fraction and Bunsen coefficient values using the real methane gas volumes in F. Din, *Thermodynamic Functions of Gases*, Butterworths, London, 1961, Vol. 3.

AUXILIARY INFORMATION

METHOD/APPARATUS/PROCEDURE:

The solubility apparatus and procedure employed were similar to that described by Ben-Naim and Baer (ref]) with modifications suggested by Wen and Hung (ref 2). The apparatus consists of a mercury manometer, a gas-volume measuring buret, a dissolution cell of about 450 cm^3 capacity, and a mercury reservoir.

The degassing apparatus and procedure used were similar to that described by Battino *et al.* (ref 3).

SOURCE AND PURITY OF MATERIALS:

(1) Methane. Matheson Co., Inc. Specified to have a purity of 99.95 percent.

(2) Water. Carbon dioxide free.

ESTIMATED ERROR:
$$\delta T/K = \pm 0.005$$
$$\delta p_1/mmHg = \pm 3$$
$$\delta L/L = \pm 0.005$$

REFERENCES:
1. Ben-Naim, A.; Baer, S. *Trans. Faraday Soc.* 1964, *60*,1736.
2. Wen, W.-Y.; Hung, J. H. *J. Phys. Chem.* 1970, *74*, 170.
3. Battino, R.; Banzhof, M.; Bogan, M.; Wilhelm, E. *Anal. Chem.* 1971, *43*, 806.

COMPONENTS:	ORIGINAL MEASUREMENTS:
(1) Methane; CH_4; [74-82-8] (2) Water; H_2O; [7732-18-5]	Cosgrove, B. A.; Walkley, J. *J. Chromatogr.* <u>1981</u>, *216*, 161 - 7.
VARIABLES: $T/K = 278.15 - 318.15$ $p_1/kPa = 101.325$	PREPARED BY: H. L. Clever

EXPERIMENTAL VALUES:

Temperature T/K	Mol Fraction $10^4 x_1$
278.15	0.3978
283.15	0.3467
288.15	0.2940
293.15	0.2720
298.15	0.2485
303.15	0.2278
308.15	0.2140
313.15	0.1943
318.15	0.1899

AUXILIARY INFORMATION

METHOD/APPARATUS/PROCEDURE:	SOURCE AND PURITY OF MATERIALS:
A 20 ml volume of degassed solvent (sublimation technique) is transferred to a previously evacuated (10^{-4} mmHg) saturation cell immersed in an insulated controlled (± 0.01 K) water bath. The gas is dispersed through the constantly stirred solution at 1 atm by a course, fritted glass disc. Saturation is obtained within a few hours. Prior to analysis the solution is allowed to sit under 1 atm gas pressure for one hour.	(1) Methane. No information. (2) Water. No information.

	ESTIMATED ERROR: $\delta x_1/x_1 = \pm 0.015$ (compiler)

A saturated sample is withdrawn from the saturation cell using a grease-less, gas tight (2.500 ± 0.001 ml) Gilmont syringe. A 0.250 ml sample is injected to "wet" the frit. It is stripped and then four 0.500 ml samples are injected sequentially into the cell. The stripped gas is dried before entering the column. The gas is analyzed on a dual filament conductivity detector. Calibrations with pure gas are made dry before and wet after each series of runs.

REFERENCES:

COMPONENTS:	ORIGINAL MEASUREMENTS:
(1) Methane; CH_4; [74-82-8] (2) Water; H_2O; [7732-18-5]	Rettich, T. R.; Handa, Y. P.; Battino, R.; Wilhelm, E. *J. Phys. Chem.* 1981, *85*, 3230-7.

VARIABLES:
    T/K = 275.46 - 328.15
    p_1/kPa = 50.772 - 118.310
	PREPARED BY:
	
	    H. L. Clever

EXPERIMENTAL VALUES:

T/K	Pressure		Henry's Constant		Mol Fraction
	p_1/atm	p_1/kPa^a	H/atm^b	H/GPa^b	$10^5 x_1$^{c}
275.46	0.8063	81.694	22984	2.3288	4.3509
278.14	0.8455	85.673	24877	2.5207	4.0197
283.95	0.8556	86.693	29000	2.9385	3.4482
288.10	0.8491	86.036	31989	3.2413	3.1261
288.15	0.9032	91.520	32010	3.2434	3.1240
293.16	0.9051	91.709	35618	3.6090	2.8076
298.14	0.5079	51.462	39166	3.9685	2.5532
298.16	0.8531	86.442	39214	3.9734	2.5501
298.16	0.9470	95.959	39186	3.9705	2.5520
303.15	0.9701	98.296	42588	4.3152	2.3481
308.15	1.0309	104.458	45876	4.6483	2.1798
313.16	1.0344	104.810	48999	4.9648	2.0409
318.15	0.5011	50.772	51691	5.2376	1.9346
318.16	1.0904	110.482	51678	5.2362	1.9351
323.16	1.1471	116.234	54157	5.4874	1.8465
328.15	1.1676	118.310	56491	5.7239	1.7702

^{a} Calculated by the compiler.

^{b} Henry's constant evaluated at saturation pressure of the solvent from:

$$H/\text{atm} = \lim_{x_1 \to 0}(f_1/x_1) \text{ where } f_1 \text{ is the fugacity.}$$

^{c} Mole fraction determined at unit fugacity of 1 atm (101.325 kPa).

AUXILIARY INFORMATION

METHOD/APPARATUS/PROCEDURE:	SOURCE AND PURITY OF MATERIALS:
The apparatus used was modelled after that of Benson, Krause, and Peterson (1). Degassed water is flowed in a thin film over the surface of a one dm^3 sphere to contact the gas. After equilibrium is attained the solution is sealed in a chamber of calibrated volume. The dissolved gas is extracted and its amount determined by a direct PVT measurement. A sample of the gas phase is analyzed in an identical manner. From the results, the saturation pressure of the solvent and Henry's constant are calculated in a thermodynamically rigorous manner, applying all non-ideal corrections.	(1) Methane. Airco. Both chemical pure grade, 99.0 minimum mole percent, and ultrahigh purity grade, 99.99 minimum mole percent were used with no detectable difference in results. (2) Water. Reverse osmosis, "house-distilled". Resistivity greater than 5 x 10^4 Ωm.

	ESTIMATED ERROR:
	
	    $\delta H/H$ = ± 0.0008
	    $\delta T/K$ = ± 0.01

The authors smoothing equation, which fits their data to 0.06 %, is:
 	
ln H = 127.174 - 155.5756/$(T/100$ K)
 	
    - 65.2553 ln $(T/100$ K)
 	
    + 6.1698 $(T/100$ K).   For H/Pa.
	REFERENCES:
	
	1. Benson, B. B.; Krause, D.;
	    Peterson, M. A.
	    J. Soln. Chem. 1979, *8*, 655.

COMPONENTS:	EVALUATOR:
(1) Methane; CH_4; [74-82-8] (2) Water; H_2O; [7732-18-5]	Rubin Battino Department of Chemistry Wright State University Dayton, OH 45435 USA 1984, December

CRITICAL EVALUATION:

THE SOLUBILITY OF METHANE IN WATER BETWEEN 298 AND 627 K AT A TOTAL PRESSURE BETWEEN 0.5 AND 200 MPa.

There were thirteen papers (1-13) that reported on the solubility of methane in water as a function of pressure. Of these, several reported additional data on methane, ethane, and propane mixtures (1), methane and butane mixtures (6), aqueous electrolyte solutions (2), and brine solutions (13) as a function of temperature and pressure.

The table below summarizes the ranges of temperature and pressure studied for the methane + water system for each paper. Also listed are the number of data points from each study and the *estimated* per cent precision of the data in terms of the mole fraction solubility of methane in water.

Reference Number	Temperature Range T/K	Total Pressure Range p/MPa	Number of Experimental Points	Estimated Precision, Percent Mol Fraction
1	311 - 344	4.1 - 34.5	8	2
2	373	15 - 154	9	5
3	298 - 518	1.3 - 6.5	7	1-2
4	298 - 444	2.3 - 68.9	71	3
5	298 - 303	0.3 - 5.2	17	1-2
6	311 - 411	0.3 - 13	Graphs	8
7	298 -423	4.1 - 46.9	39	3-5
8	325 - 398	10.1 - 61.6	18	1
9	427 - 627	3.5 - 197	71	3-5
10	298	2.4 - 5.2	6	1-2
11	423 - 633	9.9 - 113.3	58	5
12	277 -573	1.1 - 13.2	16	6
13	298	3.6 - 66.7	11	5

Although the studies covered widely varying ranges of temperature and pressure, they report data of roughly comparable precision. Initially, all the data from references 1-11 (304 data points) were fit as a function of temperature and pressure. (References 12 and 13 were found later and will

be treated separately below - this omission did not affect the analysis.)
The temperature was expressed as the function $\tau = T/100$ K, since this gives
regression coefficients of comparable magnitude. Since the papers report
total pressure and not partial pressure, the total pressures were converted
to MPa and then fit. Initially, four equations were tested and they are:

$$\ln x_1 = A_0 + A_1/\tau + B_0 \ln (p/\text{MPa}) \tag{1}$$

$$\ln x_1 = A_0 + A_1 \ln \tau + B_0 \ln (p/\text{MPa}) \tag{2}$$

$$\ln x_1 = A_0 + A_1/\tau + A_2 \ln \tau + B_0 \ln (p/\text{MPa}) \tag{3}$$

$$\ln x_1 = A_0 + A_1/\tau + A_2 \ln \tau + B_0 \ln (p/\text{MPa}) + B_1 (p/\text{MPa}) \tag{4}$$

Equation (3) gave the best fit in all tests. The precision of the fit was
not significantly improved by the addition of the linear pressure term in
equation (4).

An additional seven equations were fit in an attempt to find a form
related to the Kasarnovsky-Kritchevsky and Kritchevsky-Ilinskaya equations.
These equations follow:

$$\ln x_1 = A_0 + A_1 \ln \tau + C_0 \tau \ln (p/\text{MPa}) \tag{5}$$

$$\ln x_1 = A_0 + A_1/\tau + C_0 \tau \ln (p/\text{MPa}) \tag{6}$$

$$\ln x_1 = A_0 + A_1/\tau + A_2 \ln \tau + C_0 \tau \ln (p/\text{MPa}) \tag{7}$$

$$\ln x_1 = A_0 + A_1/\tau + A_2 \ln \tau + B_0 \ln (p/\text{MPa}) + C_0 \tau (p/\text{MPa}) \tag{8}$$

$$\ln x_1 = A_0 + A_1/\tau + A_2 \ln \tau + B_0 \ln (p/\text{MPa}) + C_0 \ln (\tau p/\text{MPa}) \tag{9}$$

$$\ln x_1 = A_0 + A_1/\tau + A_2 \ln \tau + B_0 \ln (p/\text{MPa}) + C_0 \ln (\tau p/\text{MPa}) + D_0 (1 - x_2^2) \tag{10}$$

$$\ln x_1 = A_0 + A_1/\tau + A_2 \ln \tau + B_0 \ln (p/\text{MPa}) + C_0 \tau (p/\text{MPa}) + D_0 (1 - x_2^2) \tag{11}$$

The degree of fit for equations (8-11) was of the same order as that of
equation (3), while the fit to equations (5-7) was significantly poorer.
Thus, there appears to be no reason to prefer any other equation over
equation (3).

If Henry's law is obeyed exactly, the coefficient B_0 in equation (3)
would be unity. B_0 is not unity, so methane in water solubilities are not
accurately described by Henry's law, although plots of $\ln x_1$ *vs.* $\ln (p/\text{MPa})$
are *linear*. Since we are using total pressure rather than partial pressures
(due to the difficulty in calculating the later), any discussion of Henry's
law is not meaningful.

In the evaluation procedure used all points which deviated from the
smoothed curve by about two or more standard deviations were deleted and
the linear regression repeated. This procedure was carried out three times
for equation (3) with the results shown below:

Number of points	304	275	242	192
Standard deviation in $\ln x_1$	0.27	0.15	0.11	0.081

COMPONENTS:	EVALUATOR:
(1) Methane; CH_4; [74-82-8] (2) Water; H_2O; [7732-18-5]	Rubin Battino Department of Chemistry Wright State University Dayton, OH 45435 USA 1984, December

CRITICAL EVALUATION:

The process could be continued until any desired precision were attained, but this does some violence to the original data set. Also, the further one goes, the more the pressure and temperature ranges narrow, and the larger deviations accumulate at the extremes.

After studying the individual papers and the results from combining their data, we recommend as a most reasonable choice the results obtained with 242 data points and their associated standard deviation of 0.11 in $\ln x_1$. These 242 data points fall in the 298 to 627 K and 0.6 to 192 MPa ranges. The error in mole fraction is a function of pressure at each temperature. At a represenative temperature of 479 K the per cent errors in x_1 at (p/MPa) are: 0.11 % (5 MPa); 1.3 % (95 MPa); and 12 % (192 MPa). It is reasonable to expect poorer precision as pressure increases.

Taking into consideration the discussion in the previous paragraph the *recommended* smoothing equation is:

$$\ln x_1 = -55.8111 + 74.7884/\tau + 20.6794 \ln \tau + 0.753158 \ln (p/\text{MPa}) \qquad (12)$$

where $\tau = T/100$ K and p/MPa is the *total* pressure. Smoothed values of the mole fraction solubility at 25 K intervals between 300 and 625 K are given at seven pressures in Table 1. Several isotherms of $\ln x_1$ *vs.* $\ln (p/\text{MPa})$ are shown in Figure 1. At all pressures there appears to be a minimum in the mole fraction solubility at a temperature of about 350 K.

An important characteristic to keep in mind at these elevated temperatures and pressures is the vapor pressure of the solvent water. Ambrose and Lawrenson (14) provided a smoothing equation for the vapor pressure of pure water using Chebyshev polynomials. We provide in Table 2 for reference at 10 K intervals the vapor pressure of water calculated from their equation. We have added to Figure 1 a line showing the vapor pressure of water. In Table 1 the water vapor pressure exceeds the total pressure heading of 0.5 MPa at 425 K, 1.0 MPa at 475 K, and 10 MPa at 600 K.

TABLE 1. The *tentative* mole fraction solubility of methane in water as a function of temperature between 300 and 625 K at total pressures between 0.5 and 200 MPa.

T/K	Mol Fraction Solubility, $10^3 x_1$, at a Total Pressure of:						
	0.5 MPa	1.0 MPa	10 MPa	50 MPa	100 MPa	150 MPa	200 MPa
300	0.1691	0.2850	1.614	5.43	9.14	12.41	15.41
325	0.1301	0.2192	1.242	4.17	7.03	9.55	11.86
350	0.1164	0.1962	1.111	3.73	6.29	8.54	10.61
375	0.1166	0.1966	1.114	3.74	6.31	8.56	10.63
400	0.1274	0.2147	1.216	4.09	6.89	9.35	11.61
425	0.1486	0.2504	1.418	4.77	8.03	10.90	13.54
450	-	0.3072	1.740	5.85	9.86	13.38	16.61
475	-	-	2.220	7.46	12.57	17.08	21.19
500	-	-	2.918	9.81	16.53	22.43	27.86
525	-	-	3.925	13.19	22.23	30.18	37.48
550	-	-	5.376	18.07	30.45	41.33	51.33
575	-	-	7.463	25.08	42.27	57.37	71.25
600	-	-	-	35.17	59.28	80.46	99.92
625	-	-	-	49.69	83.76	113.7	141.2

Table 2. The vapor pressure of water (14).

T/K	p/kPa	T/K	p/MPa
273.15	0.6107	470	1.4538
280	0.9912	480	1.7890
290	1.9191	490	2.1814
300	3.5352	500	2.6372
310	6.2280	510	3.1633
320	10.540	520	3.7665
330	17.202	530	4.4540
340	27.167	540	5.2336
350	41.647	550	6.1134
360	62.138	560	7.1019
370	90.451	570	8.2084
380	128.73	580	9.4427
390	179.48	590	10.816
400	245.54	600	12.339
410	330.15	610	14.026
420	436.90	620	15.892
430	569.74	630	17.958
440	733.00	640	20.256
450	931.36	647.31[a]	22.106
460	1169.9		

a critical temperature

After the above analysis was completed we found two more papers (12, 13). Their points were added to the 242 used for this analysis. Eleven out of the sixteen of the points of Cramer (13) reported were off the smoothing curve by 1.5 σ or more, while half of Culberson *et al*'s (12) points showed the same deviation. The Cramer paper (13) is important and needs further study since his reported pressures are fugacities which may account for some of the discrepency reported above.

ACKNOWLEDGMENT: The evaluator thanks Professor H. L. Clever for many helpful suggestions in the preparation of this evaluation.

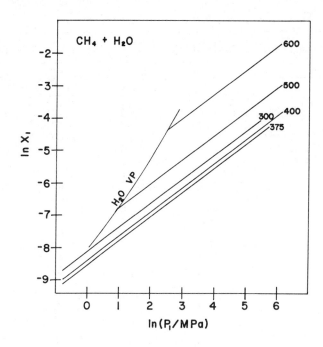

Figure 1.   Methane + Water. $\ln x_1$ vs. $\ln (p_t/\text{MPa})$ at five temperatures between 300 and 600 K.

The water vapor pressure curve is shown crossing the 500 and 600 K isotherms.

The solubility minimum shows in the order of the 300, 375 and 400 K isotherms.

REFERENCES:

1. Amirijafari, B.; Campbell, J. M. *Soc. Pet. Engnrs. J.* <u>1972</u>, *12*, 21-7.

2. Blount, C. W.; Price, L. C.; Wenger, L. M.; Tarullo, M. *DOE Contract Report DE-A508-78ET12145.*

3. Crovetto, R.; Fernandez-Prini, R.; Japas, M. L. *J. Chem. Phys.* <u>1982</u>, *76*, 1077-86.

4. Culberson, O. L.; McKetta, J. J. *J. Petrol. Tech.* <u>1951</u>, *3*, 223-6; *AIME Trans.* <u>1951</u>, *192*, 223-6.

5. Duffy, J. R.; Smith, N. O.; Nagy, B. *Geochim. Cosmochim. Acta* <u>1961</u>, *24*, 23-31.

6. McKetta, J. J.; Katz, D. L. *Ind. Eng. Chem.* <u>1948</u>, *40*, 853-62.

7. Michels, A.; Gerver, J.; Biji, A. *Physica* <u>1936</u>, *3*, 797-808.

8. O'Sullivan, T. D.; Smith, N. O. *J. Phys. Chem.* <u>1970</u>, *74*, 1460-6.

9. Price, L. C. *Am. Assn. Petr. Geol. Bull.* <u>1979</u>, *63*, 1527-33.

10. Stoessell, R. K.; Byrne, P. A. *Clays Clay Miner.* <u>1982</u>, *30*, 67-72; *Geochim. Cosmochim. Acta* <u>1982</u>, *46*, 1327-32.

11. Sultanov, R. G.; Skripka, V. G.; Namiot, A. Yu. *Zh. Fiz. Khim.* <u>1972</u>, *46*, 2160; *VINITI*, 4387-72; *Gazov. Prom.* <u>1972</u>, *17*, 6-7.

12. Culberson, O. L.; Horn, A. B.; McKetta, J. J. *J. Petrol. Tech.* <u>1950</u>, *2*, 1-6; or *AIME, Petrol. Trans.* <u>1950</u>, *189*, 1-6.

13. Cramer, S. D. *Ind. Eng. Chem. Process Des. Dev.* <u>1984</u>, *23*, 533-8.

14. Ambrose, D.; Lawrenson, I. J. *J. Chem. Thermodynamics* <u>1972</u>, *4*, 755-61.

    See also Larsen, E. R.; Prausnitz, J. M. *AIChE J.* <u>1984</u>, *30*, 732-8.

COMPONENTS:	ORIGINAL MEASUREMENTS:
1. Methane; CH_4; [74-82-8] 2. Water; H_2O; [7732-18-5]	Michels, A.; Gerver, J.; Biji, A. *Physica* 1936, *3*, 797-808.

VARIABLES:	PREPARED BY:
Temperature, pressure	C. L. Young

EXPERIMENTAL VALUES:

T/K	$P/10^5$Pa	10^3 Mole fraction of methane in liquid, $10^3 x_{CH_4}$	T/K	$P/10^5$Pa	10^3 Mole fraction of methane in liquid, $10^3 x_{CH_4}$
298.15	40.6	0.81	348.15	176.2	1.74
	46.0	0.90		208.0	1.93
	81.3	1.28	373.15	49.0	0.66
	112.0	1.58		82.2	1.01
	145.9	1.87		113.0	1.27
	176.5	2.10		148.3	1.52
	204.9	2.28		180.5	1.71
	330.8	2.68		209.2	1.84
	469.1	2.97	398.15	49.0	0.64
323.15	49.6	0.72		82.1	0.98
	82.3	1.12		113.0	1.24
	113.1	1.42		150.0	1.50
	145.6	1.69		181.1	1.66
	176.5	1.90		212.3	1.79
	208.2	2.07	423.15	47.1	0.62
348.15	44.3	0.61		81.7	0.93
	79.2	1.01		110.8	1.19
	114.5	1.33		145.4	1.42
	148.1	1.57		177.8	1.60
				206.1	1.73

AUXILIARY INFORMATION

METHOD/APPARATUS/PROCEDURE:	SOURCE AND PURITY OF MATERIALS:
Simple rocking equilibrium cell. Amount of gas absorbed calculated from volume and pressure change of charging vessel. Details in source.	No details given.
	ESTIMATED ERROR: $\delta T/K = \pm 0.1$; $\delta P/10^5$ Pa $= \pm 0.05$ to 0.5%; $\delta x_{CH_4} = \pm 3$-5% (estimated by compiler).
	REFERENCES:

COMPONENTS:	ORIGINAL MEASUREMENTS:
(1) Methane; CH_4; [74-82-8] (2) Water; H_2O; [7732-18-5]	Culberson, O. L.; Horn, A. B.; McKetta, J. J. *J. Pet. Technol.* <u>1950</u>, *2*, 1-6. or *AIME, Pet. Trans.* <u>1950</u>, *189*, 1-6.

VARIABLES:	PREPARED BY:
T/K = 298.15 p_t/MPa = 3.62 - 66.74	H. L. Clever

EXPERIMENTAL VALUES:

Temperature		Total Pressure		Mole Ratio
$t/^0F$	T/K	p/psia	p/MPa	$10^3 (n_1/n_2)$
77	298.15	525	3.62	0.770
		1000	6.89	1.10
		1450	10.00	1.80
		1845	12.72	2.02
		1930	13.31	2.27
		2535	17.48	2.31
		3615	24.92	2.88
		4435	30.58	3.28
		6342	43.72	4.07
		7935	54.71	3.91
		9680	66.74	4.51

The mole fraction solubility is $x_1 = n_1/(1 + n_1)$, when n_2 is assumed one.

AUXILIARY INFORMATION

METHOD/APPARATUS/PROCEDURE:	SOURCE AND PURITY OF MATERIALS:
The sample is equilibrated in a large rocking autoclave. Samples are analyzed by removing water and measuring the gas volumetrically. The temperature is measured with a thermocouple and the pressure with a Bourdon gage.	(1) Methane. Phillips Petroleum Co. The purity was 99.0 mole per- cent minimum. (2) Water. Distilled water was boiled to degass.
	ESTIMATED ERROR: $\delta T/K$ = ± 0.5; $\delta n_1/n_2$ = ± 5 % δp/MPa = ± 0.02;less than 10 MPa δp/MPa = ± 0.07; 10 to 35 MPa δp/MPa = ± 0.14; over 35 MPa
	REFERENCES:

COMPONENTS:	ORIGINAL MEASUREMENTS:
1. Methane; CH_4; [74-82-8] 2. Water; H_2O; [7732-18-5]	Culberson, O. L.; McKetta, J. J. *J. Petrol. Tech.* 1951, *3*, 223-226 or *AIME Petrol. Trans.* 1951, *192*, 223-226.
VARIABLES:	PREPARED BY: C. L. Young

EXPERIMENTAL VALUES:

T/K	p/psia	P/MPa	Mole fraction of methane in liquid, x_{CH_4}
298.2 (77)	341	2.35	0.000497
	459	3.16	0.000717
	659	4.54	0.001000
	934	6.44	0.001317
	1290	8.89	0.001678
	1930	13.31	0.002235
	2495	19.20	0.002585
	3515	24.24	0.003110
	4810	33.16	0.003660
	6440	44.40	0.004170
310.9 (100)	330	2.28	0.000440
	477	3.29	0.000619
	664	4.58	0.000839
	950	6.55	0.001123
	1270	8.76	0.001440
	1900	13.10	0.001890
	2575	17.75	0.002290
	3535	24.37	0.002760
	4910	33.85	0.003330
	6525	44.99	0.00391
	7870	54.26	0.00417
	9895	68.22	0.00465

(cont.)

AUXILIARY INFORMATION

METHOD/APPARATUS/PROCEDURE:	SOURCE AND PURITY OF MATERIALS:
Sample equilibrated in large rocking autoclave. Samples of liquid analysed by removing water and estimating the gas volumetrically. Temperature measured with thermocouple and pressure with Bourdon gauge. Details in ref. (1).	1. Phillips Petroleum Co. sample, purity 98.72 mole per cent. 2. Distilled and degassed.
	ESTIMATED ERROR: $\delta T/K = \pm 0.5$; $\delta P/MPa = \pm 1\%$; $\delta x_{CH_4} = \pm 3\%$ (estimated by compiler).
	REFERENCES: 1. Culberson, O. L.; Horn, A. B.; McKetta, J. J. *J. Petr. Technol. Trans AIME Pet. Div.* 1950, *189*, 1.

COMPONENTS:	ORIGINAL MEASUREMENTS:
1. Methane; CH_4; [74-82-8]	Culberson, O. L.; McKetta, J. J.
2. Water; H_2O; [7732-18-5]	*J. Petrol. Tech.* <u>1951</u>, *3*, 223-226
	or
	AIME, Petrol. Trans. <u>1951</u>, *192*,
	223-226.

EXPERIMENTAL VALUES:

T/K (T/°F)	p/psia	P/MPa	Mole fraction of methane in liquid, x_{CH_4}
344.3 (160)	331	2.28	0.000340
	467	3.22	0.000470
	659	4.54	0.000632
	943	6.50	0.000909
	1320	9.10	0.001183
	1880	12.96	0.001500
	2555	17.62	0.001924
	3535	24.37	0.002385
	4925	33.96	0.002770
	6525	44.99	0.00342
	8220	56.67	0.00375
	9865	68.02	0.00424
377.6 (220)	333	2.30	0.000323
	466	3.21	0.000432
	468	3.23	0.000472
	652	4.50	0.000611
	945	6.52	0.000886
	1310	9.03	0.001188
	1900	13.10	0.001560
	2535	17.48	0.001980
	3570	24.61	0.002510
	4965	34.23	0.00314
	6525	44.99	0.00361
	8190	56.47	0.00408
	9875	68.09	0.00451
410.9 (280)	336	2.32	0.000326
	464	3.20	0.000460
	654	4.51	0.000673
	941	6.49	0.000938
	1310	9.03	0.001326
	1900	13.10	0.001857
	2480	17.10	0.002346
	3555	24.51	0.003015
	4975	34.30	0.003805
	6525	44.99	0.00449
	8270	57.02	0.00518
	9835	67.81	0.00574
444.3 (340)	323	2.23	0.000323
	475	3.28	0.000535
	662	4.56	0.000789
	949	6.54	0.001150
	1360	9.38	0.001725
	1920	13.24	0.002355
	2580	17.79	0.003025
	3580	24.68	0.003835
	5045	34.78	0.004875
	6525	44.99	0.00595
	8210	56.61	0.00680
	9995	68.91	0.00775

COMPONENTS:	ORIGINAL MEASUREMENTS:
1. Methane; CH_4; [74-82-8] 2. Water; H_2O; [7732-18-5]	Duffy, J. R.; Smith, N. O.; Nagy, B. *Geochim. Cosmochim. Acta* 1961, *24*, 23-31.

VARIABLES:	PREPARED BY:
Temperature, pressure	C. L. Young

EXPERIMENTAL VALUES:

T/K	P/MPa	Mole fraction of methane in liquid phase, $10^4\ x_{CH_4}$
298.15	1.103	2.14
	1.482	2.73
	1.586	3.76
	2.965	7.08
	3.068	7.03
	3.544	8.00
	4.033	9.39
	4.688	9.79
	5.171	11.30
303.15	0.317	0.60
	0.552	1.15
	0.793	1.84
	0.938	2.32
	1.972	4.90
	2.048	4.93
	2.744	6.12
	3.606	7.64

AUXILIARY INFORMATION

METHOD/APPARATUS/PROCEDURE:	SOURCE AND PURITY OF MATERIALS:
Rocking equilibrium cell. Pressure measured with a Bourdon gauge. Cell charged with boiled water; gas admitted to known pressure. Cell contents allowed to equilibrate. Final pressure measured and used to calculate amounts of gas dissolved. Details in source ref.	1. C.P. grade - no other details given. 2. Degassed.

	ESTIMATED ERROR: $\delta T/K = \pm 1$; $\delta P/MPa = \pm 0.03$; $\delta x_{CH_4} = \pm 5 \times 10^{-6}$.
	REFERENCES:

COMPONENTS:	ORIGINAL MEASUREMENTS:
(1) Methane; CH_4; [74-82-8] (2) Water; H_2O; [7732-18-5]	Schröder, W. *Z. Naturforsch.* <u>1969</u>, *24b*, 500-8. *Chem.-Ing.-Tech.* <u>1973</u>, *45*, 603-8.
VARIABLES: $T/K = 303.15 - 373.15$ $p_1/MPa = 1.013 - 6.080$	PREPARED BY: H. L. Clever

EXPERIMENTAL VALUES:

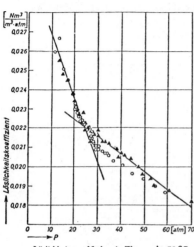

Löslichkeit von Methan in Wasser. ▲: 70 °C,
⊙: 80 °C.

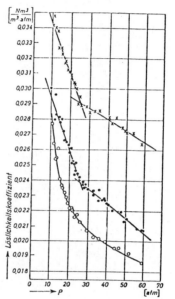

Löslichkeit von Methan in Wasser. ✕: 30 °C, ●:
50 °C, o: 100 °C.

AUXILIARY INFORMATION

METHOD/APPARATUS/PROCEDURE:	SOURCE AND PURITY OF MATERIALS:
The apparatus was a bubbling-type autoclave of the type described by Wiebe, Gaddy and Heins (ref 1). The Bunsen coefficient was calculated from the experimental data. The data were presented in the graphs above. The data were discussed briefly in an earlier paper (ref 2). The second paper gives the equation $\ln(\alpha/m^3(STP)m^{-3}atm^{-1}) = 4.211$ $- 5821/(T/K) + 1.019 \times 10^6/(T/K)^2$	
	ESTIMATED ERROR:

for the 303 - 373 temperature interval at a methane pressure of 100 atm (10.133 MPa). The temperature of minimum solubility is 350 K at this pressure, it increases to 377 K at 20 atm partial pressure methane.

REFERENCES:
1. Wiebe,R.; Gaddy,V.L.; Heins, C.
   Ind. Eng. Chem. <u>1932</u>, *24*, 823;
   J. Am. Chem. Soc. <u>1933</u>, *55*, 947.

2. Schröder, W.
   Naturwissenschaften <u>1968</u>, *55*, 542.

COMPONENTS:	ORIGINAL MEASUREMENTS:
1. Methane; CH_4; [74-82-8] 2. Water; H_2O; [7732-18-5]	O'Sullivan, T.D.; Smith, N.O. *J. Phys. Chem.*, <u>1970</u>, *74*, 1460-1466.

VARIABLES:	PREPARED BY:
Temperature, pressure	C.L. Young

EXPERIMENTAL VALUES:

T/K	P/MPa	10^3 Mole fraction of methane in liquid, $10^3 x_{CH_4}$
324.65	10.13	1.427
	20.26	2.279
	30.40	2.87
	40.53	3.34
	50.63	3.73
	60.79	4.09
375.65	10.23	1.355
	20.37	2.205
	30.60	2.87
	40.83	3.33
	50.97	3.85
	61.20	4.19
398.15	10.44	1.434
	20.67	2.321
	30.90	2.96
	41.04	3.43
	51.37	3.96
	61.61	4.30

AUXILIARY INFORMATION

METHOD/APPARATUS/PROCEDURE:	SOURCE AND PURITY OF MATERIALS:
Large steel stirred equilibrium cell. Pressure measured with Bourdon gauge. Temperature measured with iron-constantan thermocouple. Cell charged with liquid, compressed gas added. After equilibrium obtained samples removed and analysed using volumetric techniquies. Details in ref. (1).	1. Matheson Co., sample purity 99.95 mole per cent. 2. Distilled and de-ionised, air removed.

ESTIMATED ERROR:

$\delta T/K = \pm 0.5$; $\delta P/MPa = \pm 0.05\%$;
$\delta x_{CH_4} = \pm 0.4\%$.

REFERENCES:

1. O'Sullivan, T.D.; Smith, N.O.

   Geochim. Cosmochim. Acta, 1966, *30*, 617.

COMPONENTS:	ORIGINAL MEASUREMENTS:
1. Methane; CH_4; [74-82-8] 2. Water; H_2O; [7732-18-5]	Sultanov, R. G.; Skripka, V. G.; Namoit, A. Yu. *Zh. Fiz. Khim.* <u>1972</u>, *46*, 2160; <u>VINITI</u>, *4387-72*.

VARIABLES:	PREPARED BY:
	C. L. Young

EXPERIMENTAL VALUES:

T/K	P/kg cm^{-2}	P/MPa	Mole fraction of methane in liquid, x_{CH_4}	in vapor, y_{CH_4}
423.2	100	9.81	0.0010	0.9400
	200	19.61	0.0018	0.9630
	400	39.23	0.0030	0.9780
	600	58.84	0.0046	0.9830
	800	78.45	0.0056	0.9835
	1000	98.07	0.0056	0.9850
473.2	100	9.81	0.0020	0.8100
	200	19.61	0.0038	0.8915
	400	39.23	0.0067	0.9350
	600	58.84	0.0087	0.9480
	800	78.45	0.0100	0.9545
	1000	98.07	0.0104	0.9630
523.2	100	9.81	0.0025	0.5300
	200	19.61	0.0063	0.7330
	400	39.23	0.0117	0.8325
	600	58.84	0.0140	0.8720
	800	78.45	0.0146	0.8980
	1000	98.07	0.0151	0.9100
573.2	100	9.81	0.0015	0.0950
	200	19.61	0.0078	0.4360
	400	39.23	0.0185	0.6260
	600	58.84	0.0265	0.6790
	800	78.45	0.0340	0.7150
	1000	98.07	0.0407	0.7500 (cont.)

AUXILIARY INFORMATION

METHOD/APPARATUS/PROCEDURE:	SOURCE AND PURITY OF MATERIALS:
Static equilibrium cell fitted with magnetic stirrer, details in ref. (1). Samples of coexisting phases analysed by freezing out water and estimating methane volumetrically.	1. Purity 99.95 volume per cent. 2. No details given.

	ESTIMATED ERROR: $\delta T/K = \pm 0.3$; $\delta P/MPa = \pm 0.1$; δx_{CH_4}, δy_{CH_4} = ± 0.0005 (estimated by compiler).
	REFERENCES: 1. Sultanov, R. G.; Skripka, V. G.; Namoit, A. Yu. *Gazov. Prom.* <u>1971</u>, *16*, 6.

COMPONENTS: ORIGINAL MEASUREMENTS:

1. Methane; CH_4; [74-82-8]   Sultanov, R. G.; Skripka, V. G.;
                                     Namoit, A. Yu.
2. Water; H_2O; [-732-18-5]   *Zh. Fiz. Khim.*

                                     1972, *46*, 2160; VINITI, *4387-72*.

EXPERIMENTAL VALUES:

T/K	$P/kg\ cm^{-2}$	P/MPa	Mole fraction of methane in liquid, x_{CH_4}	in vapor, y_{CH_4}
603.2	200	19.61	0.0100	0.1950
	400	39.23	0.0325	0.4170
	600	58.84	0.0464	0.5040
	800	78.45	0.0572	0.5540
	1000	98.07	0.0635	0.5850
	1050	102.97	0.0650	-
623.2	200	19.61	0.0053	0.0800
	400	39.23	0.0414	0.2350
	600	58.84	0.0707	0.2980
	800	78.45	0.0955	0.3150
	1000	98.07	0.1230	0.3350
	1050	102.97	0.1300	-
	1100	113.27	0.1365	-
625.2	250	24.52	0.0135	0.1390
	300	29.42	0.0230	0.1850
	400	39.23	0.0410	0.2280
	500	49.03	0.0550	0.2490
	600	58.84	0.0660	0.2660
	700	68.65	0.0800	0.2780
	800	78.45	0.1050	0.2750
	900	88.26	0.1250	0.2310
	935*	91.69	0.1800	0.1800
	992**	97.28	0.1730	0.1730
	1000	98.07	0.1250	0.2000
	1050	102.97	0.0820	0.2400
	1100	113.27	0.0680	0.2470
628.2	250	24.52	0.0165	0.1050
	300	29.42	0.0275	0.1530
	400	39.23	0.0540	0.2050
	500	49.03	0.0830	0.2150
	600	58.84	0.1200	0.2140
	650	63.74	0.1430	0.2040
	680*	66.69	0.1720	0.1720
633.2	250	24.52	0.0160	0.0820
	300	29.42	0.0260	0.1190
	400	39.23	0.0500	0.1590
	500	49.03	0.0700	0.1710
	600	58.84	0.0960	0.1570
	620*	60.80	0.1280	0.1280

* gas-liquid critical point

** gas-gas critical point

COMPONENTS:	ORIGINAL MEASUREMENTS:
1. Methane; CH_4; [74-82-8] 2. Water; H_2O; [7732-18-5]	Amirijafari, B.; Campbell, J. M. *Soc. Pet. Engnrs. J.* 1972, 12, 21-27.
VARIABLES:	PREPARED BY: C. L. Young

EXPERIMENTAL VALUES:

T/K (T/°F)	P/psi	P/MPa	$10^3 \times$ Mole fraction of methane in water-rich phase $10^3 x_{CH_4}$
310.93 (100)	600 2000 3000 5000	4.14 13.79 20.68 34.47	0.759 1.956 2.519 3.350
344.26 (160)	600 2000 3000 5000	4.14 13.79 20.68 34.47	0.602 1.612 2.150 2.800

AUXILIARY INFORMATION

METHOD/APPARATUS/PROCEDURE:	SOURCE AND PURITY OF MATERIALS:
Static stainless steel equilibrium vessel of approximately 75 mL capacity. Pressure measured with Bourdon gauge and temperature measured with thermocouple. Samples of liquid and vapor analysed using a gas chromatograph equipped with a flame ionisation detector. Poropak R column used.	1. Pure grade sample, purity 99.9 mole per cent. 2. No details given.
	ESTIMATED ERROR: $\delta T/K = \pm 0.03$; $\delta P/MPa = \pm 1\%$; $\delta x_{CH_4} = \pm 2\%$.
	REFERENCES:

COMPONENTS:	ORIGINAL MEASUREMENTS:
1. Methane; CH_4; [74-82-8] 2. Water; H_2O; [7732-18-5]	Price, L. C. *Am. Assn. Pet. Geol. Bull.* <u>1979</u>, *63*, 1527-33.

VARIABLES:	PREPARED BY:
	C. L. Young

EXPERIMENTAL VALUES:

T/K (T/°C)	P/psi	P/MPa[a]	Solubility[c] SCF/bbl	Mole per cent of methane, [b] x_{CH_4}
427 (154)	514	3.54	5.65 ± 0.3	0.0741
	2205	15.20	21.81 ± 2.69	0.2859
	4645	32.03	34.43 ± 1.66	0.4514
	6790	46.82	42.03 ± 0.36	0.5510
	9760	67.29	46.72 ± 1.28	0.6125
	12670	87.36	49.78 ± 0.70	0.6526
	15260	105.21	58.76 ± 1.33	0.7703
	18260	125.90	67.37 ± 0.15	0.8832
	23780	163.96	78.76 ± 1.83	1.0325
479 (206)	750	5.17	9.51 ± 0.80	0.1247
	2323	16.02	30.82 ± 0.75	0.4041
	4270	29.44	48.12 ± 1.92	0.6309
	7923	54.63	72.36 ± 0.22	0.9486
	13759	94.86	98.11 ± 2.19	1.286
	18906	103.35	116.5 ± 1.5	1.527
	23652	163.07	127.0 ± 7.6	1.665
	27915	192.47	143.5 ± 0.5	1.881
494 (221)	583	4.02	9.73 ± 0.19	0.1276
	5331	36.76	62.87 ± 0.35	0.8242
	9109	62.80	101.7 ± 1.90	1.333
	12670	87.36	116.4 ± 2.9	1.526
	15020	103.56	131.4 ± 9.7	1.723

(cont.)

AUXILIARY INFORMATION

METHOD/APPARATUS/PROCEDURE:	SOURCE AND PURITY OF MATERIALS:
Static equilibrium cell filled with water, vessel sealed and brought to temperature, excess water being allowed to bleed out. Some water removed and compressed methane added. Samples for analysis removed, methane being added simultaneously so that the total pressure remained constant. Samples analysed by measuring equilibrium pressure when sample injected into an evacuated flask. Duplicate samples taken.	1. Matheson gas, purity 99.99 mole per cent. 2. Distilled and degassed.

ESTIMATED ERROR:
T/K = ±1.0.

REFERENCES:

COMPONENTS:	ORIGINAL MEASUREMENTS:
1. Methane; CH₄; [74-82-8]	Price, L. C.
	Am. Assn. Pet. Geol. Bull.
2. Water; H₂O; [7732-18-5]	1979, 63, 1527-33.

EXPERIMENTAL VALUES:

T/K (T/°C)	P/psi	P/MPa[a]	Solubility[c] SCF/bbl	Mole per cent of methane,[b] x_{CH_4}
494	17940	123.69	135.3 ± 1.6	1.774
(221)	20530	141.55	139.4 ± 4.1	1.828
507	1176	8.11	19.92 ± 0.09	0.2612
(234)	2160	14.89	34.91 ± 1.41	0.4577
	3014	20.78	54.75 ± 1.03	0.7178
	4027	27.77	63.97 ± 0.82	0.8386
	6836	47.13	108.2 ± 1.84	1.419
	8658	59.69	117.3 ± 0.3	1.538
	11330	78.12	140.0 ± 0.5	1.835
	13540	93.36	150.8 ± 4.9	1.977
	15690	108.18	161.9 ± 0.4	2.123
	15770	108.73	159.2 ± 2.6	2.087
	19230	132.59	169.3 ± 3.0	2.220
	21340	147.13	172.1 ± 1.9	2.256
	23830	164.30	181.0 ± 4.4	2.373
553	2866	19.76	65.0 ± 3.70	0.8522
(280)	4616	31.83	101.6 ± 6.4	1.332
	6953	47.94	160.0 ± 2.1	2.098
	10170	70.12	206.3 ± 9.2	2.705
	14490	99.91	252.3 ± 0.7	3.308
	18330	126.38	264.9 ± 4.6	3.473
	22020	151.82	282.8 ± 8.8	3.708
	23120	159.41	292.5 ±19.6	3.835
	27400	188.92	308.4 ± 8.9	4.043
565	1566	10.80	22.59 ± 0.08	0.2962
(292)	2770	19.10	67.26 ± 0.59	0.8818
	4337	29.90	115.2 ± 6.4	1.510
	13130	90.53	278.3 ± 1.2	3.649
	15940	109.90	293.9 ± 3.9	3.853
	22050	152.03	336.1 ± 4.6	4.406
	24500	168.92	349.9 ± 8.5	4.587
589	1632	11.25	11.2 ± 1.5	0.1468
(316)	3631	25.03	132.2 ± 3.7	1.733
	7747	53.41	321.2 ± 1.5	4.211
	10440	71.98	377.9 ± 4.5	4.954
	13390	92.32	421.1 ± 8.3	5.521
	17010	117.28	474.0 ± 6.5	6.214
	23990	165.41	509.1 ± 7.3	6.674
	27750	191.33	527.6 ± 5.7	6.917
627	2837	19.56	46.79 ± 0.71	0.6134
(354)	3631	25.03	134.7 ± 3.45	1.766
	4689	32.33	268.5 ± 2.2	3.520
	6174	42.57	422.1 ±12.0	5.534
	7688	53.01	488.7 ± 5.5	6.407
	15820	109.08	669.7 ± 8.1	8.780
	18460	127.28	700.3 ±11.3	9.181
	24650	169.96	775.9 ± 1.8	10.17
	26940	185.74	803.0 ± 1.7	10.53
	28610	197.26	828.8 ± 2.0	10.87

[a] Calculated by compiler.

[b] Calculated by compiler by multiplying solubility by conversion factor stated by author in original.

[c] Unit of standard cubic feet per barrel of water.

COMPONENTS:	ORIGINAL MEASUREMENTS:
(1) Methane; CH_4; [74-82-8] (2) Water; H_2O; [7732-18-5]	Crovetto, R.; Fernández-Prini, R.; Japas, M. L. *J. Chem. Phys.* <u>1982</u>, *76*,1077-86.
VARIABLES: T/K = 297.5 - 518.3 p/MPa = 1.327 - 6.451	PREPARED BY: R. Fernández-Prini

EXPERIMENTAL VALUES:

T/K	Total Pressure p/MPa	Methane Volume Fraction, y_1	Methane Fugacity f_1/MPa	Mol Fraction $10^4 x_1$	$\ln(H/\text{GPa})$ [1]
297.5	1.861	0.9983	1.798	4.351	1.419
333.7	1.327	0.9840	1.286	2.124	1.801
385.3	2.092	0.9226	1.908	2.985	1.855
388.4	2.156	0.9166	1.954	3.085	1.846
430.6	2.131	0.7147	1.531	3.025	1.621
473.2	3.210	0.4873	1.618	4.146	1.362
518.3	6.451	0.3875	2.697	10.337	0.959

[1] Henry's constant, $H/\text{GPa} = (f_1/\text{GPa})/x_1$.

The smoothing equation was obtained from the data and the values of H for 288, 298, and 308 K given in reference (1).

$$\ln(H/\text{GPa}) = -8.681 + 7.837/(T/1000\ \text{K}) - 1.509/(T/1000\ \text{K})^2$$
$$+ 0.0206/(T/1000\ \text{K})^3 \qquad (\sigma = 0.017)$$

Thermodynamic quantities for the process CH_4(g, 0.1 MPa, T) → CH_4(l, x_1 = 1, T) are below:

T/K	ΔG_1°/kJ mol^{-1}	ΔH_1°/kJ mol^{-1}	ΔS_1°/J(K mol)$^{-1}$	ΔC_{p1}°/J(K mol)$^{-1}$
298.2	26.28	-13.32	132.6	244
400.0	36.58	5.63	77.4	141
520.0	43.89	18.80	48.1	85

AUXILIARY INFORMATION

METHOD/APPARATUS/PROCEDURE:	SOURCE AND PURITY OF MATERIALS:
The method involved the equilibration of the gas with the liquid and the determination of the gas mole fraction by sampling the equilibrated liquid phase. Henry's constant was obtained for each temperature by employing second virial coefficients for pure components and mixture in order to correct for non-ideal behavior in the gas phase. The gas was equilibrated in a thermostated stainless steel vessel which was continuously rocked. Weighed samples of the liquid phase were withdrawn and the amount of dissolved gas determined with a gas buret. The system was then taken to a new temperature. Pressures were measured with calibrated Bourdon gauges.	(1) Methane. Matheson (UHP) 99.97 mol %. (2) Water. Conductivity water.

ESTIMATED ERROR: $\delta T/K$ = ± 0.2
(Authors') $\delta p/p$ = ± 0.003
            $\delta x_1/x_1$ = ± 0.005
($T/K \leqslant 520$) $\delta H/H$ = ± 0.01 - 0.02

REFERENCES:

1. Wilhelm, E.; Battino, R.; Wilcock, R. J.
   Chem. Rev. <u>1977</u>, *77*, 219.

COMPONENTS:	ORIGINAL MEASUREMENTS:
(1) Methane; CH_4; [74-82-8] (2) Water; H_2O; [7732-18-5]	Stoessell, R. K.; Byrne, P. A. *Clays Clay Miner.* <u>1982</u>, *30*, 67-72. *Geochim. Cosmochim. Acta* <u>1982</u>, *46*, 1327-32.

VARIABLES:	PREPARED BY:
$T/K = 298.15$ $p_1/kPa = 2410 - 5170$	H. L. Clever

EXPERIMENTAL VALUES:

Temperature		Pressure		Methane	
$t/^\circ C$	T/K	$p_1/psia$	p_1/kPa	$m_1/mol\ kg^{-1}$	Mol Fraction $10^4 x_1$
25	298.15	350	2410	0.0318[a] 0.0319[b]	5.73 5.74
		550	3790	0.0473[a] 0.0483[b]	8.51 8.69
		750	5170	0.0623[a] 0.0617[b]	11.21 11.10

[a] Values from first reference above.

[b] Values from second reference above.

The kPa pressure and the mole fraction values were calculated by the compiler.

AUXILIARY INFORMATION

METHOD/APPARATUS/PROCEDURE:

Solubility determinations were made using a titanium-lined chamber within a stainless-steel reaction vessel jacketted by a water bath for temperature control. The system pressure was set by controlling the input and output of methane within the chamber headpiece. The vessel was rocked for 3 h to allow equilibration between the methane and water.

The amount of gas in the saturated solution was measured by transfer of a sample volume to a loop at the system pressure, followed by flashing the sample in an expansion loop and measuring the gas pressure in a known volume. The total gas volume and pressure change were used to compute the moles of methane assuming ideal behavior. A correction was made for the gas not released on flashing.

SOURCE AND PURITY OF MATERIALS:

(1) Methane. Matheson Co., Inc. Ultra high purity grade. Stated to have a minimum purity of 99.97 mole percent.

(2) Water. Distilled.

ESTIMATED ERROR:

$$\delta T/K = \pm\ 0.1$$
$$\delta p_1/psia = \pm\ 1$$
$$\delta m_1/mol\ kg^{-1} = \pm\ 0.0003\ Av.$$
$$\pm\ 0.0005\ Max.$$

REFERENCES:

COMPONENTS:	ORIGINAL MEASUREMENTS:
(1) Methane; CH_4; [74-82-8] (2) Water; H_2O; [7732-18-5]	Cramer, S. D. *Ind. Eng. Chem. Process Des. Dev.* **1984**, *23*, 533-8.

| VARIABLES:
$T/K = 277.2 - 573.2$
$p_t/MPa = 1.1 - 13.2$ | PREPARED BY:

H. L. Clever |

EXPERIMENTAL VALUES:

Temperature		Total Pressure	Henry's Constant	Temperature		Total Pressure	Henry's Constant
$t/^0C$	T/K	p/MPa	k/MPa	$t/^0C$	T/K	p/MPa	k/MPa
0	*273.2*		*2460*	184.3	457.5	5.7	4050
4.0	277.2	3.0	2580	187.7	460.9	7.2	3990
12.6	285.8	3.0	3430	193.3	466.5	6.9	3120
20	*293.3*		*3950*	*200*	*473.2*		*3580*
40	*313.2*		*5370*	210.7	483.9	6.9	3080
42.0	315.2	1.1	5800	*220*	*493.2*		*2950*
60	*333.2*		*6420*	239.7	512.9	6.9	2580
61.0	334.2	1.1	6260	*240*	*513.2*		*2430*
80	*353.2*		*6940*	*260*	*533.2*		*2010*
90.4	363.6	1.1	6610	264.4	537.6	10.5	2040
100	*373.2*		*6930*	269.2	542.4	8.0	2240
111.0	384.2	1.2	6310	*280*	*553.2*		*1670*
120	*393.2*		*6510*	281.0	554.2	12.0	1780
140	*413.2*		*5850*	300.0	573.2	13.2	1130
159.3	432.5	1.1	4890	*300*	*573.2*		*1400*
160	*433.2*		*5070*				
180	*453.2*		*4290*				

The values in *italic* are the author's smoothed values. Six additional values from ref 1 were included in the smoothing.

AUXILIARY INFORMATION

METHOD/APPARATUS/PROCEDURE:	SOURCE AND PURITY OF MATERIALS:

Methane solubilities were determined from *pvT* measurements by the gas extraction technique. The measuring apparatus consisted of: (*i*) a high pressure, thermostated, stirred reactor for dissolving gas in the solvent; (*ii*) a heat exchanger for bringing the gas saturated solvent to room temperature; and (*iii*) a low pressure, thermostated gas buret for making *pvT* measurements on collected samples of vapor and liquid. The apparatus and its operation were described earlier (ref 2).
 	
Four to eight gas-saturated solution samples were taken and analyzed at 15 - 30 minute intervals after the time determined necessary for saturation. Henry's constants were computed (ref 2) and smoothed by a specially developed equation (ref 3).
 	
Henry's constant, $k^0 = f/a = (\phi p_1)/(\gamma x_1)$. $p_1 = p_t - p_2$; ϕ from reduced properties chart. See paper.
	(1) Methane.
	  No information.
	(2) Water.
	
	--- 	
	
	Henry's constants from (ref 1) corrected by author for H_2O vapor presure. (T/k)(unit as above).
	
	298.2/4550; 311.0/5360; 344.3/6900
	377.6/7320; 411.0/6320; 444.3/5040
	
	--- 	
	
	ESTIMATED ERROR:
	  $\delta k/k = \pm 0.058$ (author's est exp. error)
	Rel std error of estimate 5.1-10.5 %.
	
	--- 	
	
	REFERENCES:
	1. Culbertson,O.L.; McKetta, J.J.,Jr. *Pet. Trans. AIME* **1951**, *192*, 223.
	2. Cramer, S. D. *Ind. Eng. Chem. Process Des. Dev.* **1980**, *19*, 300.
	3. Cramer, S. D. *Ind. Eng. Chem. Process Des. Dev.* **1984**, *23*, 618.

COMPONENTS:	ORIGINAL MEASUREMENTS:
1. Methane; CH_4; [74-82-8] 2. Water; H_2O; [7732-18-5]	Yarym-Agaev, N. L.; Sinyavskaya, R. P.; Koliushko, I.L.; Levinton, L. Ya.; *Zh. Prikl. Khim.*, <u>1985</u>, *58(1)*, 165-8.

VARIABLES:
   Temperature, pressure
	PREPARED BY:
	   C. L. Young

EXPERIMENTAL VALUES:

T/K	P/MPa	Mole fraction of methane in liquid	in vapour
298.2	2.5	0.000599	0.99746
	5.0	0.00112	0.99854
	7.5	0.00146	0.999066
	10.0	0.00190	0.999180
	12.5	0.00221	0.999416
313.2	2.5	0.000490	0.99697
	5.0	0.000929	0.99813
	7.5	0.00127	0.99866
	10.0	0.00164	0.99888
	12.5	0.00187	0.999074
338.2	2.5	0.000405	0.99017
	5.0	0.000771	0.99391
	7.5	0.00110	0.99552
	10.0	0.00136	0.99652
	12.5	0.00162	0.99702

AUXILIARY INFORMATION

METHOD/APPARATUS/PROCEDURE:	SOURCE AND PURITY OF MATERIALS:
Flow method: dry methane passed through a series of six saturators containing water, each fitted with a diffuser. gas then passed through a demister fitted with packed gauze. Flow rate of methane was about 200 cm³ hr. Gas then passed through a heated needle valve to near atmospheric pressure. Samples of gas analysed either GC or by freezing out water and estimating gravi- metrically and estimating methane volumetrically.	1. Purity 99.95 mass per cent. 2. Distilled.

	ESTIMATED ERROR:
	   $\partial T/K = \pm 0.1$; $\partial p/MPa = \pm 0.05$;
	   $\partial x, \partial y = \pm 0.003$ (estimated by compiler)

	REFERENCES:

COMPONENTS:	EVALUATOR:
(1) Methane; CH$_4$; [74-82-8]	Rubin Battino
	Department of Chemistry
(2) Water-d$_2$; D$_2$O; [7789-20-0]	Wright State University
	Dayton, OH 45435 USA
	June 1984

CRITICAL EVALUATION:

The Solubility of Methane in Water-d$_2$. [CH$_4$ + D$_2$O].

Three laboratories report the solubility of methane in water-d$_2$. Two of the studies are at a methane partial pressure of about atmospheric over the 278 - 318 K temperature range (1, 2), and the third paper (3) covers the 278 - 518 K temperature interval at partial pressures up to 3.5 MPa. All three laboratories present data of similar precision.

The data have been treated to obtain three smoothing equations. The first equation is from a linear regression of the mole fraction solubility at a partial pressure of 0.101325 MPa over the 278 - 318 K temperature range from papers 1 and 2.

$$\ln x_1 = -55.6499 + 73.9115/\tau + 18.6065 \ln \tau \qquad (1)$$

where $\tau = T/100$ K. The equation has a standard deviation of 0.013 in $\ln x_1$ or about 1.0 per cent in x_1 at the middle of the temperature interval.

The second equation is the equation used to construct the table of tentative solubility values. For this equation the pressure dependent data of Crovetto et $al.$(3) were extraplolated to a methane partial pressure of 0.101325 MPa and combined with the atmospheric solubility data from papers 1 and 2 to obtain the equation

$$\ln x_1 = -120.894 + 164.161/\tau + 67.1928 \ln \tau -6.6750 \tau \qquad (2)$$

for the 278 - 518 K temperature interval for a methane partial pressure of 0.101325 MPa. The standard deviation in $\ln x_1$ is 0.017 which is about 1.7 per cent in x_1 at the middle of the temperature range.

The third equation is from a linear regression of the mole fraction methane solubility as a function of temperature and pressure from paper 3.

$$\ln x_1 = -55.6332 + 75.0373/\tau + 20.3725 \ln \tau + 0.848085 \ln (p/\text{MPa}) \quad (3)$$

where p is the methane fugacity and $\tau = T/100$ K. The standard deviation in $\ln x_1$ is 0.029 which is about 0.9 per cent in x_1 at the middle of the temperature range. The equation is for a temperature range of 278 - 518 K and a pressure (methane fugacity) up to 3.5 MPa.

Equation 2 is the evaluator's choice to represent the mole fraction solubility of methane in water-d$_2$ (heavy water) for the 278 - 518 K temperature range at a methane partial pressure of 0.101325 MPa. Values of

COMPONENTS:	EVALUATOR:
(1) Methane; CH$_4$; [74-82-8] (2) Water-d$_2$; D$_2$O; [7789-20-0]	Rubin Battino Department of Chemistr Wright State University Dayton, OH 45435 USA June 1984

CRITICAL EVALUATION:

the solubility and thermodynamic changes based on Equation 2 are given

below in Table 1.

Table 1.   The solubility of methane in water-d$_2$ at 25 K intervals over the
the 273.15 to 573.15 K temperature interval at a methane partial
pressure of 0.101325 MPa.

Thermodynamic changes for the transfer of one mole of methane
from the gas pahse at a partial pressure of 0.101325 MPa to the
infinitely dilute solution.

T/K	Mol Fraction $10^5 x_1$	Ostwald Coefficient $10^2 L/cm^3 cm^{-3}$	$\Delta \bar{G}_1^0$/ kJ mol^{-1}	$\Delta \bar{H}_1^0$/ kJ mol^{-1}	$\Delta \bar{S}_1^0$/ J K^{-1} mol^{-1}	$\Delta \bar{C}_{p\,1}^0$/ J K^{-1}mol^{-1}
273.15	5.170	5.789	22.415	−21.576	−161.1	282.7
298.15	2.636	3.213	26.137	−14.823	−137.4	257.5
323.15	1.824	2.388	29.317	− 8.702	−117.7	232.2
348.15	1.555	2.164	32.047	− 3.212	−101.3	206.0
373.15	1.527	2.239	34.405	+ 1.646	− 87.8	181.7
398.15	1.648	2.526	36.458	5.873	− 76.8	156.5
423.15	1.890	3.007	38.264	9.468	− 68.1	131.2
448.15	2.250	3.686	39.877	12.433	− 61.2	105.9
473.15	2.729	4.570	41.341	14.766	− 56.2	80.7
498.15	3.333	5.654	42.698	16.467	− 52.7	55.4
523.15	4.057	6.904	43.985	17.537	− 50.6	30.2
548.15	4.890	8.246	45.236	17.976	− 49.7	4.9
573.15	5.806	9.552	46.482	17.783	− 50.1	−20.3

The Ostwald coefficient is defined as volume of gas per volume of
pure solvent.

There is a minimum in the mole fraction solubility near a temperature
of 373.15 K.

REFERENCES.

1. Ben-Naim, A.; Wilf, J.; Yaacobi, M. *J. Phys. Chem.* 1973, *77*, 95-102.

2. Cosgrove, B. A.; Walkley, J. *J. Chromatogr.* 1981, *216*, 161-7.

3. Crovetto, R.; Fernandez-Prini, R.; Japas, M. L.
   J. Chem. Phys. 1982, *76*, 1077-86.

COMPONENTS:	ORIGINAL MEASUREMENTS:
(1) Methane; CH_4; [74-82-8] (2) Water-d$_2$; D_2O; [7789-20-0]	Ben-Naim, A.; Wilf, J.; Jaacobi, M. *J. Phys. Chem.* <u>1973</u>, *77*, 95 102.

VARIABLES:	PREPARED BY:
$T/K = 278.15 - 298.15$ $P_1/kPa = 101.325$	H. L. Clever

EXPERIMENTAL VALUES:

Temperature		Mol Fraction	Ostwald Coefficient
$t/°C$	T/K	$10^5 x_1$	$L/cm^3 \ cm^{-3}$
5	278.15	4.36	0.0549
10	283.15	3.79	0.0485
15	288.15	3.33	0.0434
20	293.15	2.97	0.0393
25	298.15	2.68	0.0360

The compiler added the Kelvin temperatures.

The compiler calculated the mole fraction solubility values at 1 atm (101.325 kPa) partial pressure using real methane gas molar volumes. The use of real molar volumes inplace of ideal volumes increases the mole fraction solubility by about 0.22 percent.

AUXILIARY INFORMATION

METHOD/APPARATUS/PROCEDURE:	SOURCE AND PURITY OF MATERIALS:
The method of Ben-Naim and Baer (ref 1) was used. The apparatus was modified by the addition of Teflon stopcocks. The degassed solvent in a volumetric container is forced by a stirrer created vortex up side arms and through tubes containing solvent vapor saturated gas. The gas uptake is measured on a buret at constant pressure.	(1) Methane. Matheson Co., Inc. Purity 99.97 percent. (2) Water-d$_2$. Darmstadt. Purity 99.75 percent. Used as received.

ESTIMATED ERROR:
$\delta L/L = \pm 0.005$ (compiler)

REFERENCES:
1. Ben-Naim, A.; Baer, S. *Trans. Faraday Soc.* <u>1963</u>, *59*, 2735.

M—C

COMPONENTS:	ORIGINAL MEASUREMENTS:
(1) Methane; CH_4; [74-82-8] (2) Water-d$_2$; D_2O; [7789-20-0]	Cosgrove, B. A.; Walkley, J. *J. Chromatogr.* <u>1981</u>, *216*, 161 - 7.

VARIABLES:	PREPARED BY:
T/K = 278.15 - 318.15 p_1/kPa = 101.325	H. L. Clever

EXPERIMENTAL VALUES:

Temperature T/K	Mol Fraction $10^4 x_1$
278.15	0.4371
283.15	0.3732
288.15	----
293.15	0.3006
298.15	0.2655
303.15	0.2382
308.15	0.2172
313.15	0.2080
318.15	0.1854

AUXILIARY INFORMATION

METHOD/APPARATUS/PROCEDURE:	SOURCE AND PURITY OF MATERIALS:
A 20 ml volume of degassed solvent (sublimation technique) is transferred to a previously evacuated (10^{-4} mmHg) saturation cell immersed in an insulated controlled (± 0.01 K) water bath. The gas is dispersed through the constantly stirred solution at 1 atm by a course, fritted glass disc. Saturation is obtained within a few hours. Prior to analysis the solution is allowed to sit under one atm gas pressure for one hour.	(1) Methane. No information. (2) Water-d$_2$. No information.

	ESTIMATED ERROR:
A saturated sample is withdrawn from the saturation cell using a greaseless, gas tight (2.500 ± 0.001 ml) Gilmont syringe. A 0.250 ml sample is injected to "wet" the frit. It is stripped and then four 0.500 ml samples are injected sequentially into the cell. The stripped gas is dried before entering the column. The gas is analyzed on a dual filament conductivity detector. Calibrations with pure gas are made dry before and wet after each series of runs.	$\delta x_1/x_1$ = ± 0.015 (compiler) REFERENCES:

COMPONENTS:	ORIGINAL MEASUREMENTS:
(1) Methane; CH_4; [74-82-8] (2) Water-d$_2$ or deuterium oxide; D_2O; [7789-20-0]	Crovetto, R.; Fernández-Prini, R.; Japas, M. L. *J. Chem. Phys.* <u>1982</u>, *76*, 1077-86.

VARIABLES:	PREPARED BY:
T/K = 298.2 - 517.5 p/MPa = 1.773 - 7.220	R. Fernández-Prini

EXPERIMENTAL VALUES:

T/K	Total Pressure p/MPa	Methane Volume Fraction, y_1	Methane Fugacity f_1/MPa	Mol Fraction $10^4 x_1$	$\ln(H$/GPa$)$ [1]
298.2	1.852	0.9984	1.789	4.515	1.377
323.2	1.803	0.9935	1.749	3.187	1.702
355.2	1.773	0.9716	1.696	2.565	1.885
393.3	2.062	0.9021	1.843	2.983	1.821
446.1	2.504	0.6479	1.639	3.562	1.527
474.4	4.115	0.5852	2.473	6.740	1.300
479.9	4.465	0.5704	2.625	7.374	1.270
517.5	7.220	0.4510	3.485	13.440	0.953

[1] Henry's constant, H/GPa = $(f_1$/GPa$)/x_1$.

$$\ln(H\text{/GPa}) = -9.062 + 8.148/(T/1000 \text{ K}) - 1.5808/(T/1000 \text{ K})^2$$
$$+ 0.0236/(T/1000 \text{ K})^3 \qquad (\sigma = 0.010)$$

Thermodynamic quantities for the process CH_4(g, 0.1 MPa, T) → CH_4(l, x_1 = 1, T) are below:

T/K	ΔG_1°/kJ mol^{-1}	ΔH_1°/kJ mol^{-1}	ΔS_1°/J (K mol)$^{-1}$	ΔC_{p1}°/J (K mol)$^{-1}$
298.2	26.20	-13.81	134.3	251
400.0	36.56	5.71	77.0	146
520.0	43.77	19.37	46.9	89

AUXILIARY INFORMATION

METHOD/APPARATUS/PROCEDURE:	SOURCE AND PURITY OF MATERIALS:
The method involved the equilibration of the gas with the liquid and the determination of the gas mole fraction by sampling the equilibrated liquid phase. Henry's constant was obtained for each temperature by employing second virial coefficients for pure components and mixture in order to correct for non-ideal behavior in the gas phase. The gas was equilibrated in a thermostated stainless steel vessel which was continuously rocked. Weighed samples of the liquid phase were withdrawn and the amount of dissolved gas determined with a gas buret. The system was then taken to a new temperature. Pressures were measured with calibrated Bourdon gauges.	(1) Methane. Matheson (UHP) 99.97 mol %. (2) Water-d$_2$. CNEA, 99.8 mol%

ESTIMATED ERROR:	$\delta T/K = \pm 0.2$
(Authors')	$\delta p/p = \pm 0.003$
	$\delta x_1/x_1 = \pm 0.005$
(T/K$\leqslant$520)	$\delta H/H = \pm 0.01 - 0.02$

REFERENCES:

COMPONENTS:	EVALUATOR:

1. Methane; CH_4; [74-82-8]
 	
 2. Sea Water
	Denis A. Wiesenburg
	 Biological & Chemical Oceanography
	  Branch - Oceanography Division
	 Naval Ocean Research & Development
	  Activity
	 NSTL Station, Mississippi 39529 U.S.A.
	 April 1983

CRITICAL EVALUATION:

There are three reports of the solubility of methane in sea water (1-3), but only one of these has an extensive amount of data. Yamamoto et al. (1) report 201 solubility measurements for distilled water and five salinities (27.738, 33.461, 33.515, 33.629, 39.379 °/oo) for many different temperatures between 273.91 and 303.16 K. Their solubility measurements are estimated to have an accuracy of 0.5%. Atkinson and Richards (2) reported seven calculated solubilities for sea water of 36°/oo salinity based on 12 unreported measurements of methane solubility in sea water having a salinity of 40°/oo. Their measurement technique was not reported. They determined a linear relationship for their data with temperature and used Winkler's (4) values for the solubility of methane in distilled water to calculate the Bunsen coefficients for 36°/oo salinity, assuming solubility to be a linear function of salinity between 0 and 40°/oo. Their calculated data in the temperature range 273.15 - 303.15 K was from 6% greater to 5% less than the results of Yamamoto et al. (1). Yamamoto et al. (1) reported that R. F. Weiss made one measurement of methane solubility at 288.24 K and salinity 36.425°/oo using the Scholander microgasometric technique (5). The value he obtained (α = 0.03023) agreed within 0.7% with the measurements of Yamamoto et al. (1) at that temperature and salinity.

Stoessell and Byrne (3) reported three averaged methane solubility measurements at 298.15 K for sea water with a salinity of 34.84°/oo at high methane pressures (2410, 3790, 5170 kPa). Their values, when converted to Bunsen coefficients at 1 atm, showed up to +10% variation with those of Yamamoto et al. (1) and were not included in the smoothing equation.

The experimental results of Yamamoto et al. (1) are considered to be of sufficient reliability to use in a smoothing equation. This decision was based not only on the quality control during their experiments and self-consistency of the measurements, but also upon the fact that these investigators had measured hydrogen solubility (6) using the same methods and equipment and obtained good agreement with other published results (7). Their methane data (1) have been fitted (8) by the method of least squares to an equation developed by Weiss (9) which expresses solubility as the natural logarithm of the Bunsen coefficient, α, and is consistent with both the integrated form of the van't Hoff equation and the Setchenow salt effect relation. The equation for methane is valid from 273.15 to 303.15 K and a salinity range, S, of 0 to 40°/oo. The smooth equation reproduced the combined hydrogen data with a root-mean-square deviation of 1.8 x 10^{-4} units ($\sim$0.39%). The equation is

$$\ln \alpha = -68.8862 + 101.4956 \ (100K/T) + 28.7314 \ \ln \ (T/100 \ K)$$

$$+ \ S \ [-0.076146 + 0.043970 \ (T/100 \ K) - 0.0068672 \ (T/100K)^2]$$

where S is the salinity in parts per thousand. Wiesenburg and Guinasso (8) give an extensive table of methane Bunsen coefficients calculated from the above equation.

Although the Bunsen solubility coefficients are well defined by the above equation, for practical purposes, oceanographers require the atmospheric equilibrium solubility values in their work. Weiss (9) has proposed an equation similar to the above which expresses the atmospheric equilibrium solubility from moist air at 1 atm total pressure, in units of volume (STP) dm^{-3}, as a function of salinity and temperature. In working with samples from the depths of the ocean, it is also advantageous to express atmospheric solubilities in terms of mol kg^{-1}, which are pressure and temperature independent (9,10). Weiss' atmospheric solubility equation is based on the assumption of a constant atmospheric concentration of methane. Since methane is variable in the atmosphere, Weiss' (9) equation has been modified (8) to include the atmospheric concentration

COMPONENTS:	EVALUATOR:

1. Methane; CH_4; [74-82-8]
 	
 2. Sea Water
	Denis A. Wiesenburg
	 Biological & Chemical Oceanography
	  Branch - Oceanography Division
	 Naval Ocean Research & Development
	  Activity
	 NSTL Station, Mississippi 39529 U.S.A.
	 April 1983

CRITICAL EVALUATION:

as a variable. The data set for methane (1) has been fitted (8) to the equations

$$\ln c_1/\text{nl dm}^{-3} = f_g - 412.1710 + 596.8104 \ (100/T)$$
$$+ 379.2599 \ \ln (T/100) - 62.0757 \ (T/100)$$
$$+ S[-0.059160 + 0.032174 \ (T/100) - 0.0048198(T/100)^2]$$

$$\ln m_1/\text{nmol kg}^{-1} = f_g - 417.5053 + 599.8626 \ (100/T)$$
$$+ 380.3636 \ \ln (T/100) - 62.0764 \ (T/100)$$
$$+ S[-0.064236 + 0.034980 \ (T/100) - 0.0052732 \ (T/100)^2]$$

where f_g is the mole fraction of methane in dry air. In these calculations methane was assumed to be an ideal gas. Vapor pressure for pure water was calculated using the equation of Bridgemen and Aldrich (11) and corrected for salinity effects using the expression of Robinson (12). Knudsen's (13) formula was used to calculate densities. These two equations can be used to calculate the atmospheric equilibrium solubility of methane under any given conditions of temperature, salinity, and atmospheric concentration. Using an atmospheric methane mole fraction of 1.41×10^{-6} (14) the equations reproduce the individual calculated atmospheric solubilities with a root-mean-square deviation of 0.48%.

References

1.   Yamamoto, S.; Alcauskas, J. B.; Crozier, T. E. *J. Chem. Eng. Data* 1976, *21*, 78.
2.   Atkinson, L. P.; Richards, F. A. *Deep-Sea Res.* 1967, *14*, 673.
3.   Stoessell, R. K.; Byrne, P. A. *Clays Clay Miner.* 1982, *30*, 67.
4.   Winkler, L. W. *Ber. dt. chem. Ges.* 1891, *24*, 3602.
5.   Scholander, P. F. *J. Biol. Chem.* 1947, *167*, 235.
6.   Crozier, T. E.; Yamamoto, S. J. *J. Chem. Eng. Data* 1974, *19*, 242.
7.   Gordon, L. I.; Cohen, Y.; Standley, D. R. *Deep Sea Res.* 1977, *24*, 937.
8.   Wiesenburg, D. A.; Guinasso, H. L., Jr. *J. Chem. Eng. Data* 1979, *24*, 356.
9.   Weiss, R. F. *Deep-Sea Res.* 1970, *17*, 721.
10.   Kester, D. in "Chemical Oceanography" v. 1 2nd Edition, J. P. Riley and G. Skirrow, eds. Academic Press, New York, 1975, pp. 497-556.
11.   Bridgeman, O. C.; Aldrich, E. W. *J. Heat Transfer* 1964, *86*, 279.
12.   Robinson, R. A. *J. Mar. Biol. Assoc. U.K.* 1954, *33*, 449.
13.   Knudsen, M. Hydrographical Tables, G. E. Gad, Copenhagen, 1901.
14.   Prabhakara, C.; Dalu, G.; Kunde, V. G. *J. Geophys. Res.* 1974, *79*, 1744.

COMPONENTS:	ORIGINAL MEASUREMENTS:
1. Methane; CH_4; [74-82-8]	Yamamoto, S.; Alcauskas, J. B.;
2. Sea Water	Crozier, T. E. *J. Chem. Eng. Data*
	1976, *21*, 78-80.

EXPERIMENTAL VALUES:

Salinity ‰

0.0		27.738		33.461	
Temp/K	Bunsen Coefficient α	Temp/K	Bunsen Coefficient α	Temp/K	Bunsen Coefficient α
273.91	0.05592	275.22	0.04431	274.75	0.04305
273.93	0.05612	275.23	0.04433	274.75	0.04299
273.94	0.05613	275.23	0.04418	274.75	0.04294
273.95	0.05600	275.24	0.04412	274.75	0.04304
273.95	0.05591	275.24	0.04410	274.80	0.04282
273.96	0.05604	275.24	0.04410	274.85	0.04301
278.07	0.04957	275.30	0.04420	274.85	0.04302
278.09	0.04963	275.55	0.04385	274.85	0.04298
278.10	0.04973	275.63	0.04394	274.88	0.04291
278.11	0.04963	282.60	0.03650	274.94	0.04312
278.12	0.04968	282.60	0.03657	274.94	0.04278
284.05	0.04235	282.60	0.03637	274.95	0.04287
284.08	0.04216	282.60	0.03626		
284.08	0.04242	282.60	0.03659		
284.09	0.04224	282.60	0.03660		
284.09	0.04220	282.60	0.03648		
284.09	0.04238	282.60	0.03669		
284.10	0.04224	282.61	0.03670		
284.10	0.04226	282.68	0.03631		
284.11	0.04252	282.69	0.03611		
291.14	0.03606	282.70	0.03640		
291.15	0.03603	282.70	0.03667		
291.16	0.03619	282.70	0.03649		
291.17	0.03607	282.71	0.03653		
291.17	0.03606	282.90	0.03620		
297.25	0.03196	288.15	0.03238		
297.26	0.03192	288.15	0.03213		
297.30	0.03189	288.15	0.03212		
297.31	0.03211	288.15	0.03204		
297.32	0.03200	288.25	0.03218		
302.67	0.02915	288.25	0.03213		
302.69	0.02913	288.25	0.03236		
302.69	0.02915	288.25	0.03200		
302.69	0.02911	288.25	0.03205		
302.70	0.02901	293.15	0.02918		
		293.15	0.02916		
		293.15	0.02918		
		293.20	0.02901		
		293.35	0.02905		
		293.36	0.02919		
		294.55	0.02902		
		294.55	0.02895		
		294.85	0.02902		
		298.30	0.02664		
		298.32	0.02645		
		298.35	0.02641		
		298.35	0.02632		
		298.40	0.02644		
		298.40	0.02621		
		298.45	0.02640		
		298.45	0.02646		
		298.50	0.02659		
		298.55	0.02658		
		298.65	0.02609		
		298.75	0.02636		

Continued on next page.

COMPONENTS:	ORIGINAL MEASUREMENTS:
1. Methane; CH_4; [74-82-8] 2. Sea Water	Yamamoto, S.; Alcauskas, J. B.; Crozier, T. E. *J. Chem. Eng. Data* <u>1976</u>, *21*, 78-80.

EXPERIMENTAL VALUES:

Salinity ‰

33.515		33.629		39.379	
Temp/K	Bunsen Coefficient α	Temp/K	Bunsen Coefficient α	Temp/K	Bunsen Coefficient α
280.00	0.03706	273.88	0.04409	283.45	0.03287
280.40	0.03705	273.93	0.04388	283.50	0.03306
280.48	0.03687	273.94	0.04396	283.50	0.03261
280.50	0.03712	273.96	0.04396	283.50	0.03307
280.50	0.03680	273.96	0.04389	283.50	0.03310
280.50	0.03682	279.80	0.03742	283.50	0.03271
280.55	0.03692	279.80	0.03771	283.50	0.03291
280.57	0.03682	279.81	0.03767	283.50	0.03315
280.57	0.03690	279.89	0.03700	287.96	0.02987
280.57	0.03681	279.90	0.03739	288.00	0.02992
280.57	0.03707	291.75	0.02902	288.14	0.02953
280.60	0.03680	291.75	0.02893	288.24	0.02996
280.60	0.03693	291.77	0.02899	288.24	0.02974
280.62	0.03677	291.80	0.02889	288.24	0.02981
280.65	0.03698	291.80	0.02898	288.24	0.02934
280.75	0.03687	291.81	0.02886	288.24	0.02973
280.79	0.03704	291.85	0.02897	293.15	0.02684
286.63	0.03218	291.85	0.02895	293.15	0.02698
286.63	0.03214	291.85	0.02880	293.20	0.02696
286.63	0.03208	291.85	0.02893	293.24	0.02673
286.68	0.03214	297.95	0.02584	293.25	0.02699
286.68	0.03207	297.95	0.02582	293.95	0.02685
286.68	0.03210	297.98	0.02580	297.85	0.02489
286.68	0.03216	298.00	0.02578	298.05	0.02478
286.69	0.03210	298.00	0.02563	298.05	0.02475
		298.00	0.02578	298.05	0.02479
		298.00	0.02585	298.10	0.02477
		298.00	0.02573	298.10	0.02459
		298.00	0.02579	298.10	0.02472
		298.03	0.02583	303.11	0.02248
		302.61	0.02420	303.11	0.02292
		302.61	0.02412	303.11	0.02288
		302.62	0.02402	303.11	0.02285
		302.64	0.02406	303.11	0.02266
		302.66	0.02409	303.14	0.02292
		302.69	0.02405	303.16	0.02260
		302.71	0.02418	303.16	0.02286

Continued on next page

COMPONENTS:	ORIGINAL MEASUREMENTS:
1. Methane; CH_4; [74-82-8] 2. Sea Water	Yamamoto, S.; Alcauskas, J. B.; Crozier, T. E. *J. Chem. Eng. Data* 1976, *21*, 78-80.

VARIABLES:	PREPARED BY:
T/K: 273.91 ~ 303.16 CH_4 P/kPa: 101.325 (1 atm) Salinity/$^o/_{oo}$: 0 - 39.379	Denis A. Wiesenburg

EXPERIMENTAL VALUES:

See preceeding pages

AUXILIARY INFORMATION

METHOD/APPARATUS/PROCEDURE:

Solubility determinations were made using the Scholander microgasometric technique (1) as modified by Douglas (2). Pure methane and degassed sea water were introduced into a reaction vessel in a constant temperature room. The vessel was shaken vigorously to allow equilibration between the methane and sea water. The amount of gas absorbed and the volume of sea water were measured volumetrically with a microburet. Bunsen solubility coefficients were calculated from the observed volumes.

SOURCE AND PURITY OF MATERIALS:

1.   Methane. Linde Specialty Gas, specified 99.99% purity.

2.   Sea Water. Passed through 0.45-μm millipore filter and poisoned with 1 mg/l of $HgCl_2$. Sea water was boiled or diluted with glass distilled water (used for 0 $^o/_{oo}$) to obtain desired salinities.

ESTIMATED ERROR:

δ T/K = 0.01

δ S/$^o/_{oo}$ = 0.003

REFERENCES:

1.   Scholander, P. F.
      J. Biol. Chem. 1947, *167*, 235-250.
2.   Douglas, E.
      J. Phys. Chem. 1964, *68*, 169-174.
      ibid. 1965, *69*, 2608-2610.

COMPONENTS:	ORIGINAL MEASUREMENTS:
(1) Methane; CH_4;[74-82-8] (2) Sea Water	Stoessell, R. K., Byrne, P. A. *Clays Clay Miner.* 1982, *30*, 67-72.

VARIABLES:	PREPARED BY:
T/K: 298.15 CH4 P/kPa: 2410 - 5170 Salinity/°‰: 34.84	Denis A. Wiesenburg

EXPERIMENTAL VALUES:

Pressure		molality/mol kg^{-1}
p_1/psia	p_1/kPa	
350	2410	0.0263
550	3790	0.0400
750	5170	0.0514

The compiler calculated the pressures
in kPa.

AUXILIARY INFORMATION

METHOD/APPARATUS/PROCEDURE:	SOURCE AND PURITY OF MATERIALS:
Solubility determinations were made using a titanium-lined chamber within a stainless steel reaction vessel jacketed by a water bath for temperature control. System pressure was set by controlling the input and output of methane within the chamber's headspace. The vessel was rocked for 3 h to allow equilibration between the methane and sea water. The amount of gas present in the sea water at equilibrium was measured by subsampling the sea water at system pressure and flash evaporating it into a second calibrated, evacuated expansion volume. The pressure of the released gas was measured with a manometer accurate to 0.7 kPa. The gas volume and pressure change were used to compute the moles of gas released and corrections were made for the small amount of dissolved methane not released during flashing. The molalities were not corrected for the volume of dissolved gas.	(1) Methane. Matheson ultra high purity, minimum 99.97%. (2) Sea Water.
	ESTIMATED ERROR: $\delta T/K = 0.1$ $\delta P/kPa = 7$
	REFERENCES:

M—C*

COMPONENTS:	EVALUATOR:
(1) Methane; CH_4; [74-82-8]	H. Lawrence Clever
	Department of Chemistry
(2) Electrolyte	Emory University
	Atlanta, GA 30322 USA
(3) Water; H_2O; [7732-18-5]	
	1985, March

CRITICAL EVALUATION:

AN EVALUATION OF THE SOLUBILITY OF
METHANE IN AQUEOUS ELECTROLYTE SOLUTIONS.

This section contains an evaluation of the solubility of methane in aqueous electrolyte solutions. Not enough workers have measured the solubility of methane in any one aqueous electrolyte system over common ranges of temperature, pressure and electrolyte concentration to recommend solubility values. A possible exception is the methane + sodium chloride + water system for which there are many papers of reasonably concordant data. For most systems the available data are classed as tentative.

In order to have a common basis for comparison the solubility data have been converted to Sechenov (Setschenow) salt effect parameters as either $k_{scc}/dm^3\ mol^{-1}$ or $k_{smm}/kg\ mol^{-1}$ values. The k_{scc} form is the most common for the older data while the k_{smm} form is more commonly used by present day workers.

Many forms of the Sechenov salt effect parameter are in use. Many of the forms and conversions among them were discussed in *NITROGEN/AIR*, Solubility Series Vol. 10 (ref 1). A briefer summary is on pp xiii-xix of this volume. A form with the gas solubility ratio given as a mole fraction ratio is coming more into use. Many of the conversions among Sechenov salt effect parameter forms require solution density data that are not available especially for systems studied at high temperatures and pressures.

The Sechenov salt effect parameters most commonly used in this evaluation are

$$k_{scc}/dm^3\ mol^{-1} = (1/(c_2/mol\ dm^{-3}))\ log\ ((c_1^0/mol\ dm^{-1})/(c_1/mol\ dm^{-3}))$$

$$k_{smm}/kg\ mol^{-1} = (1/(m_2/mol\ kg^{-1}))\ log\ ((m_1^0/mol\ kg^{-1})/(m_1/mol\ kg^{-1}))$$

$$k_{scx}/dm^3\ mol^{-1} = (1/(c_2/mol\ dm^{-3}))\ log\ (x_1^0/x_1)$$

$$k_{smx}/kg\ mol^{-1} = (1/(m_2/mol\ kg^{-1}))\ log\ (x_1^0/x_1)$$

where subscript 1 is the nonelectrolyte gas, subscript 2 the electrolyte.

The gas solubility ratios c_1^0/c_1, m_1^0/m_1 and x_1^0/x_1 represent the ratio of solubility in pure water to the solubility in the electrolyte solution. The molar (mol dm^{-3}) ratio, c_1^0/c_1, is the same as the Bunsen coefficient ratio, α^0/α, or the Ostwald coefficient ratio, L^0/L; the molal (mol kg^{-1}) ratio, m_1^0/m_1, is the same as the Kuenen coefficient ratio, S^0/S, or the solvomolality ratio, A^0/A; and the mole fraction ratio, x_1^0/x_1, is the same as the inverse of the Henry constant ratio, H_1/H_1^0 when $H/atm = (p_1/atm)/x_1$. The mole fraction is usually calculated by treating each ion of the electrolyte as an entity.

Many of the methane + electrolyte + water systems have been studied as a function of pressure as well as temperature and electrolyte concentration. Although most studies are based on solubility determinations at atmospheric pressure (0.1 MPa) there are measurements to pressures as large as 60 MPa. The studies are not in complete agreement, but it appears as if pressure has little effect on the magnitude of the salt effect parameter at a given temperature. This may be more true of k_{smm} and k_{smx} values than the k_{scc} and k_{scx} values. Several papers (3, 13, 18, 21, 23, 26) report data for either k_{scx} or k_{smx} values as a function of pressure between 400 and 600 K for the methane + sodium chloride + water system. The k_{scx} and k_{smx} values show different trends which we cannot reconcile at present because of lack of auxiliary data at these temperatures and pressures.

Other representations of the salt effect parameter are in use. A few authors prefer to use the natural logarithm instead of the base ten logarithm. Authors who want to compare the salt effect of electrolytes of different charge type use either equivalents or ionic strength for the electrolyte concentration. The ionic strength representations appear to be useful and will be used occasionally in this evaluation.

The salt effect parameters given in the evaluation were either taken from the original papers or calculated by the evaluator; the evaluator used one of two methods, which are nearly equivalent, to calculate salt effect parameters.

(i) Individual values of the salt effect parameter are calculated at each electrolyte concentration from the solubility of the gas in water and solution. If the parameter values appear statistically constant as a function of electrolyte concentration, they are averaged. If the values show a consistent change with electrolyte concentration, they are fit to an equation, usually linear, as a function of electrolyte concentration.

(ii) A graph of $\log (c_1^0/c_1)$ vs. c_2 is prepared. If linear, the slope is k_{scc}. Units of molality or mole fraction can also be used. Sometimes a better linearity is obtained with one set of units than another. If the plot is not linear a more complex function may be used to obtain the slope and k_{scc} values as a function of electrolyte concentration.

In the solubility ratio we prefer to use the author's solubility values in pure water rather than the recommended value (*this volume*, pp 1-5) in the belief that the author's systematic errors may partially cancel in the solubility ratio term. It is important to recognize that the solubility in water is not required if the solubility is known at several electrolyte concentrations. The equation can be rearranged to

$$\log c_1 = \log c_1^0 - k_{scc}\, c_2$$

and the slope of $\log c_1$ vs. c_2 is the negative of k_{scc}. The slope and intercept can be obtained without knowledge of c_1^0. However, we prefer to use the c_1^0 value and consider it especially important when solubility measurements were made only at small electrolyte concentrations (ref 1).

There has been a renewed interest in the solubility of methane in aqueous electrolyte solutions probably because of the possibility of recovery of methane from brines and sea water. Five papers containing extensive data have appeared since 1980. In all there are data on about 25 aqueous electrolyte systems plus other systems containing mixed electrolytes.

The systems are presented below in the order of the standard arrangement for inorganic compounds used in publications of the US National Bureau of Standards. The number before each system is the standard order number for the electrolyte.

10(1) Methane + Hydrochloric acid [7647-01-0] + Water

Muccitelli and Wen (ref 22) measured the solubility of methane in aqueous HCl at five temperatures between 278.15 and 298.15 K and several concentrations of HCl up to 1.28 mol dm^{-3}. The $(k_{scc}/\mathrm{dm^3\ mol^{-1}})$ values from their data are given below. The k_{scc} values decrease with electrolyte

COMPONENTS:	EVALUATOR:
(1) Methane; CH_4; [74-82-8] (2) Electrolyte (3) Water; H_2O; [7732-18-5]	H. Lawrence Clever Department of Chemistry Emory University Atlanta, GA 30322 USA 1985, March

CRITICAL EVALUATION:

concentration at three temperatures and increase at two temperatures. Thus, there is no clear trend and the values were averages at each temperature. Since the salt effect parameters at low electrolyte concentration are more likely to show errors than those at high concentration, the values above 0.7 mol dm^{-3} were weighted twice and the others once in the average. At 293.15 K the 0.074 value was not included in the average. The data are classified as tentative.

c_2/mol dm^{-3}	k_{scc} /dm^3 mol^{-1} at temperatures				
	278.15 K	283.15 K	288.15 K	293.15 K	298.15 K
0.2321	-	-	-	0.074	-
0.2822	0.065	-	-	-	-
0.2824	-	0.048	-	-	-
0.2864	-	-	0.061	-	-
0.4687	-	-	-	0.050	-
0.5330	-	0.048	-	-	-
0.5468	-	-	-	-	0.046
0.5501	0.063	-	-	-	-
0.5645	-	-	0.041	-	-
0.7324	-	-	-	-	0.043
0.7401	-	-	-	0.053	-
0.7979	-	0.051	-	-	-
0.8030	0.060	-	-	-	-
0.8164	-	-	0.033	-	-
0.8645	-	-	0.038	-	-
0.9703	0.056	-	-	-	-
1.0425	-	-	-	-	0.042
1.0914	-	0.053	-	-	-
1.127	-	-	0.044	0.049	-
1.2803	-	-	-	-	0.042
Av	0.060 ±0.004	0.051 ±0.002	0.042 ±0.009	0.051 ±0.002	0.043 ±0.002

14(1)   Methane + Sulfuric acid [7664-93-9] + Water

Christoff (ref 2) reported values from measurements in water and aqueous sulfuric acid solutions at 293.12 K and Radakov and Lutsyk (ref 19) reported values from measurements at 298.2 and 363.2 K. Both sets of data include measurements for large molalities of sulfuric acid. Results are summarized below.

T/K	H_2SO_4 $m_2/\text{mol kg}^{-1}$	Salt Effect Parameter $k_{smc}/\text{kg mol}^{-1}$
293.2	5.69	0.056
	16.4	0.026
	222.0	0.0002_5
298.2	40.8	0.014
	135.0	0.0003_3
363.2	135.0	-0.0039
	190.0	-0.0033
	433.0	-0.0018

The methane is salted out at 293.2 and 298.2 K and salted in at 363.2 K. The data are classed as tentative.

18(1)   Methane + Ammonium chloride [12125-02-9] + Water

Ben-Naim and Yaacobi (ref 15) measured the solubility of methane in water and in 1.0 mol dm^{-3} NH$_4$Cl solution at five temperatures between 283.15 K and 303.15 K. Their values are classed as tentative.

$c_2/\text{mol dm}^{-3}$	$k_{scc}/\text{dm}^3 \text{ mol}^{-1}$ at temperatures of				
	283.15 K	288.15 K	293.15 K	298.15 K	303.15 K
1.0	0.100	0.093	0.089	0.089	0.093

18(2)   Methane + Ammonium bromide [12124-97-9] + Water

Wen and Hung (ref 12) measured the solubility of methane in pure water and in one concentration of NH$_4$Br of about 0.2 M at 10 degree intervals between 278.15 and 308.15 K. Their values are classed as tentative. The smoothed values were given by Wen and Hung.

$m_2/\text{mol kg}^{-1}$	$k_{smm}/\text{kg mol}^{-1}$ at temperatures of			
	278.15 K	288.15 K	298.15 K	308.15 K
0.100 (smoothed)	0.068	0.061	0.054	0.047
0.181	–	–	–	0.044
0.196	0.071	–	–	–
0.200	–	–	0.055	–
0.202	–	0.060	–	–

18(3)   Methane + Tetramethylammonium bromide [64-20-0] + Water

Wen and Hung (ref12) measured the solubility of methane in pure water

COMPONENTS:	EVALUATOR:
(1) Methane; CH_4; [74-82-8] (2) Electrolyte (3) Water; H_2O; [7732-18-5]	H. Lawrence Clever Department of Chemistry Emory University Atlanta, GA 30322 USA 1985, March

CRITICAL EVALUATION:

and in two concentrations of tetramethylammonium bromide at 10 degree intervals between 278.15 and 308.15 K. The values are classed as tentative. Both the values calculated by the evaluator from the published data and the smoothed values given by Wen and Hung are in the table below.

m_2/mol kg^{-1}	k_{smm}/kg mol^{-1} at temperatures of			
	278.15 K	288.15 K	298.15 K	308.15 K
0.100 (smoothed)	0.007	-0.005	-0.017	-0.028
0.176	0.010	-	-	-
0.179	0.000	-	-0.022	-
0.190	-	-0.005	-	-
0.193	-	-	-0.009	-
0.194	-	-0.012	-	-
0.302	-	-	-	-0.032

18(4) Methane + Tetraethylammonium bromide [71-91-0] + H_2O

Both Wen and Hung (ref 12) and Blanco C and Smith (ref 17) have made measurements on the system. Wen and Hung's measurements were made at atmospheric pressure (0.1 MPa) while Blanco C and Smith worked between 10.1 and 50.7 MPa. It appears that pressure has relatively little effect on the salt effect parameter. The results from the two laboratories show similar order of magnitude. Although the results differ some, they are both classed as tentative. The Wen and Hung smoothed value at 298 K may be an error or a misprint. A value near -0.079 would fit the pattern better.

m_2/mol kg^{-1}	p/MPa	k_{smm}/kg mol^{-1} at temperatures of						
		278.15K	288.15K	298.15K	308.15K	311.2K	324.7K	344.2K
0.093	0.1	-0.039	-	-	-	-	-	-
0.095	0.1	-0.045	-	-	-	-	-	-
0.100 (smoothed)	0.1	-0.036	-0.049	-0.049 [sic]	-0.094	-	-	-
0.173	0.1	-	-0.052	-	-	-	-	-
0.182	0.1	-	-	-0.075	-	-	-	-
0.183	0.1	-	-	-0.074	-	-	-	-
0.209	0.1	-	-	-	-0.090	-	-	-
1.0	10.1	-	-	-0.056	-	-0.046	-0.056	-0.085
	20.3	-	-	-0.055	-	-0.067	-0.077	-0.105
	30.4	-	-	-0.044	-	-0.056	-0.073	-0.105
	40.5	-	-	-	-	-0.047	-0.070	-0.102
	50.7	-	-	-	-	-0.054	-0.077	-0.104
Blanco C and Smith (av)	-	-	-	-0.052	-	-0.054	-0.071	-0.100

18(5)   Methane + Tetrapropylammonium bromide [1941-30-6] + Water

Wen and Hung (ref12) have made extensive measurements of the solubility of methane in pure water and in tetrapropylammonium bromide solutions. The salt effect parameters show two trends. Salting in (1) increases as temperature increases and (2) decreases as the electrolyte molality increases. At 278 K methane is salted out at concentrations above 0.4 molal electrolyte. The results are classed as tentative.

m_2/mol kg^{-1}	k_{smm}/kg mol^{-1} at temperatures of			
	278.15 K	288.15 K	298.15 K	305.15 K
0.097	-	-	-0.074	-
0.098	-	-0.058	-	-
0.099	-0.051	-	-	-
0.100 (smoothed)	-0.045	-0.061	-0.082	-0.110
0.100	-	-	-	-
0.102	-	-	-0.086	-
0.103	-0.047	-	-	-
0.105	-	-	-	-0.108
0.223	-0.038	-0.059	-	-
0.224	-0.032	-	-	-
0.227	-	-0.049	-	-
0.230	-	-	-0.064	-
0.235	-	-	-0.071	-
0.245	-	-	-	-0.089
0.410	-	-	-0.059	-
0.415	-	-0.030	-	-
0.431	-0.0024	-	-	-
0.443	-	-	-	-0.079
0.587	+0.011	-	-	-
0.620	-	-0.026	-	-
0.632	-	-	-0.048	-
0.706	-	-	-	-0.080

18(6)   Methane + Tetrabutylammonium bromide [1643-19-2] + Water

Wen and Hung (ref12) measured the solubility of methane in water and 0.1 to 1.0 molal aqueous solutions of tetrabutylammonium bromide at 10 degree intervals from 278.15 to 308.15 K. The author's smoothed values for 0.1 molal solution are listed along with the individual values of the salt effect parameters.

Feillolay and Lucas (ref 14) measured the solubility in water and 1 to 4 molal aqueous solutions. Their salt effect parameters are also in the table.

The two data sets overlap for 1 molal solutions. Agreement between the salt effect parameters in one molal solution is within 10 percent at 298 K and within 2 percent at 308 K. However, the two data sets show opposite effects in that Wen and Hung's results show a decrease in salting in but Feillolay and Lucas show an increase in salting in as the electrolyte molality increases.

COMPONENTS:	EVALUATOR:
(1) Methane; CH_4; [74-82-8] (2) Electrolyte (3) Water; H_2O; [7732-18-5]	H. Lawrence Clever Department of Chemistry Emory University Atlanta, GA 30322 USA 1985, March

CRITICAL EVALUATION:

m_2/mol kg^{-1}	k_{smm}/kg mol^{-1} at temperatures of			
	278.15 K	288.15 K	298.15 K	308.15 K
0.096	-0.031	-0.043	-0.097	-
0.098	-0.033	-	-	-
0.099	-	-	-0.090	-0.150
0.100 (smoothed)	-0.030	-0.053	-0.096	-0.152
0.100	-0.028	-	-	-
0.102	-	-0.052	-	-
0.103	-	-0.051	-	-
0.185	-	-0.037	-	-
0.187	-0.019	-	-	-
0.192	-	-	-	-0.140
0.194	-	-	-0.068	-
0.201	-0.017	-	-	-
0.403	-	-0.026	-	-
0.409	+0.005	-	-	-
0.415	-	-	-0.063	-0.110
0.523	-	-0.019	-	-
0.526	0.010	-	-0.062	-
0.537	-	-	-	-0.105
0.693	-	-	-	-0.106
0.703	-	-0.019	-	-
0.704	-	-	-0.060	-
0.785	0.032	-	-	-
0.990	0.031	-	-	-
0.993	-	-	-	-0.104
1.005	-	-	-	-0.101
1.010	-	-	-0.072	-
1.018	-	-0.017	-	-
1.020	-	-	-0.078	-
1.022	-	-	-0.067	-
1.025	-	-	-	-0.103
1.981	-	-	-0.086	-
1.991	-	-'	-0.087	-
2.075	-	-	-	-0.116
2.078	-	-	-	-0.113
3.610	-	-	-	-0.128
3.623	-	-	-0.115	-
3.640	-	-	-	-0.129
3.925	-	-	-0.111	-
4.010	-	-	-	-0.128

18(7)   Methane + Tetrahydroxyethylammonium bromide [4328-04-5] + Water

Wen and Hung (ref12) have measured the solubility of methane in pure
water and in 0.1 to 0.5 molal aqueous solutions of tetrahydroxyethylammonium

bromide at 10 degree intervals between 278.15 and 308.15 K. Wen and Hung's
smoothed values for a 0.1 molal solution are included in the table below
along with the individual values of the salt effect parameter. Methane is
salted out. The salting out decreases as the temperature increases and a
small salting in is apparent at 308.15 K. There is little change in the
salt effect parameter with electrolyte concentration. The data are classed
as tentative.

m_2/mol kg^{-1}	k_{smm}/kg mol^{-1} at temperatures of			
	278.15 K	288.15 K	298.15 K	308.15 K
0.085	–	–	0.025	0.002
0.086	0.039	–	–	–
0.091	–	0.034	–	–
0.100 (smoothed)	0.042	0.033	0.022	-0.001
0.152	0.045	–	–	–
0.155	0.035	–	–	–
0.167	–	–	0.028	–
0.173	0.052	–	0.025	-0.008
0.174	–	0.030	–	–
0.177	–	0.034	–	–
0.339	–	–	–	-0.001
0.341	0.042	–	–	–
0.347	–	–	0.022	–
0.355	–	0.037	–	–
0.488	0.047	–	–	–
0.508	–	–	–	0.000
0.510	–	–	0.026	–
0.517	–	0.036	–	–

18(8) Methane + Triethylenediamine hydrochloride [2099-72-1] + Water

Muccitelli and Wen (ref 22) have measured the solubility of methane
in pure water and in aqueous solutions of 0.1 to 0.9 molar triethylenediamine hydrochloride at five degree intervals between 278.15 and 298.15 K.
The triethylenediamine hydrochloride salts out while triethylenediamine
(see data sheet) salts in. The authors did not report salt effect parameters. The evaluator calculated the values below. In general the salting
out decreases as temperature increases and as the electrolyte concentration
increases. The solubility data at 293.15 K appears to be out of line with
the data at other temperatures. All of these solutions were adjusted to a
pH of 5.40 ± 0.01 by addition of constant boiling HCl. The pure water
values were used to calculate the salt effect parameters. The values are
classed as tentative.

COMPONENTS:	EVALUATOR:
(1) Methane; CH_4; [74-82-8] (2) Electrolyte (3) Water; H_2O; [7732-18-5]	H. Lawrence Clever Department of Chemistry Emory University Atlanta, GA 30322 USA 1985, March

CRITICAL EVALUATION:

c_2 mol dm^{-3}	k_{scc}/dm^3 mol^{-1} at temperatures of				
	278.15	283.15	288.15	293.15	298.15
0.1225	0.108	-	-	-	-
0.1367	-	-	-	0.132	-
0.1469	-	-	-	-	0.014
0.1560	-	0.101	-	-	-
0.1601	-	-	0.101	-	-
0.2312	0.107	-	-	-	-
0.2340	-	-	-	0.107	-0.057
0.2538	-	0.101	-	-	-
0.2799	-	-	0.085	-	-
0.5267	-	-	-	0.086	0.014
0.5346	-	0.092	-	-	-
0.6291	0.097	-	-	-	-
0.6299	-	-	0.062	-	-
0.8029	-	-	0.073	-	-
0.8120	-	0.081	-	-	-
0.8510	-	-	-	0.084	0.039
0.9023	0.087	-	-	-	-

18(9)   Methane + Guanidine hydrochloride [50-01-1] + Water

    Wetlaufer, Malik, Stoller and Coffin (ref10) measured the solubility
of methane in water and in 4.86 mol dm^{-3} guanidine hydrochloride at three
temperatures. Values of the salt effect parameter as k_{scc} and k_{scx} are
given below. Salting out decreases as the temperature increases. The
data are classed as tentative.

T/K	c_2/mol dm^{-3}	k_{scc}/dm^3 mol^{-1}	k_{scx}/dm^3 mol^{-1}
278.2	4.86	0.058	0.035
298.2	4.86	0.034	0.008
318.2	4.86	0.025	-0.001

93(1)   Methane + Magnesium chloride [7786-30-3] + Water

    Stoessell and Byrne (ref 24) measured the solubility of methane in
water and in 0.5 to 2.16 molal $MgCl_2$ solution at 288.15 K at pressures of
2.41, 3.79 and 5.17 MPa. The salt effect parameter shows no definitive
trend with changes in either molality or pressure at this temperature.

    The authors calculated the salt effect parameter with respect to
electrolyte ionic strength, k_{sIm}/kg mol^{-1}, and the evaluator has

recalculated the values as k_{smm}/kg mol^{-1}. Both sets of values, which should differ by a factor of three, are given below. The authors recommend a value of k_{sIm} = 0.063. The values are classed as tentative.

T/K	m_2/mol kg^{-1}	k_{sIm}/kg mol^{-1} at pressures of		
		2.41 MPa	3.79 MPa	5.17 MPa
298.15	0.5	0.063	0.073	0.070
	1.0	0.068	0.069	0.066
	2.16	0.061	0.064	0.064
			Average 0.066 ± 0.004	

		k_{smm}/kg mol^{-1} at pressures of		
		2.41 MPa	3.79 MPa	5.17 MPa
298.15	0.5	0.188	0.219	0.210
	1.0	0.204	0.207	0.198
	2.16	0.183	0.192	0.192
			Average 0.199 ± 0.012	

13(2)   Methane + Magnesium sulfate [7785-87-7] + Water

Stoessel and Byrne (ref 24) measured the solubilities of methane in water and in 0.5 to 1.5 molal aqueous MgSO$_4$ solutions at 298.15 K and pressures of 2.41, 3.79 and 5.17 MPa. The salt effect parameter shows no definitive trend with changes in either molality or pressure at this temperature.

The authors report the salt effect parameter as a function of ionic strength, k_{sIm}/kg mol^{-1}, with the ionic strength in molality units. The evaluator has recalculated the salt effect parameter as k_{smm}/kg mol^{-1}. The values, which differ by a factor of 4, are given below. The authors recommend the value k_{sIm}/kg mol^{-1} = 0.066, which is the same as the numerical average of the values, 0.066 ± 0.003, of all molalities and pressures. The values are classed as tentative.

T/K	m_2/mol kg^{-1}	k_{sIm}/kg mol^{-1} at pressures of		
		2.41 MPa	3.79 MPa	5.17 MPa
298.15	0.5	0.068	0.069	0.058
	1.0	0.066	0.067	0.064
	1.5	0.066	0.067	0.065
			Average 0.066 ± 0.003	

		k_{smm}/kg mol^{-1} at pressures of		
		2.41 MPa	3.79 MPa	5.17 MPa
298.15	0.5	0.272	0.276	0.232
	1.0	0.264	0.268	0.256
	1.5	0.264	0.268	0.260
			Average 0.262 ± 0.013	

COMPONENTS:	EVALUATOR:
(1) Methane; CH_4; [74-82-8] (2) Electrolyte (3) Water; H_2O; [7732-18-5]	H. Lawrence Clever Department of Chemistry Emory University Atlanta, GA 30322 USA 1985, March

CRITICAL EVALUATION:

93(3)   Methane + Magnesium chloride [7786-30-3]

+ Magnesium sulfate [7785-87-7] + Water

Byrne and Stoessell (ref 25) report the solubility of methane in water and in two equal molal mixtures of $MgCl_2$ (m_2) and $MgSO_4$ (m_3) at 298.15 K and a pressure of 3.79 MPa. Both k_{sIm} and k_{smm} values of the salt effect parameter are given below. The mixed electrolyte salts out slightly more than one would predict from the salt effect parameter values of the individual electrolytes. The values are classed as tentative.

T/K	m_2/mol kg^{-1}	m_3/mol kg^{-1}	Salt Effect Parameter at 3.79 MPa Pressure	
			k_{sIm}/kg mol^{-1}	k_{smm}/kg mol^{-1}
298.15	0.25	0.25	0.074	0.260
	0.50	0.50	0.070	0.245

94(1)   Methane + Calcium chloride [10043-52-4] + Water

The solubility of methane in water and in aqueous $CaCl_2$ solution has been measured by Michels, Gerver and Bijl (ref 3) at 298.15 K in 2.7 mol dm^{-3} $CaCl_2$ at four pressures up to 21.0 MPa, by N. O. Smith and coworkers at 298.15 and 303.15 K at several concentrations up to saturated $CaCl_2$ and pressures up to 7.48 MPa in 1961 (ref 7) and at five temperatures between 298.2 and 398.2 K in 1.0 mol kg^{-1} $CaCl_2$ at six pressures up to 60.8 MPa in 1978 (ref 17), and by Stoessell and Byrne (ref 24) in 0.5, 1.0 and 2.0 mol kg^{-1} $CaCl_2$ at pressures of 2.41, 3.79 and 5.17 MPa.

The evaluator prepared graphs of the data of Michels *et al.* (ref 3) and Duffy *et al.* (ref 7) and calculated the k_{scx}/dm^3 mol^{-1} salt effect parameters. The solubility values from both papers showed significant scatter. There is fair agreement between salt effect parameters of the two papers at 298.15 K and 2.5 - 2.7 mol dm^{-3} $CaCl_2$, but when converted to k_{smm}/kg mol^{-1} values they are smaller than more recent data. In general, the molar scale values from the two papers are smaller and show more scatter than the data from the more recent papers (ref 17 and 24).

The data of Blanco C and Smith (ref 17) and Stoessell and Byrne (ref 24) are on a molal basis. Stoessell and Byrne report their salt effect parameters on an ionic strength (molal scale) basis. Both k_{smm}/kg mol^{-1} and k_{sIm}/kg mol^{-1} values are given in the following table. The molal scale data show no significant trend with pressure or with salt concentration. At 298.15 K the Stoessell and Byrne value is about 10 percent less than the Blanco C and Smith value. The average values at 298.15 K are 0.215 ± 0.004 and 0.235 ± 0.010 from the two papers regardless of concentration or pressure. The Blanco C and Smith values go through a minimum at 344.15 K.

Salt Effect Parameters: Methane + Calcium Chloride + water.

A. Molar (c_2/mol dm^{-3}) Scale.

T/K	c_2/mol dm^{-3}	p_1/MPa	k_{scx}/ dm^3 mol^{-1}	k_{smm}/ kg mol^{-1}	Reference
298.15	0.25	2	0.160		Duffy *et al.* (7)
		3	0.159		
	0.50	2	0.220		
		3	0.219		
	1.35	2	0.243		
		3	0.240		
	2.5	2	0.194		
		3	0.194		
	2.7	5.6	0.192	(0.165)	Michels *et al.*
		11.0	0.157	(0.132)	(3)
		15.7	0.155	(0.130)	
		21.0	0.161	(0.136)	
303.15	1.35	2	0.243		Duffy *et al.* (7)
		3	0.223		
	4.75	2	0.172		
		3	0.168		
	7.35 (sat)^a	2	0.148		
		3	0.146		

B. Molal (m_2/mol kg^{-1}) Scale.

T/K	m_2/mol kg^{-1}	p_1/MPa	k_{smm}/ kg mol^{-1}	k_{sIm}/ kg mol^{-1}	Reference
298.15	0.5	2.41	0.213	0.071	Stoessell and
		3.79	0.219	0.073	Byrne (24)
		5.17	0.210	0.070	
	1.0	2.41	0.222	0.074	
		3.79	0.213	0.071	
		5.17	0.213	0.071	
		10.1	0.244		Blanco C and
		20.3	0.225		Smith (17)
		30.4	0.225		
		40.5	0.235		
		50.7	0.247		
		Av.	0.235 ± 0.010		
	2.0	2.41	0.215	0.071	Stoessell and
		3.79	0.216	0.072	Byrne (24)
		5.17	0.210	0.070	
		Av.	0.215 ± 0.004	(All S & B data at 298 K)	
324.7	1.0	10.1	0.224		Blanco C and
		20.3	0.197		Smith (17)
		30.4	0.197		
		40.5	0.201		
		50.7	0.203		
		60.8	0.199		
		Av.	0.204 ± 0.010		
344.2	1.0	10.1	0.204		Blanco C and
		20.3	0.182		Smith (17)
		30.4	0.182		
		40.5	0.185		
		50.7	0.187		
		60.8	0.184		
		Av.	0.187 ± 0.008		
375.7	1.0	10.1	0.211		Blanco C and
		20.3	0.187		Smith (17)
		30.4	0.190		
		40.5	0.198		
		50.7	0.203		
		60.8	0.201		
		Av.	0.198 ± 0.009		
398.15	1.0	10.1	0.222		Blanco C and
		20.3	0.199		Smith (17)
		30.4	0.203		
		40.5	0.212		
		50.7	0.217		
		60.8	0.215		
		Av.	0.211 ± 0.009		

COMPONENTS:	EVALUATOR:
(1) Methane; CH_4; [74-82-8] (2) Electrolyte (3) Water; H_2O; [7732-18-5]	H. Lawrence Clever Department of Chemistry Emory University Atlanta, GA 30322 USA 1985, March

CRITICAL EVALUATION:

The Michels *et al.* and the Duffy *et al.* data are classed as doubtful and the Blanco C and Smith and the Stoessell and Byrne data are classed as tentative. Although the Stoessell and Byrne data show a better standard deviation, the Blanco C and Smith data are preferred because they cover both a large temperature and pressure range.

94(2)   Methane + Magnesium chloride [7786-30-3]

+ Calcium chloride [10043-52-4] + Water

Byrne and Stoessell (ref 25) report the solubility of methane in water and in an equal molar mixture of $MgCl_2$ (m_2) and $CaCl_2$ (m_3) in water at 298.15 K and a pressure of 3.79 MPa. Both k_{sIm} and k_{smm} values of the salt effect parameter are given below. The values are classed as tentative.

T/K			Salt Effect Parameter at 3.79 MPa	
	m_2/mol kg^{-1}	m_3/mol kg^{-1}	k_{sIm}/kg mol^{-1}	k_{smm}/kg mol^{-1}
298.15	1.0	1.0	0.074	0.223

98(1)   Methane + Lithium chloride [7447-41-8] + Water

The solubility of methane in water and aqueous LiCl solutions is reported by Michels, Gerver and Bijl (ref 3), Morrison and Billett (ref 6) and Ben-Naim and Yaacobi (ref 15).

Values of the salt effect parameter from their work are summarized below. Values in () were calculated by the evaluator.

T/K	Lithium chloride			Salt Effect Parameter				Ref
	c_2/mol dm^{-3}	m_2/mol kg^{-1}	P_1/MPa	k_{scc}	k_{scx}	k_{smm}	k_{smx}	
283.15	1.0	–	0.1	0.121	(0.129)	(0.112)	(0.127)	(15)
285.15	–	1.0	0.1	(0.139)	–	0.130	0.145	(6)
288.15	1.0	–	0.1	0.115	–	–	–	(15)
293.15	1.0	–	0.1	0.111	–	–	–	(15)
298.15	1.0	–	0.1	0.109	(0.117)	0.099	(0.114)	(15)
	2.7	–	4.80	(0.138)	0.145	–	–	(3)
	–	–	10.15	(0.077)	0.084	–	–	(3)
	–	–	14.70	(0.057)	0.064	–	–	(3)
	–	–	19.78	(0.050)	0.057	–	–	(3)
303.15	1.0	–	0.1	0.108	–	(0.098)	(0.113)	(15)
	–	1.0	0.1	(0.107)	(0.115)	0.097	0.112	(6)
322.55	–	1.0	0.1	–	–	0.082	0.097	(6)
344.85	–	1.0	0.1	–	–	0.077	0.092	(6)

At 298 and 303 K the values of Ben-Naim and Yaacobi and of Morrison and Billett agree to within one percent when converted to a common salt effect parameter as k_{smx} or k_{smm}. At 283/285 K their values differ by 15 percent. The values of Michels *et al.* cannot be compared directly with the others without knowledge of the compressibility of the aqueous LiCl solutions.

99(1)   Methane + Sodium chloride [7647-14-5] + Water

The solubility of methane in water and aqueous NaCl solution is reported in twelve papers. The salt effect parameters as a function of temperature, pressure and NaCl concentration are summarized in the following table. The table is in two parts; part A gives the results from seven papers on a salt molar (c_2/mol dm^{-3}) basis and part B was the results from five papers on a salt molal (m_2/mol kg^{-1}) basis.

Salt Effect Parameters. Methane + Sodium Chloride + water.

A. Molar Scale.

T/K	$c_2/$ $mol\ dm^{-3}$	p_1/MPa	$k_{scc}/$ $dm^3\ mol^{-1}$	$k_{scx}/$ $dm^3\ mol^{-1}$	Reference
277.15	1.06	0.1	0.162		Mishnina *et al.*
	2.10	0.1	0.157		(9)
	3.08	0.1	0.161		
	4.12	0.1	0.167		
	5.31	0.1	0.165		
283.15	1.06	0.1	0.150		Mishnina *et*
	2.10	0.1	0.156		*al.* (9)
	3.08	0.1	0.155		
	4.12	0.1	0.155		
	5.31	0.1	0.162		
	0.25	0.1	0.177		Ben-Naim,
	0.50	0.1	0.166		Yaccobi (15)
	1.0	0.1	0.167		
	2.0	0.1	0.162		
288.15	0.25	0.1	0.167		Ben-Naim,
	0.50	0.1	0.159		Yaccobi (15)
	1.0	0.1	0.160		
	2.0	0.1	0.156		
293.15	1.00	0.1	0.138		Mishnina *et*
	1.77	0.1	0.148		*al.* (9)
	2.60	0.1	0.149		
	3.90	0.1	0.147		
	5.31	0.1	0.145		
	0.25	0.1	0.160		Ben-Naim,
	0.50	0.1	0.151		Yaccobi (15)
	1.0	0.1	0.154		
	2.0	0.1	0.149		
298.15	0.25	0.1	0.158		Ben-Naim,
	0.50	0.1	0.146		Yaccobi (15)
	1.0	0.1	0.149		
	2.0	0.1	0.144		
	0.50	0.1	0.113		Yano *et al.*
	1.00	0.1	0.146		(16)
	1.50	0.1	0.163		
	2.5	10		0.117	Michaels *et*
		15		0.096	*al.* (3)
		20		0.097	
	5.4	10		0.114	
		15		0.106	
		20		0.115	
303.15	1.04	0.1	0.135		Mishnina *et*
	2.00	0.1	0.135		*al.* (9)
	2.60	0.1	0.139		
	3.90	0.1	0.137		
	5.31	0.1	0.135		
	0.25	0.1	0.158		Ben-Naim,
	0.50	0.1	0.142		Yaccobi (15)
	1.0	0.1	0.145		
	2.0	0.1	0.138		
	0.50	1,2,3,4		0.158±0.041	Duffy *et al.*
	1.0	1,2,3,4		0.154±0.025	(7)
	2.7	1,2,3,4		0.107±0.013	
	5.4	1,2,3,4		0.103±0.006	
323.15	1.02	0.1	0.131		Mishnina *et*
	1.98	0.1	0.134		*al.* (9)
	2.80	0.1	0.130		
	3.90	0.1	0.131		
	5.31	0.1	0.129		
	1.00	29.5		0.108	Namiot *et al.* (21)

Methane + Sodium Cloride + Water
A. Molar Scale (continued)

T/K	c_2/ mol dm^{-3}	p_1/MPa	k_{scc}/ dm^3 mol^{-1}	k_{scx}/ dm^3 mol^{-1}	Reference
323.15 (cont.)	2.5	10		0.100	Michaels *et al.* (3)
		15		0.097	
		20		0.093	
	5.4	10		0.105	
		15		0.103	
		20		0.101	
324.15	1.0	10.1		0.123	O'Sullivan, Smith (13)
		20.3		0.129	
		30.4		0.128	
		40.5		0.126	
		50.7		0.126	
		60.8		0.125	
	4.0	20.3		0.113	
		30.4		0.115	
		40.5		0.115	
		50.7		0.114	
		60.8		0.113	
342.15	1.02	0.1	0.129		Mishnina *et al.* (9)
	1.98	0.1	0.121		
	3.95	0.1	0.121		
	5.31	0.1	0.126		
348.15	2.5	10		0.100	Michaels *et al.* (3)
		15		0.095	
		20		0.093	
	5.4	10		0.102	
		15		0.101	
		20		0.099	
353.15	1.02	0.1	0.111		Mishnina *et al.* *et al.* (9)
	2.12	0.1	0.112		
	3.28	0.1	0.108		
	5.31	0.1	0.108		
363.15	1.06	0.1	0.108		Mishnina *et al.* (9)
	2.10	0.1	0.107		
	3.08	0.1	0.113		
	5.31	0.1	0.113		
373.15	1.0	29.5		0.114	Namiot *et al.* (21)
	2.5	10		0.105	Michaels *et al.* (3)
		15		0.098	
		20		0.092	
	5.4	10		0.103	
		15		0.100	
		20		0.098	
375.65	1.0	20.4		0.115	O'Sullivan, Smith (13)
		30.6		0.112	
		40.8		0.113	
		51.0		0.125	
		61.2		0.117	
	4.0	20.4		0.107	
		30.6		0.106	
		40.8		0.110	
		51.0		0.116	
		61.2		0.116	
398.15	2.5	10		0.115	Michaels *et al.* (3)
		15		0.105	
		20		0.099	
	5.4	10		0.105	
		15		0.101	
		20		0.099	
	1.0	10.4		0.132	O'Sullivan, Smith (13)
		20.7		0.122	
		30.9		0.124	
		41.0		0.120	
		51.4		0.129	
		61.6		0.122	

Methane + Sodium Chloride + Water
A. Molar Scale (continued)

T/K	$c_2/$ mol dm^{-3}	$p_1/$MPa	$k_{scc}/$ dm^3 mol^{-1}	$k_{scx}/$ dm^3 mol^{-1}	Reference
398.15 (cont.)	4.0	20.7		0.112	O'Sullivan, Smith (13)
		30.9		0.117	
		41.0		0.117	
		51.4		0.119	
		61.6		0.119	
423.15	2.5	10		0.145	Michaels *et al.* (3)
		15		0.109	
		20		0.106	
	5.4	10		0.104	
		15		0.102	
		20		0.099	
	1.0	29.5		0.098	Namiot *et al.*
473.15	1.0	29.5		0.084	Namiot *et al.*
523,15	1.0	29.5		0.134	Namiot *et al.*
573.15	1.0	29.5		0.205	Namiot *et al.*
623.15	1.0	29.5		0.295	Namiot *et al.* (21)

B. Molal Scale

T/K	$m_2/$ mol kg^{-1}	$p_1/$MPa	$k_{smc}/$ kg mol^{-1}	$k_{smm}/$ kg mol^{-1}	$k_{smx}/$ kg mol^{-1}	Reference
273.15	0.68	0.1	0.188			Eucken, Hertzebrg (5)
	1.37	0.1	0.178			
	2.77	0.1	0.186			
	0.81-4.70	1.9-3.3		(0.150)	0.165	Cramer (26)
285.75	1.0	0.1		0.153	(0.168)	Morrison, Billett (6)
293.15	0.795	0.1	0.163			Eucken, Hertzberg (5)
	2.63	0.1	0.157			
	0.81-4.70	2.2-4.1		(0.126)	0.141	Cramer (26)
298.15	0.5	2.41		0.101		Stoessell, Byrne (24)
		3.79		0.138		
		5.17		0.117		
	1.0	2.41		0.116		
		3.79		0.137		
		5.17		0.124		
	2.0	2.41		0.124		
		3.79		0.125		
		5.17		0.124		
	4.0	2.41		0.120		
		3.79		0.122		
		5.17		0.119		
				0.122±0.010	(0.137)	
303.15	1.0	0.1		0.127	(0.142)	Morrison, Billett (6)
313.15	0.81-4.70	1.1-4.1		(0.112)	0.127	Cramer (26)
322.55	1.0	0.1		0.111	(0.126)	Morrison, Billett (6)
333.15	0.81-4.70	1.1-3.9		(0.104)	0.119	Cramer (26)
344.85	1.0	0.1		0.102	(0.117)	Morrison, Billett (6)
353.15	0.81-4.70	1.1-4.0		(0.101)	0.116	Cramer (26)
373.15	0.81-4.70	1.1 5.5		(0.101)	0.116	Cramer (26)
	0.05-5.7	13.6-153		0.103	(0.118)	Blount *et al.* (18,23)

Methane + Sodium Chloride + Water
B. Molal Scale (continued)

T/K	$m_2/$ mol kg^{-1}	$p_1/$MPa	$k_{smc}/$ kg mol^{-1}	$k_{smm}/$ kg mol^{-1}	$k_{smx}/$ kg mol^{-1}	Reference
393.15	0.81-4.70	1.2-5.4		(0.103)	0.118	Cramer (26)
407	0.05-5.7	13.6-153		0.103	(0.118)	Blount *et al.* (18,23)
413.15	0.81-4.70	1.1-6.3		(0.107)	0.122	Cramer (26)
433.15	0.81-4.70	1.1-5.5		(0.111)	0.126	Cramer (26)
444	0.05-5.7	13.6-153		0.103	(0.118)	Blount *et al.* (18,23)
453.15	0.81-4.70	5.7-6.8		(0.115)	0.130	Cramer (26)
473.15	0.81-4.70	5.7-7.1		(0.119)	0.134	Cramer (26)
478	0.05-5.7	13.6-153		0.103	(0.118)	Blount *et al.* (18,23)
493.15	0.81-4.70	7.3-8.9		(0.121)	0.136	Cramer (26)
512	0.05-5.7	13.6-153		0.103	(0.118)	Blount *et al.* (18,23)
513.15	0.81-4.70	6.9-8.9		(0.122)	0.137	Cramer (26)
533.15	0.81-4.70	6.9-8.9		(0.121)	0.136	Cramer (26)
553.15	0.81-4.70	10.1-12.0		(0.118)	0.133	Cramer (26)
573.15	0.81-4.70	11.9-13.2		(0.113)	.128	Cramer (26)

Values in ( ) calculated by the compiler.

COMPONENTS:	EVALUATOR:
(1) Methane; CH_4; [74-82-8] (2) Electrolyte (3) Water; H_2O; [7732-18-5]	H. Lawrence Clever Department of Chemistry Emory University Atlanta, GA 30322 USA 1985, March

CRITICAL EVALUATION:

Molar Scale. All of the results are classed as tentative, but the salt effect parameters based on the work of Michels, Gerver and Bijl (ref 3), Duffy, Smith and Nagy (ref 7), and the Yano, Suetaka, Umehara and Horiuchi (ref 16) at 0.50 molar are considered less reliable than the other values. Mishnina, Avdeeva, and Bozhovskaya (ref 8,9) give salt effect parameters in two papers. The values in (ref 8) appear to be an average based on their result in (ref 9) and the results of Morrison and Billett (ref 6). Only the values based on the experimental solubility values in (ref 9) are given in the table.

The results of Mishnina *et al.* (ref 9), O'Sullivan and Smith (ref 13), Ben-Naim and Yaacobi (ref 15) and Namiot, Skripka and Ashmyan (ref 21) do not always agree well, but they do show similar trends with temperature, pressure and sodium chloride concentration.

Molal Scale. Again all of the data are classed as tentative. The solubility data of Blount *et al.* (ref 18, 23) show the most scatter. They propose an average solubility parameter of k_{smm}/kg mol^{-1} = 0.1025 ± 0.0047 for all pressure and NaCl concentrations between temperatures of 373 and 512 K, while Cramer's (ref 26) data show a steady increase from 0.101 to 0.122 over the temperature range. Blount *et al.* (ref 23) point out that Susak and McGee (ref 28) calculate a k_{smm} value of (0.120 ± 0.003) from data compiled and evaluated by Haas (ref 27) at these temperatures.

The data of Eucken and Hertzberg (ref 5), Morrison and Billett (ref 6), Stoessell and Byrne (ref 24), and Cramer (ref 26) show similar trends. Neither increasing sodium chloride molality nor pressure seem to affect the salt effect parameter at a given temperature.

The work of Namiot *et al.* (ref 21) and Cramer (ref 26) covers temperatures up to 623 and 573 K respectively. The two data sets are not directly comparable because one is on the molar scale and the other the molal scale. Both groups have taken fugacity into account. The k_{scx} values of Namiot *et al.* are much larger than the k_{smx} values of Cramer at the highest temperatures. Cramer's values actually show a slight decrease with temperature. More work is needed on salt effect parameters at the high temperatures and pressures.

99(2)   Methane + Magnesium chloride [7786-30-3]

+ Sodium chloride [7647-14-5] + Water

Byrne and Stoessell (ref 15) report the solubility of methane in water and in an equal molal mixture of $MgCl_2$ and NaCl at 298.15 K and a pressure of 3.79 MPa. The value of k_{sIm} and k_{smm} are classed as tentative.

T/K	$MgCl_2$ m_2/mol kg^{-1}	NaCl m_3/mol kg^{-1}	Salt Effect Parameter at 3.79 MPa Pressure k_{sIm}/kg mol^{-1}	k_{smm}/kg mol^{-1}
298.15	1.0	1.0	0.066	0.137

99(3)   Methane + Calcium chloride [10043-52-4]

+ Sodium chloride [7647-14-5] + Water

Duffy, Smith and Nagy (ref 7) report the solubility of methane in water and an aqueous solution that is 3.0 mol dm^{-3} CaCl$_2$ and 1.53 mol dm^{-3} NaCl at 303 K and pressures of 2.54 and 3.17 MPa. Byrne and Stoessell (ref 25) report the solubilities in equi molal solution of CaCl$_2$ and NaCl at 298.15 K and 3.79 MPa. No effort was made to directly compare the results. The values are classed as tentative.

T/K	CaCl$_2$ c_2/mol dm^{-3} or m_2/mol kg^{-1}	NaCl c_3/mol dm^{-3} or m_3/mol kg^{-1}	P_1/MPa	Salt Effect Parameter k_{scx}	k_{xIm} k_{scx}	k_{smm}
298.15	1.0 (m)	1.0 (m)	3.79		0.086 (m)	0.172
303.15	3.0 (c)	1.53 (c)	2.54	0.155	0.067 (x)	
			5.19	0.144	0.062 (x)	

99(4)   Methane + Sodium bromide [7647-15-6] + Water

Both Michels, Gerver and Bijl (ref 3) and Ben-Naim and Yaacobi (ref 15) have measured the solubility of methane in water and in aqueous NaBr solution. Salt effect parameters from their work are given below.

T/K	c_2/mol dm^{-3}	P_1/mPa	Salt Effect Parameters k_{scc}/dm^3 mol^{-1}	k_{scx}/dm^3 mol^{-1}
283.15	1.0	0.1	0.165	-
288.15	1.0	0.1	0.157	-
293.15	1.0	0.1	0.149	-
298.15	1.0	0.1	0.142	-
	2.7	52.8	-	0.152
	-	102.6	-	0.131
	-	153.6	-	0.118
	-	200.4	-	0.117
303.15	1.0	0.1	0.136	-

The two data sets are not directly comparable because of differences in NaBr concentration and methane pressure. The data are classed as tentative, but we prefer the values of Ben-Naim and Yaacobi measured at 0.1 MPa for use because they give a self-consistent set of values for use over the 283-303 K temperature interval.

99(5)   Methane + Sodium iodide [7681-82-5] + Water

Both Michels, Gerver and Bijl (ref 3) and Ben-Naim and Yaacobi (ref 15)

COMPONENTS:	EVALUATOR:
(1) Methane; CH_4; [74-82-8] (2) Electrolyte (3) Water; H_2O; [7732-18-5]	H. Lawrence Clever Department of Chemistry Emory University Atlanta, GA 30322 USA 1985, March

CRITICAL EVALUATION:

have measured the solubility of methane in water and aqueous NaI solution. The salt effect parameters calculated by the evaluator from their data are given below.

T/K	c_2/mol dm^{-3}	P_1/mPa	Salt Effect Parameter	
			k_{scc}/dm^3 mol^{-1}	k_{scx}/dm^3 mol^{-1}
283.15	1.0	0.1	0.160	–
288.15	1.0	0.1	0.152	–
293.15	1.0	0.1	0.142	–
298.15	1.0	0.1	0.130	–
	2.7	56.2	–	0.113
	–	111.7	–	0.086
	–	152.0	–	0.074
	–	204.9	–	0.064
303.15	1.0	0.1	0.118	–

It is somewhat unusual to observe a low molecular weight gas salted out more by NaI than NaBr as is the case with the Ben-Naim and Yaacobi data. However, our recommendation is the same as with the NaBr data above. The salt effect parameters are classed as tentative but we prefer the Ben-Naim and Yaacobi values measured at 0.1 MPa because they give a self-consistent set of values for use over the 283-303 K temperature interval.

99(6)   Methane + Sodium Sulfate [7757-82-6] + Water

Mishnina, Audeeva and Bozhovskaya (ref 9) and Stoessell and Byrne (ref 24) report the solubility of methane in water and aqueous Na_2SO_4 solutions at temperatures of 293 and 298 K, respectively.

T/K	c_2/mol dm^{-3} m_2/mol kg^{-1}	P_1/MPa	Salt Effect Parameters		
			k_{scc}/dm^3 mol^{-1}	k_{sIm}/kg mol^{-1}	k_{smm}/kg mol^{-1}
293.15	0.48 (c_2)	0.1	0.391	–	(0.376)
	0.90 (c_2)	0.1	0.390	–	(0.375)
298.15	0.5 (m_2)	2.41	–	0.120	0.359
	1.0 (m_2)	2.41	–	0.124	0.373
	0.5 (m_2)	3.79	–	0.127	0.381
	1.0 (m_2)	3.79	–	0.122	0.366
	0.5 (m_2)	5.17	–	0.120	0.360
	1.0 (m_2)	5.17	–	0.116	0.348
				Average 0.122 ± 0.004 (Authors)	
				Average 0.365 ± 0.012	

The Mishnina *et al.* k_{scc} values when converted to k_{smm} values agree well with the Stoessell and Byrne values, although they were measured under quite different pressures and temperatures that differed by 5 K. Both sets of values are classed as tentative.

99(7)   Methane + Sulfuric acid [7664-93-9]

+ Sodium sulfate [7757-82-6] + Water

Kobe and Kenton (ref 4) measured the solubility of methane in a solution that was 0.90 mol kg^{-1} H_2SO_4 and 1.76 mol kg^{-1} Na_2SO_4 at 298.15 K. The salt effect parameter k_{smm}/kg mol^{-1} = 0.20 is classed as tentative.

99(8)   Methane + Magnesium sulfate [7785-87-7]

+ Sodium sulfate [7757-82-6] + Water

Stoessell and Byrne (ref 24) report the following salt effect parameter for the $MgSO_4$ + Na_2SO_4 mixed electrolyte solvent at 298.15 K and 3.79 MPa. The value is classed as tentative.

T/K	m_2/mol kg^{-1}	m_3/mol kg^{-1}	P_1/MPa	Salt Effect Parameter	
				k_{sIm}/kg mol^{-1}	k_{smm}/kg mol^{-1}
298.15	0.5	0.5	3.79	0.097	0.340

19(9)   Methane + Sodium chloride [7647-14-5]

+ Sodium sulfate [7757-82-6] + Water

Stoessell and Byrne (ref 24) report the following salt effect parameters for NaCl + Na_2SO_4 at 298.15 K and 3.79 MPa. The value is classed as tentative.

T/K	m_2/mol kg^{-1}	m_3/mol kg^{-1}	P_1/MPa	Salt Effect Parameter	
				k_{sIm}/kg mol^{-1}	k_{smm}/kg mol^{-1}
298.15	1.0	1.0	3.79	0.111	0.223

99(10)   Methane + Sodium bicarbonate [144-55-8] + Water

Stoessell and Byrne (ref 24) report the solubility of methane in water and in aqueous $NaHCO_3$ solution at 298.15 K and methane partial pressures of 2.41, 3.79 and 5.17 MPa. The salt effect parameters are classed as tentative.

COMPONENTS:	EVALUATOR:
(1) Methane; CH_4; [74-82-8] (2) Electrolyte (3) Water; H_2O; [7732-18-5]	H. Lawrence Clever Department of Chemistry Emory University Atlanta, GA 30322 USA 1985, March

CRITICAL EVALUATION:

T/K	m_2/mol kg^{-1}	P_1/MPa	k_{sIm} or k_{smm}/kg mol^{-1}
298.15	0.25 0.50	2.41	0.130 0.145
	0.25 0.50	3.79	0.123 0.164
	0.25 0.50	5.17	0.132 0.129
		Authors	0.146
		Average	0.137 ± 0.015

For a 1-1 electrolyte k_{sIm} and k_{smm} will be the same.

99(11)   Methane + Sodium carbonate [497-19-8] + Water

Stoessel and Byrne (ref 24) measured the solubility of methane in water and several aqueous Na_2CO_3 solutions at 298.15 K and methane partial pressures of 2.41, 3.79 and 5.17 MPa. The solubility parameters calculated from these data are classed as tentative.

T/K	m_2/mol kg^{-1}	P_1/MPa	Salt Effect Parameters	
			k_{sIm}/kg mol^{-1}	k_{smm}/kg mol^{-1}
298.15	0.5 1.0 1.5	2.41	0.120 0.124 0.121	0.360 0.372 0.363
	0.5 1.0 1.5	3.79	0.127 0.125 0.114	0.381 0.375 0.342
	0.5 1.0 1.5	5.17	0.125 0.125 0.118	0.375 0.375 0.354
		Average	(0.122 ± 0.004)	
		Authors	0.118	

99(12)   Water + Sodium chloride [7647-14-5]

+ Sodium carbonate [497-19-8] + Water

Stoessell and Byrne (ref 24) report the following salt effect parameters for the NaCl + Na_2CO_3 mixed electrolyte solvent at 298.15 K and a methane partial pressure of 3.79 MPa. The value is classed as tentative.

T/K	NaCl m_2/mol kg^{-1}	Na_2CO_3 m_3/mol kg^{-1}	P_1/MPa	Salt Effect Parameter k_{sIm}/kg mol^{-1}	k_{smm}/kg mol^{-1}
298.15	1.0	1.0	3.79	0.113	0.225

99(13) Methane + Sodium sulfate [7757-82-6]

+ Sodium carbonate [497-19-8] + Water

Stoessell and Byrne (ref 24) report the following salt effect parameter for the Na_2SO_4 + Na_2CO_3 mixed electrolyte solvent at 298.15 K and at methane partial pressure of 3.79 MPa. The value is classed as tentative.

T/K	Na_2SO_4 m_2/mol kg^{-1}	Na_2CO_3 m_3/mol kg^{-1}	P_1/MPa	Salt Effect Parameter k_{sIm}/kg mol^{-1}	k_{smm}/kg mol^{-1}
298.15	0.5	0.5	3.79	0.120	0.360

100(1) Methane + Potassium hydroxide [1310-58-3] + Water

Shoor, Walker and Gubbins (ref 11) measured the solubility of methane in water and up to 10 molar KOH solution at four temperatures between 298 and 353 K. Their salt effect parameters as k_{scx}/dm^3 mol^{-1} values are classed as tentative.

KOH c_2/mol dm^{-3}	Salt Effect Parameters k_{scx}/dm^3 mol^{-1} at temperatures of			
	298.15 K	313.15 K	333.15 K	353.15 K
1.03	0.214	0.174	0.165	0.139
2.77	0.203	0.179	0.163	0.144
5.13	0.197	0.177	0.164	0.154
7.35	0.194	0.176	0.166	0.158
10.15	0.198	0.178	0.168	0.156
Average	(0.201 ± 0.008)	(0.177 ± 0.002)	(0.165 ± 0.002)	(0.151 ± 0.009)
Author's value	0.197	0.176	0.164	0.154

100(2) Methane + Potassium chloride [7447-40-7] + Water

Four papers report data of the solubility of methane in water and aqueous KCl solutions. Three of the papers use a molar (c_2/mol dm^{-3}) scale and one a molal (m_2/mol kg^{-1}) scale. Two report values from solubility measurements at one atm. while two report values as a function of pressure.

The data of Michels *et al.* (ref 3) show the most scatter and are the least reliable. At 298.15 K the data of Stoessell and Byrne (ref 24) average k_{smm}/kg mol^{-1} = (0.111 ± 0.010), but the authors suggest the value 0.101, based primarily on the 4.0 modal KCl solution for use. Yano *et al.*

COMPONENTS:	EVALUATOR:
(1) Methane; CH_4; [74-82-8] (2) Electrolyte (3) Water; H_2O; [7732-18-5]	H. Lawrence Clever Department of Chemistry Emory University Atlanta, GA 30322 USA 1985, March

CRITICAL EVALUATION:

(ref 16) suggest a value of $k_{scc}/dm^3\ mol^{-1} = 0.137$ which agrees well with the Ben-Naim and Yaacobi (ref 15) value of $k_{scc}/dm^3\ mol^{-1} = 0.138$. The last two values are equivalent to $k_{smm}/kg\ mol^{-1} = 0.120 - 0.121$ which is about 8 percent larger than the Stoessell and Byrne value.

All of the data are classed as tentative but the salt effect parameters of Ben-Naim and Yaacobi are preferred because they are a set of self-consistent values for use over a 20 degree temperature interval.

T/K	c_2/mol dm^{-3} m_2/mol kg^{-1}	P_1/MPa	Salt Effect Parameter				Ref
			k_{scc}	k_{scx}	k_{sIm}	k_{smm}	
283.15	1.0	0.1	0.156	-	-	-	15
288.15	1.0	0.1	0.147	-	-	-	15
293.15	1.0	0.1	0.141	-	-	-	15
298.15	1.0	0.1	0.138	-	-	(0.121)	15
298.15	0.5	0.1	0.128	-	-	-	16
	1.0	0.1	0.139	-	-	-	16
	1.5	0.1	0.140	-	-	-	16
298.15	0.5 (m)	2.41	-	-	0.107	0.107	24
	1.0 (m)	-	-	-	0.111	0.111	24
	2.0 (m)	-	-	-	0.108	0.108	24
	4.0 (m)	-	-	-	0.101	0.101	24
	0.5 (m)	3.79	-	-	0.130	0.130	24
	1.0 (m)	-	-	-	0.129	0.129	24
	2.0 (m)	-	-	-	0.112	0.112	24
	4.0 (m)	-	-	-	0.104	0.104	24
	0.5 (m)	5.17	-	-	0.108	0.108	24
	1.0 (m)	-	-	-	0.114	0.114	24
	2.0 (m)	-	-	-	0.111	0.111	24
	4.0 (m)	-	-	-	0.100	0.100	24
298.15	2.7	4.85	-	0.119	-	-	3
	-	9.85	-	0.098	-	-	3
	-	15.07	-	0.084	-	-	3
	-	20.06	-	0.110	-	-	3
303.15	1.0	0.1	0.138	-	-	-	15

100(3)   Methane + Magnesium chloride [7786-30-3]

+ Calcium chloride [10043-52-4]

+ Sodium chloride [7647-14-5]

+ Potassium chloride [7447-40-7] + Water

Stoessell and Byrne (ref 24) report the following salt effect parameter

for the $MgCl_2$ + $CaCl_2$ + NaCl + KCl mixed electrolyte solvent at 298.15 K
and a methane partial pressure of 3.79 MPa. The value is classed as tentative.

T/K	m_2	m_3	m_4	m_5	P_1/MPa	k_{sIm}/kg mol^{-1}	k_{smm}/kg mol^{-1}
298.15	0.5	0.5	0.5	0.5	3.79	0.074	0.147

100(4)   Methane + Sodium chloride [7647-14-5]

+ Potassium chloride [7447-40-7] + Water

Stoessell and Byrne (ref 24) report the following salt effect parameter for the NaCl + KCl mixed electrolyte solvent at 298.15 K and a methane
partial pressure 3.79 MPa. The value is classed as tentative.

T/K	NaCl m_2/mol kg^{-1}	KCl m_3/mol kg^{-1}	P_1/MPa	Salt Effect Parameter k_{sIm}/kg mol^{-1}	k_{smm}/kg mol^{-1}
298.15	1.0	1.0	3.79	0.114	0.114

100(5)   Methane + Potassium iodide [7681-11-0] + Water

Morrison and Billett (ref 6) measured the solubility of methane in
water and one mol kg^{-1} KI solution at four temperatures. The salt effect
parameters are classed as tentative.

T/K	KI m_2/mol kg^{-1}	P_1/MPa	Salt Effect Parameter k_{smm}/kg mol^{-1}	k_{smx}/kg mol^{-1}
285.75	1.0	0.1	0.130	0.145
303.15	1.0	0.1	0.097	0.112
322.55	1.0	0.1	0.071	0.086
344.85	1.0	0.1	0.064	0.069

100(6)   Methane + Potassium sulfate [7778-80-5] + Water

Stoessel and Byrne (ref 24) measured the solubility of methane in
water and in K_2SO_4 solution at several concentrations and methane partial
pressures. The salt effect parameters from their data are classed as
tentative.

COMPONENTS:	EVALUATOR:
(1) Methane; CH_4; [74-82-8] (2) Electrolyte (3) Water; H_2O; [7732-18-5]	H. Lawrence Clever Department of Chemistry Emory University Atlanta, GA 30322 USA 1985, March

CRITICAL EVALUATION:

T/K	m_2/mol kg^{-1}	P_1/MPa	Salt Effect Parameter	
			k_{sIm}/kg mol^{-1}	k_{smm}/kg mol^{-1}
298.15	0.25 0.50	2.41 -	0.116 0.104	0.348 0.312
	0.25 0.50	3.79 -	0.119 0.115	0.357 0.345
	0.25 0.50	5.17 -	0.105 0.104	0.315 0.312

Average 0.111 ± 0.007

Authors 0.108

100(7)   Methane + Magnesium sulfate [7785-87-7]

+ Potassium sulfate [7778-80-5] + Water

Stoessell and Byrne (ref 24) report the following salt effect parameter for the $MgSO_4$ + K_2SO_4 mixed electrolyte solvent at 298.15 K and a methane partial pressure of 3.79 MPa. The value is classed as tentative.

T/K	m_2/mol kg^{-1}	m_3/mol kg^{-1}	P_1/MPa	Salt Effect Parameter	
				k_{sIm}/kg mol^{-1}	k_{smm}/kg mol^{-1}
298.15	0.25	0.25	3.79	0.084	0.292

100(8)   Methane + Sodium sulfate [7757-82-6]

+ Potassium sulfate [7778-80-5] + Water

Stoessell and Byrne (ref 24) report the following value for the salt effect parameter for the Na_2SO_4 + K_2SO_4 mixed electrolyte solution at 298.15 K and a methane partial pressure of 3.79 MPa. The value is classed as tentative.

T/K	Na_2SO_4 m_2/mol kg^{-1}	K_2SO_4 m_3/mol kg^{-1}	P_1/MPa	Salt Effect Parameter	
				k_{sIm}/kg mol^{-1}	k_{smm}/kg mol^{-1}
298.15	0.25	0.25	3.79	0.117	0.351

100(9)   Methane + Potassium bicarbonate [298-14-6] + Water

Stoessell and Byrne (ref 24) measured the solubility of methane in
water and aqueous $KHCO_3$ solutions at 298.15 K and methane partial pressures
of 2.41, 3.79 and 5.17 MPa. The salt effect parameters calculated from
their data are classed as tentative.

T/K	m_2/mol kg^{-1}	P_1/MPa	Salt Effect Parameter	
			k_{sIm}/kg mol^{-1}	k_{smm}/kg mol^{-1}
298.15	0.25	2.41	0.130	0.130
	0.50	–	0.131	0.131
	0.25	3.79	0.096	0.096
	0.50	–	0.151	0.151
	0.25	5.17	0.125	0.125
	0.50	–	0.132	0.132
			Average (0.128 ± 0.018)	
			Authors 0.145	

100(10)   Methane + Potassium carbonate [584-08-7] + Water

Stoessell and Byrne (ref 24) measured the solubility of methane in
water and in aqueous K_2CO_3 solution of three concentrations at 298.15 K
and methane partial pressures of 2.41, 3.79 and 5.17 MPa. The salt effect
parameters calculated from their solubility data are classed as tentative.

T/K	m_2/mol kg^{-1}	P_1/MPa	Salt Effect Parameters	
			k_{sIm}/kg mol^{-1}	k_{smm}/kg mol^{-1}
298.15	0.5	2.41	0.109	0.327
	1.0	–	0.117	0.351
	2.0	–	0.107	0.321
	0.5	3.79	0.116	0.348
	1.0	–	0.116	0.348
	2.0	–	0.112	0.336
	0.5	5.17	0.107	0.321
	1.0	–	0.109	0.327
	2.0	–	0.114	0.342
			Average (0.112 ± 0.004)	
			Authors 0.111	

100(11)   Methane + Potassium chloride [7447-40-7]

+ Potassium carbonate [584-08-7] + Water

Stoessell and Byrne (ref 24) report the following value of the salt
effect parameter for the KCl + K_2CO_3 mixed electrolyte solution at 298.15 K
and a methane partial pressure of 3.79 MPa. The value is classed as
tentative.

COMPONENTS:	EVALUATOR:
(1) Methane; CH_4; [74-82-8] (2) Electrolyte (3) Water; H_2O; [7732-18-5]	H. Lawrence Clever Department of Chemistry Emory University Atlanta, GA USA 1985, March

CRITICAL EVALUATION:

T/K	KCl m_2/mol kg^{-1}	K_2CO_3 m_3/mol kg^{-1}	P_1/MPa	Salt Effect Parameter k_{sIm}/kg mol^{-1}	k_{smm}/kg mol^{-1}
298.15	1.0	1.0	3.79	0.109	0.218

101 Methane + Cesium chloride [9647-17-8] + Water

Ben-Naim and Yaacobi (ref 15) measured the solubility of methane in water and in 1.0 mol dm^{-3} CsCl solution at five degree temperature intervals between 283 and 303 K. The salt effect parameters are classed as tentative.

T/K	CsCl c_2/mol dm^{-3}	P_1/MPa	Salt Effect Parameter k_{scc}/dm^3 mol^{-1}
283.15	1.0	0.1	0.139
288.15	1.0	0.1	0.140
293.15	1.0	0.1	0.135
298.15	1.0	0.1	0.127
303.15	1.0	0.1	0.114

REFERENCES

1. Clever, H. L. *J. Chem. Eng. Data* 1983, *23*, 401-4; *NITROGEN*, Battino, R., Editor, Solubility Series, Vol. 10, xxix - xlii, Pergamon Press, Ltd., Oxford and New York, 1982.

2. Christoff, A. *Z. Phys. Chem.* 1906, *55*, 622-34.

3. Michels, A.; Gerver, J.; Bijl, A. *Physica* 1936, *3*, 797-808.

4. Kobe, K. A.; Kenton, F. H. *Ind. Eng. Chem., Anal. Ed.* 1938, *10*, 76-7.

5. Eucken, A.; Hertzberg, G. *Z. Phys. Chem.* 1950, *195*, 1-23.

6. Morrison, T. J.; Billett, F. *J. Chem. Soc.* 1952, 3819-22.

7. Duffy, J. R.; Smith, N. O.; Nagy, B. *Geochim. Cocmochim. Acta* 1961, *24*, 23-31.

8. Mishnina, T. A.; Avdeeva, O. I.; Bozhovskaya, T. K. *Materialy Vses. Nauchn.-Issled. Geol. Inst.* 1961, *No. 46*, 93-110; *Chem. Abstr.* 1962, *57*, 11916b.

9. Mishnina, A. A.; Avdeeva, O. I.; Bozhovskaya, T. K. *Inf. Sb., Vses. Nauchn-Issled. Geol. Inst.* 1962, *No. 56*, 137-45; *Chem. Abstr.* 1964, *60*, 8705g.

COMPONENTS:	EVALUATOR:
(1) Methane; CH_4; [74-82-8] (2) Electrolyte (3) Water; H_2O; [7732-18-5]	H. Lawrence Clever Department of Chemistry Emory University Atlanta, GA 30322 USA 1985, March

CRITICAL EVALUATION:

10. Wetlaufer, D. B.; Malik, S. K.; Stoller, L.; Coffin, R. L. *J. Am. Chem. Soc.* <u>1964</u>, *86*, 580-14.

11. Shoor, S. K.; Walker, R. D., Jr.; Gubbins, K. E. *J. Phys. Chem.* <u>1969</u>, *73*, 312-7.

12. Wen, W.-Y.; Hung, J. H. *J. Phys. Chem.* <u>1970</u>, *74*, 170-80.

13. O'Sullivan, T. D.; Smith, N. O. *J. Phys. Chem.* <u>1970</u>, *74*, 1460-6.

14. Feillolay, A.; Lucas, M. *J. Phys. Chem.* <u>1972</u>, *76*, 3068-72.

15. Ben-Naim, A.; Yaacobi, M. *J. Phys. Chem.* <u>1974</u>, *78*, 170-5.

16. Yano, T.; Suetaka, T.; Umehara, T.; Horiuchi, A. *Kagaku Kogaku* <u>1974</u>, *38*, 320-3.

17. Blanco C, L. H.; Smith N. O. *J. Phys. Chem.* <u>1978</u>, *82*, 186-91.

18. Blount, C. W.; Price, L. C.; Wenger, L. M.; Tarullo, M. *Proc.-US Gulf Coast Geopressured Geotherm. Energy Conf.* <u>1979</u> (Pub. <u>1980</u>), *4(3)*, 1225-62.

19. Rudakov, E. S.; Lutsyk, A. I. *Zh. Fiz. Khim.* <u>1979</u>, *53*, 1298-1300; *Russ. J. Phys. Chem.* <u>1979</u>, *53*, 731-3.

20. Barone, G.; Castronuovo, G.; Volpe, D.; Elia, V.; Grassi, L. *J. Phys. Chem.* <u>1979</u>, *83*, 2703-14.

21. Namiot, A. Yu.; Skripka,V. G.; Ashmyan, K. D. *Geokhimiya* <u>1979</u>, *(1)*, 147-8.

22. Muccitelli, J. A.; Wen, W.-Y. *J. Solution Chem.* <u>1980</u>, *9*, 141-61.

23. Blount, C. W.; Price, C. W. *REPORT* <u>1982</u> *DOE/ET*12145-1, 159 pp.; *Chem. Abstr.* <u>1983</u>, *98*, 22026p.

24. Stoessell, R. K.; Byrne, P. A. *Geochim. Cosmochim. Acta* <u>1982</u>, *46*, 1327-32.

25. Byrne, P. A.; Stoessell, R. K. *Ibid.* <u>1982</u>, *46*, 2395-7.

26. Cramer, S. D. *Ind. Eng. Chem. Process Des. Dev.* <u>1984</u>, *23*, 533-8.

27. Haas, J. L., Jr. US Geological Survey Open-file Report <u>1978</u>, No. 78-1004, 42 pp.

28. Susak, N. J.; McGee, K. A. US Geological Survey Open-file Report <u>1980</u>, No. 80-371.

COMPONENTS:	ORIGINAL MEASUREMENTS:
(1) Methane; CH_4; [74-82-8]	Muccitelli, J. A.; Wen, W.-Y.
(2) Hydrochloric Acid; HCl; [7647-01-0]	J. Solution Chem. 1980, 9, 141 - 161.
(3) Water; H_2O; [7732-18-5]	

VARIABLES:	PREPARED BY:
T/K: 278.15 - 298.15 p/kPa: 101.325 (1 atm)	H. L. Clever

EXPERIMENTAL VALUES:

T/K	HCl c_2/mol dm^{-3}	Ostwald Coefficient $10^3 L$/cm^{3}cm^{-3}
278.15	0.0000	51.12 ± 0.35
	0.2822	49.01
	0.5501	47.19
	0.8030	45.76
	0.9703	45.11
283.15	0.0000	45.04 ± 0.17
	0.2824	43.67
	0.5330	42.49
	0.7979	41.05
	1.0914	39.45
288.15	0.0000	40.39 ± 0.18
	0.2864	38.81
	0.5645	38.31
	0.8164	37.94
	0.8645	37.45
	1.127	36.04
293.15	0.0000	37.42 ± 0.10
	0.2321	35.97
	0.4687	35.46
	0.7401	34.20
	1.127	32.91
298.15	0.0000	34.26 ± 0.15
	0.5468	32.34
	0.7324	31.84
	1.0425	30.99
	1.2803	30.35

AUXILIARY INFORMATION

METHOD/APPARATUS/PROCEDURE:

The solubility apparatus and procedure employed were similar to that described by Ben-Naim and Baer (1) with modifications suggested by Wen and Hung (2). The apparatus consists mainly of a mercury manometer, a gas-volume measuring buret, a dissolution cell of about 450 cm³ capacity, and a mercury reservoir.

The degassing apparatus and procedure used were similar to that described by Battino et al. (3).

From published ionization constants the authors estimated that nearly 100 per cent of the triethylenediamine is unprotonated when the solution pH is 12 or above, and about 99.7 per cent is in the monoprotonated form when the solution pH is 5.7 to 5.9.

SOURCE AND PURITY OF MATERIALS:

(1) Methane. Matheson Co., Inc. Specified to have a purity of 99.95 per cent.

(2) Hydrochloric acid. Source not given. Reagent grade. Diluted with water and distilled at 1 atm to prepare constant boiling HCl solution.

(3) Water. Carbon dioxide free.

ESTIMATED ERROR:
$$\delta T/K = \pm 0.005$$
$$\delta P/\text{mmHg} = \pm 3$$
$$\delta L/L = \pm 0.005$$

REFERENCES:
1. Ben-Naim, A.; Baer, S. Trans. Faraday Soc. 1964,60,1736.
2. Wen, W.-Y.; Hung, J. H. J. Phys. Chem. 1970, 74, 170.
3. Battino, R.; Banzhof, M.; Bogan, M.; Wilhelm, E. Anal. Chem. 1971, 43, 806.

COMPONENTS:	ORIGINAL MEASUREMENTS:
(1) Methane; CH_4; [74-82-8] (2) Sulfuric acid; H_2SO_4; [7664-93-9] (3) Water; H_2O; [7732-18-5]	Christoff, A. Z. Phys. Chem. 1906, 55, 622-34.

VARIABLES:	PREPARED BY:
T/K = 293.15 p_1/kPa = Atmospheric H_2SO_4/wt % = 0 - 95.6	H. L. Clever

EXPERIMENTAL VALUES:

Temperature		Sulfuric Acid		Ostwald Coefficient
$t/^{\circ}C$	T/K	H_2SO_4/wt %	m_2/mol kg^{-1}	L/cm^3 cm^{-3}
20	293.15	0.0	0.0	0.03756
		35.82	5.69	0.01815
		61.62	16.37	0.01407
		95.6	222.	0.03303

The compiler calculated the acid molality values.

AUXILIARY INFORMATION

METHOD/APPARATUS/PROCEDURE:	SOURCE AND PURITY OF MATERIALS:
The apparatus was an Ostwald type (ref 1) with a lead capillary tube through which the gas flows to the absorption flask, gas buret, and gasometer. The acid solution was degassed by boiling under reflux. The author estimates a one percent change in the acid concentration due to the degassing procedure. The adsorption flask was filled with solvent, the gas was introduced, and the system shaken until equilibrium was reached.	(1) Methane. Prepared by heating soda lime and anhydrous sodium acetate in an iron dish. (2) Sulfuric acid. Merck. Specific gravity 1.271, 1.523, and 1.839 for the 35.82, 61.62, and 95.6 wt % acid, respectively. (3) Water. Distilled.

ESTIMATED ERROR:
$\delta T/K$ = ± 0.02 for solvent
         ± 0.5 for gas
Barometric fluctuations were
stated to be negligible.

REFERENCES:

1. Ostwald, W.
   Lehrbuch der allgem. Chemie
   (2 Aufl.), 1, 615.

COMPONENTS:	ORIGINAL MEASUREMENTS:
(1) Methane; CH_4; [74-82-8] (2) Sulfuric acid; H_2SO_4; [7664-93-9] (3) Water; H_2O; [7732-18-5]	Rudakov, E. S.; Lutsyk, A. I. *Zh. Fiz. Khim.* <u>1979</u>, *53*, 1298-1300. *<u></u>*Russ. J. Phys. Chem.* <u>1979</u>, *53*, 731-3.

VARIABLES:	PREPARED BY:
T/K = 298.2 , 363.2 H_2SO_4/ wt % = 80.0 97.7	H. L. Clever

EXPERIMENTAL VALUES:

Temperature		Sulfuric Acid /wt %	Partition Coefficient[a] k/cm^{3}cm^{-3}	Ostwald Coefficient[b] L/cm^{3}cm^{-3}	Bunsen Coefficient[b] α/cm^3(STP)cm^{-3}atm^{-1}
$t/^0$C	T/K				
25.0	298.2	0	29 ± 2	0.034	0.031
		80.0	110 ± 8	0.0091	0.0083
		93.0	32 ± 2	0.031	0.028
90.0	363.2	0	148 ± 10	0.0068	0.0051
		93.0	43 ± 3	0.023	0.017
		94.9	35 ± 3	0.029	0.022
		97.7	25 ± 2	0.040	0.030

[a] original data from the paper.

[b] The Ostwald and Bunsen coefficient values were calculated by the compiler on the basis that the partition coefficient is equivalent to the inverse of the Ostwald coefficient and assuming that the ideal gas law is obeyed.

The enthalpy of solution of methane in 93.0 wt % sulfuric acid is estimated by the authors to be, ΔH/kcal mol^{-1}= -(1.0 ± 0.5).

AUXILIARY INFORMATION

METHOD/APPARATUS/PROCEDURE:	SOURCE AND PURITY OF MATERIALS:
A gas chromatographic method was used to evaluate the partition coefficients. A reactor containing methane and aqueous acid solution was mechanically shaken for 10 m to establish equilibrium. Equal volumes of samples of the gas and solution phases were introduced by syringe into a special cell for stripping the methane by the carrier gas. The carrier gas entered a gas chromatograph and the partition coefficient was obtained from the ratio of areas of the peaks from each phase. The actual partial pressure of the methane was not specified.	(1) Methane. (2) Sulfuric acid. (3) Water. Sources and purities were not given.
	ESTIMATED ERROR: $\delta k/k$ = ± 0.10 (authors)
	REFERENCES:

COMPONENTS:	ORIGINAL MEASUREMENTS:
1. Methane; CH_4; [74-82-8] 2. Ammonium chloride; NH_4Cl [12125-02-9] 3. Water; H_2O; [7732-18-5]	Ben-Naim, A.; Yaacobi, M. *J. Phys. Chem.* 1974, *78*, 175-8.

VARIABLES:	PREPARED BY:
Temperature	C. L. Young

EXPERIMENTAL VALUES:

T/K	Conc. of ammonium chloride /mol ℓ^{-1}	Ostwald coefficient,[*] L
283.15	1.0	0.03556
288.15		0.03273
293.15		0.03017
298.15		0.02786
303.15		0.02577

[*] Smoothed values of Ostwald coefficient obtained from

$$kT \ln L = 1,721.5 - 9.329 \, (T/K) + 0.01195 \, (T/K)^2 \text{ cal mol}^{-1}$$

where k is in units of cal mol^{-1} K^{-1}.

AUXILIARY INFORMATION

METHOD/APPARATUS/PROCEDURE:	SOURCE AND PURITY OF MATERIALS:
The apparatus was similar to that described by Ben-Naim and Baer (1) and Wen and Hung (2). It consists of three main parts, a dissolution cell of 300 to 600 cm^3 capacity, a gas volume measuring column, and a manometer. The solvent is degassed in the dissolution cell, the gas is introduced and dissolved while the liquid is kept stirred by a magnetic stirrer immersed in the water bath. Dissolution of the gas results in the change in the height of a column of mercury which is measured by a cathetometer.	1. Matheson sample, purity 99.97 mole per cent. 2. AR grade. 3. Deionised, doubly distilled.

ESTIMATED ERROR:

$\delta T/K = \pm 0.01$; $\delta L/L = \pm 0.005$
(estimated by compiler).

REFERENCES:

1. Ben-Naim, A.; Baer, S.
   Trans. Faraday Soc. 1963, *59*,
   2735.

2. Wen, W.-Y.; Hung, J. H.
   J. Phys. Chem. 1970, *74*, 170.

COMPONENTS:	ORIGINAL MEASUREMENTS:
(1) Methane; CH_4; [74-82-8]	Wen, W.-Y.; Hung, J. H.
(2) Water; H_2O; [7732-18-5]	
	J. Phys. Chem. 1970, *74*, 170 - 180.
(3) Ammonium bromide; NH_4Br; [12124-97-9]	

VARIABLES:	PREPARED BY:
T/K: 278.15 - 308.15	
P/kPa: 101.325 (1 atm)	H. L. Clever
m_3/mol kg^{-1}: 0 - 0.202	

EXPERIMENTAL VALUES:

T/K	Ammonium bromide Molality m_3/mol kg^{-1}	Methane Solubility S_1/cm^3(STP) kg^{-1}	Setchenow Constant[1] k/kg mol^{-1}
278.15	0	49.48 ± 0.08	0.068
	0.196	47.93	
288.15	0	38.49 ± 0.11	0.061
	0.202	37.43	
298.15	0	31.35 ± 0.10	0.054
	0.200	30.56	
308.15	0	26.51 ± 0.08	0.047
	0.181	26.03	

[1] Setchenow constant, k/kg mol^{-1} = (1/(m_3/mol kg^{-1}))

log (S_1°/S_1)

The authors specify the value of the constant for m_3/mol kg^{-1} = 0.1.

AUXILIARY INFORMATION

METHOD/APPARATUS/PROCEDURE:	SOURCE AND PURITY OF MATERIALS:
The apparatus was similar to that described by Ben-Naim and Baer (1). Teflon needle valves were used in place of stopcocks.	(1) Methane. Matheson Co. Stated to be better than 99.9 per cent pure.
The apparatus consists of three main parts, a dissolution cell of 300 to 600 cm^3 capacity, a gas volume measuring column, and a manometer.	(2) Water. Distilled from an all Pyrex apparatus. Specific conductivity 1.5 x 10^{-6} (ohm cm)$^{-1}$.
The solvent is degassed in the dissolution cell, the gas is introduced and dissolved while the liquid is kept stirred by a magnetic stirrer immersed in the water bath.	(3) Ammonium bromide. Baker Chemical Co. Analyzed reagent grade. Used as received.
Dissolution of the gas results in the change in the height of a column of mercury which is measured by a cathetometer.	**ESTIMATED ERROR:** $\delta T/K = \pm 0.005$ $\delta S_1/S_1 = \pm 0.003$
	REFERENCES: 1. Ben-Naim, A.; Baer, S. *Trans. Faraday Soc.* 1963, *59*, 2735.

COMPONENTS:	ORIGINAL MEASUREMENTS:
(1) Methane; CH_4; [74-82-8] (2) Water; H_2O; [7732-18-5] (3) N,N,N-Trimethylmethanaminium bromide or tetramethylammonium bromide; $C_4H_{12}NBr$; [64-20-0]	Wen, W.-Y.; Hung, J. H. *J. Phys. Chem.* <u>1970</u>, *74*, 170 - 180.

| VARIABLES:
 T/K: 278.15 - 308.15
 P/kPa: 101.325 (1 atm)
 m_3/mol kg^{-1}: 0 - 0.302 | PREPARED BY:

 H. L. Clever |

EXPERIMENTAL VALUES:

T/K	Salt Molality m_3/mol kg^{-1}	Methane Solubility S_1/cm^3(STP) kg^{-1}	Setchenow Constant[1] k/kg mol^{-1}
278.15	0 0.179 0.176	49.48 ± 0.08 49.48 49.27	+0.007
288.15	0 0.190 0.194	38.49 ± 0.11 38.57 38.70	-0.005
298.15	0 0.179 0.193	31.35 ± 0.10 31.64 31.47	-0.017
308.15	0 0.302	26.51 ± 0.08 27.10	-0.028

[1] Setchenow constant, k/kg mol^{-1} = $(1/(m_3/\text{mol kg}^{-1}))$

log (S_1°/S_1)

The authors specify the value of the constant for

m_3/mol kg^{-1} = 0.1.

AUXILIARY INFORMATION

METHOD/APPARATUS/PROCEDURE:	SOURCE AND PURITY OF MATERIALS:
The apparatus was similar to that described by Ben-Naim and Baer (1). Teflon needle valves were used in place of stopcocks. The apparatus consists of three main parts, a dissolution cell of 300 to 600 cm^3 capacity, a gas volume measuring column, and a manometer. The solvent is degassed in the dissolution cell, the gas is introduced and dissolved while the liquid is kept stirred by a magnetic stirrer immersed in the water bath. Dissolution of the gas results in the change in the height of a column of mercury which is measured by a cathetometer.	(1) Methane. Matheson Co. Stated to be better than 99.9 per cent pure. (2) Water. Distilled from an all Pyrex apparatus. Specific conductivity 1.5 x 10^{-6} (ohm cm)$^{-1}$. (3) Tetramethylammonium bromide. Eastman Kodak Co. Recrystallized and analyzed. Better than 99.9 per cent pure.

| | ESTIMATED ERROR:

 δT/K = ±0.005
 $\delta S_1/S_1$ = ±0.003 |

| | REFERENCES:

 1. Ben-Naim, A.; Baer, S. *Trans. Faraday Soc.* <u>1963</u>, *59*, 2735. |

COMPONENTS:	ORIGINAL MEASUREMENTS:
(1) Methane; CH_4; [74-82-8]	Wen, W.-Y.; Hung, J. H.
(2) Water; H_2O; [7732-18-5]	
(3) N,N,N-Triethylethanaminium bromide or tetraethylammonium bromide; $C_8H_{20}NBr$; [71-91-0]	*J. Phys. Chem.* 1970, *74*, 170 - 180.

| VARIABLES: T/K: 278.15 - 308.15 $\quad$ P/kPa: 101.325 (1 atm) $\quad$ $m_3/mol\ kg^{-1}$: 0 - 0.209 | PREPARED BY: H. L. Clever |

EXPERIMENTAL VALUES:

T/K	Salt Molality $m_3/mol\ kg^{-1}$	Methane Solubility S_1/cm^3 (STP) kg^{-1}	Setchenow Constant[1] $k/kg\ mol^{-1}$
278.15	0	49.48 ± 0.08	-0.036
	0.093	49.89	
	0.095	49.97	
288.15	0	38.49 ± 0.11	-0.049
	0.173	39.30	
298.15	0	31.35 ± 0.10	-0.049
	0.182	32.36	
	0.183	32.34	
308.15	0	26.51 ± 0.08	-0.094
	0.209	27.68	

[1] Setchenow constant, $k/kg\ mol^{-1} = (1/(m_3/mol\ kg^{-1}))$ $\log (S_1^\circ/S_1)$

The authors specify the value of the constant for $m_3/mol\ kg^{-1} = 0.1$.

AUXILIARY INFORMATION

METHOD/APPARATUS/PROCEDURE:

The apparatus was similar to that described by Ben-Naim and Baer (1). Teflon needle valves were used in place of stopcocks.

The apparatus consists of three main parts, a dissolution cell of 300 to 600 cm^3 capacity, a gas volume measuring column, and a manometer.

The solvent is degassed in the dissolution cell, the gas is introduced and dissolved while the liquid is kept stirred by a magnetic stirrer immersed in the water bath. Dissolution of the gas results in the change in the height of a column of mercury which is measured by a cathetometer.

SOURCE AND PURITY OF MATERIALS:

(1) Methane. Matheson Co. Stated to be better than 99.9 per cent pure.

(2) Water. Distilled from an all Pyrex apparatus. Specific conductivity 1.5×10^{-6} (ohm cm)$^{-1}$.

(3) Tetraethylammonium bromide. Eastman Kodak Co. Recrystallized and analyzed. Better than 99.9 per cent pure.

ESTIMATED ERROR:

$$\delta T/K = \pm 0.005$$
$$\delta S_1/S_1 = \pm 0.003$$

REFERENCES:

1. Ben-Naim, A.; Baer, S. *Trans. Faraday Soc.* 1963, *59*, 2735.

COMPONENTS:	ORIGINAL MEASUREMENTS:
(1) Methane; CH_4; [74-82-8]	Blanco C, L. H.; Smith, N. O.
(2) N, N, N-Triethylethaninium bromide or tetraethylammonium bromide; $C_8H_{20}NBr$; [71-91-0]	*J. Phys. Chem.* 1978, *82*, 186-91.
(3) Water; H_2O; [7732-18-5]	

EXPERIMENTAL VALUES:

Temperature		$C_8H_{20}NBr$	Total Pressure	Mol Fraction	Salt Effect Parameter
$t/^0C$	T/K	$m_2/$ mol kg^{-1}	p/MPa	$10^3 x_1$	$k_{smm}/$kg mol^{-1}
25.0	298.2	1.0	10.1	2.025	-0.056
			20.3	3.117	-0.055
			30.4	3.740	-0.044
			40.5	3.914	-
38.0	311.2	1.0	10.1	1.747	-0.046
			20.3	2.759	-0.067
			30.4	3.436	-0.056
			40.5	3.994	-0.047
			50.7	4.451	-0.054
			60.8	4.496	-
51.5	324.7	1.0	10.1	1.591	-0.056
			20.3	2.602	-0.077
			30.4	3.264	-0.073
			40.5	3.817	-0.070
			50.7	4.274	-0.077
			60.8	4.448	-
55.0	328.2	1.0	20.3	2.580	-
60.0	333.2	1.0	20.3	2.550	-
65.0	338.2	1.0	20.3	2.527	-
71.0	344.2	1.0	10.1	1.541	-0.085
			20.3	2.517	-0.105
			30.4	3.304	-0.105
			40.5	3.880	-0.102
			50.7	4.350	-0.104
			60.8	4.618	-
75.0	348.2	1.0	20.3	2.433	-
80.0	353.2	1.0	20.3	2.549	-
85.0	358.2	1.0	20.3	2.579	-
89.5	362.7	1.0	20.3	2.595	-
95.0	368.2	1.0	20.3	2.625	-
100.0	373.3	1.0	20.3	2.659	-
102.5	375.7	1.0	20.3	2.636	-
115.0	388.2	1.0	20.3	2.676	-

The total pressures given in the paper were 100, 200, 300, 400, 500 and 600 atm. They are given above as 10.1, 20.3, 30.4, 40.5, 50.7, and 60.8 MPa.

The salt effect parameters were calculated from smoothed data. The solubility in water was smoothed data from an earlier paper, O'Sullivan, T. D.; Smith, N. O. *J. Phys. Chem.* 1970, *74*, 1460. For the salt effect parameter calculation the methane solubility was converted to molality, thus the salt effect parameter is $k$smm/kg mol^{-1}.

COMPONENTS:	ORIGINAL MEASUREMENTS:
(1) Methane; CH_4; [74-82-8] (2) N,N,N-Triethylethaninium bromide or tetraethylammonium bromide; $C_8H_{20}NBr$; [71-91-0] (3) Water; H_2O; [7732-18-5]	Blanco C, L. H.; Smith, N. O. J. Phys. Chem. 1978, 82, 186-91.

VARIABLES:	PREPARED BY:
T/K = 298.2 - 388.2 p/MPa = 10.1 - 60.8 $m_2/mol\ kg^{-1}$ = 1.0	C. L. Young H. L. Clever

ADDITIONAL INFORMATION:

The authors fitted the mole fraction solubility by the method of least squares to the equation

$$x_1 = a(p/atm) + b(p/atm)^2 + c(p/atm)^3$$

The constants are:

T/K	$10^5 a$	$10^8 b$	$10^{11} c$
298.2	2.6462	-6.9850	7.7333
311.2	2.0875	-4.1892	3.6007
324.7	1.8972	-3.5513	2.9404
344.2	1.7631	-2.8433	2.1206

At each temperature the solubility value at the largest pressure was not included in the curve fitting.

The solubility values of methane in one molal $(C_2H_5)_4NBr$ at 200 atm (20.3 MPa) total pressure were fitted by the method of least squares to two equations as a function of temperature. The equations are:

$\ln (f_1/x_1)$ = 154.8978 - 7445.408/(T/K) - 20.909 $\ln (T/K)$ and

$\ln (x_1$ at 200 atm)   = -135.2291 + 650.371/(T/K) + 18.892 $\ln (T/K)$

From the equations they calculated the thermodynamic changes for the transfer of one mole of gas at unit fugacity to the hypothetical dissolved state (x_1 = 1) at 298.15 K and 200 atm (20.3 MPa) are ΔH^0/kcal mol^{-1}= -2.41; ΔS^0/cal K^{-1} mol^{-1} = -29.5; ΔC_p^0/cal K^{-1} mol^{-1} = 41.6; temperature of minimum solubility is 71 ^{0}C (344.2 K).

AUXILIARY INFORMATION

METHOD/APPARATUS/PROCEDURE:	SOURCE AND PURITY OF MATERIALS:
The apparatus employed and the procedures used were described in earlier work (1, 2). Solubilities were determined in a stirred, thermostated one gallon stainless steel autoclave, samples of the liquid phase were withdrawn into a thermostated buret system for analysis when equilibrium was reached. Pressure was measured with a Bourdon gauge. Tempertaure was measured with an iron-constantan thermocouple. Salt concentration was determined by gravimetric analysis.	(1) Methane. Matheson Co., Inc. Gold label sample, purity 99.97 mole percent. (2) Tetraethylammonium bromide. Eastman Kodak Co. Recrystallized. (3) Water. Distilled and boiled.

ESTIMATED ERROR:
$\delta T/K$ = ± 0.5
$\delta p/MPa$ = ± 0.05 %
$\delta x_1/x_1$ ± 0.004

REFERENCES:
1. O'Sullivan, T. D.; Smith, N. O.
   J. Phys. Chem. 1970, 74, 1460.

2. Gardiner, G. E.; Smith, N. O.
   J. Phys. Chem. 1972, 76, 1195.
   J. Phys. Chem. 1973, 77, 2928.

COMPONENTS:	ORIGINAL MEASUREMENTS:
(1) Methane; CH_4; [74-82-8] (2) Water; H_2O; [7732-18-5] (3) N,N,N-Tripropylpropanaminium bromide or tetrapropylammonium bromide; $C_{12}H_{28}NBr$; [1941-30-6]	Wen, W.-Y.; Hung, J. H. *J. Phys. Chem.* <u>1970</u>, *74*, 170 - 180.

| VARIABLES: T/K: 278.15 - 308.15
 P/kPa: 101.325 (1 atm)
 $m_3/mol\ kg^{-1}$:   0 - 0.706 | PREPARED BY:

 H. L. Clever |

EXPERIMENTAL VALUES:

T/K	Salt Molality $m_3/mol\ kg^{-1}$	Methane Solubility S_1/cm^3 (STP) kg^{-1}	Setchenow Constant[1] $k/kg\ mol^{-1}$
278.15	0	49.48 ± 0.08	-0.045
	0.099	50.06	
	0.103	50.03	
	0.224	50.31	
	0.223	50.46	
	0.431	49.60	
	0.587	48.74	
288.15	0	38.49 ± 0.11	-0.061
	0.098	39.00	
	0.223	39.67	
	0.227	39.48	
	0.415	39.60	
	0.620	39.97	
298.15	0	31.35 ± 0.10	-0.082
	0.097	31.87	
	0.102	31.99	
	0.230	32.43	
	0.235	32.58	
	0.410	33.14	
	0.632	33.63	
308.15	0	26.51 ± 0.08	-0.110
	0.105	27.21	
	0.245	27.87	
	0.443	28.73	
	0.706	30.17	

AUXILIARY INFORMATION

METHOD/APPARATUS/PROCEDURE:	SOURCE AND PURITY OF MATERIALS:
The apparatus was similar to that described by Ben-Naim and Baer (1). Teflon needle valves were used in place of stopcocks. The apparatus consists of three main parts, a dissolution cell of 300 to 600 cm^3 capacity, a gas volume measuring column, and a manometer. The solvent is degassed in the dissolution cell, the gas is introduced and dissolved while the liquid is kept stirred by a magnetic stirrer immersed in the water bath. Dissolution of the gas results in the change in the height of a column of mercury which is measured by a cathetometer.	(1) Methane. Matheson Co. Stated to be better than 99.9 per cent pure. (2) Water. Distilled from an all Pyrex apparatus. Specific conductivity 1.5 x 10^{-6} (ohm cm)$^{-1}$. (3) Tetrapropylammonium bromide. Eastman Kodak Co. Recrystallized and analyzed. Better than 99.9 per cent pure.
	ESTIMATED ERROR: $\delta T/K = \pm 0.005$ $\delta S_1/S_1 = \pm 0.003$
[1] Setchenow constant, $k/kg\ mol^{-1} =$ $(1/(m_3/mol\ kg^{-1})) \log (S_1^o/S_1)$ The authors specify the value of the constant for $m_3/mol\ kg^{-1} = 0.1$.	REFERENCES: 1. Ben-Naim, A.; Baer, S. *Trans. Faraday Soc.* <u>1963</u>, *59*, 2735.

COMPONENTS:	ORIGINAL MEASUREMENTS:
(1) Methane; CH_4; [74-82-8]	Wen, W.-Y.; Hung, J. H.
(2) Water; H_2O; [7732-18-5]	
(3) N,N,N-Tributylbutanaminium bromide or tetrabutylammonium bromide; $C_{16}H_{36}NBr$; [1643-19-2]	*J. Phys. Chem.* <u>1970</u>, *74*, 170 - 170.

EXPERIMENTAL VALUES:

T/K	Salt Molality m_3/mol kg^{-1}	Methane Solubility S_1/cm^3(STP) kg^{-1}	Setchenow Constant[1] k/kg mol^{-1}
278.15	0	49.48 ± 0.08	-0.030
	0.096	49.82	
	0.098	49.85	
	0.100	49.80	
	0.187	49.88	
	0.201	49.86	
	0.409	49.27	
	0.526	48.91	
	0.785	46.67	
	0.990	46.10	
288.15	0	38.49 ± 0.11	-0.053
	0.096	38.86	
	0.102	38.95	
	0.103	38.96	
	0.185	39.10	
	0.403	39.44	
	0.523	39.40	
	0.703	39.69	
	1.018	40.10	
298.15	0	31.35 ± 0.10	-0.096
	0.096	32.03	
	0.099	32.00	
	0.194	32.32	
	0.415	33.30	
	0.526	33.79	
	0.704	34.53	
	1.022	36.72	
308.15	0	26.51 ± 0.08	-0.152
	0.099	27.43	
	0.192	28.20	
	0.415	29.45	
	0.537	30.18	
	0.693	31.38	
	0.993	33.60	

[1] Setchenow constant, k/kg mol^{-1} = $(1/(m_3$/mol kg^{-1})) log $(S_1^°/S_1)$

The authors specify the value of the constant for m_3/mol kg^{-1} = 0.1.

COMPONENTS:	ORIGINAL MEASUREMENTS:
1. Methane; CH_4; [74-82-8] 2. 1-Methyl-2-pyrrolidinone; C_5H_9NO ; [872-50-4] 3. Water; H_2O ; [7732-18-5]	Wu, Z.; Zeck, S.; Knapp, H. *Ber. Bunsenges. Phys. Chem.*, <u>1985</u>, *89*, 1009-1013.
VARIABLES:	PREPARED BY:
Composition of solvent	C. L. Young.

EXPERIMENTAL VALUES:

T/K	Mole fraction of water	Henry's constant /MPa	Ostwald coefficient	Mole fraction of gas x 10^4 [a]
298.15	1.000	39.06	0.0349	0.2594
	0.950	30.80	0.0367	0.3290
	0.883	22.78	0.0406	0.4448
	0.806	15.80	0.0483	0.6413
	0.645	7.192	0.0768	1.409
	0.494	3.862	0.113	2.624
	0.361	2.401	0.153	4.220
	0.193	1.581	0.193	6.409
	0.075	1.236	0.221	8.198
	0.000	1.059	0.242	9.568

[a] Calculated by compiler for a partial pressure of 1 atmosphere

---

AUXILIARY INFORMATION

METHOD APPARATUS/PROCEDURE:	SOURCE AND PURITY OF MATERIALS:
Precision volumetric apparatus described in detail in ref. (1). Pressure measured with mercury manometer. Composition of solvent estimated from density and refractive index as composition changed on degassing.	1. Purity better than 99 volume per cent. 2. Merck sample, dried with molecular sieve 4 X. Final water content less than 0.01 mass per cent, purity 99.9 mole per cent by GC. 3. Twice distilled.

ESTIMATED ERROR:

$\partial T/K = \pm 0.01$; $\partial P/Pa = \pm 50$;
∂x(solvent)$=\pm 0.003$; ∂x = 0.005

REFERENCES:

1. Zeck, S.; Dissertion,
   TU Berlin, 1985.

COMPONENTS:	ORIGINAL MEASUREMENTS:
(1) Methane; CH_4; [74-82-8] (2) N,N,N-Tributyl-1-butanaminium bromide or tetrabutyl ammonium bromide; $C_{16}H_{36}NBr$; [1643-19-2] (3) Water; H_2O; [7732-18-5]	Feillolay, A.; Lucas, M. *J. Phys. Chem.* <u>1972</u>, *76*, 3068-72.

VARIABLES:	PREPARED BY:
T/K = 298.15, 308.15 p_1/kPa = 101.325 $m_2/mol\ kg^{-1}$ = 0 - 4.010	H. L. Clever

EXPERIMENTAL VALUES:

Temperature		Tetrabutyl Ammonium Bromide $m_2/mol\ kg^{-1}$	Methane Solubility $/cm^3(STP)\ kg^{-1}$	Salt Effect Parameter $k_{smm}/kg\ mol^{-1}$
$t/^0C$	T/K			
25	298.15	0	29.87, 29.99, 30.05, Av. 29.97	-
		1.010	35.40	-0.0716
		1.020	36.00	-0.0781
		1.981	44.31	-0.0857
		1.991	44.59	-0.0867
		3.623	78.01	-0.1147
		3.925	81.68	-0.1109
35	308.15	0	25.32, 25.38 Av. 25.35	-
		1.025	32.30	-0.1027
		1.005	32.05	-0.1013
		2.025	43.52	-0.1159
		2.078	43.60	-0.1133
		3.610	73.49	-0.1280
		3.640	74.55	-0.1287
		4.010	82.40	-0.1277

The salt effect parameters were calculated by the compiler.

AUXILIARY INFORMATION

METHOD/APPARATUS/PROCEDURE:	SOURCE AND PURITY OF MATERIALS:
The apparatus is modeled after the apparatus used by Hung (ref 1). The procedure was the same as that used by Hung except that the time allowed for equilibration is longer. In the present work gas equilibration required about 16 h.	(1) Methane. l'Air Liquide. Stated to be of 99.99 percent purity. (2) Tetrabutyl ammonium bromide. Southwestern Analytical Chemical. Polarographic grade, used as received.
	ESTIMATED ERROR: Methane solubility ± 0.5 percent.
	REFERENCES: 1. Hung, J. H. <u>1968</u>, Ph. D. thesis, Clark University, Worcester, MA

COMPONENTS:	ORIGINAL MEASUREMENTS:
(1) Methane; CH_4; [74-82-8] (2) Water; H_2O; [7732-18-5] (3) 2-Hydroxy-N,N,N-tris(2-hydroxyethyl)-ethanaminium bromide or tetraethanolammonium bromide; $C_8H_{20}NO_4Br$; [4328-04-5]	Wen, W.-Y.; Hung, J. H. *J. Phys. Chem.* <u>1970</u>, *74*, 170 - 180.

VARIABLES:	PREPARED BY:
T/K: 278.15 - 308.15 P/kPa: 101.325 (1 atm) m_3/mol kg^{-1}: 0 - 0.517	H. L. Clever

EXPERIMENTAL VALUES:

T/K	Salt Molality m_3/mol kg^{-1}	Methane Solubility S_1/cm^3(STP) kg^{-1}	Setchenow Constant k/kg mol^{-1}
278.15	0	49.48 ± 0.08	0.042
	0.086	49.10	
	0.152	48.70	
	0.155	48.86	
	0.173	48.47	
	0.341	47.87	
	0.488	46.92	
288.15	0	38.49 ± 0.11	0.033
	0.091	38.22	
	0.174	38.03	
	0.177	37.96	
	0.355	37.34	
	0.517	36.87	
298.15	0	31.35 ± 0.10	0.022
	0.085	31.20	
	0.167	31.02	
	0.173	31.04	
	0.347	30.81	
	0.510	30.42	
308.15	0	26.51 ± 0.08	-0.001
	0.085	26.50	
	0.173	26.59	
	0.339	26.53	
	0.508	26.51	

AUXILIARY INFORMATION

METHOD/APPARATUS/PROCEDURE:	SOURCE AND PURITY OF MATERIALS:
The apparatus was similar to that described by Ben-Naim and Baer (1). Teflon needle valves were used in place of stopcocks. The apparatus consists of three main parts, a dissolution cell of 300 to 600 cm^3 capacity, a gas volume measuring column, and a manometer. The solvent is degassed in the dissolution cell, the gas is introduced and dissolved while the liquid is kept stirred by a magnetic stirrer immersed in the water bath. Dissolution of the gas results in the change in the height of a column of mercury which is measured by a cathetometer.	(1) Methane. Matheson Co. Stated to be better than 99.9 per cent pure. (2) Water. Distilled from an all Pyrex apparatus. Specific conductivity 1.5 x 10^{-6} (ohm cm)$^{-1}$. (3) Tetraethanolammonium bromide. Prepared and analyzed. Better than 99.9 per cent pure. m.p., t/°C 102.

ESTIMATED ERROR:
$\delta T/K = \pm0.005$ $\delta S_1/S_1 = \pm0.003$

[1] Setchenow constant, k/kg mol^{-1} = $(1/(m_3/mol\ kg^{-1}))$ log (S_1°/S_1)

The authors specify the value of the constant for m_3/mol kg^{-1} = 0.1.

REFERENCES:
1. Ben-Naim, A.; Baer, S. *Trans. Faraday Soc.* <u>1963</u>, *59*, 2735.

COMPONENTS:	ORIGINAL MEASUREMENTS:
(1) Methane; CH_4; [74-82-8] (2) N, N, N-Trimethyl-1-hexadecana-minium bromide or cetyltrimethyl ammonium bromide; $C_{19}H_{42}NBr$; [57-09-0] (3) Water; H_2O; [7732-18-5]	Prapaitrakul, W.; King, A. D. Jr. $J.$ $Coll.$ $Interface$ $Sci.$ 1985, 106 186-93.

VARIABLES:	PREPARED BY:
T/K = 299.0 p_1/kPa = 120 - 690 (Est., see ref 1)	H. L. Clever

EXPERIMENTAL VALUES:

Temperature		Cetyltrimethyl ammonium bromide	Henry's Constant[a]
t/^{0}C	T/K	m_2/mol kg^{-1}	$10^3 H$/mol kg^{-1} atm^{-1}
26	299.0	0.0	1.55[b]
		0.10	1.79
		0.20	2.08
		0.30	2.41

[a] The authors reported these results as $10^3 m_1$/mol kg^{-1} at 1 atm.

Henry's constnat is defined as

$$H/\text{mol kg}^{-1} \text{ atm}^{-1} = (m_1/\text{mol kg}^{-1})/(p_1/\text{atm}).$$

[b] The solubility value in water is from (ref 2).

AUXILIARY INFORMATION

METHOD/APPARATUS/PROCEDURE:	SOURCE AND PURITY OF MATERIALS:
The apparatus and procedure are described in detail in earlier papers (ref 1, 2). The solvent, contained in a glass-lined brass equilibrium cell resting on a magnetic stirrer, is degassed by evacuation and stirring. The gas is introduced at pressures above atmospheric and the solution stirred until equilibrium is reached. The pressure is reduced to atmospheric over the still liuqid. The liquid is stirred and the gas evolved from the super-saturated solution is collected at atmospheric pressure and ambient temperature in a Warburg manometer, and its volume is measured. Corrections are made for the gas lost during the venting procedure, for the differences in temperature and pressure, and for the water vapor pressure in the calculation of Henry's constant.	(1) Methane. Matheson Co., Inc. Stated to be 99.0 mol % or better. (2) Cetyltrimethylammonium bromide. Aldrich Chemical Co. Lot No. 5814AJ. Recrystallized once from 2-propanol and dried in $vacuo$. (3) Water. Double distilled.

	ESTIMATED ERROR:
	    δT/K = ± 0.1
	    δm_1/mol kg^{-1} = ± 0.00002
	                    (authors)

	REFERENCES: 1. Matheson, I. B. C.; King, A. D. Jr. $J.$ $Coll.$ $Interface$ $Sci.$ 1978, 66, 464. 2. Hoskins, J. C.; King, A. D. Jr. $J.$ $Coll.$ $Interface$ $Sci.$ 1981, 82, 264.

COMPONENTS:	ORIGINAL MEASUREMENTS:
(1) Methane; CH_4; [74-82-8] (2) N,N,N-Trimethyl-1-decanaminium bromide or decyltrimethylammonium bromide; $C_{13}H_{30}NBr$; [2082-84-0] (3) Water; H_2O; [7732-18-5]	Prapaitrakul, W.; King, A. D. Jr. *J. Coll. Interface Sci.* <u>1985</u>, *106*, 186-93.

VARIABLES:	PREPARED BY:
T/K = 299.0 p_1/kPa = 120 - 690 (est., see ref 1)	H. L. Clever

EXPERIMENTAL VALUES:

Temperature		Decyltrimethyl ammonium bromide	Henry's Constant[a]
$t/^0C$	T/K	$m_2/mol\ kg^{-1}$	$10^3 H/mol\ kg^{-1}\ atm^{-1}$
26	299	0.0	1.55[b]
		0.04	1.50
		0.10	1.56
		0.20	1.71
		0.30	1.93
		0.40	2.06
		0.50	2.25

[a] The authors reported these results as $10^3 m_1/mol\ kg^{-1}$ at 1 atm.

Henry's constant is defined,

$$H/mol\ kg^{-1}\ atm^{-1} = (m_1/mol\ kg^{-1})/(p_1/atm).$$

[b] The solubility value in water is from (ref 2).

AUXILIARY INFORMATION

METHOD/APPARATUS/PROCEDURE:	SOURCE AND PURITY OF MATERIALS:
The apparatus and procedure are described in detail in earlier papers (ref 1,2). The solvent, contained in a glass-lined brass equilibrium cell resting on a magnetic stirrer, is degassed by evacuation and stirring. The gas is introduced at pressures above atmospheric and the solution stirred until equilibrium is reached. The pressure is reduced to atmospheric over the still liquid. The liquid is stirred and the gas evolved from the super-saturated solution is collected at atmospheric pressure and ambient temperature in a Warburg manometer, and its volume is measured. Corrections are made for the gas lost during the venting procedure, the differences in temperature and pressure, and the water vapor pressure in the calculation of Henry's constant.	(1) Methane. Matheson Co. Inc. Stated to be 99.0 mol % or better. (2) Decyltrimethylammonium bromide. Eastman Kodak Co. Lot No. A10E and A10F. Recrystalized once from 2-propanol and dried *in vacuo*. CMC agreed well with accepted value. (3) Water. Double distilled.

ESTIMATED ERROR:

$\delta T/K$ = ± 0.1
$\delta m_1/mol\ kg^{-1}$ = ± 0.000
             (authors)

REFERENCES:

1. Matheson, I.B.C.; King, A.D.Jr. *J. Coll. Interface Sci.* <u>1978</u>, *66*, 464.
2. Hoskins, J.C.; King, A.D. Jr. *J. Coll. Interface Sci.* <u>1981</u>, *82*, 264.

COMPONENTS:	ORIGINAL MEASUREMENTS:
(1) Methane; CH_4; [74-82-8]	Muccitelli, J. A.; Wen, W.-Y.
(2) Triethylenediamine hydrochloride or 1,4-diazabicyclo[2.2.2]octane hydrochloride; $C_6H_{13}ClN_2$	*J. Solution Chem.* <u>1980</u>, *9*, 141 - 161.
(3) Water; H_2O; [7732-18-5]	

VARIABLES:	PREPARED BY:
T/K: 278.15 - 298.15 p/kPa: 101.325 (1 atm)	H. L. Clever

EXPERIMENTAL VALUES:

T/K	$C_6H_{13}ClN_2$ $c_2/mol\ dm^{-3}$	pH	Ostwald Coefficient $10^3 L/cm^3 cm^{-3}$
278.15	0.1225	5.37	49.58
	0.2312	5.53	48.29
	0.6291	5.40	44.40
	0.9023	5.59	42.65
283.15	0.1560	5.60	43.43
	0.2538	5.07	42.45
	0.5346	5.46	40.20
	0.8120	5.72	38.70
288.15	0.1601	5.79	38.92
	0.2799	5.62	38.24
	0.6299	5.87	36.91
	0.8029	5.80	35.31
293.15	0.1367	5.44	35.90
	0.2340	5.59	35.32
	0.5267	5.77	33.70
	0.8510	5.74	31.74
298.15	0.1469	5.49	34.10
	0.2340	5.59	35.32
	0.5267	5.77	33.70
	0.8510	5.74	31.74

AUXILIARY INFORMATION

METHOD/APPARATUS/PROCEDURE:

The solubility apparatus and procedure employed were similar to that described by Ben-Naim and Baer (1) with modifications suggested by Wen and Hung (2). The apparatus consists mainly of a mercury manometer, a gas-volume measuring buret, a dissolution cell of about 450 cm³ capacity, and a mercury reservoir.

The degassing apparatus and procedure used were similar to that described by Battino *et al.* (3).

From published ionization constants the authors estimated that nearly 100 per cent of the triethylenediamine is unprotonated when the solution pH is 12 or above, and about 99.7 per cent is in the monoprotonated form when the solution pH is 5.7 to 5.9.

SOURCE AND PURITY OF MATERIALS:

(1) Methane. Matheson Co., Inc. Specified to have a purity of 99.95 per cent.

(2) Triethylenediamine hydrochloride. The pH of the triethylenediamine solution was adjusted to a pH of 5.40 ± 0.01 by constant boiling HCl.

(3) Water. Carbon dioxide free.

ESTIMATED ERROR:
$\delta T/K = \pm\ 0.005$
$\delta P/mmHg = \pm\ 3$
$\delta L/L = \pm\ 0.005$

REFERENCES:
1. Ben-Naim, A.; Baer, S. *Trans. Faraday Soc.* <u>1964</u>,*60*,1736.
2. Wen, W.-Y.; Hung, J. H. *J. Phys. Chem.* <u>1970</u>, *74*, 170.
3. Battino, R.; Banzhof, M.; Bogan, M.; Wilhelm, E. *Anal. Chem.* <u>1971</u>, *43*, 806.

COMPONENTS:	ORIGINAL MEASUREMENTS:

1. Methane; CH_4; [74-82-8]
2. Water; H_2O; [7732-18-5]
3. Guanidine monohydrochloride
   (Guanidinium chloride);
   CH_6ClN_3; [50-01-1]
	Wetlaufer, D. B.; Malik, S. K.;
	Stoller, L.; Coffin, R. L.
	J. Am. Chem. Soc.
	<u>1964</u>, *86*, 508-514.

VARIABLES:	PREPARED BY:
Temperature	C. L. Young

EXPERIMENTAL VALUES:

T/K	Conc. of guanidinium chloride in soln. /mol dm^{-3}	10^3 Conc. of methane[†] in soln. /mol dm^{-3}	Mole fraction[*] of methane x_{CH_4}
278.2	4.86	1.15	0.0000280
298.2	4.86	0.96	0.0000234
318.2	4.86	0.805	0.0000196

[†] at a partial pressure of 101.3 kPa.

[*] calculated by compiler.

AUXILIARY INFORMATION

METHOD/APPARATUS/PROCEDURE:	SOURCE AND PURITY OF MATERIALS:
Modified Van Slyke-Neill apparatus fitted with a magnetic stirrer. Solution was saturated with gas and then sample transferred to the Van Slyke extraction chamber.	1. Matheson c.p. grade, purity 99 mole per cent or better. 2. Distilled. 3. Prepared from the action of reagent grade hydrochloric acid on twice or three times recrystallized guanidinium carbonate.
	ESTIMATED ERROR: $\delta T/K = \pm 0.05$; $\delta x_{CH_4} = \pm 2\%$.
	REFERENCES:

COMPONENTS:	ORIGINAL MEASUREMENTS:
(1) Methane; CH_4; [74-82-8]	Stoessell, R. K.; Byrne, P. A.
(2) Magnesium chloride; $MgCl_2$; [7786-30-3]	*Geochim. Cosmochim. Acta* <u>1982</u>, *46*, 1327-32.
(3) Water; H_2O; [7732-18-5]	

VARIABLES:	PREPARED BY:
$T/K = 298.15$ $p_1/kPa = 2410-5170$ $m_2/mol\ kg^{-1} = 0-2.16$	H. L. Clever

EXPERIMENTAL VALUES:

Temperature		Pressure		Magnesium Chloride	Methane	Salt Effect Parameter
$t/°C$	T/K	$p_1/psia$	p_1/kPa	$m_2/mol\ kg^{-1}$	$m_1/mol\ kg^{-1}$	$k_{smm}/kg\ mol^{-1}$
25	298.15	350	2410	0	0.0319	–
				0.5	0.0257	0.063
				1.0	0.0199	0.068
				2.16	0.0128	0.061
		550	3790	0	0.0483	–
				0.5	0.0376	0.073
				1.0	0.0300	0.069
				2.16	0.0185	0.064
		750	5170	0	0.0617	–
				0.5	0.0485	0.070
				1.0	0.0390	0.066
				2.16	0.0237	0.064
						0.063 (authors)

The salt effect parameter is defined as $k_{smm} = \log \gamma_1/I$

where I is the ionic strength (molality) and $\gamma_1 = (m_1^* f_1/m_1 f_1^*)$

with m_1 and f_1 the solubility and fugacity, respectively, of methane at p

and T. The "*" refers to saturation in distilled water.

AUXILIARY INFORMATION

METHOD/APPARATUS/PROCEDURE:	SOURCE AND PURITY OF MATERIALS:
Solubility determinations were made using a titanium-lined chamber within a stainless steel reaction vessel jacketed by a water bath for temperature control. The system pressure was set by controlling the input and output of methane within the chamber's headpiece. The vessel was rocked for three h to allow equilibration between the methane and solution. The amount of gas in the saturated solution was measured by transfer of a sample volume to a loop at the system pressure, followed by flashing the sample in an ansion loop and measuring the gas pressure in a known volume. The total gas volume and pressure change were used to compute the moles of released gas assuming ideal behavior. A correction was made for the gas not released on flashing.	(1) Methane. Matheson Co., Inc. Ultra high purity grade, stated to be a minimum of 99.97 mole present methane. (2) Magnesium chloride. The salt solutions were made up gravimetrically using analytical grade chemicals. (3) Water. Distilled.
	ESTIMATED ERROR: $\delta\ p_1/psia = \pm 1$ $\delta\ m_1/m_1 = \pm 0.01$
	REFERENCES:

COMPONENTS:	ORIGINAL MEASUREMENTS:
(1) Methane; CH_4; [74-82-8] (2) Magnesium sulfate; $MgSO_4$; [7785-87-7] (3) Water; H_2O; [7732-18-5]	Stoessell, R. K.; Byrne, P. A. *Geochim. Cosmochim. Acta* <u>1982</u>, *46*, 1327-32.

| VARIABLES: T/K = 298.15
 p_1/kPa = 2410-5170
 m_2/mol kg^{-1} = 0-1.5 | PREPARED BY:

 H. L. Clever |

EXPERIMENTAL VALUES:

Temperature		Pressure		Magnesium Sulfate	Methane	Salt Effect Parameter
$t/°C$	T/K	p_1/psia	p_1/kPa	m_2/mol kg^{-1}	m_1/mol kg^{-1}	k_{smm}/kg mol^{-1}
25	298.15	350	2410	0	0.0319	–
				0.5	0.0233	0.068
				1.0	0.0174	0.066
				1.5	0.0128	0.066
		550	3790	0	0.0483	–
				0.5	0.0352	0.069
				1.0	0.0261	0.067
				1.5	0.0192	0.067
		750	5170	0	0.0617	–
				0.5	0.0472	0.058
				1.0	0.0342	0.064
				1.5	0.0253	0.065
						0.066 (authors)

The salt effect parameter is defined as $k_{smm} = \log \gamma_1 / I$

where I is the ionic strength (molality) and $\gamma_1 = (m_1^* f_1 / m_1 f_1^*)_{p,T}$

with m_1 and f_1 the solubility and fugacity, respectively, of methane at p

and T. The "*" refers to saturation in distilled water.

AUXILIARY INFORMATION

METHOD/APPARATUS/PROCEDURE:	SOURCE AND PURITY OF MATERIALS:
Solubility determinations were made using a titanium-lined chamber within a stainless steel reaction vessel jacketed by a water bath for temperature control. The system pressure was set by controlling the input and output of methane within the chamber's headpiece. The vessel was rocked for three h to allow equilibration between the methane and solution. The amount of gas in the saturated solution was measured by transfer of a sample volume to a loop at the system pressure, followed by flashing the sample in an ansion loop and measuring the gas pressure in a known volume. The total gas volume and pressure change were used to compute the moles of released gas assuming ideal behavior. A correction was made for the gas not released on flashing.	(1) Methane. Matheson Co., Inc. Ultra high purity grade, stated to be a minimum of 99.97 mole present methane. (2) Magnesium sulfate. The salt solutions were made up gravimetrically using analytical grade chemicals. (3) Water. Distilled.

	ESTIMATED ERROR:
	     δp_1/psia = ± 1
	       $\delta m_1/m_1$ = ± 0.01

	REFERENCES:

COMPONENTS:	ORIGINAL MEASUREMENTS:
(1) Methane; CH_4; [74-82-8] (2) (3) Electrolytes, see below (4) Water; H_2O; [7732-18-5]	Byrne, P. A.; Stoessell, R. K. *Geochim. Cosmochim. Acta* <u>1982</u>, *46*, 2395-7.

VARIABLES:	PREPARED BY:
T/K = 298.15 p_1/kPa = 3790	H. L. Clever

EXPERIMENTAL VALUES:

Temperature		Pressure		Electrolyte		Methane	Salt Effect
$t/°C$	T/K	$p_1/psia$	p_1/kPa		$m_i/mol\ kg^{-1}$	$m_1/mol\ kg^{-1}$	Parameter
25	298.15	550	3790	–	0	0.0483^a	–
				$MgCl_2$	0.25 ⎫	0.0358	0.076
				$MgSO_4$	0.25 ⎭		
25	298.15	550	3790	–	0	0.0483^a	–
				$MgCl_2$	0.5 ⎫	0.0275	0.070
				$MgSO_4$	0.5 ⎭		

[a] Value of methane solubility in water, $m_1°$, from (ref 1)

The salt effect parameter, $k_{smm}/kg\ mol^{-1}$ =

$$(\Sigma((k_{smim}/kg\ mol^{-1})(I_i/mol\ kg^{-1}))/(I/mol\ kg^{-1})$$

where I_i and I are ionic strength due to component i and the total ionic strength, respectively, and $k_{smim}/kg\ mol^{-1}$ =

$$log((m_1°/mol\ kg^{-1})/(m_1/mol\ kg^{-1}))/(I_i/mol\ kg^{-1})$$

Magnesium chloride; $MgCl_2$; [7786-30-3]
Magnesium sulfate; $MgSO_4$; [7785-87-7]

AUXILIARY INFORMATION

METHOD/APPARATUS/PROCEDURE:	SOURCE AND PURITY OF MATERIALS:
Solubility determinations were made using a titanium-lined chamber within a stainless steel reaction vessel jacketted by a water bath for temperature control. The system pressure was set by controlling the input and output of methane within the chamber's headpiece. The vessel was rocked for 3 h to allow equilibration between the methane and the solution. The amount of gas in the saturated solution was measured by transfer of a sample volume to a loop at the system pressure, followed by flashing the sample in an expansion loop and measuring the gas pressure in a known volume. The total gas volume and pressure change were used to compute the moles of gas assuming ideal behavior. A correction was made for the gas not released on flashing. Solution densities were measured gravimetrically with pycnometers.	(1) Methane. Matheson Co., Inc. Ultra high purity grade, stated to have a minimum purity of 99.97 mole percent. (2) Electrolytes. The salt solutions (3) were made up gravimetrically using analytical grade chemicals. (4) Water. Distilled. ESTIMATED ERROR: $\delta p_1/psia$ = ± 1 $\delta m_1/mol\ kg^{-1}$ = ± 0.0003 - 0.0005 $\delta m_{2,3}/mol\ kg^{-1}$ = ± 0.0001 REFERENCES: 1. Stoessell, R. K.; Byrne, P. A. *Geochim. Cosmochim. Acta* <u>1982</u>, *46*, 1327.

COMPONENTS:	ORIGINAL MEASUREMENTS:
1. Methane; CH_4; [74-82-8] 2. Water; H_2O; [7732-18-5] 3. Calcium chloride; $CaCl_2$; [10043-52-4]	Michels A.; Gerver, J.; Bijl, A. *Physica*, <u>1936</u>, *3*, 797-808.

VARIABLES:	PREPARED BY:
Pressure	C.L. Young

EXPERIMENTAL VALUES:

T/K	Conc. of $CaCl_2$/mol l^{-1}	$p/10^5$Pa	10^3 Mole fraction of methane in liquid, $10^3 x_{CH_4}$
298.15	2.7	56.2 110.4 157.2 209.9	0.34 0.56 0.69 0.80

AUXILIARY INFORMATION

METHOD/APPARATUS/PROCEDURE:	SOURCE AND PURITY OF MATERIALS:
Simple rocking equilibrium cell. Amount of gas absorbed calculated from volume and pressure change in charging vessel. Details in source.	No details given.

ESTIMATED ERROR:
$\delta T/K = \pm 0.1$; $\delta p/10^5$Pa $= \pm 0.05$ to 0.5%; $\delta x_{CH_4} = \pm 3\text{-}5\%$. (estimated by compiler.)

REFERENCES:

COMPONENTS:	ORIGINAL MEASUREMENTS:
1. Methane; CH_4; [74-82-8] 2. Water; H_2O; [7732-18-5] 3. Calcium chloride; $CaCl_2$; [10043-52-4]	Blanco, L. H.; Smith, N. O. *J. Phys. Chem.* <u>1978</u>, *82*, 186-191.

VARIABLES:	PREPARED BY:
	C. L. Young

EXPERIMENTAL VALUES:

T/K	Conc. of $CaCl_2$ /mol dm^{-3}	P/MPa	P/atm	10^3 Mole fraction of methane in liquid, $10^3 x_{CH_4}$
298.2	1.0	10.1	100	1.032
		20.3	200	1.591
		30.4	300	1.956
		40.5	400	2.202
		50.7	500	2.471
		60.8	600	2.720
324.7	1.0	10.1	100	0.834
		20.3	200	1.360
		30.4	300	1.724
		40.5	400	2.004
		50.7	500	2.232
		60.8	600	2.455
344.2	1.0	10.1	100	0.787
		20.3	200	1.286
		30.4	300	1.665
		40.5	400	1.963
		50.7	500	2.215
		60.8	600	2.443
375.7	1.0	10.1	100	0.803
		20.3	200	1.347
		30.4	300	1.740

(cont.)

AUXILIARY INFORMATION

METHOD/APPARATUS/PROCEDURE:	SOURCE AND PURITY OF MATERIALS:
Large steel stirred equilibrium cell. Pressure measured with Bourdon gauge. Temperature measured with iron-constantan thermocouple. Cell charged with liquid, compressed gas added. After equilibrium obtained samples removed and analysed using volumetric techniques. Salt concentration determined by gravimetric analysis.	1. Matheson, gold label sample, purity 99.97 mole per cent. 2. Distilled and boiled. 3. Fisher Certified grade used without further purification.

	ESTIMATED ERROR:
	$\delta T/K = \pm 0.5$; $\delta P/MPa = \pm 0.05\%$ δx_{CH_4} $= \pm 0.4\%$.

	REFERENCES:

COMPONENTS:	ORIGINAL MEASUREMENTS:
1. Methane; CH_4; [74-82-8]	Blanco, L. H.; Smith, N. O.
2. Water; H_2O; [7732-18-5]	J. *Phys. Chem.*
3. Calcium chloride; $CaCl_2$; [10043-52-4]	1978, *82*, 186-191.

EXPERIMENTAL VALUES:

T/K	Conc. of $CaCl_2$ /mol dm^{-3}	P/MPa	P/atm	10^3 Mole fraction of methane in liquid, $10^3 x_{CH_4}$
375.7	1.0	40.5	400	2.018
		50.7	500	2.281
		60.8	600	2.477
398.2	1.0	10.1	100	0.826
		20.3	200	1.371
		30.4	300	1.709
		40.5	400	2.034
		50.7	500	2.252
		60.8	600	2.477

COMPONENTS:	ORIGINAL MEASUREMENTS:
(1) Methane; CH_4; [74-82-8] (2) Calcium chloride; $CaCl_2$; [10043-52-4] (3) Water; H_2O; [7732-18-5]	Duffy, J. R.; Smith, N. O.; Nagy, B. *Geochim. Cosmochim. Acta* <u>1961</u>, *24*, 23-31.

EXPERIMENTAL VALUES:

Temperature		Calcium Chloride	Pressure		Mol Fraction
$t/^0C$	T/K	$c_2/mol\ dm^{-1}$	$p_1/psia$	p_1/MPa	$10^4 x_1$
25	298.15	0	160	1.10	2.14
			215	1.48	2.73
			230	1.59	3.76
			430	2.96	7.08
			445	3.07	7.03
			514	3.54	8.00
			585	4.03	9.39
			680	4.69	9.79
			750	5.17	11.30
		0.25	165	1.14	2.08
			328	2.26	4.49
			493	3.40	6.72
			631	4.35	8.98
		0.50	118	0.81	1.26
			212	1.46	2.39
			281	1.94	3.30
			377	2.60	4.50
			578	3.99	6.50
			683	4.71	8.33
		1.35	163	1.12	1.15
			210	1.45	1.36
			277	1.94	2.02
			455	3.14	3.34
			555	3.83	3.89
			925	6.38	7.10
		2.50	293	2.02	1.01
			44]	3.04	2.39
			746	5.14	3.58
			1085	7.48	5.49
30	303.15	0	46	0.32	0.60
			80	0.55	1.15
			115	0.79	1.84
			136	0.94	2.32
			286	1.97	4.90
			297	2.05	4.93
			398	2.74	6.12
			523	3.61	7.64
		1.35	278	1.92	1.86
			550	3.79	4.38
		4.75	415	2.86	0.54
			574	3.96	1.04
			770	5.31	1.86
			955	6.58	2.71
		7.35 (sat)[a]	190	1.31	0.12
			460	3.17	0.43
			840	5.79	1.39

[a] The solid in equilibrium with the saturated solution is $CaCl_2 \cdot 6H_2O$.

The authors gave the $CaCl_2$ concentrations in normality, $c_2/eq\ dm^{-3}$.

At 25 ^0C the values were 0.5, 1.0, 2.7 and 5.0 N, and at 30 ^0C the normality values were 2.7, 9.5, and 14.7 N.

COMPONENTS:	ORIGINAL MEASUREMENTS:
(1) Methane; CH_4; [74-82-8] (2) Calcium chloride; $CaCl_2$; [10043-52-4] (3) Water; H_2O; [7732-18-5]	Duffy, J. R.; Smith, N. O.; Nagy, B. *Geochim. Cosmochim. Acta* <u>1961</u>, *24*, 23-31.

VARIABLES:	PREPARED BY:
T/K = 298.15, 303.15 p_1/MPa = 1.14 - 7.48 c_2/mol l^{-1} = 0.25 - 7.35	H. L. Clever C. L. Young

EXPERIMENTAL VALUES:

AUXILIARY INFORMATION

METHOD/APPARATUS/PROCEDURE:	SOURCE AND PURITY OF MATERIALS:
Rocking equilibrium cell. The pressure is measured with a Bourdon gage. The cell is charged with salt solution, the gas is admitted to a known pressure, and the cell contents allowed to equilibrate. The final pressure is measured and used to calculate the amount of gas dissolved.	(1) Methane. Source not given. Stated to be *c.p.* grade. (2) Calcium chloride. Source not given. Stated to be reagent grade of known water content. (3) Water. Distilled, degassed.
	ESTIMATED ERROR: $\delta T/K$ = ± 1; δp/MPa = ± 0.03; δx_1 = ± 5 x 10^{-6}.
	REFERENCES:

COMPONENTS:	ORIGINAL MEASUREMENTS:
(1) Methane; CH_4; [74-82-8]	Stoessell, R. K.; Byrne, P. A.
(2) Calcium chloride; $CaCl_2$; [10043-52-4]	*Geochim. Cosmochim. Acta* 1982, *46*, 1327-32.
(3) Water; H_2; [7732-18-5]	

VARIABLES: T/K = 298.15 $\quad$ p_1/kPa = 2410-5170 $\quad$ m_2/mol kg^{-1} = 0-2.0	PREPARED BY: H. L. Clever

EXPERIMENTAL VALUES:

Temperature		Pressure		Calcium Chloride	Methane	Salt Effect Parameter
t/°C	T/K	p_1/psia	p_1/kPa	m_2/mol kg^{-1}	m_1/mol kg^{-1}	k_{smm}/kg mol^{-1}
25	298.15	350	2410	0	0.0319	-
				0.5	0.0250	0.071
				1.0	0.0191	0.074
				2.0	0.0120	0.071
		550	3790	0	0.0483	-
				0.5	0.0375	0.073
				1.0	0.0295	0.071
				2.0	0.0179	0.072
		750	5170	0	0.0617	-
				0.5	0.0485	0.070
				1.0	0.0379	0.071
				2.0	0.0236	0.070
						0.071 (authors)

The salt effect parameter is defined as $k_{smm} = \log \gamma_1 / I$

where I is the ionic strength (molality) and $\gamma_1 = (m_1^* f_1 / m_1 f_1^*)_{p,T}$

with m_1 and f_1 the solubility and fugacity, respectively, of methane at p

and T. The "*" refers to saturation in distilled water.

AUXILIARY INFORMATION

METHOD/APPARATUS/PROCEDURE:	SOURCE AND PURITY OF MATERIALS:
Solubility determinations were made using a titanium-lined chamber within a stainless steel reaction vessel jacketed by a water bath for temperature control. The system pressure was set by controlling the input and output of methane within the chamber's headpiece. The vessel was rocked for three h to allow equilibration between the methane and solution.	(1) Methane. Matheson Co., Inc. Ultra high purity grade, stated to be a minimum of 99.97 mole present methane.
	(2) Calcium chloride. The salt solutions were made up gravimetrically using analytical grade chemicals.
	(3) Water. Distilled.
The amount of gas in the saturated solution was measured by transfer of a sample volume to a loop at the system pressure, followed by flashing the sample in an ansion loop and measuring the gas pressure in a known volume. The total gas volume and	ESTIMATED ERROR: δp_1/psia = ± 1 $\quad$ $\delta m_1/m_1$ = ± 0.01
pressure change were used to compute the moles of released gas assuming ideal behavior. A correction was made for the gas not released on flashing.	REFERENCES:

COMPONENTS:	ORIGINAL MEASUREMENTS:
(1) Methane; CH_4; [74-82-8] (2), (3), (4), (5) Electrolytes 　　see below (6) Water; H_2O; [7732-18-5]	Byrne, P. A.; Stoessell, R. K. *Geochim. Cosmochim. Acta* <u>1982</u>, *46*, 2395-7.

VARIABLES:	PREPARED BY:
$T/K = 298.15$ $p_1/kPa = 3790$	H. L. Clever

EXPERIMENTAL VALUES:

Temperature		Pressure		Electrolyte		Methane	Salt Effect Parameter
$t/°C$	T/K	p_1/psia	p_1/kPa		m_i/mol kg^{-1}	m_1/mol kg^{-1}	
25	298.15	550	3790	–	0	0.0483[a]	–
				$MgCl_2$	1.0 ⎫	0.0173	0.074
				$CaCl_2$	1.0 ⎭		
25	298.15	550	3790	–	0	0.0483[a]	–
				NaCl	0.5 ⎫		
				KCl	0.5 ⎪	0.0245	0.074
				$MgCl_2$	0.5 ⎪		
				$CaCl_2$	0.5 ⎭		

[a] Value of methane solubility in water, $m_1°$, from (ref 1).

The salt effect parameter, k_{smm}/kg mol^{-1} =

$(\Sigma((k_{smim}$/kg mol$^{-1})(I_i$/mol kg$^{-1})))/(I$/mol kg$^{-1})$

where I_i and I are ionic strength due to component i and the total
ionic strength, respectively, and k_{smim}/kg mol^{-1} =

$\log((m_1°$/mol kg$^{-1})/(m_1$/mol kg$^{-1}))/(I_i$/mol kg$^{-1})$

Sodium chloride; NaCl; [7647-14-5]   Magnesium chloride; $MgCl_2$; [7786-30-3]
Potassium chloride; KCl; [7447-40-7] Calcium chloride; $CaCl_2$; [10043-52-4]

AUXILIARY INFORMATION

METHOD/APPARATUS/PROCEDURE:	SOURCE AND PURITY OF MATERIALS:
Solubility determinations were made using a titanium-lined chamber within a stainless steel reaction vessel jacketted by a water bath for temperature control. The system pressure was set by controlling the input and output of methane within the chamber's headpiece. The vessel was rocked for 3 h to allow equilibration between the methane and the solution. 　The amount of gas in the saturated solution was measured by transfer of a sample volume to a loop at the system pressure, followed by flashing the sample in an expansion loop and measuring the gas pressure in a known volume. The total gas volume and pressure change were used to compute the moles of gas assuming ideal behavior. A correction was made for the gas not released on flashing. 　Solution densities were measured gravimetrically with pycnometers.	(1) Methane. Matheson Co., Inc. 　Ultra high purity grade, stated 　to have a minimum purity of 　99.97 mole percent. (2), (3), (4), (5) Electrolytes. 　The salt solutions were made up 　gravimetrically using analytical 　grade chemicals. (6) Water. Distilled.

ESTIMATED ERROR:
　　　　δp_1/psia = ± 1
　　δm_1/mol kg^{-1} = ± 0.0003-0.0005
　$\delta m_{2,3}$/mol kg^{-1} = ± 0.0001

REFERENCES:
1. Stoessell, R. K.; Byrne, P. A.
　Geochim. Cosmochim. Acta <u>1982</u>,
　46, 1327.

COMPONENTS:	ORIGINAL MEASUREMENTS:
1. Methane; CH_4; [74-82-8] 2. Water; H_2O; [7732-18-5] 3. Lithium chloride; LiCl; [7447-41-8]	Michels, A.; Gerver, J.; Bijl, A. *Physica*, <u>1936</u>, *3*, 797-808.

VARIABLES:	PREPARED BY:
Pressure	C.L. Young

EXPERIMENTAL VALUES:

T/K	Conc. of LiCl/mol 1^{-1}	$p/10^5$Pa	10^3 Mole fraction of methane in liquid, 10^3 x_{CH_4}
298.15	2.7	48.0	0.43
		101.5	0.85
		147.0	1.17
		197.8	1.46

AUXILIARY INFORMATION

METHOD/APPARATUS/PROCEDURE:	SOURCE AND PURITY OF MATERIALS:
Simple rocking equilibrium cell. Amount of gas absorbed calculated from volume and pressure change in charging vessel. Details in source.	No details given

ESTIMATED ERROR:

$\delta T/K = \pm 0.1$; $\delta p/10^5$Pa $= \pm 0.05$ to 0.5%; $\delta x_{CH_4} = \pm 3$-5%. (estimated by compiler).

REFERENCES:

COMPONENTS:	ORIGINAL MEASUREMENTS:
(1) Methane; CH_4; [74-82-8] (2) Lithium chloride; LiCl; [7447-41-8] (3) Water; H_2O; [7732-18-5]	Morrison, T. J.; Billett, F. *J. Chem. Soc.* <u>1952</u>, 3819 - 3822.

VARIABLES:	PREPARED BY:
T/K: 285.75 - 344.85 p/kPa: 101.325 (1 atm)	H. L. Clever

EXPERIMENTAL VALUES:

Temperature			Salt Effect Parameters	
t/°C	T/K	$1/(T/K)$	$(1/m_2)\log(S°/S)$[1]	$(1/m_2)\log(x°/x)$
12.6	285.75	0.0035	0.130	0.145
30.0	303.15	0.0033	0.097	0.112
49.4	322.55	0.0031	0.082	0.097
71.7	344.85	0.0029	0.077	0.092

[1] The authors used $(1/c)\log(S°/S)$ with c defined as g eq salt per kg of water. For the 1-1 electrolyte the compiler changed the c to an m for m_2/mol kg^{-1}. The methane solubility S is cm^3(STP) kg^{-1}.

The salt effect parameters were calculated from two measurements. The solubility of methane in water, S°, and in the one molal salt solution, S. Only the solubility of the methane in water, and the value of the salt effect parameter are given in the paper. The solubility values in the salt solution are not given.

The compiler calculated the values of the salt effect parameter using the mole fraction gas solubility ratio.

AUXILIARY INFORMATION

METHOD/APPARATUS/PROCEDURE:	SOURCE AND PURITY OF MATERIALS:
The degassed solvent flows in a thin film down an absorption helix containing the methane gas plus solvent vapor at a total pressure of one atmosphere. The volume of gas absorbed is measured in an attached buret system (1).	(1) Methane. Prepared from Grignard reagent. (2) Lithium chloride. "AnalaR" material. (3) Water. No information given.
	ESTIMATED ERROR: $\delta k/kg^{-1}$ mol = 0.010
	REFERENCES: 1. Morrison, T. J.; Billett, F. *J. Chem. Soc.* <u>1948</u>, 2033.

COMPONENTS:	ORIGINAL MEASUREMENTS:
1. Methane; CH_4; [74-82-8] 2. Lithium chloride; LiCl; [7447-41-8] 3. Water; H_2O; [7732-18-5]	Ben-Naim, A.; Yaacobi, M. *J. Phys. Chem.* 1974, *78*, 175-8.

VARIABLES:	PREPARED BY:
Temperature	C. L. Young

EXPERIMENTAL VALUES:

T/K	Conc. of lithium chloride /mol ℓ^{-1}	Ostwald coefficient,[*] L
283.15	0.1	0.03394
288.15		0.03106
293.15		0.02864
298.15		0.02661
303.15		0.02488

[*] Smoothed values of Ostwald coefficient obtained from

$$kT \ln L = 7,264.1 - 47.609\,(T/K) + 0.05380\,(T/K)^2 \text{ cal mol}^{-1}$$

where k is in units of cal mol^{-1} K^{-1}.

AUXILIARY INFORMATION

METHOD/APPARATUS/PROCEDURE:	SOURCE AND PURITY OF MATERIALS:
The apparatus was similar to that described by Ben-Naim and Baer (1) and Wen and Hung (2). It consists of three main parts, a dissolution cell of 300 to 600 cm^3 capacity, a gas volume measuring column, and a manometer. The solvent is degassed in the dissolution cell, the gas is introduced and dissolved while the liquid is kept stirred by a magnetic stirrer immersed in the water bath. Dissolution of the gas results in the change in the height of a column of mercury which is measured by a cathetometer.	1. Matheson sample, purity 99.97 mole per cent. 2. AR grade. 3. Deionised, doubly distilled.

	ESTIMATED ERROR:
	$\delta T/K = \pm 0.01$; $\delta L/L = \pm 0.005$ (estimated by compiler).

REFERENCES:

1. Ben-Naim, A.; Baer, S. *Trans. Faraday Soc.* 1963, *59*, 2735.

2. Wen, W.-Y.; Hung, J. H. *J. Phys. Chem.* 1970, *74*, 170.

COMPONENTS:	ORIGINAL MEASUREMENTS:
1. Methane; CH_4; [74-82-8] 2. Water; H_2O; [7732-18-5] 3. Sodium chloride; NaCl; [7647-14-5]	Michels, A.; Gerver, J.; Bijl, A. *Physica*, <u>1936</u>, *3*, 797-808.
VARIABLES: Temperature, pressure, concentration	PREPARED BY: C.L. Young

EXPERIMENTAL VALUES:

T/K	Conc. of NaCl / mol l^{-1}	$p/10^5$Pa	10^3 Mole fraction of methane in liquid, $10^3 \, x_{CH_4}$
298.15	1.0	41.8	0.58
		102.6	1.14
		142.0	1.42
		177.0	1.62
		206.8	1.73
	1.7	50.8	0.55
		105.9	1.00
		151.7	1.26
		201.6	1.47
	2.5	53.0	0.49
		87.8	0.72
		117.8	0.91
		144.4	1.06
		182.9	1.22
		224.5	1.36
		325.3	1.55
		456.0	1.67
	2.7	53.4	0.47
		105.8	0.77
		152.6	0.98
		195.7	1.12
	3.2	48.5	0.35
		99.2	0.62
		151.8	0.82

AUXILIARY INFORMATION

METHOD/APPARATUS/PROCEDURE:	SOURCE AND PURITY OF MATERIALS:
Simple rocking equilibrium cell. Amount of gas absorbed calculated from volume and pressure change in charging vessel. Details in source.	No details given.
	ESTIMATED ERROR: $\delta T/K = \pm 0.1$; $\delta p/10^5$Pa $= \pm 0.05$ to 0.5%; $\delta x_{CH_4} = \pm 3$-5%. (estimated by compiler)
	REFERENCES:

COMPONENTS:

1. Methane; CH_4; [74-82-8]
2. Water; H_2O; [7732-18-5]
3. Sodium chloride; NaCl;
   [7647-14-5]

ORIGINAL MEASUREMENTS:

Michels, A.; Gerver, J.;
Bijl, A.

Physica, 1936, *3*, 797-808.

EXPERIMENTAL VALUES:

T/K	Conc. of NaCl/mol 1^{-1}	$p/10^5$ Pa	10^3 Mole fraction of methane in liquid, 10^3 x_{CH_4}
298.15	3.2	200.8	0.95
	4.0	59.8	0.31
		109.3	0.49
		160.9	0.63
		206.1	0.70
	5.4	47.6	0.21
		82.1	0.32
		111.2	0.41
		140.9	0.49
		177.8	0.56
		216.4	0.62
		324.2	0.75
		445.3	0.85
323.15	2.5	53.6	0.44
		88.5	0.67
		118.4	0.85
		142.7	0.95
		182.9	1.13
		224.0	1.26
348.15	2.5	55.9	0.39
		87.6	0.62
		117.7	0.78
		145.9	0.90
		182.0	1.05
		224.1	1.18
373.15	2.5	51.9	0.33
		85.2	0.55
		115.8	0.74
		144.1	0.85
		177.2	0.99
		224.5	1.12
398.15	2.5	56.6	0.36
		87.2	0.52
		117.8	0.69
		147.5	0.83
		182.2	0.94
		224.2	1.04
423.15	2.5	56.9	0.33
		87.2	0.50
		117.4	0.66
		147.5	0.77
		182.6	0.88
		224.4	0.99
323.15	5.4	49.1	0.195
		83.1	0.295
		113.4	0.383
		142.6	0.461
		179.9	0.536
		217.1	0.601
348.15	5.4	50.9	0.189
		83.0	0.281
		113.1	0.373
		141.8	0.444
		175.2	0.510
		223.4	0.585
373.15	5.4	50.4	0.186
		85.2	0.281
		114.2	0.365
		143.4	0.425
		181.3	0.495
		217.2	0.565

COMPONENTS:	ORIGINAL MEASUREMENTS:
1. Methane; CH_4; [74-82-8]	Michels, A.; Gerver, J.; Bijl, A.
2. Water; H_2O; [7732-18-5]	*Physica*, 1936, *3*, 797-808.
3. Sodium chloride; NaCl; [7647-14-5]	

EXPERIMENTAL VALUES:

T/K	Conc. of NaCl/mol l^{-1}	$p/10^5$Pa	10^3 Mole fraction of methane in liquid, $10^3 \, x_{CH_4}$
398.15	5.4	50.5	0.177
		82.4	0.262
		113.5	0.351
		142.0	0.409
		175.8	0.475
		223.8	0.548
423.15	5.4	49.5	0.173
		81.9	0.254
		115.0	0.337
		142.8	0.395
		176.8	0.455
		222.7	0.539

COMPONENTS:	ORIGINAL MEASUREMENTS:
(1) Methane; CH_4; [74-82-8] (2) Sodium chloride; NaCl; [7647-14-5] (3) Water; H_2O; [7732-18-5]	Eucken, A.; Hertzberg, G. Z. Phys. Chem. <u>1950</u>, 195, 1-23.

| VARIABLES:
$T/K = 273.15, 293.15$
$p_1/kPa = 101.325$
$m_2/mol\ kg^{-1} = 0 - 2.77$ | PREPARED BY:

H. L. Clever |

EXPERIMENTAL VALUES:

Temperature		Sodium Chloride	Ostwald Coefficient	Salt Effect Parameter
$t/^0C$	T/K	$m_2/mol\ kg^{-1}$	L/cm^3cm^{-3}	$k_{smc}/kg\ mol^{-1}$
0	273.15	0	0.0550	–
		0.68	0.0409	0.188
		1.37	0.0313	0.178
		2.77	0.0167	0.186
				Av. 0.184
20	293.15	0	0.0359	–
		0.795	0.0266	0.163
		2.63	0.0139	0.157
				Av. 0.160

Salt effect parameter, $k_{smc}/kg\ mol^{-1} =$

$(1/(m_2/mol\ kg^{-1}))\log((L^0\ cm^3cm^{-3})/(L/cm^3cm^{-3}))$

AUXILIARY INFORMATION

METHOD/APPARATUS/PROCEDURE:	SOURCE AND PURITY OF MATERIALS:
The apparatus consists of a gas buret and an absorption flask connected by a capillary tube. The whole apparatus is shaken. The capillary tube is a two m long glass helix. An amount of gas is measured at STP and placed in the gas buret. After shaking, the difference from the original amount of gas placed in the buret is determined.	(1) Methane. (2) Sodium chloride. (3) Water. No information.
	ESTIMATED ERROR: $\delta L/L = \pm\ 0.01$ (authors)
	REFERENCES:

COMPONENTS:	ORIGINAL MEASUREMENTS:
(1) Methane; CH_4; [74-82-8]	Morrison, T. J.; Billett, F.
(2) Sodium chloride; NaCl; [7647-14-5]	*J. Chem. Soc.* <u>1952</u>, 3819 - 3822.
(3) Water; H_2O; [7732-18-5]	

VARIABLES:	PREPARED BY:
T/K: 285.75 - 344.85 p/kPa: 101.325 (1 atm)	H. L. Clever

EXPERIMENTAL VALUES:

Temperature			Salt Effect Parameters	
$t/°C$	T/K	$1/(T/K)$	$(1/m_2)\log(S°/S)$ [1]	$(1/m_2)\log(x°/x)$
12.6	285.75	0.0035	0.153	0.168
30.0	303.15	0.0033	0.127	0.142
49.4	322.55	0.0031	0.111	0.126
71.7	344.85	0.0029	0.102	0.117

[1] The authors used $(1/c)\log(S°/S)$ with c defined as g eq salt per kg of water. For the 1-1 electrolyte the compiler[3] changed the c to an m for m_2/mol kg^{-1}. The methane solubility S is cm^3(STP) kg^{-1}.

The salt effect parameters were calculated from two measurements. The solubility of methane in water, S°, and in the one molal salt solution, S. Only the solubility of the methane in water, and the value of the salt effect parameter are given in the paper. The solubility values in the salt solution are not given.

The compiler calculated the values of the salt effect parameter using the mole fraction gas solubility ratio.

AUXILIARY INFORMATION

METHOD/APPARATUS/PROCEDURE:	SOURCE AND PURITY OF MATERIALS:
The degassed solvent flows in a thin film down an absorption helix containing the methane gas plus solvent vapor at a total pressure of one atmosphere. The volume of gas absorbed is measured in an attached buret system (1).	(1) Methane. Prepared from Grignard reagent. (2) Sodium chloride. "AnalaR" material. (3) Water. No information given.

	ESTIMATED ERROR:
	$\delta k/kg^{-1}$ mol = 0.010

	REFERENCES:
	1. Morrison, T. J.; Billett, F. *J. Chem. Soc.* <u>1948</u>, 2033.

COMPONENTS:	ORIGINAL MEASUREMENTS:
1. Methane; CH_4; [74-82-8] 2. Water; H_2O; [7732-18-5] 3. Sodium chloride; NaCl; [7647-14-5]	Duffy, J.R.; Smith, N.O.; Nagy, B. *Geochim. Cosmochim. Acta,* <u>1961</u>, *24,* 23-31.

VARIABLES:	PREPARED BY:
Concentration, pressure	C.L. Young

EXPERIMENTAL VALUES:

T/K	Conc. /mol 1^{-1}	p/MPa	10^4 Mole fraction of methane, $10^4 x_{CH_4}$
303.15	0.5	21.48	2.54
		41.64	5.36
		52.99	6.53
		60.29	7.72
		64.14	8.53
	1.0	36.68	4.00
		53.70	6.22
		65.05	6.82
	2.7	23.51	1.41
		33.64	2.73
		46.41	3.30
		54.11	4.44
		77.21	6.00
		77.31	6.34
	5.4	22.29	0.73
		32.42	1.28
		38.00	1.11
		42.05	1.56
		57.25	2.58
		67.38	2.78
		69.41	2.95
		71.33	3.25
		95.75	4.26

AUXILIARY INFORMATION

METHOD/APPARATUS/PROCEDURE:	SOURCE AND PURITY OF MATERIALS:

Rocking equilibrium cell. Pressure measured with a Bourdon gauge. Cell charged with salt solution; gas admitted to known pressure cell contents allowed to equilibriate. Final pressure measured and used to calculate amount of gas dissolved. Details in source ref.	1. C.P. grade. No other details
	    given.
	
	2. Degassed.
	
	3. Reagent grade of known water
	    content.

	ESTIMATED ERROR:
	REFERENCES:

COMPONENTS:	ORIGINAL MEASUREMENTS:
(1) Methane; CH_4; [74-82-8]	Mishnina, T. A.; Avdeeva, O. I. Bozhovskaya, T. K.
(2) Sodium chloride; NaCl; [7647-14-5]	*Inf. Sb., Vses. Nauchn-Issled. Geol. Inst.* 1962, No. 56, 137-45.
(3) Water; H_2O; [7732-18-5]	*Chem. Abstr.* 1964, 60, 8705g

EXPERIMENTAL VALUES:

Temperature		Sodium Chloride	Solubility	Salt Effect Parameter
$t/^0C$	T/K	$c_2/mol\ dm^{-3}$	$cm^3(STP)dm^{-3}$	k_{scc}/dm^3mol^{-1}
4	277.15	0	50.6	–
		1.06	34.1	0.162
		2.10	23.7	0.157
		3.08	16.2	0.161
		4.12	10.4	0.167
		5.31	6.7	0.165
				0.162 av.
10	283.15	0	43.8	–
		1.06	30.4	0.150
		2.10	20.6	0.156
		3.08	14.6	0.155
		4.12	10.1	0.155
		5.31	6.3	0.162
				0.155 av.
20	293.15	0	33.9	–
		1.00	24.7	0.138
		1.77	18.6	0.148
		2.60	13.9	0.149
		3.90	9.1	0.147
		5.31	5.7	0.145
				0.145 av.
30	303.15	0	28.5	–
		1.04	20.6	0.135
		2.00	15.3	0.135
		2.60	12.4	0.139
		3.90	8.3	0.137
		5.31	5.4	0.135
				0.136 av.
50	323.15	0	23.4	–
		1.02	17.2	0.131
		1.98	12.7	0.134
		2.80	10.1	0.130
		3.90	7.2	0.131
		5.31	4.8	0.129
				0.131 av.
70	343.15	0	20.7	–
		1.02	15.3	0.129
		1.98	11.9	0.121
		3.95	6.7	0.121
		5.31	4.4	0.126
				0.124 av.
80	353.15	0	20.4	–
		1.02	15.7	0.111
		2.12	11.8	0.112
		3.28	9.0	0.108
		5.31	5.4	0.108
				0.110 av.
90	363.15	0	22.1	–
		1.06	17.0	0.108
		2.10	13.2	0.107
		3.08	9.9	0.113
		5.31	5.6	0.111
				0.110 av.

COMPONENTS:	ORIGINAL MEASUREMENTS:
(1) Methane; CH_4; [74-82-8] (2) Sodium chloride; NaCl; [7647-14-5] (3) Water; H_2O; [7732-18-5]	Mishnina, T. A.; Avdeeva, O. I. Bozhovskaya, T. K. *Inf. Sb., Vses. Nauchn-Issled. Geol. Inst.* <u>1962</u>, No. *56*, 137-45. *Chem. Abstr.* <u>1964</u>, *60*, 8705g.

| VARIABLES:
 T/K = 277.15 - 363.15
 p_1/kPa = 101.3
 $c_2/mol\ dm^{-3}$ = 0 - 5.31 | PREPARED BY:

 H. L. Clever |

EXPERIMENTAL VALUES:

See preceeding page.

The values at 50, 70, 80, and 90 C were first published in the authors' earlier paper (1). In (1) the description of the table says "methane solubility in solutions of sodium chloride at various temperatures and atmospheric pressure above the solution.

Reference (1) also gives a table of smoothed values. The experimental values are about 7 percent greater than the smoothed values, and the salt effect parameter values are about 20 percent larger than the smoothed values.

AUXILIARY INFORMATION

METHOD/APPARATUS/PROCEDURE:	SOURCE AND PURITY OF MATERIALS:
See authors' earlier paper (1).	(1) Methane. Source not given. Stated to contain one percent air.

COMMENTS:	(2) Sodium chloride.
The compiler estimated the values of the solubility of methane in water from a figure in the paper and values of the salt effect parameter. The other values were tabulated in the paper. The methane solubility is equivalent to the Bunsen coefficient, $10^2\alpha/cm^3$ (STP) $cm^{-3}\ atm^{-1}$. The Kelven temperature values were added by the compiler.	(3) Water.

| | ESTIMATED ERROR:

 $\delta T/K$ = ± 0.1
 ± 0.5 at 50 °C and above. |

| | REFERENCES:

 1. Mishnina, T. A.; Avdeeva, O. I.; Bozhovskaya, T. K. *Materialy Vses. Nauchn. Issled. Geol. Inst.* <u>1961</u>, *46*, 93. |

COMPONENTS:	ORIGINAL MEASUREMENTS:
1. Methane; CH_4; [74-82-8]	O'Sullivan, T.D.; Smith, N.O.
2. Water; H_2O; [7732-18-5]	*J. Phys. Chem.*, 1970, *74*, 1460-1466
3. Sodium chloride; NaCl; [7647-14-5]	

VARIABLES:	PREPARED BY:
Temperature, pressure, concentration	C.L. Young

EXPERIMENTAL VALUES:

T/K	Conc./mol l^{-1}	P/MPa	10^3 Mole fraction of methane in liquid, $10^3\ x_{CH_4}$
324.65	1.000	10.13	1.076
		20.26	1.695
		30.40	2.138
		40.53	2.50
		50.66	2.79
		60.79	3.07
375.65		20.37	1.693
		30.60	2.219
		40.83	2.57
		50.97	2.89
		61.20	3.20
398.15		10.44	1.058
		20.67	1.752
		30.90	2.223
		41.04	2.60
		51.37	2.94
		61.61	3.25
324.65	4.000	20.26	0.805
		30.40	0.997
		40.53	1.154
		50.66	1.303
		60.79	1.444

AUXILIARY INFORMATION

METHOD/APPARATUS/PROCEDURE:	SOURCE AND PURITY OF MATERIALS:
Large steel stirred equilibrium cell. Pressure measured with Bourdon gauge. Temperature measured with iron-constantan thermocouple. Cell charged with liquid, compressed gas added. After equilibrium obtained samples removed and analysed using volumetric techniques. Details in ref. (1).	1. Matheson Co. sample purity 99.95 mole per cent. 2. Distilled and de-ionised air removed. 3. Baker analysed reagent dried at 388K.

ESTIMATED ERROR:
$\delta T/K = \pm 0.5$; $\delta P/MPa = \pm 0.05\%$;
$\delta x_{CH_4} = \pm 0.4\%$

REFERENCES:

1. O'Sullivan, T.D.; Smith, N.O.

   Geochim. Cosmochim. Acta, 1966, *30*, 617.

COMPONENTS:	ORIGINAL MEASUREMENTS:

COMPONENTS:

1. Methane; CH_4; [74-82-8]
2. Water; H_2O; [7732-18-5]
3. Sodium chloride; NaCl; [7647-14-5]

ORIGINAL MEASUREMENTS:

O'Sullivan, T.D.; Smith, N.O.

J. Phys. Chem. 1970, *74*, 1460-1466.

EXPERIMENTAL VALUES:

T/K	Conc. /mol l^{-1}	P/MPa	10^3 Mole fraction of methane in liquid, $10^3 x_{CH_4}$
375.65	4.000	20.37	0.826
		30.60	1.079
		40.83	1.211
		50.97	1.319
		61.20	1.433
398.15		20.67	0.825
		30.90	1.005
		41.04	1.164
		51.37	1.322
		61.61	1.438

COMPONENTS:	ORIGINAL MEASUREMENTS:
1. Methane; CH_4; [74-82-8] 2. Sodium chloride; NaCl; [7647-14-5] 3. Water; H_2O; [7732-18-5]	Ben-Naim, A.; Yaacobi, M. *J. Phys. Chem.* 1974, *78*, 170-5.
VARIABLES: Temperature, concentration	PREPARED BY: C. L. Young

EXPERIMENTAL VALUES:

T/K	Conc. of salt /mol ℓ^{-1}	Ostwald coefficient,* L	T/K	Conc. of salt /mol ℓ^{-1}	Ostwald coefficient,* L
283.15	0.25	0.04047	283.15	1.0	0.03048
288.15		0.03679	288.15		0.02800
293.15		0.03375	293.15		0.02594
298.15		0.03123	298.15		0.02424
303.15		0.02914	303.15		0.02285
283.15	0.50	0.03700	283.15	2.0	0.02123
288.15		0.03374	288.15		0.01977
293.15		0.03108	293.15		0.01860
298.15		0.02890	298.15		0.01765
303.15		0.02711	303.15		0.01691

* Smoothed values of Ostwald coefficient obtained from

$$kT \ln L = 8{,}677.1 - 56.405 \, (T/K) + 0.06846 \, (T/K)^2 \text{ cal mol}^{-1}$$

$$kT \ln L = 9{,}392.2 - 61.966 \, (T/K) + 0.07855 \, (T/K)^2 \text{ cal mol}^{-1}$$

$$kT \ln L = 8{,}645.6 - 57.885 \, (T/K) + 0.07209 \, (T/K)^2 \text{ cal mol}^{-1}$$

$$kT \ln L = 9{,}327.1 - 64.964 \, (T/K) + 0.08605 \, (T/K)^2 \text{ cal mol}^{-1}$$

(where k is in units of cal mol^{-1} K^{-1}) for concentrations of 0.25, 0.50, 1.0 and 2.0 mol ℓ^{-1}, respectively.

AUXILIARY INFORMATION

METHOD/APPARATUS/PROCEDURE:	SOURCE AND PURITY OF MATERIALS:
The apparatus was similar to that described by Ben-Naim and Baer (1) and Wen and Hung (2). It consists of three main parts, a dissolution cell of 300 to 600 cm^3 capacity, a gas volume measuring column, and a manometer. The solvent is degassed in the dissolution cell, the gas is introduced and dissolved while the liquid is kept stirred by a magnetic stirrer immersed in the water bath. Dissolution of the gas results in the change in the height of a column of mercury which is measured by a cathetometer.	1. Matheson sample, purity 99.97 mole per cent. 2. AR grade. 3. Deionised, doubly distilled.
	ESTIMATED ERROR: $\delta T/K = \pm 0.01$; $\delta L/L = \pm 0.005$ (estimated by compiler).
	REFERENCES: 1. Ben-Naim, A.; Baer, S. *Trans. Faraday Soc.* 1963, *59*, 2735. 2. Wen, W.-Y.; Hung, J. H. *J. Phys. Chem.* 1970, *74*, 170.

COMPONENTS:	ORIGINAL MEASUREMENTS:
(1) Methane; CH_4; [74-82-8] (2) Sodium chloride; NaCl; [7647-14-5] (3) Water; H_2O; [7732-18-5]	Yano, T.; Suetaka, T.; Umehara, T.; Horiuchi, A. *Kagaku Kogaku* <u>1974</u>, *38*, 320-3.

VARIABLES: $T/K = 298.15$ $p_1/kPa = 101.325$ $c_2/mol\ dm^{-3} = 0 - 1.500$	PREPARED BY: H. L. Clever C. L. Young

EXPERIMENTAL VALUES:

Temperature		Sodium Chloride	Methane Solubility	Salt Effect Parameter
$t/^0C$	T/K	$c_2/mol\ dm^{-3}$	$10^3 c_1/mol\ dm^{-3}$	$k_{scc}/dm^3\ mol^{-1}$
25	298.15	0	1.31	–
		0.500	1.15	0.113
		1.000	0.935	0.146
		1.500	0.745	0.163
				0.149 (authors)

$k_{scc}/dm^3\ mol^{-1} = (1/(c_2/mol\ dm^{-3}))\ \log((c_1^0/mol\ dm^{-3})/(c_1/mol\ dm^{-3}))$

The compiler added the salt effect parameter values at the individual salt concentrations.

The authors defined the salt effect parameter in terms of the electrolyte ionic strength.

AUXILIARY INFORMATION

METHOD/APPARATUS/PROCEDURE:	SOURCE AND PURITY OF MATERIALS:
Volumetric apparatus. Salt solution allowed to enter stirred absorption chamber. Pressure within absorption chamber adjusted to be as near atmospheric pressure as possible. Details in source and ref. 1.	1. High purity sample, purity better than 99.5 mole per cent. 2. Special grade. 3. Distilled.
	ESTIMATED ERROR:
	REFERENCES: 1. Yano, T.; Suetaka, T.; Umehara, T. *Nippon Kagaku Kaishi* <u>1972</u>, *11*, 2194.

COMPONENTS:	ORIGINAL MEASUREMENTS:
(1) Methane; CH_4; [74-82-8] (2) Sodium chloride; NaCl; [7647-14-5] (3) Water; H_2O; [7732-18-5]	Namiot, A. Yu.; Skripka, V. G. 　　Ashmyan, K. D. *Geokhimiya* <u>1979</u>, *(1)*, 147-9.

VARIABLES:	PREPARED BY:
T/K = 323 - 623 p_t/MPa = 29.5 $c_2/mol\ dm^{-3}$ = 0 - 1.11	H. L. Clever

EXPERIMENTAL VALUES:

Temperature		Sodium Chloride	Methane Solubility	Salt Effect Parameter
$t/^{0}C$	T/K	$c_2/mol\ dm^{-3}$	S_1/cm^3 (STP) g^{-1}	$k_{scx}/dm^3\ mol^{-1}$
50	323	0.0 1.00	3.59 2.80	0.108
100	373	0.0 1.00	3.66 2.81	0.114
150	423	0.0 1.00	4.76 3.79	0.098
200	473	0.0 1.00	7.83 6.48	0.084
250	523	0.0 1.02	12.68 8.95	0.134
300	573	0.0 1.08	21.78 13.58	0.205
350	623	0.0 1.11	31.97 16.66	0.295

AUXILIARY INFORMATION

METHOD/APPARATUS/PROCEDURE:	SOURCE AND PURITY OF MATERIALS:
See earlier paper on methane + water system (1, 2). The salt effect parameter was calculated from a Henry's constant ratio which included the fugacity of methane. The fugacity ratio was corrected for the amount of water in pure water and in the salt solution. The correction is very important at temperatures of 523 K and above.	(1) Methane. Stated to be 99.9 percent. (2) Sodium chloride. (3) Water. Distilled.
	ESTIMATED ERROR: 　　$\delta T/K$ = ± 0.1 　$\delta p_t/MPa$ = ± 0.15
	REFERENCES: 1. Sultanov, R. G.; Skripka, V. G.; Namiot, A. Yu. 　*Gazov. Prom.* <u>1972</u>, *17 (5)*, 6-7. 2. Sultanov, R. G.; Skripka, V. G.; Namiot, A. Yu. 　*Zh. Fiz. Khim.* <u>1972</u>, *46*, 2160.

COMPONENTS:	ORIGINAL MEASUREMENTS:
1. Methane; CH₄: [74-82-8] 2. Sodium chloride; NaCl; [7647-14-5] 3. Water; H₂O; [7732-18-5]	Blount, C. W.; Price, L. C.; Wenger, L. M.; Tarullo, M. DOE Contract report DE-A508- 78ET12145.

VARIABLES:	PREPARED BY:
Temperature, pressure, concentration of component 2.	C. L. Young

EXPERIMENTAL VALUES:

T/°F	T/K	P/psi	P/MPa	Conc. of NaCl /g dm⁻³	Methane solubility /mol kg⁻¹	Mole fraction x_{CH_4}
212.36	373.35	22365	154.2	0.0	0.3563	0.00638
212.36	373.35	22364	154.2	0.0	0.3636	0.00651
212.36	373.35	22336	154.0	0.0	0.3833	0.00686
211.10	372.65	19102	131.7	0.0	0.3556	0.00636
211.10	372.65	19102	131.7	0.0	0.3483	0.00623
211.10	372.65	19102	131.7	0.0	0.3505	0.00627
211.10	372.65	19102	131.7	0.0	0.3636	0.00651
211.10	372.65	19102	131.7	0.0	0.3600	0.00644
212.36	373.35	16027	110.5	0.0	0.3169	0.00568
212.36	373.35	16027	110.5	0.0	0.3183	0.00570
212.36	373.35	16085	110.9	0.0	0.3322	0.00595
212.36	373.35	16085	110.9	0.0	0.3154	0.00565
212.00	373.15	16128	111.2	0.0	0.3242	0.00581
212.00	373.15	16128	111.2	0.0	0.3300	0.00591
213.80	374.15	13169	90.8	0.0	0.2986	0.00535
213.80	374.15	13169	90.8	0.0	0.2994	0.00536
212.00	373.15	10240	70.6	0.0	0.2585	0.00463
212.00	373.15	10240	70.6	0.0	0.2592	0.00465
212.36	373.35	7107	49.0	0.0	0.2227	0.00400
212.36	373.35	5004	34.5	0.0	0.1884	0.00338
212.36	373.35	5004	34.5	0.0	0.1774	0.00319
212.36	373.35	3524	24.3	0.0	0.1438	0.00258
212.36	373.35	3524	24.3	0.0	0.1365	0.00245

(cont.)

AUXILIARY INFORMATION

METHOD/APPARATUS/PROCEDURE:	SOURCE AND PURITY OF MATERIALS:
Teflon-lined stainless steel equilibrium cell. Methane added to saline solution under pressure. Cell equilibrated over a period of up to two days. Samples of solution taken while additional methane was being added to cell to keep the pressure constant. Sample sizes varied from 3 to 10 ml. Weight of sample measured and methane dissolved determined by volumetric method. Details in ref. (1) and (2).	1. Purity 99.99 mole per cent. 2. Reagent grade. 3. Distilled.

	ESTIMATED ERROR:
	$\delta T/K = \pm 2$; $\delta P/MPa = \pm 0.3$; $\delta x/x = \pm 0.05$.

	REFERENCES:
	1. Price, L. C. *Am. Ass. Petr. Geol.* 1979, *63*, 1527. 2. Blount, C. W.; Price, L. C.; Wenger, L. M.; Tarullo, M. *Proc. U.S. Gulf Coast Geopressured Geotherm. Energy Conf.* 1980, *4*, 1225.

COMPONENTS:	ORIGINAL MEASUREMENTS:
1. Methane; CH₄; [74-82-8]	Blount, C. W.; Price, L.C.;
2. Sodium chloride; NaCl; [7647-14-5]	Wenger, L. M.; Tarullo, M. DOE Contract report DE-A508-
3. Water; H₂O; [7732-18-5]	78ET12145.

EXPERIMENTAL VALUES:

T/°F	T/K	P/psi	P/MPa	Conc. of NaCl /g dm^{-3}	Methane solubility /mol kg^{-1}	Mole fraction x_{CH_4}
212.36	373.35	2176	15.0	0.0	0.1081	0.00194
212.36	373.35	2176	15.0	0.0	0.1095	0.00197
213.35	373.90	22539	155.4	3.19	0.3533	0.00632
213.35	373.90	19000	131.0	3.19	0.3489	0.00624
213.35	373.90	19000	131.0	3.19	0.3366	0.00602
212.45	373.40	16056	110.7	3.19	0.3074	0.00550
212.45	373.40	16056	110.7	3.19	0.3133	0.00561
212.45	373.40	13227	91.2	3.19	0.2848	0.00510
212.45	373.40	13227	91.2	3.19	0.2856	0.00511
212.45	373.40	10008	69.0	3.19	0.2462	0.00441
212.45	373.40	10008	69.0	3.19	0.2331	0.00418
213.35	373.90	7063	48.7	3.19	0.2215	0.00397
213.35	373.90	5105	35.2	3.19	0.1719	0.00308
213.35	373.90	5105	35.2	3.19	0.1719	0.00308
212.90	373.65	3640	25.1	3.19	0.1377	0.00247
212.90	373.65	3640	25.1	3.19	0.1464	0.00263
212.90	373.65	2147	14.8	3.19	0.1209	0.00217
212.90	373.65	2147	14.8	3.19	0.0998	0.00179
212.90	373.65	2219	15.3	3.19	0.1034	0.00186
212.90	373.65	2219	15.3	3.19	0.1100	0.00198
212.45	373.40	19421	133.9	51.1	0.2682	0.00473
212.45	373.40	19421	133.9	51.1	0.2491	0.00440
212.00	373.15	19421	133.9	51.1	0.2534	0.00447
212.00	373.15	19421	133.9	51.1	0.2654	0.00468
212.00	373.15	16027	110.5	51.1	0.2357	0.00416
212.00	373.15	13111	90.4	51.1	0.2280	0.00402
212.00	373.15	13111	90.4	51.1	0.2301	0.00406
212.00	373.15	13111	90.4	51.1	0.2322	0.00410
211.55	372.90	10182	70.2	51.1	0.2011	0.00355
211.55	372.90	10182	70.2	51.1	0.1842	0.00325
211.10	372.65	7223	49.8	51.1	0.1687	0.00298
211.10	372.65	7223	49.8	51.1	0.1701	0.00301
212.00	373.15	5033	34.7	51.1	0.1454	0.00257
212.00	373.15	5033	34.7	51.1	0.1461	0.00258
212.00	373.15	3597	24.8	51.1	0.1200	0.00212
212.00	373.15	3597	24.8	51.1	0.1228	0.00217
213.80	374.15	16186	111.6	106.0	0.1960	0.00340
213.80	374.15	13213	91.1	106.0	0.1735	0.00301
213.80	374.15	13213	91.1	106.0	0.1755	0.00305
212.00	373.15	22510	155.2	106.5	0.2349	0.00407
212.00	373.15	22510	155.2	106.5	0.2308	0.00400
212.00	373.15	22510	155.2	106.5	0.2329	0.00404
212.00	373.15	22510	155.2	106.5	0.2308	0.00400
212.00	373.15	22394	154.4	106.5	0.23626	0.00401
212.00	373.15	22394	154.4	106.5	0.23831	0.00413
212.90	373.65	19043	131.3	106.5	0.21919	0.00380
212.90	373.65	19043	131.3	106.5	0.21646	0.00375
212.36	373.35	15360	105.9	106.5	0.20007	0.00347
212.36	373.35	15360	105.9	106.5	0.20485	0.00355
212.90	373.65	15273	105.3	106.5	0.19529	0.00339
212.90	373.65	15273	105.3	106.5	0.20076	0.00348
212.90	373.65	13242	91.3	106.5	0.17617	0.00306
212.90	373.65	13242	91.3	106.5	0.17822	0.00309
212.90	373.65	12908	89.0	106.5	0.17959	0.00312
212.90	373.65	12908	89.0	106.5	0.18505	0.00321
212.00	373.15	10182	70.2	106.5	0.16457	0.00286

(cont.)

COMPONENTS:	ORIGINAL MEASUREMENTS:
1. Methane; CH_4; [74-82-8]	Blount, C. W.; Price, L. C.;
2. Sodium chloride; NaCl; [7647-14-5]	Wenger, L. M.; Tarullo, M. DOE Contract report DE-A508- 78ET12145.
3. Water; H_2O; [7732-18-5]	

EXPERIMENTAL VALUES:

T/°F	T/K	P/psi	P/MPa	Conc. of NaCl /g dm^{-3}	Methane solubility /mol kg^{-1}	Mole fraction x_{CH_4}
212.00	373.15	10182	70.2	106.5	0.16183	0.00281
212.00	373.15	10095	69.6	106.5	0.1489	0.00258
212.00	373.15	10095	69.6	106.5	0.1550	0.00269
212.90	373.65	7165	49.4	106.5	0.1407	0.00244
212.90	373.65	7165	49.4	106.5	0.1284	0.00223
212.90	373.65	7078	48.8	106.5	0.1338	0.00232
212.90	373.65	7078	48.8	106.5	0.1359	0.00236
212.36	373.35	5163	35.6	104.0	0.1245	0.00216
212.36	373.35	5163	35.6	104.0	0.1197	0.00208
212.36	373.35	5120	35.6	105.0	0.1080	0.00188
212.36	373.35	5120	35.3	105.0	0.1230	0.00214
212.90	373.65	3669	35.3	105.0	0.0936	0.00163
212.00	373.15	3698	25.5	105.0	0.0977	0.00170
212.36	373.35	3640	25.1	105.0	0.0916	0.00159
212.36	373.35	3640	25.1	105.0	0.0889	0.00154
212.00	373.15	2060	14.2	105.0	0.0663	0.00115
212.00	373.15	2060	14.2	105.0	0.0622	0.00108
211.55	372.90	21843	150.6	166.3	0.1772	0.00301
211.55	372.90	21843	150.6	166.3	0.1726	0.00293
211.10	372.65	22104	152.4	166.3	0.1806	0.00307
211.10	372.65	19087	131.6	166.3	0.1620	0.00276
211.10	372.65	19087	131.6	166.3	0.1614	0.00274
210.65	372.40	16157	111.4	166.3	0.1528	0.00260
211.55	372.90	13314	91.8	166.3	0.1343	0.00228
211.55	372.90	13314	91.8	166.3	0.1369	0.00233
211.55	372.90	13198	91.0	163.5	0.1497	0.00255
211.55	372.90	13198	91.0	163.5	0.1358	0.00231
211.55	372.90	13373	92.2	163.5	0.1497	0.00255
211.55	372.90	13373	92.2	163.5	0.1411	0.00240
210.65	372.40	9993	68.9	164.6	0.1165	0.00198
210.65	372.40	9993	68.9	164.6	0.1198	0.00204
212.00	373.15	9834	67.8	163.5	0.1358	0.00231
212.00	373.15	9834	67.8	163.5	0.1166	0.00199
212.45	373.40	9935	68.5	163.5	0.1159	0.00197
212.45	373.40	9935	68.5	163.5	0.1311	0.00223
210.92	372.55	7208	49.7	164.6	0.1019	0.00174
209.75	371.90	7107	49.0	163.5	0.1146	0.00195
209.75	371.90	7107	49.0	163.5	0.1099	0.00187
209.75	371.90	7107	49.0	163.5	0.1126	0.00192
209.75	371.90	7107	49.0	163.5	0.1126	0.00192
211.55	372.90	7034	48.5	163.5	0.1086	0.00185
211.55	372.90	7034	48.5	163.5	0.1060	0.00181
211.55	372.90	5018	34.6	163.5	0.0960	0.00164
211.55	372.90	5018	34.6	163.5	0.0960	0.00164
210.92	372.55	3205	22.1	163.5	0.0656	0.00112
210.92	372.55	3205	22.1	163.5	0.0642	0.00110
211.10	372.65	1915	13.2	163.5	0.0570	0.00097
211.10	372.65	1915	13.2	163.5	0.0563	0.00096
211.10	372.65	1973	13.6	163.5	0.0570	0.00097
211.10	372.65	1973	13.6	163.5	0.0570	0.00097
211.55	372.90	22466	154.9	227.6	0.1149	0.00191
211.55	372.90	22466	154.9	227.6	0.1194	0.00199
211.55	372.90	22466	154.9	227.6	0.1342	0.00223
211.55	372.90	22466	154.9	227.6	0.1258	0.00210
211.55	372.90	22437	154.7	227.6	0.1329	0.00221
211.55	372.90	22437	154.7	227.6	0.1246	0.00207
211.55	372.90	17463	120.4	227.6	0.1169	0.00195

(cont.)

COMPONENTS:

1. Methane; CH_4; [74-82-8]

2. Sodium chloride; NaCl; [7647-14-5]

3. Water; H_2O; [7732-18-5]

ORIGINAL MEASUREMENTS:

Blount, C. W.; Price, L. C.;
Wenger, L. M.; Tarullo, M.
DOE Contract report DE-A508-78ET12145.

EXPERIMENTAL VALUES:

T/°F	T/K	P/psi	P/MPa	Conc. of NaCl /g dm^{-3}	Methane solubility /mol kg^{-1}	Mole fraction x_{CH_4}
211.55	372.90	17463	120.4	227.6	0.1162	0.00194
211.55	372.90	15345	105.8	227.6	0.1136	0.00189
210.65	372.40	15519	107.0	227.6	0.1181	0.00197
210.65	372.40	15519	107.0	227.6	0.1226	0.00204
211.10	372.65	16302	112.4	227.6	0.1085	0.00181
210.65	372.40	16360	112.8	227.6	0.10272	0.00171
210.65	372.40	16360	112.8	227.6	0.10786	0.00180
210.65	372.40	13285	91.6	227.6	0.09438	0.00157
210.65	372.40	13518	93.2	227.6	0.09053	0.00151
210.65	372.40	13518	93.2	227.6	0.09951	0.00166
210.20	372.15	9747	67.2	227.6	0.08346	0.00139
210.65	372.40	10182	70.2	227.6	0.07833	0.00131
210.65	372.40	10182	70.2	227.6	0.08796	0.00147
210.20	372.15	7281	50.2	225.1	0.07328	0.00122
210.20	372.15	7281	50.2	225.1	0.07649	0.00128
210.65	372.40	5047	34.8	225.1	0.07713	0.00129
210.20	372.15	5366	37.0	223.7	0.06432	0.00107
210.65	372.40	5395	37.2	223.7	0.05506	0.00092
210.65	372.40	5395	37.2	223.7	0.06110	0.00102
210.65	372.40	3597	24.8	223.7	0.05210	0.00087
210.65	372.40	3597	24.8	223.7	0.05467	0.00091
210.65	372.40	3597	24.8	223.7	0.06753	0.00113
210.65	372.40	3568	24.6	223.7	0.05210	0.00087
210.20	372.15	2118	14.6	223.7	0.04181	0.00070
210.20	372.15	2118	14.6	223.7	0.03988	0.00067
213.44	373.95	22220	153.2	293.5	0.07734	0.00126
212.90	373.65	19000	131.0	293.5	0.08545	0.00139
212.90	373.65	19000	131.0	293.5	0.08920	0.00145
212.90	372.65	19000	131.0	293.5	0.08732	0.00142
212.90	373.65	19000	131.0	293.5	0.08670	0.00141
212.90	373.65	19000	131.0	293.5	0.09169	0.00149
212.90	373.65	15998	110.3	293.5	0.06487	0.00106
212.90	373.65	15998	110.3	293.5	0.08233	0.00134
212.90	373.65	13068	90.1	293.5	0.06924	0.00113
212.90	373.65	13068	90.1	293.5	0.07610	0.00124
214.16	374.35	13169	90.8	294.6	0.06671	0.00109
214.16	374.35	13169	90.8	294.6	0.07107	0.00116
213.44	373.95	13111	90.4	294.6	0.07357	0.00120
213.44	373.95	13111	90.4	294.6	0.07793	0.00127
213.80	374.15	10240	70.6	294.6	0.06983	0.00114
213.80	374.15	10240	70.6	294.6	0.06796	0.00111
213.44	373.95	7194	49.6	294.6	0.05424	0.00088
213.44	373.95	7194	49.6	294.6	0.05673	0.00092
213.80	374.15	5178	35.7	294.6	0.04801	0.00078
213.80	374.15	5178	35.7	294.6	0.04551	0.00074
214.16	374.35	3626	25.0	294.6	0.03678	0.00060
214.16	374.35	3626	25.0	294.6	0.04177	0.00068
213.44	373.95	2045	14.1	294.6	0.03180	0.00052
213.44	373.95	2045	14.1	294.6	0.03055	0.00050
213.44	373.95	2045	14.1	294.6	0.02806	0.00046
213.44	373.95	2045	14.1	294.6	0.03055	0.00050
272.75	406.90	22495	155.1	3.19	0.44146	0.00788
272.75	406.90	22495	155.1	3.19	0.44000	0.00786
272.75	406.90	19203	132.4	3.19	0.43563	0.00778
272.75	406.90	19203	132.4	3.19	0.38828	0.00694
273.20	407.15	19218	132.5	3.19	0.41086	0.00734
273.20	407.15	19218	132.5	3.19	0.41086	0.00734

(cont.)

COMPONENTS:	ORIGINAL MEASUREMENTS:
1. Methane; CH_4; [74-82-8]	Blount, C. W.; Price, L. C.;
2. Sodium chloride; NaCl; [7647-14-5]	Wenger, L. M.; Tarullo, M. DOE Contract report DE-A508-
3. Water; H_2O; [7732-18-5	78ET12145.

EXPERIMENTAL VALUES:

T/°F	T/K	P/psi	P/MPa	Conc. of NaCl /g dm^{-3}	Methane solubility /mol kg^{-1}	Mole fraction x_{CH_4}
272.75	406.90	16186	111.6	3.19	0.39047	0.00698
272.75	406.90	16186	111.6	3.19	0.35696	0.00638
273.20	407.15	13169	90.8	3.19	0.34603	0.00619
273.20	407.15	13169	90.8	3.19	0.35914	0.00642
273.20	407.15	10399	71.7	3.19	0.32636	0.00584
273.20	407.15	10399	71.7	3.19	0.29576	0.00529
272.30	406.65	10385	71.6	3.19	0.31470	0.00563
272.30	406.65	10385	71.6	3.19	0.29576	0.00529
273.65	407.40	7223	49.8	3.19	0.24695	0.00442
272.75	406.90	7266	50.1	3.19	0.24040	0.00431
272.75	406.90	7266	50.1	3.19	0.25715	0.00461
273.20	407.15	7629	52.6	3.19	0.26080	0.00467
273.20	407.15	7629	52.6	3.19	0.25351	0.00454
274.10	407.65	5149	35.5	3.19	0.22292	0.00400
273.20	407.15	5163	35.6	3.19	0.20543	0.00368
273.65	407.40	5134	35.4	3.19	0.20835	0.00374
273.65	407.40	5134	35.4	3.19	0.22656	0.00406
273.20	407.15	3626	25.0	3.19	0.17119	0.00307
273.20	407.15	3626	25.0	3.19	0.18358	0.00329
274.10	407.65	2248	15.5	3.19	0.13404	0.00241
274.10	407.65	2248	15.5	3.19	0.15007	0.00269
275.00	408.15	16128	111.2	51.1	0.34511	0.00608
274.55	407.90	16186	111.6	51.1	0.32252	0.00568
274.55	407.90	16186	111.6	51.1	0.34017	0.00599
275.00	408.15	13155	90.7	51.1	0.28441	0.00502
275.00	408.15	13242	91.3	51.1	0.29147	0.00514
275.00	408.15	13242	91.3	51.1	0.27383	0.00483
274.55	407.90	10327	71.2	51.1	0.26042	0.00460
274.55	407.90	10327	71.2	51.1	0.27171	0.00479
272.75	406.90	7034	48.5	51.1	0.20749	0.00366
272.75	406.90	7034	48.5	51.1	0.19831	0.00350
271.99	406.48	7034	48.5	51.1	0.20325	0.00359
271.99	406.48	7034	48.5	51.1	0.21031	0.00371
272.75	406.90	5062	34.9	51.1	0.16797	0.00297
272.75	406.90	5062	34.9	51.1	0.16867	0.00298
272.75	406.90	3568	24.6	51.1	0.14115	0.00250
272.75	406.90	3568	24.6	51.1	0.14185	0.00251
272.30	406.65	2016	13.9	51.1	0.09316	0.00165
272.30	406.65	2016	13.9	51.1	0.08963	0.00159
275.00	408.15	19014	131.1	106.0	0.24931	0.00432
275.00	408.15	19014	131.1	106.0	0.25136	0.00436
275.00	408.15	16273	112.2	106.0	0.22130	0.00384
275.00	408.15	16273	112.2	106.0	0.21789	0.00378
275.00	408.15	16157	111.4	106.0	0.23292	0.00404
275.00	408.15	16157	111.4	106.0	0.25409	0.00440
275.00	408.15	16128	111.2	106.0	0.26229	0.00455
275.00	408.15	16128	111.2	106.0	0.25887	0.00449
275.45	408.40	13169	90.8	106.0	0.21584	0.00374
273.20	407.15	13373	92.2	106.0	0.20218	0.00351
273.20	407.15	13373	92.2	106.0	0.20901	0.00363
274.55	407.90	13097	90.3	106.0	0.20423	0.00354
274.55	407.90	13097	90.3	106.0	0.20286	0.00352
274.55	407.90	13256	91.4	106.0	0.21994	0.00381
276.80	409.15	13082	90.2	106.0	0.21379	0.00371
276.80	409.15	13082	90.2	106.0	0.21789	0.00378
276.80	409.15	10312	71.1	106.0	0.20969	0.00364

(cont.)

COMPONENTS:	ORIGINAL MEASUREMENTS:
1. Methane; CH$_4$; [74-82-8]	Blount, C. W.; Price, L. C.;
2. Sodium chloride; NaCl; [7647-14-5]	Wenger, L. M.; Tarullo, M.
3. Water; H$_2$O; [7732-18-5]	DOE Contract report DE-A508-78ET12145.

EXPERIMENTAL VALUES:

T/°C	T/K	P/psi	P/MPa	Conc. of NaCl /g dm^{-3}	Methane solubility /mol kg^{-1}	Mole fraction x_{CH_4}
276.80	409.15	10312	71.1	106.0	0.20628	0.00358
274.10	407.65	7223	49.8	106.0	0.17213	0.00299
274.10	407.65	7223	49.8	106.0	0.17076	0.00296
274.55	407.90	7426	51.2	106.0	0.17691	0.00307
274.55	407.90	7426	51.2	106.0	0.18237	0.00316
274.55	407.90	5279	36.4	106.0	0.12978	0.00225
274.10	407.65	3579	24.8	106.0	0.11407	0.00198
274.10	407.65	3597	24.8	106.0	0.10519	0.00183
274.10	407.65	3626	25.0	106.0	0.11680	0.00203
274.10	407.65	3626	25.0	106.0	0.10929	0.00190
275.90	408.65	19043	131.3	105.0	0.25082	0.00435
276.44	408.95	19131	131.9	105.0	0.25423	0.00441
276.44	408.95	19131	131.9	105.0	0.25082	0.00435
276.80	409.15	15940	109.9	105.0	0.23168	0.00402
276.80	409.15	15940	109.9	105.0	0.23988	0.00416
276.98	409.25	13024	89.8	105.0	0.20161	0.00350
276.98	409.25	13024	89.8	105.0	0.20298	0.00352
277.16	409.35	13155	90.7	105.0	0.20708	0.00359
277.16	409.35	13155	90.7	105.0	0.20161	0.00350
276.80	409.15	13126	90.5	105.0	0.20571	0.00357
277.16	409.35	7020	48.4	107.0	0.15496	0.00269
277.16	409.35	7063	48.7	107.0	0.16657	0.00289
277.16	409.35	7063	48.7	107.0	0.15564	0.00270
277.16	409.35	5018	34.6	107.0	0.13243	0.00230
277.16	409.35	5018	34.6	107.0	0.13448	0.00233
276.80	409.15	5149	35.5	107.0	0.14199	0.00247
276.80	409.15	5149	35.5	107.0	0.14131	0.00245
276.80	409.15	5149	35.5	107.0	0.14131	0.00245
276.80	409.15	3684	25.4	107.0	0.10581	0.00184
276.98	409.25	3365	23.2	107.0	0.10786	0.00187
276.98	409.25	3365	23.2	107.0	0.10991	0.00191
276.98	409.25	2016	13.9	106.8	0.07647	0.00133
276.98	409.25	2016	13.9	106.8	0.07647	0.00133
271.04	405.95	2016	13.9	106.8	0.07305	0.00127
271.04	405.95	2016	13.9	106.8	0.06964	0.00121
275.00	408.15	22539	155.4	163.5	0.23644	0.00402
275.00	408.15	22539	155.4	163.5	0.23047	0.00392
275.00	408.15	22278	153.6	161.7	0.22538	0.00383
275.00	408.15	22350	154.1	161.7	0.23864	0.00406
275.00	408.15	22437	154.7	161.7	0.23599	0.00401
275.00	408.15	19029	131.2	161.7	0.22804	0.00388
275.00	408.15	19203	132.4	161.7	0.20682	0.00352
275.00	408.15	19203	132.4	161.7	0.21213	0.00361
275.00	408.15	16157	111.4	160.7	0.20560	0.00350
275.00	408.15	16157	111.4	160.7	0.21688	0.00369
275.90	408.65	13227	91.2	163.5	0.17219	0.00293
275.45	408.40	12850	88.6	160.7	0.18239	0.00311
275.00	408.15	13024	89.8	160.7	0.19300	0.00329
275.00	408.15	13024	89.8	160.7	0.18504	0.00315
275.90	408.65	10211	70.4	163.5	0.16425	0.00280
275.00	408.15	10182	70.2	163.5	0.16160	0.00275
275.00	408.15	10182	70.2	163.5	0.17948	0.00305
275.45	408.40	10153	70.0	160.7	0.18040	0.00307
275.45	408.40	10153	70.0	160.7	0.15918	0.00271
275.00	408.15	7136	49.2	160.7	0.12734	0.00217
275.00	408.15	7136	49.2	160.7	0.12734	0.00217

(cont.)

COMPONENTS:	ORIGINAL MEASUREMENTS:
1. Methane; CH_4; [74-82-8]	Blount, C. W.; Price, L. C.;
2. Sodium chloride; NaCl; [7647-14-5]	Wenger, L. M.; Tarullo, M. DOE Contract report DE-A508- 78ET12145.
3. Water; H_2O; [7732-18-5]	

EXPERIMENTAL VALUES:

T/°C	T/K	P/psi	P/MPa	Conc. of NaCl /g dm^{-3}	Methane solubility /mol kg^{-1}	Mole fraction x_{CH_4}
275.45	408.40	5163	35.6	160.3	0.10813	0.00184
275.45	408.40	5091	35.1	160.3	0.10548	0.00180
275.45	408.40	5091	35.1	160.3	0.10481	0.00179
275.00	408.15	3568	24.6	159.7	0.08162	0.00139
274.55	407.90	3626	25.0	163.9	0.08211	0.00140
274.55	407.90	3655	25.2	163.9	0.10330	0.00176
275.00	408.15	2147	14.8	163.9	0.06224	0.00106
275.00	408.15	2147	14.8	163.9	0.05761	0.00098
274.55	407.90	2292	15.8	163.9	0.05694	0.00097
274.55	407.90	2292	15.8	163.9	0.07019	0.00120
275.00	408.15	19087	131.6	224.3	0.15754	0.00263
275.00	408.15	19087	131.6	224.3	0.15818	0.00264
275.00	408.15	19232	132.6	224.3	0.14468	0.00241
274.55	407.90	19174	132.2	224.3	0.15175	0.00253
273.65	407.40	19072	131.5	227.8	0.15086	0.00251
273.65	407.40	19072	131.5	227.8	0.14509	0.00242
273.65	407.40	19072	131.5	227.6	0.14189	0.00236
273.65	407.40	19072	131.5	227.6	0.15794	0.00263
273.65	407.40	16041	110.6	227.1	0.14385	0.00240
273.65	407.40	16070	110.8	225.7	0.15294	0.00255
273.65	407.40	16070	110.8	225.7	0.14908	0.00248
273.65	407.40	13024	89.8	225.4	0.12532	0.00209
273.65	407.40	13024	89.8	225.4	0.13882	0.00231
273.65	407.40	10066	69.4	223.7	0.13764	0.00229
273.65	407.40	10066	69.4	223.7	0.14150	0.00236
273.65	407.40	10080	69.5	222.6	0.13449	0.00224
273.65	407.40	10080	69.5	222.6	0.12162	0.00203
273.65	407.40	10095	69.6	221.3	0.13651	0.00228
273.65	407.40	10109	69.7	219.0	0.12178	0.00203
273.65	407.40	10109	69.7	219.9	0.12178	0.00203
273.65	407.40	7136	49.2	218.6	0.10573	0.00177
272.30	406.65	7194	49.6	224.3	0.10224	0.00171
272.30	406.65	7194	49.6	224.3	0.10674	0.00178
273.65	407.40	4873	33.6	224.3	0.09131	0.00152
272.20	407.15	4931	34.0	224.3	0.07973	0.00133
272.20	407.15	4931	34.0	224.3	0.07780	0.00130
273.65	407.40	3597	24.8	223.7	0.05853	0.00098
273.65	407.40	3597	24.8	223.7	0.06110	0.00102
273.65	407.40	2176	15.0	223.2	0.04568	0.00076
273.65	407.40	2176	15.0	222.6	0.05792	0.00097
273.65	407.40	2176	15.0	222.6	0.05406	0.00090
273.65	407.40	2161	14.9	222.6	0.04633	0.00077
273.65	407.40	2161	14.9	222.6	0.04440	0.00074
275.90	408.65	22640	156.1	295.2	0.13463	0.00219
275.90	408.65	22640	156.1	295.1	0.13464	0.00219
275.90	408.65	19218	132.5	295.1	0.12030	0.00196
275.90	408.65	19218	132.5	295.1	0.11906	0.00194
275.90	408.65	16273	112.2	294.4	0.11784	0.00192
275.90	408.65	16273	112.2	294.4	0.11036	0.00179
277.25	409.40	13068	90.1	293.6	0.09730	0.00158
277.25	409.40	13068	90.1	293.6	0.09480	0.00154
277.70	409.65	10167	70.1	293.0	0.08609	0.00140
277.70	409.65	10167	70.1	293.0	0.08734	0.00142
276.80	409.15	7150	49.3	292.1	0.07240	0.00118
276.80	409.15	7150	49.3	292.1	0.07115	0.00116
276.35	408.90	5163	35.6	291.5	0.05931	0.00097

(cont.)

COMPONENTS:	ORIGINAL MEASUREMENTS:
1. Methane; CH$_4$; [74-82-8]	Blount, C. W.; Price, L. C.;
2. Sodium chloride; NaCl; [7647-14-5]	Wenger, L. M.; Tarullo, M.
3. Water; H$_2$O; [7732-18-5]	DOE Contract report DE-A508-78ET12145.

EXPERIMENTAL VALUES:

T/°C	T/K	P/psi	P/MPa	Conc. of NaCl /g dm^{-3}	Methane solubility /mol kg^{-1}	Mole fraction x_{CH_4}
276.35	408.90	5163	35.6	291.5	0.05931	0.00097
276.80	409.15	2263	15.6	286.0	0.04005	0.00065
276.80	409.15	2263	15.6	286.0	0.03629	0.00059
335.12	441.55	22539	155.4	3.0	0.58942	0.01050
335.12	441.55	22539	155.4	3.0	0.66665	0.01186
335.75	441.90	22278	153.6	3.0	0.65500	0.01165
335.75	441.90	22278	153.6	3.0	0.60399	0.01075
335.75	441.90	21843	150.6	3.0	0.65135	0.01159
335.75	441.90	22597	155.8	3.0	0.58724	0.01046
335.75	441.90	22597	155.8	3.0	0.62585	0.01114
336.20	442.15	22800	157.2	3.0	0.60909	0.01084
336.20	442.15	22771	157.0	3.0	0.69142	0.01229
336.20	442.15	22771	157.0	3.0	0.67685	0.01204
336.20	442.15	18332	126.4	3.0	0.58432	0.01041
336.20	442.15	18332	126.4	3.0	0.62512	0.01113
336.20	442.15	17927	123.6	3.0	0.63678	0.01113
335.75	441.90	19218	132.5	3.0	0.61711	0.01099
335.75	441.90	16360	112.8	3.0	0.53187	0.00948
335.75	441.90	16360	112.8	3.0	0.57412	0.01023
336.65	442.40	13169	90.8	3.0	0.53041	0.00946
336.65	442.40	13169	90.8	3.0	0.52021	0.00928
336.65	442.40	10225	70.5	3.0	0.43132	0.00770
335.84	441.95	10269	70.8	3.0	0.42695	0.00763
336.20	442.15	10167	70.1	3.0	0.46119	0.00823
337.55	442.90	5076	35.0	3.0	0.33733	0.00603
337.55	442.90	5076	35.0	3.0	0.32713	0.00585
336.20	442.15	3626	25.0	3.1	0.23167	0.00415
336.20	442.15	3626	25.0	3.1	0.23677	0.00424
336.20	442.15	3640	25.1	3.1	0.25717	0.00461
336.20	442.15	3640	25.1	3.1	0.27247	0.00488
340.25	444.40	18347	126.5	51.1	0.44956	0.00791
340.25	444.40	18347	126.5	51.1	0.45026	0.00792
340.25	444.40	17985	124.0	51.1	0.44744	0.00787
340.25	444.40	17985	124.0	51.1	0.46720	0.00821
339.98	444.25	15650	107.9	51.0	0.42771	0.00753
339.98	444.25	15650	107.9	51.0	0.43759	0.00770
340.70	444.65	12821	88.4	50.7	0.37414	0.00659
340.70	444.65	12821	88.4	50.7	0.39602	0.00697
339.98	444.25	9776	67.4	50.4	0.34949	0.00616
339.98	444.25	9776	67.4	50.4	0.31913	0.00563
339.98	444.25	6773	46.7	50.0	0.26978	0.00476
339.98	444.25	6773	46.7	50.0	0.28320	0.00500
339.08	443.75	4801	33.1	49.7	0.25076	0.00443
339.08	443.75	4801	33.1	49.7	0.23451	0.00414
339.08	443.75	4801	33.1	49.7	0.23592	0.00417
338.90	443.65	3553	24.5	49.5	0.19357	0.00342
338.90	443.65	3553	24.5	49.5	0.19922	0.00352
339.35	443.90	1958	13.5	49.5	0.13281	0.00235
339.35	443.90	1958	13.5	49.5	0.12716	0.00225
339.44	443.95	19043	131.3	104.6	0.32060	0.00555
339.80	444.15	19072	131.5	104.6	0.35204	0.00609
339.80	444.15	19072	131.5	104.6	0.35819	0.00620
339.80	444.15	16157	111.4	104.6	0.31991	0.00554
339.80	444.15	16157	111.4	104.6	0.31171	0.00540
339.80	444.15	16027	110.5	104.6	0.31308	0.00542
339.80	444.15	16027	110.5	104.6	0.30693	0.00532

(cont.)

COMPONENTS:	ORIGINAL MEASUREMENTS:
1. Methane; CH_4; [74-82-8] 2. Sodium chloride; NaCl; [7647-14-5] 3. Water; H_2O; [7732-18-5]	Blount, C. W.; Price, L. C.; Wenger, L. M.; Tarullo, M. DOE Contract report DE-A508- 78ET12145.

EXPERIMENTAL VALUES:

T/°C	T/K	P/psi	P/MPa	Conc. of NaCl /g dm⁻³	Methane solubility /mol kg⁻¹	Mole fraction x_{CH_4}
339.80	444.15	16027	110.5	104.6	0.32538	0.00564
339.80	444.15	16027	110.5	104.6	0.32538	0.00564
339.80	444.15	13024	89.8	104.6	0.27822	0.00482
339.80	444.15	13024	89.8	104.6	0.28505	0.00494
338.36	443.35	13024	89.8	104.6	0.28505	0.00494
338.36	443.35	13024	89.8	104.6	0.29599	0.00513
339.80	444.15	10008	69.0	106.0	0.25819	0.00447
339.80	444.15	10008	69.0	106.0	0.26638	0.00462
339.80	444.15	10037	69.2	106.0	0.24863	0.00431
339.80	444.15	7049	48.6	106.0	0.20218	0.00351
339.80	444.15	7020	48.4	106.0	0.19603	0.00340
339.80	444.15	7020	48.4	106.0	0.20355	0.00353
339.80	444.15	6991	48.2	106.0	0.19945	0.00346
339.80	444.15	6991	48.2	106.0	0.19808	0.00344
339.80	444.15	5047	34.8	106.0	0.16529	0.00287
339.80	444.15	5047	34.8	106.0	0.17691	0.00307
339.80	444.15	5250	36.2	106.0	0.17486	0.00303
339.80	444.15	5250	36.2	106.0	0.16325	0.00283
339.80	444.15	3553	24.5	106.0	0.14412	0.00250
339.80	444.15	3553	24.5	106.0	0.13388	0.00233
339.80	444.15	3568	24.6	104.0	0.15386	0.00267
339.80	444.15	3568	24.6	104.0	0.13676	0.00238
339.80	444.15	2002	13.8	104.0	0.11762	0.00204
339.80	444.15	2002	13.8	104.0	0.11420	0.00199
339.80	444.15	2002	13.8	104.0	0.09505	0.00165
339.80	444.15	2002	13.8	104.0	0.10667	0.00185
339.80	444.15	22568	155.6	163.9	0.30657	0.00521
338.00	443.15	22568	155.6	163.9	0.30194	0.00513
338.00	443.15	22582	155.7	162.1	0.30487	0.00518
338.00	443.15	22582	155.7	162.1	0.29426	0.00500
338.36	443.35	19072	131.5	163.4	0.27884	0.00474
338.36	443.35	19072	131.5	163.4	0.27155	0.00461
336.20	442.15	16099	111.0	159.8	0.26409	0.00449
336.20	442.15	16099	111.0	159.8	0.27736	0.00472
338.36	443.35	13097	90.3	157.8	0.21853	0.00372
338.36	443.35	13097	90.3	157.8	0.21986	0.00375
336.20	442.15	10298	71.0	159.5	0.19777	0.00337
336.20	442.15	10298	71.0	159.5	0.18250	0.00311
336.20	442.15	7165	49.4	159.5	0.17786	0.00303
336.20	442.15	7165	49.4	159.5	0.17454	0.00297
336.20	442.15	5105	35.2	159.5	0.14401	0.00246
336.20	442.15	5105	35.2	159.5	0.14268	0.00243
336.20	442.15	3553	24.5	159.5	0.11348	0.00194
336.20	442.15	3553	24.5	159.5	0.11680	0.00199
337.10	442.65	2060	14.2	159.5	0.08760	0.00150
337.64	442.95	2016	13.9	159.5	0.08229	0.00140
337.64	442.95	2016	13.9	159.5	0.08229	0.00140
338.36	443.35	22539	155.4	224.3	0.24563	0.00409
338.36	443.35	22539	155.4	224.3	0.24177	0.00402
338.36	443.35	22336	155.4	224.3	0.25077	0.00417
338.36	443.35	22336	155.4	224.3	0.25206	0.00419
338.00	443.15	19087	131.6	223.7	0.22061	0.00367
338.00	443.15	19087	131.6	223.7	0.20518	0.00342
337.64	442.95	16027	110.5	223.5	0.19362	0.00323
338.00	443.15	16070	110.8	223.5	0.18911	0.00315
338.00	443.15	16070	110.8	223.5	0.20262	0.00337

(cont.)

COMPONENTS:	ORIGINAL MEASUREMENTS:
1. Methane; CH₄; [74-82-8]	Blount, C. W.; Price, L. C.;
2. Sodium chloride; NaCl; [7647-14-5]	Wenger, L. M.; Tarullo, M.
	DOE Contract report DE-A508-
3. Water; H₂O; [7732-18-5]	78ET12145.

EXPERIMENTAL VALUES:

T/°F	T/K	P/psi	P/MPa	Conc. of NaCl /g dm^{-3}	Methane solubility /mol kg^{-1}	Mole fraction x_{CH_4}
338.00	443.15	13082	90.2	223.7	0.18074	0.00301
338.00	443.15	13097	90.3	223.7	0.18652	0.00311
338.00	443.15	13097	90.3	223.7	0.19553	0.00326
337.64	442.95	10066	69.4	223.5	0.14988	0.00250
338.36	443.35	10051	69.3	223.5	0.14988	0.00250
338.36	443.35	10051	69.3	223.5	0.15052	0.00251
335.84	441.95	7034	48.5	220.4	0.12497	0.00209
338.00	443.15	7092	48.9	220.4	0.13656	0.00228
338.00	443.15	7092	48.9	220.4	0.13334	0.00223
338.00	443.15	5178	35.7	226.5	0.09764	0.00163
338.00	443.15	5033	34.7	226.5	0.09378	0.00156
338.00	443.15	5033	34.7	226.5	0.09700	0.00162
338.00	443.15	3626	25.0	220.4	0.08245	0.00138
338.00	443.15	3626	27.1	220.4	0.07666	0.00128
338.40	443.35	2263	15.6	226.5	0.06167	0.00103
338.40	443.35	2277	15.7	226.5	0.06167	0.00103
338.00	443.15	22480	155.0	289.9	0.18303	0.00298
338.00	443.15	22480	155.0	289.9	0.17803	0.00290
340.00	444.15	21553	148.6	289.9	0.18178	0.00296
338.00	443.15	19232	132.6	287.2	0.16823	0.00274
338.00	443.15	19232	132.6	287.2	0.16948	0.00276
338.00	443.15	16258	112.1	284.5	0.16467	0.00268
338.00	443.15	16258	112.1	284.5	0.15778	0.00257
338.00	443.15	13256	91.4	283.0	0.14472	0.00236
338.00	443.15	13256	91.4	283.1	0.13407	0.00219
338.00	443.15	13256	91.4	283.1	0.13595	0.00222
338.00	443.15	13256	91.4	283.1	0.14096	0.00230
338.90	443.65	10211	70.4	283.1	0.12655	0.00207
338.90	443.65	10211	70.4	283.1	0.12091	0.00197
339.44	443.95	7252	50.0	283.1	0.09397	0.00153
339.44	443.95	7252	50.0	283.1	0.09836	0.00161
339.44	443.95	5163	35.6	282.1	0.08085	0.00132
339.44	443.95	5163	35.6	282.1	0.08524	0.00139
338.90	443.65	3539	24.4	282.1	0.06581	0.00108
338.90	443.65	3539	24.4	282.1	0.06769	0.00111
401.00	478.15	22452	154.8	3.1	0.91139	0.01614
401.45	478.40	22742	156.8	3.5	0.89293	0.01581
401.45	478.40	22742	156.8	3.5	0.91696	0.01623
401.45	478.40	19406	133.8	3.5	0.83321	0.01477
401.45	478.40	19406	133.8	3.5	0.90094	0.01595
400.55	477.90	19319	133.2	3.5	0.86962	0.01541
400.55	477.90	19319	133.2	3.5	0.88856	0.01574
401.90	478.65	16418	113.2	3.5	0.78732	0.01397
401.90	478.65	16360	112.8	3.5	0.83685	0.01484
401.90	478.65	16360	112.8	3.6	0.81281	0.01442
401.90	478.65	13140	90.6	3.5	0.70138	0.01246
401.90	478.65	13140	90.6	3.5	0.71740	0.01275
401.45	478.40	10356	71.4	3.5	0.62637	0.01115
401.18	478.25	10269	70.8	3.5	0.66642	0.01185
400.82	478.05	10327	71.2	3.5	0.63656	0.01133
400.82	478.05	10327	71.2	3.5	0.61908	0.01102
398.75	476.90	7310	50.4	3.5	0.57684	0.01027
398.75	476.90	7310	50.4	3.5	0.58266	0.01038
398.30	476.65	5207	35.9	3.5	0.41369	0.00739
398.30	476.76	5207	35.9	3.5	0.41223	0.00736
398.12	476.55	3655	25.2	3.5	0.31755	0.00568

(cont.)

COMPONENTS: ORIGINAL MEASUREMENTS:

1. Methane; CH_4; [74-82-8]   Blount, C. W.; Price, L. C.;

                                               Wenger, L. M.; Tarullo, M.
2. Sodium chloride; NaCl;
   [7647-14-5]   DOE Contract report DE-A508-
                                               78ET12145.
3. Water; H_2O; [7732-18-5]

EXPERIMENTAL VALUES:

$T/°F$	T/K	P/psi	P/MPa	Conc. of NaCl $/\text{g dm}^{-3}$	Methane solubility $/\text{mol kg}^{-1}$	Mole fraction x_{CH_4}
398.12	476.55	3655	25.2	3.5	0.31901	0.00571
397.22	476.05	2060	14.2	3.5	0.26802	0.00480
397.22	476.05	2060	14.2	3.5	0.24108	0.00432
401.90	478.65	19247	132.7	49.8	0.68301	0.01197
401.90	478.65	19377	133.6	49.4	0.67894	0.01190
401.45	478.40	19522	134.6	49.3	0.65355	0.01146
401.45	478.40	19522	134.6	49.3	0.66062	0.01158
401.45	478.40	16070	110.8	49.2	0.61685	0.01082
401.45	478.40	16244	112.0	49.0	0.67063	0.01176
401.45	478.40	16244	112.0	49.0	0.64307	0.01128
401.45	478.40	13227	91.2	49.0	0.64025	0.01123
401.45	478.40	13227	91.2	49.0	0.59785	0.01049
401.45	478.40	9979	68.8	49.0	0.53283	0.00936
401.45	478.40	9979	68.8	49.0	0.52223	0.00918
401.90	478.65	7194	49.6	48.8	0.37883	0.00668
401.90	478.65	7194	49.6	48.8	0.39296	0.00692
402.35	478.90	7513	51.8	48.8	0.40710	0.00717
401.00	478.15	4873	33.6	46.8	0.33755	0.00596
401.00	478.15	4917	33.9	46.8	0.36091	0.00637
401.00	478.15	4960	34.2	46.8	0.37364	0.00659
401.00	478.15	4960	34.2	46.8	0.37577	0.00663
401.36	478.35	3553	24.5	46.3	0.29023	0.00513
401.36	478.35	3553	24.5	46.3	0.30439	0.00538
401.36	478.35	2161	14.9	46.3	0.19254	0.00341
401.36	478.35	2161	14.9	46.3	0.19679	0.00348
404.60	480.15	22278	153.6	161.7	0.39508	0.00670
404.60	480.15	22278	153.6	161.7	0.40105	0.00680
404.60	480.15	22292	153.7	161.7	0.40304	0.00684
404.60	480.15	22292	153.7	161.7	0.40569	0.00688
404.60	480.15	18855	130.0	161.7	0.38050	0.00646
404.60	480.15	18855	130.0	161.7	0.38647	0.00656
404.60	480.15	15969	110.1	161.7	0.36525	0.00620
404.60	480.15	15969	110.1	161.7	0.35067	0.00595
404.24	479.95	12966	89.4	161.7	0.34537	0.00586
403.70	479.65	12952	89.3	161.7	0.32681	0.00555
404.24	479.95	10124	69.8	161.7	0.30029	0.00510
404.24	479.95	10124	69.8	161.7	0.28968	0.00492
404.24	479.95	7005	48.3	161.7	0.25654	0.00436
404.24	479.95	7005	48.3	161.7	0.24660	0.00419
404.24	479.95	5076	35.0	161.7	0.21146	0.00360
404.24	479.95	5076	35.0	161.7	0.23334	0.00397
404.24	479.95	3568	24.6	161.7	0.16837	0.00287
404.24	479.95	3568	24.6	161.7	0.16572	0.00282
404.24	479.95	2074	14.3	161.7	0.12926	0.00220
404.24	479.95	2074	14.3	161.7	0.11402	0.00194
402.44	478.95	22481	155.0	221.0	0.30525	0.00508
402.44	478.95	22481	155.0	221.0	0.32457	0.00540
401.90	478.65	19218	132.5	221.0	0.30268	0.00504
401.90	478.65	19087	131.6	216.0	0.30274	0.00505
401.90	478.65	19087	131.6	216.0	0.29757	0.00496
402.44	478.95	15635	107.8	212.6	0.27284	0.00456
402.44	478.95	15635	107.8	212.6	0.29676	0.00495
402.44	478.95	13169	90.8	212.3	0.23149	0.00387
402.44	478.95	13169	90.8	212.3	0.23357	0.00393
401.90	478.65	13068	90.1	211.5	0.23546	0.00394
401.90	478.65	13068	90.1	211.5	0.23934	0.00400

(cont.)

COMPONENTS:	ORIGINAL MEASUREMENTS:
1. Methane; CH_4; [74-82-8]	Blount, C. W.; Price, L. C.;
2. Sodium chloride; NaCl; [7647-14-5]	Wenger, L. M.; Tarullo, M. DOE Contract report DE-A508-
3. Water; H_2O; [7732-18-5]	78ET12145.

EXPERIMENTAL VALUES:

T/°F	T/K	P/psi	P/MPa	Conc. of NaCl /g dm^{-3}	Methane solubility /mol kg^{-1}	Mole fraction CH_4
401.90	478.65	10385	71.6	219.7	0.22360	0.00373
402.44	478.93	10428	71.9	218.6	0.23274	0.00388
402.44	478.95	10428	71.9	218.6	0.23210	0.00387
402.44	478.95	7092	48.9	216.0	0.18139	0.00303
402.44	478.95	7092	48.9	216.0	0.16202	0.00271
402.44	478.95	4989	34.4	211.5	0.14878	0.00249
402.44	478.95	4989	34.4	211.5	0.13325	0.00223
401.90	478.65	3510	24.2	207.0	0.11604	0.00195
401.90	478.65	3510	24.2	207.0	0.11279	0.00189
398.75	476.90	22684	156.4	281.7	0.27832	0.00453
398.75	476.90	22568	155.6	281.7	0.25826	0.00421
398.75	476.90	22568	155.6	281.7	0.26077	0.00425
399.20	477.15	22452	154.8	281.7	0.26704	0.00435
398.75	476.90	22220	153.2	281.7	0.28146	0.00458
398.75	476.90	22220	153.2	281.7	0.28835	0.00470
398.75	476.90	22597	155.8	281.7	0.26892	0.00438
398.75	476.90	22597	155.8	281.7	0.28020	0.00456
399.20	477.15	19363	133.5	281.7	0.24071	0.00392
399.20	477.15	19363	133.5	281.7	0.26140	0.00426
399.20	477.15	16273	112.2	281.7	0.24071	0.00392
399.20	477.15	16273	112.2	281.7	0.22379	0.00365
398.75	476.90	13198	91.0	281.7	0.20749	0.00338
398.75	476.90	13198	91.0	281.7	0.20498	0.00334
398.30	476.65	10269	70.8	273.7	0.18618	0.00305
398.30	476.65	10269	70.8	273.7	0.18430	0.00302
398.30	476.65	7107	49.0	273.7	0.15159	0.00248
398.30	476.65	5163	35.6	255.8	0.11664	0.00192
398.30	476.65	5163	35.6	255.8	0.10840	0.00179
398.75	476.90	5453	37.6	249.1	0.12271	0.00203
398.75	476.90	5453	37.6	249.1	0.12207	0.00202
398.30	476.65	3684	25.4	243.6	0.09688	0.00160
398.30	476.65	3684	25.4	243.6	0.09114	0.00151
462.20	512.15	18942	130.6	47.7	0.95621	0.01669
462.20	512.15	19145	132.0	47.6	1.06307	0.01852
462.20	512.15	19145	132.0	47.6	0.99941	0.01743
462.20	512.15	19218	132.5	47.5	0.97401	0.01699
462.20	512.15	15969	110.1	47.4	0.89556	0.01565
462.20	512.15	15969	110.1	47.4	0.86302	0.01509
462.20	512.15	13329	91.9	46.3	0.89972	0.01572
462.20	512.15	13329	91.9	46.3	0.88202	0.01542
462.20	512.15	13576	93.6	46.1	0.75966	0.01331
462.20	512.15	14214	98.0	45.8	0.84194	0.01473
462.20	512.15	14504	100.0	45.8	0.83415	0.01460
462.20	512.15	14504	100.0	45.8	0.84548	0.01479
462.20	512.15	9573	66.0	45.8	0.71873	0.01260
462.20	512.15	9573	66.0	45.8	0.71519	0.01254
462.20	512.15	7397	51.0	45.3	0.65450	0.01149
462.20	512.15	7397	51.0	45.3	0.59996	0.01054
462.20	512.15	5018	34.6	45.1	0.52919	0.00931
462.20	512.15	5221	36.0	44.0	0.47923	0.00844
462.20	512.15	5308	36.6	43.7	0.48074	0.00847
462.20	512.15	5308	36.6	43.7	0.48145	0.00848
460.40	511.15	3568	24.6	43.7	0.44174	0.00779
460.40	511.15	3568	24.6	43.7	0.38289	0.00676
460.40	511.15	2031	14.0	43.2	0.29364	0.00519
460.40	511.15	2031	14.0	43.2	0.26243	0.00464

(cont.)

COMPONENTS:	ORIGINAL MEASUREMENTS:
1. Methane; CH_4; [74-82-8]	Blount, C. W.; Price, L. C.;
2. Sodium chloride; NaCl; [7647-14-5]	Wenger, L. M.; Tarullo, M.
	DOE Contract report DE-A508-
3. Water; H_2O; [7732-18-5]	78ET12145.

EXPERIMENTAL VALUES:

$T/°F$	T/K	P/psi	P/MPa	Conc. of NaCl $/g\ dm^{-3}$	Methane solubility $/mol\ kg^{-1}$	Mole fraction x_{CH_4}
460.40	511.15	2045	14.1	43.2	0.24896	0.00441
461.84	511.95	19203	132.4	158.4	0.52258	0.00886
461.84	511.95	19203	132.4	158.4	0.51926	0.00880
462.20	512.15	16172	111.5	158.4	0.49270	0.00835
462.20	512.15	16172	111.5	158.4	0.49403	0.00838
461.30	511.65	13169	90.8	158.4	0.43759	0.00743
461.30	511.65	13169	90.8	158.4	0.46282	0.00785
461.84	511.95	9964	68.7	161.9	0.38510	0.00653
461.84	511.95	9964	68.7	161.9	0.38245	0.00649
461.84	511.95	7078	48.8	161.9	0.31153	0.00529
461.84	511.95	7078	48.8	161.9	0.29297	0.00498
459.50	510.65	5004	34.5	161.9	0.28369	0.00482
459.50	510.65	5004	34.5	161.9	0.28568	0.00486
461.84	511.95	3510	24.2	161.9	0.21012	0.00358
461.84	511.95	3510	24.2	161.9	0.21542	0.00367
461.84	511.95	3452	23.8	161.9	0.21210	0.00361
461.84	511.95	3452	23.8	161.9	0.21012	0.00358
462.20	512.15	19232	132.6	253.8	0.37688	0.00619
462.20	512.15	19232	132.6	253.8	0.37054	0.00609
464.00	513.15	16331	112.6	253.2	0.33446	0.00550
464.00	513.15	16331	112.6	253.2	0.37889	0.00623
464.00	513.15	15867	109.4	250.4	0.34441	0.00567
464.00	513.15	15867	109.4	250.4	0.33297	0.00548
464.00	513.15	13140	90.6	248.3	0.32947	0.00543
464.00	513.15	13140	90.6	248.3	0.33583	0.00553
464.00	513.15	13256	91.4	243.6	0.32442	0.00535
464.00	513.15	13256	91.4	243.6	0.32506	0.00536
464.00	513.15	10298	71.0	229.4	0.28226	0.00469
464.00	513.15	10298	71.0	229.4	0.27392	0.00455

COMPONENTS:	ORIGINAL MEASUREMENTS:
(1) Methane; CH_4; [74-82-8] (2) Sodium chloride; NaCl; [7647-14-5] (3) Water; H_2O; [7732-18-5]	Stoessell, R. K.; Byrne, P. A. *Geochim. Cosmochim. Acta* <u>1982</u>, *46*, 1327-32.

VARIABLES:	PREPARED BY:
$T/K = 298.15$ $p_1/kPa = 2410-5170$ $m_2/mol\ kg^{-1} = 0 - 4.0$	H. L. Clever

EXPERIMENTAL VALUES:

Temperature		Pressure		Sodium Chloride	Methane	Salt Effect Parameter
$t/°C$	T/K	$p_1/psia$	p_1/kPa	$m_2/mol\ kg^{-1}$	$m_1/mol\ kg^{-1}$	$k_{smm}/kg\ mol^{-1}$
25	298.15	350	2410	0	0.0319	–
				0.5	0.0284	0.101
				1.0	0.0244	0.116
				2.0	0.0180	0.124
				4.0	0.0106	0.120
		550	3790	0	0.0483	–
				0.5	0.0412	0.138
				1.0	0.0352	0.137
				2.0	0.0272	0.125
				4.0	0.0157	0.122
		750	5170	0	0.0617	–
				0.5	0.0539	0.117
				1.0	0.0464	0.124
				2.0	0.0347	0.124
				4.0	0.0206	0.119
						0.121 (authors)

The salt effect parameter is defined as $k_{smm} = \log \gamma_1/I$

where I is the ionic strength (molality) and $\gamma_1 = (m_1^* f_1/m_1 f_1^*)_{p,T}$ with m_1 and f_1 the solubility and fugacity, repectively, of methane at p and T. The "*" refers to saturation in distilled water.

AUXILIARY INFORMATION

METHOD/APPARATUS/PROCEDURE:
Solubility determinations were made using a titanium-lined chamber within a stainless steel reaction vessel jacketed by a water bath for temperature control. The system pressure was set by controlling the input and output of methane within the chamber's headpiece. The vessel was rocked for three h to allow equilibration between the methane and solution.
The amount of gas in the saturated solution was measured by transfer of a sample volume to a loop at the system pressure, followed by flashing the sample in an expansion loop and measuring the gas pressure in a known volume. The total gas volume and pressure change were used to compute the moles of released gas assuming ideal behavior. A correction was made for the gas not released on flashing.

SOURCE AND PURITY OF MATERIALS:

(1) Methane. Matheson Co., Inc. Ultra high purity grade, stated to be a minimum of 99.97 mole percent methane.

(2) Sodium chloride. The salt solutions were made up gravimetrically using analytical grade chemicals.

(3) Water. Distilled.

ESTIMATED ERROR:
$\delta p_1/psia = \pm 1$
$\delta m_1/m_1 = \pm 0.01$

REFERENCES:

M—F

COMPONENTS:	ORIGINAL MEASUREMENTS:
(1) Methane; CH_4; [74-82-8] (2) Sodium chloride; NaCl; [7647-14-5] (3) Water; H_2O; [7732-18-5] Also Salton sea geothermal brine.	Cramer, S. D. *Ind. Eng. Chem. Processs Des. Dev.* <u>1984</u>, *23*, 533-8.

EXPERIMENTAL VALUES:

Temperature		Total Pressure	Henry's Constant	Temperature		Total Pressure	Henry's Constant
$t/^0C$	T/K	p_t/MPa	k/MPa	$t/^0C$	T/K	p_t/MPa	k/MPa

NaCl-1, 0.81m, 0.79 M, 0.81 I.

3.0	276.2	1.9	3620
13.0	286.2	2.2	4190
20.0	293.2	3.3	5640
41.6	314.8	4.1	7120
73.3	346.5	2.7	8400
100.4	373.6	3.0	9260
124.6	397.8	3.0	7870
146.0	419.2	6.3	8150
174.5	447.7	5.0	6440
191.6	464.8	5.5	5350
240.6	513.8	8.7	3400
264.0	537.2	10.3	2450

NaCl-4, 4.70 m, 4.26 M, 4.70 I.

3.5	276.7	2.4	17560
13.3	286.5	2.3	14650
31.7	304.9	3.7	23560
61.6	334.8	3.9	23360
89.5	362.7	5.5	28150
112.6	385.8	5.4	21110
115.7	388.9	4.7	27470
146.9	420.1	4.7	20110
159.4	432.6	3.7	18710
204.8	478.0	5.8	15400
241.1	514.3	8.2	10240
287.9	561.1	10.9	6740

NaCl-2, 1.95 m, 1.86 M, 1.95 I.

0.5	273.7	2.3	5530
5.0	278.2	2.7	5870
12.6	285.8	2.3	7080
13.0	286.2	2.6	6060
26.3	299.5	2.3	8230
47.5	320.7	1.9	10530
78.0	351.2	2.3	11770
102.0	375.2	4.6	12260
131.9	405.1	3.5	10450
161.8	435.0	4.4	8620
205.2	478.4	7.1	6770
224.6	497.8	8.9	5330
244.9	518.1	8.1	3820
271.4	544.6	10.1	3180
301.1	574.3	11.9	2070

SSGB, Synthetic Salton sea geothermal brine, 4.05 m, 3.55 M, 6.18 I.

4.5	277.7	2.6	12490
12.5	285.7	2.7	13160
16.0	289.2	2.7	12920
45.3	318.5	1.1	17990
60.2	333.4	1.1	17630
88.1	361.3	1.1	18000
121.7	394.9	1.1	16790
166.8	440.0	2.7	16450
204.3	477.5	5.3	13180
233.0	506.2	6.4	9200
268.0	541.2	10.5	6320
301.0	574.2	12.7	4830

Composition of the synthetic Salton sea geothermal brine.

Constituent	ppm by wt
barium	207
boron	324
calcium	23 900
cesium	17
chlorine	129 000
iron	1 660
lead	66
lithium	174
magnesium	8
manganese	1 140
potassium	13 700
rubidium	58
silica	332
sodium	44 000
strontium	365
sulfur	25
zinc	415

NaCl-3, 3.18 m, 2.97 M, 3.18 I.

3.5	276.7	3.3	8160
12.5	285.7	2.4	8860
12.5	285.7	2.5	9090
24.5	297.7	4.1	13390
45.7	318.9	3.1	15590
60.2	333.4	1.1	17330
61.8	335.0	3.4	14290
61.9	335.1	3.3	15350
75.4	348.6	2.4	17160
88.8	362.0	4.0	16300
111.9	385.1	5.2	15920
134.4	407.6	4.3	14870
162.2	435.4	5.5	12870
180.7	453.9	6.8	11200
203.7	476.9	5.7	7430
205.3	478.5	6.8	10870
225.1	498.3	7.3	7160
256.0	529.2	9.4	5670
269.5	542.7	11.3	5250
271.5	544.7	10.8	5050
295.6	568.8	12.4	3410

$m = m_2$/mol kg^{-1}; $M = c_2$/mol dm^{-3};
$I = \Sigma m_i z_i^2$, ionic strength in molality.

COMPONENTS:	ORIGINAL MEASUREMENTS:
(1) Methane; CH_4; [74-82-8]	Cramer, S. D.
(2) Sodium chloride; NaCl; [7647-14-5	*Ind. Eng. Chem. Process Des. Dev.*
(3) Water; H_2O; [7732-18-5]	**1984**, *23*, 533-8.
Also Salton sea geothermal brine.	

VARIABLES:	PREPARED BY:
T/K = 273.7 - 301.1 p_t/MPa = 1.1 - 12.7 m_2/mol kg^{-1} = 0.81 - 4.70	H. L. Clever

ADDITIONAL INFORMATION: Author's smoothed values of Henry's constant and NaCl salt effect parameters.

Temperature		Henry's constant, k/MPa = $(f_1$/MPa$)/x_1$					Salt Effect Parameter
$t/^0$C	T/K	NaCl-1 0.81 *m*	NaCl-2 1.95 *m*	NaCl-3 3.18 *m*	NaCl-4 4.70 *m*	SSGB 4.05 *m*	k_{smx}/kg mol^{-1}
0	273.2	3400	5320	7550	14970	11680	0.165
20	293.2	5140	7540	10840	19170	14090	0.141
40	313.2	6810	9540	13760	22500	16120	0.127
60	333.2	8120	11020	15830	24620	17610	0.119
80	353.2	8880	11810	16830	25440	18470	0.116
100	373.2	9050	11900	16810	25090	18670	0.116
120	393.2	8700	11380	15970	23810	18260	0.118
140	413.2	7990	10420	14570	21890	17310	0.122
160	433.2	7060	9210	12890	19600	15960	0.126
180	453.2	6040	7910	11120	17190	14340	0.130
200	473.2	5040	6620	9410	14810	12560	0.134
220	493.2	4120	5430	7860	12580	10760	0.136
240	513.2	3310	4370	6500	10570	9010	0.137
260	533.2	2630	3480	5340	8810	7390	0.136
280	553.2	2070	2730	4380	7290	5940	0.133
300	573.2	1620	2130	3590	6000	4690	0.128

The methane fugacity is estimated by subtracting water vapor pressure from total pressure and taking fugacity coefficient reduced properties chart.

AUXILIARY INFORMATION

METHOD/APPARATUS/PROCEDURE:

Methane solubilities were determined from pvT measurements by the gas extraction technique. The measuring apparatus consisted of : (*i*) a high pressure, thermostated, stirred reactor for dissolving the gas in the solvent; (*ii*) a heat exchanger for bringing the gas saturated solvent to room temperature; and (*iii*) a low pressure, thermostated gas buret for making pvT measurements on collected samples of vapor and liquid. The apparatus and its operation were described earlier (ref 1).

Four to eight gas-saturated solution samples were taken and analyzed at 15-30 minute intervals after the time determined necessary for saturation. Henry's constants were computed (ref 1), and smoothed by a specially developed equation (ref 2).

Henry's constant:

$k^0 = f/a = (\phi p_1)/(\gamma x_1)$ see paper.
Salt solution, $k = \gamma k^0$

SOURCE AND PURITY OF MATERIALS:

1. Methane.

2. Sodium chloride   No information.

3. Water.

ESTIMATED ERROR:
$\delta k/k$ = ± 0.058 is author's estimated experimental error. The relative std error of estimate is 5.1-10.5 %.

REFERENCES:

1. Cramer, S. D.
   Ind. Eng. Chem. Process Des. Dev.
   1980, *19*, 300.

2. Cramer, S. D.
   ibid **1984**, *23*, 618.

COMPONENTS:	ORIGINAL MEASUREMENTS:
(1) Methane; CH_4; [74-82-8] (2) (3) Electrolytes, see below (4) Water; H_2O; [7732-18-5]	Byrne, P. A.; Stoessell, R. K. *Geochim. Cosmochim. Acta* 1982, *46*, 2395-7.
VARIABLES: $T/K = 298.15$ $p_1/kPa = 3790$	PREPARED BY: H. L. Clever

EXPERIMENTAL VALUES:

Temperature		Pressure		Electrolyte		Methane	Salt Effect
$t/°C$	T/K	$p_1/psia$	p_1/kPa		$m_i/mol\ kg^{-1}$	$m_1/mol\ kg^{-1}$	Parameter
25	298.15	550	3790	–	0	0.0483^a	–
				NaCl	1.0		
				$MgCl_2$	1.0	0.0264	0.066
25	298.15	550	3790	–	0	0.0483^a	–
				NaCl	1.0		
				$CaCl_2$	1.0	0.0219	0.086

[a] Value of methane solubility in water, $m_1°$, from (ref 1).

The salt effect parameter, $k_{smm}/kg\ mol^{-1} =$

$(\Sigma ((k_{smim}/kg\ mol^{-1})(I_i/mol\ kg^{-1}))/(I/mol\ kg^{-1})$

where I_i and I are ionic strength due to component i and the total ionic strength, respectively, and $k_{smim}/kg\ mol^{-1} =$

$log((m_1°/mol\ kg^{-1})/(m_1/mol\ kg^{-1}))/(I_i/mol\ kg^{-1})$

Sodium chloride; NaCl; [7647-14-5]

Magnesium chloride; $MgCl_2$; [7786-30-3]

Calcium chloride; $CaCl_2$; [10043-52-4]

AUXILIARY INFORMATION

METHOD/APPARATUS/PROCEDURE:

Solubility determinations were made using a titanium-lined chamber within a stainless steel reaction vessel jacketted by a water bath for temperature control. The system pressure was set by controlling the input and output of methane within the chamber's headpiece. The vessel was rocked for 3 h to allow equilibration between the methane and the solution.

The amount of gas in the saturated solution was measured by transfer of a sample volume to a loop at the system pressure, followed by flashing the sample in an expansion loop and measuring the gas pressure in a known volume. The total gas volume and pressure change were used to compute the moles of gas assuming ideal behavior. A correction was made for the gas not released on flashing.

Solution densities were measured gravimetrically with pycnometers.

SOURCE AND PURITY OF MATERIALS:

(1) Methane. Matheson Co., Inc. Ultra high purity grade, stated to have a minimum purity of 99.97 mole percent.

(2) Electrolytes. The salt solutions
(3) were made up gravimetrically using analytical grade chemicals.

(4) Water. Distilled.

ESTIMATED ERROR:
$\delta p_1/psia = \pm 1$
$\delta m_1/mol\ kg^{-1} = \pm 0.0003-0.0005$
$\delta m_{2,3}/mol\ kg^{-1} = \pm 0.0001$

REFERENCES:
1. Stoessell, R. K.; Byrne, P. A. *Geochim. Cosmochim. Acta* 1982, *46*, 1327.

COMPONENTS:	ORIGINAL MEASUREMENTS:
(1) Methane; CH_4; [74-82-8]	Duffy, J. R.; Smith, N. O.; Nagy, B.
(2) Calcium chloride; $CaCl_2$; [10043-52-4]	*Geochim. Cosmochim. Acta* <u>1961</u>, *24*, 23-31.
(3) Sodium chloride; NaCl; [7647-14-5]	
(4) Water; H_2O; [7732-18-5]	

VARIABLES:	PREPARED BY:
T/K = 303.15 p_1/MPa = 0.32 - 5.19 c_2/mol l^{-1} = 0, 3.0 c_3/mol l^{-1} = 0, 1.53	H. L. Clever

EXPERIMENTAL VALUES:

Temperature		Calcium Chloride	Sodium Chloride	Pressure		Mol Fraction
$t/°C$	T/K	c_2/mol l^{-1}	c_3/mol l^{-1}	p_1/psia	p_1/MPa	$10^4 x_1$
30	303.15	0	0	46	0.32	0.60
				80	0.55	1.15
				115	0.79	1.84
				136	0.94	2.32
				286	1.97	4.90
				297	2.05	4.93
				398	2.74	6.12
				523	3.61	7.64
		3.0	1.53	368	2.54	1.16
		3.0	1.53	753	5.19	2.32

The above solution is nearly saturated with NaCl, but not with the $CaCl_2 \cdot 6H_2O$.

The authors describe the solution as 1.53 N NaCl and 6.0 N $CaCl_2$.

AUXILIARY INFORMATION

METHOD/APPARATUS/PROCEDURE:	SOURCE AND PURITY OF MATERIALS:
Rocking equilibrium cell. Pressure measured with a Bourdon gage. The cell is charged with salt solution, the gas is admitted to a known pressure, and the cell contents allowed to equilibrate. The final pressure is measured and used to calculate the amount of gas dissolved.	(1) Methane. Source not given. Stated to be *c. p.* grade. (2, 3) Electrolytes. Reagent grade of known water constent. (4) Water. Distilled, degassed.
	ESTIMATED ERROR: $\delta T/K$ = ± 1; δp/MPa = ± 0.03; δx_1 = ± 5 x 10^{-6}
	REFERENCES:

COMPONENTS:	ORIGINAL MEASUREMENTS:
1. Methane; CH₄; [74-82-8]	Michels, A.; Gerver, J.; Bijl, A.
2. Sodium bromide; NaBr; [7647-15-6]	*Physica*
3. Water; H₂O; [7732-18-5]	1936, *3*, 797-808.

VARIABLES:	PREPARED BY:
Pressure	C. L. Young

EXPERIMENTAL VALUES:

T/K	Conc/mol ℓ^{-1}	$P/10^5$ Pa	10^3 Mole fraction of methane in liquid, $10^3 \, x_{CH_4}$
298.15	2.7	52.8	0.38
		102.6	0.67
		153.6	0.93
		200.4	1.09

AUXILIARY INFORMATION

METHOD/APPARATUS/PROCEDURE:	SOURCE AND PURITY OF MATERIALS:
Simple rocking equilibrium cell. Amount of gas absorbed calculated from volume and pressure change of charging vessel. Details in source.	No details given.

ESTIMATED ERROR:

δT/K = ±0.1; $\delta P/10^5$Pa = ±0.05 to 0.5%; δx_{CH_4} = ±3-5% (estimated by compiler).

REFERENCES:

COMPONENTS:	ORIGINAL MEASUREMENTS:
1. Methane; CH_4; [74-82-8] 2. Sodium bromide; NaBr; [7647-15-6] 3. Water; H_2O; [7732-18-5]	Ben-Naim, A.; Yaacobi, M. *J. Phys. Chem.* <u>1974</u>, *78*, 170-5.

VARIABLES:	PREPARED BY:
Temperature	C. L. Young

EXPERIMENTAL VALUES:

T/K	Conc. of sodium bromide /mol ℓ^{-1}	Ostwald coefficient,[*] L
283.15	1.0	0.03062
288.15		0.02823
293.15		0.02626
298.15		0.02465
303.15		0.02332

[*] Smoothed values of Ostwald coefficient obtained from

$$kT \ln L = 8{,}631.7 - 58.231(T/K) + 0.07352(T/K)^2 \text{ cal mol}^{-1}$$

where k is in units of cal mol^{-1} K^{-1}.

AUXILIARY INFORMATION

METHOD/APPARATUS/PROCEDURE:	SOURCE AND PURITY OF MATERIALS:
The apparatus was similar to that described by Ben-Naim and Baer (1) and Wen and Hung (2). It consists of three main parts, a dissolution cell of 300 to 600 cm^3 capacity, a gas volume measuring column, and a manometer. The solvent is degassed in the dissolution cell, the gas is introduced and dissolved while the liquid is kept stirred by a magnetic stirrer immersed in the water bath. Dissolution of the gas results in the change in the height of a column of mercury which is measured by a cathetometer.	1. Matheson sample, purity 99.97 mole per cent. 2. AR grade. 3. Deionised, doubly distilled.

ESTIMATED ERROR:
$\delta T/K = \pm 0.01$; $\delta L/L = \pm 0.005$ (estimated by compiler).

REFERENCES:
1. Ben-Naim, A.; Baer, S. *Trans. Faraday Soc.* <u>1963</u>, *59*, 2735.   2. Wen, W.-Y.; Hung, J. H. *J. Phys. Chem.* <u>1970</u>, *74*, 170.

COMPONENTS:	ORIGINAL MEASUREMENTS:
1. Methane; CH_4; [74-82-8]	Michels A.; Gerver, J.; Bijl, A.
2. Water; H_2O; [7732-18-5]	*Physica*, 1936, *3*, 797-808.
3. Sodium iodide; NaI; [7681-82-5]	

VARIABLES:	PREPARED BY:
Pressure	C.L. Young

EXPERIMENTAL VALUES:

T/K	Conc. of NaI /mol 1^{-1}	$p/10^5$Pa	10^3Mole fraction of methane in liquid, $10^3 x_{CH_4}$
298.15	2.7	56.2	0.50
		111.7	0.94
		152.0	1.22
		204.9	1.52

AUXILIARY INFORMATION

METHOD/APPARATUS/PROCEDURE:	SOURCE AND PURITY OF MATERIALS:
Simple rocking equilibrium cell. Amount of gas absorbed calculated from volume and pressure change in charging vessel. Details in source.	No details given.

ESTIMATED ERROR:
$\delta T/K = \pm 0.1$; $\delta p/10^5$Pa $= \pm 0.05$ to 0.5%; $\delta x_{CH_4} = \pm 3-5\%$.
(estimated by compiler).

REFERENCES:

COMPONENTS:	ORIGINAL MEASUREMENTS:
1. Methane; CH_4; [74-82-8] 2. Sodium iodide; NaI; [7681-82-5] 3. Water; H_2O; [7732-18-5]	Ben-Naim, A.; Yaacobi, M. *J. Phys. Chem.* <u>1974</u>, *78*, 170-5.

VARIABLES:	PREPARED BY:
Temperature	C. L. Young

EXPERIMENTAL VALUES:

T/K	Conc. of sodium iodide /mol ℓ^{-1}	Ostwald coefficient,[*] L
283.15	1.0	0.03102
288.15		0.02857
293.15		0.02671
298.15		0.02533
303.15		0.02435

[*] Smoothed values of Ostwald coefficient obtained from

$$kT \ln L = 14{,}243.5 - 97.387 \, (T/K) + 0.14190 \, (T/K)^2 \text{ cal mol}^{-1}$$

where k is in units of cal mol^{-1} K^{-1}.

AUXILIARY INFORMATION

METHOD/APPARATUS/PROCEDURE:	SOURCE AND PURITY OF MATERIALS:
The apparatus was similar to that described by Ben-Naim and Baer (1) and Wen and Hung (2). It consists of three main parts, a dissolution cell of 300 to 600 cm^3 capacity, a gas volume measuring column, and a manometer. The solvent is degassed in the dissolution cell, the gas is introduced and dissolved while the liquid is kept stirred by a magnetic stirrer immersed in the water bath. Dissolution of the gas results in the change in the height of a column of mercury which is measured by a cathetometer.	1. Matheson sample, purity 99.97 mole per cent. 2. AR grade. 3. Deionised, doubly distilled.

ESTIMATED ERROR:

$\delta T/K = \pm 0.01$; $\delta L/L = \pm 0.005$
(estimated by compiler).

REFERENCES:
1. Ben-Naim, A.; Baer, S. *Trans. Faraday Soc.* <u>1963</u>, *59*, 2735.
2. Wen, W.-Y.; Hung, J. H. *J. Phys. Chem.* <u>1970</u>, *74*, 170.

COMPONENTS:	ORIGINAL MEASUREMENTS:
(1) Methane; CH_4; [74-82-8] (2) Sodium sulfate; Na_2SO_4; [7757-82-6] (3) Water; H_2O; [7732-18-5]	Mishnina, T. A.; Avdeeva, O. I.; Bozhovskaya, T. K. *Inf. Sb., Vses. Nauchn-Issled. Geol. Inst.* No. 56, <u>1962</u>, 137-45. *Chem. Abstr.* <u>1964</u>, *60*, 8705g

VARIABLES:	PREPARED BY:
T/K = 293.15 p_1/kPa = 101.3 c_2/mol dm$^-$ = 0.48, 0.90	H. L. Clever

EXPERIMENTAL VALUES:

Temperature		Sodium Sulfate		Solubility	Salt Effect Parameter
$t/^0$C	T/K	$c_2/$ mol dm^{-3}	$c_2/$ eq dm^{-3}	cm^3(STP) dm^{-3}	k_{scc}/dm^{3}mol^{-1}
20	293.15	0	0	33.9[a]	–
		0.48	0.96	22.0	0.391
		0.90	1.80	15.1	0.390

[a] The compiler estimated the solubility of methane in water from a figure in the paper and values of the salt effect parameter.

The values of Kelvin temperature, sodium sulfate molar concentration, and salt effect parameter were calculated by the compiler.

The solubility is equivalent to the Bunsen coefficient, $10^2\alpha$/cm^3(STP) cm^{-3} atm^{-1}.

AUXILIARY INFORMATION

METHOD/APPARATUS/PROCEDURE:	SOURCE AND PURITY OF MATERIALS:
See author's earlier paper (1).	(1) Methane. Source not given. Contained one per cent air. (2) Sodium sulfate. (3) Water.

ESTIMATED ERROR:

δT/K = ± 0.1

REFERENCES:

1. Mishnina, T. A.; Avdeeva, O. I.; Bozhovskaya, T. K.
Materialy Vses. Nauchn. Issled. Geol. Inst. <u>1961</u>, *46*, 93.

COMPONENTS:	ORIGINAL MEASUREMENTS:
(1) Methane; CH_4; [74-82-8] (2) Sodium sulfate; Na_2SO_4; [7757-82-6] (3) Water; H_2O; [7732-18-5]	Stoessell, R. K.; Byrne, P. A. *Geochim. Cosmochim. Acta* <u>1982</u>, *46*, 1327-32.
VARIABLES: $T/K = 298.15$ $p_1/kPa = 2410-5170$ $m_2/mol\ kg^{-1} = 0-1.0$	PREPARED BY: H. L. Clever

EXPERIMENTAL VALUES:

Temperature		Pressure		Sodium Sulfate	Methane	Salt Effect Parameter
$t/°C$	T/K	$p_1/psia$	p_1/kPa	$m_2/mol\ kg^{-1}$	$m_1/mol\ kg^{-1}$	$_{smm}/kg\ mol^{-1}$
25	298.15	350	2410	0	0.0319	-
				0.5	0.0211	0.120
				1.0	0.0135	0.124
		550	3790	0	0.0483	-
				0.5	0.0311	0.127
				1.0	0.0208	0.122
		750	5170	0	0.0617	-
				0.5	0.0407	0.120
				1.0	0.0277	0.116
						0.121 (authors)

The salt effect parameter is defined as $k_{smm} = \log \gamma_1/I$

where I is the ionic strength (molality) and $\gamma_1 = (m_1^* f_1/m_1 f_1^*)_{p,T}$

with m_1 and f_1 the solubility and fugacity, respectively, of methane at p and T. The "*" refers to saturation in distilled water.

AUXILIARY INFORMATION

METHOD/APPARATUS/PROCEDURE:	SOURCE AND PURITY OF MATERIALS:
Solubility determinations were made using a titanium-lined chamber within a stainless steel reaction vessel jacketed by a water bath for temperature control. The system pressure was set by controlling the input and output of methane within the chamber's headpiece. The vessel was rocked for three h to allow equilibration between the methane and solution. The amount of gas in the saturated solution was measured by transfer of a sample volume to a loop at the system pressure, followed by flashing the sample in an ansion loop and measuring the gas pressure in a known volume. The total gas volume and pressure change were used to compute the moles of released gas assuming ideal behavior. A correction was made for the gas not released on flashing.	(1) Methane. Matheson Co., Inc. Ultra high purity grade, stated to be a minimum of 99.97 mole present methane. (2) Sodium sulfate. The salt solutions were made up gravimetrically using analytical grade chemicals. (3) Water. Distilled.
	ESTIMATED ERROR: $\delta p_1/psia = \pm 1$ $\delta m_1/m_1 = \pm 0.01$
	REFERENCES:

COMPONENTS:	ORIGINAL MEASUREMENTS:
(1) Methane; CH_4; [74-82-8] (2) Sulfuric acid; H_2SO_4; [7664-93-9] (3) Sodium sulfate; Na_2SO_4; [7757-82-6] (4) Water; H_2O; [7732-18-5]	Kobe, K. A.; Kenton, F. H. *Ind. Eng. Chem., Anal. Ed.* 1938, *10*, 76 - 77.

VARIABLES:	PREPARED BY:
T/K: 298.15 p_1/kPa: 101.325 (1 atm)	P. L. Long H. L. Clever

EXPERIMENTAL VALUES:

Temperature		Solvent Volume V/cm^3	Methane Volume Absorbed v_1/cm	Bunsen Coefficient $\alpha/cm^3(STP)cm^{-3}atm^{-1}$	Ostwald Coefficient L/cm^3cm^{-3}
$t/°C$	T/K				
25	298.15	49.54	0.47		
		49.54	0.45	0.0085	0.0093

The solvent is a mixture of 800 g H_2O

$\qquad\qquad\qquad\qquad$ 200 g Na_2SO_4 (anhydrous)

$\qquad\qquad\qquad\qquad$ 40 ml H_2SO_4 (Conc., 36 normal)

Thus the molality of the solution is

$$m_2/mol\ kg^{-1} = 0.90\ (H_2SO_4)$$

$$m_3/mol\ kg^{-1} = 1.76\ (Na_2SO_4)$$

AUXILIARY INFORMATION

METHOD/APPARATUS/PROCEDURE:	SOURCE AND PURITY OF MATERIALS:
The apparatus is described in detail in an earlier paper (1). The apparatus consists of a gas buret, a pressure compensator, and a 200 cm^3 absorption bulb and mercury leveling bulb. The absorption bulb is attached to a shaking mechanism. The solvent and the gas are placed in the absorption bulb. The bulb is shaken until equilibrium is reached. The remaining gas is returned to the buret. The difference in the final and initial volumes is taken as the volume of gas absorbed.	(1) Methane. Source not given. Purity stated to be 99+ per cent. (2, 3) Sulfuric acid and sodium sulfate. Sources not given. Analytical grade. (4) Water. Distilled.
	ESTIMATED ERROR: $\qquad\delta\alpha/cm^3 = \pm0.001$ (authors)
	REFERENCES: 1. Kobe, K. A.; Williams, J. S. *Ind. Eng. Chem., Anal. Ed.* 1935, *7*, 37.

COMPONENTS:	ORIGINAL MEASUREMENTS:
(1) Methane; CH_4; [74-82-8] (2) (3) Electrolytes, see below (4) Water; H_2O; [7732-18-5]	Byrne, P. A.; Stoessell, R. K. *Geochim. Cosmochim. Acta* <u>1982</u>, *46*, 2395-7.

VARIABLES:	PREPARED BY:
T/K = 298.15 p_1/kPa = 3790	H. L. Clever

EXPERIMENTAL VALUES:

Temperature		Pressure		Electrolyte		Methane	Salt Effect Parameter
$t/°C$	T/K	p_1/psia	p_1/kPa		m_i/mol kg^{-1}	m_1/mol kg^{-1}	
25	298.15	550	3790	–	0	0.0438[a]	–
				Na_2SO_4	0.5 ⎫		
				$MgSO_4$	0.5 ⎭	0.0221	0.097
25	298.15	550	3790	–	0	0.0483	–
				K_2SO_4	0.25 ⎫		
				$MgSO_4$	0.25 ⎭	0.0345	0.084

[a] Value of methane solubility in water, $m_1°$, from (ref 1).

The salt effect parameter, k_{smm}/kg mol^{-1} =

$(\Sigma(k_{smim}$/kg mol$^{-1})(I_i$/mol kg$^{-1}))/(I$/mol kg$^{-1})$

where I_i and I are ionic strength due to component i and the total ionic strength, respectively, and k_{smim}/kg mol^{-1} =

$\log((m_1°$/mol kg$^{-1})/(m_1$/mol kg$^{-1}))/(I_i$/mol kg$^{-1})$

Sodium sulfate; Na_2SO_4; [7757-82-6]
Potassium sulfate; K_2SO_4; [7778-80-5]
Magnesium sulfate; $MgSO_4$; [7785-87-7]

AUXILIARY INFORMATION

METHOD/APPARATUS/PROCEDURE:	SOURCE AND PURITY OF MATERIALS:
Solubility determinations were made suing a titanium-lined chamber within a stainless steel reaction vessel jacketted by a water bath for temperature control. The system pressure was set by controlling the input and output of methane within the chamber's headpiece. The vessel was rocked for 3 h to allow equilibration between the methane and the solution. The amount of gas in the saturated solution was measured by transfer of a sample volume to a loop at the system pressure, followed by flashing the sample in an expansion loop and measuring the gas pressure in a known volume. The total gas volume and pressure change were used to compute the moles of gas assuming ideal behavior. A correction was made for the gas not released on flashing. Solution densities were measured gravimetrically with pycnometers.	(1) Methane. Matheson Co., Inc. Ultra high purity grade, stated to have a minimum purity of 99.97 mole percent. (2) Electrolytes. The salt solutions (3) were made up gravimetrically using analytical grade chemicals. (4) Water. Distilled. ESTIMATED ERROR: δp_1/psia = ± 1 δm_1/mol kg^{-1} = ± 0.0003 - 0.0005 $\delta m_{2,3}$/mol kg^{-1} = ± 0.0001 REFERENCES: 1. Stoessell, R. K.; Byrne, P. A. *Geochim. Cosmochim. Acta* <u>1982</u>, *46*, 1327.

COMPONENTS:	ORIGINAL MEASUREMENTS:
(1) Methane; CH_4; [74-82-8] (2) (3) Electrolytes, see below (4) Water; H_2O; [7732-18-5]	Byrne, P. A.; Stoessell, R. K. *Geochim. Cosmochim. Acta* <u>1982</u>, *46*, 2395-7.

VARIABLES:	PREPARED BY:
$T/K = 298.15$ $p_1/kPa = 3790$	H. L. Clever

EXPERIMENTAL VALUES:

Temperature		Pressure		Electrolyte		Methane	Salt Effect Parameter
$t/°C$	T/K	p_1/psia	p_1/kpa		m_i/mol kg^{-1}	m_1/mol kg^{-1}	
25	298.15	550	3790	–	0	0.0483^a	–
				NaCl	1.0 ⎫	0.0173	0.111
				Na_2SO_4	1.0 ⎭		
25	298.15	550	3790	–	0	0.0483^a	–
				NaCl	1.0 ⎫	0.0171	0.113
				Na_2CO_3	1.0 ⎭		

^{a} Value of methane solubility in water, $m_1°$, from (ref 1).

The salt effect parameter, k_{smm}/kg mol^{-1} =

$(\Sigma((k_{smim}/kg\ mol^{-1})(I_i/mol\ kg^{-1}))/(I/mol\ kg^{-1})$

where I_i and I are ionic strength due to component i and the total ionic strength, respectively, and k_{smim}/kg mol^{-1} =

$log((m_1°/mol\ kg^{-1})/(m_1/mol\ kg^{-1}))/(I_i/mol\ kg^{-1})$

Sodium chloride; NaCl; [7647-14-5]

Sodium sulfate; Na_2SO_4; [7757-82-6]

Sodium carbonate; Na_2CO_3; [497-19-8]

AUXILIARY INFORMATION

METHOD/APPARATUS/PROCEDURE:	SOURCE AND PURITY OF MATERIALS:
Solubility determinations were made using a titanium-lined chamber within a stainless steel reaction vessel jacketted by a water bath for temperature control. The system pressure was set by controlling the input and output of methane within the chamber's headpiece. The vessel was rocked for 3 h to allow equilibration between the methane and the solution. The amount of gas in the saturated solution was measured by transfer of a sample volume to a loop at the system pressure, followed by flashing the sample in an expansion loop and measuring the gas pressure in a known volume. The total gas volume and pressure change were used to compute the moles of gas assuming ideal behavior. A correction was made for the gas not released on flashing. Solution densities were measured gravimetrically with pycnometers.	(1) Methane. Matheson Co., Inc. Ultra high purity grade, stated to have a minimum purity of 99.97 mole percent. (2) (3) Electrolytes. The salt solutions were made up gravimetrically using analytical grade chemicals. (4) Water. Distilled.

ESTIMATED ERROR:

δp_1/psia = ± 1
δm_1/mol kg^{-1} = ± 0.0003-0.0005
$\delta m_{2,3}$/mol kg^{-1} = ± 0.0001

REFERENCES:

1. Stoessell, R. K.; Byrne, P. A. *Geochim. Cosmochim. Acta* <u>1982</u>, *46*, 1327.

COMPONENTS:	ORIGINAL MEASUREMENTS:
(1) Methane; CH_4; [74-82-8] (2) Sodium bicarbonate; $NaHCO_3$; [144-55-8] (3) Water; H_2O; [7732-18-5]	Stoessell, R. K.; Byrne, P. A. *Geochim. Cosmochim. Acta* <u>1982</u>, *46*, 1327-32.

| VARIABLES: T/K = 298.15
 p_1/kPa = 2410-5170
 $m_2/mol\ kg^{-1}$ = 0-0.5 | PREPARED BY:

 H. L. Clever |

EXPERIMENTAL VALUES:

Temperature		Pressure		Sodium Bicarbonate	Methane	Salt Effect Parameter
$t/°C$	T/K	$p_1/psia$	p_1/kPa	$m_2/mol\ kg^{-1}$	$m_1/mol\ kg^{-1}$	$k_{smm}/kg\ mol^{-1}$
25	298.15	350	2410	0	0.0319	-
				0.25	0.0296	0.130
				0.5	0.0270	0.145
		550	3790	0	0.0483	-
				0.25	0.0450	0.123
				0.5	0.0400	0.164
		750	5170	0	0.0617	-
				0.25	0.0572	0.132
				0.5	0.0532	0.129
						0.146 (authors)

The salt effect parameter is defined as $k_{smm} = \log \gamma_1 / I$

where I is the ionic strength (molality) and $\gamma_1 = (m_1^* f_1 / m_1 f_1^*)_{p,T}$

with m_1 and f_1 the solubility and fugacity, respectively, of methane at p and T. The "*" refers to saturation in distilled water.

AUXILIARY INFORMATION

METHOD/APPARATUS/PROCEDURE:	SOURCE AND PURITY OF MATERIALS:
Solubility determinations were made using a titanium-lined chamber within a stainless steel reaction vessel jacketed by a water bath for temperature control. The system pressure was set by controlling the input and output of methane within the chamber's headpiece. The vessel was rocked for three h to allow equilibration between the methane and solution. The amount of gas in the saturated solution was measured by transfer of a sample volume to a loop at the system pressure, followed by flashing the sample in an ansion loop and measuring the gas pressure in a known volume. The total gas volume and pressure change were used to compute the moles of released gas assuming ideal behavior. A correction was made for the gas not released on flashing.	(1) Methane. Matheson Co., Inc. Ultra high purity grade, stated to be a minimum of 99.97 mole present methane. (2) Sodium bicarbonate. The salt solutions were made up gravimetrically using analytical grade chemicals. (3) Water. Distilled.

| | ESTIMATED ERROR:
 $\delta p_1/psia$ = ± 1
 $\delta m_1/m_1$ = ± 0.01 |
| | REFERENCES: |

COMPONENTS:	ORIGINAL MEASUREMENTS:
(1) Methane; CH_4; [74-82-8] (2) Sodium carbonate; Na_2CO_3; [497-19-8] (3) Water; H_2O; [7732-18-5]	Stoessell, R. K.; Byrne, P. A. *Geochim. Cosmochim. Acta* <u>1982</u>, *46*, 1327-32.

VARIABLES:	PREPARED BY:
T/K = 298.15 p_1/kPa = 2410-5170 m_2/mol kg^{-1} = 0-1.5	H. L. Clever

EXPERIMENTAL VALUES:

Temperature		Pressure		Sodium Carbonate	Methane	Salt Effect Parameter
t/°C	T/K	p_1/psia	p_1/kPa	m_2/mol kg^{-1}	m_1/mol kg^{-1}	k_{smm}/kg mol^{-1}
25	298.15	350	2410	0	0.0319	–
				0.5	0.0211	0.120
				1.0	0.0135	0.124
				1.5	0.0091	0.121
		550	3790	0	0.0483	–
				0.5	0.0311	0.127
				1.0	0.0204	0.125
				1.5	0.0149	0.114
		750	5170	0	0.0617	–
				0.5	0.0400	0.125
				1.0	0.0260	0.125
				1.5	0.0182	0.118
						0.118 (authors)

The salt effect parameter is defined as $k_{smm} = \log \gamma_1 / I$

where I is the ionic strength (molality) and $\gamma_1 = (m_1^* f_1 / m_1 f_1^*)_{p,T}$

with m_1 and f_1 the solubility and fugacity, respectively, of methane at p and T. The "*" refers to saturation in distilled water.

AUXILIARY INFORMATION

METHOD/APPARATUS/PROCEDURE:	SOURCE AND PURITY OF MATERIALS:
Solubility determinations were made using a titanium-lined chamber within a stainless steel reaction vessel jacketed by a water bath for temperature control. The system pressure was set by controlling the input and output of methane within the chamber's headpiece. The vessel was rocked for three h to allow equilibration between the methane and solution. The amount of gas in the saturated solution was measured by transfer of a sample volume to a loop at the system pressure, followed by flashing the sample in an ansion loop and measuring the gas pressure in a known volume. The total gas volume and pressure change were used to compute the moles of released gas assuming ideal behavior. A correction was made for the gas not released on flashing.	(1) Methane. Matheson Co., Inc. Ultra high purity grade, stated to be a minimum of 99.97 mole present methane. (2) Sodium carbonate. The salt solutions were made up gravimetrically using analytical grade chemicals. (3) Water. Distilled.
	ESTIMATED ERROR: δp_1/psia = ± 1 $\delta m_1/m_1$ = ± 0.01
	REFERENCES:

COMPONENTS:	ORIGINAL MEASUREMENTS:
(1) Methane; CH_4; [74-82-8] (2) (3) Electrolytes, see below (4) Water; H_2O; [7732-18-5]	Byrne, P. A.; Stoessell, R. K. *Geochim. Cosmochim. Acta* <u>1982</u>, *46*, 2395-7.

VARIABLES:	PREPARED BY:
$T/K = 298.15$ $p_1/kPa = 3790$	H. L. Clever

EXPERIMENTAL VALUES:

Temperature		Pressure		Electrolyte		Methane	Salt Effect Parameter
$t/°C$	T/K	$p_1/psia$	p_1/kPa		$m_i/mol\ kg^{-1}$	$m_1/mol\ kg^{-1}$	
25	298.15	550	3790	-	0	0.0483^a	-
				Na_2SO_4	0.5		
				Na_2CO_3	0.5	0.0211	0.120
25	298.15	550	3790	-	0	0.0483^a	-
				Na_2SO_4	0.25		
				K_2SO_4	0.25	0.0322	0.119

[a] Value of methane solubility in water, $m_i°$, from (ref 1).

The salt effect parameter, $k_{smm}/kg\ mol^{-1} =$

$(\ ((k_{smim}/kg\ mol^{-1})(I_i/mol\ kg^{-1}))/(I/mol\ kg^{-1})$

where I_i and I are ionic strength due to component i and the total ionic strength, respectively, and $k_{smim}/kg\ mol^{-1} =$

$\log((m_1°/mol\ kg^{-1})/(m_1/mol\ kg^{-1}))/(I_i/mol\ kg^{-1})$

Sodium sulfate; Na_2SO_4; [7757-82-6]
Sodium carbonate; Na_2CO_3; [497-19-8]
Potassium sulfate; K_2SO_4; [7778-80-5]

AUXILIARY INFORMATION

METHOD/APPARATUS/PROCEDURE:
Solubility determinations were made using a titanium-lined chamber within a stainless steel reaction vessel jacketted by a water bath for temperature control. The system pressure was set by controlling the input and output of methane within the chamber's headpiece. The vessel was rocked for 3 h to allow equilibration between the methane and the solution.

The amount of gas in the saturated solution was measured by transfer of a sample volume to a loop at the system pressure, followed by flashing the sample in an expansion loop and measuring the gas pressure in a known volume. The total gas volume and pressure change were used to compute the moles of gas assuming ideal behavior. A correction was made for the gas not released on flashing.

Solution densities were measured gravimetrically with pycnometers.

SOURCE AND PURITY OF MATERIALS:
(1) Methane. Matheson Co., Inc. Ultra high purity grade, stated to have a minimum purity of 99.97 mole percent.

(2) Electrolytes. The salt solutions
(3) were made up gravimetrically using analytical grade chemicals.

(4) Water. Distilled.

ESTIMATED ERROR:
$\delta p_1/psia = \pm 1$
$\delta m_1/mol\ kg^{-1} = \pm 0.0003-0.0005$
$\delta m_{2,3}/mol\ kg^{-1} = \pm 0.0001$

REFERENCES:
1. Stoessell, R. K.; Byrne, P. A. *Geochim. Cosmochim. Acta* <u>1982</u>, *46*, 1327.

COMPONENTS:	ORIGINAL MEASUREMENTS:
(1) Methane; CH_4; [74-82-8]	Shoor, S. K.; Walker, R. D., Jr.; Gubbins, K. E.
(2) Potassium hydroxide; KOH; [1310-58-3]	
(3) Water; H_2O; [7732-18-5]	*J. Phys. Chem.* <u>1969</u>, *73*, 312-7.

EXPERIMENTAL VALUES:

T/K	Potassium Hydroxide KOH/wt %	Potassium Hydroxide $c_2/mol\ dm^{-3}$	Gas Mol Fraction $10^5 x_1$	Solubility Ratio $\gamma = x_1^0/x_1$	Salt Effect Parameter $k_{scx}/dm^3\ mol^{-1}$
298.15	0.0	0.0	2.48	1.00	–
	5.61	1.03		1.66	0.214
	13.90	2.77		3.64	0.203
	23.50	5.13		10.2	0.197
	31.61	7.35		26.7	0.194
	40.70	10.12		100	0.198
					0.197 (authors)
313.15	0.0	0.0	1.90	1.00	–
	5.61	1.03		1.51	0.174
	13.90	2.77		3.14	0.179
	23.50	5.13		8.08	0.177
	31.61	7.35		19.8	0.176
	40.70	10.12		63.4	0.178
					0.176 (authors)
333.15	0.0	0.0	1.62	1.00	–
	5.61	1.03		1.48	0.165
	13.90	2.77		2.83	0.163
	23.50	5.13		6.90	0.164
	31.61	7.35		16.7	0.166
	40.70	10.12		49.6	0.168
					0.164 (authors)
353.15	0.0	0.0	1.44	1.00	–
	5.61	1.03		1.39	0.139
	13.90	2.77		2.50	0.144
	23.50	5.13		6.22	0.154
	31.61	7.35		14.4	0.158
	40.70	10.12		38.0	0.156
					0.154 (authors)

The KOH concentrations were measured at 298.15 K.

The compiler calculated the salt effect parameter values at the individual KOH concentrations. The author's values are also given.

The salt effect parameter, $k_{scx}/dm^3 mol^{-1} = (1/(c_2/mol\ dm^{-3}))\log\ (x_1^0/x_1)$.

COMPONENTS:	ORIGINAL MEASUREMENTS:
(1) Methane; CH_4; [74-82-8] (2) Potassium hydroxide; KOH; [1310-58-3] (3) Water; H_2O; [7732-18-5]	Shoor, S. K.; Walker, R. D., Jr.; Gubbins, K. E. *J. Phys. Chem.* <u>1969</u>, *73*, 312-7.

VARIABLES:	PREPARED BY:
T/K = 298.15 - 353.15 p_1/kPa = 101.325 c_2/mol dm^{-3} = 0 - 10.12	H. L. Clever

EXPERIMENTAL VALUES:

See preceeding page.

AUXILIARY INFORMATION

METHOD/APPARATUS/PROCEDURE:

A gas chromatographic method was used (ref 1). The gas saturated solutions were prepared by bubbling the g gas through presaturators and then the KOH solution. Samples were drawn over a 48 h period to determine whether or not equilibrium was established. Samples were transferred from the saturator to the gas chromatograph in gas tight syringes. All analyses were made with a thermal conductivity cell, and with nitrogen as the carrier gas.

The results are reported as activity coefficients, which are the mole fraction solubility ratio x_1^0/x_1. x_1^0 is the mole fraction solubility in water, and x_1 is the mole fraction solubility in the KOH solution. Both mole fractions were adjusted to a gas partial pressure of one atm assuming Henry's law is obeyed. The γ's are the average of at least four measurements.

SOURCE AND PURITY OF MATERIALS:

(1) Methane. Source not given. Minimum purity stated to be 99.0 %.

(2) Potassium hydroxide. Baker Analyzed Reagent Grade. Contained a maximum of 1 percent K_2CO_3. The KOH solutions were protected from atm CO_2 by Ascarite.

(3) Water. Specially distilled and degassed from an all glass-Teflon still.

ESTIMATED ERROR:

δT/K = ± 0.05
$\delta\gamma/\gamma$ = ± 0.01

REFERENCES:

1. Gubbins, K. E.; Carden, S. N.; Walker, R. D., Jr.
   J. Gas Chromatog. <u>1965</u>, *3*, 98.

COMPONENTS:	ORIGINAL MEASUREMENTS:
1. Methane; CH_4; [74-82-8] 2. Water; H_2O; [7732-18-5] 3. Potassium chloride; KCl; [7447-40-7]	Michels, A.; Gerver, J.; Bijl, A. *Physica*, 1936, *3*, 797-808.

VARIABLES:	PREPARED BY:
Pressure	C.L. Young

EXPERIMENTAL VALUES:

T/K	Conc. of KCl /mol l^{-1}	$p/10^5 Pa$	10^3 Mole fraction of methane in liquid, $10^3 x_{CH_4}$
298.15	2.7	48.5	0.44
		98.5	0.80
		150.7	1.14
		200.6	1.39

AUXILIARY INFORMATION

METHOD/APPARATUS/PROCEDURE:	SOURCE AND PURITY OF MATERIALS:
Simple rocking equilibrium cell. Amount of gas absorbed calculated from volume and pressure change in charging vessel. Details in source.	No details given.
	ESTIMATED ERROR:
	$\delta T/K = \pm 0.1$; $\delta p/10^5 Pa = \pm 0.05$ to 0.5%; $\delta x_{CH_4} = \pm 3-5\%$. (estimated by compiler)
	REFERENCES:

COMPONENTS:	ORIGINAL MEASUREMENTS:
1. Methane; CH_4; [74-82-8] 2. Potassium chloride; KCl; [7447-40-7] 3. Water; H_2O; [7732-18-5]	Ben-Naim, A.; Yaacobi, M. *J. Phys. Chem.* 1974, *78*, 175-8.

VARIABLES:	PREPARED BY:
Temperature	C. L. Young

EXPERIMENTAL VALUES:

T/K	Conc. of potassium chloride /mol ℓ^{-1}	Ostwald coefficient,[*] L
283.15	1.0	0.03129
288.15		0.02888
293.15		0.02676
298.15		0.02488
303.15		0.02321

[*] Smoothed values of Ostwald coefficient obtained from

$$kT \ln L = 3{,}595.6 - 23.037 \, (T/K) + 0.01219 \, (T/K)^2 \text{ cal mol}^{-1}$$

where k is in units of cal mol^{-1} K^{-1}.

AUXILIARY INFORMATION

METHOD/APPARATUS/PROCEDURE:	SOURCE AND PURITY OF MATERIALS:
The apparatus was similar to that described by Ben-Naim and Baer (1) and Wen and Hung (2). It consists of three main parts, a dissolution cell of 300 to 600 cm^3 capacity, a gas volume measuring column, and a manometer. The solvent is degassed in the dissolution cell, the gas is introduced and dissolved while the liquid is kept stirred by a magnetic stirrer immersed in the water bath. Dissolution of the gas results in the change in the height of a column of mercury which is measured by a cathetometer.	1. Matheson sample, purity 99.97 mole per cent. 2. AR grade. 3. Deionised, doubly distilled.

ESTIMATED ERROR:

$\delta T/K = \pm 0.01$; $\delta L/L = \pm 0.005$

(estimated by compiler).

REFERENCES:

1. Ben-Naim, A.; Baer, S.
Trans. Faraday Soc. 1963, *59*, 2735.

2. Wen, W.-Y.; Hung, J. H.
J. Phys. Chem. 1970, *74*, 170.

COMPONENTS:	ORIGINAL MEASUREMENTS:

(1) Methane; CH_4; [74-82-8]
 	
(2) Potassium chloride; KCl;
  [7447-40-7]
 	
(3) Water; H_2O; [7732-18-5]
	Yano, T.; Suetaka, T.; Umehara, T.;
	  Horiuchi, A.
	
	Kagaku Kogaku <u>1974</u>, *38*, 320-3.

VARIABLES:	PREPARED BY:
T/K = 298.15 p_1/kPa = 101.325 $c_2/mol\ dm^{-3}$ = 0 - 1.500	H. L. Clever C. L. Young

EXPERIMENTAL VALUES:

Temperature		Potassium Chloride	Methane Solubility	Salt Effect Parameter
$t/^0C$	T/K	$c_2/mol\ dm^{-3}$	$10^3 c_1/mol\ dm^{-3}$	$k_{scc}/dm^3\ mol^{-1}$
25	298.15	0	1.31	–
		0.500	1.13	0.128
		1.000	0.951	0.139
		1.500	0.809	0.140
				0.137 (authors)

$$k_{scc}/dm^3\ mol^{-1} = (1/(c_2/mol\ dm^{-3}))\log\ ((c_1^0/mol\ dm^{-3})/(c_1/mol\ dm^{-3}))$$

The compiler added the salt effect parameter values at the individual salt concentrations.

The authors defined the salt effect parameter in terms of the electrolyte ionic strength.

AUXILIARY INFORMATION

METHOD/APPARATUS/PROCEDURE:	SOURCE AND PURITY OF MATERIALS:
Volumetric apparatus. Salt solution allowed to enter stirred absorption chamber. Pressure within absorption chamber adjusted to be as near atmospheric pressure as possible. Details in source and ref. 1.	1. High purity sample, purity better than 99.5 mole per cent. 2. Special grade. 3. Distilled.
	ESTIMATED ERROR:

	REFERENCES:
	
	1. Yano, T.; Suetaka, T.;
	   Umehara, T.
	   Nippon Kagaku Kaishi
	   <u>1972</u>, *11*, 2194.

COMPONENTS:	ORIGINAL MEASUREMENTS:
(1) Methane; CH_4; [74-82-8] (2) Potassium chloride; KCl; [7447-40-7] (3) Water; H_2O; [7732-18-5]	Stoessell, R. K.; Byrne, P. A. *Geochim. Cosmochim. Acta* 1982, *46*, 1327-32.
VARIABLES: T/K = 298.15 p_1/kPa = 2410-5170 $m_2/mol\ kg^{-1}$ = 0 - 4.0	PREPARED BY: H. L. Clever

EXPERIMENTAL VALUES:

Temperature		Pressure		Potassium Chloride	Methane	Salt Effect Parameter
$t/°C$	T/K	$p_1/psia$	p_1/kPa	$m_2/mol\ kg^{-1}$	$m_1/mol\ kg^{-1}$	$k_{smm}/kg\ mol^{-1}$
25	298.15	350	2410	0	0.0319	–
				0.5	0.0282	0.107
				1.0	0.0247	0.111
				2.0	0.0194	0.108
				4.0	0.0126	0.101
		550	3790	0	0.0483	–
				0.5	0.0416	0.130
				1.0	0.0359	0.129
				2.0	0.0289	0.112
				4.0	0.0186	0.104
		750	5170	0	0.0617	–
				0.5	0.0545	0.108
				1.0	0.0475	0.114
				2.0	0.0370	0.111
				4.0	0.0245	0.100
						0.101 (authors)

The salt effect parameter is defined as $k_{smm} = \log \gamma_1/I$

where I is the ionic strength (molality) and $\gamma_1 = (m_1^* f_1/m_1 f_1^*)_{p,T}$
with m_1 and f_1 the solubility and fugacity, respectively, of methane at
and T. The "*" refers to saturation in distilled water.

AUXILIARY INFORMATION

METHOD/APPARATUS/PROCEDURE:	SOURCE AND PURITY OF MATERIALS:
Solubility determinations were made using a titanium-lined chamber within a stainless steel reaction vessel jacketed by a water bath for temperature control. The system pressure was set by controlling the input and output of methane within the chamber's headpiece. The vessel was rocked for three h to allow equilibration between the methane and solution. The amount of gas in the saturated solution was measured by transfer of a sample volume to a loop at the system pressure, followed by flashing the sample in an ansion loop and measuring the gas pressure in a known volume. The total gas volume and pressure change were used to compute the moles of released gas assuming ideal behavior. A correction was made for the gas not released on flashing.	(1) Methane. Matheson Co., Inc. Ultra high purity grade, stated to be a minimum of 99.97 mole present methane. (2) Potassium chloride. The salt solutions were made up gravimetrically using analytical grade chemicals. (3) Water. Distilled.
	ESTIMATED ERROR: $\delta p_1/psia$ = ± 1 $\delta m_1/m_1$ = ± 0.01
	REFERENCES:

COMPONENTS:	ORIGINAL MEASUREMENTS:
(1) Methane; CH_4; [74-82-8] (2) (3) Electrolytes, see below (4) Water; H_2O; [7732-18-5]	Byrne, P. A.; Stoessell, R. K. *Geochim. Cosmochim. Acta* <u>1982</u>, *46*, 2395-7.

| VARIABLES:
$T/K = 298.15$
$p_1/kPa = 3790$ | PREPARED BY:

H. L. Clever |

EXPERIMENTAL VALUES:

Temperature		Pressure		Electrolyte		Methane	Salt Effect Parameter
$t/°C$	T/K	$p_1/psia$	p_1/kPa		$m_i/mol\ kg^{-1}$	$m_1/mol\ kg^{-1}$	
25	298.15	550	3790	–	0	0.0483^a	–
				NaCl	1.0 ⎫		
				KCl	1.0 ⎭	0.0286	0.114
25	298.15	550	3790	–	0	0.0483^a	–
				KCl	1.0 ⎫		
				K_2CO_4	1.0 ⎭	0.0177	0.109

The salt effect parameter, $k_{smm}/kg\ mol^{-1} =$

$(\Sigma((k_{sm_im}/kg\ mol^{-1})(I_i/mol\ kg^{-1}))/(I/mol\ kg^{-1})$

where I_i and I are ionic strength due to component i and the total ionic strength, respectively, and $k_{sm_im}/kg\ mol^{-1} =$

$\log((m_1^0/mol\ kg^{-1})/(m_1/mol\ kg^{-1}))/(I_i/mol\ kg^{-1})$

Sodium chloride; NaCl; [7647-14-5]
Potassium chloride; KCl; [7447-40-7]
Potassium carbonate; K_2CO_3; [584-08-7]

[a] Value of methane solubility in water, m_1^0, from (ref 1)

AUXILIARY INFORMATION

METHOD/APPARATUS/PROCEDURE:	SOURCE AND PURITY OF MATERIALS:
Solubility determinations were made using a titanium-lined chamber within a stainless steel reaction vessel jacketted by a water bath for temperature control. The system pressure was set by controlling the input and output of methane within the chamber's headpiece. The vessel was rocked for 3 h to allow equilibration between the methane and the solution. The amount of gas in the saturated solution was measured by transfer of a sample volume to a loop at the system pressure, followed by flashing the sample in an expansion loop and measuring the gas pressure in a known volume. The total gas volume and pressure change were used to compute the moles of gas assuming ideal behavior. A correction was made for the gas not released on flashing. Solution densities were measured gravimetrically with pycnometers.	(1) Methane. Matheson Co., Inc. Ultra high purity grade, stated to have a minimum purity of 99.97 mole percent. (2) Electrolytes. The salt solutions (3) were made up gravimetrically using analytical grade chemicals. (4) Water. Distilled.
	ESTIMATED ERROR: $\delta p_1/psia = \pm 1$ $\delta m_1/mol\ kg^{-1} = \pm 0.0003-0.0005$ $\delta m_{2,3}/mol\ kg^{-1} = \pm 0.0001$
	REFERENCES: 1. Stoessell, R. K.; Byrne, P. A. *Geochim. Cosmochim. Acta* <u>1982</u>, *46*, 1327.

COMPONENTS:	ORIGINAL MEASUREMENTS:
(1) Methane; CH_4; [74-82-8] (2) Potassium iodide; KI; [7681-11-0] (3) Water; H_2O; [7732-18-5]	Morrison, T. J.; Billett, F. *J. Chem. Soc.* <u>1952</u>, 3819 - 3822.

VARIABLES:	PREPARED BY:
T/K: 285.75 - 344.85 p/kPa: 101.325 (1 atm)	H. L. Clever

EXPERIMENTAL VALUES:

Temperature			Salt Effect Parameters	
t/°C	T/K	$1/(T/K)$	$(1/m_2)\log(S°/S)$ [1]	$(1/m_2)\log(x°/x)$
12.6	285.75	0.0035	0.130	0.145
30.0	303.15	0.0033	0.097	0.112
49.4	322.55	0.0031	0.071	0.086
71.7	344.85	0.0029	0.054	0.069

[1] The authors used $(1/c)\log(S°/S)$ with c defined as g eq salt per kg of water. For the 1-1 electrolyte the compiler changed the c to an m for m_2/mol kg^{-1}. The methane solubility S is cm^3(STP) kg^{-1}.

The salt effect parameters were calculated from two measurements. The solubility of methane in water, S°, and in the one molal salt solution, S. Only the solubility of the methane in water, and the value of the salt effect parameter are given in the paper. The solubility values in the salt solution are not given.

The compiler calculated the values of the salt effect parameter using the mole fraction gas solubility ratio.

AUXILIARY INFORMATION

METHOD/APPARATUS/PROCEDURE:	SOURCE AND PURITY OF MATERIALS:
The degassed solvent flows in a thin film down an absorption helix containing the methane gas plus solvent vapor at a total pressure of one atmosphere. The volume of gas absorbed is measured in an attached buret system (1).	(1) Methane. Prepared from Grignard reagent. (2) Potassium iodide. "AnalaR" material. (3) Water. No information given.

ESTIMATED ERROR:

$$\delta k/kg^{-1}\ mol = 0.010$$

REFERENCES:
1. Morrison, T. J.; Billett, F. *J. Chem. Soc.* <u>1948</u>, 2033.

COMPONENTS:	ORIGINAL MEASUREMENTS:
(1) Methane; CH_4; [74-82-8] (2) Potassium sulfate; K_2SO_4; [7778-80-5] (3) Water; H_2O; [7732-18-5]	Stoessell, R. K.; Byrne, P. A. *Geochim. Cosmochim. Acta* <u>1982</u>, *46*, 1327-32.

| VARIABLES: T/K = 298.15
       p_1/kPa = 2410-5170
   m_2/mol kg^{-1} = 0-0.5 | PREPARED BY:

         H. L. Clever |

EXPERIMENTAL VALUES:

Temperature		Pressure		Potassium Sulfate	Methane	Salt Effect Parameter
t/°C	T/K	p_1/psia	p_1/kPa	m_2/mol kg^{-1}	m_1/mol kg^{-1}	k_{smm}/kg mol^{-1}
25	298.15	350	2410	0	0.0319	-
				0.25	0.0261	0.116
				0.5	0.0223	0.104
		550	3790	0	0.0483	-
				0.25	0.0393	0.119
				0.5	0.0325	0.115
		750	5170	0	0.0617	-
				0.25	0.0515	0.105
				0.5	0.0431	0.104
						0.108 (authors)

The salt effect parameter is defined as $k_{smm} = \log \gamma_1 / I$

where I is the ionic strength (molality) and $\gamma_1 = (m_1^* f_1 / m_1 f_1^*)_{p,T}$

with m_1 and f_1 the solubility and fugacity, respectively, of methane at p and T. The "*" refers to saturation in distilled water.

AUXILIARY INFORMATION

METHOD/APPARATUS/PROCEDURE:	SOURCE AND PURITY OF MATERIALS:
Solubility determinations were made using a titanium-lined chamber within a stainless steel reaction vessel jacketed by a water bath for temperature control. The system pressure was set by controlling the input and output of methane within the chamber's headpiece. The vessel was rocked for three h to allow equilibration between the methane and solution. The amount of gas in the saturated solution was measured by transfer of a sample volume to a loop at the system pressure, followed by flashing the sample in an ansion loop and measuring the gas pressure in a known volume. The total gas volume and pressure change were used to compute the moles of released gas assuming ideal behavior. A correction was made for the gas not released on flashing.	(1) Methane. Matheson Co., Inc. Ultra high purity grade, stated to be a minimum of 99.97 mole present methane. (2) Potassium sulfate. The salt solutions were made up gravimetrically using analytical grade chemicals. (3) Water. Distilled.

	ESTIMATED ERROR:
	         δp_1/psia = ± 1
	           $\delta m_1/m_1$ = ± 0.01

	REFERENCES:

COMPONENTS:	ORIGINAL MEASUREMENTS:
(1) Methane; CH_4; [74-82-8] (2) Potassium bicarbonate; $KHCO_3$; [298-14-6] (3) Water; H_2O; [7732-18-5]	Stoessell, R. K.; Byrne, P. A. *Geochim. Cosmochim. Acta* <u>1982</u>, *46*, 1327-32.

VARIABLES:	PREPARED BY:
$T/K = 298.15$ $p_1/kPa = 2410-5170$ $m_2/mol\ kg^{-1} = 0-0.5$	H. L. Clever

EXPERIMENTAL VALUES:

Temperature		Pressure		Potassium Bicarbonate m_2/mol kg^{-1}	Methane m_1/mol kg^{-1}	Salt Effect Parameter k_{smm}/kg mol^{-1}
$t/°C$	T/K	p_1/psia	p_1/kPa			
25	298.15	350	2410	0	0.0319	-
				0.25	0.0296	0.130
				0.5	0.0268	0.151
		550	3790	0	0.0483	-
				0.25	0.0457	0.096
				0.5	0.0406	0.151
		750	5170	0	0.0617	-
				0.25	0.0574	0.125
				0.5	0.0530	0.132
						0.145 (authors)

The salt effect parameter is defined as $k_{smm} = \log \gamma_1/I$

where I is the ionic strength (molality) and $\gamma_1 = (m_1^* f_1/m_1 f_1^*)_{p,T}$

with m_1 and f_1 the solubility and fugacity, respectively, of methane at p and T. The "*" refers to saturation in distilled water.

AUXILIARY INFORMATION

METHOD/APPARATUS/PROCEDURE:	SOURCE AND PURITY OF MATERIALS:
Solubility determinations were made using a titanium-lined chamber within a stainless steel reaction vessel jacketed by a water bath for temperature control. The system pressure was set by controlling the input and output of methane within the chamber's headpiece. The vessel was rocked for three h to allow equilibration between the methane and solution. The amount of gas in the saturated solution was measured by transfer of a sample volume to a loop at the system pressure, followed by flashing the sample in an ansion loop and measuring the gas pressure in a known volume. The total gas volume and pressure change were used to compute the moles of released gas assuming ideal behavior. A correction was made for the gas not released on flashing.	(1) Methane. Matheson Co., Inc. Ultra high purity grade, stated to be a minimum of 99.97 mole present methane. (2) Potassium bicarbonate. The salt solutions were made up gravimetrically using analytical grade chemicals. (3) Water. Distilled.
	ESTIMATED ERROR: $\delta p_1/psia = \pm 1$ $\delta m_1/m_1 = \pm 0.01$
	REFERENCES:

COMPONENTS:	ORIGINAL MEASUREMENTS:
(1) Methane; CH_4; [74-82-8] (2) Potassium carbonate; K_2CO_3; [584-08-7] (3) Water; H_2O; [7732-18-5]	Stoessell, R. K.; Byrne, P. A. *Geochim. Cosmochim. Acta* <u>1982</u>, *46*, 1327-32.

| VARIABLES: T/K = 298.15
 p_1/kPa = 2410-5170
 m_2/mol kg^{-1} = 0-2.0 | PREPARED BY:

 H. L. Clever |

EXPERIMENTAL VALUES:

Temperature		Pressure		Potassium Carbonate	Methane	Salt Effect Parameter
$t/°C$	T/K	p_1/psia	p_1/kPa	m_2/mol kg^{-1}	m_1/mol kg^{-1}	$_{smm}$/kg mol^{-1}
25	298.15	350	2410	0	0.0319	–
				0.5	0.0219	0.109
				1.0	0.0142	0.117
				2.0	0.0073	0.107
		550	3790	0	0.0483	–
				0.5	0.0324	0.116
				1.0	0.0217	0.116
				2.0	0.0103	0.112
		750	5170	0	0.0617	–
				0.5	0.0427	0.107
				1.0	0.0290	0.109
				2.0	0.0128	0.114
						0.111 (authors)

The salt effect parameter is defined as $k_{smm} = \log \gamma_1 / I$

where I is the ionic strength (molality) and $\gamma_1 = (m_1^* f_1 / m_1 f_1^*)_{p,T}$

with m_1 and f_1 the solubility and fugacity, respectively, of methane at p

and T. The "*" refers to saturation in distilled water.

AUXILIARY INFORMATION

METHOD/APPARATUS/PROCEDURE:	SOURCE AND PURITY OF MATERIALS:
Solubility determinations were made using a titanium-lined chamber within a stainless steel reaction vessel jacketed by a water bath for temperature control. The system pressure was set by controlling the input and output of methane within the chamber's headpiece. The vessel was rocked for three h to allow equilibration between the methane and solution. The amount of gas in the saturated solution was measured by transfer of a sample volume to a loop at the system pressure, followed by flashing the sample in an ansion loop and measuring the gas pressure in a known volume. The total gas volume and pressure change were used to compute the moles of released gas assuming ideal behavior. A correction was made for the gas not released on flashing.	(1) Methane. Matheson Co., Inc. Ultra high purity grade, stated to be a minimum of 99.97 mole present methane. (2) Potassium carbonate. The salt solutions were made up gravimetrically using analytical grade chemicals. (3) Water. Distilled.
	ESTIMATED ERROR: δp_1/psia = ± 1 $\delta m_1/m_1$ = ± 0.01
	REFERENCES:

COMPONENTS:	ORIGINAL MEASUREMENTS:
1. Methane; CH_4; [74-82-8] 2. Cesium chloride; CsCl; [7647-17-8] 3. Water; H_2O; [7732-18-5]	Ben-Naim, A.; Yaacobi, M. *J. Phys. Chem.* 1974, *78*, 175-8.

VARIABLES:	PREPARED BY:
Temperature	C. L. Young

EXPERIMENTAL VALUES:

T/K	Conc. of cesium chloride / mol ℓ^{-1}	Ostwald coefficient,* L
283.15	1.0	0.03250
288.15		0.02936
293.15		0.02710
298.15		0.02553
303.15		0.02453

* Smoothed values of Ostwald coefficient obtained from

$$kT \ln L = 20,487.3 - 138.831 \, (T/K) + 0.21072 \, (T/K)^2 \text{ cal mol}^{-1}$$

where k is in units of cal mol^{-1} K^{-1}.

AUXILIARY INFORMATION

METHOD/APPARATUS/PROCEDURE:	SOURCE AND PURITY OF MATERIALS:
The apparatus was similar to that described by Ben-Naim and Baer (1) and Wen and Hung (2). It consists of three main parts, a dissolution cell of 300 to 600 cm^3 capacity, a gas volume measuring column, and a manometer. The solvent is degassed in the dissolution cell, the gas is introduced and dissolved while the liquid is kept stirred by a magnetic stirrer immersed in the water bath. Dissolution of the gas results in the change in the height of a column of mercury which is measured by a cathetometer.	1. Matheson sample, purity 99.97 mole per cent. 2. AR grade. 3. Deionised, doubly distilled.

ESTIMATED ERROR:
$\delta T/K = \pm 0.01$; $\delta L/L = \pm 0.005$ (estimated by compiler).

REFERENCES:
1. Ben-Naim, A.; Baer, S. *Trans. Faraday Soc.* 1963, *59*, 2735.   2. Wen, W.-Y.; Hung, J. H. *J. Phys. Chem.* 1970, *74*, 170.

COMPONENTS:	ORIGINAL MEASUREMENTS:
1. Methane; CH_4; [74-82-8] 2. Water; H_2O; [7732-18-5] 3. Formaldehyde; CH_2O; [50-00-0]	Michels, A.; Gerver, J.; Bijl, A. *Physica*, 1936, *3*, 797-808.

VARIABLES:	PREPARED BY:
Pressure	C.L. Young

EXPERIMENTAL VALUES:

T/K	Conc. of formaldehyde /mol l^{-1}	$p/10^5$Pa	10^3 Mole fraction of methane in liquid, $10^3 x_{CH_4}$
298.15	1.0	49.6	0.97
		100.7	1.62
		151.8	2.02
		198.7	2.29

AUXILIARY INFORMATION

METHOD/APPARATUS/PROCEDURE:	SOURCE AND PURITY OF MATERIALS:
Simple rocking equilibrium cell. Amount of gas absorbed calculated from volume and pressure change in charging vessel. Details in source.	No details given.

ESTIMATED ERROR:

$\delta T/K = \pm 0.1$; $\delta p/10^5$Pa $= \pm 0.05$ to 0.5%; $\delta x_{CH_4} = \pm 3-5\%$
(estimated by compiler)

REFERENCES:

COMPONENTS:	ORIGINAL MEASUREMENTS:
1. Methane; CH_4; [74-82-8] 2. Methanol; CH_4O; [67-56-1] 3. Water; H_2O; [7732-18-5]	Tokunaga, J.; Kawai, M. *J. Chem. Eng. Japan* <u>1975</u>, *8*, 326-327.

VARIABLES:	PREPARED BY:
	C. L. Young

EXPERIMENTAL VALUES:

$$T/K = 293.2$$

Mole fraction of methanol	Ostwald coefficient, [a] L	Henry's law constant [a] /atm	Mole fraction of methane, [a,b] x_{CH_4}
0	0.0353	37800	0.0000265
0.0650	0.0461	27000	0.0000370
0.1386	0.0513	22600	0.0000442
0.2016	0.0555	19800	0.0000505
0.2983	0.0743	13600	0.0000735
0.4871	0.131	6590	0.000152
0.6011	0.187	4200	0.000238
0.6762	0.237	3140	0.000318
1.0000	0.525	1130	0.000885

[a] At a partial pressure of methane of 101.3 kPa.

[b] Calculated by compiler.

AUXILIARY INFORMATION

METHOD/APPARATUS/PROCEDURE:	SOURCE AND PURITY OF MATERIALS:
Volumetric apparatus with multibulb buret and magnetic stirrer. Amount of solution and gas absorbed determined volumetrically. Partial pressure determined from total pressure of solvent solution. Details in ref. (1).	1. Obtained from Seitetsu Kagaku Co., purity better than 99 mole per cent. 2. Guaranteed reagent obtained from Wako Pure Chemical Ind. Fractionated.
	ESTIMATED ERROR: $\delta T/K = \pm 0.5$; $\delta L/L = \pm 0.01$.
	REFERENCES: 1. Tokunaga, J. *J. Chem. Eng. Data* <u>1975</u>, *20*, 41.

COMPONENTS:	ORIGINAL MEASUREMENTS:
1. Methane; CH_4; [74-82-8] 2. Ethanol; C_2H_6O; [64-17-5] 3. Water; H_2O; [7732-18-5]	Tokunaga, J.; Kawai, M. *J. Chem. Eng. Japan* 1975, *8*, 326-327.
VARIABLES:	PREPARED BY: C. L. Young

EXPERIMENTAL VALUES: T/K = 293.2

Mole fraction of ethanol	Ostwald coefficient, [a] L	Henry's law constant [a] /atm	Mole fraction of methane, [a,b] x_{CH_4}
0	0.0353	37800	0.0000265
0.0260	0.0415	30600	0.0000327
0.0609	0.0462	25800	0.0000388
0.1170	0.0501	21700	0.0000461
0.2432	0.0840	10700	0.0000935
0.3112	0.106	7680	0.000130
0.5285	0.216	2820	0.000355
0.7601	0.367	1370	0.000730
1.0000	0.540	763	0.00131

[a] At a partial pressure of methane of 101.3 kPa.

[b] Calculated by compiler.

AUXILIARY INFORMATION

METHOD/APPARATUS/PROCEDURE:	SOURCE AND PURITY OF MATERIALS:
Volumetric apparatus with multibulb buret and magnetic stirrer. Amount of solution and gas absorbed determined volumetrically. Partial pressure determined from total pressure of solvent solution. Details in ref. (1).	1. Obtained from Seitetsu Kagaku Co., purity better than 99 mole per cent. 2. Guaranteed reagent obtained from Wako Pure Chemical Ind. Fractionated.
	ESTIMATED ERROR: $\delta T/K = \pm 0.5$; $\delta L/L = \pm 0.01$.

	REFERENCES:
	1. Tokunaga, J.
	   J. Chem. Eng. Data
	   1975, *20*, 41.

COMPONENTS:	ORIGINAL MEASUREMENTS:
1. Methane; CH_4; [74-82-8] 2. 1-Propanol; C_3H_8O; [71-23-8] 3. Water; H_2O; [7732-18-5]	Ben-Naim, A.; Yaacobi, M. *J. Phys. Chem.* <u>1974</u>, *78*, 170-5.

VARIABLES:	PREPARED BY:
Temperature	C. L. Young

EXPERIMENTAL VALUES:

T/K	Mole[#] fraction of 1-propanol, $x_{C_3H_8O}$	Ostwald coefficient[*], L
283.15	0.03	0.04594
288.15		0.04248
293.15		0.03951
298.15		0.03696
303.15		0.03475

[*] Smoothed values of Ostwald coefficient obtained from

$$kT \ln L = 5{,}537.5 - 36.091 \, (T/K) + 0.03677 \, (T/K)^2 \text{ cal mol}^{-1}$$

where k is in units of cal mol^{-1} K^{-1}.

[#] Mole fraction before saturation with methane which is virtually the same as the mole fraction after saturation.

AUXILIARY INFORMATION

METHOD/APPARATUS/PROCEDURE:	SOURCE AND PURITY OF MATERIALS:
The apparatus was similar to that described by Ben-Naim and Baer (1) and Wen and Hung (2). It consists of three main parts, a dissolution cell of 300 to 600 cm^3 capacity, a gas volume measuring column, and a manometer. The solvent is degassed in the dissolution cell, the gas is introduced and dissolved while the liquid is kept stirred by a magnetic stirrer immersed in the water bath. Dissolution of the gas results in the change in the height of a column of mercury which is measured by a cathetometer.	1. Matheson sample, purity 99.97 mole per cent. 2. CP grade. 3. Deionised, doubly distilled.

ESTIMATED ERROR:
$\delta T/K = \pm 0.01$; $\delta L/L = \pm 0.005$ (estimated by compiler).

REFERENCES:

1. Ben-Naim, A.; Baer, S.
   Trans. Faraday Soc. <u>1963</u>, *59*,
   2735.
 	
2. Wen, W.-Y.; Hung, J. H.
   J. Phys. Chem. <u>1970</u>, *74*, 170.

COMPONENTS:	ORIGINAL MEASUREMENTS:
1. Methane; CH_4; [74-82-8] 2. 1,4-Dioxane; $C_4H_8O_2$; [123-91-1] 3. Water; H_2O; [7732-18-5]	Ben-Naim, A.; Yaacobi, M. *J. Phys. Chem.* 1974, *78*, 170-5.

VARIABLES:	PREPARED BY:
Temperature	C. L. Young

EXPERIMENTAL VALUES:

T/K	#Mole fraction of dioxane, $x_{C_4H_8O_2}$	Ostwald coefficient,[*] L
283.15	0.03	0.04516
288.15		0.04211
293.15		0.03949
298.15		0.03724
303.15		0.03530

[*] Smoothed values of Ostwalt coefficient obtained from

$$kT \ln L = 11,689.2 - 72.532(T/K) + 0.09180(T/K)^2 \text{ cal mol}^{-1}$$

where k is in units of cal mol^{-1} K^{-1}.

[#] Mole fraction before saturation with methane which is virtually the same as the mole fraction after saturation.

AUXILIARY INFORMATION

METHOD/APPARATUS/PROCEDURE:	SOURCE AND PURITY OF MATERIALS:
The apparatus was similar to that described by Ben-Naim and Baer (1) and Wen and Hung (2). It consists of three main parts, a dissolution cell of 300 to 600 cm^3 capacity, a gas volume measuring column, and a manometer. The solvent is degassed in the dissolution cell, the gas is introduced and dissolved while the liquid is kept stirred by a magnetic stirrer immersed in the water bath. Dissolution of the gas results in the change in the height of a column of mercury which is measured by a cathetometer.	1. Matheson sample, purity 99.97 mole per cent. 2. AR grade. 3. Deionised, doubly distilled.
	ESTIMATED ERROR: $\delta T/K = \pm 0.01$; $\delta L/L = \pm 0.005$ (estimated by compiler).
	REFERENCES: 1. Ben-Naim, A.; Baer, S. *Trans. Faraday Soc.* 1963, *59*, 2735. 2. Wen, W.-Y.; Hung, J. H. *J. Phys. Chem.* 1970, *74*, 170.

COMPONENTS:	ORIGINAL MEASUREMENTS:
1. Methane; CH_4; [74-82-8] 2. Water; H_2O; [7732-18-5] 3. Urea; CH_4N_2O; [57-13-6]	Wetlaufer, D. B.; Malik, S. K.; Stoller, L.; Coffin, R. L. *J. Am. Chem. Soc.* 1964, *86*, 508-514.

VARIABLES:
   Temperature
	PREPARED BY:
	   C. L. Young

EXPERIMENTAL VALUES:

T/K	Conc. of urea in soln. / mol dm^{-3}	10^3 Conc. of methane[†] in soln. / mol dm^{-3}	Mole fraction[*] of methane x_{CH_4}
278.2	6.96	0.00131	0.0000291
298.2	6.96	0.00102	0.0000227
318.2	6.96	0.00086	0.0000198

[†] at a partial pressure of 101.3 kPa.

[*] calculated by compiler.

AUXILIARY INFORMATION

METHOD/APPARATUS/PROCEDURE:	SOURCE AND PURITY OF MATERIALS:
Modified Van Slyke-Neill apparatus fitted with a magnetic stirrer. Solution was saturated with gas and then sample transferred to the Van Slyke extraction chamber.	1. Matheson c.p. grade, purity 99 mole per cent or better. 2. Distilled. 3. Commercial sample, purified by two recrystallizations from 65% ethanol.
	ESTIMATED ERROR: $\delta T/K = \pm 0.05$; $\delta x_{CH_4} = \pm 2\%$.
	REFERENCES:

COMPONENTS:	ORIGINAL MEASUREMENTS:
1. Methane; CH_4; [74-82-8] 2. Urea; CH_4N_2O; [57-13-6] 3. Water; H_2O; [7732-18-5]	Ben-Naim, A.; Yaacobi, M. *J. Phys. Chem.* <u>1974</u>, *78*, 170-5.

VARIABLES:	PREPARED BY:
Temperature, concentration	C. L. Young

EXPERIMENTAL VALUES:

T/K	Conc. of urea /mol ℓ^{-1}	Ostwald* Coefficient, L	T/K	Conc. of urea /mol ℓ^{-1}	Ostwald* Coefficient, L
283.15	1.0	0.04117	283.15	4.0	0.03228
288.15		0.03757	288.15		0.03066
293.15		0.03471	293.15		0.02894
298.15		0.03244	298.15		0.02715
303.15		0.03065	303.15		0.02532
283.15	2.0	0.03782	283.15	7.0	0.02676
288.15		0.03480	288.15		0.02579
293.15		0.03249	293.15		0.02482
298.15		0.03078	298.15		0.02387
303.15		0.02955	303.15		0.02294

* Smoothed values of Ostwald coefficient obtained from

$$kT \ln L = 11,562.0 - 77.00(T/K) + 0.10534(T/K)^2 \text{ cal mol}^{-1}$$
$$kT \ln L = 14,255.0 - 96.928(T/K) + 0.14153(T/K)^2 \text{ cal mol}^{-1}$$
$$kT \ln L = -6,004.4 + 41.018(T/K) + 0.09407(T/K)^2 \text{ cal mol}^{-1}$$
$$kT \ln L = -855.0 + 2.980(T/K) + 0.02528(T/K)^2 \text{ cal mol}^{-1}$$

(where k is in units of cal $mol^{-1} K^{-1}$) for concentrations of 1.0, 2.0, 4.0 and 7.0 mol ℓ^{-1}, respectively.

AUXILIARY INFORMATION

METHOD/APPARATUS/PROCEDURE:	SOURCE AND PURITY OF MATERIALS:
The apparatus was similar to that described by Ben-Naim and Baer (1) and Wen and Hung (2). It consists of three main parts, a dissolution cell of 300 to 600 cm^3 capacity, a gas volume measuring column, and a manometer. The solvent is degassed in the dissolution cell, the gas is introduced and dissolved while the liquid is kept stirred by a magnetic stirrer immersed in the water bath. Dissolution of the gas results in the change in the height of a column of mercury which is measured by a cathetometer.	1. Matheson sample, purity 99.97 mole per cent. 2. AR grade. 3. Deionised, doubly distilled.
	ESTIMATED ERROR: $\delta T/K = \pm 0.01$; $\delta L/L = \pm 0.005$ (estimated by compiler).

	REFERENCES:
	1. Ben-Naim, A.; Baer, S.
	  Trans. Faraday Soc. <u>1963</u>, *59*, 2735.
	
	2. Wen, W.-Y.; Hung, J. H.
	  J. Phys. Chem. <u>1970</u>, *74*, 170.

COMPONENTS:	ORIGINAL MEASUREMENTS:
1. Methane; CH_4; [74-82-8] 2. 1,3,5,7-Tetraazatricyclo[3,3,- 1,1^{3,7}] decane; $C_6H_{12}N_4$; [100-97-0] 3. Water; H_2O; [7732-18-5]	Barone, G.; Castronuovo, G.; Volpe, D.; Elia, V.; Grassi, L. *J. Phys. Chem.* 1979, *83*, 2703-2714.
VARIABLES:	PREPARED BY: C. L. Young

EXPERIMENTAL VALUES:

t/°C	T/K	Conc. of component 2[a] /mol dm^{-3}	Ostwald coefficient, L
15	288.15	0.0	0.03918
		0.5	0.03936
		1.0	0.03932
		1.5	0.03905
		2.0	0.03856
		2.5	0.03785
		3.0	0.03692
25	298.15	0.0	0.03371
		0.5	0.03434
		1.0	0.03445
		1.5	0.03405
		2.0	0.03314
		2.5	0.03172
		3.0	0.02979
35	308.15	0.0	0.02986
		0.5	0.03010
		1.0	0.03026
		1.5	0.03034
		2.0	0.03034
		2.5	0.03026
		3.0	0.03010

[a] Units not explicitly given in original.

AUXILIARY INFORMATION

METHOD/APPARATUS/PROCEDURE:	SOURCE AND PURITY OF MATERIALS:
The apparatus was similar to that described by Ben-Naim and Baer (1). Teflon needle valves were used in place of stopcocks. The apparatus consists of three main parts, a dissolution cell of 300 to 600 cm^3 capacity, a gas volume measuring column and a manometer. The solvent is degassed in the dissolution cell, the gas is introduced and dissolved while the liquid is kept stirred by a magnetic stirrer. Dissolution of the gas results in the change in the height of a column of mercury which is measured with a cathetometer.	1. Matheson sample, purity 99.97 mole per cent. 2. Fluka sample, recrystallized from ethanol. 3. Doubly distilled.
	ESTIMATED ERROR: $\delta T/K = \pm 0.03$; $\delta L/L = \pm 0.005$ (estimated by compiler).
	REFERENCES: 1. Ben-Naim, A.; Baer, S. *Trans. Faraday Soc.* 1963, *59*, 2735.

COMPONENTS:	ORIGINAL MEASUREMENTS:

1. Methane; CH_4; [74-82-8]
2. β-D-Fructofuranosyl-α-D-gluco-
   pyranoside, (Sucrose); $C_{12}H_{22}O_{11}$;
   [57-50-1]
3. Water; H_2O; [7732-18-5]
	Ben-Naim, A.; Yaacobi, M.
	J. Phys. Chem.
	<u>1974</u>, *78*, 170-5.

VARIABLES:	PREPARED BY:
Temperature	C. L. Young

EXPERIMENTAL VALUES:

T/K	Conc. of sucrose /mol l^{-1}	Ostwald coefficient,[*] L
283.15	0.5	0.03592
288.15		0.03276
293.15		0.03020
298.15		0.02813
303.15		0.02646

[*] Smoothed values of Ostwald coefficient obtained from

$$kT \ln L = 10,386.5 - 68.957 \ (T/K) + 0.09063 \ (T/K)^2 \ \text{cal mol}^{-1}$$

where k is in units of cal mol^{-1} K^{-1}.

AUXILIARY INFORMATION

METHOD/APPARATUS/PROCEDURE:	SOURCE AND PURITY OF MATERIALS:
The apparatus was similar to that described by Ben-Naim and Baer (1) and Wen and Hung (2). It consists of three main parts, a dissolution cell of 300 to 600 cm^3 capacity, a gas volume measuring column, and a manometer. The solvent is degassed in the dissolution cell, the gas is introduced and dissolved while the liquid is kept stirred by a magnetic stirrer immersed in the water bath. Dissolution of the gas results in the change in the height of a column of mercury which is measured by a cathetometer.	1. Matheson sample, purity 99.97 mole per cent. 2. AR grade. 3. Deionised, doubly distilled.

ESTIMATED ERROR:
$\delta T/K = \pm 0.01$; $\delta L/L = \pm 0.005$ (estimated by compiler).

REFERENCES:

1. Ben-Naim, A.; Baer, S.
   Trans. Faraday Soc. <u>1963</u>, *59*, 2735.

2. Wen, W.-Y.; Hung, J. H.
   J. Phys. Chem. <u>1970</u>, *74*, 170.

COMPONENTS:	ORIGINAL MEASUREMENTS:
1. Methane; CH_4; [74-82-8] 2. Water; H_2O; [7732-18-5] 3. α-D-Glucopyronoside, β-D-fructofuranosyl (Sucrose); $C_{12}H_{22}O_{11}$; [57-50-1]	Michels, A.; Gerver, J.; Bijl, A. *Physica*, <u>1936</u>, *3*, 797-808.

VARIABLES:	PREPARED BY:
Pressure, concentration	C.L. Young

EXPERIMENTAL VALUES:

T/K	$p/10^5 Pa$	Conc. of sucrose /mol l^{-1}	10^3 Mole fraction of methane in liquid, $10^3\ x_{CH_4}$
298.15	60.2	1.0	0.88
	109.4		1.35
	157.5		1.70
	191.5		1.94
	60.3	2.0	0.83
	146.4		1.53
	265.8		1.96
	448		2.23

AUXILIARY INFORMATION

METHOD/APPARATUS/PROCEDURE:	SOURCE AND PURITY OF MATERIALS:
Simple rocking equilibrium cell. Amount of gas absorbed calculated from volume and pressure change in charging vessel. Details in source.	No details given.

ESTIMATED ERROR:
$\delta T/K = \pm 0.1$; $\delta p/10^5 Pa = \pm 0.05$ to 0.5%; $\delta x_{CH_4} = \pm 3-5\%$

(estimated by compiler).

REFERENCES:

COMPONENTS:	ORIGINAL MEASUREMENTS:
1. Methane; CH_4; [74-82-8] 2. Sulfinylbismethane; (Dimethyl-sulfoxide, DMSO); C_2H_6OS; [67-68-5] 3. Water; H_2O; [7732-18-5]	Ben-Naim, A.; Yaacobi, M. *J. Phys. Chem.* 1974, *78*, 170-5.

VARIABLES:	PREPARED BY:
Temperature	C. L. Young

EXPERIMENTAL VALUES:

T/K	[#]Mole fraction of DMSO, x_{DMSO}	Ostwald coefficient,[*] L
283.15	0.03	0.04329
288.15		0.03977
293.15		0.03684
298.15		0.03440
303.15		0.03236

[*] Smoothed values of Ostwald coefficient obtained from

$$kT \ln L = 7,904.5 - 52.042(T/K) + 0.06316(T/K)^2 \text{ cal mol}^{-1}$$

where k is in units of cal $mol^{-1} K^{-1}$.

[#] Mole fraction before saturation with methane which is virtually the same as the mole fraction after saturation.

AUXILIARY INFORMATION

METHOD/APPARATUS/PROCEDURE:	SOURCE AND PURITY OF MATERIALS:
The apparatus was similar to that described by Ben-Naim and Baer (1) and Wen and Hung (2). It consists of three main parts, a dissolution cell of 300 to 600 cm³ capacity, a gas volume measuring column, and a manometer. The solvent is degassed in the dissolution cell, the gas is introduced and dissolved while the liquid is kept stirred by a magnetic stirrer immersed in the water bath. Dissolution of the gas results in the change in the height of a column of mercury which is measured by a cathetometer.	1. Matheson sample, purity 99.97 mole per cent. 2. CP grade. 3. Deionised, doubly distilled.

METHOD/APPARATUS/PROCEDURE continued:

ESTIMATED ERROR:
$\delta T/K = \pm 0.01$; $\delta L/L = \pm 0.005$
(estimated by compiler).

REFERENCES:

1. Ben-Naim, A.; Baer, S. *Trans. Faraday Soc.* 1963, *59*, 2735.

2. Wen, W.-Y.; Hung, J. H. *J. Phys. Chem.* 1970, *74*, 170.

COMPONENTS:	ORIGINAL MEASUREMENTS:
1. Methane; CH_4; [74-82-8] 2. Water; H_2O; [7732-18-5] 3. Glucose; $C_6H_{12}O_6$; [50-99-7]	Michels, A.; Gerver, J.; Bijl, A. *Physica*, 1936, *3*, 797-808.

VARIABLES:	PREPARED BY:
Pressure, concentration	C.L. Young

EXPERIMENTAL VALUES:

T/K	$p/10^5 Pa$	Conc. of glucose /mol l^{-1}	10^3 Mole fraction of methane in liquid, $10^3 x_{CH_4}$
298.15	53.0	1.0	0.76
	108.1		1.32
	149.8		1.59
	192.8		1.77
	54.4	2.0	0.67
	102.1		1.15
	154.5		1.55
	204.4		1.74
	414.4		2.66

AUXILIARY INFORMATION

METHOD/APPARATUS/PROCEDURE:	SOURCE AND PURITY OF MATERIALS:
Simple rocking equilibrium cell. Amount of gas absorbed calculated from volume and pressure change in charging vessel. Details in source.	No details given.

ESTIMATED ERROR:
$\delta T/K = \pm 0.1$; $\delta p/10^5 Pa = \pm 0.05$ to 0.5%; $\delta x_{CH_4} = \pm 3-5\%$.
(estimated by compiler).

REFERENCES:

M—G*

COMPONENTS:	ORIGINAL MEASUREMENTS:
1. Methane; CH_4; [74-82-8] 2. Water; H_2O; [7732-18-5] 3. 2-Aminoethanol, (Monoethanol- amine); C_2H_7NO; [141-43-5]	Lawson, J.D.; Garst, A.W. *J. Chem. Engng. Data.* <u>1976</u>, *21*, 30-2

VARIABLES:	PREPARED BY:
Temperature, pressure, composition	C.L. Young

EXPERIMENTAL VALUES:

T/K	P/MPa	Conc. Wt. % amine	Mole fraction of methane in liquid, x_{CH_4}	$10^5 x$ Solubility / mol g^{-1} (soln)
310.93	6.578	15	0.00132	6.55
	3.447	40	0.000835	3.48
	6.578		0.00157	6.26
338.71	3.433	15	0.000618	3.07
	6.846		0.00152	5.80
	3.433	40	0.000835	3.33
	6.578		0.00155	6.20
366.48	3.378	40	0.00230	9.16
	6.715		0.00416	16.60
394.26	3.503	40	0.00105	4.19
	6.550		0.00197	7.85

AUXILIARY INFORMATION

METHOD/APPARATUS/PROCEDURE:	SOURCE AND PURITY OF MATERIALS:

Rocking equilibrium cell fitted
with liquid sampling valve.
Pressure measured with Bourdon gauge.
Cell charged with amine and then
methane added. Liquid phase
samples analysed volumetrically.
	1. Purity 99 mole per cent minimum.
	2. Distilled.
	3. Commercial sample, purity better
	    than 99 mole per cent as
	    determined by acid titration.

ESTIMATED ERROR:
$\delta T/K = \pm 0.15$; $\delta P/MPa = \pm 0.5\%$
$\delta x_{CH_4} = \pm 3\%$.

REFERENCES:

COMPONENTS:	ORIGINAL MEASUREMENTS:
1. Methane; CH_4; [74-82-8] 2. Water; H_2O; [7732-18-5] 3. 2,2´-Iminobisethanol,(Diethanol- amine); $C_4H_{11}NO$; [111-42-2]	Lawson, J.D.; Garst, A.W. *J. Chem. Engng. Data.* <u>1976</u>, *21*, 30-2

VARIABLES:	PREPARED BY:
Temperature, pressure, composition	C.L. Young

EXPERIMENTAL VALUES:

T/K	P/MPa	Conc. Wt. %	Mole fraction of methane in liquid, x_{CH_4}	$10^5 x$ Solubility / mol g^{-1}(soln)
310.93	3.530	5	0.000653	3.48
	6.640		0.00121	6.46
	3.515	25	0.000727	3.20
	6.674		0.00136	5.00
	3.627	40	0.000872	3.24
	6.433		0.00149	5.52
338.71	3.558	5	0.000520	2.77
	6.743		0.000976	5.20
	3.523	25	0.000656	2.89
	6.771		0.00123	5.40
	3.571	40	0.000824	3.06
	6.460		0.00145	5.38
366.48	3.558	25	0.000586	2.58
	6.343		0.00114	5.02
	3.654	40	0.000819	3.04
	6.640		0.000155	5.74
394.26	3.454	25	0.000674	2.97
	6.343		0.001321	5.82
	3.434	40	0.000959	3.56
	6.260		0.00170	6.30

AUXILIARY INFORMATION

METHOD/APPARATUS/PROCEDURE:	SOURCE AND PURITY OF MATERIALS:
Rocking equilibrium cell fitted with liquid sampling valve. Pressure measured with Bourdon gauge. Cell charged with amine then methane added. Liquid phase samples analysed volumetrically.	1. Purity 99 mole per cent minimum. 2. Distilled. 3. Commercial sample, purity better than 99 mole per cent as determined by acid titration

ESTIMATED ERROR:
$$\delta T/K = \pm 0.15; \quad \delta P/MPa = \pm 0.5\%$$
$$\delta x_{CH_4} = \pm 3\%.$$

REFERENCES:

COMPONENTS:	ORIGINAL MEASUREMENTS:
(1) Methane; CH_4; [74-82-8] (2) Triethylenediamine or 1,4-Diaza-bicyclo[2.2.2]octane; $C_6H_{12}N_2$; [280-57-9] (3) Sodium hydroxide; NaOH; [7646-69-7] (4) Water; H_2O; [7732-18-5]	Muccitelli, J. A.; Wen, W.-Y. *J. Solution Chem.* <u>1980</u>, *9*, 141 - 161.

VARIABLES:	PREPARED BY:
T/K: 278.15 - 298.15 p/kPa: 101.325 (1 atm) c_2/mol dm^{-3}: 0 - 1.157	H. L. Clever

EXPERIMENTAL VALUES:

T/K	$C_6H_{12}N_2$ c_2/mol dm^{-3}	NaOH c_3/mol dm^{-3}	pH	Ostwald Coefficient $10^3 L$/cm^{3}cm^{-3}
278.15	0.0	0.01	12.02	49.82 ± 0.29
	0.291	0.01	12.03	52.83
	0.3785	0.01	12.06	52.31
	0.8554	0.01	12.06	51.80
	1.010	0.01	11.90	52.01
283.15	0.0	0.01	12.08	44.42 ± 0.25
	0.1782	0.01	12.01	45.41
	0.2778	0.01	12.02	45.42
	0.5629	0.01	12.07	46.48
	0.8753	0.01	12.02	46.44
	1.039	0.01	12.08	47.15
288.15	0.0	0.01	12.02	39.46 ± 0.25
	0.1110	0.01	12.10	40.07
	0.1722	0.01	12.00	40.02
	0.2182	0.01	12.22	40.49
	0.3835	0.01	12.01	41.98
	0.4763	0.01	12.09	48.81 [*sic*]
	0.6238	0.01	12.23	42.33
	0.8941	0.01	12.03	41.74

Table continued on next page.

AUXILIARY INFORMATION

METHOD/APPARATUS/PROCEDURE:	SOURCE AND PURITY OF MATERIALS:
The solubility apparatus and procedure employed were similar to that described by Ben-Naim and Baer (1) with modifications suggested by Wen and Hung (2). The apparatus consists mainly of a mercury manometer, a gas-volume measuring buret, a dissolution cell of about 450 cm^3 capacity, and a mercury reservoir. The degassing apparatus and procedure used were similar to that described by Battino *et al.* (3). From published ionization constants the authors estimated that nearly 100 per cent of the triethylenediamine is unprotonated when the solution pH is 12 or above, and about 99.7 per cent is in the monoprotonated form when the solution pH is 5.7 to 5.9.	(1) Methane. Matheson Co., Inc. Specified to have a purity of 99.95 per cent. (2) Triethylenediamine. Aldrich Chemical Co. Recrytallized twice from diethylether, triturated and dried *in vacuo* over P_2O_5 for eight days at 50 ^{0}C. (3) Sodium hydroxide. Carbonate free. (4) Water. Carbon dioxide free.

ESTIMATED ERROR:
δT/K = ± 0.005
δP/mmHg = ± 3
$\delta L/L$ = ± 0.005

REFERENCES:
1. Ben-Naim, A.; Baer, S.
   Trans. Faraday Soc. <u>1964</u>,*60*,1736.
2. Wen, W.-Y.; Hung, J. H.
   J. Phys. Chem. <u>1970</u>, *74*, 170.
3. Battino, R.; Banzhof, M.;
   Bogan, M.; Wilhelm, E.
   Anal. Chem. <u>1971</u>, *43*, 806.

COMPONENTS:	ORIGINAL MEASUREMENTS:
(1) Methane; CH_4; [74-82-8] (2) Triethylenediamine or 1,4-Diazabicyclo[2.2.2]octane; $C_6H_{12}N_2$; [280-57-9] (3) Sodium hydroxide; NaOH; [7646-69-7] (4) Water; H_2O; [7732-18-5]	Muccitelli, J. A.; Wen, W.-Y. *J. Solution Chem.* <u>1980</u>, *9*, 141 - 161.

VARIABLES:	PREPARED BY:
T/K: 278.15 - 298.15 P/kPa: 101.325 (1 atm) c_2/mol dm^{-3}: 0 - 1.157	H. L. Clever

EXPERIMENTAL VALUES:

T/K	$C_6H_{12}N_2$ c_2/mol dm^{-3}	NaOH c_3/mol dm^{-3}	pH	Ostwald Coefficient $10^3 L$/cm^{3}cm^{-3}
293.15	0.0	0.01	12.02	36.63 ± 0.37
	0.1395	0.01	12.02	36.75
	0.2440	0.01	12.02	37.32
	0.6772	0.01	12.08	38.29
	1.157	0.01	12.01	38.71
298.15	0.0	0.01	11.88	33.48 ± 0.16
	0.1807	0.01	11.74	34.10
	0.4836	0.01	11.94	35.55
	0.7682	0.01	11.93	35.71
	0.9947	0.01	12.03	36.38

AUXILIARY INFORMATION

METHOD/APPARATUS/PROCEDURE:	SOURCE AND PURITY OF MATERIALS:
See preceding page.	See preceding page.

	ESTIMATED ERROR:
	See preceding page.

	REFERENCES:
	See preceding page.

COMPONENTS:	ORIGINAL MEASUREMENTS:
(1) Methane; CH_4; [74-82-8] (2) Sulfuric acid monohexyl ester, sodium salt or sodium hexyl-sulfate; $C_6H_{14}O_4S.Na$; [2207-98-9] (3) Water; H_2O; [7732-18-5]	Bolden, P. L.; Hoskins, J. C.; King, A. D. Jr. *J. Colloid Interface Sci.* **1983**, *91*, 454-463.

| VARIABLES: $T/K = 298.15$
$p_1/kPa =$
$m_2/mol\ kg^{-1} = 0 - 1.10$ | PREPARED BY:

H. L. Clever |

EXPERIMENTAL VALUES:

Temperature		Sodium Hexyl-sulfate	Henry's Constant Methane
$t/^0C$	T/K	$m_2/mol\ kg^{-1}$	$K/mol\ kg^{-1}\ atm^{-1}$
25	298.15	0	1.55[a]
		0.10	1.62
		0.30	1.61
		0.50	1.51
		0.60	1.43
		0.70	1.60
		0.80	1.71
		0.90	1.78
		1.00	1.84
		1.10	1.90

[a] Solubility value in water from reference 1.

Henry's constant, $K/mol\ kg^{-1}atm^{-1} = (m_1/mol\ kg^{-1})/(p_1/atm)$.

AUXILIARY INFORMATION

METHOD/APPARATUS/PROCEDURE:	SOURCE AND PURITY OF MATERIALS:
The solution of surface active material is contained in a glass-lined brass equilibrium cell resting on a magnetic stirrer. The solution is degassed by evacuation and stirring. Gas is introduced at pressures above atmospheric and equilibration is continued for at least five hours. Subsequently as the pressure is released to a lower pressure, the gas evolved from the supersaturated solution is collected at atmospheric pressure and ambient temperature in a Warburg manometer and its volume measured. Corrections are made for the gas lost during the venting procedure, the differences in temperature and pressure, and the water vapor pressure in the calculation of Henry's constant. Details are in an earlier paper (ref 1).	(1) Methane. Source not given. Stated to be c. p. grade, 99 % or better. (2) Sodium hexylsulfate. Eastman Kodak Co. Recrytallized from 2-propanol and dried *in vacuo*. (3) Water. Double distilled.

ESTIMATED ERROR:

$\delta K/mol\ kg^{-1}\ atm^{-1} = \pm\ 0.04$

REFERENCES:

1. Matheson, I. B. C.; King, A. D. *J. Coll. Interface Sci.* **1978**, *66*, 464

COMPONENTS:	ORIGINAL MEASUREMENTS:
(1) Methane; CH_4; [74-82-8] (2) Sulfuric acid monododecyl ester sodium salt (sodium dodecyl sulfate or SDS); $C_{12}H_{26}O_4S.Na$; [151-21-3] (3) Water; H_2O; [7732-18-5]	Matheson, I. B. C.; King, A. D. *J. Coll. Interface Sci.* <u>1978</u>, *66*, 464 - 469.

| VARIABLES: T/K: 298.15
p/kPa: 255.1-820.5(37.0-119.0 psig)
SDS/mol kg^{-1} H_2O: 0 - 0.300 | PREPARED BY:

    H. L. Clever |

EXPERIMENTAL VALUES:

T/K	Sulfuric acid monododecyl ester sodium salt m_2/mol kg^{-1}	Pressure pounds per square inch,gauge p/psig	Volume gas evolved V_1/cm^3	Ambient Pressure p/mmHg	Ambient Temperature t/°C	Henry's constant 10^3K/mol kg^{-1}atm^{-1}
298.15	0	42.7	11.6	744.6	24.5	
		68.2	18.8	749.9	23.0	
		119.0	31.1	744.7	23.9	1.55±0.03
	0.150	37.0	12.1	755.4	21.9	
		70.5	22.7	758.6	21.2	
		94.5	29.6	756.7	22.0	1.88±0.03
	0.300	53.2	19.6	749.6	21.9	
		75.4	26.7	752.8	21.2	
		112.0	39.2	752.0	22.0	2.09±0.03

AUXILIARY INFORMATION

METHOD/APPARATUS/PROCEDURE:	SOURCE AND PURITY OF MATERIALS:
The apparatus consists of a jacketed thermostated thick-walled cylindrical brass bomb which rests on a variable speed magnetic stirrer. An inlet line to the bomb is connected to a gas manifold, and an exit line is connected to a Warburg manometer. Bourden gauges are used to record the pressure.	(1) Methane. Source not given. Chemically pure or equivalent of 99.0 mole percent purity. (2) Sulfuric acid monododecyl ester sodium salt. Aldrich Chemical Co., Inc. Recrystallized from ethanol and dried *in vacuo*. (3) Water. Laboratory distilled.

The solution, consisting of 100 g of water and the colloidal electrolyte, is contained in a glass liner inside of the bomb. The solution is degassed by evacuation to just above water vapor pressure and then stirring for several hours. The gas is introduced over the solution at the desired pressure and the solution is stirred for a minimum of five hours.

| | ESTIMATED ERROR:

    $\delta K/K = 0.02$ |

The gas is vented to atmospheric pressure. The gas from the supersaturated solution is collected in the Warburg manometer and its volume measured at atmospheric pressure and ambient temperature. Corrections for gas lost during venting and thermal equilibration and for water vapor pressure are made.

The solubility is reported as Henry's constant, K/mol gas kg^{-1} atm^{-1} = gas molality/pressure = $(m_1$/mol kg^{-1} $)/(p$/atm$)$.

COMPONENTS:	ORIGINAL MEASUREMENTS:
(1) Methane; CH_4; [74-82-8] (2) Sodium dodecylsulfate or SDS; $C_{12}H_{25}SO_4Na$; [151-21-8] (3) Water; H_2O; [7732-18-5]	Ben-Naim, A.; Battino, R. *J. Solution Chem.* <u>1985</u>, *14*, 245-53.

EXPERIMENTAL VALUES:

Temperature		Sodium Dodecyl sulfate	Ostwald Coefficient
$t/^0C$	T/K	$c_2/mol\ dm^{-3}$	$10^3 L/cm\ cm^{-3}$
15	288.15	0.0	40.94
			40.60
		0.002	42.0
		0.005	42.2
		0.01	41.6
		0.03	42.2
		0.05	42.2
		0.08	44.1
18	291.15	0.0	39.1
			38.6
		0.008	39.2
		0.05	40.5
		0.08	41.8
21	294.15	0.0	37.0
			36.7
		0.005	36.4
		0.008	37.64
		0.01	37.0
		0.03	38.3
		0.05	38.0
		0.08	39.6
24	297.15	0.0	35.5
			34.85
		0.005	36.0
		0.008	34.9
		0.05	36.2
		0.08	37.8
27	300.15	0.0	32.7
			33.1
		0.002	33.5
		0.004	33.5
		0.005	33.8
			33.0
		0.008	33.45
		0.01	33.5
		0.03	34.5
		0.05	35.5
		0.08	36.1

The values of $\Delta\mu_g^0$ calculated from the experimental Ostwald absorption coefficients were fit to the following equation by the authors.

$$\Delta\mu_g^0 = -\ RT\ \ln L$$

$$= -\ 2986.19 + 16.7082\ (T/K) - 523.534\ (c_2/mol\ dm^{-3})$$

with R in cal K^{-1} mol^{-1}.

COMPONENTS:	ORIGINAL MEASUREMENTS:
(1) Methane; CH_4; [74-82-8]	Ben-Naim, A.; Battino, R.
(2) Sodium dodecylsulfate; [151-21-3] $C_{12}H_{25}SO_4Na$	*J. Solution Chem.* <u>1985</u>, *14*, 245-53.
(3) Water; H_2O; [7732-18-5]	

VARIABLES: T/K = 288.15 - 300.15 p_1/kPa = 101.325 c_2/mol dm^{-3} = 0 - 0.1	PREPARED BY: H. L. Clever

EXPERIMENTAL VALUES:

AUXILIARY INFORMATION

METHOD/APPARATUS/PROCEDURE:

The apparatus is similar to that described by Ben-Naim and Baer (ref 1). It consists of three main parts, a dissolution cell of 300 to 600 cm^3 capacity, a gas volume measuring buret, and a manometer. The solvent is degassed in the dissolution cell, and the gas is introduced and dissolved while the liquid is stirred by a magnetic stirrer. Dissolution of the gas results in the change in the height of a column of mercury which is measured by a cathetometer.

SOURCE AND PURITY OF MATERIALS:

(1) Methane. Matheson Co., Inc. Stated to be 99.95 percent minimum purity.

(2) Sodium dodecylsulfate. British Drug House. Stated to be 'especially pure'. Used without further purification.

(3) Water. Distilled. Specific conductivity < 1 x 10^{-6} (ohm cm)$^{-1}$.

ESTIMATED ERROR:
$$\delta T/K = \pm 0.05$$
$$\delta c_2/\text{mol dm}^{-3} = \pm 0.0005$$
$$\delta L/L = \pm 0.01$$

REFERENCES:
1. Ben-Naim, A.; Baer, S. *Trans. Faraday Soc.* <u>1963</u>, *59*, 2735.

COMPONENTS:	ORIGINAL MEASUREMENTS:
(1) Methane; CH_4; [74-82-8]	Stoessell, R. K.; Byrne, P.A.
(2) Clay and sediment slurries	*Clays Clay Miner*. <u>1982</u>, *30*, 67-72.
(3) Water; H_2O; [7732-18-5]	
Sea water	

EXPERIMENTAL VALUES:

Temperature		Pressure		Clay or Sediment /wt%	Methane
t/°C	T/K	p_1/psia	p_1/kPa		m_1/mol kg^{-1}
Na saturated SAz-1 dispersed into distilled water					
25	298.15	350	2410	0	0.0318
				1.09	0.0300
				1.99	0.0298
				5.52	0.0299
				10.32	0.0296
		550	3790	0	0.0473
				1.09	0.0457
				1.99	0.0457
				5.52	0.0466
				10.32	0.0472
		750	5170	0	0.0623
				1.09	0.0596
				1.99	0.0610
				5.52	0.0620
				10.32	0.0608
Na saturated SWy-1 dispersed in distilled water					
25	298.15	350	2410	0	0.0318
				1.01	0.0306
				2.01	_a
		550	3790	0	0.0473
				1.01	0.0467
				2.01	0.0465
		750	5170	0	0.0623
				1.01	0.0621
				2.01	0.0613
Sediment dispersed in distilled water					
25	298.15	350	2410	0	0.0318
				7.1	0.0312
		550	3790	0	0.0473
				7.1	0.0468
		750	5170	0	0.0623
				7.1	0.0616
Sediment dispersed in sea water of 34.84% salinity					
25	298.15	350	2410	0	0.0263
				9.7	0.0250
		550	3790	0	0.0400
				9.7	0.0396
		750	5170	0	0.0514
				9.7	0.0517

[a] High viscosity gel that would not release gas at these conditions.

COMPONENTS:	ORIGINAL MEASUREMENTS:
(1) Methane; CH_4; [74-82-8] (2) Clay and sediment slurries (3) Water; H_2O; [7732-18-5] Sea water	Stoessell, R. K.; Byrne, P.A. *Clays Clay Miner* <u>1982</u>, *30*, 67-72.

VARIABLES:	PREPARED BY:
$T/K = 298.15$ $p_1/kPa = 2410 - 5170$ slurries/wt % = 0 - 10.32	H. L. Clever

SOURCE AND PURITY OF MATERIALS:
(1) Methane. Matheson ultra high purity. Minimum purity 99.97 percent.

(2) SAz-1. Cheto montmorillonite from Arizona.

SWy-1. Bentonite montmorillonite from Wyoming.

Both were obtained from The Clay Minerals Society. Data on these clays are reported by Van Olphen, H.; Fripiant, J. J., Editors, *Data Handbook for Clay Minerals and Other Non-Metallic Minerals*, Pergamon Press, Ltd., Oxford and New York, 1979, 346pp.

Samples of each clay were treated with H_2O_2 to remove organic material. Following settling to remove non-clay minerals, the samples were centrifuged and washed to remove soluble salts. X-ray powder photographs showed no crystalline impurities in SAz-1, but showed the presence of minor quartz in SWy-1. The exchange sites were saturated with Na by mixing 10 g of clay with one liter of 1 mol dm^{-3} NaCl solution. The slurries were allowed to set for one week with occasional shaking. They were then washed and centrifuged until chloride ion could not be detected by silver nitrate. The washed clay was then dispersed into distilled water to make the slurries.

Marine sediment. Obtained from a core off the present Mississippi delta in 60 m of water and 10 m below the bottom. The argillaceous sediment was stored for 2 months in a brine containing 150,000 ppm NaCl. It was then centrifuged and washed until chloride ion could not be detected. The organic carbon content was 1.2 ± 0.2 wt %. X-ray diffraction patterns of the sediment showed the following major
(continued below)

AUXILIARY INFORMATION

METHOD/APPARATUS/PROCEDURE:	SOURCE AND PURITY OF MATERIALS:(continued)
Solubility determinations were made using a titanium-lined chamber within a stainless-steel reaction vessel jacketed by a water bath for temperature control. The system pressure was set by controlling the input and output of methane within the chamber headpiece. The vessel was rocked for 3 h to allow equilibration between the methane and slurry.	components: quartz, feldspar, dioctahedral smecite, and well-crystallized mica and kaolinite. (3) Water. Distilled. Sea water. 34.85% salinity.

The amount of gas in the saturated solution was measured by transfer of a sample volume to a loop at the system pressure, followed by flashing the sample in an expansion loop and measuring the gas pressure in a known volume. The total gas volume and pressure change were used to compute the moles of methane assuming ideal behavior. A correction was made for the gas not released on flashing.

ESTIMATED ERROR:
$\delta T/K = \pm 0.1$ $\delta p_1/psia = \pm 1$ $\delta m_1/mol\ kg^{-} = \pm 0.0003$ Av. ± 0.0005 Max.

REFERENCES:

COMPONENTS:	ORIGINAL MEASUREMENTS:
1. Methane; CH_4; [74-82-8] 2. Ethane; C_2H_6; [74-84-0] 3. Propane; C_3H_8; [74-98-6] 4. Water; H_2O; [7732-18-5]	Amirijafari, B.; Campbell, J. M. *Soc. Pet. Engnrs. J.* <u>1972</u>, 12, 21-27.

VARIABLES:	PREPARED BY:
	C. L. Young

EXPERIMENTAL VALUES:

T/K (T/°F)	P/psi	P/MPa	$10^3 \times$ Mole fraction in liquid			Mole fraction in vapor[a]		
			$10^3 x_{CH_4}$	$10^3 x_{C_2H_6}$	$10^3 x_{C_3H_8}$	y'_{CH_4}	$y'_{C_2H_6}$	$y'_{C_3H_8}$
377.59 (220)	665	4.59	1.178	0.398	0.019	0.5023	0.2527	0.2450
	2065	14.24	2.068	0.700	0.032	0.5023	0.2527	0.2450
	3015	20.79	2.370	0.803	0.041	0.5023	0.2527	0.2450
	4015	27.68	2.676	0.906	0.042	0.5023	0.2527	0.2450
	5015	34.58	2.942	0.998	0.039	0.5023	0.2527	0.2450
	665	4.59	1.276	0.271	0.048	0.6110	0.1780	0.2110
	2065	14.24	2.244	0.476	0.086	0.6110	0.1780	0.2110
	3015	20.79	2.570	0.546	0.102	0.6110	0.1780	0.2110
	4015	27.68	2.882	0.613	0.111	0.6110	0.1780	0.2110
	5015	34.58	3.162	0.672	0.123	0.6110	0.1780	0.2110
344.26 (160)	5000	34.47	2.780	0.323	0.198	0.7015	0.1065	0.1920
360.93 (190)	5000	34.47	2.830	0.329	0.206	0.7015	0.1065	0.1920
377.59 (220)	5000	34.47	3.158	0.367	0.225	0.7015	0.1065	0.1920
	665	4.59	1.175	0.137	0.083	0.7015	0.1065	0.1920
	2065	14.24	2.234	0.260	0.161	0.7015	0.1065	0.1920
	3015	20.79	2.614	0.304	0.161	0.7015	0.1065	0.1920
	4015	27.68	2.882	0.328	0.205	0.7015	0.1065	0.1920
	665	4.59	1.230	0.119	0.040	0.8218	0.0945	0.0837

(cont.)

[a] Mole fraction of hydrocarbon in water-free gas phase.

AUXILIARY INFORMATION

METHOD/APPARATUS/PROCEDURE:	SOURCE AND PURITY OF MATERIALS:
Static stainless steel equilibrium vessel of approximately 75 mL capacity. Pressure measured with Bourdon gauge and temperature measured with thermocouple. Samples of liquid and vapor analysed using a gas chromatograph equipped with a flame ionisation detector. Poropak R column used.	1, 2 and 3. Pure grade samples, purity 99.9 mole per cent. 4. No details given.
	ESTIMATED ERROR: $\delta T/K = \pm 0.03$; $\delta P/MPa = \pm 1\%$; δx, $\delta y = \pm 2\%$.
	REFERENCES:

COMPONENTS:	ORIGINAL MEASUREMENTS:
1. Methane; CH_4; [74-82-8]	Amirijafari, B.; Campbell, J. M.
2. Ethane; C_2H_6; [74-84-0]	*Soc. Pet. Engnrs. J.*
3. Propane; C_3H_8; [74-98-6]	<u>1972</u>, 12, 21-27.
4. Water; H_2O; [7732-18-5]	

EXPERIMENTAL VALUES:

T/K $(T/°F)$	P/psi	P/MPa	$10^3 \times$ Mole fraction in liquid			Mole fraction in vapor[a]		
			$10^3 x_{CH_4}$	$10^3 x_{C_2H_6}$	$10^3 x_{C_3H_8}$	y'_{CH_4}	$y'_{C_2H_6}$	$y'_{C_3H_8}$
377.59 (220)	2065	14.24	2.000	0.194	0.056	0.8218	0.0945	0.0837
	3015	20.79	2.668	0.258	0.079	0.8218	0.0945	0.0837
	4015	27.68	2.842	0.275	0.087	0.8218	0.0945	0.0837
	5015	34.58	3.276	0.317	0.099	0.8218	0.0945	0.0837
	665	4.59	0.768	0.596	0.216	0.2594	0.3558	0.3848
	2065	14.24	1.305	1.002	0.348	0.2594	0.3558	0.3848
	3015	20.79	1.508	1.170	0.427	0.2594	0.3558	0.3848
	4015	27.68	1.635	1.269	0.461	0.2594	0.3558	0.3848
	5015	34.58	1.800	1.396	0.607	0.2594	0.3558	0.3848

[a] Mole fraction of hydrocarbon in water-free gas phase.

COMPONENTS:	ORIGINAL MEASUREMENTS:
1. Methane; CH_4; [74-82-8] 2. Butane; C_4H_{10}; [106-97-8] 3. Water; H_2O; [7732-18-5]	McKetta, J. J.; Katz, D. L. *Ind. Eng. Chem.* 1948, *40*, 853-862.

VARIABLES:	PREPARED BY:
	C. L. Young

EXPERIMENTAL VALUES: Experimental data in the three-phase region.

T/K (T/°F)	P/MPa (P/psi)	Phase[a]	Mole fraction[b] x_{CH_4}	$x_{C_4H_{10}}$	x_{H_2O}
310.9 (100)	4.35 (631)	V Lh Lw	0.8710 0.2076 0.000649	0.1290 0.7924 0.0000594	0.00184 0.000838 0.99929
	1.39 (202)	V Lh Lw	0.7850 0.0694 0.0002635	0.2150 0.9306 0.000087	0.00508 0.000654 0.9996495
	9.69 (1406)	V Lh Lw	0.8580 0.4505 0.001505	0.1420 0.5495 0.000045	0.001028 0.000864 0.99845
	6.75 (979)	V Lh Lw	0.8650 0.3070 0.001051	0.1350 0.6930 0.000065	0.001315 0.000871 0.998884
	3.27 (474)	V Lh Lw	0.8470 0.1479 0.00072	0.1530 0.8521 0.000080	0.002315 0.000809 0.9992
	1.46 (212)	V Lh Lw	0.7125 0.0480 0.000299	0.2875 0.9520 0.0000507	0.00485 0.000692 0.99965
	14.13 (1838)	V Lh Lw	0.8140 0.6240 0.001805	0.1860 0.3760 0.000033	0.000858 0.000824 0.998162
(cont.)	13.10 (1900)	V Lh Lw	0.7770 0.6690 0.001818	0.2230 0.3310 0.0000295	0.0008405 0.000831 0.99815

AUXILIARY INFORMATION

METHOD/APPARATUS/PROCEDURE:	SOURCE AND PURITY OF MATERIALS:
Equilibrium cell of approx. 1 dm³ capacity fitted with glass window, 3 sampling ports, mercury injection port and stirring mechanism. Temperature measured with copper-constantan thermocouples. Pressure measured with Bourdon gauge. Methane then butane and water charged into cell. Samples taken after equilibrium established. Samples analysed by removing water and weighing the hydrocarbon and determining amounts from knowledge of mass, volume and pressure of gas. Details in source.	1. Phillips Petroleum Co. sample, purity 99.9 mole per cent, dried. 2. Phillips Petroleum Co. sample, purity 99.9 mole per cent, dried. 3. Redistilled in atmosphere of methane.

ESTIMATED ERROR:
$\delta T/K = \pm 0.12$; $\delta P/MPa = \pm 0.03$;
δx_{CH_4}, $\delta x_{C_4H_{10}}$, δy_{CH_4}, $\delta y_{C_4H_{10}} = \pm 1\%$
(estimated by compiler).

REFERENCES:

COMPONENTS:	ORIGINAL MEASUREMENTS:
1. Methane; CH_4; [74-82-8]	McKetta, J. J.; Katz, D. L.
2. Butane; C_4H_{10}; [106-97-8]	*Ind. Eng. Chem.*
3. Water; H_2O; [7732-18-5]	1948, *40*, 853-862.

EXPERIMENTAL VALUES:

Experimental data in the three-phase region.

T/K (T/°F)	P/MPa (P/psi)	Phase[a]	Mole fraction[b]		
			x_{CH_4}	$x_{C_4H_{10}}$	x_{H_2O}
310.9 (100)	13.17 (1910)	V	0.7490	0.2510	0.000831
		Lh	0.6975	0.3025	0.000832
		Lw	0.01852	0.000028	0.99812
	13.18 (1912)	V	0.7390	0.2610	0.000835
		Lh	0.7130	0.2870	0.000835
		Lw	0.001852	0.000028	0.99812
	12.96 (1880)	V	0.8040	0.1960	0.000841
		Lh	0.6420	0.3580	0.000828
		Lw	0.001822	0.0000314	0.99815
	8.41 (1220)	V	0.8790	0.1210	0.00106
		Lh	0.3860	0.6140	0.00082
		Lw	0.00129	0.00006	0.99865
344.3 (160)	11.60 (1683)	V	0.7425	0.2575	0.0038
		Lh	0.5105	0.4895	0.0035
		Lw	0.00149	0.00072	0.99844
	10.20 (1479)	V	0.7810	0.2190	0.004295
		Lh	0.4380	0.5620	0.0035
		Lw	0.00135	0.00001	0.99858
	7.05 (1022)	V	0.7960	0.2040	0.00539
		Lh	0.2938	0.7062	0.00309
		Lw	0.00099	0.000091	0.99892
	3.69 (535)	V			
		Lh	0.1348	0.8652	0.0030
		Lw	0.00052	0.000098	0.99938
	1.32 (192)	V	0.3360	0.6640	0.02210
		Lh	0.0266	0.9734	0.00242
		Lw	0.000015	0.00021	0.99977
	11.27 (1635)	V	0.7695	0.2305	0.00398
		Lh	0.4965	0.5035	0.00352
		Lw	0.001462	0.000064	0.99847
	11.92 (1729)	V	0.7155	0.2845	0.00366
		Lh	0.5330	0.4670	0.00355
		Lw	0.001535	0.000064	0.9984
	7.25 (1051)	V	0.7990	0.2010	0.00538
		Lh	0.2975	0.7025	0.00327
		Lw	0.000975	0.000133	0.99889
	12.38 (1796)	V	0.6645	0.3355	0.00354
		Lh	0.5895	0.4205	0.00370
		Lw	0.001585	0.000048	0.99863
	4.16 (604)	V	0.7445	0.2554	0.00782
		Lh	0.1568	0.8432	0.00280
		Lw	0.000555	0.000165	0.99928
	12.48 (1810)	V	0.6475	0.3525	0.00359
		Lh	0.60000	0.4000	0.00360
		Lw	0.001602	0.000046	0.99835
377.6 (220)	8.56 (1241)	V	0.5790	0.4210	0.01535
		Lh	0.3202	0.6898	0.01238
		Lw	0.000998	0.000205	0.9988
	9.07 (1316)	V	0.5770	0.4230	0.01485
		Lh	0.3476	0.6524	0.01255
		Lw	0.001091	0.000156	0.99875
	7.35 (1066)	V			
		Lh	0.2620	0.7380	0.01148
		Lw	0.000929	0.000119	0.99895
	5.90 (855)	V	0.5875	0.4125	0.02145
		Lh	0.1918	0.8082	0.01108
		Lw	0.000626	0.000256	0.99912

(cont.)

COMPONENTS: ORIGINAL MEASUREMENTS:

1. Methane; CH_4; [74-82-8]   McKetta, J. J.; Katz, D. L.

2. Butane; C_4H_{10}; [106-97-8]   *Ind. Eng. Chem.*

3. Water; H_2O; [7732-18-5]   1948, *40*, 853-862.

EXPERIMENTAL VALUES:

Experimental data in the three-phase region.

T/K (T/°F)	P/MPa (P/psi)	Phase[a]	Mole fraction[b]		
			x_{CH_4}	$x_{C_4H_{10}}$	x_{H_2O}
377.6 (220)	3.45 (500)	V	0.4268	0.5732	0.0324
		Lh	0.0775	0.9225	0.0098
		Lw	0.00027	0.000266	0.99946
	2.33 (338)	V	0.2380	0.7620	0.04365
		Lh	0.0267	0.9733	0.00882
		Lw	0.000185	0.000175	0.99964
	10.48 (1520)	V	0.5245	0.4755	0.01298
		Lh	0.4545	0.5455	0.01288
		Lw	0.001304	0.000098	0.9986
	10.07 (1460)	V	0.5550	0.4450	0.0136
		Lh	0.4105	0.5895	0.01172
		Lw	0.00117	0.000137	0.9987
	4.49 (651)	V	0.5335	0.4665	0.0287
		Lh	0.1264	0.8736	0.102
		Lw	0.00052	0.00019	0.99929
	2.08 (301)	V	0.1618	0.8382	0.0494
		Lh	0.01555	0.9845	0.0088
		Lw	0.00005	0.000286	0.99966
	10.34 (1499)	V	0.5350	0.4650	0.01348
		Lh	0.4360	0.5640	0.01276
		Lw	0.001258	0.0001305	0.99861
410.9 (280)	4.59 (665)	V	0.2220	0.7780	0.0606
		Lh	0.0906	0.9094	0.0341
		Lw	0.000286	0.00037	0.999344
	6.17 (895)	V	0.2406	0.7594	0.0485
		Lh	0.1680	0.8320	0.04285
		Lw	0.000455	0.000405	0.99914
	4.96 (720)	V	0.2358	0.7642	0.05725
		Lh	0.1067	0.8933	0.03705
		Lw	0.000375	0.000325	0.99930
	3.45 (500)	V	0.0836	0.9164	0.0718
		Lh	0.0053	0.9947	0.0274
		Lw	0.00006	0.00043	0.99951
	6.54 (948)	V	0.2348	0.7652	0.0439
		Lh	0.1880	0.8120	0.0456
		Lw	0.00055	0.00037	0.99908
	5.55 (805)	V	0.2428	0.7572	0.0536
		Lh	0.1378	0.8622	0.03815
		Lw	0.00046	0.00033	0.99921
	4.19 (608)	V	0.1977	0.8023	0.066
		Lh	0.0624	0.9376	0.0318
		Lw	0.00006	0.00052	0.99942

(cont.)

[a] *V* - vapor; *Lh* - liquid (hydrocarbon-rich); *Lw* - liquid (water-rich).

[b] In *V* and *Lh* phases the mole fraction of methane and butane are given on the dry basis.

COMPONENTS:	ORIGINAL MEASUREMENTS:
1. Methane; CH_4; [74-82-8]	McKetta, J. J.; Katz, D. L.
2. Butane; C_4H_{10}; [106-97-8]	*Ind. Eng. Chem.*
3. Water; H_2O; [7732-18-5]	<u>1948</u>, *40*, 853-862.

EXPERIMENTAL VALUES:

Experimental data in the two-phase region.

T/K (T/°F)	P/MPa (P/psi)	Phase[c]	Mole fraction[d] x_{CH_4}	$x_{C_4H_{10}}$	x_{H_2O}
310.9	24.27	*Fh*	0.7360	0.2640	0.000656
(100)	(3520)	*Lw*	0.00259	0.00011	0.9973
	20.66	*Fh*	0.7325	0.2675	0.00065
	(2996)	*Lw*	0.002414	0.000086	0.9975
	13.78	*Fh*	0.7310	0.2690	0.000804
	(1998)	*Lw*	0.01868	0.000032	0.9981
	20.66	*Fh*	0.0896	0.9104	0.000639
	(2996)	*Lw*	0.00081	0.00169	0.9975
344.3	20.81	*Fh*	0.6300	0.3700	0.00274
(160)	(3018)				
	14.12	*Fh*	0.6302	0.3698	0.00328
	(2048)	*Lw*	0.001682	0.000068	0.99825
	20.33	*Fh*	0.795	0.2005	0.00270
	(2948)	*Lw*	0.002158	0.000062	0.99778
	13.71	*Fh*	0.7996	0.2004	0.00339
	(1988)	*Lw*	0.001724	0.000026	0.99825
	20.67	*Fh*	0.1905	0.8095	0.00254
	(2998)	*Lw*	0.00132	0.0009	0.99778
	13.72	*Fh*	0.1874	0.8126	0.00316
	(1990)	*Lw*	0.00107	0.00068	0.99825
377.6	20.67	*Fh*	0.4915	0.5085	0.00795
(220)	(2998)	*Lw*	0.00208	0.00024	0.99768
	13.78	*Fh*	0.4945	0.5055	0.01056
	(1999)	*Lw*	0.00158	0.00158	0.99825
	20.66	*Fh*	0.0908	0.9092	0.00718
	(2996)	*Lw*	0.00073	0.00173	0.99754
410.9	20.67	*Fh*	0.2190	0.7810	0.0193
(280)	(2998)	*Lw*	0.00157	0.00083	0.9976
	13.78	*Fh*	0.2190	0.7810	0.0253
	(1999)	*Lw*	0.00113	0.00063	0.99824
	20.67	*Fh*	0.0765	0.9235	0.0177
	(2998)	*Lw*	0.00061	0.00184	0.99755

[c] *Fh* - fluid phase.

[d] In the fluid phase the mole fraction of methane
and butane are given on the dry basis.

COMPONENTS:	ORIGINAL MEASUREMENTS:
(1) Methane; CH_4; [74-82-8] (2) Carbon dioxide; CO_2; [124-38-9] (3) Hydrogen sulfide; H_2S; [7783-06-4] (4) Water; H_2O; [7732-18-5]	Froning, H. R.; Jacoby, R. H.; Richards, W. L. *Proc., Ann. Conv., Nat. Gasoline Assn. Am., Tech. Papers* 1963, 42, 32-9. *Chem. Abstr.* 1963, 59, 10812c.

EXPERIMENTAL VALUES:

Temperature		Total Pressure		Vapor Composition			Liquid Composition		
				CH_4	CO_2	H_2S	CH_4	CO_2	H_2S
$t/^0F$	T/K	p/psia	p/MPa	y_1	y_2	y_3	$10^4 x_1$	$10^3 x_2$	$10^3 x_3$
85	302.6	60	0.41	0.6026	0.1928	0.1945	0.55	0.406	1.221
		252	1.74	0.0	0.5320	0.4656	0.0	4.690	11.54
		256	1.79	0.6145	0.1058	0.2774	2.41	0.960	7.448
		256	1.79	0.6340	0.1793	0.1844	2.48	1.570	4.878
		1002	6.91	0.9435	0.01498	0.04088	12.63	0.416	3.068
		1003	6.92	0.6324	0.08595	0.2810	9.43	2.275	19.99
		1006	6.94	0.9489	0.02538	0.02508	12.51	0.689	1.819
		1008	6.95	0.5985	0.1984	0.2025	8.36	5.035	13.77
100	310.9	15.2	0.105	0.6340	0.1926	0.1109	0.13	0.074	0.119
		17.3	0.119	0.0	0.4658	0.4794	0.0	0.252	0.635
		63	0.43	0.6183	0.2725	0.09406	0.58	0.553	0.555
		251	1.73	0.6158	0.1974	0.1831	2.15	1.473	4.138
		252	1.74	0.0	0.5331	0.4632	0.0	3.787	10.09
		256	1.77	0.0	0.2994	0.6969	0.0	2.215	15.48
		258	1.78	0.5979	0.05191	0.3465	2.13	0.408	7.727
		258	1.78	0.6130	0.09286	0.2904	2.25	0.732	7.265
		263	1.81	0.6392	0.04374	0.3135	2.40	0.393	8.276
		268	1.85	0.6171	0.08221	0.2971	2.20	0.732	7.662
		991	6.83	0.8421	0.08582	0.07113	9.90	1.935	4.852
		994	6.85	0.9381	0.01539	0.04546	12.83	0.395	3.140
		1002	6.91	0.9330	0.05185	0.01409	10.92	1.300	1.060
		1005	6.93	0.6093	0.04486	0.3449	8.04	1.048	21.181
		1006	6.94	0.5999	0.1894	0.2097	7.68	4.205	12.663
		1006	6.94	0.9482	0.02608	0.02478	11.74	0.615	1.696
		1011	6.97	0.5922	0.3075	0.09931	7.23	6.471	6.327
		1012	6.98	0.6103	0.08742	0.3013	7.91	1.947	18.475
		1014	6.99	0.9482	0.00879	0.04026	11.98	0.229	3.079
		1015	7.00	0.8284	0.05461	0.1161	10.22	1.274	7.506
115	319.3	64	0.44	0.6047	0.1939	0.1784	0.45	0.304	0.895
		254	1.75	0.0	0.5162	0.4780	0.0	3.119	9.037
		256	1.77	0.6384	0.09117	0.2647	2.09	0.589	5.340
		257	1.77	0.6820	0.1751	0.1372	2.20	1.103	2.719
		989	6.82	0.9477	0.01508	0.03575	10.34	0.294	2.209
		994	6.85	0.8605	0.06820	0.06980	9.42	1.544	4.057
		999	6.89	0.6152	0.08567	0.2977	7.22	1.638	16.68
		1012	6.98	0.6092	0.1895	0.1997	6.88	3.613	11.02
		1014	6.99	0.5987	0.04504	0.3548	7.06	0.861	18.74

The compositions are mole fraction. The water mole fraction in each phase can be obtained by difference.

The paper also gave values of the equilibrium ratio, $K = y_i/x_i$, for each gas at each temperature and pressure of measurment, and values of the total acid gas in the gas phase, $AG = y_2 + y_3$, and the gas phase ratio, $R = y_3/(y_2 + y_3)$.

The equilibrium ratios for each gas were correlated by equations which were a function of $t/^0F, p$/psia, AG, and R. The equations are given on the following page.

COMPONENTS:	ORIGINAL MEASUREMENTS:
(1) Methane; CH_4; [74-82-8]	Froning, H. R.; Jacoby, R. H.; Richards, W. L.
(2) Carbon dioxide; CO_2; [124-38-9]	
(3) Hydrogen sulfide; H_2S; [7783-06-4]	Proc., Ann. Conv., Nat. Gasoline Assn. Am., Tech. Papers 1963, 42, 32-9.
(4) Water; H_2O; [7732-18-5]	Chem. Abstr. 1963, 59, 10812c.

VARIABLES: $T/K = 302.6 - 319.3$ $p_t/MPa = 0.105 - 7.00$	PREPARED BY: H. L. Clever

ADDITIONAL DATA AND COMMENTS:

The data were correlated to obtain the following equations for the equilibrium ratios as functions of $t/°F$, $p_t/psia$, AG, and R. AG and R were defined on the previous page.

$$K_{CH_4} = y_1/x_1 = 306,000/(p_t/psia) + 2.19(t/°F) + 3910(t/°F)/(p_t/psia)$$
$$-145.0 \text{ AG} - 121.6 \text{ R}$$

$$K_{CO_2} = y_2/x_2 = -3500/(p_t/psia) + 0.12(t/°F) + 360.0(t/°F)/(p_t/psia)$$
$$+ 8.30 \text{ AG} - 5825 \text{ R}/(p_t/psia)$$

$$K_{H_2S} = y_3/x_3 = 4.53 - 1087/(p_t/psia) + 110.4(t/°F)/(p_t/psia) + 4.65 \text{ AG}$$

The authors calculate equilibrium ratios, K, for the gas + water systems. To obtain the values the following AG and R values are used.

Methane + water   AG infinitesimal, R = 0.5

Carbon dioxide   AG = 1, R = 0
+ Water
Hydrogen sulfide AG = 1
+ Water

The objective of the study was to evaluate composition effects on K values. For methane the prescence of the acid gases increases the methane solubility in water. At 1000 psia K decreases 6 % as AG increases from 2.5 to 40 % and decreases 9 % as R changes from 0.2 to 0.8.

AUXILIARY INFORMATION

METHOD/APPARATUS/PROCEDURE:	SOURCE AND PURITY OF MATERIALS:
Equilibrium was attained in a 6.6 liter stainless steel bomb equiped with ports to sample both the gas and liquid phases.	(1) Methane. Described as "pure grade".
The systems was evacuated, 1.6 liters of distilled water was drawn in. The gases were added in the order hydrogen sulfide, carbon dioxide and methane. Each gas was added to a predetermined pressure to obatin the desired concentration.	(2) Carbon dioxide. Described as "bone dry".
	(3) Hydrogen sulfide. Described as "purified'.
	(4) Water. Distilled.
The bomb was gently rocked at 7 oscillations per minute in a thermostated bath. Equilibrium was attained in about 30 minutes, the usual mixing time was 60 minutes.	ESTIMATED ERROR: $\delta t/°F = \pm 0.1$ The correlating equations reproduce the data within ± 4 percent.
Both vapor and liquid phases were sampled. The vapor phase was analyzed on a dry basis by mass spectrometry. The liquid phase was extracted and analyzed by mass spectrometry.	REFERENCES:

COMPONENTS:	EVALUATOR:
(1) Methane; CH_4; [74-82-8] (2) Alkanes	H. Lawrence Clever Chemistry Department Emory University Atlanta, GA 30322 USA 1984, January

CRITICAL EVALUATION:

The Solubility of Methane in Alkanes at Partial

Pressures up to 200 kPa (*ca*. 2 atm).

I. The solubility of methane in normal alkanes.

The two methods most commonly used to determine the solubility of methane in alkanes are volumetric methods used at a total pressure of about one atmosphere and gas liquid chromatography methods in which retention times or volumes are measured when the solvent is the stationary phase. Also used was a gas pressure change method at very low methane partial pressure and a gas stripping method with the GLC used as a detector.

With two exceptions the volumetric methods used since 1960 give consistent results which appear to be reliable within several percent. The GLC methods used since 1974 also appear to give reliable results. Earlier work by both methods gave solubility values that appear to be too small sometimes by as much as 40 percent.

The volumetric methods used cover a variety of degassing and equilibration techniques. The solubility values from the early work of McDaniel (ref. 1) are consistently too small. This may be because of poor equilibration in the hand shaken apparatus or incomplete degassing of the solvent.

Guerry (ref. 2) used a modified van Slyke method. The small solvent volumes and large solvent vapor pressure gave problems which resulted in too small solubility values. Tilquin *et al*. (ref. 5) measured pressure changes at low methane partial pressures when degassed solvent and gas were contacted. When the methane solubility is calculated for 101.325 kPa partial pressure, the value appears to be too small for the normal alkane and too large for the branched alkane. It is possible that Henry's Law is not obeyed between the low partial pressure of the measurement and atmospheric pressure, but Henry's law is supported by the GLC results on solvents of higher carbon number. The results of Makranczy *et al*. (ref. 11) are often too large. Their volumetric technique appears to have problems when solvents of relatively large vapor pressure are studied. Both Lannung and Gjaldbaek (ref. 3) and Wilcock *et al*. (ref. 12) usually report reliable solubility values. However, both find a smaller temperature coefficient of solubility and enthalpy of solution for methane in hexane and octane than do most other workers which casts some doubts on their results for these systems. The other solubility values by volumetric methods (ref. 7, 9, 10, and 13) appear to show a consistent and reliable pattern of results.

The GLC retention time studies of Ng *et al*. (ref. 6) and Lenoir *et al*. (ref. 8) give methane solubility values that appear to be too small. The results obtained by Lin and Parcher (ref. 15) from GLC retention volume studies and by Richon and Renon (ref. 14) by gas stripping and GLC detector method appear to give reasonable methane solubility values.

The mole fraction solubility values at 298.15 K and 101.325 kPa methane pressure are given as a function of normal alkane carbon number in Fig. 1. Although the values show considerable scatter, we believe that there are enough reliable values measured by traditional volumetric methods near atmospheric pressure to allow a reliable line to be placed through the data. Octadecane melts at 301.33 K, thus the normal alkanes of C_{18} and larger melt at temperatures greater than 298.15 K. The solubility values of methane in hydrocarbons of carbon number 18 and greater are for a hypothetical liquid hydrocarbon. The values were estimated from solubility data at higher temperatures by assuming $\ln x_1$ vs $1/(T/K)$ is linear. The

COMPONENTS:	EVALUATOR:
(1) Methane; CH_4; [74-82-8] (2) Alkanes	H. Lawrence Clever Chemistry Department Emory University Atlanta, GA 30322 USA 1984, January

CRITICAL EVALUATION:

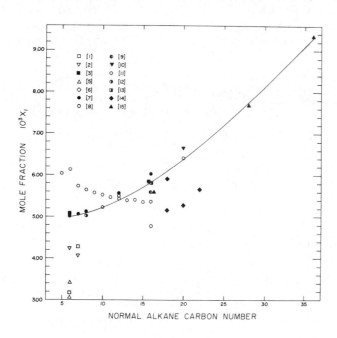

Figure 1. The mole fraction solubility of methane at 298.15 K and 0.1013 MPa partial pressure in the normal alkanes.

The C_{18} and above hydrocarbons are solids at 298.15 K. The solubility values are extrapolated from experimental values in the liquid state at higher temperatures.

The numbers refer to the references at the end of the evaluation.

solubility of Tilquin *et al.* (ref. 5) in hexane was adjusted assuming Henry's law and an average enthalpy of solution to estimate a 298 K value from the experimental value at 288 K.

The line drawn in Fig. 1 must be considered as only an approximation as to how the methane mole fraction solubility at 298.15 K and 101.325 kPa methane pressure changes with carbon number. The steady increase in solubility as the carbon number increases seems reasonable. The mole fraction solubility does increase more rapidly with carbon number than does the contact surface of the hydrocarbon.

The partial molal enthalpy of solution of methane in normal alkanes averages (-4.1 ± 0.5) kJ mol^{-1} at 298.15 K for hydrocarbons C_6 through C_{20}. The enthalpy values in heptane and hexadecane indicate the methane solubility values in these solvents are not quite consistent with the data on the other systems.

The work of Cukor and Prausnitz (ref. 9) and Chappelow and Prausnitz (ref. 10) indicate the enthalpy of solution decreases in magnitude as the temperature increases. Values of the partial molal enthalpies of solution of methane in normal alkanes from a three constant equation fitted to the data are about

T/K	298	323	373	473
$\Delta \overline{H}_1^{\circ}$/kJ mol^{-1}	-4.1	-3.1	-2.2	-0.4

The individual systems are discussed in more detail below.

Methane + Pentane; C_5H_{12}; [109-66-0]

The mole fraction solubility of 6.02×10^{-3} at 298.15 K and 101.32 101.325 kPa methane pressure of Makranczy *et al.* (ref. 11) is the only experimental value for the system. Although it is classed as tentative, we believe it may be as much as 15-20 percent too large.

Methane + Hexane; C_6H_{14}; [110-54-3]

The solubility of methane in hexane is reported from six laboratories. The early work of McDaniel (ref. 1) and Guerry (ref. 2) is rejected. Their solubility values at temperatures between 293 and 333 K are much too small. The single value of Tilquin *et al.* (ref. 5) at 288.15 is also rejected, however, it may have some validity as a distribution coefficient at small methane partial pressures. The single value of Makranczy *et al.* (ref. 11) at 298.15 K is classed as doubtful. It may be 15-20 percent too large.

The values of Lunnung and Gjaldbaek (ref. 3) at 291.15, 298.15, and 310.15 K and of Hayduk and Buckley (ref. 7) over the 273.15 to 323.15 K interval are classed as tentative. There is an inconsistency in the two data sets in that the Lannung and Gjaldbaek data gives an enthalpy of solution of -2.1 kJ mol^{-1} while the Hayduk and Buckley data gives a value of -4.7 kJ mol^{-1}. The Lannung, Gjaldbaek data extends over a much shorter temperature interval. The two studies appear internally consistent. Although the temperature coefficients of solubility differences casts some doubts on the results, the mole fraction solubility values from the two papers were combined in a linear regression to obtain the tentative equation for the mole fraction solubility over the 273.15 to 323.15 K interval.

$$\ln x_1 = -6.92496 + 4.85402/(T/100 \text{ K})$$

With a standard error about the regression line of 1.60×10^{-4}. From the equation, the temperature independent thermodynamic quantities are

$$\Delta \overline{H}_1^{\circ}/ \quad \text{kJ mol}^{-1} = -4.04 \text{ and } \Delta \overline{S}_1^{\circ}/ \quad \text{J K}^{-1} \text{ mol}^{-1} = -57.6$$

COMPONENTS:	EVALUATOR:
(1) Methane; CH_4; [74-82-8] (2) Alkanes	H. Lawrence Clever Chemistry Department Emory University Atlanta, GA 30322 USA 1984, January

CRITICAL EVALUATION:

Smoothed values of the mole fraction solubility and partial molar Gibbs energy of solution are in Table 1.

Table 1. Solubility of methane in hexane at a methane partial pressure of 101.325 kPa (1 atm). Tentative mole fraction solubility and partial molal Gibbs energy of solution as a function of temperature.

T/K	Mol Fraction $10^3 x_1$	$\Delta \bar{G}_1^\circ/kJ\ mol^{-1}$
273.15	5.81	11.691
283.15	5.46	12.267
293.15	5.15	12.843
298.15	5.01	13.131
303.15	4.87	13.418
313.15	4.63	13.994
323.15	4.41	14.570

Methane + Heptane; C_7H_{16}; [142-82-5]

McDaniel (ref. 1) reports the solubility of methane in heptane at 295.35, 303.25, and 313.15 K, Guerry (ref. 2) at 293.15 and 298.15 K, Hayduk and Buckley (ref. 7) at 298.15, 323.15 and 348.15 K, and Makranczy et al. (ref. 11) at 298.15 K. Again the McDaniel and the Guerry data are rejected as too small. The other data are classed as tentative, but the data of Hayduk and Buckley are preferred.

The tentative data below are based entirely on the three measurements of Hayduk and Buckley. The mole fraction solubility values were fitted by a linear regression to obtain the tentative equation for the 298.15 to 348.15 K temperature interval

$$\ln x_1 = -7.28446 + 5.95159/(T/100\ K)$$

with standard error about the regression line of 2.18×10^{-5}. From the equation the temperature independent thermodynamic changes are

$$\Delta \bar{H}_1^\circ/kJ\ mol^{-1} = -4.95 \text{ and } \Delta \bar{S}_1^\circ/J\ K^{-1}\ mol^{-1} = -60.6$$

Smoothed values of the mole fraction solubility and partial molar Gibbs energy of solution are in Table 2.

Table 2.   Solubility of methane in heptane at a partial pressure of
            101.325 kPa (1 atm) methane. Tentative mole fraction
            solubility and partial molal Gibbs energy as a function
            of temperature.

T/K	Mol Fraction $10^3 x_1$	$\Delta \bar{G}_1^{\circ}/kJ\ mol^{-1}$
298.15	5.05	13.109
303.15	4.89	13.412
313.15	4.59	14.017
323.15	4.33	14.623
333.15	4.09	15.229
343.15	3.89	15.835

Methane + Octane; C_8H_{18}; [111-65-9]

    Hayduk and Buckley (ref. 7) report the solubility of methane in
octane at four temperatures between 273.15 and 348.15 K, Makranczy *et al.*
(ref. 11) report one solubility value at 298.15 K and Wilcock *et al.*
(ref. 12) report values 298.15 and 313.35 K. All are classed as tentative but the data from (ref. 7 and 12) are preferred.

    The temperature coefficients of solubility differ in the two studies.
Wilcock *et al.* data are consistent with an enthalpy of solution of
-2.97 kJ mol^{-1} and the Hayduk and Buckley a value of -4.16 kJ mol^{-1}.   The
Wilcock *et al.* enthalpy is based on only two experimental points
15 degrees apart and thus subject to some error.

    The six experimental values from Hayduk and Buckley and from Wilcock
et al. were combined in a linear regression to obtain the tentative
equation

    $\ln x_1 = -6.94961 + 4.95903/(T/100\ K)$

with a standard error around the regression line of 5.37 x 10^{-5}. The
temperature independent thermodynamic changes from the equation are

    $\Delta \bar{H}_1^{\circ}/kJ\ mol^{-1} = -4.12$ and $\Delta \bar{S}_1^{\circ}/J\ K^{-1}\ mol^{-1} = -57.8$

Smoothed values of the mole fraction solubility and partial molar Gibbs
energy of solution are in Table 3.

Table 3.   Solubility of methane in octane. Tentative values of the
            mole fraction solubility at 101.325 kPa (1 atm) partial
            pressure methane and partial molal Gibbs energy as a func-
            tion of temperature.

T/K	Mol Fraction $10^3 x_1$	$\Delta \bar{G}_1^{\circ}/kJ\ mol^{-1}$
273.15	5.89	11.660
283.15	5.53	12.238
293.15	5.21	12.815
298.15	5.06	13.104
303.15	4.92	13.393
313.15	4.67	13.971
323.15	4.45	14.549
333.15	4.25	15.127
343.15	4.07	15.705

COMPONENTS:	EVALUATOR:
(1) Methane; CH_4; [74-82-8] (2) Alkanes	H. Lawrence Clever Chemistry Department Emory University Atlanta, GA 30322 USA 1984, January

CRITICAL EVALUATION:

Methane + Nonane; C_9H_{20}; [111-84-2]

The mole fraction solubility of 5.57×10^{-3} at 298.15 K and 101.325 kPa methane pressure of Makranczy *et al.* (ref. 11) is the only experimental value reported for the system. Although it is classed as tentative, the evaluator believes it may be as much as 10 percent too large.

Methane + Decane; $C_{10}H_{22}$; [124-18-5]

Makranczy *et al.* (ref. 11) report the solubility of methane in decane at 298.15 K, Wilcock *et al.* (ref. 12) report the solubility at 282.80 and 313.15 K. Both sets of data are classed as tentative but the Wilcock *et al.* data are preferred.

The tentative values are based on the Wilcock *et al.* data. The equation for the mole fraction solubility between 282.80 and 313.15 K is

$$\ln x_1 = -7.0050 + 5.2154/(T/100 \text{ K})$$

and the corresponding temperature independent thermodynamic changes are

$$\Delta \overline{H}_1^\circ/\text{kJ mol}^{-1} = -4.34 \quad \text{and} \quad \Delta \overline{S}_1^\circ/\text{J K}^{-1} \text{ mol}^{-1} = -58.2$$

Tentative values of the mole fraction solubility and partial molal Gibbs energy are given in Table 4.

Table 4. The solubility of methane in decane. Tentative values of the mole fraction solubility at 101.325 kPa (1 atm) partial pressure methane and partial molal Gibbs energy of solution as a function of temperature.

T/K	Mol Fraction $10^3 x_1$	$\Delta \overline{G}_1^\circ/\text{kJ mol}^{-1}$
283.15	5.72	12.135
293.15	5.37	12.737
298.15	5.22	13.029
303.15	5.07	13.320
313.15	4.80	13.902

Methane + Undecane; $C_{11}H_{24}$; [1120-21-4]

The mole fraction solubility of 5.46×10^{-3} at 298.15 K and 101 101.325 kPa (1 atm) methane pressure of Makranczy *et al.* (ref. 11) is the only experimental value reported for the system. It is classed as tentative.

Methane + Dodecane; $C_{12}H_{26}$; [112-40-3]

Hayduk and Buckley (ref. 7) report four solubility values at temperatures between 273.15 and 348.15 K. Makranczy *et al.* (ref. 11) report one value at 298.15 K. All of the data are classed as tentative.

The data were fitted by a linear regression to obtain the equation for the 273.15 to 348.15 K temperature interval

$$\ln x_1 = -6.76819 + 4.63437/(T/100 \text{ K})$$

with a standard deviation about the regression line of 4.3×10^{-5}. The temperature independent thermodynamic changes from the equation are

$$\Delta \overline{H}_1^\circ / \text{kJ mol}^{-1} = -3.85 \quad \text{and} \quad \Delta \overline{S}_1^\circ / \text{J K}^{-1} \text{mol}^{-1} = -56.3$$

Smoothed values of the mole fraction solubility and partial molal Gibbs energy of solution are in Table 5.

Table 5. Solubility of methane in dodecane. Tentative values of the mole fraction solubility at 101.325 kPa (1 atm) methane partial pressure and partial molal Gibbs energy of solution as a function of temperature.

T/K	Mol Fraction $10^3 x_1$	$\Delta \overline{G}_1^\circ / \text{kJ mol}^{-1}$
273.15	6.27	11.518
283.15	5.91	12.080
293.15	5.59	12.643
298.15	5.44	12.925
303.15	5.30	13.206
313.15	5.05	13.769
323.15	4.82	14.331
333.15	4.62	14.894
343.15	4.44	15.457

Methane + Tridecane; $C_{13}H_{28}$; [629-50-5]

Methane + Tetradecane; $C_{14}H_{30}$; [629-59-4]

Methane + Pentadecane; $C_{15}H_{32}$; [629-62-9]

Only Makranczy *et al.* (ref. 11) have reported data on these systems. They report the solubility at 298.15 K. The mole fraction solubility at 298.15 K and 101.325 kPa (1 atm) methane pressure from their measurements is as follows:

Tridecane	5.39×10^{-3}
Tetradecane	5.40×10^{-3}
Pentadecane	5.35×10^{-3}

The data are classed as tentative.

Methane + Hexadecane; $C_{16}H_{34}$; [544-76-3]

Seven papers report solubility data on the $CH_4 + C_{16}H_{34}$ system. The mole fraction solubility values at 298.15 K and 101.325 kPa methane pressure are

COMPONENTS:	EVALUATOR:
(1) Methane; CH_4; [74-82-8] (2) Alkanes	H. Lawrence Clever Chemistry Department Emory University Atlanta, GA 30322 USA 1984, January

CRITICAL EVALUATION:

6.02×10^{-3} Hayduk, Buckley (ref. 7)

5.841×10^{-3} Richon, Renon (ref. 14)

5.824×10^{-3} Rivas, Prausnitz (ref. 13)

5.75×10^{-3} (300K)   Cukor, Prausnitz (ref. 9)

5.59×10^{-3} Lin, Parcher (ref. 15)

5.36×10^{-3} Makranczy et $al.$ (ref. 11)

4.78×10^{-3} Lenoir et $al.$ (ref. 8)

The larger five values average a mole fraction solubility of 5.81×10^{-3} with standard deviation 0.16×10^{-3}.

The data of Hayduk and Buckley, Richon and Renon, Rivas and Prausnitz, and Lin and Parcher were combined in a linear regression to obtain the equation

$$\ln x_1 = -10.68231 + 9.91533/(T/100 \text{ K}) + 2.02051 \ln (T/100 \text{ K})$$

for the 298.15 to 473.15 K temperature interval.

The equation gives the following thermodynamic changes for the solution process:

T/K	$\Delta \bar{H}_1^\circ$/kJ mol^{-1}	$\Delta \bar{S}_1^\circ$/J K^{-1} mol^{-1}	$\Delta \bar{C}_{p_1}^\circ$ /J K^{-1} mol^{-1}
298.15	-3.24	-53.7	16.8
323.15	-2.82	-52.3	16.8
348.15	-2.40	-51.1	16.8
373.15	-1.98	-49.9	16.8
398.15	-1.56	-48.8	16.8
423.15	-1.14	-47.8	16.8
448.15	-0.72	-46.8	16.8
473.15	-0.30	-45.9	16.8

The smoothed mole fraction solubility and partial molar Gibbs energy values are in Table 6.

Table 6. Solubility of methane in hexadecane. Tentative values of the mole fraction solubility at 101.325 kPa (1 atm) methane partial pressure partial molal Gibbs energy of solution as a function of temperature.

T/K	Mol Fraction $10^3 x_1$	$\Delta \bar{G}_1^\circ$/kJ mol^{-1}
298.15	5.80	12.765
323.15	5.28	14.089
348.15	4.92	15.381
373.15	4.68	16.643
398.15	4.52	17.877
423.15	4.41	19.084
448.15	4.34	20.267
473.15	4.31	21.425

The smoothed solubility values differ only a few percent from the values of Cukor and Prausnitz and the enthalpies of solution are 3.5 to 7 percent less negative than the Cukor and Prausnitz values up to 400 K. The present evaluation does not show the change to a positive enthalpy of solution at 475 K. The Cukor and Prausnitz data are a good alternative to the tentative data presented here.

Methane + Octadecane; $C_{18}H_{38}$; [593-45-3]

Ng, Harris and Prausnitz (ref. 6) used gas liquid chromatography retention times to estimate the solubility of methane in octadecane at six temperatures between 308.2 and 423.2 K. Richon and Renon (ref. 14) report a mole fraction solubility value of 5.079×10^{-3} at 323.15 K and 101.325 kPa methane pressure. The Richon and Renon value is classed as tentative and the Ng *et al.* data as doubtful. Unfortunately there is some problem with the Ng *et al.* experiment and their results for this system as well as the methane + eicosane and docosane systems appear to be too small mole fraction solubilities by 15 to 30 percent.

Methane + Eicosane; $C_{20}H_{42}$; [112-95-8]

Both Ng *et al.* (ref. 6) and Chappelow and Prausnitz (ref. 10) report solubility data as a function of temperature. As discussed for the previous system the Ng *et al.* data are doubtful. The Chappelow and Prausnitz data are classed as tentative.

The Chappelow and Prausnitz data were treated by a linear regression to obtain the equation

$$\ln x_1 = -11.65137 + 11.01303/(T/100 \text{ K}) + 2.41842 \ln (T/100 \text{ K})$$

with a standard error about the regression line of 9.1×10^{-6} for the 323.15 to 473.15 K temperature interval.

The equation gives the following thermodynamic changes for the solution process:

T/K	$\Delta \bar{H}_1^\circ/\text{kJ mol}^{-1}$	$\Delta \bar{S}_1^\circ/\text{J K}^{-1} \text{ mol}^{-1}$	$\Delta \bar{C}_{p_1}^\circ /\text{J K}^{-1} \text{ mol}^{-1}$
298.15	-3.91[a]	-54.8[a]	20.1[a]
323.15	-3.41	-53.2	20.1
348.15	-2.90	-51.7	20.1
373.15	-2.40	-50.3	20.1
398.15	-1.90	-49.0	20.1
423.15	-1.40	-47.8	20.1
448.15	-0.88	-46.6	20.1
473.15	-0.39	-45.5	20.1

[a] Hypothetical liquid state. Eicosane melts at 310.0 K.

The smoothed mole fraction solubilities and partial molal Gibbs energy values are in Table 7.

Table 7. Solubility of methane in eicosane. Tentative mole fraction solubility at 101.325 kPa (1 atm) methane pressure and partial molal Gibbs energy of solution as a function of temperature.

T/K	Mol Fraction $10^3 x_1$	$\Delta \bar{G}_1^\circ/\text{kJ mol}^{-1}$
298.15	6.65[a]	12.429[a]
323.15	5.93	13.778
348.15	5.45	15.089
373.15	5.12	16.363
398.15	4.90	17.358
423.15	4.76	18.813
448.15	4.67	19.992
473.15	4.63	21.143

[a] For hypothetical liquid state. Eicosane melts at 310.0 K.

COMPONENTS:	EVALUATOR:
(1) Methane; CH_4; [74-82-8] (2) Alkanes	H. Lawrence Clever Chemistry Department Emory University Atlanta, GA 30322 USA 1984, January

CRITICAL EVALUATION:

Methane + Docosane; $C_{22}H_{46}$; [629-98-0]

The only data on this system is from the work of Ng *et al.* (ref. 6) which appears to give mole fraction values 15 to 30 percent too small with the octadecane and eicosane systems discussed earlier. Although the data are doubtful, they are presented because they are the only data on the system.

The data were fitted to a two constant equation by a linear regression

$$\ln x_1 = -7.29746 + 6.32687/(T/100 \text{ K})$$

with a standard error about the regression line at 1.50×10^{-4}.

The corresponding temperature independent thermodynamic changes are

$$\Delta H_1^\circ/\text{kJ mol}^{-1} = -5.26 \quad \text{and} \quad \Delta S_1^\circ/\text{J K}^{-1} \text{ mol}^{-1} = -60.8$$

Smoothed values of the mole fraction solubility and partial molal Gibbs energy are given in Table 8. The values are probably 15 to 30 percent smaller than the true values.

Table 8.   The solubility of methane in docosane. Smoothed values of the mole fraction solubility at 101.3 kPa (1 atm) and partial molal Gibbs energy of solution as a function of temperature.

T/K	Mol Fraction $10^3 x_1$	$\Delta \bar{G}_1^\circ/\text{kJ mol}^{-1}$
298.15	5.65[a]	12.829[a]
323.15	4.80	14.346
348.15	4.17	15.863
373.15	3.69	17.380
398.15	3.32	18.897
423.15	3.02	20.414
448.15	2.78	21.930
473.15	2.58	23.447

[a] For hypothetical liquid state. Docosane melts at 317.6 K.

Methane + Octacosane; $C_{28}H_{58}$; [630-02-4]

Only Lin and Parcher (ref. 15) have reported solubility data on the system. They report three solubility values between 353.2 and 393.2 K by a GLC method. The three points were fitted to the two constant equation by linear regression

$$\ln x_1 = -6.06004 + 3.55025/(T/100 \text{ K})$$

with a standard error about the regression line of 2.00×10^{-5}. The equation gives the temperature independent thermodynamic changes

$$\Delta \bar{H}_1^\circ/\text{kJ mol}^{-1} = -2.95 \quad \text{and} \quad \Delta \bar{S}_1^\circ/\text{J K}^{-1} \text{ mol}^{-1} = -50.4$$

The smoothed values of the mole fraction solubility and partial molal Gibbs energy are given in Table 9.

Table 9.   Solubility of methane in octacosane. Tentative values of the
           mole fraction solubility at 101.325 kPa (1 atm) and partial
           molal Gibbs energy of solution as a function of temperature.

T/K	Mol Fraction $10^3 x_1$	$\Delta \bar{G}_1^o/kJ\ mol^{-1}$
298.15	7.68[a]	12.071[a]
353.15	6.38	14.842
363.15	6.20	15.346
373.15	6.04	15.849
383.15	5.90	16.353
393.15	5.76	16.857

[a] Extrapolated hypothetical liquid state. Octacosane melts at 337.7 K.

Methane + Hexatriacontane; $C_{36}H_{74}$; [630-06-8]

The system at temperatures between 353.2 and 413.2 K by a GLC method.
The results are classed as tentative. The data of Lin and Parcher were
treated by a linear regression to obtain the equation.

$$\ln x_1 = -5.93818 + 3.77226/(T/100\ K)$$

with a standard error about the regression line of 5.5×10^{-5}. The
temperature independent thermodynamic changes are

$$\Delta H_1^o/kJ\ mol^{-1} = -3.14 \quad \text{and} \quad \Delta S_1^o/J\ K^{-1}\ mol^{-1} = -49.4.$$

Smoothed values of the mole fraction solubility and partial molal Gibbs
energy of solution are in Table 10.

Table 10. Solubility of methane in hexatriacontane. Tentative mole
           fraction solubility at 101.325 kPa (1 atm) methane partial
           pressure and partial molal Gibbs energy of solution as a
           function of temperature.

T/K	Mol Fraction $10^3 x_1$	$\Delta \bar{G}_1^o/kJ\ mol^{-1}$
298.15	9.34[a]	11.584
353.15	7.67	14.299
363.15	7.45	14.793
373.15	7.25	15.287
383.15	7.06	15.780
393.15	6.88	16.274
403.15	6.72	16.768
413.15	6.57	17.202

[a] Extrapolated hypothetical liquid state. Hexatriacontane melts at 349.

II. The solubility of methane in branched alkanes.

The solubility of methane is reported for only four branched chain
alkanes. All of the available data are consistent with a larger solubility of methane in the branched chain alkane than in the linear alkane
of the same number of carbon atoms. The data for the C_6 and C_{16} branched
hydrocarbon gives a branched/normal methane solubility ratio of greater
than 2 while the C_8 and C_{30} solubility branched/normal ratio is 1.06-1.12.
The smaller ratio appears to be more reasonable.

COMPONENTS:	EVALUATOR:
(1) Methane; CH_4; [74-82-8] (2) Alkanes	H. Lawrence Clever Chemistry Department Emory University Atlanta, GA 30322 USA 1984, January

CRITICAL EVALUATION:

Methane + 2,2-Dimethylbutane; C_6H_{14}; [75-83-2]

Tilquin *et al.* (ref. 5) report measurements at low methane partial pressures from which we calculate a mole fraction solubility of 12.7×10^{-3} at 101.325 kPa methane pressure at 288.15 K. The C_6 branched/normal solubility ratio is 2.4 using the tentative hexane solubility value from this evaluation. Since there is no other value to compare with the present value, it is classed as tentative, but we believe the value is probably too large and that it should be used with caution.

Methane + 2,2,4-Trimethylpentane; C_8H_{18}; [540-84-1]

Hiroka and Hildebrand report the solubility of methane at four temperatures between 277.51 and 308.22 K. The data are classed as tentative. At 298.15 K the branched/normal solubility ratio is 1.06.

The solubility data were fitted by a linear regression to the equation

$$\ln x_1 = -7.29405 + 6.14129/(T/100 \text{ K})$$

with a standard error about the regression line of 3.3×10^{-5}.

The temperature independent thermodynamic changes are

$$\Delta \overline{H}_1^\circ / \text{kJ mol}^{-1} = 5.11 \quad \text{and} \quad \Delta \overline{S}_1^\circ / \text{J K}^{-1} \text{ mol}^{-1} = -60.6$$

Table 11. Solubility of methane in 2,2,4-trimethylpentane. Tentative values of the mole fraction solubility at 101.325 kPa methane partial pressure and partial molal Gibbs energy of solution as a function of temperature.

T/K	Mol Fraction $10^3 x_1$	$\Delta \overline{G}_1^\circ / \text{kJ mol}^{-1}$
283.15	5.95	12.066
293.15	5.52	12.672
298.15	5.33	12.975
303.15	5.15	13.278

Methane + 2,2,4,4,6,8,8-Heptamethylnonane; $C_{16}H_{34}$; [4390-04-9]

Richon and Renon (ref. 14) report measurements from which a mole fraction solubility of 30.5×10^{-3} at 101.325 kPa methane pressure at 298.15 K can be calculated. The branched/normal C_{16} methane solubility ratio is 5.3. Since there is no other value with which to compare the solubility value it is classed as tentative. However, it is suspected that the value is too large and it should be used with caution.

Methane + 2,6,10,15,19,23-Hexamethyltetracosane; $C_{30}H_{62}$; [111-01-3]

Chappelow and Prausnitz (ref. 10) report eight solubility values over the 300 to 475 K temperature interval for this system. Taking the linear C_{30} hydrocarbon solubility from Fig. 1, the branched/normal methane solubility ratio is 1.12. The data are classed as tentative.

The data were fitted to a three constant equation by a linear regression

$$\ln x_1 = -10.32638 + 10.42056/(T/100\ K) + 1.9508 \ln (T/100\ K)$$

The equation reproduces the experimental data so closely that there is reason to believe the data are smoothed data rather than experimental points.

The equation gives the following values of the thermodynamic functions:

T/K	$\Delta \bar{H}_1^\circ/kJ\ mol^{-1}$	$\Delta \bar{S}_1^\circ/J\ K^{-1}\ mol^{-1}$	$\Delta \bar{C}_{p_1}^\circ /J\ K^{-1}\ mol^{-1}$
298.15	-3.83	-51.9	16.2
323.15	-3.42	-50.6	16.2
348.15	-3.02	-49.4	16.2
373.15	-2.61	-48.2	16.2
398.15	-2.21	-47.2	16.2
423.15	-1.80	-46.2	16.2
448.15	-1.40	-45.3	16.2
473.15	-0.99	-44.4	16.2

The smoothed mole fraction solubility and partial molal Gibbs energy of solution are in Table 12.

Table 12. Solubility of methane in 2,6,10,15,19,23-Hexamethylnonane. Tentative values of the mole fraction solubility at 101.325 kPa methane pressure and the partial molal Gibbs energy of solution as a function of temperature.

T/K	Mol Fraction $10^3 x_1$	$\Delta \bar{G}_1^\circ/kJ\ mol^{-1}$
298.15	9.09	11.651
323.15	8.12	12.933
348.15	7.45	14.183
373.15	6.98	15.404
398.15	6.65	16.597
423.15	6.41	17.766
448.15	6.25	18.910
473.15	6.15	20.031

REFERENCES:

1. McDaniel, A. S. J. Phys. Chem. 1911, 15, 587.

2. Guerry, D. Jr. Ph. D. Thesis, 1944, Vanderbilt University, Nashville, TN.

3. Lannung, A.; Gjaldbaek, J. C. Acta Chem. Scand. 1960, 14, 1124.

4. Hiraoka, H.; Hildebrand, J. H. J. Phys. Chem. 1964, 68, 213.

5. Tilquin, B.; Decannière, L.; Fontaine, R.; Claes, P. Ann. Soc. Sc. Bruxelles (Belgium) 1967, 81, 191.

6. Ng, S.; Harris, H. G.; Prausnitz, J. M. J. Chem. Eng. Data, 1969, 14, 482.

7. Hayduk, W.; Buckley, W. D. Can. J. Chem. Eng. 1971, 49, 667.

8. Lenoir, J-Y.; Renault, P.; Renon, H. J. Chem. Eng. Data 1971, 16, 340.

9. Cukor, P. M.; Prausnitz, J. M. J. Phys. Chem. 1972, 76, 598.

COMPONENTS:	EVALUATOR:
(1) Methane; CH_4; [74-82-8] (2) Alkanes	H. Lawrence Clever Chemistry Department Emory University Atlanta, GA 30322 USA 1984, January

CRITICAL EVALUATION:

10. Chappelow, C. C.; Prausnitz, J. M. *Am. Inst. Chem. Engnrs. J.*
     1974, *20*, 1097.

11. Makranczy, J.; Megyery-Balog, K.; Rusz, L.; Patyi, L.
     Hung. J. Ind. Chem. 1976, *4*, 269.

12. Wilcock, R. J.; Battino, R.; Danforth, W. F.; Wilhelm, E.
     J. Chem. Thermodyn. 1978, *10*, 817.

13. Rivas, O. R.; Prausnitz, J. M. *Ind. Eng. Chem. Fundam.*
     1979, *18*, 289.

14. Richon, D.; Renon, H. *J. Chem. Eng. Data* 1980, *25*, 59.

15. Lin, P. J.; Parcher, J. F. *J. Chromatog. Sci.* 1982, *20*, 33.

The evaluation of the solubility of methane in hydrocarbons at high
pressure is given in separate sections later in the volume.

A useful paper to consult for additional information on methane +
hydrocarbons systems is

    Legret, D.; Richon, D.; Renon, H. *Fluid Phase Equilib.* 1984, *17*,
    323-50.

COMPONENTS:	ORIGINAL MEASUREMENTS:
(1) Methane; CH_4; [74-82-8] (2) Pentane; C_5H_{12}; [109-66-0] Hexane; C_6H_{14}; [110-54-3]	Makranczy, J.; Megyery-Balog, K.; Rusz, L.; Patyi, L. *Hung. J. Ind. Chem.* <u>1976</u>, *4*, 269 - 280.

VARIABLES:	PREPARED BY:
T/K: 298.15 p/kPa: 101.325 (1 atm)	S. A. Johnson H. L. Clever

EXPERIMENTAL VALUES:

T/K	Mol Fraction $10^3 x_1$	Bunsen Coefficient α/cm^3(STP)cm^{-3}atm^{-1}	Ostwald Coefficient L/cm^{3}cm^{-3}
Pentane			
298.15	6.03	1.172	1.279
Hexane			
298.15	6.13	1.050	1.146

The Bunsen coefficient and mole fraction values were calculated by the compiler assuming that the gas is ideal and that Henry's law is obeyed.

AUXILIARY INFORMATION

METHOD/APPARATUS/PROCEDURE:	SOURCE AND PURITY OF MATERIALS:
Volumetric method. The apparatus described by Bodor, Bor, Mohai and Sipos was used (1).	Both the gas and the liquid were analytical grade reagents of Hungarian or foreign origin. No further information.
	ESTIMATED ERROR: $\delta L/L$ = ± 0.03
	REFERENCES: 1. Bodor, E.; Bor, Gy.; Mohai, B.; Sipos, G. *Veszpremi Vegyip. Egy. Kozl.* <u>1957</u>, *1*, 55. *Chem. Abstr.* <u>1961</u>, *55*, 3175h.

COMPONENTS:	ORIGINAL MEASUREMENTS:
(1) Methane; CH_4; [74-82-8] (2) Hexane; C_6H_{14}; [110-54-3]	McDaniel, A. S. J. Phys. Chem. <u>1911</u>, 15, 587-610.
VARIABLES: T/K = 295.35 - 333.15 p_1/kPa = 101.3 (1 atm)	PREPARED BY: H. L. Clever

EXPERIMENTAL VALUES:

Temperature		Mol Fraction	Bunsen Coefficient[a]	Ostwald Coefficient[b]
t/°C	T/K	$10^3 x_1$	α	L/cm^3 cm^{-3}
22.2	295.35	3.25	0.5585	0.6035
25.0	298.15	3.17	0.5422	0.5918[c]
40.2	313.35	2.77	0.4639	0.5320
49.7	322.85	2.66	0.4380	0.5180
60.0	333.15	2.51	0.4068	0.4964

[a] Bunsen coefficient, α/cm^3(STP) cm^{-3} atm^{-1}.

[b] Listed as absorption coefficient in the original paper.
Interpreted to be equivalent to Ostwald coefficient by compiler.

[c] Ostwald coefficient (absorption coefficient) estimated as
298.15 K value by author.

[d] Mole fraction and Bunsen coefficient values calculated by
compiler assuming ideal gas behavior.

EVALUATOR'S COMMENT: McDaniel's data should be used with caution.
His values are often 20 percent or more too small when compared
with more reliable data.

AUXILIARY INFORMATION

METHOD/APPARATUS/PROCEDURE:	SOURCE AND PURITY OF MATERIALS:
The apparatus is all glass. It consists of a gas buret connected to a contacting vessel. The solvent is degassed by boiling under reduced pressure. Gas pressure or volume is adjusted using mercury displacement. Equilibration is achieved at atm pressure by hand shaking, and incrementally adding gas to the contacting chamber. Solubility measured by obtaining total uptake of gas by known volume of the solvent.	(1) Methane. Prepared by reaction of methyl iodide with zinc-copper. Passed through water and sulfuric acid. (2) Hexane.
	ESTIMATED ERROR: $\delta L/L \geq -0.20$
	REFERENCES:

COMPONENTS:	ORIGINAL MEASUREMENTS:
(1) Methane; CH_4; [74-82-8] (2) Hexane; C_6H_{14}; [110-54-3]	Lannung, A.; Gjaldbaek, J. C. *Acta Chem. Scand.* 1960, *14*, 1124 - 1128.
VARIABLES: $T/K = 291.15 - 310.15$ $p_1/kPa = 101.325$ (1 atm)	PREPARED BY: J. Chr. Gjaldbaek

EXPERIMENTAL VALUES:

T/K	Mol Fraction $10^3 x_1$	Bunsen Coefficient $\alpha/cm^3(STP)cm^{-3}atm^{-1}$	Ostwald Coefficient L/cm^3cm^{-3}
291.15	5.14	0.884	0.942
291.15	5.15	0.886	0.944
298.15	5.06	0.865	0.944
298.15	5.10	0.872	0.952
310.15	4.87	0.813	0.923
310.15	4.90	0.818	0.929

Smoothed Data: For use between 291.15 and 310.15 K.

$$\ln x_1 = -6.1312 + 2.5167/(T/100 \text{ K})$$

The standard error about the regression line is 2.75×10^{-5}.

T/K	Mol Fraction $10^3 x_1$
298.15	5.06
308.15	4.92

AUXILIARY INFORMATION

METHOD/APPARATUS/PROCEDURE:	SOURCE AND PURITY OF MATERIALS:
A calibrated all-glass combined manometer and bulb containing degassed solvent and the gas was placed in an air thermostat and shaken until equilibrium (1). The absorbed volume of gas is calculated from the initial and final amounts, both saturated with solvent vapor. The amount of solvent vapor. The amount of solvent is determined by the weight of displaced mercury. The values are at 101.325 kPa (1 atm) pressure assuming Henry's law is obeyed.	(1) Methane. Generated from magnesium methyl iodide. Purified by fractional distillation. Specific gravity corresponds with mol wt 16.08. (2) Hexane. Kahlbaum. "Hexan aus petroleum". Fractionated by distillation. B.p. (760 mmHg)/°C = 68.85, vapor pressure (25°C)/mmHg = 154.
	ESTIMATED ERROR: $\delta T/K = \pm 0.05$ $\delta x_1/x_1 = \pm 0.015$
	REFERENCES: 1. Lannung, A. *J. Am. Chem. Soc.* 1930, *52*, 68.

COMPONENTS:	ORIGINAL MEASUREMENTS:
(1) Methane; CH_4; [74-82-8] (2) Alkanes; C_6H_{14} and C_7H_{16}	Guerry, D. Jr. Ph.D. thesis, <u>1944</u> Vanderbilt University Nashville, TN Thesis Director: L. J. Bircher
VARIABLES: 　　　　T/K: 293.15, 298.15 　　　P/kPa: 101.325 (1 atm)	PREPARED BY: 　　　　　H. L. Clever

EXPERIMENTAL VALUES:

T/K	Mol Fraction x_1 x 10^4	Bunsen Coefficient α	Ostwald Coefficient L
Hexane; C_6H_{14}; [110-54-3]			
293.15	43.4	0.748	0.803
298.15	42.4	0.726	0.792
Heptane; C_7H_{16}; [142-82-5]			
293.15	46.7	0.718	0.771
298.15	40.7	0.622	0.679

The Ostwald coefficients were calculated by the compiler.

AUXILIARY INFORMATION

METHOD/APPARATUS/PROCEDURE:	SOURCE AND PURITY OF MATERIALS:
A Van Slyke-Neill Manometric Appara-tus manufactured by the Eimer and Amend Co. was used. The procedure of Van Slyke (1) for pure liquids was modified (2) so that small solvent samples (2 cm^3) could be used with almost complete recovery of the sample. An improved temperature control system was used.	Hexane. Eastman Kodak Co. B.p. (760.3 mmHg) t/°C 68.85 - 68.90. Heptane. B.p. (753.9 mmHg) t/°C 98.27 - 98.28 (corr.).

SOURCE AND PURITY OF MATERIALS:	
(1) Methane. Prepared by hydrolysis of crystaline methyl Grignard reagent. Passed through conc. H_2SO_4, solid KOH, and Dririte.	ESTIMATED ERROR: 　　　　　$\delta T/K = 0.05$
(2) Alkanes. Distilled from sodium in air. In addition to the solubility data the thesis con-tains data of the refractive index, density, vapor pressure, and b.p.	REFERENCES: 　1. Van Slyke, D. D. 　　　*J. Biol. Chem.* <u>1939</u>, *130*, 545. 　2. Ijams, C. C. 　　　Ph.D. thesis, <u>1941</u> 　　　Vanderbilt University

COMPONENTS:	ORIGINAL MEASUREMENTS:
1. Methane; CH_4;]74-82-8] 2. Hexane; C_6H_{14}; [110-54-3]	Tilquin, B.; Decannière, L.; Fontaine, R.; Claes, P. *Ann. Soc. Sc. Bruxelles (Belgium)* <u>1967</u>, *81*, 191-199.

VARIABLES:	PREPARED BY:
T/K: 288.15 P/kPa: 4.11-8.13	C. L. Young

EXPERIMENTAL VALUES:

t/C	T/K	Ostwald coefficient,[a] L	Mole fraction[b] $/x_1$	Henry's constant[b] H/atm
15.0	288.15	0.56	0.00306	326

[a] Original data at low pressure reported as distribution co-efficient; but if Henry's law and ideal gas law apply, distribution coefficient is equivalent to Ostwald coefficient as shown here.

[b] Calculated by compiler for a gas partial pressure of 101.325 kPa assuming that Henry's law and ideal gas law apply.

AUXILIARY INFORMATION

METHOD/APPARATUS/PROCEDURE:	SOURCE AND PURITY OF MATERIALS:
All glass apparatus used at very low gas partial pressures, containing a replaceable degassed solvent ampule equipped with a breakable point which could be broken by means of a magnetically activated plunger. Quantity of gas fed into system determined by measuring the pressure change in a known volume. Quantity of liquid measured by weight. Pressure change observed after solvent released. Experimental details described by Rzad and Claes, ref. (1).	1. Source not given; minimum purity specified as 99.0 mole per cent. 2. Fluka pure grade; minimum purity specified as 99.0 mole per cent.

ESTIMATED ERROR:
δT/K = 0.05; $\delta x_1/x_1$ = 0.01   (estimated by compiler).

REFERENCES:
1. Rzad, S.; Claes, P.   *Bull. Soc. Chim. Belges*   <u>1964</u>, *73*, 689.

COMPONENTS:	ORIGINAL MEASUREMENTS:
(1) Methane; CH_4; [74-82-8] (2) Hexane; C_6H_{14}; [110-54-3]	Hayduk, W.; Buckley, W.D. *Can. J. Chem. Eng.* <u>1971</u>, *49*, 667-671.

VARIABLES:	PREPARED BY:
T/K: 273.15-323.15 P/kPa: 101.325	W. Hayduk

EXPERIMENTAL VALUES:

T/K	Ostwald Coefficient[1] L/cm^3 cm^{-3}	Bunsen Coefficient[2] α/cm^3(STP) cm^{-3}atm^{-1}	Mole Fraction[1] $10^4\ x_1$
273.15	1.004	1.004	57.1 (58.1)[3]
298.15	0.935	0.857	50.2 (48.7)
323.15	0.801	0.677	41.3 (41.9)

[1] Original data

[2] Calculated by compiler

[3] The original mole fraction solubility data were used to determine the following equations for $\Delta G°$ and ln x_1 and table of smoothed values:

$\Delta G°$/J mol^{-1} =-RT ln x_1 = 60.361 T - 4795.3
ln x_1 = 576.8/T - 7.260

Std. deviation for $\Delta G°$ = 66.9 J mol^{-1}; Correlation coefficient = 0.9991

T/K	$10^{-4}\Delta G°$/J mol^{-1}	$10^4 x_1$
273.15	1.169	58.1
283.15	1.230	53.9
293.15	1.290	50.3
298.15	1.320	48.7
303.15	1.350	47.1
313.15	1.411	44.3
323.15	1.471	41.9

AUXILIARY INFORMATION

METHOD/APPARATUS/PROCEDURE:	SOURCE AND PURITY OF MATERIALS:
A volumetric method using a glass apparatus was employed. Degassed solvent contacted the gas while flowing as a thin film, at a constant rate, through an absorption spiral into a solution buret. A constant solvent flow was obtained by means of a calibrated syringe pump. The solution at the end of the spiral was considered saturated. Dry gas was maintained at atmospheric pressure in a gas buret by mechanically raising the mercury level in the buret at an adjustable rate. Volumes of solvent injected and residual gas were obtained for regular time intervals. The solubility was calculated from the constant slope of volume of gas dissolved and volume of solvent injected. Degassing was accomplished using a two stage vacuum process described by Clever et al. (1).	1. Matheson Co. Specified as ultra high purity grade of 99.97 per cent. 2. Canlab. Chromatoquality grade of specified minimum purity of 99.0 per cent.
	ESTIMATED ERROR: δT/K = 0.1 $\delta x_1/x_1$ = 0.01
	REFERENCES: 1. Clever, H.L.; Battino, R.; Saylor, J.H.; Gross, P.M. *J. Phys. Chem.* <u>1957</u>, *61*, 1078.

COMPONENTS:	ORIGINAL MEASUREMENTS:
(1) Methane; CH_4; [74-82-8] (2) Hexane; C_6H_{14}; [110-54-3]	Yokoyama, C.; Masuoka, H. Aral, K.; Saito, S. *J. Chem. Eng. Data* <u>1985</u>, *30*, 177-9.

VARIABLES:	PREPARED BY:
$T/K = 311.0$ $p_t/MPa = 0.57 - 1.98$	H. L. Clever

EXPERIMENTAL VALUES:

Temperature		Total Pressure	Mol Fraction	
			Liquid	Vapor
$t/^0C$	T/K	p_t/MPa	x_1	y_1
37.8	311.0	0.57		0.9264
		0.59		0.9256
		0.83		0.9468
		1.06		0.9468
		1.20	0.0567	0.9607
		1.30	0.0616	0.9627
		1.53	0.0734	0.9680
		1.58	0.0757	
		1.77	0.0860	0.9696
		1.79	0.0892	
		1.81	0.0880	0.9724
		1.98	0.0965	

AUXILIARY INFORMATION

METHOD/APPARATUS/PROCEDURE:	SOURCE AND PURITY OF MATERIALS:
The equipment consists of an equil-ibration system and an analysis sys-tem. The procedures are essentially the same as those used by King *et al.* (ref 1) and Kubota *et al.* (ref 2). The equilibration system is in a thermostated water bath. The analysis system is in an air bath at 100 C to avoid condensation problems. Details of the degassing, equili-bration, and sampling procedures are not given in the paper. The composi-tion analysis was made by a gas chro-matograph and digital integrator. Calibration curves were obtained from mixtures of known composition.	(1) Methane. Takachio Kagaku Co., Ltd. Used as received. (2) Hexane. Takachio Kagaku Co., Ltd. Used as received. A trace anaylsis of the components found no measurable impurities. The samples were used without further purification.

| | ESTIMATED ERROR:
$\delta T/K = \pm 0.05$
$\delta p_t/MPa = \pm 0.01$
$\delta x_1/x_1 = \pm 0.015$ |

REFERENCES:	REFERENCES: (continued)
1. King,M.B.;Alderson,D.A.;Fallah,F.; Kassim,D.M.;Sheldon,J.R.;Mahmud,R. *"Chemical Engineering at Super- critical Conditions"* Paulatis,M.E.; Penninger,J.M.L.;Gray,R.D.,Jr.; Davidson,P, Editors; Ann Arbor Sci.; Ann Arbor, MI <u>1983</u>, p. 31.	2. Kubota, H.; Inatome, H.; Tanaka, Y.; Makita, T. *J. Chem. Eng. Jpn.* <u>1983</u>, *16*, 99.

COMPONENTS:	ORIGINAL MEASUREMENTS:
1. Methane; CH_4; [74-82-8] 2. 2,2'-Dimethylbutane (Neo-hexane); C_6H_{14}; [75-83-2]	Tilquin, B.; Decannière, L.; Fontaine, R.; Claes, P. *Ann. Soc. Sc. Bruxelles (Belgium)* <u>1967</u>, *81*, 191-199.

VARIABLES:	PREPARED BY:
T/K: 288.15 P/kPa: 2.05-2.11	C. L. Young

EXPERIMENTAL VALUES:

t/C	T/K	Ostwald coefficient,[a] L	Mole fraction[b] /x_1	Henry's constant[b] H/atm
15.0	288.15	2.30	0.01266	79.0

[a] Original data at low pressure reported as distribution coefficient; but if Henry's law and ideal gas law apply, distribution coefficient is equivalent to Ostwald coefficient as shown here.

[b] Calculated by compiler for a gas partial pressure of 101.325 kPa assuming that Henry's law and ideal gas law apply.

AUXILIARY INFORMATION

METHOD/APPARATUS/PROCEDURE:	SOURCE AND PURITY OF MATERIALS:
All glass apparatus used at very low gas partial pressures, containing a replaceable degassed solvent ampule equipped with a breakable point which could be broken by means of a magnetically activated plunger. Quantity of gas fed into system determined by measuring the pressure change in a known volume. Quantity of liquid measured by weight. Pressure change observed after solvent released. Experimental details described by Rzad and Claes, ref. (1).	1. Source not given; minimum purity specified as 99.0 mole per cent. 2. Fluka pure grade; minimum purity specified as 99.0 mole per cent.

	ESTIMATED ERROR:
	δT/K = 0.05; $\delta x_1/x_1$ = 0.01 (estimated by compiler).

	REFERENCES:
	1. Rzad, S.; Claes, P. *Bull. Soc. Chim. Belges* <u>1964</u>, *73*, 689.

COMPONENTS:	ORIGINAL MEASUREMENTS:
(1) Methane; CH_4; [74-82-8] (2) Heptane; C_7H_{16}; [142-82-5]	McDaniel, A. S. *J. Phys. Chem.* <u>1911</u>, *15*, 587-610.

VARIABLES:	PREPARED BY:
$T/K = 295.35 - 313.15$ $p_1/kPa = 101.3$ (1 atm)	H. L. Clever

EXPERIMENTAL VALUES:

Temperature		Mol Fraction	Bunsen Coefficient[a]	Ostwald Coefficient[b]
$t/°C$	T/K	$10^3 x_1$	α	$L/cm^3\ cm^{-3}$
22.2	295.35	4.37	0.6720	0.7242
25.0	298.15	4.27	0.6519	0.7116[c]
30.1	303.25	4.10	0.6221	0.6906
40.0	313.15	3.89	0.5820	0.6675

[a] Bunsen coefficient, α/cm^3 (STP) cm^{-3} atm^{-1}.

[b] Listed as absorption coefficient in the original paper. Interpreted to be equivalent to Ostwald coefficient by compiler.

[c] Ostwald coefficient (absorption coefficient) estimated as 298.15 K value by author.

[d] Mole fraction and Bunsen coefficient values calculated by compiler assuming ideal gas behavior.

EVALUATOR'S COMMENT: McDaniel's data should be used with caution. His values are often 20 percent or more too small when compared with more reliable data.

AUXILIARY INFORMATION

METHOD/APPARATUS/PROCEDURE:	SOURCE AND PURITY OF MATERIALS:
The apparatus is all glass. It consists of a gas buret connected to a contacting vessel. The solvent is degassed by boiling under reduced pressure. Gas pressure or volume is adjusted using mercury displacement. Equilibration is achieved at atm pressure by hand shaking, and incrementally adding gas to the contacting chamber. Solubility measured by obtaining total uptake of gas by known volume of the solvent.	(1) Methane. Prepared by reaction of methyl iodide with zinc-copper. Passed through water and sulfuric acid. (2) Heptane.
	ESTIMATED ERROR: $\delta L/L \geq -0.20$
	REFERENCES:

COMPONENTS:	ORIGINAL MEASUREMENTS:
(1) Methane; CH_4; [74-82-8] (2) Heptane; C_7H_{16}; [142-82-5] Octane; C_8H_{18}; [111-65-9]	Makranczy, J.; Megyery-Balog, K.; Rusz, L.; Patyi, L. *Hung. J. Ind. Chem.* 1976, *4*, 269 - 280.

VARIABLES:	PREPARED BY:
T/K: 298.15 p/kPa: 101.325 (1 atm)	S. A. Johnson H. L. Clever

EXPERIMENTAL VALUES:

T/K	Mol Fraction $10^3 x_1$	Bunsen Coefficient $\alpha/cm^3 (STP) cm^{-3} atm^{-1}$	Ostwald Coefficient $L/cm^3 cm^{-3}$
Heptane			
298.15	5.73	0.876	0.956
Octane			
298.15	5.64	0.778	0.849

The Bunsen coefficient and mole fraction values were calculated by the compiler assuming that the gas is ideal and that Henry's law is obeyed.

AUXILIARY INFORMATION

METHOD/APPARATUS/PROCEDURE:	SOURCE AND PURITY OF MATERIALS:
Volumetric method. The apparatus described by Bodor, Bor, Mohai and Sipos was used (1).	Both the gas and the liquid were analytical grade reagents of Hungarian or foreign origin. No further information.

ESTIMATED ERROR:

$$\delta L/L = \pm 0.03$$

REFERENCES:
1. Bodor, E.; Bor, Gy.;
   Mohai, B.; Sipos, G.
   Veszpremi Vegyip. Egy. Kozl.
   1957, *1*, 55.
   Chem. Abstr. 1961, *55*, 3175h.

COMPONENTS:	ORIGINAL MEASUREMENTS:
(1) Methane; CH_4; [74-82-8] (2) Heptane; C_7H_{16}; [142-82-5]	Hayduk, W.; Buckley, W.D. *Can. J. Chem. Eng.* <u>1971</u>, *49*, 667-671.

VARIABLES:	PREPARED BY:
T/K: 273.15-348.15 P/kPa: 101.325	W. Hayduk

EXPERIMENTAL VALUES:

T/K	Ostwald Coefficient[1] L/cm^3 cm^{-3}	Bunsen Coefficient[2] α/cm^3 (STP)cm^{-3}atm^{-1}	Mole Fraction[1] 10^4 x_1
273.15	0.894	0.894	57.1 (58.1)[3]
298.15	0.840	0.770	50.6 (49.4)
323.15	0.752	0.636	43.1 (43.1)
348.15	0.688	0.540	38.0 (38.3)

[1]Original data.

[2]Calculated by compiler

[3]The mole fraction solubility of the original data was used to determine the following equations for $\Delta G°$ and ln x_1 and table of smoothed values:

$$\Delta G°/J\ mol^{-1} = -RT\ ln\ x_1 = 58.916\ T - 4401.9$$

$$ln\ x_1 = 529.5/T - 7.086$$

Std. deviation for $\Delta G°$ = 43.2 J mol^{-1}; Correlation coefficient = 0.9997

T/K	$10^{-4}\Delta G°$/J mol^{-1}	10^4 x_1	T/K	$10^{-4}\Delta G°$/J mol^{-1}	10^4 x_1
273.15	1.169	58.1	313.15	1.404	45.4
283.15	1.228	54.2	323.15	1.464	43.1
293.15	1.287	50.9	333.15	1.523	41.0
298.15	1.316	49.4	343.15	1.582	39.1
303.15	1.346	48.0	348.15	1.611	38.3

AUXILIARY INFORMATION

METHOD/APPARATUS/PROCEDURE:	SOURCE AND PURITY OF MATERIALS:
A volumetric method using a glass apparatus was employed. Degassed solvent contacted the gas while flowing as a thin film, at a constant rate, through an absorption spiral into a solution buret. A constant solvent flow was obtained by means of a calibrated syringe pump. The solution at the end of the spiral was considered saturated. Dry gas was maintained at atmospheric pressure in a gas buret by mechanically raising the mercury level in the buret at an adjustable rate. The solubility was calculated from the constant slope of volume of gas dissolved and volume of solvent injected. Degassing was accomplished using a two stage vacuum process described by Clever et al. (1).	1. Matheson Co. Specified as ultra high purity grade of 99.97 per cent. 2. Canlab. Chromatoquality grade of specified minimum purity of 99.0 per cent.

<table>
<tr><td></td><td>ESTIMATED ERROR:

δT/K   = 0.1
$\delta x_1/x_1$ = 0.01</td></tr>
</table>

REFERENCES:

1. Clever, H.L.; Battino, R.; Saylor, J.H.; Gross, P.M.
 J. Phys. Chem. <u>1957</u>, *61*, 1078.

COMPONENTS:	ORIGINAL MEASUREMENTS:
(1) Methane; CH_4; [74-82-8] (2) Octane; C_8H_{18}; [111-65-9]	Wilcock, R. J.; Battino, R.; Danforth, W. F.; Wilhelm, E. *J. Chem. Thermodyn.* 1978, *10*, 817 - 822.

VARIABLES:
    T/K: 298.25, 313.35
    p/kPa: 101.325 (1 atm)
	PREPARED BY:
	
	        H. L. Clever

EXPERIMENTAL VALUES:

T/K	Mol Fraction $10^3 x_1$	Bunsen Coefficient α/cm^3 (STP) cm^{-3} atm^{-1}	Ostwald Coefficient L/cm^3 cm^{-3}
298.25	5.026	0.6913	0.7548
313.35	4.744	0.6410	0.7353

The Bunsen coefficients were calculated by the compiler.

It is assumed that the gas is ideal and that Henry's law is obeyed.

AUXILIARY INFORMATION

METHOD/APPARATUS/PROCEDURE:

The solubility apparatus is based on the design of Morrison and Billett (1) and the version used is described by Battino, Evans, and Danforth (2). The degassing apparatus is that described by Battino, Banzhof, Bogan, and Wilhelm (3).

Degassing. Up to 500 cm^3 of solvent is placed in a flask of such size that the liquid is about 4 cm deep. The liquid is rapidly stirred, and vacuum is intermittently applied through a liquid N_2 trap until the permanent gas residual pressure drops to 5 microns.

Solubility Determination. The degassed solvent is passed in a thin film down a glass helical tube containing solute gas plus the solvent vapor at a total pressure of one atm. The volume of gas absorbed is found by difference between the initial and final volumes in the buret system. The solvent is collected in a tared flask and weighed.

SOURCE AND PURITY OF MATERIALS:

(1) Methane. Matheson Co., Inc. Minimum mole percent purity stated to be 99.97.

(2) Octane. Phillips Petroleum Co. 99 mole per cent, distilled, density at 298.15 K, ρ/g cm^{-3} 0.6988.

ESTIMATED ERROR:

$$\delta T/K = 0.02$$
$$\delta P/\text{mmHg} = 0.5$$
$$\delta x_1/x_1 = 0.01$$

REFERENCES:

1. Morrison, T. J.; Billett, F. *J. Chem. Soc.* 1948, 2033.

2. Battino, R.; Evans, F. D.; Danforth, W. F. *J. Am. Oil Chem. Soc.* 1968, *45*, 830.

3. Battino, R.; Banzhof, M.; Bogan, M.; Wilhelm, E. *Anal. Chem.* 1971, *43*, 806.

COMPONENTS:	ORIGINAL MEASUREMENTS:
(1) Methane; CH_4; [74-82-8] (2) Octane; C_8H_{18}; [111-65-9]	Hayduk, W.; Buckley, W.D. *Can. J. Chem. Eng.* <u>1971</u>, *49*, 667-671.

VARIABLES:	PREPARED BY:
T/K: 273.15-348.15 P/kPa: 101.325	W. Hayduk

EXPERIMENTAL VALUES:

T/K	Ostwald Coefficient[1] L/cm^3 cm^{-3}	Bunsen Coefficient[2] α/cm^3 (STP)cm^{-3}atm^{-1}	Mole Fraction[1] $10^4\ x_1$
273.15	0.826	0.826	58.6 (58.9)[3]
298.15	0.767	0.703	51.1 (50.5)
323.15	0.696	0.588	44.1 (44.4)
348.15	0.653	0.512	39.7 (39.7)

[1]Original data.

[2]Calculated by compiler

[3]The mole fraction solubility of the original data was used to determine the following equations for $\Delta G°$ and $\ln x_1$ and table of smoothed values:

$$\Delta G°/\text{J mol}^{-1} = -RT \ln x_1 = 57.958\ T - 4172.4$$

$$\ln x_1 = 501.9/T - 6.971$$

Std. deviation for $\Delta G° = 20.0$ J mol^{-1}; Correlation coefficient = 1.000

T/K	$10^{-4}\Delta G°$/J mol^{-1}	$10^4\ x_1$	T/K	$10^{-4}\Delta G°$/J mol^{-1}	$10^4\ x_1$
273.15	1.166	58.9	303.15	1.340	49.1
283.15	1.224	55.2	313.15	1.398	46.6
293.15	1.282	52.0	323.15	1.456	44.4
298.15	1.311	50.5	348.15	1.601	39.7

AUXILIARY INFORMATION

METHOD/APPARATUS/PROCEDURE:	SOURCE AND PURITY OF MATERIALS:
A volumetric method using a glass apparatus was employed. Degassed solvent contacted the gas while flowing as a thin film, at a constant rate, through an absorption spiral into a solution buret. A constant solvent flow was obtained by means of a cabibrated syringe pump. The solution at the end of the spiral was considered saturated. Dry gas was maintained at atmospheric pressure in a gas buret by mechanically raising the mercury level in the buret at an adjustable rate. The solubility was calculated from the constant slope of volume of gas dissolved and volume of solvent injected. Degassing was accomplished using a two stage vacuum process described by Clever et al. (1).	1. Matheson Co. Specified as ultra high purity grade of 99.97 per cent. 2. Canlab. Chromatoquality grade of specified minimum purity of 99.0 per cent.

ESTIMATED ERROR:
δT/K   = 0.1 $\delta x_1/x_1$ = 0.01

REFERENCES:
1. Clever, H.L.; Battino, R.; Saylor, J.H.; Gross, P.M.   *J. Phys. Chem.* <u>1957</u>, *61*, 1078.

COMPONENTS:	ORIGINAL MEASUREMENTS:
(1) Methane; CH_4; [74-82-8] (2) 2,2,4-Trimethylpentane or isooctane; C_8H_{18}; [540-84-1]	Hiraoka, H.; Hildebrand, J. H. *J. Phys. Chem.* 1964, *68*, 213-214.
VARIABLES: $T/K = 277.51 - 308.22$ $p_1/kPa = 101.325$ (1 atm)	PREPARED BY: M. E. Derrick H. L. Clever

EXPERIMENTAL VALUES:

Temperature		Mol Fraction	Bunsen Coefficient	Ostwald Coefficient
$t/°C$	T/K	$10^3 x_1$	$\alpha/cm^3 (STP) cm^{-3} atm^{-1}$	$L/cm^3 cm^{-3}$
4.36	277.51	6.236	0.868	0.882
14.97	288.12	5.688	0.781	0.824
24.77	297.92	5.351	0.726	0.792
35.07	308.22	4.989	0.669	0.755

The Bunsen and Ostwald coefficients were calculated by the compiler assuming ideal gas behavior.

Smoothed Data: For use between 277.51 and 308.22 K.

$$\ln x_1 = -7.2940 + 6.1413/(T/100 \text{ K})$$

The standard error about the regression line is 3.32×10^{-4}.

T/K	Mol Fraction $10^2 x_1$
278.15	6.182
288.15	5.726
298.15	5.331
308.15	4.986

AUXILIARY INFORMATION

METHOD/APPARATUS/PROCEDURE:	SOURCE AND PURITY OF MATERIALS:
The apparatus consists of a gas measuring buret, an absorption pipet, and a reservoir for the solvent. The buret is thermostated at 25°C, the pipet at any temperature from 5 to 30°C. The pipet contains an iron bar in glass for magnetic stirring. The pure solvent is degassed by freezing with liquid nitrogen, evacuating, then boiling with a heat lamp. The degassing process is repeated three times. The solvent is flowed into the pipet where it is again boiled for final degassing. Manipulation of the apparatus is such that the solvent never comes in contact with stopcock grease. The liquid in the pipet is sealed off by mercury. Its volume is the difference between the capacity of the pipet and the volume of mercury that confines it. Gas is admitted into the pipet. Its exact amount is determined by P-V measurements in the buret before and after	(1) Methane. Phillips Petroleum Co. Gas passed through a cold trap. (2) Isooctane. Source not given. Distilled, purity checked by ultraviolet absorbance.
	ESTIMATED ERROR: $\delta T/K = 0.02$ $\delta x_1/x_1 = 0.003$
	REFERENCES: 1. Kobatake, Y.; Hildebrand, J. H. *J. Phys. Chem.* 1961, *65*, 331.

introduction of the gas into the pipet. The stirrer is set in motion. Equilibrium is attained within 24 hours.

COMPONENTS:	ORIGINAL MEASUREMENTS:
(1) Methane; CH_4; [74-82-8] (2) Nonane; C_9H_{20}; [111-84-2] Decane; $C_{10}H_{22}$; [124-18-5]	Makranczy, J.; Megyery-Balog, K.; Rusz, L.; Patyi, L. *Hung. J. Ind. Chem.* 1976, *4*, 269 - 280.

VARIABLES:
  T/K: 298.15
  p/kPa: 101.325 (1 atm)
	PREPARED BY:
	  S. A. Johnson
	  H. L. Clever

EXPERIMENTAL VALUES:

T/K	Mol Fraction $10^3 x_1$	Bunsen Coefficient $\alpha/cm^3 (STP) cm^{-3} atm^{-1}$	Ostwald Coefficient $L/cm^3 cm^{-3}$
Nonane			
298.15	5.57	0.698	0.762
Decane			
298.15	5.52	0.635	0.693

The Bunsen coefficient and mole fraction values were calculated by the compiler assuming that the gas is ideal and that Henry's law is obeyed.

AUXILIARY INFORMATION

METHOD/APPARATUS/PROCEDURE: Volumetric method. The apparatus described by Bodor, Bor, Mohai and Sipos was used (1).	SOURCE AND PURITY OF MATERIALS: Both the gas and the liquid were analytical grade reagents of Hungarian or foreign origin. No further information.
	ESTIMATED ERROR: $\delta L/L = \pm 0.03$

	REFERENCES:
	1. Bodor, E.; Bor, Gy.;
	  Mohai, B.; Sipos, G.
	  Veszpremi Vegyip. Egy. Kozl.
	  1957, *1*, 55.
	  Chem. Abstr. 1961, *55*, 3175h.

COMPONENTS:	ORIGINAL MEASUREMENTS:
(1) Methane; CH_4; [74-82-8] (2) Decane; $C_{10}H_{22}$; [124-18-5]	Wilcock, R. J.; Battino, R.; Danforth, W. F.; Wilhelm, E. *J. Chem. Thermodyn.* <u>1978</u>, *10*, 817 - 822.

VARIABLES:
    T/K: 282.80, 313.35
    p/kPa: 101.325 (1 atm)
	PREPARED BY:
	
	        H. L. Clever

EXPERIMENTAL VALUES:

T/K	Mol Fraction $10^3 x_1$	Bunsen Coefficient α/cm^3 (STP) cm^{-3} atm^{-1}	Ostwald Coefficient L/cm^3 cm^{-3}
282.80	5.737	0.6693	0.6929
313.35	4.798	0.5420	0.6218

The Bunsen coefficients were calculated by the compiler.

It is assumed that the gas is ideal and that Henry's law is obeyed.

AUXILIARY INFORMATION

METHOD/APPARATUS/PROCEDURE:	SOURCE AND PURITY OF MATERIALS:
The solubility apparatus is based on the design of Morrison and Billett (1) and the version used is described by Battino, Evans, and Danforth (2). The degassing apparatus is that described by Battino, Banzhof, Bogan, and Wilhelm (3). Degassing. Up to 500 cm^3 of solvent is placed in a flask of such size that the liquid is about 4 cm deep. The liquid is rapidly stirred, and vacuum is intermittently applied through a liquid N_2 trap until the permanent gas residual pressure drops to 5 microns. Solubility Determination. The degassed solvent is passed in a thin film down a glass helical tube containing solute gas plus the solvent vapor at a total pressure of one atm. The volume of gas absorbed is found by difference between the initial and final volumes in the buret system. The solvent is collected in a tared flask and weighed.	(1) Methane. Matheson Co., Inc. Minimum mole percent purity stated to be 99.97. (2) Decane. Phillips Petroleum Co. 99 mol %, distilled, density at 298.15 K, ρ/g cm^{-3} 0.7264.

	ESTIMATED ERROR:
	        δT/K = 0.02
	      δP/mmHg = 0.5
	      $\delta x_1/x_1$ = 0.01

	REFERENCES:
	
	1. Morrison, T. J.; Billett, F.
	   J. Chem. Soc. <u>1948</u>, 2033.
	
	2. Battino, R.; Evans, F. D.;
	   Danforth, W. F.
	   J. Am. Oil Chem. Soc. <u>1968</u>,
	   45, 830.
	
	3. Battino, R.; Banzhof, M.;
	   Bogan, M.; Wilhelm, E.
	   Anal. Chem. <u>1971</u>, *43*, 806.

COMPONENTS:	ORIGINAL MEASUREMENTS:
(1) Methane; CH_4; [74-82-8] (2) Dodecane; $C_{12}H_{26}$; [112-40-3]	Hayduk, W.; Buckley, W.D. *Can. J. Chem. Eng.* <u>1971</u>, *49*, 667-671.

VARIABLES:	PREPARED BY:
T/K: 273.15-348.15 P/kPa: 101.325	W. Hayduk

EXPERIMENTAL VALUES:

T/K	Ostwald Coefficient[1] L/cm^3 cm^{-3}	Bunsen Coefficient[2] α/cm^3 (STP)$cm^{-3}atm^{-1}$	Mole Fraction[1] 10^4 x_1
273.15	0.627	0.627	62.4 (62.8)[3]
298.15	0.590	0.540	54.9 (54.4)
323.15	0.547	0.462	48.2 (48.2)
348.15	0.517	0.406	43.4 (43.5)

[1]Original data.

[2]Calculated by compiler

[3]The mole fraction solubility of the original data was used to determine the following equations for $\Delta G°$ and $\ln x_1$ and table of smoothed values:

$$\Delta G°/J\ mol^{-1} = -RT\ \ln x_1 = 56.325\ T - 3869.4$$
$$\ln x_1 = 465.4/T - 6.775$$

Std. deviation for $\Delta G°$ = 15.3 J mol^{-1}; Correlation coefficient = 0.9999

T/K	$10^{-4}\Delta G°/J\ mol^{-1}$	10^4 x_1	T/K	$10^{-4}\Delta G°/mol^{-1}$	10^4 x_1
273.15	1.152	62.8	303.15	1.321	53.0
283.15	1.208	59.1	313.15	1.377	50.5
293.15	1.264	55.9	323.15	1.433	48.2
298.15	1.292	54.4	348.15	1.574	43.5

AUXILIARY INFORMATION

METHOD/APPARATUS/PROCEDURE:	SOURCE AND PURITY OF MATERIALS:
A volumetric method using a glass apparatus was employed. Degassed solvent contacted the gas while flowing as a thin film, at a constant rate, through an absorption spiral into a solution buret. A constant solvent flow was obtained by means of a calibrated syringe pump. The solution at the end of the spiral was considered saturated. Dry gas was maintained at atmospheric pressure in a gas buret by mechanically raising the mercury level in the buret at an adjustable rate. The solubility was calculated from the constant slope of volume of gas dissolved and volume of solvent injected. Degassing was accomplished using a two stage vacuum process described by Clever et al. (1).	1. Matheson Co. Specified as ultra high purity grade of 99.98 per cent. 2. Canlab. Chromatoquality grade of specified minimum purity of 99.0 per cent.

METHOD/APPARATUS/PROCEDURE section continues with:

	ESTIMATED ERROR: $\delta T/K$ = 0.1 $\delta x_1/x_1$ = 0.01
	REFERENCES: 1. Clever, H.L.; Battino, R.; Saylor, J.H.; Gross, P.M. *J. Phys. Chem.* <u>1957</u>, *61*, 1078.

COMPONENTS:	ORIGINAL MEASUREMENTS:
(1) Methane; CH_4; [74-82-8] (2) Undecane; $C_{11}H_{24}$; [1120-21-4] Dodecane; $C_{12}H_{26}$; [112-40-3]	Makranczy, J.; Megyery-Balog, K.; Rusz, L.; Patyi, L. *Hung. J. Ind. Chem.* <u>1976</u>, *4*, 269 - 280.

VARIABLES:	PREPARED BY:
T/K: 298.15 p/kPa: 101.325 (1 atm)	S. A. Johnson H. L. Clever

EXPERIMENTAL VALUES:

T/K	Mol Fraction $10^3 x_1$	Bunsen Coefficient $\alpha/cm^3(STP)cm^{-3}atm^{-1}$	Ostwald Coefficient $L/cm^3 cm^{-3}$
Undecane			
298.15	5.46	0.580	0.633
Dodecane			
298.15	5.45	0.537	0.586

The Bunsen coefficient and mole fraction values were calculated by the compiler assuming that the gas is ideal and that Henry's law is obeyed.

AUXILIARY INFORMATION

METHOD/APPARATUS/PROCEDURE:	SOURCE AND PURITY OF MATERIALS:
Volumetric method. The apparatus described by Bodor, Bor, Mohai and Sipos was used (1).	Both the gas and the liquid were analytical grade reagents of Hungarian or foreign origin. No further information.

ESTIMATED ERROR:

$$\delta L/L = \pm\ 0.03$$

REFERENCES:

1. Bodor, E.; Bor, Gy.; Mohai, B.;
   Sipos, G.
   Veszpremi Vegyip. Egy. Kozl.
   <u>1957</u>, *1*, 55.
   Chem. Abstr. <u>1961</u>, *55*, 3175h.

COMPONENTS:	ORIGINAL MEASUREMENTS:
(1) Methane; CH_4; [74-82-8] (2) Tridecane; $C_{13}H_{28}$; [629-50-5] Tetradecane; $C_{14}H_{30}$; [629-59-4]	Makranczy, J.; Megyery-Balog, K.; Rusz, L.; Patyi, L. *Hung. J. Ind. Chem.* <u>1976</u>, *4*, 269 - 280.

VARIABLES:	PREPARED BY:
T/K: 298.15 p/kPa: 101.325 (1 atm)	S. A. Johnson H. L. Clever

EXPERIMENTAL VALUES:

T/K	Mol Fraction $10^3 x_1$	Bunsen Coefficient α/cm^3(STP)cm^{-3}atm^{-1}	Ostwald Coefficient L/cm^{3}cm^{-3}
Tridecane			
298.15	5.39	0.496	0.541
Tetradecane			
298.15	5.40	0.465	0.508

The Bunsen coefficient and mole fraction values were calculated by the compiler assuming that the gas is ideal and that Henry's law is obeyed.

AUXILIARY INFORMATION

METHOD/APPARATUS/PROCEDURE:	SOURCE AND PURITY OF MATERIALS:
Volumetric method. The apparatus described by Bodor, Bor, Mohai and Sipos was used (1).	Both the gas and the liquid were analytical grade reagents of Hungarian or foreign origin. No further information.
	ESTIMATED ERROR: $\delta L/L$ = ± 0.03
	REFERENCES: 1. Bodor, E.; Bor, Gy.; Mohai, B.; Sipos, G. *Veszpremi Vegyip. Egy. Kozl.* <u>1957</u>, *1*, 55. *Chem. Abstr.* <u>1961</u>, *55*, 3175h.

COMPONENTS:	ORIGINAL MEASUREMENTS:
(1) Methane; CH_4; [74-82-8] (2) Pentadecane; $C_{15}H_{32}$; [629-62-9] Hexadecane; $C_{16}H_{34}$; [544-76-3]	Makranczy, J.; Megyery-Balog, K.; Rusz, L.; Patyi, L. *Hung. J. Ind. Chem.* <u>1976</u>, *4*, 269 - 280.

VARIABLES:	PREPARED BY:
T/K: 298.15 p/kPa: 101.325 (1 atm)	S. A. Johnson H. L. Clever

EXPERIMENTAL VALUES:

T/K	Mol Fraction $10^3 x_1$	Bunsen Coefficient $\alpha/cm^3(STP)cm^{-3}atm^{-1}$	Ostwald Coefficient $L/cm^3 cm^{-3}$
Pentadecane			
298.15	5.35	0.434	0.474
Hexadecane			
298.15	5.36	0.410	0.448

The Bunsen coefficient and mole fraction values were calculated by the
compiler assuming that the gas is ideal and that Henry's law is obeyed.

AUXILIARY INFORMATION

METHOD/APPARATUS/PROCEDURE:	SOURCE AND PURITY OF MATERIALS:
Volumetric method. The apparatus described by Bodor, Bor, Mohai and Sipos was used (1).	Both the gas and the liquid were analytical grade reagents of Hungarian or foreign origin. No further information.
	ESTIMATED ERROR: $\delta L/L = \pm 0.03$

	REFERENCES:
	1. Bodor, E.; Bor, Gy.;
	   Mohai, B.; Sipos, G.
	   Veszpremi Vegyip. Egy. Kozl.
	   <u>1957</u>, *1*, 55.
	   Chem. Abstr. <u>1961</u>, *55*, 3175h.

COMPONENTS:	ORIGINAL MEASUREMENTS:
1. Methane; CH_4; [74-82-8]	Lenoir, J-Y.; Renault, P.; Renon, H.
2. Hexadecane; $C_{16}H_{34}$; [544-76-3]	*J. Chem. Eng. Data*, <u>1971</u>, *16*, 340-2.

VARIABLES:	PREPARED BY:
	C. L. Young

EXPERIMENTAL VALUES:

T/K	Henry's constant H_{CH_4}/atm	Mole fraction at 1 atm* x_{CH_4}
298.2	209	0.00478

* Calculated by compiler assuming a linear function of P_{CH_4} vs x_{CH_4}, i.e., x_{CH_4} (1 atm) = $1/H_{CH_4}$.

AUXILIARY INFORMATION

METHOD/APPARATUS/PROCEDURE:	SOURCE AND PURITY OF MATERIALS:
A conventional gas-liquid chromatographic unit fitted with a thermal conductivity detector was used. The carrier gas was helium. The value of Henry's law constant was calculated from the retention time. The value applies to very low partial pressures of gas and there may be a substantial difference from that measured at 1 atm. pressure. There is also considerable uncertainty in the value of Henry's constant since surface adsorption was not allowed for although its possible existence was noted.	(1) L'Air Liquide sample, minimum purity 99.9 mole per cent. (2) Touzart and Matignon or Serlabo sample, purity 99 mole per cent.
	ESTIMATED ERROR: $\delta T/K = \pm0.1$; δH/atm = ±6% (estimated by compiler).
	REFERENCES:

COMPONENTS:	ORIGINAL MEASUREMENTS:
(1) Methane; CH_4; [74-82-8] (2) Hexadecane; $C_{16}H_{34}$; [544-76-3]	Hayduk, W.; Buckley, W.D. *Can. J. Chem. Eng.* <u>1971</u>, *49*, 667-671.

VARIABLES:	PREPARED BY:
T/K: 298.15-348.15 P/kPa: 101.325	W. Hayduk

EXPERIMENTAL VALUES:

T/K	Ostwald Coefficient[1] L/cm^3 cm^{-3}	Bunsen Coefficient[2] α/cm^3 (STP)cm^{-3}atm^{-1}	Mole Fraction[1] 10^4 x_1
298.15	0.500	0.458	60.2 (59.9) [3]
323.15	0.475	0.401	53.7 (55.2)
348.15	0.464	0.364	49.9 (49.7)

[1]Original data.

[2]Calculated by compiler.

[3]The mole fraction solubility of the original data was used to determine the following equations for $\Delta G°$ and ln x_1 and table of smoothed values:

$$\Delta G°/J\ mol^{-1} = -RT\ \ln x_1 = 53.380\ T - 3229.7$$

$$\ln x_1 = 388.5/T - 6.420$$

Std. deviation for $\Delta G° = 20.4$ J mol^{-1}; Correlation coefficient = 0.9999

T/K	$10^{-4}\Delta G°$/mol^{-1}	10^4 x_1
298.15	1.269	59.9
303.15	1.295	58.6
313.15	1.349	56.3
323.15	1.375	55.2
333.15	1.455	52.3
348.15	1.535	49.7

AUXILIARY INFORMATION

METHOD/APPARATUS/PROCEDURE:	SOURCE AND PURITY OF MATERIALS:
A volumetric method using a glass apparatus was employed. Degassed solvent contacted the gas while flowing as a thin film, at a constant rate, through an absorption spiral into a solution buret. A constant solvent flow was obtained by means of a calibrated syringe pump. The solution at the end of the spiral was considered saturated. Dry gas was maintained at atmospheric pressure in a gas buret by mechanically raising the mercury level in the buret at an adjustable rate. The solubility was calculated from the constant slope of volume of gas dissolved and volume of solvent injected. Degassing was accomplished using a two stage vacuum process described by Clever et al. (1).	1. Matheson Co. Specified as ultra high purity grade of 99.97 per cent. 2. Canlab. Olefin-free grade of specified minimum purity of 99.0 per cent.

ESTIMATED ERROR:

δT/K = 0.1

$\delta x_1/x_1$ = 0.01

REFERENCES:

1. Clever, H.L.; Battino, R.; Saylor, J.H.; Gross, P.M.

    J. Phys. Chem. <u>1957</u>, *61*, 1078.

COMPONENTS:	ORIGINAL MEASUREMENTS:
1. Methane; CH_4; [74-82-8] 2. Hexadecane; $C_{16}H_{34}$; [544-76-3]	Cukor, P.M.; Prausnitz, J.M. *J. Phys. Chem.* <u>1972</u>, *76*, 598-601

VARIABLES:	PREPARED BY:
Temperature	C.L. Young

EXPERIMENTAL VALUES:

T/K	Henry's Constant[a] /atm	Mole fraction of methane[b] in liquid, x_{CH_4}
300	174	0.00575
325	191	0.00524
350	206	0.00485
375	218	0.00459
400	228	0.00439
425	233	0.00429
450	235	0.00426
475	233	0.00429

a. Quoted in supplementary material for original paper

b. Calculated by compiler for a partial pressure of 1 atmosphere

AUXILIARY INFORMATION

METHOD/APPARATUS/PROCEDURE:	SOURCE AND PURITY OF MATERIALS:
Volumetric apparatus similar to that described by Dymond and Hildebrand (1). Pressure measured with a null detector and precision gauge. Details in ref. (2).	No details given

ESTIMATED ERROR:

$\delta T/K = \pm 0.05$; $\delta x_{CH_4} = \pm 2\%$

REFERENCES:

1. Dymond, J.; Hildebrand, J.H.
    Ind. Eng. Chem. Fundam. <u>1967</u>, *6*,
    130.

2. Cukor, P.M.; Prausnitz, J.M.
    Ind. Eng. Chem. Fundam. <u>1971</u>, *10*
    638.

COMPONENTS:	ORIGINAL MEASUREMENTS:
1. Methane; CH_4; [74-82-8] 2. Hexadecane; $C_{16}H_{34}$; [544-76-3]	Rivas, O.R.; Prausnitz, J.M. *Ind. Eng. Chem. Fundam.* <u>1979</u>,*18*, 289-292.

VARIABLES:	PREPARED BY:
Temperature	C.L. Young

EXPERIMENTAL VALUES:

T/K	Henry's constant / atm	Mole fraction at 1 atm partial pressure*, x_{CH_4}
298.15	171.7	0.005824
373.15	216.1	0.004627
473.15	231.7	0.004316

* Calculated by compiler assuming mole fraction
  solubility linear with pressure.

AUXILIARY INFORMATION

METHOD/APPARATUS/PROCEDURE:	SOURCE AND PURITY OF MATERIALS:
Volumetric apparatus with a fused quartz precision bourdon pressure gauge. Solubility apparatus carefully thermostatted. Solvent degassed *in situ*. Apparatus described in source and similar to that described in ref (1).	No details given.

ESTIMATED ERROR:

$\delta T/K = \pm 0.05$; $\delta x_{CH_4} = \pm 1\%$.

REFERENCES:

1. Cukor, P.M.; Prausnitz, J.M.

   Ind. Eng. Chem. Fundam. <u>1971</u>, *10*, 638.

COMPONENTS:	ORIGINAL MEASUREMENTS:

1. Methane; CH_4; [74-82-8]
 	
2. Hexadecane; $C_{16}H_{34}$; [544-76-3]
   or
   Octadecane; $C_{18}H_{38}$; [593-45-3]
	Richon, D.; Renon, H.
	
	J. Chem. Eng. Data <u>1980</u>, *25*, 59-60.

VARIABLES:	PREPARED BY:
	C. L. Young

EXPERIMENTAL VALUES:

T/K	Limiting value of Henry's constant, H^{∞} /atm	Mole fraction of methane, * x_{CH_4}
	Hexadecane	
298.15	171.2	0.005841
	Octadecane	
323.15	196.9	0.005079

* Calculated by compiler assuming mole fraction is a linear function of pressure up to 1 atm.

AUXILIARY INFORMATION

METHOD/APPARATUS/PROCEDURE:	SOURCE AND PURITY OF MATERIALS:
Inert gas stripping plus gas chromatographic method. Details given in ref. (1). Method based on passing constant stream of inert gas through dissolved gas-solvent mixture and periodically injecting mixture into gas chromatograph. Henry's law constant determined from variation of gas peak area with time.	1. L'Air Liquide sample, purity 99.95 mole per cent. 2. Hexadecane was a Merck sample, Octadecane was a Fluka sample, both had purities of not less than 99 mole per cent.
	ESTIMATED ERROR: $\delta T/K = \pm 0.05$; $\delta H^{\infty} = \pm 4\%$ (estimated by compiler).

	REFERENCES:
	1. Leroi, J. C.; Masson, J. C.;
	   Renon, H.; Fabries, J. F.;
	   Sannier, H. *Ind. Eng. Chem.*
	   Process. Des. Develop. <u>1977</u>, *16*,
	   139.

COMPONENTS:	ORIGINAL MEASUREMENTS:
1. Methane; CH_4; [74-82-8] 2. Alkanes	Lin, P. J.; Parcher, J. F. *J. Chromatog. Sci.* 1982, *20*, 33-38.

VARIABLES:	PREPARED BY:
	C. L. Young

EXPERIMENTAL VALUES:

T/K	Henry's law constant, H /atm	Mole fraction at a partial pressure of 1 atmosphere x_{CH_4}
	Hexadecane; $C_{16}H_{34}$; [544-76-3]	
298.2	179	0.00559
313.2	189	0.00529
328.2	192	0.00521
	Octacosane; $C_{28}H_{58}$; [630-02-4]	
353.2	157	0.00637
373.2	165	0.00606
393.2	174	0.00575
	Hexatriacontane; $C_{36}H_{74}$; [630-06-8]	
353.2	131	0.00763
373.2	137	0.00730
393.2	145	0.00690
413.2	153	0.00654

AUXILIARY INFORMATION

METHOD/APPARATUS/PROCEDURE:	SOURCE AND PURITY OF MATERIALS:
Henry's law constant determined from retention volume of gas on a chromatographic column. Helium was used as a carrier gas and a mass spectrometer was used as a detector. The measured Henry's law constants were independent of sample size, flow rate and composition of injected sample. The dead volume was determined by two independent methods and the values agreed within experimental error.	No details given.
	ESTIMATED ERROR: $\delta T/K = \pm 0.1$; $\delta x_{CH_4} = \pm 5\%$ (estimated by compiler).
	REFERENCES:

COMPONENTS:	ORIGINAL MEASUREMENTS:
1. Methane; CH_4; [74-82-8] 2. 2,2,4,4,6,8,8-Heptamethyl nonane; $C_{16}H_{34}$; [4390-04-9]	Richon, D.; Renon, H. *J. Chem. Eng. Data* 1980, *25*, 59-60.

VARIABLES:	PREPARED BY:
	C. L. Young

EXPERIMENTAL VALUES:

T/K	Limiting value of Henry's constant, H^∞/atm	Mole fraction of methane, * x_{CH_4}
298.15	32.8	0.0305

* Calculated by compiler assuming mole fraction is a linear
  function of pressure up to 1 atm.

AUXILIARY INFORMATION

METHOD/APPARATUS/PROCEDURE:	SOURCE AND PURITY OF MATERIALS:
Inert gas stripping plus gas chromatographic method. Details given in ref. (1). Method based on passing constant stream of inert gas through dissolved gas-solvent mixture and periodically injecting mixture into gas chromatograph. Henry's law constant determined from variation of gas peak area with time.	1. L'Air Liquide sample, purity 99.95 mole per cent. 2. Sigma sample, purity not less than 99 mole per cent.
	ESTIMATED ERROR: $\delta T/K = \pm 0.05$; $\delta x_{CH_4} = \pm 4\%$ (estimated by compiler).
	REFERENCES: 1. Leroi, J. C.; Masson, J. C.; Renon, H.; Fabries, J. F.; Sannier, H. *Ind. Eng. Chem. Process. Des. Develop.* 1977, *16*, 139.

COMPONENTS:	ORIGINAL MEASUREMENTS:
1. Methane; CH_4; [74-82-8] 2. Octadecane; $C_{18}H_{38}$; [593-45-3]	Ng, S.; Harris, H.G.; Prausnitz, J.M.; *J. Chem. Engng. Data,* <u>1969</u>, *14*, 482-3.

VARIABLES:	PREPARED BY:
Temperature	C.L. Young

EXPERIMENTAL VALUES:

T/K	Henry's Constant, H /atm.	Mole fraction[+] of methane in liquid, x_{CH_4}
308.2	209	0.00478
323.2	239	0.00418
343.2	272	0.00368
363.2	255	0.00392
373.2	306	0.00327
423.2	395	0.00253

+ at 1 atmosphere partial pressure, calculated by compiler assuming mole fraction equals $1/H$.

AUXILIARY INFORMATION

METHOD/APPARATUS/PROCEDURE:	SOURCE AND PURITY OF MATERIALS:
Gas chromatographic method. Solvent supported on Chromosorb P in 6 m column. Gas injected as sample, helium used as carrier gas. Henry's law constant calculated from knowledge of retention time and flow rate.	1. Matheson sample purity greater than 99 mole per cent. 2. Matheson, Coleman and Bell sample, m.pt. 27-28.5°C.
	ESTIMATED ERROR: $\delta T/K = \pm 0.1$; $\delta H/atm = \pm 5\%$.
	REFERENCES:

COMPONENTS:	ORIGINAL MEASUREMENTS:
1. Methane; CH_4; [74-82-8] 2. Eicosane; $C_{20}H_{42}$; [112-95-8]	Ng. S.; Harris, H.G.; Prausnitz, J.M. *J. Chem. Engng. Data*, <u>1969</u>, *14*, 482-3

VARIABLES:	PREPARED BY:
Temperature	C. L. Young

EXPERIMENTAL VALUES:

T/K	Henry's Constant, H /atm	Mole fraction of methane in liquid, x_{CH_4} [*]
323.2	226	0.00442
373.2	286	0.00350
393.2	301	0.00332
413.2	356	0.00281

* At 1 atmosphere partial pressure, calculated by compiler assuming mole fraction equals $1/H$

AUXILIARY INFORMATION

METHOD/APPARATUS/PROCEDURE:	SOURCE AND PURITY OF MATERIALS:
Gas chromatographic method. Solvent supported on Chromosorb P in 6 m column. Gas injected as sample, helium used as carrier gas. Henry's law constant claculated from knowledge of retention time and flow rate.	1. Matheson sample, purity greater than 99 mole per cent. 2. Matheson, Coleman and Bell sample, m.pt. 35-36.5°C.
	ESTIMATED ERROR: $\delta T/K = \pm 0.1$; $\delta H/atm = \pm 5\%$
	REFERENCES:

COMPONENTS:	ORIGINAL MEASUREMENTS:
1. Methane; CH_4; [74-82-8] 2. Eicosane; $C_{20}H_{42}$; [112-95-8]	Chappelow, C.C.; Prausnitz, J.M. *Am. Inst. Chem. Engnrs. J.* <u>1974</u>, *20*, 1097-1104.

VARIABLES:
   Temperature
	PREPARED BY:
	   C.L. Young

EXPERIMENTAL VALUES:

T/K	Henry's Constant[a] /atm	Mole fraction[b] of methane at 1 atm. partial pressure, x_{CH_4}
325	170	0.00588
350	184	0.00543
375	196	0.00510
400	205	0.00488
425	210	0.00476
450	214	0.00467
475	216	0.00463

a.   Authors stated measurements were made at several pressures
     and values of solubility used were all within the Henry's
     Law region.

b.   Calculated by compiler assuming linear relationship between
     mole fraction and pressure.

AUXILIARY INFORMATION

METHOD/APPARATUS/PROCEDURE:	SOURCE AND PURITY OF MATERIALS:
Volumetric apparatus similar to that described by Dymond and Hildebrand (1). Pressure measured with a null detector and precision gauge. Details in ref. (2).	Solvent degassed, no other details given.
	ESTIMATED ERROR: $\delta T/K = \pm 0.1$; $\delta x_{CH_4} = \pm 1\%$
	REFERENCES: 1. Dymond, J.; Hildebrand, J.H. *Ind.Eng.Chem.Fundam.*<u>1967</u>,*6*,130. 2. Cukor, P.M.; Prausnitz, J.M. *Ind.Eng.Chem.Fundam.*<u>1971</u>,*10*,638.

COMPONENTS:	ORIGINAL MEASUREMENTS:
1. Methane; CH_4; [74-82-8] 2. Docosane; $C_{22}H_{46}$ [629-97-0]	Ng, S.; Harris, H.G.; Prausnitz, J.M. *J. Chem. Engng. Data*, <u>1969</u>, *14*, 482-3

VARIABLES:	PREPARED BY:
Temperature	C. L. Young

EXPERIMENTAL VALUES:

T/K	Henry's Constant, H /atm	Mole fraction of methane * in liquid, x_{CH_4}
333.2	229	0.00437
383.2	269	0.00372
408.2	314	0.00318
433.2	338	0.00296
453.2	355	0.00282
473.2	411	0.00243

*   At 1 atmosphere partial pressure, calculated by compiler assuming
    mole fraction equals $1/H$

AUXILIARY INFORMATION

METHOD/APPARATUS/PROCEDURE:	SOURCE AND PURITY OF MATERIALS:
Gas chromatographic method. Solvent supported on Chromosorb P in 6 m column. Gas injected as sample, helium used as carrier gas. Henry's law constant calculated from knowledge of retention time and flow rate.	1. Matheson sample, purity greater than 99 mole per cent. 2. Matheson, Coleman and Bell sample, m.pt . 43-45°C.
	ESTIMATED ERROR: $\delta T/K = \pm 0.1$; $\delta H/atm = \pm 5\%$
	REFERENCES:

COMPONENTS:	ORIGINAL MEASUREMENTS:
1. Methane; CH_4; [74-82-8] 2. 2,6,10,15,19,23,-Hexamethyl-tetracosane,(Squalane); $C_{30}H_{62}$; [111-01-3]	Chappelow, C.C.; Prausnitz, J.M. *Am. Inst. Chem. Engnrs. J.* <u>1974</u>, *20*, 1097-1104.

VARIABLES:
 	
    Temperature
	PREPARED BY:
	
	    C.L. Young

EXPERIMENTAL VALUES:

T/K	Henry's Constant[a] /atm	Mole fraction[b] of methane at 1 atm. partial pressure, x_{CH_4}
300	111	0.00901
325	124	0.00806
350	135	0.00741
375	144	0.00694
400	151	0.00662
425	156	0.00641
450	160	0.00625
475	163	0.00613

a. Authors stated measurements were made at several pressures and values of solubility used were all within the Henry's Law region.

b. Calculated by compiler assuming linear relationship between mole fraction and pressure.

AUXILIARY INFORMATION

METHOD/APPARATUS/PROCEDURE:	SOURCE AND PURITY OF MATERIALS:
Volumetric apparatus similar to that described by Dymond and Hildebrand (1). Pressure measured with a null detector and precision gauge. Details in ref. (2).	

ESTIMATED ERROR:

$$\delta T/K = \pm 0.1; \quad \delta x_{CH_4} = \pm 1\%$$

REFERENCES:
1. Dymond, J.; Hildebrand, J.H.
Ind.Chem.Eng.Fundam. <u>1967</u>,*6*,130.
2. Cukor, P.M.; Prausnitz, J.M.
Ind.Chem.Eng.Fundam. <u>1971</u>,*10*,638.

COMPONENTS:	EVALUATOR:

1. Methane; CH_4; [74-82-8]

2. Ethane; C_2H_6; [74-84-0]

Colin L. Young,
School of Chemistry,
University of Melbourne,
Parkville, Victoria 3052,
Australia.

February 1984

EVALUATION:

Measurements on this system have recently been evaluated by Hiza, Miller and Kidnay (1). Some of the data on this system are perhaps more correctly considered as vapor-liquid equilibrium data rather than gas-liquid solubility since the gas-liquid critical temperatures of methane and ethane are 190.6 K and 305.3 K, respectively.

The data of Uehara (2) were determined using a static experimental technique. Six solubility measurements were made between 140 and 155 K at a total pressure of 10^5 Pa. These data are rejected in the light of more recent data in which the partial pressures have been determined. The data of Ruhemann (3) and Guter, Newitt and Ruhemann (4) are not presented here. This early work appears to be in disagreement with more recent work and the experimental data show considerable scatter. These workers used a flow method and determined the solubility at 169 K, 185 K, 195 K and 273 K at a series of pressures between 10^5 and 8.3×10^6 Pa.

Michels and Nederbragt (5) made a very brief study of this system at 273 K and their data are rejected. Levitskaya (6) used a recirculating flow method to study this system at 178 K and 188 K; only three data points were given and the data are also rejected because of their limited nature.

The most extensive study of this system is that of Wichterle and Kobayashi (7) which covers the temperature range from 139 K to 200 K. Their data are believed to be accurate and are classified as recommended. Their measurements were carried out using a recirculating vapor flow apparatus. The data of Bloomer and coworkers (8) were determined using a dew-point/bubble-point apparatus over the temperature range 140 K to 300 K. This type of apparatus in general is not the most accurate for determining gas-liquid solubilities and the data are classified as tentative.

Wilson (9) and Miller and Staveley (10) investigated this system at 111 K and 116 K, respectively; their measurements are believed to be of high quality but are more correctly considered as vapor-liquid equilibrium. Therefore, these data are not considered further. The data of Kidnay and coworkers (11), (12), (13) are thought to be reliable. The apparatus used was of proven design and high purity materials were used in each study. The data of Kidnay and coworkers are in good agreement with those of Wichterle and Kobayashi (7) and are classified as recommended.

The limited data of Price and Kobayashi (14) are of moderate precision and their data at 227.95 K are classified as tentative. The limited data of Chang and Lu (15) are rejected, being superseded by the more accurate and more extensive data of Wichterle and Kobayashi (7).

The data of Hsi and Lu (16) deviate somewhat from values obtained by extrapolation of the data of Wichterle and Kobayashi (7) and the former data are classified as doubtful. The data of Skripka *et al*. (17) are in fair agreement with the data of Wichterle and Kobayashi (7) and are classified as tentative.

References

1. Hiza, M. J.; Miller, R. C.; Kidnay, A. J.
    J. Phys. Chem. Ref. Data, 1979, *8*, 799.

2. Uehara, K. *Nippon Kagaku Zasshi*, 1932, *53*, 931.

3. Ruhemann, M. *Proc. Roy. Soc. A.*, 1939, *171*, 121.

(cont.)

COMPONENTS:	EVALUATOR:

1. Methane; CH_4 ; [74-82-8]

2. Ethane; C_2H_6 ; [74-84-0]

Colin L. Young,
School of Chemistry,
University of Melbourne,
Parkville, Victoria 3052,
Australia.

February 1982

EVALUATION:

References (cont.)

4.   Guter, M.; Newitt, D. M.; Ruhemann, M.
     Proc. Roy. Soc. A., <u>1940</u>, *176*, 140.

5.   Michels, A.; Nederbragt, G. W.
     Physica, <u>1939</u>, *6*, 656.

6.   Levitskaya, E. P.
     Zh. Tekh. Fiz., <u>1941</u>, *11*, 197.

7.   Wichterle, I.; Kobayashi, R.
     J. Chem. Engng. Data, <u>1972</u>, *17*, 9.

8.   Ellington, R. T.; Eakin, B. E.; Parent, J. D.; Gami, D. C.;
     Bloomer, O. T.
     Thermodynamic and Transport Properties of Gases, Liquids and Solids.
     Am. Soc. Mech. Eng. Heat Transfer Div. McGraw-Hill, New York, 180
     (1959).

     Bloomer, O. T.; Gami, D. C.; Parent, J. D.
     Inst. Gas Technol. Res. Bull. No. 22 (1953).

9.   Wilson, G. M.
     Adv. Gryog. Engng., <u>1975</u>, *20*, 164.

10. Miller, R. C.; Staveley, L. A. K.
     Adv. Cryog. Engng., <u>1976</u>, *21*, 493.

11. Miller, R. C.; Kidnay, A. J.; Hiza, M. J.
     J. Chem. Thermodyn., <u>1977</u>, *9*, 167.

12. Gupta, M. K.; Gardner, G. C.; Hegarty, M. J.; Kidnay, A. J.
     J. Chem. Engng. Data, <u>1980</u>, *25*, 313.

13. Davalos, J.; Anderson, W. R.; Phelps, R. E.; Kidnay, A. J.
     J. Chem. Engng. Data, <u>1976</u>, *21*, 81.

14. Price, A. R.; Kobayashi, R.
     J. Chem. Engng. Data, <u>1959</u>, *4*, 40.

15. Chang, S.-D.; Lu, B. C.-Y.
     Chem. Eng. Prog. Symp. Ser., <u>1967</u>, *63* (81), 18.

16. Hsi, C.; Lu, B. C.-Y.
     Can. J. Chem. Eng., <u>1971</u>, *49*, 140.

17. Skripka, V. G.; Nikitina, I. E.; Zhdanovich, L. A.; Sirotin, A. G.;
     Benyaminovich, O. A.
     Gazov. Prom., <u>1970</u>, *15*, 35.

COMPONENTS:	ORIGINAL MEASUREMENTS:
1. Methane; CH_4; [74-82-8] 2. Ethane; C_2H_6; [74-84-0]	Bloomer, O. T.; Gami, D. C.; Parent, J. D. *Inst. Gas Tech. Res. Bull. no. 22,* <u>1953</u>.

VARIABLES:	PREPARED BY:
	C. L. Young

EXPERIMENTAL VALUES:

T/K	T/°F	P/psia	P/MPa	Mole fraction of methane in liquid, x_{CH_4}	in vapor, y_{CH_4}
149.82	-190.00	54.1	0.373	-	0.9750
159.82	-172.00	109.0	0.751	-	0.9750
168.71	-156.00	185.8	1.281	-	0.9750
177.04	-141.00	293.1	2.021	-	0.9750
183.15	-130.00	393.5	2.713	-	0.9750
192.59	-113.00	579.5	3.996	-	0.9746
		667.5	4.602	0.9746	-
209.26	-115.00	558.0	3.847	-	0.9746
		646.5	4.457	0.9746	-
188.15	-121.00	482.0	3.323	-	0.9746
		585.5	4.037	0.9746	-
183.15	-130.00	387.0	2.668	-	0.9746
		503.1	3.469	0.9746	-
169.82	-154.00	323.2	2.228	0.9746	-
161.48	-169.00	237.2	1.635	0.9746	-
150.93	-188.00	152.9	1.054	0.9746	-
138.71	-210.00	85.0	0.586	0.9746	-
194.26	-110.00	623.3	4.298	-	0.9746
		694.5	4.789	0.9746	-
195.37	-108.00	669.4	4.615	-	0.9746
		713.3	4.918	0.9746	-
213.68	-107.04	690.5	4.761	-	0.9746

(cont.)

AUXILIARY INFORMATION

METHOD/APPARATUS/PROCEDURE:	SOURCE AND PURITY OF MATERIALS:
Bubble point-dew point apparatus with glass equilibrium cell. Temperature measured with copper-constantan thermocouple. Pressure measured using a Bourdon pressure gauge. Dew point and bubble points observed visually.	1. Phillips Petroleum Co. pure grade sample, distilled; final purity better than 99.9 mole per cent; major impurity nitrogen. 2. Phillips Petroleum Co. sample; no impurities detected by mass spectrometry.
	ESTIMATED ERROR: $\delta T/K = \pm 0.15$; $\delta P/MPa = \pm 0.03$ or less.
	REFERENCES:

COMPONENTS:	ORIGINAL MEASUREMENTS:
1. Methane; CH₄; [74-82-8]	Bloomer, O. T.; Gami, D. C.;
2. Ethane; C₂H₆; [74-84-0]	Parent, J. D.
	Inst. Gas Tech. Res. Bull. no. 22,
	<u>1953</u>.

EXPERIMENTAL VALUES:

T/K	T/°F	P/psia	P/MPa	Mole fraction of methane in liquid, x_{CH_4}	in vapor, y_{CH_4}
213.68	-107.04	715.5	4.933	0.9746	-
205.37	-90.00	603.6	4.162	-	0.9250
		792.2[a]	5.462	0.9250	-
204.82	-91.00	788.1[a]	5.434	0.9250	-
204.26	-92.00	784.3[a]	5.408	0.9250	-
202.59	-95.00	768.3	5.297	0.9250	-
199.82	-100.00	728.7	5.024	0.9250	-
196.48	-106.00	672.7	4.638	0.9250	-
192.04	-114.00	602.1	4.151	0.9250	-
185.93	-125.00	508.4	3.505	0.9250	-
178.71	-138.00	407.5	2.810	0.9250	-
171.48	-151.00	321.5	2.217	0.9250	-
162.04	-168.00	228.6	1.576	0.9250	-
151.48	-187.00	148.2	1.022	0.9250	-
140.43	-206.89	88.2	0.608	0.9250	-
160.93	-170.00	47.0	0.324	-	0.9250
175.37	-144.00	116.0	0.800	-	0.9250
185.37	-126.00	206.4	1.423	-	0.9250
193.15	-112.00	312.8	2.157	-	0.9250
198.71	-102.00	420.0	2.896	-	0.9250
202.59	-95.00	519.1	3.579	-	0.9250
207.04	-87.00	689.7	4.755	-	0.9250
207.37	-86.40	712.7	4.914	-	0.9250
207.34	-86.46	780.3[b]	5.380	-	0.9250
207.07	-86.94	785.9[b]	5.405	-	0.9250
206.26	-88.40	793.8[b]	5.473	-	0.9250
218.71	-66.00	888.6[a]	6.127	0.8516	-
217.59	-68.00	881.1[a]	6.075	0.8516	-
216.48	-70.00	874.7	6.031	0.8516	-
215.37	-72.00	867.1	5.978	0.8516	-
213.71	-75.00	849.8	5.859	0.8516	-
210.93	-80.00	814.0	5.612	0.8516	-
205.37	-90.00	731.1	5.041	0.8516	-
198.71	-102.00	630.9	4.350	0.8516	-
190.93	-116.00	519.8	3.584	0.8516	-
183.15	-130.00	417.1	2.876	0.8516	-
172.04	-150.00	49.4	0.341	-	0.8516
185.93	-125.00	109.3	0.754	-	0.8516
198.71	-102.00	209.0	1.441	-	0.8516
174.26	-146.00	320.6	2.211	0.8516	-
164.26	-164.00	226.9	1.564	0.8516	-
153.15	-184.00	146.6	1.011	0.8516	-
142.04	-204.00	88.6	0.611	0.8516	-
207.04	-87.00	313.5	2.162	-	0.8516
213.15	-76.00	420.1	2.896	-	0.8516
217.04	-69.00	508.3	3.505	-	0.8516
220.37	-63.00	607.4	4.188	-	0.8516
222.59	-59.00	703.3	4.849	-	0.8516
223.43	-57.50	761.6	5.251	-	0.8516
222.87	-58.50	865.7[b]	5.969	0.8516	-
222.04	-60.00	878.4[b]	6.056	0.8516	-
220.93	-62.00	886.6[b]	6.113	0.8516	-
219.82	-64.00	887.8[b]	6.121	0.8516	-
238.71	-30.00	553.8	3.818	-	0.7000
240.93	-26.00	610.8	4.211	-	0.7000
244.26	-20.00	718.9	4.957	-	0.7000

(cont.)

COMPONENTS:

1. Methane; CH₄; [74-82-8]

2. Ethane; C₂H₆; [74-84-0]

ORIGINAL MEASUREMENTS:

Bloomer, O. T.; Gami, D. C.;
Parent, J. D.

Inst. Gas Tech. Res. Bull. no. 22,
1953.

EXPERIMENTAL VALUES:

T/K	T/°F	P/psia	P/MPa	Mole fraction of methane in liquid, x_{CH_4}	in vapor, y_{CH_4}
245.93	-17.00	802.9	5.536	–	0.7000
246.32	-16.30	834.9	5.756	–	0.7000
246.32	-16.30	894.7[b]	6.169	–	0.7000
245.37	-18.00	941.9[b]	6.494	–	0.7000
244.26	-20.00	963.3[b]	6.642	–	0.7000
242.04	-24.00	978.2[b]	6.745	–	0.7000
237.04	-33.00	973.4[a]	6.711	0.7000	–
234.82	-37.00	965.3	6.656	0.7000	–
232.04	-42.00	948.1	6.537	0.7000	–
227.59	-50.00	899.9	6.205	0.7000	–
219.26	-65.00	788.9	5.439	0.7000	–
210.93	-80.00	673.9	4.646	0.7000	–
202.59	-95.00	565.4	2.519	0.7000	–
183.15	-130.00	49.2	0.339	–	0.7000
199.82	-100.00	109.7	0.756	–	0.7000
214.82	-73.00	210.2	1.449	–	0.7000
225.93	-53.00	325.5	2.244	–	0.7000
233.15	-40.00	439.3	3.029	–	0.7000
194.26	-110.00	464.3	3.201	0.7000	–
184.26	-128.00	356.3	2.457	0.7000	–
173.71	-147.00	260.3	1.795	0.7000	–
170.21	-153.29	232.3	1.602	0.7000	–
160.19	-171.32	163.3	1.126	0.7000	–
145.47	-197.82	88.3	0.609	0.7000	–
244.26	-20.00	346.2	2.387	–	0.5002
244.26	-50.00	850.7	5.865	0.5002	–
227.59	-50.00	660.0	4.551	0.5002	–
216.48	-70.00	541.0	3.370	0.5002	–
205.37	-90.00	433.3	2.987	0.5002	–
194.26	-110.00	334.3	2.305	0.5002	–
252.59	-5.00	456.2	3.145	–	0.5002
258.15	+5.00	548.7	3.783	–	0.5002
266.43	+19.92	750.5	5.175	–	0.5002
264.71	+24.00	844.7[b]	5.824	–	0.5002
268.62	+23.84	926.2[b]	6.385	–	0.5002
265.98	+19.10	970.0[b]	6.688	–	0.5002
197.04	-105.00	55.5	0.383	–	0.5002
210.93	-80.00	101.5	0.700	–	0.5002
222.04	-60.00	157.8	1.088	–	0.5002
233.15	-40.00	236.0	1.627	–	0.5002
180.37	-135.00	231.0	1.593	0.5002	–
169.26	-155.00	165.6	1.142	0.5002	–
153.73	-182.96	96.8	0.667	0.5002	–
263.15	+14.00	649.4	4.477	–	0.5002
266.48	+20.00	744.4	5.132	–	0.5002
258.15	+5.00	548.7	3.783	–	0.5002
263.15	+14.00	981.4[a]	6.767	0.5002	–
264.26	+16.00	978.2[a]	6.744	0.5002	–
261.48	+11.00	983.9	6.784	0.5002	–
259.26	+7.00	981.9	6.770	0.5002	–
254.84	-0.95	952.6	6.568	0.5002	–
205.37	-10.00	901.4	6.215	0.5002	–
244.26	-20.00	845.0	5.826	0.5002	–
235.93	-35.00	751.6	5.182	0.5002	–
230.37	-45.00	149.1	1.028	–	0.3002
		442.0	3.047	0.3002	–
216.48	-70.00	338.7	2.335	0.3002	–

(cont.)

COMPONENTS:	ORIGINAL MEASUREMENTS:
1. Methane; CH₄; [74-82-8]	Bloomer, O. T.; Gami, D. C.;
2. Ethane; C₂H₆; [74-84-0]	Parent, J. D.
	Inst. Gas Tech. Res. Bull. no. 22,
	1953.

EXPERIMENTAL VALUES:

T/K	T/°F	P/psia	P/MPa	Mole fraction of methane in liquid, x_{CH_4}	in vapor, y_{CH_4}
202.59	-95.00	250.9	1.729	0.3002	-
188.71	-120.00	178.6	1.231	0.3002	-
172.40	-149.35	112.9	0.778	0.3002	-
202.59	-95.00	50.7	0.350	-	0.3002
216.48	-70.00	90.2	0.622	-	0.3002
230.37	-45.00	148.9	1.027	-	0.3002
244.26	-20.00	234.2	1.615	-	0.3002
255.37	0.00	326.9	2.254	-	0.3002
266.48	+20.00	449.1	3.096	-	0.3002
273.15	+32.00	543.4	3.747	-	0.3002
278.71	+42.00	638.3	4.401	-	0.3002
283.15	+50.00	738.4	5.091	-	0.3002
285.37	+54.00	815.4	5.622	-	0.3002
285.11	+53.53	870.4[b]	6.001	0.3002	-
283.71	+51.00	888.8[a]	6.128	-	0.3002
280.93	+46.00	685.3	4.725	0.3002	-
280.23	+44.75	899.8	6.204	0.3002	-
279.09	+42.70	894.3	6.166	0.3002	-
274.55	+34.52	858.8	5.921	0.3002	-
266.48	+20.00	777.3	5.359	0.3002	-
252.59	-5.00	638.8	4.404	0.3002	-
242.04	-24.00	538.8[b]	3.715	0.3002	-
285.03	+53.39	872.3[b]	6.014	-	0.2998
284.26	+52.00	888.3[a]	6.125	0.2998	-
282.59	+49.00	898.3[a]	6.821	0.2998	-
204.26	-92.00	45.8	0.316	-	0.1507
216.48	-70.00	74.5	0.514	-	0.1507
232.04	-42.00	129.9	0.896	-	0.1507
245.93	-17.00	200.6	1.383	-	0.1507
255.37	0.00	263.5	1.817	-	0.1507
266.48	+20.00	356.2	2.456	-	0.1507
275.93	+37.00	454.9	3.136	-	0.1507
283.15	+50.00	546.6	3.769	-	0.1507
289.82	+62.00	647.5	4.464	-	0.1507
294.26	+70.00	734.0	5.061	-	0.1507
295.65	+72.50	774.0	5.337	-	0.1507
293.71	+69.00	720.1	4.965	-	0.1507
		809.3	5.580	0.1507	-
286.94	+56.83	749.6	5.168	0.1507	-
280.93	+46.00	684.9	4.722	0.1507	-
271.48	+29.00	587.6	4.051	0.1507	-
263.15	+14.00	508.1	3.503	0.1507	-
254.82	-1.00	437.9	3.019	0.1507	-
245.93	-17.00	370.6	2.555	0.1507	-
216.48	-70.00	201.6	1.390	0.1507	-
202.59	-95.00	145.2	1.001	0.1507	-
185.93	-125.00	93.9	0.647	0.1507	-
172.04	-150.00	61.9	0.427	0.1507	-
160.93	-170.00	40.6	0.280	0.1507	-
202.59	-95.00	42.4	0.292	-	0.1498
216.48	-70.00	74.5	0.514	-	0.1498
232.04	-42.00	129.7	0.894	-	0.1498
245.93	-17.00	200.4	1.382	-	0.1498
		370.6	2.555	0.1498	-
230.37	-45.00	270.9	1.868	0.1498	-
216.48	-70.00	201.2	1.387	0.1498	-

(cont.)

COMPONENTS: ORIGINAL MEASUREMENTS:

1. Methane; CH₄; [74-82-8]   Bloomer, O. T.; Gami, D. C.;
                                         Parent, J. D.
2. Ethane; C₂H₆; [74-84-0]
                                         Inst. Gas Tech. Res. Bull. no. 22,
                                         1953.

EXPERIMENTAL VALUES:

				Mole fraction of methane	
				in liquid,	in vapor,
T/K	$T/°F$	P/psia	P/MPa	x_{CH_4}	y_{CH_4}
202.59	-95.00	144.7	0.998	0.1498	–
289.82	+62.00	648.9	4.474	–	0.1498
294.26	+70.00	735.0	5.068	–	0.1498
295.65	+72.50	769.1	5.302	–	0.1498
295.54	+72.30	798.4[b]	5.505	–	0.1498
295.37	+72.00	802.1[a]	5.530	0.1498	–
294.82	+71.00	805.9[a]	5.556	0.1498	–
293.71	+69.00	810.1	5.585	0.1498	–
292.04	+66.00	799.4	5.512	0.1498	–
287.99	+58.71	760.8	5.246	0.1498	–
208.15	-85.00	46.9	0.323	–	0.0500
222.04	-60.00	80.7	0.556	–	0.0500
235.93	-35.00	129.6	0.894	–	0.0500
249.82	-10.00	197.4	1.361	–	0.0500
263.71	+15.00	289.1	1.993	–	0.0500
273.93	+33.40	374.7	2.583	–	0.0500
284.26	+52.00	380.1	2.621	–	0.0500
293.15	+68.00	589.6	4.065	–	0.0500
299.82	+80.00	687.9	4.743	–	0.0500
296.59	+74.20	638.7	4.404	–	0.0500
296.79	+74.55	687.0	4.737	0.0500	–
289.82	+62.00	601.2	4.145	0.0500	–
282.04	+48.00	520.4	3.588	0.0500	–
272.04	+30.00	424.6	2.928	0.0500	–
260.93	+10.00	333.9	2.302	0.0500	–
247.04	-15.00	242.4	1.671	0.0500	–
230.37	-45.00	160.0	1.103	0.0500	–
210.93	-80.00	93.5	0.645	0.0500	–
188.71	-120.00	46.7	0.322	0.0500	–
302.12	+84.15	734.9[a]	5.067	–	0.0500
		739.4[a]	5.098	0.0500	–
301.48	+83.00	743.4	5.126	0.0500	–
300.37	+81.00	731.9	5.046	0.0500	–
298.71	+78.00	712.4	4.912	0.0500	–

[a] Phase boundary disappeared.

[b] Second (upper) dew point.

COMPONENTS:	ORIGINAL MEASUREMENTS:
1. Methane; CH_4; [74-82-8] 2. Ethane; C_2H_6; [74-84-0]	Price, A. R.; Kobayashi, R. *J. Chem. Engng. Data* 1959, *4*, 40-52.
VARIABLES: Temperature, pressure	PREPARED BY: C. L. Young

EXPERIMENTAL VALUES:

T/K	P/MPa	Mole fraction of methane in liquid, x_{CH_4}	in vapor, y_{CH_4}
227.59	1.38	0.0968	0.505
	2.76	0.282	0.724
	4.14	0.438	0.800
	5.52	0.622	0.832
199.82	0.689	0.101	0.680
	1.38	0.250	0.814
	2.76	0.528	0.9165
	4.14	0.784	0.9498
172.04	0.689	0.251	0.9312
	1.38	0.561	0.970
144.26	0.689	0.833	0.9965

AUXILIARY INFORMATION

METHOD/APPARATUS/PROCEDURE:	SOURCE AND PURITY OF MATERIALS:
Recirculating vapor flow apparatus with modified Jerguson sight gauge for equilibrium cell. Vapor recycled with magnetic pump. Pressure measured with Bourdon pressure gauge and temperature measured with thermocouple. Details in source.	1. Phillips Petroleum Co. research grade, purity 99.5 mole per cent. 2. Phillips Petroleum Co. research grade, purity 99.9 mole per cent.

ESTIMATED ERROR:

$\delta T/K = \pm 0.06$; $\delta P/MPa = \pm 1\%$;

δx_{CH_4}, $\delta y_{CH_4} = \pm 2\%$ (estimated by compiler).

REFERENCES:

COMPONENTS:	ORIGINAL MEASUREMENTS:
1. Methane; CH_4; [74-82-8] 2. Ethane; C_2H_6; [74-84-0]	Skripka, V. G.; Nikitina, I. E.; Zhdanovich, L. A.; Sirotin, A. G.; Benyaminovich, O. A. *Gazov. Prom.* <u>1970</u>, *15*, 35-36.

VARIABLES:	PREPARED BY:
	C. L. Young

EXPERIMENTAL VALUES:

T/K (T/°C)	Bubble pt. pressure		Dew pt. pressure		Mole fraction x_{CH_4}
	P/kg f cm^{-2}	P/MPa	P/kg f cm^{-2}	P/MPa	
123.2 (-150)	0.005	0.0005	0.005	0.0005	0.0
	0.11	0.0108	–	–	0.05
	0.48	0.047	–	–	0.20
	0.99	0.097	–	–	0.40
	1.52	0.149	–	–	0.60
	2.02	0.198	–	–	0.80
	2.23	0.219	–	–	0.95
	2.43	0.238	2.43	0.238	1.00
133.2 (-140)	0.02	0.002	0.02	0.002	0.0
	0.30	0.029	–	–	0.05
	1.10	0.108	–	–	0.20
	2.06	0.202	–	–	0.40
	2.92	0.286	–	–	0.60
	3.74	0.367	–	–	0.80
	4.32	0.424	–	–	0.95
	4.51	0.442	4.51	0.442	1.00
143.2 (-130)	0.05	0.005	0.05	0.005	0.00
	0.50	0.049	–	–	0.05
	1.79	0.176	–	–	0.20
	3.43	0.336	–	–	0.40
	4.89	0.480	–	–	0.60
	6.24	0.612	–	–	0.80
	7.36	0.722	–	–	0.95
	7.67	0.752	7.67	0.752	1.00

(cont.)

AUXILIARY INFORMATION

METHOD/APPARATUS/PROCEDURE:	SOURCE AND PURITY OF MATERIALS:
Recirculating vapor flow apparatus fitted with magnetic stirrer. Temperature measured with platinum resistance thermometer. Liquid and gas analysed by gas chromatography. Details of apparatus in ref. (1)	1 and 2. Purity not less than 99.9 per cent by volume.

	ESTIMATED ERROR:

REFERENCES:

1. Skripka, V. G.; Barsuk, S. D.; Nikitina, I. E.; Ben'yaminovic, O. A.

   Gazov. Prom.

   <u>1964</u>, *14*, 11.

COMPONENTS:

1. Methane; CH_4; [74-82-8]

2. Ethane; C_2H_6; [74-84-0]

ORIGINAL MEASUREMENTS:

Skripka, V. G.; Nikitina, I. E.;
Zhdanovich, L. A.; Sirotin, A. G.;
Benyaminovich, O. A.

Gazov. Prom.

1970, *15*, 35-36.

EXPERIMENTAL VALUES:

T/K (T/°C)	Bubble pt. pressure $P/\text{kg f cm}^{-2}$	P/MPa	Dew pt. pressure $P/\text{kg f cm}^{-2}$	P/MPa	Mole fraction x_{CH_4}
153.2	0.13	0.013	0.13	0.013	0.00
(-120)	0.73	0.072	-	-	0.05
	2.55	0.250	-	-	0.20
	4.95	0.485	-	-	0.40
	7.36	0.722	-	-	0.60
	9.77	0.958	-	-	0.80
	11.57	1.135	2.46	0.241	0.95
	12.18	1.194	12.18	-	1.00

COMPONENTS:	ORIGINAL MEASUREMENTS:
1. Methane; CH₄; [74-82-8] 2. Ethane; C₂H₆; [74-84-0]	Hsi, C.; Lu, B. C.-Y. *Can. J. Chem. Eng.* <u>1971</u>, *49*, 140-143. (Supplementary data)

VARIABLES:	PREPARED BY:
	C. L. Young

EXPERIMENTAL VALUES:

T/K (T/°F)	P/psi	P/MPa	Mole fraction of methane	
			in liquid, x_{CH_4}	in vapor, y_{CH_4}
159.2 (-173.1)	32.54	0.2244	0.1106	0.9130
	48.25	0.3327	0.1668	0.9400
	72.05	0.4968	0.2658	0.9611
	87.10	0.6005	0.3234	0.9676
	102.00	0.7033	0.3990	0.9747
	113.60	0.7832	0.4592	0.9791
	133.00	0.9170	0.5471	0.9831
	145.50	0.9997	0.6214	0.9863
	164.35	1.1332	0.7135	0.9893
	172.22	1.1874	0.7779	0.9916
	193.00	1.3307	0.8766	0.9943
	201.4	1.3886	0.9173	0.9963

AUXILIARY INFORMATION

METHOD/APPARATUS/PROCEDURE:	SOURCE AND PURITY OF MATERIALS:
Recirculating vapor flow apparatus constructed of 100 ml Jerguson gauge. Temperature measured using copper-constantan thermocouples. Pressure measured with Bourdon gauge. Cell charged and vapor recirculated with magnetic pump for 2 or more hours. Samples of vapor and liquid removed at constant pressure and analysed using gas chromatography. Helium was used as a carrier gas.	1. Matheson research grade sample, 99.99 mole per cent. 2. Matheson research grade sample, purity 99.9 mole per cent.
	ESTIMATED ERROR: δT/K = ±0.02; δP/MPa ∿ ±0.005; δx_{CH_4}, δy_{CH_4} = ±1% (estimated by compiler).
	REFERENCES:

COMPONENTS:	ORIGINAL MEASUREMENTS:
1. Methane; CH_4; [74-82-8] 2. Ethane; C_2H_6; [74-84-0]	Wichterle, I.; Kobayashi, R. *J. Chem. Engng. Data.* <u>1972</u>, *17*, 9-12.

VARIABLES:	PREPARED BY:
Temperature, pressure	C.L. Young

EXPERIMENTAL VALUES:

T/K	P/psi	P/MPa	Mole fraction of methane in liquid, x_{CH_4}	in vapor, y_{CH_4}
199.92	31.45a	0.2168a	0.0000	0.0000
	45.0	0.367	0.0214	0.3005
	65.0	0.448	0.0512	0.5098
	100.0	0.689	0.1039	0.6800
	160.0	1.104	0.1875	0.7957
	250.0	1.723	0.3100	0.8679
	350.0	2.413	0.4526	0.9052
	500.0	3.447	0.6601	0.9337
	600.0	4.137	0.7852	0.9461
	700.0	4.826	0.8942	0.9562
	719.0	4.957	0.9126	0.9584
	726.0	5.006	0.9175	0.9578
	732.0	5.047	0.9222	0.9575
	740.0	5.102	0.9319	0.9577
	748.0b	5.157b	0.9520	0.9520
195.44	522.0	3.599	0.7648	0.9546
	591.0	4.075	0.8539	0.9627
	634.0	4.371	0.9007	0.9680
	680.0	4.688	0.9437	0.9755
	693.0	4.778	0.9529	0.9764
	704.0	4.854	0.9613	0.9770
	708.0	4.881	0.9641	0.9756
	713.0b	4.916b	0.9706	0.9706

AUXILIARY INFORMATION

METHOD/APPARATUS/PROCEDURE:	SOURCE AND PURITY OF MATERIALS:
Recirculating vapor flow apparatus with magnetic vapor pump. Pressure measured with Bourdon gauge and temperature with platinum resistance thermometer. Samples of both phases analysed using gas chromatography with flame ionisation detector. Details in source and ref. (1).	1. Matheson Gas Products sample purity 99.97 mole per cent; purified by passing through molecular sieve. 2. Phillips Petroleum research grade. sample purity 99.99 mole per cent.

ESTIMATED ERROR:

$\delta T/K = \pm 0.1$; $\delta P/MPa = \pm 0.015$ or less; $\delta x_{CH_4}, \delta y_{CH_4} = \pm 2\%$ or less.

REFERENCES:

1. Chang, H.L.; Hurt, L.J.; Kobayashi, R.

   Am. Inst. Chem. Engnrs. J. <u>1966</u>, *11*, 1212.

COMPONENTS:		ORIGINAL MEASUREMENTS:
1. Methane; CH₄; [74-82-8]		Wichterle, I.; Kobayashi, R.
2. Ethane; C₂H₆; [74-84-0]		J. Chem. Engng. Data. 1972, 17, 9-12.

EXPERIMENTAL VALUES:

T/K	P/psi	P/MPa	Mole fraction of methane in liquid, x_{CH_4}	in vapor, y_{CH_4}
193.44	612.0	4.220	0.9007	0.9715
	660.0	4.551	0.9469	0.9787
	675.0	4.654	0.9590	0.98145
	685.0	4.723	0.9671	0.98316
	692.0	4.771	0.9724	0.98352
	696.0	4.799	0.9756	0.98371
	698.0b	4.813b	0.98214	0.98214
192.39	22.05b	0.1520a	0.0000	0.0000
	39.0	0.269	0.0308	0.4339
	61.5	0.424	0.0681	0.6358
	100.0	0.689	0.1364	0.7755
	160.0	1.104	0.2299	0.8605
	250.0	1.723	0.3854	0.9087
	350.0	2.413	0.5512	0.9372
	500.0	3.447	0.7885	0.9607
	600.0	4.137	0.9154	0.9755
	640.0	4.413	0.9529	0.9822
	662.0	4.564	0.9715	0.9862
	675.0	4.654	0.9813	0.98973
	682.0	4.702	0.9858	0.99100
	685.0b	4.723b	0.99125	0.99125
190.94	20.41a	0.1407a	0.0000	0.0000
	39.5	0.273	0.0340	0.4724
	60.5	0.417	0.0714	0.6513
	100.0	0.689	0.1386	0.7907
	160.0	1.104	0.2384	0.8709
	250.0	1.723	0.3955	0.9181
	350.0	2.413	0.5687	0.9425
	500.0	3.447	0.8150	0.9647
	563.0	3.882	0.9007	0.9750
	612.0	4.220	0.9474	0.9832
	630.0	4.344	0.9646	0.9869
	642.0	4.426	0.9749	0.9898
	651.0	4.488	0.9819	0.99187
	660.0	4.551	0.9882	0.99445
	664.0	4.578	0.99160	0.99579
	668.0	4.606	0.99383	0.99693
	670.0	4.619	0.99561	0.99765
	671.0c	4.626c	1.0000	1.0000
189.65	19.22a	0.1325a	0.0000	0.0000
	35.8	0.247	0.0320	0.4600
	60.0	0.413	0.0752	0.6741
	95.5	0.659	0.1380	0.7964
	140.0	0.966	0.2138	0.8626
	200.0	1.378	0.3202	0.9026
	300.0	2.068	0.4975	0.9369
	400.0	2.758	0.6816	0.9550
	500.0	3.447	0.8403	0.9706
	562.0	3.875	0.9214	0.9803
	600.0	4.137	0.9580	0.9876
	615.0	4.240	0.9729	0.99084
	625.0	4.309	0.9819	0.99388
	634.0	4.371	0.9893	0.99626
	648.0c	4.468c	1.0000	1.0000
188.04	477.0	3.288	0.8403	0.9719
	518.0	3.571	0.8970	0.9789
	556.0	3.833	0.9413	0.9849

COMPONENTS:		ORIGINAL MEASUREMENTS:
1. Methane; CH$_4$; [74-82-8]		Wichterle, I.; Kobayashi, R.
		J. Chem. Engng. Data. <u>1972</u>,
2. Ethane; C$_2$H$_6$; [74-84-0]		*17*, 9-12.

EXPERIMENTAL VALUES:

T/K	P/psi	P/MPa	Mole fraction of methane in liquid, x_{CH_4}	in vapor, y_{CH_4}
188.04	580.0	3.999	0.9643	0.9898
	598.0	4.123	0.9819	0.99448
	606.0	4.178	0.9887	0.99661
	616.0c	4.247c	1.0000	1.0000
186.11	16.00a	0.1103a	0.0000	0.0000
	36.7	0.253	0.0417	0.5585
	61.5	0.424	0.0897	0.7358
	104.0	0.717	0.1707	0.8471
	180.0	1.241	0.3100	0.9116
	275.0	1.896	0.5019	0.9435
	400.0	2.758	0.7486	0.9674
	482.0	3.323	0.8844	0.9792
	524.0	3.613	0.9395	0.9866
	550.0	3.792	0.9684	0.99187
	562.0	3.875	0.9802	0.99495
	568.0	3.916	0.9877	0.99730
	579.0c	3.992c	1.0000	1.0000
172.04	7.099a	0.04895a	0.0000	0.0000
	30.8	0.213	0.0685	0.7681
	45.5	0.314	0.1087	0.8469
	81.0	0.558	0.2050	0.9161
	120.0	0.828	0.3164	0.9434
	180.0	1.241	0.5042	0.9656
	247.5	1.706	0.7082	0.9788
	299.0	2.062	0.8609	0.9878
	324.0	2.234	0.9175	0.99209
	339.5	2.341	0.9513	0.99531
	361.5c	2.492c	1.0000	1.0000
158.15	2.715a	0.01872a	0.0000	0.0000
	25.8	0.178	0.1090	0.8990
	28.8	0.199	0.1230	0.9089
	40.0	0.276	0.1640	0.9354
	50.0	0.345	0.2186	0.9485
	70.0	0.482	0.2953	0.9645
	100.0	0.689	0.4382	0.9767
	140.0	0.965	0.6528	0.9864
	170.0	1.172	0.8107	0.99221
	181.0	1.248	0.8650	0.99420
	199.0	1.372	0.9407	0.99740
	213.5c	1.472c	1.0000	1.0000
144.26	0.841a	0.00580a	0.0000	0.0000
	27.3	0.188	0.1965	0.9716
	37.0	0.255	0.2702	0.9796
	43.0	0.297	0.3241	0.9834
	56.0	0.386	0.4385	0.9880
	66.0	0.455	0.5314	0.99100
	81.0	0.558	0.6882	0.99433
	98.0	0.676	0.8581	0.99728
	114.0c	0.786c	1.0000	1.0000
130.37	0.186a	0.00128a	0.0000	0.0000
	28.0	0.193	0.4319	0.99479
	35.0	0.241	0.5989	0.99654
	43.3	0.299	0.7788	0.99845
	48.6	0.334	0.8935	0.99919
	54.0c	0.372c	1.0000	1.0000

a) saturated vapor pressure of ethane. b) gas-liquid critical pressure.
c) saturated vapor pressure of methane.

COMPONENTS:	ORIGINAL MEASUREMENTS:
1. Methane; CH_4; [74-82-8] 2. Ethane; C_2H_6; [74-84-0]	Davalos, J.; Anderson, W. R.; Phelps, R. E.; Kidnay, A. J. *J. Chem. Eng. Data* <u>1976</u>, *21*, 81-84.

VARIABLES:	PREPARED BY:
	C. L. Young

EXPERIMENTAL VALUES:

T/K	P/atm	P/MPa	Mole fraction of methane in liquid, x_{CH_4}	in vapor, y_{CH_4}
250.00	15.10		0.024	0.134
	22.00		0.089	0.365
	22.50		0.097	0.383
	32.50		0.196	0.540
	34.00		–	0.554
	45.00		0.320	0.643
	55.20		0.426	0.673
	65.70		0.546	0.673
	67.50		single phase	

AUXILIARY INFORMATION

METHOD/APPARATUS/PROCEDURE:	SOURCE AND PURITY OF MATERIALS:
Recirculating vapor flow apparatus. Temperature measured with platinum resistance thermometer. Pressure measured with Bourdon gauge. Gas and liquid samples analysed by gas chromatography using a thermal conductivity detector. Details in source and ref. 1.	1. Matheson ultra high purity sample, maximum impurity 0.03 mole per cent. 2. Purity probably better than 99.9 mole per cent.

ESTIMATED ERROR:
$\delta T/K = \pm0.01$; $\delta P/MPa = \pm0.003$ up to 3.5 MPa, ±0.005 above 3.5 MPa; δx_{CH_4}, $\delta y_{CH_4} = \pm1.5\%$.

REFERENCES:
1. Miller, R. C.; Kidnay, A. J.; Hiza, M. J.   *J. Chem. Thermodyn.*   <u>1972</u>, *4*, 807.

COMPONENTS:	ORIGINAL MEASUREMENTS:
1. Methane; CH_4; [74-82-8] 2. Ethane; C_2H_6; [74-84-0]	Miller, R. C.; Kidnay, A. J.; Hiza, M. J. *J. Chem. Thermodynamics* 1977, *9*, 167-178.
VARIABLES: Temperature, pressure	PREPARED BY: C. L. Young

EXPERIMENTAL VALUES:

T/K	P/MPa	Mole fraction of methane in liquid, x_{CH_4}	in vapor, y_{CH_4}
160.00	0.1233	0.0602	0.8270
	0.2026	0.1118	0.8970
	0.2037	0.1127	0.8986
	0.3050	0.1720	0.9344
	0.3995	0.2225	0.9511
	0.4116	0.2292	0.9534
	0.5294	0.3035	0.9652
	0.5984	0.3504	0.9704
	0.793	0.4795	0.9794
	0.993	0.6203	0.9854
	1.195	0.7596	0.9909
	1.338	0.8559	–
180.00	0.184	0.0327	0.5734
	0.313	0.0745	0.7493
	0.471	0.1241	0.8335
	0.567	0.1567	0.8633
	0.953	0.2825	0.9213
	1.567	0.4928	0.9551
	2.128	0.6898	0.9712
	2.482	0.8025	0.9792
	2.828	0.9002	0.9874

AUXILIARY INFORMATION

METHOD/APPARATUS/PROCEDURE:	SOURCE AND PURITY OF MATERIALS:
Vapor-recirculation system similar to that in refs. 1 and 2. Pressure measured with Bourdon gauge. Temperature measured with platinum resistance thermometer. Samples of liquid and vapor analysed by gas chromatography. Details in source.	1. Purity 99.99 mole per cent. 2. Purity 99.99 mole per cent.

ESTIMATED ERROR:
$\delta T/K = \pm 0.02$; $\delta P/MPa = \pm 0.001$ up to 0.6 MPa; ± 0.005 above 0.6 MPa; δx_{CH_4}, $\delta y_{CH_4} = \pm 0.001$.

REFERENCES:

1. Duncan, A. G.; Hiza, M. J.
   Adv. Cryogen. Engng. 1970, *15*, 42.

2. Hiza, M. J.; Duncan, A. G.
   Rev. Sci. Instr. 1969, *40*, 513.

COMPONENTS:	ORIGINAL MEASUREMENTS:
1. Methane; CH_4; [74-82-8] 2. Ethane; C_2H_6; [74-84-0]	Gupta, M.K.; Gardner, G.C.; Hegarty, M.J.; Kidnay, A.J. *J. Chem. Engng. Data* <u>1980,</u> 25, 313-318.

VARIABLES:	PREPARED BY:
Temperature, pressure	C.L. Young

EXPERIMENTAL VALUES:

	Total pressure		Mole fraction of methane	
			in liquid,	in gas
T/K	p/atm	p/MPa	x_{CH_4}	y_{CH_4}
260.00	16.80	1.702	0.0000	0.0000
	17.80	1.804	0.0088	0.0467
	18.88	1.913	-	0.0942
	19.42	1.968	0.0239	0.1150
	21.60	2.189	0.0445	0.1922
	23.22	2.353	0.0592	0.2387
	25.26	2.559	0.0788	0.2924
	29.98	3.038	0.1220	0.3810
	34.69	3.515	0.1652	0.4483
	40.59	4.113	0.2189	0.5052
	50.30	5.097	0.3087	0.5664
	55.18	5.591	0.3545	0.5823
	59.40	6.019	0.3952	0.5896
	62.65	6.348	0.4297	0.5897
	65.18	6.604	0.4615	0.5823
	69.90	7.083		single phase
270.00	21.68	2.197	0.0000	0.0000
	22.53	2.283	0.0070	0.0298
	23.39	2.421	0.0188	0.0762
	26.10	2.645	0.0357	0.1332
	30.69	3.110	0.0739	0.2361
	35.07	3.553	0.1127	0.3140
	40.00	4.053	0.1550	0.3768

AUXILIARY INFORMATION

METHOD/APPARATUS/PROCEDURE:	SOURCE AND PURITY OF MATERIALS:
Recirculating vapor flow apparatus with diaphragm pump. Temperature measured with platinum resistance thermometer and pressure with Bourdon gauges Cell stirred with double propeller stirrer. Vapor and liquid samples analysed by gas chromatography using a thermal conductivity detector. Details in ref. (1).	1 and 2. Purity at least 99 mole per cent. No extraneous peaks were found when samples analysed by gas chromatography.

ESTIMATED ERROR:

$\delta T/K = \pm 0.02$; $\delta p/MPa = \pm 0.03$ up to 3.4 MPa, ± 0.1 above 3.4 MPa, δx_{CH_4}, $\delta y_{CH_4} = \pm 0.002$.

REFERENCES:

1. Somait, F.; Kidnay, A.J.
    J. Chem. Engng.Data <u>1978</u>, *23*, 301.

COMPONENTS:	ORIGINAL MEASUREMENTS:
1. Methane; CH_4; [74-82-8] 2. Ethane; C_2H_6; [74-84-0]	Gupta, M. K.; Gardner, G. C.; Hegarty, M. J.; Kidnay, A. J. *J. Chem. Engng. Data* 1980, *25*, 313-318.

T/K	Total pressure		Mole fraction of methane	
	p/atm	p/MPa	in liquid, x_{CH_4}	in gas y_{CH_4}
270.00	45.30	4.590	0.1984	0.4237
	45.50	4.610	-	0.4251
	49.06	4.971	0.2315	0.4484
	50.26	5.093	0.2419	0.4558
	55.02	5.575	0.2819	-
	60.22	6.102	0.3321	0.4889
	62.96	6.379	-	0.4896
	64.53	6.539	0.3862	-
	65.14	6.600	0.3980	-
	65.44	6.631	-	0.4856
	66.50	6.738	single phase	
280.00	27.60	2.797	0.0000	0.0000
	28.50	2.888	0.0081	0.0279
	30.35	3.075	0.0231	0.0744
	31.62	3.204	0.0335	0.1036
	36.08	3.656	0.0693	0.1864
	39.85	4.038	0.0996	0.2404
	43.58	4.416	0.1292	0.2818
	46.50	4.712	0.1528	0.3091
	49.75	5.041	0.1798	0.3336
	52.60	5.330	0.2044	0.3512
	55.72	5.646	0.2333	0.3654
	59.06	5.984	0.2668	0.3719
	60.07	6.087	0.2774	0.3711
	60.51	6.131	-	0.3699
	61.10	6.191	0.2913	0.3675
	61.84	6.266	0.3046	0.3609
	62.10	6.292	0.3109	0.3557
	62.40	6.323	single phase	

COMPONENTS:	EVALUATOR:
1. Methane; CH₄; [74-82-8] 2. Propane; C₃H₈; [74-98-6]	Colin L. Young, School of Chemistry, University of Melbourne, Parkville, Victoria 3052, Australia. February 1984

EVALUATION:

The most extensive study on this system is due to Wichterle and Kobayashi (1) who studied the system at temperatures from 130 K to 214 K. This work extended earlier work by Price and Kobayashi (2) who studied this system in the temperature range 144 K to 278 K. The data of Price and Kobayashi (2) at 213.7 K and at lower temperatures are less accurate than those of Wichterle and Kobayashi (1). The data of Sage and coworkers (3), (4) cover the temperature range 277 K to 363 K. Their data are only of moderate accuracy due to the techniques available at that time. The data of Akers, Burns and Fairchild (5) are also only of moderate accuracy and cover the temperature range 158 K to 273 K. The data of Poon and Lu (6) are thought to be of fairly high accuracy.

The data of Wichterle and Kobayashi (1) and Poon and Lu (8) are classified as recommended, whereas those of Price and Kobayashi (2) and Sage and coworkers (3), (4) are of lower accuracy. The data of Akers, Burns and Fairchild (5) are superseded by the more recent and more accurate data of Wichterle and Kobayashi (1).

The limited data of Cheung and Wang (7) are restricted to pressures below 2 atmospheres and are classified as tentative. The four measurements by Kalra and Robinson (8) were determined to test the reliability of their apparatus and agree well with the data tiven in ref. (1) at 213.7 K. The data of Frolich *et al.* (9) at 298 K which were presented in graphical form are rejected.

References

1.   Wichterle, I.; Kobayashi, R. *J. Chem. Eng. Data*, <u>1972</u>, *17*, 4.

2.   Price, R. A.; Kobayashi, R. *J. Chem. Eng. Data*, <u>1959</u>, *4*, 40.

3.   Sage, B. H.; Lacey, W. N.; Schassfma, J. G. *Ind. Eng. Chem.*, <u>1934</u>, *26*, 214.

4.   Reamer, H. H.; Sage, B. H.; Lacey, W. N. *Ind. Eng. Chem.*, <u>1950</u>, *42*, 534.

5.   Akers. W. W.; Burns, J. F.; Fairchild, W. R. *Ind. Eng. Chem.*, <u>1954</u>, *46*, 2531.

6.   Poon, D. P. L.; Lu, B. C.-Y. *Adv. Cryog. Eng.*, <u>1973</u>, *19*, 292.

7.   Cheung, H.; Wang, D. I. J. *Ind. Eng. Chem. Fundam.*, <u>1964</u>, *3*, 355.

8.   Kalra, H.; Robinson, D. B. *Cryogenics*, <u>1975</u>, *15*, 409.

9.   Frolich, P. K.; Tauch, E. J.; Hogan, J. J.; Peer, A. A. *Ind. Eng. Chem.*, <u>1931</u>, *23*, 548.

COMPONENTS:	ORIGINAL MEASUREMENTS:
1. Methane; CH_4; [74-82-8] 2. Propane; C_3H_8; [74-98-6]	Frolich, P.K.; Tauch, E.J.; Hogan, J.J.; Peer, A.A. *Ind. Eng. Chem.* <u>1931</u>,*23*,548-550.

VARIABLES:
 	
    Pressure
	PREPARED BY:
	
	    C.L. Young

EXPERIMENTAL VALUES:

T/K	P/MPa	Solubility*,S	Mole fraction of methane in liquid,[+] x_{CH_4}
298.15	1.0	19.4	0.0660
	2.0	38.7	0.124
	3.0	58.1	0.175
	4.0	80.4	0.227
	5.0	102	0.271
	6.0	124	0.311
	7.0	146	0.347
	8.0	168.5	0.380
	9.07	208	0.431

*   Data taken from graph in original article.

+   calculated by compiler.

AUXILIARY INFORMATION

METHOD/APPARATUS/PROCEDURE:	SOURCE AND PURITY OF MATERIALS:
Static equilibrium cell. Liquid saturated with gas and after equilibrium established samples removed and analysed by volumetric method. Allowance was made for vapor pressure of liquid and the solubility of the gas at atmospheric pressure. Details in source.	Stated that the materials were the highest purity available.
	ESTIMATED ERROR: $\delta T/K = \pm 0.1$; $\delta x_{CH_4} = \pm 5\%$
	REFERENCES:

COMPONENTS:	ORIGINAL MEASUREMENTS:
1. Methane; CH_4; [74-82-8] 2. Propane; C_3H_8; [74-98-6]	Sage, B. H.; Lacey, W. N.; Schaafsma, J. G. *Ind. Eng. Chem.* <u>1934</u>, *26*, 214-217.
VARIABLES: Temperature, pressure	PREPARED BY: C. L. Young

EXPERIMENTAL VALUES:

T/K	$P/10^5$Pa	Mole fraction of methane in liquid, x_{CH_4}	in vapor, y_{CH_4}
293.15	10.1	0.008	0.132
	15.2	-	0.386
	20.3	0.062	0.505
	25.4	-	0.575
	30.4	0.116	0.619
	35.5	-	0.654
	40.5	0.176	0.681
	45.6	-	0.701
	50.7	0.236	0.714
	60.8	0.298	0.728
	70.9	0.363	0.736
	81.1	0.436	0.734
	91.19	0.524	0.714
	96.26	0.583	0.690
313.15	15.2	0.007	0.079
	20.3	0.035	0.273
	25.4	-	0.379
	30.4	0.092	0.451
	35.5	-	0.503
	40.5	0.149	0.543
	45.6	-	0.571
	50.7	0.208	0.592
	55.7	-	0.606
	60.8	0.260 (cont.)	0.615

AUXILIARY INFORMATION

METHOD/APPARATUS/PROCEDURE:	SOURCE AND PURITY OF MATERIALS:
PVT cell charged with mixture of known composition. Dew point or bubble point determined from PVT data and vapor-liquid equilibrium data obtained by graphical means. Pressure measured with pressure balance and temperature measured with a copper-constantan thermo-couple. Details in source and ref. (1).	1. Prepared from natural gas, carbon dioxide, water and hydrocarbons removed. Distilled. Final purity 99.47 mole per cent; major impurities, nitrogen (0.5 mole per cent) and ethane and higher hydrocarbons (0.03 mole per cent). 2. Phillips Petroleum Co. sample.
	ESTIMATED ERROR: $\delta T/K = \pm 0.1$; $\delta P/MPa = \pm 0.01$; δx_{CH_4}, $\delta y_{CH_4} = \pm 0.001$.
	REFERENCES: 1. Sage, B. H.; Lacey, W. N. *Ind. Eng. Chem.* <u>1934</u>, *26*, 103.

COMPONENTS:	ORIGINAL MEASUREMENTS:
1. Methane; CH_4; [74-82-8]	Sage, B. H.; Lacey, W. N.;
2. Propane; C_3H_8; [74-98-6]	Schaafsma, J. G.
	Ind. Eng. Chem.
	<u>1934</u>, *26*, 214-217.

EXPERIMENTAL VALUES:

T/K	$P/10^5$Pa	Mole fraction of methane in liquid, x_{CH_4}	in vapor, y_{CH_4}
313.15	70.9	0.329	0.622
	81.1	0.395	0.621
	86.1	0.439	0.614
	91.2	0.506	0.593
328.15	20.3	0.005	0.037
	25.4	–	0.196
	30.4	0.055	0.292
	35.5	–	0.355
	40.5	0.108	0.401
	45.6	–	0.439
	50.7	0.167	0.470
	60.8	0.228	0.511
	70.9	0.292	0.524
	76.0	0.326	0.521
	81.1	0.375	0.501
343.15	30.4	0.021	0.103
	35.5	–	0.189
	40.5	0.074	0.252
	45.6	–	0.299
	50.7	0.130	0.336
	60.8	0.199	0.388
	65.9	0.241	0.392
353.15	35.5	0.020	0.085
	40.5	0.046	0.149
	45.6	–	0.202
	50.7	0.106	0.243
	55.7	0.028	0.274
	60.8	0.183	0.273
363.15	40.5	0.010	0.044
	45.6	0.038	0.121
	50.7	0.069	0.152

COMPONENTS:	ORIGINAL MEASUREMENTS:
1. Methane; CH_4; [74-82-8] 2. Propane; C_3H_8; [74-98-6]	Reamer, H. H.; Sage, B. H.; Lacey, W. N. *Ind. Engng. Chem.* 1950, *42*, 534-539.

VARIABLES:	PREPARED BY:
Temperature, pressure	C. L. Young

EXPERIMENTAL VALUES:

T/K	$P/10^5$Pa	Mole fraction of methane in liquid, x_{CH_4}	in vapor, y_{CH_4}	T/K	$P/10^5$Pa	Mole fraction of methane in liquid, x_{CH_4}	in vapor, y_{CH_4}
277.59	6.89	0.0099	0.2034	277.59	86.18	0.5492	0.8222
	10.34	0.0324	0.4432		89.63	0.5773	0.8222
	13.79	0.0549	0.5627		93.08	0.6080	0.8217
	17.24	0.0779	0.6382		96.53	0.6434	0.8173
	20.68	0.1008	0.6875		99.97	0.6891	0.7924
	24.13	0.1242	0.7235		101.63	0.7459	0.7459
	27.58	0.1471	0.7505	294.26	10.34	0.0106	0.1513
	31.03	0.1695	0.7677		13.79	0.0321	0.3435
	34.47	0.1923	0.7819		17.24	0.0535	0.4575
	37.92	0.2171	0.7966		20.68	0.0749	0.5338
	41.37	0.2378	0.8042		24.13	0.0959	0.5853
	44.82	0.2607	0.8099		27.58	0.1168	0.6231
	48.26	0.2834	0.8135		31.03	0.1372	0.6501
	51.71	0.3060	0.8159		34.47	0.1580	0.6721
	55.16	0.3289	0.8180		37.92	0.1782	0.6908
	58.61	0.3517	0.8188		41.37	0.1987	0.7038
	62.05	0.3769	0.8199		44.82	0.2196	0.7141
	65.50	0.3986	0.8205		48.26	0.2407	0.7228
	68.95	0.4226	0.8208		51.71	0.2616	0.7300
	72.39	0.4473	0.8211		55.16	0.2828	0.7357
	75.84	0.4719	0.8214		58.61	0.3042	0.7403
	79.29	0.4968	0.8217		62.05	0.3261	0.7442
	82.74	0.5225	0.8220		65.50	0.3481	0.7471

(cont.)

AUXILIARY INFORMATION

METHOD/APPARATUS/PROCEDURE:	SOURCE AND PURITY OF MATERIALS:
PVT cell charged with mixture of known composition. Pressure measured with pressure balance. Temperature measured using resistance thermometer. Bubble point and dew point determined for various compositions. Co-existing liquid and gas phase properties determined by graphical means. Details in source and ref. (1).	1. Crude sample treated for removal of alkanes, CO_2 and H_2O; final purity 99.9 mole per cent. 2. Phillips Petroleum Co. sample distilled; initial purity 99.9 mole per cent.

	ESTIMATED ERROR: $\delta T/K = \pm 0.1$; $\delta/Pa = \pm 0.05\%$; δx_{CH_4}, $\delta y_{CH_4} = \pm 0.002$.

REFERENCES:
1. Sage, B. H.; Lacey, W. N.
   Trans. Am. Inst. Mining and Met. Engnrs.
   1940, *136*, 136.

COMPONENTS:	ORIGINAL MEASUREMENTS:
1. Methane; CH_4; [74-82-8] 2. Propane; C_3H_8; [74-98-6]	Reamer, H. H.; Sage, B. H.; Lacey, W. N. *Ind. Engng. Chem.* <u>1950</u>, *42*, 534-539.

EXPERIMENTAL VALUES:

T/K	$P/10^5$Pa	Mole fraction of methane in liquid, x_{CH_4}	in vapor, y_{CH_4}	T/K	$P/10^5$Pa	Mole fraction of methane in liquid, x_{CH_4}	in vapor, y_{CH_4}
294.26	68.95	0.3707	0.7497	327.59	31.03	0.0664	0.2964
	72.39	0.3938	0.7520		34.47	0.0852	0.3418
	75.84	0.4179	0.7539		37.92	0.1040	0.3797
	79.29	0.4425	0.7553		41.37	0.1227	0.4109
	82.74	0.4679	0.7567		44.82	0.1419	0.4361
	86.18	0.4954	0.7570		48.26	0.1612	0.4582
	89.63	0.5244	0.7561		51.71	0.1810	0.4768
	93.08	0.5670	0.7503		55.16	0.2008	0.4938
	96.53	0.6046	0.7309		58.61	0.2213	0.5086
	99.97	0.6772	0.6772		62.05	0.2430	0.5224
310.93	13.79	0.0049	0.0521		65.50	0.2652	0.5351
	17.24	0.0257	0.2184		68.95	0.2885	0.5459
	20.68	0.0460	0.3255		72.39	0.3118	0.5532
	24.13	0.0652	0.3949		75.84	0.3361	0.5546
	27.58	0.0845	0.4472		79.29	0.3654	0.5473
	31.03	0.1040	0.4884		82.74	0.4101	0.5130
	34.47	0.1235	0.5209		83.98	0.4691	0.4691
	37.92	0.1432	0.5481	344.26	27.47	0.0063	0.0276
	41.37	0.1630	0.5714		31.03	0.0249	0.0981
	44.82	0.1821	0.5911		34.47	0.0433	0.1550
	48.26	0.2019	0.6073		37.92	0.0622	0.2007
	51.71	0.2216	0.6210		41.37	0.0813	0.2392
	55.16	0.2418	0.6321		44.82	0.1002	0.2712
	58.61	0.2611	0.6420		48.26	0.1199	0.2983
	62.05	0.2836	0.6503		51.71	0.1402	0.3215
	65.50	0.3051	0.6572		55.16	0.1618	0.3414
	68.95	0.3271	0.6635		58.61	0.1820	0.3566
	72.39	0.3498	0.6691		62.05	0.2081	0.3656
	75.84	0.3731	0.6738		65.50	0.2375	0.3678
	79.29	0.3969	0.6767		68.95	0.2800	0.3558
	82.74	0.4226	0.6779		70.33	0.3228	0.3228
	86.18	0.4511	0.6766	360.93	37.92	0.0107	0.0280
	89.63	0.4889	0.6643		41.37	0.0333	0.0798
	93.08	0.5610	0.6087		44.82	0.0555	0.1208
	93.29	0.5882	0.5882		48.26	0.0786	0.1489
327.59	20.68	0.0104	0.0699		49.99	0.0926	0.1570
	24.13	0.0289	0.1663		51.71	0.1120	0.1601
	27.57	0.0480	0.2414		52.88	0.1400	0.1400

COMPONENTS:	ORIGINAL MEASUREMENTS:
1. Methane; CH_4; [74-82-8] 2. Propane; C_3H_8; [74-98-6]	Akers, W. W.; Burns, J. F.; Fairchild, W. R. *Ind. Eng. Chem.* <u>1954</u>, *46*, 2531-2534.

VARIABLES:	PREPARED BY:
Temperature, pressure	C. L. Young

EXPERIMENTAL VALUES:

T/K	P/MPa	Mole fraction of methane in liquid, x_{CH_4}	in vapor, y_{CH_4}
273.15	0.689	0.012	0.230
	1.38	0.059	0.566
	2.07	0.106	0.715
	2.76	0.152	0.780
	3.45	0.200	0.808
	4.14	0.248	0.830
	4.83	0.296	0.843
	5.52	0.347	0.852
	6.21	0.399	0.856
	6.89	0.451	0.854
	7.58	0.508	0.850
	8.27	0.568	0.833
	8.96	0.628	0.812
	9.65	0.700	0.781
	10.00	0.745	0.745
256.48	0.689	0.034	0.560
	1.38	0.089	0.767
	2.07	0.142	0.832
	2.76	0.197	0.861
	3.45	0.249	0.880
	4.14	0.303	0.888
	4.83	0.357	0.890
	5.52	0.410	0.892 (cont.)

AUXILIARY INFORMATION

METHOD/APPARATUS/PROCEDURE:	SOURCE AND PURITY OF MATERIALS:
Equilibrium cell containing liquid and vapor phases. Vapor portion recirculated via external line and re-entered the cell through liquid phase. Equilibrium established with a fixed quantity of vapor and liquid. Details of apparatus and procedure in source.	1. Phillips Petroleum Co. sample, purity better than or equal to 99 mole per cent. Major impurities: nitrogen (0.3 mole per cent) and ethane (0.5 mole per cent). 2. Phillips Petroleum Co. sample, purity better than or equal to 99 mole per cent.

ESTIMATED ERROR:

$\delta T/K = \pm 0.5$; $\delta P/MPa = \pm 0.007$;

δx_{CH_4}, $\delta y_{CH_4} = \pm 0.005$ (estimated by compiler).

REFERENCES:

COMPONENTS:

1. Methane; CH_4; [74-82-8]

2. Propane; C_3H_8; [74-98-6]

ORIGINAL MEASUREMENTS:

Akers, W. W.; Burns, J. F.;
Fairchild, W. R.
Ind. Eng. Chem.
<u>1954</u>, *46*, 2531-2534.

EXPERIMENTAL VALUES:

T/K	P/MPa	Mole fraction of methane in liquid, x_{CH_4}	in vapor, y_{CH_4}
256.48	6.21	0.464	0.891
	6.89	0.518	0.889
	7.58	0.572	0.882
	8.27	0.636	0.869
	8.96	0.718	0.845
	9.52	0.80	0.80
241.48	0.689	0.050	0.765
	1.38	0.112	0.868
	2.07	0.175	0.902
	2.76	0.237	0.920
	3.45	0.300	0.930
	4.14	0.361	0.933
	4.83	0.422	0.936
	5.52	0.485	0.935
	6.21	0.548	0.933
	6.89	0.609	0.930
	7.58	0.671	0.919
	8.27	0.734	0.910
	8.96	0.796	0.872
	9.45	0.835	0.835
226.48	0.689	0.061	0.850
	1.38	0.136	0.915
	2.07	0.208	0.937
	2.76	0.284	0.946
	3.45	0.361	0.952
	4.14	0.440	0.960
	4.83	0.522	0.964
	5.52	0.605	0.963
	6.21	0.696	0.960
	6.89	0.792	0.952
	7.58	0.897	0.925
	7.67	0.921	0.921
213.15	0.689	0.100	0.920
	1.38	0.198	0.955
	2.07	0.290	0.966
	2.76	0.382	0.970
	3.45	0.469	0.972
	4.14	0.552	0.972
	4.83	0.638	0.970
	5.52	0.720	0.968
	6.21	0.804	0.961
	6.89	0.890	0.951
	7.31	0.945	0.945
194.82	0.689	0.140	0.975
	1.38	0.280	0.985
	2.07	0.420	0.991
	2.76	0.560	0.995
	3.45	0.700	0.990
	4.14	0.840	0.975
	4.76	0.960	0.960
174.26	0.689	0.237	0.997
	1.38	0.499	0.998
	2.07	0.769	0.999
	2.69	0.999	0.999
157.59	0.345	0.170	1.00
	0.689	0.355	1.00
	1.38	0.907	1.00

COMPONENTS:	ORIGINAL MEASUREMENTS:
1. Methane; CH₄; [74-82-8]	Price, A. R.; Kobayashi, R.
2. Propane; C₃H₈; [74-98-6]	*J. Chem. Engng. Data*
	<u>1959</u>, *4*, 40-52.

VARIABLES:	PREPARED BY:
Temperature, pressure	C. L. Young

EXPERIMENTAL VALUES:

T/K	P/MPa	Mole fraction of methane in liquid, x_{CH_4}	in vapor, y_{CH_4}	T/K	P/MPa	Mole fraction of methane in liquid, x_{CH_4}	in vapor, y_{CH_4}
283.15	2.76	0.128	0.685	255.37	8.96	0.708	0.845
	4.14	0.216	0.762	227.59	0.689	0.0769	0.840
	5.52	0.300	0.788		1.38	0.146	0.9216
	6.89	0.413	0.805		2.76	0.296	0.9493
	7.58	0.451	0.803		4.14	0.438	0.9585
	8.27	0.498	0.784		5.52	0.581	0.959
255.37	0.689	0.0358	0.573		6.89	0.736	0.9458
	1.38	0.0904	0.768	199.82	0.689	0.125	0.9591
	2.76	0.199	0.862		1.38	0.222	0.9792
	4.14	0.311	0.891		2.76	0.477	0.9855
	5.52	0.415	0.899		5.52	0.744	0.9852
	6.89	0.522	0.895	172.04	0.689	0.238	0.9932
	7.58	0.582	0.888		1.38	0.502	0.9955
	8.27	0.637	0.873	144.26	0.689	0.802	0.9993

AUXILIARY INFORMATION

METHOD/APPARATUS/PROCEDURE:	SOURCE AND PURITY OF MATERIALS:
Recirculating vapor flow apparatus with modified Jerguson sight gauge for equilibrium cell. Vapor recycled with magnetic pump. Pressure measured with Bourdon pressure gauge and temperature measured with thermocouple. Details in source.	1. Phillips Petroleum Co. research grade, purity 99.5 mole per cent. 2. Phillips Petroleum Co. pure grade, purity 99.0 mole per cent.

ESTIMATED ERROR:

$\delta T/K = \pm 0.06$; $\delta P/MPa = \pm 1\%$;

δx_{CH_4}, $\delta y_{CH_4} = \pm 2\%$ (estimated by compiler).

REFERENCES:

COMPONENTS:	ORIGINAL MEASUREMENTS:
1. Methane; CH_4; [74-82-6] 2. Propane; C_3H_8; [74-98-6]	Cheung, H.; Wang, D. I. J. *Ind. Eng. Chem. Fundam.* *1964, 3,* 355.

VARIABLES:	PREPARED BY:
	C. L. Young

EXPERIMENTAL VALUES:

T/K	P/cmHg	P/kPa	Mole fraction of methane in liquid, x_{CH_4}
91.7	0.8	1.1	0.0272
91.7	4.7	6.3	0.234
91.8	6.4	8.5	0.374
91.7	7.0	9.3	0.473
112.5	4.3	5.7	0.0268
112.7	30.8	41.0	0.232
112.5	43.1	57.4	0.372
128.4	10.4	13.9	0.0264
128.4	78.1	104.0	0.230
128.3	112.7	163.5	0.371

AUXILIARY INFORMATION

METHOD/APPARATUS/PROCEDURE:	SOURCE AND PURITY OF MATERIALS:
Static equilibrium cell of accurately known volume. Solvent added then solute gas added. Liquid composition determined from known volume of cell and liquid and amounts of solvent and solute present. Pressure measured with mercury manometer and temperature measured with thermocouple.	No details given.

ESTIMATED ERROR:

$\delta T/K = \pm 0.1$; δP/cmHg = ± 0.1;

$\delta x_{CH_4} = \pm 7\%$ (estimated by compiler).

REFERENCES:

COMPONENTS:	ORIGINAL MEASUREMENTS:
1. Methane; CH_4; [74-82-8] 2. Propane; C_3H_8; [74-98-6]	Skripka, V. G.; Nikitina, I. E.; Zhdanovich, L. A.; Sirotin, A. G.; Benyaminovich, O. A. *Gazov. Prom.* 1970, *15*, 35-36.

VARIABLES:	PREPARED BY:
	C. L. Young

EXPERIMENTAL VALUES:

T/K (T/°C)	Bubble pt. pressure		Mole fraction, x_{CH_4}	T/K (T/°C)	Bubble pt. pressure		Mole fraction, x_{CH_4}
	P/kg f cm^{-2}	P/MPa			P/kg f cm^{-2}	P/MPa	
123.2	0.00005	0.000005	0.00	143.2	0.001	0.0001	0.00
(-150)	0.17	0.017	0.05	(-130)	0.50	0.049	0.05
	0.65	0.064	0.20		1.94	0.190	0.20
	1.24	0.122	0.40		3.65	0.358	0.40
	1.72	0.169	0.60		5.13	0.503	0.60
	2.08	0.204	0.80		6.46	0.634	0.80
	2.34	0.229	0.95		7.37	0.723	0.95
	2.43	0.238	1.00		7.67	0.752	1.00
133.2	0.0003	0.00003	0.00	153.2	0.004	0.0004	0.00
(-140)	0.38	0.037	0.05	(-120)	0.86	0.084	0.05
	1.35	0.132	0.20		3.15	0.309	0.20
	2.31	0.227	0.40		5.80	0.569	0.40
	3.09	0.303	0.60		8.14	0.798	0.60
	3.80	0.373	0.80		10.24	1.004	0.80
	4.33	0.425	0.95		11.72	1.149	0.95
	4.51	0.442	1.00		12.18	1.194	1.00

AUXILIARY INFORMATION

METHOD/APPARATUS/PROCEDURE:	SOURCE AND PURITY OF MATERIALS:
Recirculating vapor flow apparatus fitted with magnetic stirrer. Temperature measured with platinum resistance thermometer. Liquid analysed by gas chromatography. Details of apparatus in ref. (1).	1. Purity 99.9 per cent by volume. 2. Purity 99.5 per cent by volume.
	ESTIMATED ERROR:
	REFERENCES: 1. Skripka, V. G.; Barsuk, S. D.; Nikitina, I. E.; Ben'yaminovic, O.A. *Gazov. Prom.* 1964, *14*, 11.

COMPONENTS:	ORIGINAL MEASUREMENTS:
1. Methane; CH_4; [74-82-8] 2. Propane; C_3H_8; [74-98-6]	Wichterle, I.; Kobayashi, R. *J. Chem. Eng. Data* <u>1972</u>, *17*, 4-9.

VARIABLES:	PREPARED BY:
Temperature, pressure	C. L. Young

EXPERIMENTAL VALUES:

T/K	P/MPa	Mole fraction of methane in liquid, x_{CH_4}	in vapor, y_{CH_4}
213.71	0.189	0.0205	0.7669
	0.355	0.0443	0.8706
	0.689	0.0899	0.9288
	1.034	0.1358	0.9505
	2.067	0.2709	0.9698
	2.756	0.3656	0.9741
	3.445	0.4580	0.9760
	4.139	0.5563	0.9767
	4.828	0.6555	0.9755
	5.517	0.7573	0.9726
	6.206	0.8600	0.9646
	6.475	0.9053	0.9519
	6.510	0.9469	0.9469
195.2	0.211	0.0377	0.9244
	0.362	0.0677	0.9541
	0.517	0.0958	0.9670
	0.683	0.1263	0.9742
	1.378	0.2545	0.9845
	2.067	0.3969	0.9882
	2.756	0.5392	0.9898
	3.445	0.6947	0.9904
	3.795	0.7734	0.9905
	4.139	0.8454	0.9905

(cont.)

AUXILIARY INFORMATION

METHOD/APPARATUS/PROCEDURE:	SOURCE AND PURITY OF MATERIALS:
Recirculating vapor flow apparatus with magnetic vapor pump. Pressure measured with Bourdon gauge and temperature with a platinum resistance thermometer. Samples of both phases analysed using gas chromatography with flame ionisation detector. Details in source and ref. (1).	1. Matheson Gas Products sample, purity 99.97 mole per cent; purified by passage through molecular sieve. 2. Phillips Petroleum Co. research grade sample, purity 99.99 mole per cent.

ESTIMATED ERROR:

$\delta T/K = \pm 0.1$; $\delta P/MPa = \pm 0.015$ or less;

$\delta x_{CH_4} \simeq \delta y_{CH_4} = \pm 2\%$ (details in source).

REFERENCES:

1. Chang, H. L.; Hunt, L. J.;
   Kobayashi, R.
   Am. Inst. Chem. Engnrs. J.
   <u>1966</u>, *11*, 1212.

COMPONENTS:	ORIGINAL MEASUREMENTS:
1. Methane; CH₄; [74-82-8]	Wichterle, I.; Kobayashi, R.
2. Propane; C₃H₈; [74-98-6]	J. Chem. Eng. Data
	1972, 17, 4-9.

EXPERIMENTAL VALUES:

T/K	P/MPa	Mole fraction of methane in liquid, x_{CH_4}	in vapor, y_{CH_4}
195.2	4.484	0.9061	0.99075
	4.742	0.9546	0.9912
	4.884	0.9719	0.9911
	4.990	0.9856	0.9856
192.3	0.207	0.0409	0.9355
	0.345	0.0692	0.9601
	0.517	0.1052	0.9718
	0.689	0.1379	0.9783
	1.378	0.2737	0.9874
	2.067	0.4207	0.9899
	2.756	0.5819	0.9913
	3.445	0.7529	0.9919
	3.967	0.8728	0.9924
	4.509	0.9578	0.9941
	4.590	0.9782	0.9953
	4.646	0.9844	0.9957
	4.747	0.9926	0.9926
187.54	0.283	0.0629	0.9656
	0.689	0.1506	0.9839
	1.378	0.3042	0.9907
172.04	0.213	0.0692	0.9862
	0.362	0.1196	0.9915
	0.689	0.2270	0.99505
158.15	0.172	0.0873	0.9958
	0.355	0.1791	0.99793
	0.689	0.3510	0.99888
144.26	0.214	0.2109	0.99940
	0.331	0.3005	0.99959
130.37	0.186	0.3924	0.99921

Additional vapor-liquid equilibrium data in which
the mole fraction is greater than 0.30 are given
in source.

COMPONENTS:	ORIGINAL MEASUREMENTS:
1. Methane; CH_4; [74-82-8] 2. Propane; C_3H_8; [74-98-6]	Poon, D.P.L.; Lu, B.C.Y. *Advan. Cryog. Engng.* <u>1973</u>, *19*, 292-299.

VARIABLES:	PREPARED BY:
Temperature, pressure	C.L. Young

EXPERIMENTAL VALUES:

T/K	P/psia	P/MPa	Mole fraction of methane in liquid x_{CH_4}	in vapor y_{CH_4}
114.1	6.1	0.042	0.1812	0.9990
	8.6	0.059	0.2911	0.9995
	11.2	0.077	0.4102	0.9997
	13.0	0.090	0.5488	1.0
	14.0	0.097	0.6647	0.9998
	16.2	0.112	0.8812	1.0
	17.8	0.123	1.0	1.0
118.3	7.9	0.054	0.1775	0.9986
	11.0	0.076	0.2717	0.9993
	14.1	0.097	0.3909	0.9997
	17.6	0.121	0.5714	0.9992
	19.6	0.135	0.6540	0.9999
	20.6	0.142	0.7399	1.0
	22.9	0.158	0.9031	1.0
	24.7	0.170	1.0	1.0
122.2	7.1	0.049	0.1130	0.9976
	9.1	0.063	0.1409	0.9986
	13.1	0.090	0.2219	0.9996
	13.5	0.093	0.2253	0.9996
	18.8	0.130	0.3701	0.9999
	23.1	0.159	0.5297	0.9999
	26.9	0.185	0.7090	0.9999
	28.8	0.199	0.8095	1.0
	31.0	0.214	0.8910	1.0
	32.3	0.223	1.0	1.0

AUXILIARY INFORMATION

METHOD/APPARATUS/PROCEDURE:	SOURCE AND PURITY OF MATERIALS:
Recirculating vapor flow apparatus constructed from 100 ml. Jerguson gauge with stainless steel body. Temperature measured using copper-constantan thermocouples. Pressure measured using Bourdon gauges. Magnetic circulating pump. Cell charged vapour recirculated for 2 or more hours. Samples of vapor and liquid removed at constant pressure and analysed using gas chromatography. Details in source.	1. Matheson research grade, purity 99.99 mole per cent. 2. Phillips Petroleum Co. sample, research grade purity 99.99 mole per cent.

ESTIMATED ERROR:

$\delta T/K = \pm 0.05$; $\delta P/MPa \sim \pm 0.005$;
$\delta x_{CH_4} + \pm 1\%$; $\delta y_{CH_4} = \pm 0.0001$

REFERENCES:

COMPONENTS:	ORIGINAL MEASUREMENTS:
1. Methane; CH_4; [74-82-8] 2. Propane; C_3H_8; [74-98-6]	Kalra, H.; Robinson, D. B. *Cryogenics* <u>1975</u>, *15*, 409.

VARIABLES:	PREPARED BY:
	C. L. Young

EXPERIMENTAL VALUES:

T/K (T/°F)	P/psia	P/MPa	Mole fraction of methane in liquid, x_{CH_4}	in vapor, y_{CH_4}
213.8 (-74.9)	109.1 295.5 494 686	0.7522 2.037 3.406 4.730	0.0949 0.271 0.450 0.642	0.941 0.972 0.977 0.978

AUXILIARY INFORMATION

METHOD/APPARATUS/PROCEDURE:	SOURCE AND PURITY OF MATERIALS:
Windowed equilibrium cell con- structed of stainless steel fitted with specially made sampling valves. Contents of cell mixed with a high speed magnetic stirrer rotating at more than 500 rpm. Temperature measured with a copper-constantan thermocouple and pressure measured with Bourdon gauges. Details in source.	No details given.

ESTIMATED ERROR:
$\delta T/K = \pm 0.06$; $\delta P/\text{lbs in}^{-2} = \pm 1.0$;
δx_{CH_4}, $\delta y_{CH_4} = \pm 0.005$.

REFERENCES:

COMPONENTS:	EVALUATOR:
1. Methane; CH_4; [74-82-8]	Colin L. Young,
	School of Chemistry,
2. Butane; C_4H_{10}; [106-97-8]	University of Melbourne,
	Parkville, Victoria 3052,
	Australia.
	May 1982

EVALUATION:

This system has been extensively studied. The early data of Nederbragt (1) are rejected because of their limited nature and the low precision of the analytical techniques used in that work. The data of Frolich *et al.* are also rejected since the results were presented in the form of a small graph and the measurements have been superseded by more recent data.

The data of Wang and McKetta (3) and Roberts *et al.* (4) are classified as doubtful. These data show a fair degree of scatter in the reported solubilities. This probably arose because of the sampling and analyzing techniques employed.

The data of Wiese *et al.* (5) are not in good agreement with the data of Elliott *et al.* (6) at the overlapping temperature of 277.6 K. There is excellent agreement between the data of Wiese *et al.* (5) and Sage *et al.* (7) but in fact it appears that the two sets of data are derived from the same set of raw experimental measurements.

Although the data of Kahre (8) and Elliott *et al.* (6) agree more closely than do the data of Wiese *et al.* (5) and Elliott *et al.* (6), there are still significant discrepancies between the two sets of data. The precision, and probably the accuracy, of the data of Elliott *et al.* (6) is greater than that of the data of Kahre (8).

The data of Elliott *et al.* (6) are classified as tentative for the temperature range 144 K to 278 K and the data of Sage *et al.* (7) are classified as tentative for the range 294 K to 394 K although the accuracy of the later work is considerably less than the former.

In another paper Sage and coworkers (9) have made a detailed evaluation of phase behavior of this system.

References

1.   Nederbragt, G. W. *Ing. Eng. Chem.*, <u>1938</u>, *30*, 587.

2.   Frolich, P. K.; Tauch, E. J.; Hogan, J. J.; Peer, A. A. *Ind. Eng. Chem.*, <u>1931</u>, *23*, 548.

3.   Wang, R. H.; McKetta, J. J. *J. Chem. Eng. Data*, <u>1964</u>, *9*, 30.

4.   Roberts, L. R.; Wang, R. H.; Azarnoosh, A.; McKetta, J. J. *J. Chem. Eng. Data*, <u>1962</u>, *7*, 484.

5.   Wiese, H. C.; Jacobs, J.; Sage, B. H. *J. Chem. Eng. Data*, <u>1970</u>, *15*, 82.

6.   Elliott, D. G.; Chen, R. J. J.; Chappelear, P. S.; Kobayashi, R. *J. Chem. Eng. Data*, <u>1974</u>, *19*, 71.

7.   Sage, B. H.; Hicks, B. L.; Lacey, W. N. *Ind. Eng. Chem.*, <u>1940</u>, *32*, 1085.

8.   Kahre, L. *J. Chem. Eng. Data*, <u>1974</u>, *19*, 67.

9.   Sage, B. H.; Budenholzer, R. A.; Lacey, W. N. *Ind. Eng. Chem.*, <u>1940</u>, *32*, 1262.

COMPONENTS:	ORIGINAL MEASUREMENTS:
1. Methane; CH_4; [74-82-8] 2. Butane; C_4H_{10}; [106-97-8]	Frolich, P.K.; Tauch, E.J.; Hogan, J.J.; Peer, A.A. *Ind. Eng. Chem.* 1931, 23, 548-550.

VARIABLES:
 	
         Pressure
	PREPARED BY:
	
	          C.L. Young

EXPERIMENTAL VALUES:

T/K	P/MPa	Solubility[*], S	Mole fraction of methane in liquid, x_{CH_4}
298.15	1.0	18	0.0718
	2.0	35.5	0.132
	3.0	52	0.183
	4.0	70	0.231
	5.0	88	0.274
	6.0	106	0.313
	7.0	123	0.346

*   Data taken from graph in original article.

+   calculated by compiler.

AUXILIARY INFORMATION

METHOD/APPARATUS/PROCEDURE:	SOURCE AND PURITY OF MATERIALS:
Static equilibrium cell. Liquid saturated with gas and after equilibrium established samples removed and analysed by volumetric method. Allowance was made for vapor pressure of liquid and the solubility of the gas at atmospheric pressure. Details in source.	Stated that the materials were the highest purity available (98 to 99 mole per cent.)
	ESTIMATED ERROR: $\delta T/K = \pm 0.1$; $\delta x_{CH_4} = \pm 5\%$
	REFERENCES:

COMPONENTS:	ORIGINAL MEASUREMENTS:
1. Methane; CH_4; [74-82-8] 2. Butane; C_4H_{10}; [106-97-8]	Sage, B. H.; Hicks, B. L.; Lacey, W. N. *Ind. Eng. Chem.* <u>1940</u>, *32*, 1085-1092.

VARIABLES:	PREPARED BY:
Temperature, pressure	C. L. Young

EXPERIMENTAL VALUES:

T/K	P/kPa	Wt. fraction of methane in liquid,	in gas,	Mole fraction of methane in liquid, x_{CH_4}	in gas, y_{CH_4}
294.25	0.276	0.0010	0.06845	0.0036	0.2103
	0.414	0.0031	0.1899	0.0111	0.4593
	0.552	0.0052	0.2795	0.0186	0.5843
	0.689	0.0074	0.3489	0.0263	0.6601
	1.034	0.0129	0.4651	0.0452	0.7591
	1.379	0.0185	0.5387	0.0639	0.8089
	2.068	0.0301	0.6265	0.1011	0.8587
	2.758	0.0423	0.6758	0.1380	0.8831
	3.447	0.0551	0.7081	0.1745	0.8979
	4.137	0.0686	0.7300	0.2107	0.9074
	5.516	0.0971	0.7540	0.2804	0.9174
	6.895	0.1289	0.7610	0.3491	0.9203
	8.274	0.1644	0.7550	0.4162	0.9178
	8.618	0.1726	0.7510	0.4305	0.9162
	9.653	0.2014	0.7331	0.4775	0.9087
	10.34	0.2232	0.7190	0.5101	0.9027
	11.03	0.2485	0.7000	0.5451	0.8942
	11.72	0.2812	0.6750	0.5864	0.8827
	12.07	0.3020	0.6590	0.6106	0.8751
	12.41	0.3268	0.6392	0.6376	0.8652
	12.76	0.3591	0.6120	0.6700	0.8511
	13.10	0.4094	0.5659	0.7153	0.8253
	13.26	0.482	0.482	0.7713 (cont.)	0.7713

AUXILIARY INFORMATION

METHOD/APPARATUS/PROCEDURE:	SOURCE AND PURITY OF MATERIALS:
PVT cell charged with mixture of known composition. Pressure measured with pressure balance. Temperature measured with resistance thermometer. Bubble point and dew point determined for various compositions from discontinuity in PV isotherm. Coexisting liquid and gas phase properties determined by graphical means.	1. Crude sample treated for removal of higher alkanes, carbon dioxide and water vapor. Final purity 99.9 mole per cent. 2. Phillips petroleum sample, distilled, final purity better than 99.96 mole per cent.

ESTIMATED ERROR:
$\delta T/K = \pm 0.1$; $\delta P/kPa = \pm 0.007$;
$\delta x_{CH_4} = \pm 0.0005$; $\delta y_{CH_4} = \pm 0.003$

(estimated by compiler).

REFERENCES:

1. Sage, B. H.; Lacey, W. N.
   Trans. Am. Inst. Mining Met. Engnrs.
   <u>1940</u>, *136*, 136.

COMPONENTS:	ORIGINAL MEASUREMENTS:
1. Methane; CH₄; [74-82-8]	Sage, B. H.; Hicks, B. L.; Lacey, W. N.
2. Butane; C₄H₁₀; [106-97-8]	*Ind. Eng. Chem.* <u>1940</u>, *32*, 1085-1092.

EXPERIMENTAL VALUES:

T/K	P/kPa	Wt. fraction of methane in liquid,	in gas,	Mole fraction of methane in liquid, x_{CH_4}	in gas, y_{CH_4}
310.93	0.414	0.0008	0.0408	0.0029	0.1336
	0.552	0.0029	0.1230	0.0104	0.3370
	0.689	0.0049	0.1901	0.0175	0.4596
	1.034	0.0100	0.3128	0.0353	0.6226
	1.379	0.0152	0.3948	0.0530	0.7027
	2.068	0.0257	0.4985	0.0873	0.7827
	2.758	0.0368	0.5636	0.1216	0.8240
	3.447	0.0484	0.6050	0.1556	0.8473
	4.137	0.0604	0.6335	0.1889	0.8623
	5.516	0.0859	0.6640	0.2540	0.8775
	6.895	0.1136	0.6712	0.3172	0.8809
	8.274	0.1452	0.6690	0.3810	0.8799
	8.618	0.1540	0.6664	0.3975	0.8786
	9.653	0.1821	0.6550	0.4467	0.8731
	10.34	0.2030	0.6418	0.4800	0.8665
	11.03	0.2279	0.6232	0.5168	0.8570
	11.72	0.2589	0.5990	0.5587	0.8441
	12.07	0.2790	0.5825	0.5837	0.8349
	12.41	0.3024	0.5616	0.6110	0.8228
	12.76	0.3325	0.5330	0.6435	0.8053
	13.10	0.3954	0.4815	0.7033	0.7709
	13.18	0.4195	0.4195	0.7237	0.7237
327.59	0.689	0.0018	0.05675	0.0065	0.1790
	1.034	0.0066	0.1724	0.0235	0.4302
	1.379	0.0115	0.2546	0.0405	0.5531
	2.068	0.0214	0.3672	0.0734	0.6777
	2.758	0.0317	0.4387	0.1061	0.7391
	3.447	0.0424	0.4867	0.1383	0.7746
	4.137	0.0534	0.5194	0.1697	0.7966
	5.516	0.0769	0.5559	0.2319	0.8194
	6.895	0.1028	0.5688	0.2934	0.8270
	8.274	0.1322	0.5711	0.3557	0.8283
	8.618	0.1403	0.5697	0.3716	0.8275
	9.653	0.1662	0.5600	0.4194	0.8218
	10.34	0.1855	0.5481	0.4522	0.8147
	11.03	0.2088	0.5313	0.4889	0.8042
	11.72	0.2362	0.5040	0.5285	0.7864
	12.07	0.2560	0.4858	0.5550	0.7740
	12.41	0.2756	0.4617	0.5796	0.7566
	12.76	0.3131	0.4266	0.6229	0.7295
	12.93	0.3610	0.3610	0.6719	0.6719
344.26	1.034	0.0025	0.0543	0.0090	0.1722
	1.379	0.0068	0.1303	0.0242	0.3519
	2.068	0.0165	0.2387	0.0573	0.5319
	2.758	0.0261	0.3135	0.0885	0.6233
	3.447	0.0363	0.3661	0.1201	0.6767
	4.137	0.0469	0.4030	0.1513	0.7098
	5.516	0.0691	0.4465	0.2120	0.7451
	6.895	0.0943	0.4628	0.2740	0.7574
	8.274	0.1210	0.4655	0.3328	0.7594
	8.618	0.1285	0.4648	0.3483	0.7589
	9.653	0.1542	0.4567	0.3978	0.7529
	10.34	0.1740	0.4450	0.4329	0.7440
	11.03	0.1964	0.4255	0.4697	0.7286
	11.72	0.2234	0.3959	0.5104	0.7037
	12.07	0.2466	0.3730	0.5426	0.6831
	12.41	0.2665	0.3330	0.5684	0.6440
	12.48	0.3073	0.3073	0.6165	0.6165

(cont.)

COMPONENTS:	ORIGINAL MEASUREMENTS:
1. Methane; CH$_4$; [74-82-8]	Sage, B. H.; Hicks, B. L.; Lacey, W. N.
2. Butane; C$_4$H$_{10}$; [106-97-8]	*Ind. Eng. Chem.* <u>1940</u>, *32*, 1085-1092.

EXPERIMENTAL VALUES:

T/K	P/kPa	Wt. fraction of methane in liquid,	in gas,	Mole fraction of methane in liquid, x_{CH_4}	in gas, y_{CH_4}
360.93	1.379	0.0022	0.0333	0.0079	0.1110
	2.068	0.0110	0.1357	0.0387	0.3627
	2.758	0.0202	0.2062	0.0695	0.4849
	3.447	0.0299	0.2584	0.1005	0.5581
	4.137	0.0399	0.2960	0.1309	0.6038
	5.516	0.0614	0.3417	0.1916	0.6529
	6.895	0.0851	0.3610	0.2521	0.6719
	8.274	0.1128	0.3638	0.3155	0.6745
	8.618	0.1206	0.3627	0.3320	0.6735
	9.653	0.1479	0.3534	0.3861	0.6645
	10.34	0.1702	0.3406	0.4264	0.6518
	11.03	0.1957	0.3165	0.4685	0.6266
	11.38	0.2108	0.2975	0.4919	0.6055
	11.71	0.2525	0.2525	0.5504	0.5504
383.15	2.068	0.0049	0.0499	0.0175	0.1599
	2.758	0.0136	0.1169	0.0476	0.3242
	3.447	0.0229	0.1666	0.0783	0.4201
	4.137	0.0328	0.2030	0.1094	0.4800
	5.516	0.0543	0.2468	0.1722	0.5429
	6.895	0.0784	0.2623	0.2356	0.5630
	8.274	0.1063	0.2610	0.3012	0.5614
	8.618	0.1145	0.2586	0.3191	0.5583
	8.963	0.1227	0.2545	0.3364	0.5530
	9.653	0.1457	0.2425	0.3820	0.5371
	10.00	0.1575	0.2322	0.4039	0.5229
	10.34	0.1768	0.2154	0.4377	0.4987
	10.48	0.1980	0.1980	0.4722	0.4722
394.26	2.758	0.0061	0.0436	0.0218	0.1418
	3.447	0.0152	0.0905	0.0530	0.2650
	4.137	0.0250	0.1250	0.0850	0.3411
	5.516	0.0471	0.1636	0.1519	0.4148
	6.895	0.0719	0.1739	0.2192	0.4328
	7.584	0.0900	0.1720	0.2639	0.4295
	8.274	0.1067	0.1610	0.3021	0.4102
	8.618	0.1243	0.1474	0.3397	0.3852
	8.715	0.1345	0.1345	0.3603	0.3603

COMPONENTS:	ORIGINAL MEASUREMENTS:
1. Methane; CH_4; [74-82-8] 2. Butane; C_4H_{10}; [106-97-8]	Sage, B. H.; Budenholzer, R. A.; Lacey, W. N. *Ind. Eng. Chem.* <u>1940</u>, *32*, 1262-1277.
VARIABLES: Temperature, pressure	PREPARED BY: C. L. Young

EXPERIMENTAL VALUES:

T/K (T/°F)	p/psi	P/MPa	Wt. fraction of methane	Mole fraction of methane, x_{CH_4}
294.3 (70)	257.0	1.772	0.025	0.0849
	460.5	3.175	0.050	0.160
	645.4	4.450	0.075	0.227
	819	5.647	0.100	0.287
	976	6.729	0.125	0.341
	1122	7.736	0.150	0.390
	1393	9.604	0.200	0.475
	1604	11.06	0.250	0.547
	1745	12.03	0.300	0.608
	1893	13.05	0.400	0.707
	1924	13.27	0.500	0.784
	1867	12.87	0.600	0.844
	1600	11.03	0.700	0.894
310.9 (100)	275.0	1.896	0.025	0.0849
	514.2	3.545	0.050	0.160
	716.1	4.937	0.075	0.227
	903.3	6.228	0.100	0.287
	1074	7.405	0.125	0.341
	1228	8.467	0.150	0.390
	1485	10.24	0.200	0.475
	1672	11.53	0.250	0.547
	1796	12.38	0.300	0.608
	1906	13.14	0.400	0.707

(cont.)

AUXILIARY INFORMATION

METHOD/APPARATUS/PROCEDURE:	SOURCE AND PURITY OF MATERIALS:
PVT cell charged with mixture of known composition. Pressure measured with pressure balance. Temperature measured with resistance thermometer. Bubble point and dew point determined from discontinuity in PV isotherm. Coexisting liquid and gas phase properties determined by graphical means. Details of apparatus in ref. (1). NOTE: Source contains extensive PVT data.	1. Crude sample, treated for re-moval of higher alkanes, carbon dioxide and water vapor. Final purity 99.9 mole per cent. 2. Phillips Petroleum sample, dis-tilled, final purity better than 99.96 mole per cent.

	ESTIMATED ERROR: $\delta T/K = \pm 0.1$; $\delta P/MPa = \pm 0.007$; $\delta x_{CH_4} = \pm 0.002$ (estimated by compiler).
	REFERENCES: 1. Sage, B. H.; Lacey, W. N. *Trans. Am. Inst. Mining Met. Engnrs.* <u>1940</u>, *136*, 136.

COMPONENTS:	ORIGINAL MEASUREMENTS:
1. Methane; CH_4; [74-82-8]	Sage, B. H.; Budenholzer, R. A.; Lacey, W. N.
2. Butane; C_4H_{10}; [106-97-8]	*Ind. Eng. Chem.*
	<u>1940</u>, *32*, 1262-1277.

EXPERIMENTAL VALUES:

T/K (T/°F)	p/psi	P/MPa	Wt. fraction of methane	Mole fraction of methane, x_{CH_4}
310.9 (100)	1888	13.02	0.500	0.784
	1696	11.69	0.600	0.844
327.6 (130)	335.0	2.310	0.025	0.0849
	569.6	3.927	0.050	0.160
	784.0	5.403	0.075	0.227
	979.5	6.753	0.100	0.287
	1153	7.950	0.125	0.341
	1308	9.018	0.150	0.390
	1565	10.79	0.200	0.475
	1736	11.97	0.250	0.547
	1833	12.64	0.300	0.608
	1870	12.89	0.400	0.707
	1712	11.80	0.500	0.784
344.3 (160)	378.2	2.608	0.025	0.0849
	628.3	3.142	0.050	0.160
	848.5	5.850	0.075	0.227
	1049	7.233	0.100	0.287
	1228	8.467	0.125	0.341
	1377	9.494	0.150	0.390
	1611	11.11	0.200	0.475
	1757	12.11	0.250	0.547
	1810	12.48	0.300	0.608
	1689	11.64	0.400	0.707
360.0 (190)	449.8	3.101	0.025	0.0849
	696.5	4.802	0.050	0.160
	848.6	5.851	0.075	0.227
	1111	7.660	0.100	0.287
	1276	8.798	0.125	0.341
	1409	9.715	0.150	0.390
	1602	11.05	0.200	0.475
	1698	11.71	0.250	0.547
	1645	11.34	0.300	0.608
377.6 (220)	521.8	3.598	0.025	0.0849
	762.0	5.254	0.050	0.160
	973.1	6.709	0.075	0.227
	1167	8.046	0.100	0.287
	1307	9.011	0.125	0.341
	1422	9.804	0.150	0.390
	1517	10.46	0.200	0.475
	1349	9.301	0.250	0.547
394.3 (250)	660	4.551	0.025	0.0840
	824.8	5.687	0.050	0.160
	1024	7.060	0.075	0.227
	1173	8.088	0.100	0.287
	1255	8.653	0.125	0.341
	1242	8.563	0.150	0.390

COMPONENTS:	ORIGINAL MEASUREMENTS:
1. Methane; CH$_4$; [74-82-8] 2. Butane; C$_4$H$_{10}$; [106-97-8]	Roberts, L. R.; Wang, R. H.; Azarnoosh, A.; McKetta, J. J. *J. Chem. Eng. Data* 1962, *7*, 484-5.

VARIABLES:	PREPARED BY:
Temperature, pressure	C. L. Young

EXPERIMENTAL VALUES:

T/K (T/°F)	P/psi	P/MPa	Mole fraction of methane in liquid, x_{CH_4}	in vapor, y_{CH_4}
410.9 (280)	535	3.69	0.038	0.113
	669	5.68	0.087	0.202
	824	5.68	0.158	0.227
	831	5.73	0.158	–
	798	5.50	0.143	–
	787	5.43	0.127	0.231
	747	5.15	0.117	0.242
	735	5.07	0.115	0.234
377.6 (220)	1348	9.29	–	0.452
	1342	9.25	–	0.478
	1339	9.23	0.389	0.496
	1336	9.21	0.382	0.506
	1125	7.76	0.287	0.552
	878	6.05	0.201	0.533
277.6 (40)	53	0.36	0.0330	0.6213
	74	0.51	0.0317	0.7053
	102	0.70	0.0547	0.7969
	152	1.05	0.0768	0.8633
	192	1.32	0.0887	0.8867
	253	1.74	0.0914	0.9039
	298	2.05	0.1157	0.9140
	341	2.35	0.1484	0.9200
	447	3.08	0.1806	0.9420

(cont.)

AUXILIARY INFORMATION

METHOD/APPARATUS/PROCEDURE:	SOURCE AND PURITY OF MATERIALS:
Windowed stainless steel equilibrium cell. Vapor recirculated with magnetic pump. Temperature measured with thermocouple and pressure measured with Bourdon type gauge. Details of apparatus in source. Samples of liquid and gas analyzed by gas chromatography.	1. Phillips Petroleum Co., research grade sample, purity better than 99.5 mole per cent, major impurity nitrogen. 2. Phillips Petroleum Co., research grade sample, purity better than 99.9 mole per cent.

	ESTIMATED ERROR:
	$\delta T/K = \pm 0.1$; $\delta P/psi = \pm 2$; δx_{CH_4}, $\delta y_{CH_4} = \pm 0.002$.

	REFERENCES:

COMPONENTS:	ORIGINAL MEASUREMENTS:
1. Methane; CH₄; [74-82-8]	Roberts, L. R.; Wang, R. H.;
2. Butane; C₄H₁₀; [106-97-8]	Azarnoosh, A.; McKetta, J. J.
	J. Chem. Eng. Data
	1962, 7, 484-5.

EXPERIMENTAL VALUES:

T/K (T/°F)	P/psi	P/MPa	Mole fraction of methane in liquid, x_{CH_4}	in vapor, y_{CH_4}
277.6 (40)	449	3.10	0.2061	0.9282
	515	3.56	0.1979	0.9312
	584	4.03	0.2232	0.9424
	640	4.41	0.2424	0.9432
	735	5.07	0.2887	0.9510
	835	5.76	0.3139	0.9464
	840	5.79	0.3139	0.9463
	930	6.41	0.3453	0.9479
	1060	7.31	0.3674	0.9456
	1155	7.96	0.4245	0.9437
	1285	8.86	0.4795	0.9381
	1370	9.45	0.4842	0.9351
	1480	10.20	0.5227	0.9321
	1615	11.14	0.5641	0.9177
	1685	11.62	0.5888	0.8937
	1750	12.07	0.6369	0.9015
	1770	12.20	0.6275	0.8862
	1835	12.65	0.6898	0.8545
	1905	13.13	0.7749	0.8171
	1915	13.20	0.7953	0.7953
244.3 (-20)	26	0.18	0.015	0.457
	49	0.34	0.023	0.785
	78	0.54	0.043	0.875
	120	0.83	0.064	0.925
	149	1.03	0.077	0.930
	177	1.22	0.095	0.941
	251	1.73	0.116	0.971
	348	2.40	0.174	0.973
	429	2.96	0.205	0.978
	506	3.49	0.246	0.980
	613	4.23	0.306	0.978
	720	4.96	0.334	0.975
	845	5.83	0.403	0.977
	910	6.27	0.412	0.982
	930	6.41	0.422	0.977
	1075	7.41	0.504	0.973
	1225	8.44	0.552	0.970
	1235	8.52	0.563	0.968
	1290	8.89	0.578	0.967
	1295	8.93	0.580	0.970
	1380	9.51	0.608	0.957
	1590	10.96	0.719	0.938
	1645	11.34	0.793	0.903
	1724	11.89	0.863	0.863
210.9 (-80)	27	0.19	0.0350	0.8782
	57	0.39	0.0728	0.9437
	110	0.76	0.1058	0.9758
	169	1.17	0.1759	0.9839
	207	1.43	0.1796	0.9940
	263	1.81	0.2376	0.9918
	359	2.48	0.3165	0.9883
	518	3.57	0.4133	0.9917
	725	5.00	0.5986	0.9918
	785	5.41	0.6554	0.9948
	890	6.14	0.7412	0.9840
	975	6.72	0.8112	0.9670
	1041	7.18	0.9214	0.9214

COMPONENTS:	ORIGINAL MEASUREMENTS:
1. Methane; CH_4; [74-82-8] 2. Butane; C_4H_{10}; [106-97-8]	Wang, R. H.; McKetta, J. J. *J. Chem. Engng. Data* <u>1964</u>, *9*, 30-35.

VARIABLES:
 	
            Pressure
	PREPARED BY:
	
	           C. L. Young

EXPERIMENTAL VALUES:

T/K (T/°F)	P/MPa	P/psi	Mole fraction of methane in liquid, x_{CH_4}	in vapor, y_{CH_4}
177.6 (-140)	0.503	73	0.1579	0.9732
	0.841	122	0.2652	0.9924
	1.18	171	0.3582	0.9945
	1.50	217	0.4601	0.9868
	1.77	256	0.4913	0.9925
	2.28	330	0.7037	0.9942
	2.66	386	0.8241	0.9901
	2.93	425	0.9086	0.9942
	3.12	453	1.000	1.000

AUXILIARY INFORMATION

METHOD/APPARATUS/PROCEDURE:	SOURCE AND PURITY OF MATERIALS:
Stainless steel windowed equilibrium cell with magnetic pump for re-circulating vapor. Samples analysed by gas chromatography and mass spectrometry. Some details given in source and ref. (1).	1 and 2. Phillips Petroleum Co. research grade samples, purity at least 99.9 mole per cent.
	ESTIMATED ERROR: $\delta T/K = \pm 0.3$; $\delta P/MPa = \pm 0.2\%$; δx_{CH_4}, $\delta y_{CH_4} = \pm 0.001$.

	REFERENCES:
	 1. Wang, R. H.
	    Ph.D. thesis, University of Texas,
	    Austin, <u>1963</u>.

COMPONENTS:	ORIGINAL MEASUREMENTS:
1. Methane; CH$_4$; [74-82-8] 2. Butane; C$_4$H$_{10}$; [106-97-8]	Wiese, H. C.; Jacobs, J.; Sage, B. H. *J. Chem. Engng. Data* <u>1970</u>, *15*, 82-91.
VARIABLES:	PREPARED BY:
	C. L. Young

EXPERIMENTAL VALUES:

T/K	T/°F	P/MPa	P/psia	Mole fraction of methane in liquid, x_{CH_4}	in vapor, y_{CH_4}
277.6	40	1.38	200	0.0808	0.8888
		3.45	500	0.1913	0.9369
		6.89	1000	0.3651	0.9461
		8.62	1250	0.4513	0.9407
		10.34	1500	0.5390	0.9262
		11.72	1700	0.6194	0.9044
310.9	100	1.38	200	0.0530	0.7027
		3.45	500	0.1556	0.8473
		6.89	1000	0.3171	0.8809
		8.62	1250	0.3974	0.8786
		10.34	1500	0.4799	0.8665
		11.72	1700	0.5586	0.8440
344.3	160	1.38	200	0.0091	0.1171
		3.45	500	0.1201	0.6796
		6.89	1000	0.2717	0.7567
		8.62	1250	0.3482	0.7588
		10.34	1500	0.4329	0.7439
		11.72	1700	0.5103	0.7036
377.6	220	3.45	500	0.0783	0.4200
		6.89	1000	0.2361	0.5630
		8.62	1250	0.3200	0.5338

AUXILIARY INFORMATION

METHOD/APPARATUS/PROCEDURE:	SOURCE AND PURITY OF MATERIALS:
PVT cell charged with mixture of known composition. Pressure measured with pressure balance and temperature measured using a platinum resistance thermometer. Details in ref. (1). Samples of coexisting phases analysed by GC.	1. Texaco sample, passed over calcium chloride, activated charcoal, Ascarite and anhydrous calcium sulfate at pressures in excess of 3 MPa, purity 99.99 mole per cent. 2. Phillips Petroleum Co. samples, degassed, purity 99.95 mole per cent.
	ESTIMATED ERROR: $\delta T/K = \pm 0.01$; $\delta P/MPa = \pm 0.1\%$; δx_{CH_4}, $\delta y_{CH_4} = 0.005$ or better.
	REFERENCES: 1. Sage, B. H.; Lacey, W. N. *Trans. Am. Inst. Mining Met.* <u>1940</u>, *136*, 136.

COMPONENTS:	ORIGINAL MEASUREMENTS:
1. Methane; CH_4; [74-82-8] 2. Butane; C_4H_{10}; [106-97-8]	Elliott, D. G.; Chen, R. J. J.; Chappelear, P. S.; Kobayashi, R. *J. Chem. Eng. Data* <u>1974</u>, *19*, 71-7.

VARIABLES:	PREPARED BY:
Temperature, pressure	C. L. Young

EXPERIMENTAL VALUES:

T/K	P/psi	P/MPa	Mole fraction of methane in liquid, x_{CH_4}	in gas, y_{CH_4}
277.59	17.66	0.1218	0.0000	0.0000
	100.00	0.692	0.04258	0.796
	200.00	1.382	0.08986	0.890
	300.4	2.071	0.1339	0.9176
	400	2.758	0.1759	0.9313
	500	3.447	0.2152	0.9385
	600	4.137	0.2536	0.9425
	800	5.516	0.3262	0.9469
	1000	6.895	0.3976	0.9459
	1200	8.274	0.4651	0.9390
	1400	9.653	0.5331	0.9294
	1600	11.03	0.6078	0.9100
	1700	11.72	0.6558[a]	0.8967
	1800[b]	12.41	0.7278[a]	0.8460
	1822[b]	12.56	0.7828[a]	0.7828
255.38	7.25	0.050	0.0000	0.0000
	50.3	0.347	0.02570	0.837
	100.0	0.692	0.05591	0.9161
	200.3	1.381	0.1124	0.9516
	300.4	2.071	0.1643	0.9639
	400	2.758	0.2135	0.9689
	500	3.447	0.2580	0.9716
	700	4.826	0.3455	0.9729
	800	5.516	0.3905	0.9728

(cont.)

AUXILIARY INFORMATION

METHOD/APPARATUS/PROCEDURE:	SOURCE AND PURITY OF MATERIALS:
Recirculating vapor flow apparatus with windowed equilibrium cell. Temperature measured with platinum resistance thermometer, pressure measured with Bourdon gauge. Butane added to cell, air removed, methane added and recirculated for at least 0.5 hour. Sample analysed by G.C. Details in source and ref. (1).	1. Ultra high purity Matheson sample, purity 99.97 mole per cent. 2. Matheson research grade sample, purity 99.93 mole per cent.

ESTIMATED ERROR:
   $\delta T/K = \pm 0.02$; $\delta P/MPa = \pm 0.013$;
   $\delta x_{CH_4} < \pm 2\%$, $\delta(1-y_{CH_4}) = \pm 2\%$ or
0.00001 whichever is larger.

REFERENCES:
1. Wichterle, I.; Kobayashi, R.
   J. Chem. Eng. Data
   <u>1972</u>, *17*, 4.

1. Methane; CH_4; [74-82-8]	Elliott, D. G.; Chen, R. J. J.; Chappelear, P. S.; Kobayashi, R.
2. Butane; C_4H_{10}; [106-97-8]	*J. Chem. Eng. Data* 1974, *19*, 71-7.

EXPERIMENTAL VALUES:

T/K	P/psi	P/MPa	Mole fraction of methane in liquid, x_{CH_4}	in gas, y_{CH_4}
255.38	1000	6.896	0.4651	0.9696
	1200	8.274	0.5466	0.9625
	1400	9.653	0.6326	0.9499
	1600[b]	11.03	0.7498	0.9175
	1652	11.39	0.8543[a]	0.8543
244.28	4.326	0.0298	0.0000	0.0000
	100.2	0.691	0.06304	0.9488
	200.3	1.381	0.1237	0.9703
	400	2.758	0.2335	0.9801
	600	4.137	0.3325	0.9818
	800	5.516	0.4223	0.9809
	1000	6.895	0.5101	0.9772
	1200	8.274	0.6062	0.9688
	1400	9.653	0.7189	0.9498
	1500[b]	10.34	0.8059[a]	0.9159
	1515	10.45	0.8605[a]	0.8605
233.18	2.439	0.0168	0.0000	0.0000
	100.0	0.692	0.07208	0.9703
	200.3	1.381	0.1400	0.9827
	400	2.758	0.2655	0.9878
	600	4.137	0.3739	0.9882
	800	5.516	0.4804	0.9868
	1000	6.895	0.5875	0.9822
	1200	8.274	0.6948	0.9705
	1300	8.963	0.7886	0.9608
	1350[b]	9.308	0.8549[a]	0.9318
	1355	9.342	0.9097[a]	0.9097
222.07	1.285	0.00886	0.0000	0.0000
	100.2	0.691	0.08202	0.9840
	200.3	1.381	0.1586	0.99034
	400	2.758	0.2981	0.99298
	600	4.137	0.4284	0.99262
	800	5.516	0.5564	0.99046
	1000	6.895	0.7056	0.9843
	1100	7.584	0.8001	0.9774
	1150[b]	7.929	0.8637[a]	0.9648
	1169	8.060	0.9326[a]	0.9326
210.94	0.625	0.00431	0.0000	0.0000
	200.0	1.379	0.1880	0.99506
	400	2.758	0.3523	0.99608
	600	4.137	0.5104	0.99546
	800	5.516	0.6954	0.99271
	900	6.205	0.8232	0.9896
	950[b]	6.550	0.9036	0.9862
	973	6.709	0.9591[a]	0.9591
199.88	0.276	0.00190	0.0000	0.0000
	200.1	1.380	0.2267	0.99757
	300.3	2.070	0.3261	0.99794
	400	2.758	0.4267	0.99795
	500	3.447	0.5322	0.99770
	600	4.137	0.6591	0.99715
	700	4.826	0.8296	0.99584
	750[b]	5.171	0.9257	0.99445
	792	5.461	0.9829[a]	0.9829
190.58	0.12506	0.00086	0.0000	0.0000
	100.1	0.690	0.1466	0.99814
	199.7	1.377	0.2773	0.99880
	299.7	2.066	0.3988	0.99895

(cont.)

1. Methane; CH_4; [74-82-8]

2. Butane; C_4H_{10}; [106-97-8]

Elliott, D. G.; Chen, R. J. J.;
Chappelear, P. S.; Kobayashi, R.
J. Chem. Eng. Data
<u>1974</u>, *19*, 71-7.

EXPERIMENTAL VALUES:

T/K	P/psi	P/MPa	Mole fraction of methane in liquid, x_{CH_4}	in gas, y_{CH_4}
190.58	400	2.758	0.5314	0.99889
	500	3.447	0.7031	0.99866
	600	4.137	0.9469	0.99823
	671	4.626	1.000	1.000
189.06	0.109	0.00075	0.0000	0.0000
	101.1	0.697	0.1526	0.99835
	201.0	1.386	0.2860	0.99896
	300.3	2.070	0.4150	0.999083
	400	2.758	0.5521	0.999049
	501	3.454	0.7511	0.99875
	550	3.792	0.9009	0.99861
	600	4.137	0.9808	0.99873
	636	4.385	1.000	1.000
177.62	0.0360	0.000248	0.0000	0.0000
	50.0	0.345	0.09796	0.999001
	100.1	0.690	0.1879	0.999391
	149.8	1.033	0.2804	0.999531
	199.9	1.378	0.3716	0.999597
	299.9	2.068	0.5812	0.999651
	350	2.413	0.7288	0.999651
	400	2.758	0.9370	0.999671
	420	2.896	0.9793	0.999767
	420	2.896	0.9841	0.999767
	440	3.034	1.000	1.000
166.50	0.0106	0.000073	0.0000	0.0000
	50.0	0.345	0.1370	0.999656
	100.0	0.689	0.2640	0.999801
	150.0	1.034	0.3930	0.999831
	200.0	1.379	0.5451	0.999866
	250.0	1.724	0.7910	0.999911
	296	2.041	1.000	1.000
155.38	0.00264	0.000018	0.0000	0.0000
	20.1	0.139	0.08016	0.999828
	50.1	0.345	0.1925	0.999901
	100.1	0.690	0.3860	0.999940
	150.1	1.035	0.6678	0.999948
	187	1.289	1.000	1.000
144.26	0.0005	0.000003	0.0000	0.0000
	25.2	0.174	0.1492	0.999960
	49.9	0.344	0.3006	0.999971
	99.9	0.689	0.8173	0.999983
	115	0.793	1.0000	1.0000

[a] Bubble point analysis by gas chromatography.

[b] Critical point of mixture.

COMPONENTS:	ORIGINAL MEASUREMENTS:
1. Methane; CH_4; [74-82-8] 2. Butane; C_4H_{10}; [106-97-8]	Kahre, L. C. *J. Chem. Eng. Data* 1974, *19*, 67.

VARIABLES:	PREPARED BY:
Temperature, pressure	C. L. Young

EXPERIMENTAL VALUES:

T/K	P/psi	P/MPa	Mole fraction of methane in liquid, x_{CH_4}	in vapor, y_{CH_4}
283.15	21.6	0.1489	0.00	0.00
	51	0.352	0.013	0.566
	100	0.689	0.035	0.775
	201	1.386	0.076	0.875
	400	2.758	0.152	0.925
	600	4.137	0.232	0.939
	800	5.516	0.304	0.941
	1000	6.895	0.377	0.941
	1200	8.274	0.442	0.933
	1400	9.653	0.514	0.924
255.35	7.26	0.0501	0.00	0.00
	20	0.138	0.063[b]	0.623
	50	0.345	0.0212[b]	0.846
	100	0.689	0.0461[b]	0.9174
	400	2.758	0.195	0.9704
	599	4.130	0.2925	0.9746
	998	6.881	0.470	0.9710
	1397	9.632	0.651	0.9531
	1597	11.011	0.758	0.9317
227.55	1.77	0.0122	0.00	0.00
	50	0.345	0.034[b]	0.962
	100	0.689	0.069[b]	0.980
	200	1.379	0.139[b]	0.988

(cont.)

AUXILIARY INFORMATION

METHOD/APPARATUS/PROCEDURE:	SOURCE AND PURITY OF MATERIALS:
Apparatus for isotherm at 283.15 described in ref. (1). Other isotherm determined with a re-circulating vapor flow apparatus described in ref. (2). Liquid sample added to windowed equilibrium cell, air removed. Methane added to cell and re-circulated for at least ½ hour. Samples analysed by G.C.	1. Phillips Petroleum Co. research grade methane. 2. Phillips Petroleum Co. research grade butane.
	ESTIMATED ERROR: $\delta T/K = \pm 0.05$; $\delta P/MPa = \pm 0.015$; $\delta x_{CH_4} = \pm 2\%$; $\delta(1-y_{CH_4}) = 2\%$.

	REFERENCES:
	1. Kahre, L.
	    J. Chem. Eng. Data 1973, *18*, 267.
	
	2. Wichterle, I.; Kobayashi, R. J.
	    J. Chem. Eng. Data 1972, *17*, 4.

1. Methane; CH_4; [74-82-8]   Kahre, L. C.

2. Butane; C_4H_{10}; [106-97-8]   *J. Chem. Eng. Data*

                                         <u>1974</u>, *19*, 67.

EXPERIMENTAL VALUES:

T/K	P/psi	P/MPa	Mole fraction of methane in liquid, x_{CH_4}	in vapor, y_{CH_4}
227.55	299	2.062	0.209	0.9907
	499	3.440	0.350	0.9918
	798	5.502	0.534	0.990
	998	6.881	0.656	0.985
	1197	8.253	0.805	0.972
210.95	0.62	0.0043	0.00	0.00
	20	0.138	0.019[b]	0.969
	40	0.276	0.038[b]	0.984
	80	0.552	0.077[b]	0.9913
	120	0.827	0.111[b]	0.9936
	160	1.103	0.148[b]	0.9948
	200	1.379	0.184	0.9955
	399	2.751	0.351	0.9965
	599	4.130	0.532	0.9959
	798	5.502	0.721	0.9934
	936	6.453	...	0.9867
	973	6.709	0.933	0.9802
194.10	0.17	0.0012	0.00	0.00
	20	0.138	0.025[b]	0.9917
	40	0.276	0.051[b]	0.9955
	81	0.558	0.103	0.9975
	100	0.689	0.130	0.9979
	200	1.379	0.248	0.9986
	399	2.751	0.500	0.9988
	595	4.102	0.830	0.9980
	627	4.323	0.896	0.9977
	649	4.475	0.930	0.9975
	677	4.668	0.968	0.9972
185.95	0.084	0.00058	0.00	0.00
	20	0.138	0.028[b]	0.9956
	50	0.345	0.069	0.9981
	100	0.689	0.144	0.9989
	200	1.379	0.290	0.99927
	299	2.062	0.444	0.99934
	399	2.751	0.608	0.99926
	449	3.096	0.728	0.99920
	499	3.440	0.871	0.99914
	549	3.785	0.972	0.99931
	578	3.985	1.00	1.00
177.55	0.035	0.00024	0.00	0.00
	20	0.138	0.036[b]	0.9981
	50	0.345	0.091	0.99918
	100	0.689	0.180	0.99950
	200	1.379	0.360	0.99963
	299	2.062	0.573	0.99967
	354	2.441	0.732	0.99966
	386	2.661	0.875	0.99967
	404	2.785	0.934	0.99970
	441	3.041	1.00	1.00
166.45	0.010	0.000069	0.00	0.00
	20	0.138	0.047[b]	0.99940
	49	0.338	0.116	0.99972
	100	0.689	0.251	0.99984
	148	1.020	0.379	0.99987
	199	1.372	0.545	0.99989
	249	1.717	0.753	0.999902
	272	1.875	0.897	0.999920
	283	1.951	0.950	0.999943
	298	2.055	1.00	1.00

COMPONENTS:	EVALUATOR:
1. Methane; CH_4; [74-82-8] 2. 2-Methylpropane (*isobutane*); C_4H_{10}; [75-28-5]	Colin L. Young, School of Chemistry, University of Melbourne, Parkville, Victoria 3052, Australia. March 1982

CRITICAL EVALUATION:

The most extensive sets of data on this system are those of Barsuk *et al.* (1). These data which cover the temperature range 198 to 377 K and are in reasonable agreement with those of Olds *et al.* (2) at 310.9 K and 344.25 K. There are significant discrepancies between the two sets of data at 377.6 K and near the critical region at the low temperatures.

References

1. Barsuk, S. D.; Skripka, V. G.; Benyaminovich, O. A.
    Gazov. Prom. 1970, *15*, 38.

2. Olds, R. H.; Sage, B. H.; Lacey, W. N.
    Ind. Eng. Chem. 1942, *34*, 1008.

COMPONENTS:	ORIGINAL MEASUREMENTS:
1. Methane; CH_4; [74-82-8] 2. 2-Methylpropane (isobutane); C_4H_{10}; [75-28-5]	Olds, R. H.; Sage, B. H.; Lacey, W. N. *Ind. Eng. Chem.* <u>1942</u>, *34*, 1008-1013.

VARIABLES:	PREPARED BY:
	C. L. Young

EXPERIMENTAL VALUES:

T/K (T/°F)	P/MPa	p/psi	Wt. fraction of methane in liquid,	Wt. fraction of methane in vapor,	Mole fraction of methane in liquid, x_{CH_4}	Mole fraction of methane in vapor, y_{CH_4}
310.9	0.55	80	0.00077	0.0248	0.00278	0.0843
(100)	0.69	100	0.00311	0.0929	0.01117	0.2704
	1.03	150	0.00843	0.2049	0.02985	0.4826
	1.38	200	0.01400	0.2869	0.04888	0.5929
	2.07	300	0.02586	0.4031	0.08766	0.7097
	2.76	400	0.03793	0.4646	0.1249	0.7585
	3.45	500	0.05076	0.5076	0.1622	0.7886
	4.14	600	0.06434	0.5380	0.1993	0.8082
	4.83	700	0.07870	0.5594	0.2362	0.8213
	5.52	800	0.09390	0.5740	0.2728	0.8298
	6.21	900	0.1098	0.5826	0.4273	0.8348
	6.89	1000	0.1266	0.5859	0.3441	0.8366
	7.58	1100	0.1449	0.5845	0.3802	0.8358
	8.27	1200	0.1648	0.5785	0.4166	0.8324
	8.96	1300	0.1870	0.5673	0.4543	0.8259
	9.65	1400	0.2123	0.5498	0.4938	0.8155
	10.34	1500	0.2430	0.5233	0.5374	0.7989
	11.03	1600	0.2858	0.4810	0.5916	0.7703
	11.58	1679	0.3800	0.3800	0.6893	0.6893
344.3	1.38	200	0.00359	0.0492	0.01287	0.1577
(160)	2.07	300	0.01345	0.1468	0.04702	0.3838

(cont.)

AUXILIARY INFORMATION

METHOD/APPARATUS/PROCEDURE:	SOURCE AND PURITY OF MATERIALS:
PVT cell charged with mixture of known composition. Pressure measured with pressure balance. Temperature measured with resistance thermometer. Bubble point and dew point determined for various compositions from discontinuity in PV isotherm. Coexisting liquid and gas phase properties determined by graphical means. Details in ref. (1).	1. Crude sample treated for removal of higher alkanes, carbon dioxide and water vapor. Final purity 99.9 mole per cent. 2. Phillips Petroleum sample, purity at least 99.97 mole per cent.

	ESTIMATED ERROR: $\delta T/K = \pm 0.1$; $\delta P/MPa = \pm 0.007$; $\delta x_{CH_4} = \pm 0.001$; $\delta y_{CH_4} = \pm 0.005$ (estimated by compiler).

REFERENCES:

1. Sage, B. H.;   Lacey, W. N.
   Trans. Am. Inst. Mining Met. Engnrs.
   <u>1940</u>, *136*, 136.

COMPONENTS: ORIGINAL MEASUREMENTS:

1. Methane; CH$_4$; [74-82-8]   Olds, R. H.; Sage, B. H.;
                                              Lacey, W. N.
2. 2-Methylpropane (*iso*butane)
   C$_4$H$_{10}$; [75-28-5]   *Ind. Eng. Chem.*

                                              1942, *34*, 1008-1013.

EXPERIMENTAL VALUES:

T/K (T/°F)	P/MPa	p/psi	Wt. fraction of methane in liquid,	Wt. fraction of methane in vapor,	Mole fraction of methane in liquid, x_{CH_4}	Mole fraction of methane in vapor, y_{CH_4}
344.3	2.76	400	0.02381	0.2139	0.08112	0.4962
(160)	3.45	500	0.03481	0.2633	0.1155	0.5640
	4.14	600	0.04645	0.2998	0.1499	0.6078
	4.83	700	0.05875	0.3262	0.1843	0.6367
	5.52	800	0.07200	0.3454	0.2192	0.6563
	6.21	900	0.08624	0.3587	0.2546	0.6694
	6.89	1000	0.1020	0.3677	0.2913	0.6779
	7.58	1100	0.1189	0.3683	0.3281	0.6785
	8.27	1200	0.1381	0.3631	0.3671	0.6736
	8.96	1300	0.1623	0.3477	0.4122	0.6586
	9.65	1400	0.1980	0.3156	0.4719	0.6253
	10.05	1457	0.2580	0.2580	0.5572	0.5572
377.6	2.76	400	0.00871	0.0472	0.03082	0.1520
(220)	3.45	500	0.01912	0.0876	0.06590	0.2579
	4.14	600	0.03030	0.1182	0.1016	0.3267
	4.83	700	0.04160	0.1387	0.1358	0.3682
	5.52	800	0.05460	0.1521	0.1729	0.3937
	6.21	900	0.07030	0.1580	0.2149	0.4045
	6.89	1000	0.09370	0.1488	0.2723	0.3875
	7.14	1035	0.1230	0.1230	0.3367	0.3367

COMPONENTS:	ORIGINAL MEASUREMENTS:
1. Methane; CH_4; [74-82-8] 2. 2-Methylpropane; C_4H_{10}; [75-28-5]	Barsuk, S.D.; Skripka, V.G.; Benyaminovich, O.A. *Gazov. Prom.* 1970, *15*, 38-41.

VARIABLES:	PREPARED BY:
Temperature, pressure	C.L. Young

EXPERIMENTAL VALUES:

T/K	$P/10^5$Pa	Mole fraction of methane in liquid, x_{CH_4}	in vapor, y_{CH_4}
198.15	4.9	0.092	0.993
	9.8	0.184	0.996
	19.6	0.354	0.998
	29.4	0.523	0.998
	39.2	0.698	0.997
	49.0	0.826	0.995
	53.4	0.981	0.981
213.15	4.9	0.074	0.981
	9.8	0.141	0.988
	19.6	0.269	0.993
	29.4	0.394	0.996
	39.2	0.516	0.995
	49.0	0.635	0.993
	58.8	0.755	0.991
	68.6	0.890	0.973
	70.6	0.960	0.960
233.15	4.9	0.058	0.948
	9.8	0.111	0.966
	19.6	0.201	0.980
	29.4	0.293	0.985
	39.4	0.382	0.986
	49.0	0.472	0.985
	58.8	0.562	0.983

AUXILIARY INFORMATION

METHOD/APPARATUS/PROCEDURE:	SOURCE AND PURITY OF MATERIALS:
Recirculating vapor flow apparatus fitted with magnetic stirrer. Temperature measured with platinum resistance thermometer. Liquid and gas phases analysed by gas chromatography. Details in source and ref. (1).	Both samples had purity of 99.5 mole per cent.

ESTIMATED ERROR:

$\delta T/K = \pm 0.1$; $\delta P/10^5$Pa $= \pm 0.4$;
$\delta x_{CH_4}, \delta y_{CH_4} = \pm 3\%$

REFERENCES:

1. Skripka, V.G.; Barsuk, S.D.; Nikitina, I.E.; Benyaminovich, O.A.

Gazov. Prom. 1964, *14*, 41.

COMPONENTS:	ORIGINAL MEASUREMENTS:
1. Methane; CH_4; [74-82-8]	Barsuk, S.D.; Skripka, V.G.; Benyaminovich, O.A.
2. 2-Methylpropane; C_4H_{10}; [75-28-5]	*Gazov. Prom.* <u>1970</u>, *15*, 38-41.

EXPERIMENTAL VALUES:

T/K	$P/10^5$Pa	Mole fraction of methane in liquid, x_{CH_4}	in vapor, y_{CH_4}
233.15	68.6	0.651	0.978
	78.5	0.745	0.971
	88.3	0.850	0.952
	91.2	0.908	0.908
253.15	4.9	0.038	0.670
	9.8	0.080	0.894
	19.6	0.155	0.946
	29.4	0.230	0.963
	39.2	0.306	0.967
	49.0	0.377	0.966
	58.8	0.449	0.964
	68.6	0.520	0.962
	78.5	0.593	0.957
	88.3	0.670	0.948
	98.1	0.754	0.924
	106.5	0.856	0.856
273.15	4.9	0.024	-
	9.8	0.060	0.829
	19.6	0.132	0.899
	29.4	0.200	0.921
	39.2	0.268	0.930
	49.0	0.331	0.935
	58.8	0.393	0.938
	68.6	0.457	0.939
	78.5	0.521	0.937
	88.3	0.586	0.931
	98.1	0.651	0.918
	107.9	0.714	0.889
	114.9	0.810	0.810
293.15	4.9	0.012	0.340
	9.8	0.043	0.741
	19.6	0.102	0.827
	29.4	0.163	0.860
	39.2	0.222	0.880
	49.0	0.281	0.890
	58.8	0.339	0.898
	68.6	0.396	0.899
	78.5	0.454	0.896
	88.3	0.512	0.889
	98.1	0.574	0.877
	107.9	0.638	0.854
	117.7	0.725	0.778
	117.9	0.750	0.750
310.95	4.9	0.002	0.030
	9.8	0.030	0.485
	19.6	0.084	0.707
	29.4	0.140	0.770
	39.2	0.194	0.805
	49.0	0.249	0.824
	58.8	0.303	0.834
	68.6	0.357	0.835
	78.5	0.412	0.833
	88.3	0.467	0.824
	98.1	0.524	0.807
	107.9	0.603	0.760
	112.0	0.690	0.690

COMPONENTS:

1. Methane; CH_4; [74-82-8]
2. 2-Methylpropane; C_4H_{10}; [75-28-5]

ORIGINAL MEASUREMENTS:

Barsuk, S.D.; Skripka, V.G.; Benyaminovich, O.A.

Gazov. Prom. 1970, *15*, 38-41.

EXPERIMENTAL VALUES:

T/K	$P/10^5$Pa	Mole fraction of methane in liquid, x_{CH_4}	Mole fraction of methane in vapor, y_{CH_4}
344.25	14.7	0.018	0.231
	19.6	0.045	0.377
	29.4	0.096	0.531
	39.2	0.145	0.605
	49.0	0.196	0.645
	58.8	0.248	0.670
	68.6	0.301	0.680
	78.5	0.358	0.676
	88.3	0.425	0.654
	97.0	0.558	0.558
377.55	24.5	0.018	0.111
	29.4	0.044	0.200
	39.2	0.096	0.306
	49.0	0.150	0.377
	58.8	0.208	0.405
	68.6	0.316	0.358
	68.9	0.337	0.337

COMPONENTS:	EVALUATOR:
1. Methane; CH_4; [74-82-8] 2. Pentane; C_5H_{12}; [109-66-0]	Colin L. Young, School of Chemistry, University of Melbourne, Parkville, Victoria 3052, Australia. March 1982

EVALUATION:

This system has been studied over the temperature range 176.2 K to 444.3 K. The data of Frolich *et al.* (1) are classified as doubtful on account of their low precision and graphical presentation. The data of Boomer *et al.* (2) are also classified as doubtful in view of the fact that significant amounts of nitrogen were present in the system. The more limited data of Velikovskii *et al.* (3) are rejected since they are restricted to 273.2 K and the methane used contained about 1.5 mole per cent nitrogen.

The data of Prodany and Williams (4) and Sage and coworkers (5), (6) are classified as tentative. The earlier data of Sage *et al.* (5) are more limited than the latter data (6) although they cover part of the same temperature range. The later data covers the temperature range 310 K - 344 K (100 °F - 340 °F). The data published by Prodany and Williams (4) are probably more accurate than those of Sage and coworkers (5) and (6) but are restricted to 377.6 K (220 °F).

The data of Kahre (7) and Chu *et al.* (8) cover a similar temperature (176 K to 283 K) and pressure range. However, recommendation of either set of data is unwarranted since there are some discrepancies between the two sets of data. Both sets of data are therefore classified as tentative.

Dew point data for this system has been obtained by Chen *et al.* (9) but are not compiled nor evaluated here.

References

1.   Frolich, P. K.; Tauch, E. J.; Hogan, J. J.; Peer, A. A.
      Ind. Eng. Chem., 1931, *23*, 548.

2.   Boomer, E. H.; Johnson, C. A.; Piercey, A. G. A.
      Can. J. Res., 1938, *B16*, 319.

3.   Velikovskii, A. S.; Stepanova, G. S.; Vybornova, Ya. I.
      Gazov. Prom., 1964, *9* (2), 1.

4.   Prodany, N. W.; Williams, B.
      J. Chem. Eng. Data, 1971, *16*, 1.

5.   Sage, B. H.; Webster, D. C.; Lacey, W. N.
      Ind. Eng. Chem., 1936, *28*, 1045.

6.   Sage, B. H.; Reamer, H. H.; Olds, R. H.; Lacey, W. N.
      Ind. Eng. Chem., 1942, *34*, 1108.

7.   Kahre, L. C.
      J. Chem. Eng. Data, 1975, *20*, 363.

8.   Chu, T. C.; Chen, R. J. J.; Chappelear, P. S.; Kobayashi, R.
      J. Chem. Eng. Data, 1976, *21*, 41.

9.   Chen, R. J. J.; Chappelear, P. S.; Kobayashi, R.
      J. Chem. Eng. Data, 1974, *19*, 58.

COMPONENTS:	ORIGINAL MEASUREMENTS:
1. Methane; CH_4; [74-82-8] 2. Pentane; C_5H_{12}; [109-66-0]	Frolich, P.K.; Tauch, E.J.; Hogan, J.J.; Peer, A.A. *Ind. Eng. Chem.* <u>1931</u>, *23*, 548-550.

VARIABLES:	PREPARED BY:
Pressure	C.L. Young

EXPERIMENTAL VALUES:

T/K	P/MPa	Solubility[*]	Mole fraction of methane in liquid[+], x_{CH_4}
298.15	1.0	15	0.066
	2.0	31	0.128
	3.0	45	0.176
	4.0	61	0.225
	5.0	77	0.268
	6.0	95	0.311
	7.0	113	0.349
	8.0	129	0.380
	9.0	147	0.411
	10.0	166	0.441

* Data taken from graph in original article. Volume of gas
  measured at 101.325 kPa pressure and 298.15 K dissolved
  by unit volume of liquid measured under the same conditions.

+ Calculated by compiler.

AUXILIARY INFORMATION

METHOD/APPARATUS/PROCEDURE:	SOURCE AND PURITY OF MATERIALS:
Static equilibrium cell. Liquid saturated with gas and after equilibrium established samples removed and analysed by volumetric method. Allowance was made for the vapor pressure of the liquid and the solubility of the gas at atmospheric pressure. Details in source.	Stated that the materials were the highest purity available. Purity 98 to 99 mole per cent.
	ESTIMATED ERROR: $\delta T/K = \pm 0.1$; $\delta x_{CH_4} = \pm 5\%$
	REFERENCES:

COMPONENTS:	ORIGINAL MEASUREMENTS:
1. Methane; CH_4; [74-82-8] 2. Pentane; C_5H_{12}; [109-66-0]	Sage, B. H.; Webster, D. C.; Lacey, W. N. *Ind. Eng. Chem.* 1936, *28*, 1045-1047.

VARIABLES:	PREPARED BY:
	C. L. Young

EXPERIMENTAL VALUES:

T/K (T/°F)	p/psi	P/MPa[†]	Mass fraction of methane	Mole fraction[†] of methane, x_{CH_4}
310.9 (100)	854 1945 2228	5.89 13.41 15.36	0.0715 0.2031 0.2706	0.257 0.534 0.625
344.3 (160)	968 2064 2327	6.67 14.23 16.04	0.0715 0.2031 0.2706	0.257 0.534 0.625
377.6 (220)	1043 2026 2152	7.19 13.97 14.84	0.0715 0.2031 0.2706	0.257 0.534 0.625

[†] calculated by compiler.

AUXILIARY INFORMATION

METHOD/APPARATUS/PROCEDURE:	SOURCE AND PURITY OF MATERIALS:
PVT cell charged with mixture of known composition. Pressure measured with pressure balance. Bubble point determined from the discontinuity in the pressure, volume isotherm. Details of apparatus in ref. (1).	1. Prepared from natural gas, treated for removal of higher alkanes, carbon dioxide and water vapor. Final purity 99.9 mole per cent. 2. Phillips petroleum sample, fractionally distilled, final purity probably better than 99.8 mole per cent.

ESTIMATED ERROR:

$\delta T/K = \pm 0.1$; $\delta P/MPa = \pm 0.02$;

$\delta x_{CH_4} = \pm 0.002$ (estimated by compiler).

REFERENCES:

1. Sage, B. H.; Lacey, W. N.
   Ind. Eng. Chem.
   1934, *26*, 103.

COMPONENTS:	ORIGINAL MEASUREMENTS:
1. Methane; CH₄; [74-82-8] 2. Pentane; C₅H₁₂; [109-66-0]	Sage, B.H.; Reamer, H.H.; Olds, R.H. Lacey, W.N. *Ind. Eng. Chem.* 1942, *34*, 1108-1117

VARIABLES:	PREPARED BY:
Temperature, pressure	C.L. Young

EXPERIMENTAL VALUES:

T/K	P/10⁵Pa	Mole fraction of methane	
		in liquid, x_{CH_4}	in gas, y_{CH_4}
310.93	1.38	0.0015	0.2090
	2.76	0.0085	0.5893
	4.14	0.0154	0.7160
	5.52	0.0221	0.7797
	6.89	0.0288	0.8178
	10.34	0.0458	0.8696
	13.79	0.0626	0.8940
	20.68	0.0957	0.9195
	27.58	0.1282	0.9320
	41.37	0.1911	0.9430
	55.16	0.2508	0.9460
	68.95	0.3077	0.9470
	86.18	0.3748	0.9460
	103.4	0.4390	0.9410
	120.7	0.5041	0.9330
	137.9	0.5788	0.9204
	155.1	0.6770	0.8972
	169.3	0.8236	0.8236
344.26	4.14	0.0054	0.2805
	5.52	0.0115	0.4505
	6.89	0.0176	0.5524
	10.34	0.0329	0.6894
	13.79	0.0480	0.7568
	20.68	0.0777	0.8186
	27.58	0.1070	0.8485

AUXILIARY INFORMATION

METHOD/APPARATUS/PROCEDURE:	SOURCE AND PURITY OF MATERIALS:
PVT cell charged with mixture of known composition. Pressure measured with pressure balance. Temperature measured using resistance thermometer. Bubble point and dew point determined for various compositions. Co-existing liquid and gas phase properties determined by graphical means. Details in ref. (1).	1. Crude sample purified by removal of CO_2 and hydrocarbons. Final purity of 99.9 mole per cent. 2. Phillips petroleum sample purified fractionated purity better than 99.9 mole per cent.

ESTIMATED ERROR:
$\delta T/K = \pm0.03$; $\delta P/10$ Pa $= \pm0.1$
δx_{CH_4}, $\delta y_{CH_4} = \pm0.002$.

(estimated by compiler)

REFERENCES:
1. Sage, B.H.; Lacey, W.N.
   Trans. Am. Inst. Mining and Met. Engnrs. 1940, *136*, 136.

COMPONENTS:	ORIGINAL MEASUREMENTS
1. Methane; CH₄; [74-82-8]	Sage, B.H.; Reamer, H.H.; Olds, R.H. Lacey, W.N.
2. Pentane; C₅H₁₂; [109-66-0]	*Ind. Eng. Chem.* 1942,*34*, 1108-1117.

EXPERIMENTAL VALUES:

T/K	$P/10^5$Pa	Mole fraction of methane in liquid, x_{CH_4}	in gas, y_{CH_4}
344.26	41.37	0.1655	0.8785
	55.16	0.2213	0.8900
	68.95	0.2743	0.8937
	86.18	0.3381	0.8929
	103.4	0.4002	0.8875
	120.7	0.4670	0.8772
	137.9	0.5460	0.8558
	155.1	0.6654	0.8142
	161.2	0.7665	0.7665
377.59	6.89	0.0015	0.0458
	10.34	0.0159	0.3304
	13.79	0.0301	0.4722
	20.68	0.0587	0.6138
	27.58	0.0870	0.6846
	41.37	0.1435	0.7566
	55.16	0.1984	0.7880
	68.95	0.2509	0.7981
	86.18	0.3156	0.8009
	103.4	0.3817	0.7940
	120.7	0.4564	0.7584
	137.9	0.5659	0.7420
	143.5	0.6705	0.6705
410.93	13.79	0.0043	0.0578
	20.68	0.0338	0.3051
	27.58	0.0623	0.4289
	41.37	0.1178	0.5532
	55.16	0.1728	0.6134
	68.95	0.2297	0.6429
	86.18	0.3068	0.6420
	103.4	0.4076	0.6010
	111.0	0.5211	0.5211
444.26	27.58	0.0231	0.0938
	41.37	0.0853	0.2795
	55.16	0.1534	0.3561
	68.95	0.2569	0.3364
	70.67	0.2950	0.2950

COMPONENTS:	ORIGINAL MEASUREMENTS:
1. Methane; CH_4; [74-82-8] 2. Pentane; C_5H_{12}; [109-66-0]	Prodany, N.W.; Williams, B. *J. Chem. Engng. Data.* <u>1971</u>, *16*, 1-6.
VARIABLES: Pressure	PREPARED BY: C.L. Young

EXPERIMENTAL VALUES:

T/K	$p/10^5 Pa$	Mole fraction of methane in liquid x_{CH_4}	in vapor, y_{CH_4}
377.59	69.02	0.247	0.805
	69.29	0.248	0.814
	70.53	0.253	0.806
	84.87	0.306	0.810
	86.87	0.310	0.816
	87.22	0.324	0.812
	103.49	0.382	0.808
	103.56	0.380	0.808
	122.52	0.456	0.788
	137.83	0.532	0.740

AUXILIARY INFORMATION

METHOD/APPARATUS/PROCEDURE:	SOURCE AND PURITY OF MATERIALS:
Stirred equilibrium cell fitted with vapor and liquid sampling valves. Temperature measured with mercury in glass thermometer. Pressure measured with Bourdon gauge. Cell charged with components and contents equilibriated. Vapor and liquid samples withdrawn through pressure lock systems. Analysed using gas chromatography. Details in source.	1. Phillips Petroleum Co., sample purity 99.3 mole per cent (0.6 mole per cent nitrogen, 0.1 mole per cent ethane). 2. Phillips Petroleum Co. sample purity 99.9 mole per cent.
	ESTIMATED ERROR: $\delta T/K = \pm 0.3$; $\delta p/MPa = \pm 0.02$; $\delta x_{CH_4} = \pm 0.75\%$.
	REFERENCES:

COMPONENTS:	ORIGINAL MEASUREMENTS:
1. Methane; CH₄; [74-82-8] 2. Pentane; C₅H₁₂; [109-66-0]	Chu, T.C.; Chen, R.J.J.; Chappelear, P.S.; Kobayashi, R. *J. Chem. Eng.ng. Data.* <u>1976</u>, *21*, 41-4.

VARIABLES:	PREPARED BY:
Temperature, pressure	C.L. Young

EXPERIMENTAL VALUES:

T/K	P/MPa	Mole fraction of methane in liquid x_{CH_4}	in vapor y_{CH_4}
273.16	1.3803	0.09091	0.9758
	2.7593	0.1653	0.9839
	4.1369	0.2320	0.9855
	5.5158	0.2920	0.9856
	6.8948	0.3481	0.9839
	8.2737	0.4005	0.9818
	9.6527	0.4480	0.9782
	11.0316	0.4980	0.9722
	12.4106	0.5501	0.9623
	13.7895	0.6117	0.9450
	14.48	0.661	-
	14.82	0.695	-
	15.10	0.695	-
	15.1685	0.9089	0.9089
248.34	0.6909	0.04943	0.9876
	1.3803	0.1119	0.99281
	2.7593	0.2089	0.99475
	4.1369	0.2958	0.99478
	5.5158	0.3695	0.99400
	6.8948	0.4309	0.99223
	8.2737	0.4765	0.9900
	9.6527	0.5708	0.9845
	11.0316	0.6491	0.9753
	12.4106	0.7279	0.9587
	12.76	0.759	-

AUXILIARY INFORMATION

METHOD/APPARATUS/PROCEDURE:	SOURCE AND PURITY OF MATERIALS:
Recirculating vapor flow apparatus. Temperature measured with Platinum resistance thermometer. Pressure measured with Bourdon gauge. Liquid added to windowed cell and air remov- ed. Methane added to cell and vapor recirculated until equilibrium established. (average time ~ 4 hours) Samples analysed by gas chromatog- raphy.	1. Ultra high purity sample from Matheson; purity 99.97 mole per cent. 2. Phillips Petroleum Co. sample purity 99.93 mole per cent.

ESTIMATED ERROR: $\delta T/K = \pm 0.02$; $\delta P/MPa = \pm 0.007$; $\delta x_{CH_4} \leq \pm 2\%$; $\delta(1-y_{CH_4}) = \pm 2\%$ or 0.00001 whichever is largest.

REFERENCES:

1. Chen, R.J.J.; Chappelear, P.S.; Kobayashi, R.,

   J. Chem. Engng. Data. <u>1974</u>, *19*, 58.

COMPONENTS:	ORIGINAL MEASUREMENTS:
1.　Methane; CH_4; [74-82-8] 2.　Pentane; C_5H_{12}; [109-66-0]	Chu, T.C.; Chen, R.J.J.; Chappelear, P.S.; Kobayashi, R. *J. Chem. Engng. Data.* <u>1976</u>, *21*, 41-4.

EXPERIMENTAL VALUES:

T/K	P/MPa	Mole fraction of methane in liquid, x_{CH_4}	in vapor, y_{CH_4}
248.34	12.96	0.811	–
	13.03	0.859	–
223.92	0.6909	0.08592	0.99738
	1.3803	0.1667	0.99842
	2.7593	0.2878	0.99848
	4.1369	0.3888	0.99815
	5.5158	0.4737	0.99738
	6.8948	0.5652	0.99522
	8.2737	0.6850	0.99071
	8.963	0.750	–
	9.170	0.785	–
	9.446	0.841	–
	9.653	0.9437	–
199.86	0.3461	0.0566	0.99933
	0.6909	0.1166	0.999576
	1.3803	0.2212	0.999667
	2.7593	0.3758	0.999581
	4.1369	0.5265	0.999324
	4.826	0.6552	–
	5.171	0.7333	–
	5.378	0.799	–
	5.447	0.863	–
194.17	0.6902	0.1251	0.999735
	1.3794	0.2378	0.999781
	2.7586	0.4041	0.999732
	4.1369	0.6226	0.999404
	4.482	0.7386	–
	4.619	0.8438	–
	4.688	0.90431	–
192.62	0.6909	0.1297	0.999767
	1.3803	0.2320	0.999811
	2.7593	0.4083	0.999775
	4.1369	0.6667	0.999425
	4.413	0.835	–
	4.488	0.9057	–
	4.551	0.9538	–
176.21	0.1386	0.03195	0.999844
	0.3461	0.08509	0.999915
	0.6909	0.1681	0.999939
	1.0356	0.2504	0.999946
	1.3803	0.3316	0.999949
	1.724	0.403	–
	2.0698	0.4819	0.999951
	2.415	0.6262	–
	2.551	0.759	–

+　vapor phase composition quoted here and in
　　original were interpolated from data given
　　in reference 1.

COMPONENTS:	ORIGINAL MEASUREMENTS:
1. Methane; CH_4; [74-8208] 2. Pentane; C_5H_{12}; [109-66-0]	Kahre, L.C. *J. Chem. Engng. Data.* <u>1975</u>, *20*, 20, 363-7

VARIABLES:
 	
     Temperature, pressure
	PREPARED BY:
	
	         C.L. Young

EXPERIMENTAL VALUES:

T/K	P/atm	P/MPa	Mole fraction of methane in liquid, x_{CH_4}	in vapor y_{CH_4}
177.6	3.40	0.345	0.0785 a	0.999937
	6.80	0.689	0.157	0.999961
	13.61	1.379	0.321	0.999970
	20.41	2.068	0.510	0.999956
	23.81	2.413	0.623	0.999949
	27.08	2.744	0.829	0.999938
	27.90	2.827	0.897	0.999935
	30.01	3.041	1.000	1.0000
186.0	3.40	0.345	0.064 a	0.99983
	6.80	0.689	0.128	0.999903
	13.61	1.379	0.260	0.999921
	20.41	2.068	0.392	0.999917
	27.22	2.758	0.540	0.999885
	30.62	3.103	0.618	0.99986
	34.02	3.447	0.740	0.99979
	36.06	3.654	0.865	0.99973
	37.01	3.750	0.9545	0.99971
	38.03	3.853	0.9715	0.99978
	39.33	3.985	1.000	1.000
191.0	3.40	0.345	0.058 a	0.99976
	6.80	0.689	0.116	0.99984
	13.61	1.379	0.228	0.99988
	20.41	2.068	0.343 a	0.99988

AUXILIARY INFORMATION

METHOD/APPARATUS/PROCEDURE:	SOURCE AND PURITY OF MATERIALS:
Recirculating vapor flow apparatus. Temperature measured with platinum resistance thermometer. Pressure measured with Bourdon gauge. Liquid sample added to windowed equilibrium cell, air removed. Methane added to cell and recirculated for at least half an hour. Samples of both phases analysed by GC.	1. Phillips research grade, purity 99.98 mole per cent. 2. Phillips research grade, purity 99.99 mole per cent.

ESTIMATED ERROR:
$$\delta T/K = \pm 0.6; \quad \delta P/MPa = \pm 0.013;$$
$$\delta x_{CH_4} = \pm 2\%; \quad \delta(1-y_{CH_4}) = \pm 5\%$$

REFERENCES:

COMPONENTS:		ORIGINAL MEASUREMENTS:

1. Methane; CH_4; [74-82-8]   Kahre, L.C.

2. Pentane; C_5H_{12}; [109-66-0]   *J. Chem. Engng. Data*, 1975,20, 363-7.

EXPERIMENTAL VALUES:

T/K	P/atm	P/MPa	Mole fraction of methane in liquid x_{CH_4}	in vapor y_{CH_4}
191.0	27.22	2.758	0.465	0.99984
	34.02	3.447	0.609 a	0.99976
	40.82	4.136	0.817	0.99946
	42.46	4.302	0.9311	0.99928
	43.68	4.426	0.9732	0.99922
	44.29	4.488	-	0.99923
	44.97	4.557	-	0.99931
	45.86	4.647	Critical opalescence observed.	
198.2	3.40	0.345	0.051 a	0.99950
	6.80	0.689	0.1015	0.99971
	13.61	1.379	0.205	0.99978
	20.41	2.068	0.298 a	0.99978
	27.22	2.758	0.391	0.99976
	34.02	3.447	0.485 a	0.99967
	40.82	4.136	0.595	0.99947
	47.63	4.826	0.737	0.99890
	51.03	5.171	0.866	0.9980
	52.39	5.308	0.956	0.9970
	53.55	5.426	Critical opalescence observed.	
227.6	3.40	0.345	0.031 a	0.9938
	6.80	0.689	0.062 a	0.9966
	27.22	2.758	0.248	0.9983
	47.63	4.826	0.411	0.9976
	68.04	6.894	0.564	0.9951
	81.65	8.273	0.657	0.9906
	95.26	9.652	0.770	0.9833
	100.70	10.203	0.831	0.9702
	102.60	10.396	Critical opalescence observed.	
255.4	3.40	0.345	0.023 a	0.9686
	6.80	0.689	0.046 a	0.9832
	13.61	1.379	0.095	0.9902
	27.22	2.758	0.183	0.9932
	40.82	4.136	0.267	0.9935
	54.43	5.515	0.348	0.9924
	68.04	6.894	0.424	0.9909
	102.06	10.341	0.597	0.9807
	122.47	12.409	0.712	0.9690
	136.08	13.788	0.811	0.9449
	136.83	13.864	Critical opalescence observed.	
283.2	6.80	0.689	0.036 a	0.9376
	13.61	1.379	0.075 a	0.9650
	20.41	2.068	0.113	0.9735
	61.24	6.205	0.320	0.9810
	88.45	8.962	0.438	0.9751
	115.67	11.720	0.558	0.9637
	136.08	13.788	0.649	0.9436
	149.67	15.165	0.725	0.9306
	156.49	15.856	Critical opalescence observed.	

a   =   Values estimated in original paper.

COMPONENTS:	ORIGINAL MEASUREMENTS:
1. Methane; CH_4; [74-82-8] 2. Nitrogen; N_2; [7727-37-9] 3. Pentane; C_5H_{12}; [109-66-0]	Boomer, E. H.; Johnson, C. A.; Piercey, A. G. A. *Can. J. Res. B* <u>1938</u>, *16*, 319-327.
VARIABLES: Temperature, pressure	PREPARED BY: C. L. Young

EXPERIMENTAL VALUES:

Mole fractions

T/K	P/atm	P/MPa	in liquid x_{CH_4}	x_{N_2}	$x_{C_5H_{12}}$	in vapor y_{CH_4}	y_{N_2}	$y_{C_5H_{12}}$
298.15	35.5	3.60	0.156	0.005	0.839	0.848	0.069	0.083
			0.161	0.003	0.836	0.884	0.033	0.083
	68.1	6.90	0.298	0.004	0.698	0.880	0.069	0.051
	101.4	10.27	0.421	0.012	0.567	0.889	0.064	0.047
			0.426	0.007	0.567	0.892	0.056	0.052
	134	13.6	0.536	0.014	0.450	0.874	0.068	0.058
	167.6	16.98	0.683	0.036	0.281	0.831	0.047	0.122
			0.677	0.041	0.282	0.820	0.058	0.122
	188	19.0	0.729	0.046	0.225	0.734	0.041	0.225
328.15	35.5	3.60	0.139	0.003	0.858	0.815	0.047	0.138
	101.4	10.27	0.386	0.011	0.603	0.862	0.057	0.081
			0.380	0.014	0.606	0.851	0.065	0.084
	134	13.6	0.495	0.023	0.482	0.841	0.056	0.103
	167.6	16.98	0.667	0.033	0.300	0.772	0.043	0.185
			0.661	0.039	0.300	0.762	0.047	0.191
	174.4	17.67	0.734	0.041	0.225	0.738	0.039	0.223
358.15	35.2	3.57	0.121	0.003	0.876	0.770	0.040	0.190
	100.7	10.20	0.353	0.014	0.633	0.791	0.064	0.145
	133.7	13.55	0.478	0.025	0.497	0.773	0.050	0.177
			0.485	0.020	0.494	0.766	0.057	0.177
	147.2	14.92	0.543	0.029	0.428	-	-	-
			0.538	0.031	0.431	0.745	0.047	0.208
			0.542	0.028	0.430	0.748	0.040	0.212
	160.1	16.22	0.750	0.042	0.208	0.750	0.044	0.206

AUXILIARY INFORMATION

METHOD/APPARATUS/PROCEDURE:	SOURCE AND PURITY OF MATERIALS:
Rocking autoclave stirred by steel piston falling under gravity. Samples of vapor and liquid trapped in two auxiliary high pressure cells. Equilibrium samples analysed in complicated volumetric and combustion apparatus. Details in ref. (1). NOTE: The source reference also contains data on a mixture of pentanes + methane + nitrogen. Since the isomeric composition of the pentane mixture is not known, the data have not been included here.	1. and 2. Natural gas sample containing 94.4 mole per cent of methane and 5.6 mole per cent of nitrogen. Impurities may have been present amounting to 0.1 mole per cent. 3. Commercial product, chemically purified, dried and fractionated.
	ESTIMATED ERROR: $\delta T/K = \pm 0.1$; $\delta P/MPa = \pm 0.02$; δx, $\delta y = \pm 1\%$ (estimated by compiler).
	REFERENCES: 1. Boomer, E. H.; Johnson, C. A.; Argue, G. H. *Can. J. Res. B* <u>1937</u>, *15*, 367.

COMPONENTS:	EVALUATOR:
1. Methane; CH_4; [74-82-8]	Colin L. Young,
	School of Chemistry,
2. 2,2-Dimethylpropane (neopentane);	University of Melbourne,
C_5H_{12}; [463-82-1]	Parkville, Victoria 3052,
	Australia.
	March 1982

EVALUATION:

        This system has been investigated by Prodany and Williams (1) over the temperature range 344 K to 411 K and by Rogers and Prausnitz (2) at 298 K.   Because of the different temperatures studied a detailed comparison between the data is impossible.   Both sets of data are classified as tentative.

References

1.   Prodany, N. W.; Williams, B. *J. Chem. Engng. Data*, <u>1971</u>, *16*, 1.

2.   Rogers, B. L.; Prausnitz, J. M. *J. Chem. Thermodyn.*, <u>1971</u>, *3*, 211.

COMPONENTS:	ORIGINAL MEASUREMENTS:
1. Methane; CH_4; [74-82-8] 2. 2,2-Dimethylpropane, (Neopentane) C_5H_{12}; [463-82-1]	Rogers, B.L.; Prausnitz, J.M. *J. Chem. Thermodynamics*, 1971, *3*, 211-6.

VARIABLES:	PREPARED BY:
Pressure	C.L. Young

EXPERIMENTAL VALUES:

T/K	P/MPa	10^2 Mole fraction of methane in liquid, $10^2\ x_{CH_4}$	in gas $10^2\ y_{CH_4}$
298.15	1.213	6.96	89.23
	2.354	13.88	92.54
	3.450	20.55	94.07
	4.371	24.22	95.19
	5.446	30.97	95.54
	6.778	37.31	95.22
	7.832	41.74	95.32
	8.998	46.87	94.81
	9.487	49.78	94.54
	10.417	52.76	94.51
	10.915	55.42	94.55
	11.990	61.28	93.54
	13.039	65.36	91.15
	13.913	69.50	89.93
	14.500	75.45	88.71
	15.100	82.87	86.32

AUXILIARY INFORMATION

METHOD/APPARATUS/PROCEDURE:	SOURCE AND PURITY OF MATERIALS:
Stainless steel equilibrium cell fitted with pistons which enabled sample of gas and liquid to be taken without a change in pressure. Pressure measured with floating piston gauge and temperature with four thermocouples. Cell charged with components and magnetically stirred. Samples removed and analysed using gas chromatography. Details in ref. (1).	1. Matheson, ultra high purity grade purity 99.97 mole per cent. 2. Phillips Petroleum Co. research grade sample, purity 99.97 mole per cent.

ESTIMATED ERROR:
$\delta T/K = \pm 0.5$; $\delta P/MPa = \pm 0.007$;
δx_{CH_4}, $\delta y_{CH_4} = \pm 1\%$

REFERENCES:

1. Rogers, B.L.; Prausnitz, J.M. *Ind.Eng. Chem. Fundam.* 1970, *9*, 1974.

COMPONENTS:	ORIGINAL MEASUREMENTS:
1. Methane; CH_4; [74-82-8] 2. 2,2-Dimethylpropane; (neopentane) C_5H_{12}; [463-82-1]	Prodany, N.W.; Williams, B. *J. Chem. Engng. Data.* <u>1971</u>, *16*, 1-6.

VARIABLES:	PREPARED BY:
Temperature, pressure	C.L. Young

EXPERIMENTAL VALUES:

T/K	$p/10^5$Pa	Mole fraction of methane in liquid, x_{CH_4}	in vapor, y_{CH_4}
344.26	21.37	0.085	0.667
	35.23	0.153	0.761
	52.61	0.232	0.797
	69.29	0.312	0.819
	87.77	0.391	0.813
	88.32	0.398	0.813
	104.87	0.482	0.784
	117.83	0.560	0.727
	120.52	0.603	0.685
377.59	21.23	0.051	0.395
	34.68	0.117	0.563
	51.57	0.197	0.639
	69.50	0.282	0.670
	86.25	0.377	0.654
	98.87	0.471	0.585
410.93	34.89	0.068	0.280
	52.06	0.163	0.407
	69.22	0.281	0.416

AUXILIARY INFORMATION

METHOD/APPARATUS/PROCEDURE:	SOURCE AND PURITY OF MATERIALS:
Stirred equilibrium cell fitted with vapor and liquid sampling valves. Temperature measured with mercury in glass thermometer. Pressure measured with Bourdon gauge. Cell charged with components and contents equilibriated. Vapor and liquid samples withdrawn through pressure lock systems. Analysed using gas chromatography. Details in source.	1. Phillips Petroleum Co. sample 99.3 mole per cent (0.6 mole per cent nitrogen, 0.1 mole per cent ethane). 2. Phillips Petroleum Co. sample purity 99.8 mole per cent.

ESTIMATED ERROR:

$\delta T/K = \pm 0.3$; $\delta p/MPa = \pm 0.02$;
$\delta x_{CH_4} = \pm 0.75\%$.

REFERENCES:

COMPONENTS:	ORIGINAL MEASUREMENTS:
1. Methane; CH_4; [74-82-8] 2. 2-Methylbutane (*isopentane*); C_5H_{12}; [78-78-4]	Prodany, N. W.; Williams, B. *J. Chem. Engng. Data* 1971, *16*, 1-6.

VARIABLES:	PREPARED BY:
	C. L. Young

EXPERIMENTAL VALUES:

T/K (T/°F)	P/MPa	p/psi	Mole fraction of methane in liquid, x_{CH_4}	in vapor, y_{CH_4}
344.26	3.46	502	0.142	0.841
(160)	5.21	755	0.218	0.872
	6.90	1001	0.283	0.885
	8.64	1253	0.351	0.879
	10.38	1505	0.418	0.869
	12.13	1759	0.489	0.853
	13.73	1992	0.545	0.821
	15.11	2191	0.633	0.741
377.59	3.44	499	0.118	0.710
(220)	5.23	759	0.192	0.765
	6.90	1001	0.262	0.741
	8.66	1256	0.331	0.788
	10.37	1503	0.396	0.774
	11.87	1721	0.454	0.746
	13.09	1899	0.566	0.686
410.93	3.52	511	0.092	0.520
(280)	5.23	759	0.161	0.603
	6.90	1001	0.231	0.636
	8.74	1267	0.315	0.651
	8.80	1277	0.330	0.643
	10.46	1517	0.488	0.581

AUXILIARY INFORMATION

METHOD/APPARATUS/PROCEDURE:	SOURCE AND PURITY OF MATERIALS:
Stirred equilibrium cell fitted with vapor and liquid sampling valves. Temperature measured with mercury in glass thermometer. Pressure measured with Bourdon gauge. Cell charged with components and contents equilibrated. Vapor and liquid samples withdrawn through pressure lock systems. Analysed using gas chromatography. Details in source.	1. Phillips Petroleum Co. sample, 99.3 mole per cent (0.6 mole per cent nitrogen, 0.1 mole per cent ethane). 2. Phillips Petroleum Co. sample, purity 99.9 mole per cent.

ESTIMATED ERROR:

$\delta T/K = \pm0.3$; $\delta P/MPa = \pm0.02$;

$\delta x_{CH_4} = \pm0.75\%$.

REFERENCES:

COMPONENTS:	ORIGINAL MEASUREMENTS:
(1) Methane; CH_4; [74-82-8] (2) 2-methylbutane or isopentane; C_5H_{12}; [78-78-4]	Pomeroy, R. D.; Lacey, W. N.; Scudder, N. F.; Stapp, F. P. *Ind. Eng. Chem.* 1933, *25*, 1014-1019.

VARIABLES:	PREPARED BY:
T/K = 303.15 p_1/MPa = 0.293 - 1.327 (2.89 - 13.10 atm)	H. L. Clever

EXPERIMENTAL VALUES:

Temperature		Pressure		Methane Dissolved[1,2] V/cm³	Solvent Volume Increase ΔV/cm³	Solubility Gas in Sat. Solu.[2] C_s/cm³ cm⁻³
t/°C	T/K	p_1/atm	p_1/MPa			
30	303.15	2.89	0.293	65.6	0.172	2.46
		6.29	0.637	142.8	0.392	5.30
		8.62	0.873	195.6	0.531	7.23
		10.06	1.019	228.0	0.614	8.40
		13.10	1.318	524.0	---	10.95

[1] Isopentane volume 26.53 cm³ except the last value for which the volume is 47.87 cm³.

[2] Gas volumes at 303.15 K (30°C) and 101.325 kPa (1 atm).

The diffusion coefficient of methane in 2-methylbutane was measured to be 10^5 D/cm² s⁻¹ = 14.00.

AUXILIARY INFORMATION

METHOD/APPARATUS/PROCEDURE:	SOURCE AND PURITY OF MATERIALS:
Measurements were carried out in a brass absorption cell designed for diffusion measurements.	(1) Methane. Gas obtained from a natural gas sample which was treated with activated charcoal at pressures up to 70 atm. The gas contained up to 2 per cent ethane and a small amount of nitrogen. (2) 2-methylbutane. Obtained by repeated fractionation of casinghead gasoline. B.p. (760 mmHg), t/°C = 27.3 - 28.2.
	ESTIMATED ERROR: $\delta T/K$ = ± 0.05 $\delta C_s/C_s$ = ± 0.05 (compiler)

COMPONENTS:	ORIGINAL MEASUREMENTS:
1. Methane; CH_4; [74-82-8] 2. 2-Methylbutane; C_5H_{12}; [78-78-4]	Amick, E. H.; Johnson, W. B.; Dodge, B. F. *Chem. Eng. Progr. Symp. Ser.,* 1952, *48*, 65-71.

VARIABLES:	PREPARED BY:
Temperature, pressure	C. L. Young

EXPERIMENTAL VALUES:

T/°F	T/K	P/Mpa (P/psi)	Mole fraction of methane in liquid	in vapor
160	344	2.76	0.089	0.811
190	361	(400)	0.072	0.726
220	378		0.057	0.618
250	394		0.043	0.494
280	405		0.030	0.354
310	428		0.017	0.017
340	444		0.004	0.034
160	344	3.45	0.141	0.832
190	361	(500)	0.115	0.766
220	378		0.093	0.677
250	394		0.072	0.563
280	405		0.054	0.440
310	428		0.037	0.296
340	444		0.021	0.130
350	450		0.015	0.083
160	344	4.14	0.210	0.842
190	361	(600)	0.180	0.792
220	378		0.155	0.718
250	394		0.134	0.622
280	405		0.117	0.507
310	428		0.102	0.372
340	444		0.088	0.222
350	450		0.084	0.160
160	344	6.89	0.300	0.870
190	361	(1000)	0.296	0.830
220	378		0.292	0.770
250	394		0.288	0.680
280	405		0.285	0.570
310	428		0.282	0.415

AUXILIARY INFORMATION

METHOD/APPARATUS/PROCEDURE:	SOURCE AND PURITY OF MATERIALS:
Bubble-point dew-point apparatus. Sample confined over mercury, pressure measured with dead-weight piston gauge and temperature measured with mercury-in-glass thermometer.	1. Sample subjected to fractional distillation. 2. Phillips Petroleum sample, stated purity 99.5 mole per cent dried and distilled.
	ESTIMATED ERROR:
	REFERENCES:

COMPONENTS:	EVALUATOR:
1. Methane; CH_4 ; [74-82-8]	Colin L. Young,
2. Hexane; C_6H_{14} ; [110-54-3]	School of Chemistry, University of Melbourne, Parkville, Victoria 3052, Australia.
	March 1982

EVALUATION:

The most extensive study of this system has been undertaken by Lin *et al.* (1).   Their data cover the temperature range 182.5 K to 273.2 K and are classified as recommended.

The data of Frolich *et al.* (2) at 298.2 K are rejected on the grounds that the data were presented in graphical form and have been superseded by more recent data.   Gunn *et al.* (3) studied the vapor composition in this system but used literature values for the coexisting liquid phase compositions.   Chen *et al.* (4) also studied the dew point loci for the methane + hexane system at temperatures between 182.5 K and 273.2 K.   The data in refs. (3) and (4) are not considered further.

The data of Boomer and Johnson (5) are classified as tentative.   The methane in their work contained a significant proportion of nitrogen. Sage, Webster and Lacey (6) reported the solubility of methane in hexane at three temperatures, 37.78 °C (100 °F), 71.11 °C (160 °F) and 104.4 °C (220 °F), but only three compositions were studied.   These data are classified as tentative but limited in scope.

The remaining three studies of Poston and McKetta (7), Shim and Kohn (8) and Schoch *et al.* (9) are all classified as tentative.   There is reasonable agreement between the three sets of data.   The data of Shim and Kohn (8) deviates slightly from the data of Lin *et al.* (1) in the temperature range where the two sets of data overlap, the deviations being greatest at the lowest temperature.   The data of Lin *et al.* (1) are superior in this region.

References

1.   Lin, Y. N.;   Chen, R. J. J.;   Chappelear, P. S.;   Kobayashi, R.
      J. Chem. Eng. Data, 1977, *22*, 402.

2.   Frolich, P. K.;   Tauch, E. J.;   Hogan, J. J.;   Peer, A. A.
      Ind. Eng. Chem., 1931, *23*, 548.

3.   Gunn, R. D.;   McKetta, J. J.;   Ata, N.
      Am. Inst. Chem. Engnrs. J., 1974, *20*, 347.

4.   Chen, R. J. J.;   Chappelear, P. S.;   Kobayashi, R.
      J. Chem. Eng. Data, 1976, *21*, 213.

5.   Boomer, E. H.;   Johnson, C. A.
      Can. J. Res., 1938, *16B*, 328.

6.   Sage, B. H.;   Webster, D. C.;   Lacey, W. N.
      Ind. Eng. Chem., 1936, *28*, 1045.

7.   Poston, R. S.;   McKetta, J. J.
      J. Chem. Eng. Data, 1966, *11*, 362.

8.   Shim, J.;   Kohn, J. P.
      J. Chem. Eng. Data, 1972, *7*, 3.

9.   Schoch, E. P.;   Hoffmann, A. E.;   Mayfield, F. D.
      Ind. Eng. Chem., 1941, *33*, 688.

COMPONENTS:	ORIGINAL MEASUREMENTS:
1. Methane; CH_4; [74-82-8] 2. Hexane; C_6H_{14}; [110-54-3]	Sage, B. H.; Webster, D. C.; Lacey, W. N. *Ind. Eng. Chem.* 1936, *28*, 1045-1047.

VARIABLES:	PREPARED BY:
	C. L. Young

EXPERIMENTAL VALUES:

T/K (T/°F)	p/psi	P/MPa[†]	Mass fraction of methane	Mole fraction[†] of methane, x_{CH_4}
310.9 (100)	655 1623 2412	4.52 11.19 16.63	0.0424 0.1233 0.1920	0.1920 0.4301 0.5605
344.3 (160)	739 1775 2528	5.10 12.24 17.43	0.0424 0.1233 0.1920	0.1920 0.4301 0.5605
377.6 (220)	795 1855 2508	5.48 12.79 17.29	0.0424 0.1233 0.1920	0.1920 0.4301 0.5605

[†] calculated by compiler.

AUXILIARY INFORMATION

METHOD/APPARATUS/PROCEDURE:	SOURCE AND PURITY OF MATERIALS:
PVT cell charged with mixture of known composition. Pressure measured with pressure balance. Bubble point determined from the discontinuity in the pressure, volume isotherm. Details of apparatus in ref. (1).	1. Prepared from natural gas, treated for removal of higher alkanes, carbon dioxide and water vapor. Final purity 99.9 mole per cent. 2. Eastman Kodak Co. sample, used without further purification.

ESTIMATED ERROR:

$\delta T/K = \pm 0.1$; $\delta P/MPa = \pm 0.02$;

$\delta x_{CH} = \pm 0.002$ (estimated by compiler).

REFERENCES:

1. Sage, B. H.; Lacey, W. N.
   Ind. Eng. Chem.
   1934, *26*, 103.

COMPONENTS:	ORIGINAL MEASUREMENTS:
1. Methane; CH_4; [74-82-8] 2. Hexane; C_6H_{14}; [110-54-3]	Schoch, E.P.; Hoffmann, A.E.; Mayfield, F.D. *Ind. Eng. Chem.* 1941, *33*,688-691.

VARIABLES:	PREPARED BY:
Temperature, pressure	C.L. Young

EXPERIMENTAL VALUES:

T/K	$p/10^5\,Pa$	Mole fraction of methane in liquid, x_{CH_4}
311.08	42.7	0.1829
	52.2	0.2205
	75.01	0.3030
	94.73	0.3743
	117.7	0.4517
	139.7	0.5218
	156.7	0.5758
	175.8	0.6411
	189.2	0.6997
	194.8	0.7401
	198.0	0.8822
344.26	40.6	0.1568
	59.1	0.2203
	83.08	0.3037
	104.3	0.3719
	129.2	0.4491
	150.7	0.5167
	167.3	0.5716
	184.2	0.6373
	195.5	0.6987
	197.7	0.7351
	200.9	0.7653
377.59	46.4	0.1646
	82.39	0.2875

AUXILIARY INFORMATION

METHOD/APPARATUS/PROCEDURE:	SOURCE AND PURITY OF MATERIALS:
Rocking equilibrium cell fitted with stirring paddles. Temperature measured with Beckmann thermometer calibrated against standard Pt resistance thermometer. Pressure measured with Bourdon gauge. Samples injected into cell using mercury displacement. Equilibrium pressure measured. Bubble point determined from change in slope of pressure volume isotherms. Details in ref. (1).	1. Crude sample treated for removal of oxygen, carbon dioxide water vapor and liquids condensible at 200K. Distilled. 2. Eastman Kodak Co. sample.

	ESTIMATED ERROR:
	$\delta T/K = \pm0.03$; $\delta p/10^5\,Pa = \pm0.1$; $\delta x_{CH_4} + \pm0.001$. (estimated by compiler)

	REFERENCES:
	1. Schoch, E.P., Hoffmann, A.E. Kasperik, A.S.; Lightfoot, J.H. Mayfield, F.D. *Ind. Eng. Chem.* 1940, *32*, 788.

COMPONENTS:

1. Methane; CH$_4$; [74-82-8]

2. Hexane; C$_6$H$_{14}$; [110-54-3]

ORIGINAL MEASUREMENTS:

Schoch, E.P.; Hoffmann, A.E.; Mayfield, F.D.

Ind. Eng. Chem. 1941, *33*, 688-691.

EXPERIMENTAL VALUES:

T/K	$p/10^5$Pa	Mole fraction of methane in liquid, x_{CH_4}
377.59	106.9	0.3662
	131.0	0.4407
	151.8	0.5180
	172.2	0.5964
	184.9	0.6805
	189.8	0.7549

COMPONENTS:	ORIGINAL MEASUREMENTS:
1. Methane; CH_4; [74-82-8] 2. Hexane; C_6H_{14}; [110-54-3]	Shim, J.; Kohn, J.P. *J. Chem. Engng. Data*, 1962, *7*, 3-8

VARIABLES:	PREPARED BY:
Temperature, pressure	C.L. Young

EXPERIMENTAL VALUES:

		Mole fraction of methane	
T/K	P/MPa	in liquid, x_{CH_4}	in gas, y_{CH_4}
183.15	0.51	0.0920	–
	1.01	0.1820	–
	1.52	0.2740	–
	2.03	0.3668	–
	2.53	0.4550	–
	3.04	0.5675	–
	3.55	0.7120	–
	3.634	1.0000	–
198.15	0.51	0.0630	–
	1.01	0.1240	–
	1.52	0.1844	–
	2.03	0.2405	–
	2.53	0.2930	–
	3.04	0.3480	–
	3.55	0.3970	–
	4.05	0.4455	–
	4.56	0.4953	–
	5.07	0.5610	–
	5.57	0.6225	–
	6.08	0.6870	–
223.15	1.01	0.0880	–
	2.03	0.1705	–
	3.04	0.2473	–

AUXILIARY INFORMATION

METHOD/APPARATUS/PROCEDURE:	SOURCE AND PURITY OF MATERIALS:
Borosilicate glass cell. Temperature measured with Platinum resistance thermometer. Pressure measured on Bourdon gauge. Details in ref. (1) and source ref. Samples of methane added to hexane, equilibrated. Liquid phase composition estimated from known overall composition and volume of both phases.	1. Phillips Petroleum Co. sample, pure grade purified by passing through silica gel and activated charcoal. Final purity better than 99.5 mole per cent. 2. Phillips Petroleum Co. sample purity 99 mole per cent.

ESTIMATED ERROR:

$\delta T/K = \pm 0.07$; $\delta P/MPa = \pm 0.01$;
δx_{CH_4}, $\delta y_{CH_4} = \pm 0.10$;

REFERENCES:

1. Kohn, J.P.; Kurata, F.; *Petrol Process*, 1956, *11*, 57.

COMPONENTS:	ORIGINAL MEASUREMENTS:
1. Methane; CH_4; [74-82-8]	Shim, J.; Kohn, J.P.
2. Hexane; C_6H_{14}; [110-54-3]	*J. Chem. Engng. Data*, <u>1962</u>, *7*, 3-8

EXPERIMENTAL VALUES:

T/K	P/MPa	Mole fraction of methane in liquid, x_{CH_4}	in gas y_{CH_4}
223.15	4.05	0.3197	–
	5.07	0.3865	–
	6.08	0.4560	–
	7.09	0.5225	–
	8.11	0.5880	–
248.15	1.01	0.0695	–
	2.03	0.1366	–
	3.04	0.1995	–
	4.05	0.2570	–
	5.07	0.3120	–
	6.08	0.3643	–
	7.09	0.4140	–
	8.11	0.4660	–
	9.12	0.5245	–
	10.13	0.5825	–
273.15	1.01	0.0560	–
	2.03	0.1108	–
	3.04	0.1637	–
	4.05	0.2135	–
	5.07	0.2590	–
	6.08	0.3045	–
	7.09	0.3493	–
	8.11	0.3910	–
	9.12	0.4300	–
	10.13	0.4720	–
298.15	1.01	0.0490	0.9530
	2.03	0.0978	0.9728
	3.04	0.1450	0.9795
	4.05	0.1890	0.9830
	5.07	0.2316	0.9850
	6.08	0.2710	0.9863
	7.09	0.3090	0.9871
	8.11	0.3447	0.9868
	9.12	0.3810	0.9854
	10.13	0.4125	0.9833
	12.16	0.4740	–
	14.19	0.5370	–
	16.21	0.6090	–
323.15	1.01	0.0422	0.9021
	2.03	0.0860	0.9465
	3.04	0.1290	0.9614
	4.05	0.1690	0.9680
	5.07	0.2080	0.9719
	6.08	0.2458	0.9747
	7.09	0.2820	0.9765
	8.11	0.3186	0.9770
	9.12	0.3540	0.9768
	10.13	0.3850	0.9751
	12.16	0.4475	–
	14.19	0.5070	–
	16.21	0.5720	–
348.15	1.01	0.0364	0.8460
	2.03	0.0767	0.9028
	3.04	0.1165	0.9267
	4.05	0.1543	0.9385
	5.07	0.1920	0.9457
	6.08	0.2280	0.9492

COMPONENTS:	ORIGINAL MEASUREMENTS:
1. Methane; CH₄; [74-82-8]	Shim, J.; Kohn, J.P.
2. Hexane; C₆H₁₄; [110-54-3]	J. Chem. Engng. Data, 1962, 7, 3-8.

EXPERIMENTAL VALUES:

T/K	P/MPa	Mole fraction of methane in liquid, x_{CH_4}	in gas, y_{CH_4}
348.15	7.09	0.2625	0.9524
	8.11	0.2963	0.9555
	9.12	0.3295	0.9580
	10.13	0.3600	0.9605
	12.16	0.4192	–
	14.19	0.4830	–
	16.21	0.5530	–
373.15	1.01	0.0300	0.7875
	2.03	0.0688	0.8316
	3.04	0.1065	0.8760
	4.05	0.1440	0.8970
	5.07	0.1810	0.9082
	6.08	0.2160	0.9158
	7.09	0.2490	0.9224
	8.11	0.2807	0.9280
	9.12	0.3135	0.9321
	10.13	0.3434	0.9340
	12.16	0.4033	–
	14.19	0.4700	–
	16.21	0.5560	–
423.15	1.01	0.0110	0.4420
	2.03	0.0464	0.5625
	3.04	0.0826	0.6695
	4.05	0.1187	0.7220
	5.07	0.1553	0.7520
	6.08	0.1915	0.7675
	7.09	0.2278	0.7780
	8.11	0.2645	0.780
	9.12	0.3040	0.760
	10.13	0.3440	0.738

COMPONENTS:	ORIGINAL MEASUREMENTS:
1. Methane; CH_4; [74-82-8] 2. Hexane; C_6H_{14}; [110-54-3]	Poston, R. S.; McKetta, J. J. *J. Chem. Eng. Data* <u>1966</u>, *11*, 362-3.

VARIABLES:	PREPARED BY:
Temperature, pressure	C. L. Young

EXPERIMENTAL VALUES:

T/K	P/atm	P/MPa	Mole fraction of methane in liquid, x_{CH_4}	in vapor, y_{CH_4}
310.93	34.01	3.446	0.153	0.969
	68.03	6.893	0.286	0.980
	102.04	10.339	0.404	0.979
	136.05	13.785	0.511	0.970
	170.07	17.232	0.638	0.969
	183.67	18.610	0.694	0.970
	188.10	19.059	0.715	0.968
	193.54	19.610	0.756	0.964
	192.18	19.473	0.740	0.962
	195.24	19.783	0.764	0.957
	197.41	20.003	Critical point	
344.26	34.01	3.446	0.135	0.937
	68.03	6.893	0.258	0.953
	102.04	10.339	0.371	0.962
	136.05	13.785	0.485	0.946
	170.07	17.232	0.594	0.928
	183.67	18.610	0.668	0.913
	187.07	18.955	0.701	0.911
	190.14	19.266	0.703	0.885
	191.63	19.417	Critical point	

(cont.)

AUXILIARY INFORMATION

METHOD/APPARATUS/PROCEDURE:	SOURCE AND PURITY OF MATERIALS:
Stainless steel glass windowed cell. Vapor recycled using high pressure magnetic pump. Pressure measured using Bourdon gauge and temperature measured using thermocouples. Samples of both phases withdrawn at constant pressure and analysed by gas chromatography. Details of apparatus in source and ref. (1).	1. Phillips Petroleum Co. sample, purity better than 99.9 mole per cent. 2. Phillips Petroleum Co. sample, purity better than 99.9 mole per cent.

ESTIMATED ERROR:

$\delta T/K = \pm 0.1$; $\delta P/MPa = \pm 0.014$;
δx_{CH_4}, δy_{CH_4} $= \pm 0.002$.

REFERENCES:

1. Roberts, L. R.; Azarnoosh, A.; Wong, R.; McKetta, J. J.
J. Chem. Eng. Data
<u>1962</u>, *7*, 484.

COMPONENTS:	ORIGINAL MEASUREMENTS:
1. Methane; CH_4; [74-82-8]	Poston, R. S.; Mcketta, J. J.
2. Hexane; C_6H_{14}; [110-54-3]	J. Chem. Eng. Data
	1966, 11, 362-3.

EXPERIMENTAL VALUES:

			Mole fraction of methane	
T/K	P/atm	P/MPa	in liquid, x_{CH_4}	in vapor, y_{CH_4}
377.55	34.01	3.446	0.119	0.871
	68.03	6.893	0.240	0.911
	102.04	10.339	0.351	0.908
	136.05	13.785	0.477	0.895
	170.07	17.232	0.604	0.861
	172.79	17.508	0.618	0.822
	175.51	17.784	0.650	0.817
	177.89	18.025	0.675	0.784
	178.37	18.073	Critical point	
410.95	34.01	3.446	0.103	0.738
	68.03	6.893	0.223	0.811
	102.04	10.339	0.345	0.830
	136.05	13.785	0.477	0.803
	144.90	14.682	0.528	0.768
	148.30	15.026	0.556	0.751
	151.16	15.316	0.598	0.725
	152.86	15.489	Critical point	
444.25	24.76	2.509	0.025	0.438
	34.01	3.446	0.079	0.522
	68.03	6.893	0.208	0.654
	102.04	10.339	0.363	0.660
	107.14	10.856	0.388	0.658
	111.09	11.256	0.422	0.641
	114.76	11.628	Critical point	

COMPONENTS:	ORIGINAL MEASUREMENTS:
1. Methane; CH_4; [74-82-8] 2. Hexane; C_6H_{14}; [110-54-3]	Lin, Y-N; Chen, R.J.J.; Chappelear, P.S.; Kobayashi, R. *J. Chem. Engng. Data*, <u>1977</u>,*22*, 402-8.

VARIABLES:	PREPARED BY:
Temperature, pressure	C.L. Young

EXPERIMENTAL VALUES:

T/K	P/MPa	Mole fraction of methane in liquid x_{CH_4}	in vapor y_{CH_4}
273.16	0.1731	0.0097	0.9643
	0.3454	0.0196	0.9815
	0.6902	0.0387	0.9898
	1.0349	0.0567	0.99266
	1.3796	0.0744	0.99396
	2.0691	0.1116	0.99528
	2.7586	0.1469	0.99566
	4.1368	0.2127	0.99578
	5.5158	0.2742	0.99549
	6.8947	0.3328	0.99465
	8.2737	0.3886	0.99343
	9.6526	0.4435	0.99166
	11.032	0.4924	0.9891
	12.410	0.5400	0.9849
	13.789	0.5933	0.9797
	15.168	0.6401	0.9692
	15.857	0.6564	-
	16.547	0.6949	0.9565
	17.237	0.7195	-
	17.582	0.7462	0.9400
	17.926	0.7758	0.9366
	18.271	0.8025	0.9348
	18.443 (a)	0.9290	0.9290

AUXILIARY INFORMATION

METHOD/APPARATUS/PROCEDURE:	SOURCE AND PURITY OF MATERIALS:
Recirculating vapor flow apparatus. Temperature measured with platinum resistance thermometer. Pressure measured with Bourdon gauge. Liquid sample added to windowed cell, air removed. Methane added to cell and recirculated for at least 1/2 hour. Samples analysed by gas chromatography.	1. Ultra-high purity sample from Union Carbide Chemicals Corp. purity 99.97 mole per cent. 2. Research grade sample from Phillips Petroleum Co. Purity 99.99 mole per cent.

ESTIMATED ERROR: $\delta T/K = \pm 0.2$; $\delta P/MPa = \pm 0.007$; $\delta x_{max} = \pm 2\%$ or 0.005 (whichever is greater); $(1-\delta y)_{max} = \pm 2\%$. See source for fuller details of errors.

REFERENCES:

1. Elliot, D.G.; Chen, R.J.J.; Chappelear, P.S.; Kobayashi, R., *J. Chem. Engng. Data.* <u>1974</u>, *19*, 71.

COMPONENTS:	ORIGINAL MEASUREMENTS:
1. Methane; CH₄; [74-82-8] 2. Hexane; C₆H₁₄; [110-54-4]	Lin, Y-N; Chen, R.J.J.; Chappelear, P.S.; Kobayashi, R. *J. Chem. Engng. Data.* <u>1977</u>, *22*, 402-8.

EXPERIMENTAL VALUES:

T/K	P/MPa	Mole fraction of methane, in liquid, x_{CH_4}	in vapor y_{CH_4}
248.14	0.13858	0.009486	0.98990
	0.17306	0.01152	0.99179
	0.34543	0.02320	0.99557
	0.69016	0.04638	0.99757
	1.0349	0.06884	0.99817
	1.3796	0.08998	0.99847
	2.0691	0.1340	0.99872
	2.7586	0.1819	0.99879
	4.1368	0.2595	0.99863
	5.5158	0.3347	0.99820
	6.8947	0.3998	0.99759
	8.2737	0.4672	0.99632
	9.6526	0.5252	0.99416
	11.032	0.5804	0.99052
	11.721	0.6123	0.9889
	12.410	0.6382	0.9844
	13.789	0.6937	0.9752
	15.168	0.7601	0.9644
	15.858	0.8191	0.9593
	16.113 (a)	0.9564	0.9564
223.15	0.13927	0.01415	0.99841
	0.17375	0.01710	0.99872
	0.34612	0.03370	0.999343
	0.69085	0.06372	0.999632
	1.0356	0.09213	0.999713
	1.3803	0.1218	0.999757
	2.0698	0.1771	0.999775
	2.7593	0.2394	0.999795
	4.1368	0.3282	0.999594
	5.5158	0.4207	0.999256
	6.8947	0.4952	0.99851
	8.2737	0.5893	0.99613
	9.6526	0.6518	0.99013
	10.342	0.7011	0.9871
	11.032	0.7423	0.9841
	11.721	0.7774	0.9816
	12.066	0.8022	-
	12.238	0.8189	-
	12.342	0.8444	-
	12.410	0.8522	-
	12.431	0.8922	-
	12.438 (a)	0.9784	0.9784
210.15	0.13927	0.01563	0.999495
	0.17375	0.01867	0.999589
	0.34612	0.03782	0.999781
	0.69085	0.07307	0.999868
	1.0356	0.1079	0.999888
	1.3803	0.1411	0.999903
	2.0698	0.2071	0.999894
	2.7593	0.2700	0.999878
	4.1368	0.3881	0.999798
	5.5158	0.5139	0.999402
	6.8947	0.6162	0.99730
	8.2737	0.7156	0.99177
	8.9631	0.7575	-
	9.6526	0.8091	0.9882
	9.7905	0.8279	-
	9.8939	0.8568	-
	9.9284	0.8713	-

COMPONENTS:	ORIGINAL MEASUREMENTS:
1. Methane; CH_4; [74-82-8] 2. Hexane; C_6H_{14}; [110-54-3]	Lin, Y-N.; Chen, R.J.J.; Chappelear, P.S.; Kobayashi, R. *J. Chem. Engng. Data.* <u>1977</u>, *22*, 402-8.

EXPERIMENTAL VALUES:

T/K	P/MPa	Mole fraction of methane in liquid, x_{CH_4}	in vapor, y_{CH_4}
210.15	9.9491	0.9872	0.9872
198.05	0.13721	0.01950	0.999848
	0.17237	0.02438	0.999885
	0.34474	0.04742	0.999927
	0.68947	0.09603	0.999949
	1.0342	0.1390	0.999955
	1.3789	0.1813	0.999952
	2.0684	0.2660	0.999939
	2.7579	0.3412	0.999919
	3.4474	0.4225	0.999893
	4.1368	0.5076	0.999832
	4.8263	0.5936	0.999640
	5.5158	0.6872	0.99789
	6.2053	0.7596	0.99572
	6.8947	0.8117	0.99435
	7.2395	0.8468	–
	7.2739	0.8932	–
	7.2877 (a)	0.99364	0.99364
195.91	0.13789	0.02082	–
	0.34474	0.04831	–
	0.68947	0.09528	–
	1.0342	0.1407	–
	1.3789	0.1849	–
	2.0684	0.2698	–
	2.7579	0.3555	–
	3.4474	0.4429	–
	4.1368	0.5323	–
	4.8263	0.6422	–
	5.2055 (b)	0.7177	0.99763
193.15	0.13858	0.02210	0.999905
	0.17306	0.02728	0.999932
	0.69016	0.09964	0.999964
	2.0691	0.2851	0.999970
	3.4474	0.4697	0.999932
	4.1368	0.5802	0.999830
	4.6539	0.6902	0.999567
	4.7712 (b)	0.7239	0.999393
	4.7712 (b)	0.9864 (L_1)	0.999393
	4.8263	0.9914 (L_1)	0.99898
	4.8815 (a)	0.9988 (L_1)	0.99882
190.50	0.13789	0.02256	0.999924
	0.17237	0.02803	0.999960
	0.68947	0.1047	0.999970
	2.0684	0.3004	0.999988
	3.4446	0.5041	0.999907
	4.1368	0.6487	0.999814
	4.3988 (b)	0.7438	0.999611
	4.3788 (b)	0.9695 (L_1)	0.999611
	4.4471 (b)	0.9724 (L_1)	0.999577
	4.4816	0.9745 (L_1)	0.999587
	4.5161	0.9897 (L_1)	0.999609
	4.5505	0.9919 (L_1)	0.999687
	4.5850 (a)	0.999847 (L_1)	0.99847
186.23	0.34474	0.06129	–
	0.68947	0.1189	–
	1.0342	0.1813	–

COMPONENTS:	ORIGINAL MEASUREMENTS:
1. Methane; CH_4; [74-82-8]	Lin, Y-N.; Chen, R.J.J.; Chappelear, P.S.; Kobayashi, R.
2. Hexane; C_6H_{14}; [110-54-3]	J. Chem. Engng. Data. 1977, 22, 402-8.

EXPERIMENTAL VALUES:

T/K	P/MPa	Mole fraction of methane in liquid, x_{CH_4}	in vapor, y_{CH_4}
186.23	1.3789	0.2357	-
	2.0684	0.3412	-
	2.7579	0.4593	-
	3.4474	0.6010	-
	3.7232	0.6826	-
	3.8541 (b)	0.8057	0.999837
	3.8541 (b)	0.9505 (L$_1$)	0.999837
	3.9576	0.9776 (L$_1$)	-
	3.9920	0.9871 (L$_1$)	-
182.46	0.13789	0.02768	-
	0.17237	0.03463	-
	0.34474	0.06623	-
	0.68947	0.1308	-
	1.0342	0.1907	-
	1.3789	0.2560	-
	2.0684	0.3822	-
	2.7579	0.5178	-
	3.1026	0.6100	-
	3.3095	0.7426	-
	3.4149	0.9286	0.999946
	3.5163	0.9640 (L$_1$)	-
	3.5508	0.9876 (L$_1$)	-

(a)   critical pressure.

(b)   pressure at which liquid-liquid-gas equilibrium exists.

(L$_1$)   light liquid in equilibrium with vapor.

COMPONENTS:	ORIGINAL MEASUREMENTS:
1. Methane; CH_4; [74-82-8] 2. Hexane; C_6H_{14}; [110-54-3]	Merrill, R. C.; Luks, K. D.; Kohn, J. P. *J. Chem. Eng. Data* <u>1983</u>, *28*, 210-215.
VARIABLES:	PREPARED BY: C. L. Young

EXPERIMENTAL VALUES:

Phases in equilibrium	T/K	P/atm	P/MPa	Mole fraction of methane, x_{CH_4} [a,b]	
				in L_1	in L_2
$L_1, L_2 \equiv V$	195.72	51.33	5.201	0.7677	0.9833
L_1, L_2, V	194.00	48.67	4.931	0.7475	0.9821
	192.00	45.81	4.642	0.7542	0.9714
	190.00	43.08	4.365	0.7698	0.9647
	188.00	40.51	4.105	0.7855	0.9566
	186.00	37.98	3.848	0.8018	0.9434
	184.00	35.61	3.608	0.8159	--
$L_1 \equiv L_2, V$	182.73	34.05	3.450	0.8521	0.8521

[a] Each point given is the average of several data points.

[b] The original article gave the mole fraction of hexane.

AUXILIARY INFORMATION

METHOD/APPARATUS/PROCEDURE:	SOURCE AND PURITY OF MATERIALS:
Glass equilibrium cell. Temperature measured with platinum resistance thermometer and pressure with Bourdon gauge. Stoichiometry and volumetric measurements were used to obtain liquid phase compositions and molar volumes. Gas phase assumed to be pure methane. Molar volume data in source. Details of apparatus in ref. (1).	1. Linde, Ultra Pure grade sample, purity 99.97 moles per cent. 2. Humphrey Chemical Co. sample, purity 99 moles per cent.
	ESTIMATED ERROR: $\delta T/K = \pm 0.03$; $\delta P/MPa = \pm 0.007$; $\delta x/x$ (hexane) = ± 0.006.
	REFERENCES: 1. Hottovy, J. D.; Kohn, J. P.; Luks, K. D. *J. Chem. Eng. Data* <u>1981</u>, *26*, 135.

COMPONENTS:	ORIGINAL MEASUREMENTS:
1. Methane; CH₄; [74-82-8] 2. Nitrogen; N₂; [7727-37-9] 3. Hexane; C₆H₁₄; [110-54-3]	Boomer, E. H.; Johnson, C. A. *Can. J. Res. B* <u>1938</u>, *16*, 328-335.

VARIABLES:	PREPARED BY:
Temperature, pressure	C. L. Young

EXPERIMENTAL VALUES:

| | | | Mole fractions | | | | | |
| | | | in liquid | | | in vapor | | |
T/K	P/atm	P/MPa	x_{CH_4}	x_{N_2}	$x_{C_6H_{14}}$	y_{CH_4}	y_{N_2}	$y_{C_6H_{14}}$
298.15	1	0.1	–	–	–	0.804	–	0.196
	36.2	3.67	0.159	0.002	0.839	0.897	0.070	0.033
			0.162	0.003	0.835	0.920	0.054	0.026
	68.4	6.93	0.278	0.009	0.713	0.934	0.048	0.018
	101.7	10.30	0.392	0.012	0.596	0.909	0.070	0.021
			0.390	0.013	0.597	0.916	0.063	0.021
	134.7	13.65	0.485	0.021	0.494	0.888	0.082	0.030
	167.9	17.01	0.578	0.030	0.392	0.883	0.071	0.046
			0.583	0.027	0.390	0.896	0.057	0.047
	202.0	20.47	0.690	0.031	0.279	0.844	0.075	0.081
			0.689	0.037	0.274	0.838	0.080	0.082
	208.2	21.10	0.734	0.029	0.237	0.838	0.057	0.105
			0.726	0.038	0.236	0.833	0.063	0.104
	229.3	23.23	0.771	0.049	0.180	0.779	0.046	0.175
328.15	1	0.1	–	–	–	0.3685	–	0.6315
	36.2	3.67	0.142	0.003	0.855	0.873	0.078	0.049
	101.7	10.30	0.355	0.017	0.628	0.899	0.067	0.034
	167.9	17.01	0.553	0.026	0.421	0.882	0.060	0.057
	202.0	20.47	0.662	0.035	0.303	0.827	0.064	0.109
			0.654	0.040	0.306	0.828	0.063	0.109
	208.2	21.10	0.710	0.041	0.249	0.808	0.046	0.146
			0.715	0.035	0.250	0.808	0.047	0.145
	219.1	22.20	0.768	0.043	0.189	0.768	0.046	0.186

(cont.)

AUXILIARY INFORMATION

METHOD/APPARATUS/PROCEDURE:	SOURCE AND PURITY OF MATERIALS:
Rocking autoclave stirred by steel piston falling under gravity. Samples of vapor and liquid trapped in two auxiliary high pressure cells. Equilibrium samples analysed in complicated volumetric and combustion apparatus. Details in ref. (1). <u>NOTE</u>: The source reference also contains data on a mixture of hexanes + methane + nitrogen. Since the isomeric composition of the hexane mixture is not known, the data have not been included here.	1. and 2. Natural gas sample containing 94.4 mole per cent of methane and 5.6 mole per cent of nitrogen. Impurities may have been present amounting to 0.1 mole per cent. 3. Synthesized from propyl bromide; product fractionated.

	ESTIMATED ERROR:
	δT/K = ±0.1; δP/MPa = ±0.02; δx, δy = ±1% (estimated by compiler).

REFERENCES:

1. Boomer, E. H.; Johnson, C. A.; Argue, G. H.
   Can. J. Res. B
   <u>1937</u>, *15*, 367.

COMPONENTS:

1. Methane; CH$_4$; [74-82-8]
2. Nitrogen; N$_2$; [7727-37-9]
3. Hexane; C$_6$H$_{14}$; [110-54-3]

ORIGINAL MEASUREMENTS:

Boomer, E. H.; Johnson, C. A.

Can. J. Res. B

<u>1938</u>, *16*, 328-335.

EXPERIMENTAL VALUES:

			Mole fractions					
			in liquid				in vapor	
T/K	P/atm	P/MPa	x_{CH_4}	x_{N_2}	$x_{C_6H_{14}}$	y_{CH_4}	y_{N_2}	$y_{C_6H_{14}}$
358.15	35.5	3.60	0.126	0.003	0.871	0.829	0.058	0.113
	101.4	10.27	0.333	0.016	0.651	0.893	0.047	0.060
	167.6	16.98	0.537	0.020	0.443	0.850	0.052	0.098
	187.9	19.04	0.611	0.037	0.352	0.801	0.060	0.139
			0.603	0.033	0.364	0.807	0.058	0.135
	201.0	20.37	0.714	0.036	0.250	0.703	0.047	0.250

COMPONENTS:	ORIGINAL MEASUREMENTS:
1. Methane; CH_4; [74-82-8] 2. 3-Methylpentane; C_6H_{14}; [96-14-0]	Kohn, J. P.; Haggin, J. H. S. *J. Chem. Engng. Data* <u>1967</u>, *12*, 313-5.

VARIABLES:	PREPARED BY:
	C. L. Young

EXPERIMENTAL VALUES:

			Mole fractions	
T/K	p/atm	p/MPa	in liquid, x_{CH_4}	in vapor, y_{CH_4}
298.15	0.2225	0.02254	0.000	0.000
	5	0.51	0.0234	-
	10	1.01	0.0480	0.9510
	15	1.52	0.0726	0.9639
	20	2.03	0.0965	0.9712
	25	2.53	0.1202	0.9758
	30	3.04	0.1440	0.9787
323.15	0.646	0.655	0.000	0.000
	5	0.51	0.0191	-
	10	1.01	0.0411	0.9001
	15	1.52	0.0630	0.9277
	20	2.03	0.0847	0.9451
	25	2.53	0.1061	0.9552
	30	3.04	0.1273	0.9602
348.15	1.427	1.446	0.000	0.000
	5	0.51	0.0149	-
	10	1.01	0.0357	0.8410
	15	1.52	0.0559	0.8760
	20	2.03	0.0757	0.8984
	25	2.53	0.0949	0.9132
	30	3.04	0.1142	0.9233

(cont.)

AUXILIARY INFORMATION

METHOD/APPARATUS/PROCEDURE:	SOURCE AND PURITY OF MATERIALS:
Borosilicate glass static equilibrium cell. Temperature measured with platinum resistance thermometer. Pressure measured with dead weight gauge. Samples of methane added to 3-methylpentane. Liquid phase composition estimated from known overall composition and volume of both phases. Dew point composition determined using similar procedure but with a cell fitted with a capillary tube at lower end so very small amounts of liquid could be measured.	1. Phillips Petroleum Co. sample, purity 99 mole per cent. Purified by passage over silica gel and activated carbon. Final purity about 99.5 mole per cent. 2. Phillips Petroleum Co. sample. Degassed. Purity at least 99 mole per cent.
	ESTIMATED ERROR: $\delta T/K = \pm 0.02$; $\delta P/MPa = \pm 0.006$; $\delta x_{CH_4} = \pm 0.001$; $\delta y_{CH_4} = \pm 0.002$.
	REFERENCES:

COMPONENTS:

1. Methane; CH_4; [74-82-8]

2. 3-Methylpentane; C_6H_{14}; [96-14-0]

ORIGINAL MEASUREMENTS:

Kohn, J. P.; Haggin, J. H. S.

J. Chem. Engng. Data

<u>1967</u>, *12*, 313-5.

EXPERIMENTAL VALUES:

T/K	p/atm	p/MPa	Mole fractions in liquid x_{CH_4}	in vapor, y_{CH_4}
373.15	2.786	0.2823	0.000	0.000
	5	0.51	0.0089	-
	10	1.01	0.0297	0.7292
	15	1.52	0.0480	0.7810
	20	2.03	0.0669	0.8190
	25	2.53	0.0860	0.8460
	30	3.04	0.1049	0.8655

COMPONENTS:	EVALUATOR:
1. Methane; CH_4; [74-82-8] 2. Heptane; C_7H_{16}; [142-82-5]	Colin L. Young, School of Chemistry, University of Melbourne, Parkville, Victoria 3052, Australia. March 1984

EVALUATION:

The solubility of methane in heptane has been studied at high pressures over the temperature range 183 K to 511 K.

The data of Boomer *et al.* (1) are classified as doubtful since these workers used methane which contained over 5 mole per cent of nitrogen. The data of Koonce and Kobayashi (2) were over a very limited range of experimental variables and are not considered further.

The data of Reamer *et al.* (3) cover the temperature range 277.6 K to 510.9 K and overlap with the data of Kohn (4) at 277.6 K. The two sets of data agree well at the common temperature. Both sets of data are classified as tentative. There is also fair agreement between the data of Kohn (4) and Chang *et al.* (5) for temperature in the range 200 K to 255 K, the later data giving a slightly greater mole solubility of methane. The data of Chang *et al.* (5) are also classified as tentative.

References

1. Boomer, E. H.; Johnson, C. A.; Piercey, A. G. A.
    Can. J. Res., <u>1938</u>, *B16*, 396.

2. Koonce, K. T.; Kobayashi, R.
    J. Chem. Eng. Data, <u>1964</u>, *9*, 494.

3. Reamer, H. H.; Sage, B. H.; Lacey, W. N.
    J. Chem. Eng. Data, <u>1956</u>, *1*, 29.

4. Kohn, J. P.
    Am. Inst. Chem. Engnrs. J., <u>1961</u>, *7*, 514.

5. Chang, H . L.; Hunt, L. J.; Kobayashi, R.
    Am. Inst. Chem. Engnrs. J., <u>1966</u>, *12*, 1212.

COMPONENTS:	ORIGINAL MEASUREMENTS:
1. Methane; CH_4; [74-82-8] 2. Heptane; C_7H_{16}; [142-82-5]	Reamer, H.H.; Sage, B.H.; Lacey, W.N. *J. Chem. Engng. Data*, 1956,1,29-42

VARIABLES:	PREPARED BY:
Temperature, pressure	C.L. Young

EXPERIMENTAL VALUES:

T/°F	T/K	p/psi	P/MPa	Mole fraction of methane in liquid, x_{CH_4}	in vapor, y_{CH_4}
40	277.6	200	1.38	0.0753	0.9964
		400	2.76	0.1445	0.9974
		600	4.14	0.2084	0.9971
		800	5.52	0.2670	0.9970
		1000	6.89	0.3215	0.9966
		1250	8.62	0.3840	0.9957
		1500	10.34	0.4410	0.9940
		1750	12.07	0.4930	0.9920
		2000	13.79	0.5425	0.9890
		2250	15.51	0.5900	0.9850
		2500	17.24	0.6373	0.9780
		2750	18.96	0.6870	0.9690
		3000	20.68	0.7400	0.9530
		3328	22.95	0.894	0.894
100	310.9	200	1.38	0.0640	0.9866
		400	2.76	0.1240	0.9905
		600	4.14	0.1810	0.9911
		800	5.52	0.2340	0.9911
		1000	6.89	0.2842	0.9910
		1250	8.62	0.3425	0.9900
		1500	10.34	0.3963	0.9881
		1750	12.07	0.4470	0.9850
		2000	13.79	0.4950	0.9801
		2250	15.51	0.5435	0.9748
		2500	17.24	0.5905	0.9690

AUXILIARY INFORMATION

METHOD/APPARATUS/PROCEDURE:	SOURCE AND PURITY OF MATERIALS:
PVT cell charged with mixture of known composition. Bubble and dew points determined for various compositions. Temperature measured with platinum resistance thermometer, pressure measured with pressure balance. Coexisting liquid and gas compositions determined by graphical means. Details in source and ref (1).	1. Crude sample treated for removal of higher alkanes, CO_2 and H_2O. Final purity 99.9 mole per cent, less than 0.02 mole per cent of other hydrocarbons. 2. Purity better than 99.95 mole per cent.

ESTIMATED ERROR:

$$\delta T/K = \pm 0.02; \quad \delta P/psi = \pm 0.1;$$
$$\delta x_{CH_4} = \pm 0.0005; \quad \delta y_{CH_4} = \pm 0.0003$$

REFERENCES:
1. Sage, G.B.; Lacey, W.N.
    *Trans. Am. Inst. Mining. Met.
    Engnrs.* 1940,*136*,136.

COMPONENTS: ORIGINAL MEASUREMENTS:

1. Methane; CH₄; [74-82-8]   Reamer, H.H.; Sage, B.G.; Lacey,
2. Heptane; C₇H₁₆; [142-82-5]   W.N.

                                         J. Chem. Engng. Data. 1956,1, 29-
                                         42.

EXPERIMENTAL VALUES:

T/°F	T/K	p/psi	P/MPa	Mole fraction of methane in liquid, x_{CH_4}	in vapor, y_{CH_4}
100	310.9	2750	18.96	0.6400	0.9620
		3000	20.68	0.6910	0.9530
		3500	24.13	0.8030	0.9050
		3609	24.88	0.8550	0.8550
160	344.3	200	1.38	0.0565	0.9597
		400	2.76	0.1110	0.9733
		600	4.14	0.1623	0.9780
		800	5.52	0.2107	0.9798
		1000	6.89	0.2567	0.9804
		1250	8.62	0.3108	0.9795
		1500	10.34	0.3620	0.9770
		1750	12.07	0.4125	0.9742
		2000	13.79	0.4620	0.9705
		2250	15.51	0.5090	0.9656
		2500	17.24	0.5580	0.9590
		2750	18.96	0.6070	0.9490
		3000	20.68	0.6610	0.9360
		3500	24.13	0.7870	0.8595
		3549	24.47	0.817	0.817
220	377.6	200	1.38	0.0494	0.8942
		400	2.76	0.1003	0.9305
		600	4.14	0.1492	0.9449
		800	5.52	0.1960	0.9517
		1000	6.89	0.2410	0.9558
		1250	8.62	0.2940	0.9566
		1500	10.34	0.3450	0.9564
		1750	12.07	0.3957	0.9532
		2000	13.79	0.4457	0.9474
		2250	15.51	0.4944	0.9392
		2500	17.24	0.5450	0.9280
		2750	18.96	0.5995	0.9120
		3000	20.68	0.6615	0.8864
		3298	22.74	0.778	0.778
280	410.9	200	1.38	0.0405	0.7481
		400	2.76	0.0918	0.8628
		600	4.14	0.1390	0.8894
		800	5.52	0.1850	0.9037
		1000	6.89	0.2290	0.9100
		1250	8.62	0.2810	0.9120
		1500	10.34	0.3308	0.9170
		1750	12.07	0.3810	0.9073
		2000	13.79	0.4329	0.9000
		2250	15.51	0.4880	0.8900
		2500	17.24	0.5446	0.8660
		2750	18.96	0.6150	0.8280
		2927	20.18	0.732	0.732
340		200	1.38	0.0305	0.5100
		400	2.76	0.0840	0.7220
		600	4.14	0.1317	0.7750
		800	5.52	0.1786	0.8060
		1000	6.89	0.2240	0.8260
		1250	8.62	0.2780	0.8369
		1500	10.34	0.3316	0.8360
		1750	12.07	0.3850	0.8250
		2000	13.79	0.4431	0.8040
		2250	15.51	0.5165	0.7730
		2469	17.02	0.672	0.672
400	477.6	200	1.38	0.0132	0.2060
		400	2.76	0.0670	0.5223
		600	4.14	0.1200	0.6210
		800	5.52	0.1740	0.6700

COMPONENTS:

1. Methane; CH_4; [74-82-8]
2. Heptane; C_7H_{16}; [142-82-5]

ORIGINAL MEASUREMENTS

Reamer, H.H.; Sage, B.H.;
Lacey, W.N.
J. Chem. Engng. Data. 1956,1,
29-42.

EXPERIMENTAL VALUES:

T/°F	T/K	p/psi	P/MPA	Mole fraction of methane in liquid x_{CH_4}	in vapor y_{CH_4}
400	477.6	1000	6.89	0.2290	0.6930
		1250	8.62	0.2980	0.6990
		1500	10.34	0.3725	0.6940
		1750	12.07	0.4620	0.6690
		1906	13.14	0.585	0.585
460	510.9	400	2.76	0.0485	0.2640
		600	4.14	0.1205	0.4199
		800	5.52	0.1918	0.4670
		1000	6.89	0.2730	0.4750
		1206	8.32	0.441	0.441

COMPONENTS:	ORIGINAL MEASUREMENTS:
1. Methane; CH_4; [74-82-8] 2. Heptane; C_7H_{16}; [142-82-5]	Kohn, J. P. *Am. Inst. Chem. Engnrs. J.* <u>1961</u>, *7*, 514-8.

VARIABLES:	PREPARED BY:
Temperature, pressure	C. L. Young

EXPERIMENTAL VALUES:

T/K	$P/10^5$Pa	Mole fraction of methane in liquid, x_{CH_4}	T/K	$P/10^5$Pa	Mole fraction of methane in liquid, x_{CH_4}
277.59	6.89	0.036	266.48	68.95	0.342
	13.79	0.072		75.84	0.370
	20.68	0.108		82.74	0.394
	27.58	0.141		89.63	0.416
	34.47	0.174		96.53	0.438
	41.37	0.204		103.42	0.462
	48.26	0.235	255.37	6.89	0.044
	55.16	0.264		13.79	0.089
	62.05	0.291		20.68	0.131
	68.95	0.318		27.58	0.171
	75.84	0.346		34.47	0.209
	82.74	0.372		41.37	0.244
	89.63	0.395		48.26	0.278
	96.53	0.416		55.16	0.312
	103.42	0.440		62.05	0.340
266.48	6.89	0.040		68.95	0.369
	13.79	0.081		75.84	0.398
	20.68	0.119		82.74	0.422
	27.58	0.156		89.63	0.444
	34.47	0.192		96.53	0.466
	41.37	0.223		103.42	0.489
	48.26	0.257	244.26	6.89	0.049
	55.16	0.288		13.79	0.098
	62.05	0.315		(cont.)	

AUXILIARY INFORMATION

METHOD/APPARATUS/PROCEDURE:	SOURCE AND PURITY OF MATERIALS:
Pyrex glass cell. Temperature measured with platinum resistance thermometer and pressure with Bourdon gauge. Bubble points of mixtures of known composition determined. Experimental data quoted obtained by smoothing.	1. Pure grade material, purity better than 99 mole per cent. Dried and pressed over activated charcoal; final purity 99.5 mole per cent or better. 2. Pure grade material, degassed.

	ESTIMATED ERROR: $\delta T/K = \pm 0.02$; $\delta P/MPa = \pm 0.1\%$ or ± 0.007 (whichever is greater); $\delta x_{CH_4} = \pm 1\%$ (compiler).
	REFERENCES:

1. Methane; CH$_4$; [74-82-8]	Kohn, J. P.	
2. Heptane; C$_7$H$_{16}$; [142-82-5]	*Am. Inst. Chem. Engnrs. J.*	
	1961, *7*, 514-8.	

EXPERIMENTAL VALUES:

T/K	P/10^{5}Pa	Mole fraction of methane in liquid, x_{CH_4}	T/K	P/10^{5}Pa	Mole fraction of methane in liquid, x_{CH_4}
244.26	20.68	0.144	222.04	89.63	0.585
	27.58	0.187		96.53	0.610
	34.47	0.228		103.42	0.640
	41.37	0.266	210.93	6.89	0.072
	48.26	0.302		13.79	0.140
	55.16	0.336		20.68	0.203
	62.05	0.370		27.58	0.262
	68.95	0.402		34.47	0.323
	75.84	0.431		41.37	0.375
	82.74	0.457		48.26	0.433
	89.63	0.480		55.16	0.476
	96.53	0.504		62.05	0.515
	103.42	0.527		68.95	0.552
233.15	6.89	0.055		75.84	0.584
	13.79	0.110	199.82	6.89	0.084
	20.68	0.159		13.79	0.163
	27.58	0.207		20.68	0.240
	34.47	0.249		27.58	0.310
	41.37	0.294		34.47	0.387
	48.26	0.330		41.37	0.460
	55.16	0.366		48.26	0.530
	62.05	0.406		55.16	0.565
	68.95	0.443	194.26	6.89	0.091
	75.84	0.473		13.79	0.181
	82.74	0.502		20.68	0.267
	89.63	0.527		27.58	0.351
	96.53	0.554		34.47	0.435
	103.42	0.578		41.37	0.520
222.04	6.89	0.063	188.71	6.89	0.100
	13.79	0.124		13.79	0.206
	20.68	0.179		20.68	0.306
	27.58	0.232		27.58	0.403
	34.47	0.279		34.47	0.498
	41.37	0.329		41.37	0.582
	48.26	0.370	183.15	6.89	0.132
	55.16	0.410		13.79	0.250
	62.05	0.450		20.68	0.350
	68.95	0.492		27.58	0.459
	75.84	0.524		34.47	0.578
	82.74	0.556			

COMPONENTS:	ORIGINAL MEASUREMENTS:
1. Methane; CH₄; [74-82-8] 2. Heptane; C₇H₁₆; [142-82-5]	Koonce, K. T.; Kobayashi, R. *J. Chem. Engng. Data* <u>1964</u>, *9*, 494-501.

VARIABLES:	PREPARED BY:
Temperature, Pressure	C. L. Young

EXPERIMENTAL VALUES:

T/K (T/°F)	P/psi	P/MPa	Mole fractions		
			x_{CH_4}	$x_{C_7H_{16}}$	y_{CH_4}
233.15 (40)	100	0.689	0.0667	0.9333	1.0
	200	1.38	0.120	0.880	1.0
	396	2.73	0.216	0.784	1.0
	605	4.17	0.311	0.689	1.0
	805	5.55	0.389	0.611	1.0
	1008	6.95	0.455	0.545	1.0
244.26 (-20)	96.0	0.662	0.0568	0.9432	1.0
	212	1.46	0.114	0.886	1.0
	400	2.76	0.196	0.804	1.0
	608	4.19	0.280	0.720	1.0
	807	5.56	0.353	0.647	1.0
	990	6.83	0.412	0.588	1.0

AUXILIARY INFORMATION

METHOD/APPARATUS/PROCEDURE:	SOURCE AND PURITY OF MATERIALS:
The solubilities were determined by measurement of retention volumes using gas chromatography. The method uses methane as a carrier gas, radioactively tagged methane as a sample and heptane as the stationary liquid. The technique is described in the source and in ref. (1).	1. Sample dried, purity 99.7 mole per cent; 0.2 mole per cent nitrogen and 0.1 mole per cent ethane. 2. Phillips Petroleum research grade sample, purity 99.90 mole per cent.

ESTIMATED ERROR:

$\delta T/K = \pm 0.1$; $\delta P/MPa = \pm 2\%$;
δx, $\delta y = \pm 6\%$ (estimated by compiler).

REFERENCES:

1. Koonce, K. T.

   Ph.D. thesis, Rice University,
   Houston, <u>1963</u>.

COMPONENTS:	ORIGINAL MEASUREMENTS:
1. Methane; CH_4; [74-82-8] 2. Heptane; C_7H_{16}; [142-82-5]	Chang, H. L.; Hunt, L. J.; Kobayashi, R. *Am. Inst. Chem. Engnr. J.* <u>1966</u>, *11*, 1212-1216.
VARIABLES: Temperature, pressure	PREPARED BY: C. L. Young

EXPERIMENTAL VALUES:

T/K	P/MPa	Mole fraction of methane in liquid, x_{CH_4}	in gas, y_{CH_4}	T/K	P/MPa	Mole fraction of methane in liquid, x_{CH_4}	in gas, y_{CH_4}
255.4	0.690	0.04450	0.99857	244.3	15.51	0.72190	0.99304
	1.379	0.09012	0.99877		17.24	0.78127	0.98833
	2.758	0.17441	0.99896		18.96	0.86340	0.97560
	4.137	0.24904	0.99873	233.2	0.690	0.05555	0.99970
	5.516	0.31600	0.99860		1.379	0.11632	0.99974
	6.895	0.37720	0.99838		2.758	0.20810	0.99968
	8.618	0.44521	0.99778		4.137	0.29901	0.99958
	10.34	0.49657	0.99639		5.516	0.37280	0.99948
	12.07	0.55700	0.99584		6.895	0.44809	0.99933
	13.79	0.61693	0.99427		8.618	0.51596	0.99907
	15.51	0.66590	0.99265		10.34	0.57895	0.99869
	17.24	0.71831	0.98535		12.07	0.65488	0.99818
	18.96	0.77888	0.97340		13.79	0.73746	0.99746
	20.68	0.85850	0.93999		15.51	0.81499	0.99600
244.3	0.690	0.04903	0.99936		17.24	0.88799	0.99200
	1.379	0.10330	0.99945	222.0	0.690	0.06280	0.99985
	2.758	0.18830	0.99934		1.379	0.12350	0.99987
	4.137	0.26685	0.99924		2.758	0.23220	0.99985
	5.516	0.34125	0.99909		4.137	0.32550	0.99980
	6.895	0.40781	0.99893		5.516	0.41170	0.99973
	8.618	0.47841	0.99849		6.895	0.49210	0.99964
	10.34	0.53885	0.99789		8.618	0.57200	0.99951
	12.07	0.60585	0.99707		10.34	0.65420	0.99929
	13.79	0.66583	0.99596			(cont.)	

AUXILIARY INFORMATION

METHOD/APPARATUS/PROCEDURE:	SOURCE AND PURITY OF MATERIALS:
Recirculating vapor flow apparatus with magnetic pump. Pressure measured with Bourdon gauge and temperature measured with thermo-pile. Samples of both phases analysed using gas chromatography with flame ionisation detector. Details in source and ref. (1).	1. High purity sample from Associated Oil and Gas Company. 2. Phillips Petroleum Co. research grade sample.

ESTIMATED ERROR:
$\delta T/K = \pm 0.01$; $\delta P/MPa = \pm 0.015$;
δx_{CH_4}, $\delta y_{CH_4} = \pm 0.00001$ (claimed by authors).

REFERENCES:
1. Price, A. R.; Kobayashi, R.
   J. Chem. Eng. Data
   <u>1958</u>, *4*, 40.

COMPONENTS:	ORIGINAL MEASUREMENTS:
1. Methane; CH_4; [74-82-8]	Chang, H. L.; Hunt, L. J.
2. Heptane; C_7H_{16}; [142-82-5]	Kobayashi, R.
	Am. Inst. Chem. Engnr. J.
	<u>1966</u>, *11*, 1212-1216.

EXPERIMENTAL VALUES:

T/K	P/MPa	Mole fraction of methane in liquid, x_{CH}	in gas, y_{CH}	T/K	P/MPa	Mole fraction of methane in liquid, x_{CH}	in gas, y_{CH}
222.0	12.07	0.72800	0.99905	210.9	12.07	0.80300	0.999220
	13.79	0.79490	0.99854		13.79	0.89110	0.998590
	15.51	0.88190	0.99659	199.8	0.690	0.08470	0.999964
210.9	0.690	0.07260	0.999930		1.379	0.16350	0.999964
	1.379	0.14090	0.999932		2.758	0.31100	0.999955
	2.758	0.26300	0.999932		4.137	0.46120	0.999940
	4.137	0.37800	0.999900		5.516	0.56520	0.999910
	5.516	0.47650	0.999854		6.895	0.62400	0.999860
	6.895	0.55490	0.999794		8.618	0.71500	0.999800
	8.618	0.65800	0.999720		10.34	0.80560	0.999700
	10.34	0.72100	0.999590		12.07	0.90000	0.999300

COMPONENTS:	ORIGINAL MEASUREMENTS:
1. Methane; CH_4; [74-82-8] 2. Nitrogen; N_2; [7727-37-9] 3. Heptane; C_7H_{16}; [142-82-5]	Boomer, E. H.; Johnson, C. A.; Piercey, A. G. A. *Can. J. Res. B* 1938, *16*, 396-410.

VARIABLES:
 	
   Temperature, pressure
	PREPARED BY:
	
	   C. L. Young

EXPERIMENTAL VALUES:

			Mole fractions in liquid			Mole fractions in vapor		
T/K	P/atm	P/MPa	x_{CH_4}	x_{N_2}	$x_{C_7H_{16}}$	y_{CH_4}	y_{N_2}	$y_{C_7H_{16}}$
298.15	1	0.1	-	-	-	0.945	-	0.045
	36.2	3.67	0.163	0.002	0.835	0.9285	0.054	0.0175
			0.159	0.003	0.838	-	-	-
	68.4	6.93	0.276	0.008	0.716	0.932	0.0579	0.0101
			-	-	-	0.928	0.0591	0.0129
	101.7	10.30	0.376	0.009	0.615	0.925	0.0641	0.0109
			0.387	0.005	0.608	0.936	0.0531	0.0109
	135	13.7	0.470	0.012	0.518	0.908	0.0787	0.0133
			-	-	-	0.910	0.0781	0.0119
	167.9	17.01	0.539	0.026	0.435	0.910	0.068	0.0220
			0.540	0.022	0.438	0.910	0.0685	0.0215
	202.2	20.49	0.622	0.024	0.354	0.894	0.069	0.037
			0.619	0.027	0.354	0.899	0.066	0.035
	236.0	23.91	0.705	0.035	0.260	-	-	-
	236.8	23.99	0.705	0.040	0.255	0.871	0.065	0.064
	243.6	24.68	0.731	0.043	0.226	0.858	0.079	0.083
	250.0	25.33	0.761	0.047	0.192	0.849	0.055	0.096
328.15	1	0.1	-	-	-	0.769	-	0.231
	34.8	3.53	-	-	-	0.904	0.064	0.032
			0.141	0.004	0.855	-	-	-
	100.9	10.22	0.352	0.009	0.639	0.928	0.051	0.021

(cont.)

AUXILIARY INFORMATION

METHOD/APPARATUS/PROCEDURE:	SOURCE AND PURITY OF MATERIALS:
Rocking autoclave stirred by steel piston falling under gravity. Samples of vapor and liquid trapped in two auxiliary high pressure cells. Equilibrium samples analysed in complicated volumetric and combustion apparatus. Details in ref. (1). NOTE: The source reference also contains data on impure heptane samples.	1 and 2. Natural gas sample containing 94.4 mole per cent of methane and 5.6 mole per cent of nitrogen. Impurities may have been present amounting to 0.1 mole per cent. 3. Jeffrey Pine Oil sample, fractionated.

ESTIMATED ERROR:

$\delta T/K = \pm 0.1$; $\delta P/MPa = \pm 0.02$;

δx, $\delta y = \pm 1\%$ (estimated by compiler).

REFERENCES:

1. Boomer, E. H.; Johnson, C. A.; Argue, G. H.
Can. J. Res. B
1937, *15*, 367.

COMPONENTS: ORIGINAL MEASUREMENTS:

1. Methane; CH$_4$; [74-82-8]   Boomer, E. H.; Johnson, C. A.;
2. Nitrogen; N$_2$; [7727-3709]   Piercey, A. G. A.
3. Heptane; C$_7$H$_{16}$; [142-82-5]   *Can. J. Res. B*
                                               1938, *16*, 396-410.

EXPERIMENTAL VALUES:

| | | | | | | Mole fractions | | |
| | | | | in liquid | | | in vapor | |
T/K	P/atm	P/MPa	x_{CH_4}	x_{N_2}	$x_{C_7H_{16}}$	y_{CH_4}	y_{N_2}	$y_{C_7H_{16}}$
328.15	167.2	16.94	0.513	0.019	0.468	0.906	0.062	0.032
	236	23.9	0.685	0.031	0.284	0.860	0.061	0.079
			0.683	0.033	0.284	0.864	0.058	0.078
	249.3	25.3	0.747	0.037	0.216	0.822	0.055	0.123
			0.743	0.041	0.216	0.818	0.065	0.117
	252.8	25.61	0.759	0.043	0.198	–	–	–
			–	–	–	0.768	0.044	0.188
358.15	1	0.1	–	–	–	0.323	–	0.677
	34.8	3.53	0.130	0.005	0.865	0.918	0.042	0.040
	100.9	10.22	0.339	0.010	0.651	0.911	0.058	0.031
	167.2	16.94	0.494	0.020	0.486	0.887	0.061	0.052
	236	23.9	–	–	–	0.819	0.054	0.127
			0.695	0.041	0.264	0.821	0.049	0.130
	242.5	24.57	0.735	0.037	0.228	0.752	0.041	0.207
	249.3	25.26	0.763	0.036	0.201	0.762	0.039	0.199

COMPONENTS:	EVALUATOR:

1. Methane; CH_4; [74-82-8]
 	
2. Octane; C_8H_{18}; [111-65-9]
   or
   Nonane; C_9H_{20}; [111-84-2]
	Colin L. Young,
	School of Chemistry,
	University of Melbourne,
	Parkville, Victoria 3052,
	Australia.
	
	March 1984.

EVALUATION:

Methane + Octane

There are only two sets of data on this system at elevated pressure and there are fairly large discrepancies between the two sets. The data of Frolich *et al.* (1) are thought to be less reliable and were originally presented in graphical form. These data are classified as doubtful. The data of Kohn and Bradish (2) are more extensive and are believed to be more reliable and hence are classified as tentative.

Methane + Nonane

Shipman and Kohn (3) have studied this system at eight temperatures between 223 K and 423 K and their data are classified as tentative. The only other study on this system appears to be that of Savvina *et al.* (4). These workers present data at temperatures between 313 K and 423 K but only in graphical form on a scale which lacks sufficient accuracy to be considered here.

References

1. Frolich, P. K.; Tauch, E. J.; Hogan, J. J.; Peer, A. A.
   Ind. Eng. Chem., 1931, *23*, 548.

2. Kohn, J. P.; Bradish, W. F.
   J. Chem. Eng. Data, 1964, *9*, 5.

3. Shipman, L. M.; Kohn, J. P.
   J. Chem. Eng. Data, 1966, *11*, 176.

4. Savvina, Ya. D.; Velikovskii, A. S.
   J. Phys. Chem. (USSR), 1956, *30*, 1597.

COMPONENTS:	ORIGINAL MEASUREMENTS:
1. Methane; CH_4; [74-82-8] 2. Octane; C_8H_{18}; [111-65-9]	Frolich, P.K.; Tauch, E.J.; Hogan, J.J.; Peer, A.A. *Ind. Eng. Chem.* <u>1931</u>, *23*, 548-550

VARIABLES:	PREPARED BY:
Pressure	C.L. Young

EXPERIMENTAL VALUES:

T/K	P/MPa	Solubility[*]	Mole fraction of methane in liquid,[+] x_{CH_2}
298.15	1.0	11	0.068
	2.0	23	0.133
	3.0	36	0.194
	4.0	50	0.251
	5.0	64	0.300
	6.0	79	0.346
	7.0	94	0.386
	8.0	109	0.422
	9.0	123	0.451
	10.0	138	0.480

*   Data taken from graph in original article. Volume of gas
    measured at 101.325 kPa and 298.15 K dissolved by unit
    volume of liquid measured under the same conditions.

+   Calculated by compiler.

AUXILIARY INFORMATION

METHOD/APPARATUS/PROCEDURE:	SOURCE AND PURITY OF MATERIALS:
Static equilibrium cell. Liquid saturated with gas and after equilibrium established samples removed and analysed by volumetric method. Allowance was made for the vapor pressure of the liquid and the solubility of the gas at atmospheric pressure. Details in source.	Stated that the materials were the highest purity available. Purity 98 to 99 mole per cent.
	ESTIMATED ERROR: $\delta T/K = \pm 0.1$; $\delta x_{CH_4} = \pm 5\%$
	REFERENCES:

COMPONENTS:	ORIGINAL MEASUREMENTS:
1. Methane; CH₄; [74-82-8] 2. Octane; C₈H₁₈; [111-65-9]	Kohn, J. P.; Bradish, W. F. *J. Chem. Engng. Data* 1964, *9*, 5-8.

VARIABLES:	PREPARED BY:
Temperature, pressure	C. L. Young

EXPERIMENTAL VALUES:

T/K	P/MPa	Mole fraction of methane in liquid, x_{CH_4}	in vapor, y_{CH_4}
223.15	1.01	0.086	–
	2.03	0.168	–
	3.04	0.244	–
	4.05	0.312	–
	5.07	0.370	–
	6.08	0.424	–
	7.09	0.472	–
248.15	1.01	0.065	–
	2.03	0.128	–
	3.04	0.187	–
	4.05	0.238	–
	5.07	0.288	–
	6.08	0.337	–
	7.09	0.373	–
273.15	1.01	0.054	–
	2.03	0.106	–
	3.04	0.155	–
	4.05	0.201	–
	5.07	0.244	–
	6.08	0.285	–
	7.09	0.319	–
298.15	1.01	0.047	0.998
	2.03	0.094	0.999

(cont.)

AUXILIARY INFORMATION

METHOD/APPARATUS/PROCEDURE:	SOURCE AND PURITY OF MATERIALS:
Borosilicate glass equilibrium cell. Temperature measured with platinum resistance thermometer, pressure measured on Bourdon gauge. Samples of methane added to octane. Dew and bubble point pressure measured. Data presented in original as smoothed values (as function of pressure in atm). Details of method in source and ref. 1.	1. Phillips Petroleum Co. sample fractionated, purity > 99.5 mole per cent. 2. Phillips Petroleum Co. sample, "pure" grade, purity at least 99 mole per cent.

	ESTIMATED ERROR:
	$\delta T/K = \pm 0.07$; $\delta P/MPa = \pm 0.01$; δx_{CH_4}, $\delta y_{CH_4} = \pm 0.0015$.

	REFERENCES:
	1. Kohn, J. P.; Kurata, F. *Petrol. Process.* 1956, *11*, 57.

COMPONENTS:	ORIGINAL MEASUREMENTS:
1. Methane; CH_4; [74-82-8]	Kohn, J. P.; Bradish, W. F.
2. Octane; C_8H_{18}; [111-65-9]	*J. Chem. Engng. Data*
	1964, *9*, 5-8.

EXPERIMENTAL VALUES:

T/K	P/MPa	Mole fraction of methane in liquid, x_{CH_4}	in vapor, y_{CH_4}
298.15	3.04	0.136	0.999
	4.05	0.178	0.999
	5.07	0.220	0.999
	6.08	0.255	0.999
	7.09	0.287	0.999
323.15	1.01	0.043	0.991
	2.03	0.083	0.995
	3.04	0.122	0.996
	4.05	0.161	0.997
	5.07	0.188	0.997
	6.08	0.234	0.997
	7.09	0.265	0.997
348.15	1.01	0.038	–
	2.03	0.075	0.985
	3.04	0.112	0.990
	4.05	0.148	0.992
	5.07	0.183	0.993
	6.08	0.218	0.994
	7.09	0.251	0.994
373.15	1.01	0.033	0.947
	2.03	0.069	0.971
	3.04	0.104	0.979
	4.05	0.139	0.982
	5.07	0.173	0.984
	6.08	0.207	0.985
	7.09	0.240	0.986
423.15	1.01	0.028	–
	2.03	0.063	0.881
	3.04	0.097	0.914
	4.05	0.131	0.930
	5.07	0.164	0.939
	6.08	0.196	0.945
	7.09	0.229	0.949

COMPONENTS:	ORIGINAL MEASUREMENTS:
1. Methane; CH₄ ; [74-82-8] 2. 2,2,3-Trimethylpentane; C_8H_{18} ; [564-02-3]	Savvina, Ya. D. *Tr. Vses. Nauchno-Issled. Inst.* *Prirodn. Gazov.*,<u>1962</u>,*17/25* , 185-196.

VARIABLES:	PREPARED BY:
Temperature, pressure	C. L. Young

EXPERIMENTAL VALUES:

T/K (t/ C)	P/kgcm^{-3}	P/Mpa	K-value methane	K-value 2,2,3-trimethyl- pentane
313.2	20	1.96	10.30	0.080
(40)	40	3.92	5.87	0.045
	60	5.88	4.14	0.039
	100	9.81	2.70	0.045
	150	14.7	1.95	0.063
	180	17.7	1.69	0.101
	200	19.6	1.52	0.516
	210	20.6	1.41	0.218
	220	21.6	1.25	0.400
	225	22.1	1.10	0.685
333.2	40	3.92	6.02	0.058
(60)	60	5.88	4.25	0.052
	100	9.81	2.74	0.058
	150	14.7	1.99	0.089
	180	17.7	1.70	0.123
	200	19.6	1.51	0.188
	210	20.6	1.37	0.272
	220	21.6	1.18	0.516
	224	22.0	1.01	0.935
353.2	20	1.96	11.56	0.148
(80)	40	3.92	6.26	0.070
	60	5.88	4.42	0.065
	100	9.81	2.83	0.075
	150	14.7	2.03	0.105
	180	17.7	1.80	0.142
	200	19.6	1.48	0.229

AUXILIARY INFORMATION

METHOD APPARATUS/PROCEDURE:	SOURCE AND PURITY OF MATERIALS:
Values appear to be determined using apparatus descibed in ref.(1).	No details given.

ESTIMATED ERROR:

REFERENCES:

1. Savvina, Ya. D.;Velikovskii, A. S.;
   Tr. Vses. Nauchno-Issed. Inst.
   Prirodn. Gazov.,<u>1962</u>, *17/25*, 163.

COMPONENTS:	ORIGINAL MEASUREMENTS:
1. Methane; CH$_4$; [74-82-8] 2. 2,2,3-Trimethylpentane; C$_8$H$_{18}$; 　　[564-02-3]	Savvina, Ya. D. *Tr. Vses. Nauchno-Issled. Inst.* *Prirodn. Gazov.* <u>1962</u>, *17/25*, 185-196.

EXPERIMENTAL VALUES:

T/K (t/ C)	P/kgcm^{-3}	P/Mpa	K-value	
			methane	2,2,3-trimethyl- pentane
353.2	210	20.6	1.34	0.309
(80)	220	21.6	1.10	0.680
373.2	20	1.96	12.23	0.180
(100)	40	3.92	6.62	0.091
	60	5.88	4.48	0.080
	100	9.81	2.86	0.093
	150	14.7	2.05	0.128
	180	17.7	1.69	0.184
	200	19.6	1.43	0.278
	210	20.6	1.25	0.429
	216	21.2	1.05	0.827
393.2	40	3.92	6.71	0.131
(120)	60	5.88	4.52	0.102
	100	9.81	2.92	0.113
	150	14.7	2.04	0.143
	180	17.7	1.66	0.214
	200	19.6	1.35	0.370
	210	20.6	1.05	0.823
423.2	20	1.96	12.53	0.316
(140)	40	3.92	6.81	0.246
	60	5.88	4.60	0.134
	100	9.81	2.91	0.139
	150	14.7	2.99	0.182
	180	17.7	1.49	0.327
	190	18.6	1.27	0.516
	195	19.1	1.06	0.826

COMPONENTS:	ORIGINAL MEASUREMENTS:
1. Methane; CH_4; [74-82-8] 2. Nonane; C_9H_{20}; [111-84-2]	Shipman, L.M.; Kohn, J.P. *J. Chem. Engng. Data.* <u>1966</u>,*11*, 176-180.

VARIABLES:	PREPARED BY:
Temperature, pressure	C.L. Young

EXPERIMENTAL VALUES:

T/K	P/atm	P/MPa	Mole fraction of methane in liquid, x_{CH_4}	in gas, y_{CH_4}
218.47	5.00	0.51	0.046	
217.43	10.00	1.01	0.093	
215.56	20.00	2.03	0.181	
213.86	30.0	3.04	0.263	data along solid-
212.35	40.0	4.05	0.343	liquid-vapor line.
209.85	60.0	6.08	0.470	
208.35	100.0	10.13	0.570	
223.15	10.00	1.013	0.0989	0.9999
	20.00	2.027	0.1737	0.9999
	30.00	3.040	0.2418	0.9999
	40.00	4.053	0.3033	0.9999
	50.00	5.066	0.3582	0.9999
	60.00	6.080	0.4065	0.9999
	70.00	7.093	0.4482	0.9999
	80.00	8.106	0.4883	0.9999
	90.00	9.119	0.5118	0.9999
	100.00	10.133	0.5336	0.9999
248.15	10.00	1.013	0.0678	0.9999
	20.00	2.027	0.1288	0.9999
	30.00	3.040	0.1856	0.9999
	40.00	4.053	0.2381	0.9999
	50.00	5.066	0.2863	0.9999
	60.00	6.080	0.3302	0.9999
	70.00	7.093	0.3698	0.9999
	80.00	8.106	0.4051	0.9999
	90.00	9.119	0.4361	0.9999

AUXILIARY INFORMATION

METHOD/APPARATUS/PROCEDURE:	SOURCE AND PURITY OF MATERIALS:
Borosilicate static equilibrium cell. Temperature measured with platinum resistance thermometer. Pressure measured with Bourdon gauge. Methane added to nonane and equilibriated. Temperature-pressure diagram constructed from bubble and dew points of mixtures of known composition. Smoothed data, as given above, reported in source. Details of apparatus in source and ref. (1) and (2).	1. Phillips Petroleum Co. sample, purified as in ref. (1). Final purity better than 99.5 mole per cent. 2. Phillips Petroleum Co. sample, purity better than 99 mole per cent.

ESTIMATED ERROR:

$\delta T/K = \pm 0.1$; $\delta P/MPa = \pm 0.007$; δx_{CH_4}; $\delta y_{CH_4} = \pm 0.003$.

REFERENCES:

1. Kohn, J.P.; *Am.Inst.Chem.Engnrs.J.* <u>1961</u>, *7*, 514.

2. Kohn, J.P.; Kurata, F. *Petrol. Process.*<u>1956</u>,*11*, 57.

COMPONENTS:	ORIGINAL MEASUREMENTS:
1. Methane; CH₄; [74-82-8]	Shipman, L.M.; Kohn, J.P.
2. Nonane; C₉H₂₀; ; [111-84-2]	J. Chem. Eng. Data. 1966,11, 176-180.

EXPERIMENTAL VALUES:

			Mole fraction of methane	
T/K	P/atm	P/MPa	in liquid,x_{CH_4}	in gas,y_{CH_4}
248.15	100.00	10.133	0.4629	0.9999
273.15	10.00	1.013	0.0540	0.9999
	20.00	2.027	0.1069	0.9999
	30.00	3.040	0.1563	0.9999
	40.00	4.053	0.2023	0.9999
	50.00	5.066	0.2448	0.9999
	60.00	6.080	0.2838	0.9999
	70.00	7.093	0.3194	0.9999
	80.00	8.106	0.3515	0.9999
	90.00	9.119	0.3801	0.9999
	100.00	10.133	0.4087	0.9999
	120.00	12.159	0.4639	0.9999
	140.00	14.186	0.5131	0.9999
	160.00	16.212	0.5538	0.9999
	180.00	18.239	0.5922	0.9999
	200.00	20.265	0.6234	0.9999
	220.00	22.292	0.6512	0.9990
	240.00	24.318	0.6781	0.9980
	260.00	26.345	0.7075	0.9931
	280.00	28.371	0.7434	0.9788
	300.00	30.398	0.7908	0.9572
	310.00	31.411	0.8200	0.9410
	315.00	31.917	0.8450	0.9250
	318.00	32.221	0.8850	0.8850
298.15	10.00	1.013	0.0509	0.9999
	20.00	2.027	0.0986	0.9999
	30.00	3.040	0.1426	0.9999
	40.00	4.053	0.1832	0.9999
	50.00	5.066	0.2205	0.9999
	60.00	6.080	0.2550	0.9999
	70.00	7.093	0.2870	0.9999
	80.00	8.106	0.3166	0.9999
	90.00	9.119	0.3442	0.9999
	100.00	10.133	0.3664	0.9999
	120.00	12.159	0.4160	0.9999
	140.00	14.186	0.4637	0.9999
	160.00	16.212	0.5094	0.9999
	180.00	18.239	0.5533	0.9929
	200.00	20.265	0.5952	0.9883
	220.00	22.292	0.6352	0.9867
	240.00	24.318	0.6733	0.9851
	260.00	26.345	0.7094	0.9803
	280.00	28.371	0.7437	0.9695
	300.00	30.398	0.7760	0.9495
	310.00	31.411	0.8090	0.9351
	315.00	31.917	0.8230	0.9200
	319.00	32.323	0.8800	0.8800
323.15	10.00	1.013	0.0449	0.9999
	20.00	2.027	0.0860	0.9999
	30.00	3.040	0.1251	0.9999
	40.00	4.054	0.1622	0.9999
	50.00	5.066	0.1974	0.9999
	60.00	6.080	0.2306	0.9999
	70.00	7.093	0.2618	0.9999
	80.00	8.106	0.2911	0.9999
	90.00	9.119	0.3184	0.9999
	100.00	10.133	0.3471	0.9999
	120.00	12.159	0.3994	0.9999
	140.00	14.186	0.4480	0.9999
	160.00	16.212	0.4935	0.9970
	180.00	18.239	0.5365	0.9906
	200.00	20.265	0.5775	0.9870
	220.00	22.292	0.6173	0.9857

COMPONENTS:	ORIGINAL MEASUREMENTS:
1. Methane; CH_4; [74-82-8]	Shipman, L.M.; Kohn, J.P.
2. Nonane; C_9H_{20}; [111-84-2]	*J. Chem. Eng. Data.* <u>1966</u>,*11*, 176-180.

T/K	P/atm	P/MPa	Mole fraction of methane in liquid, x_{CH_4}	in gas, y_{CH_4}
323.15	240.00	24.318	0.6562	0.9837
	260.00	26.345	0.6950	0.9783
	280.00	28.371	0.7343	0.9662
	300.00	30.398	0.7745	0.9445
	310.00	31.411	0.8003	0.9292
	315.00	31.917	0.8230	0.9120
	318.00	32.221	0.8720	0.8720
348.15	10.00	1.013	0.0448	0.9999
	20.00	2.027	0.0826	0.9999
	30.00	3.040	0.1187	0.9999
	40.00	4.053	0.1532	0.9999
	50.00	5.066	0.1861	0.9999
	60.00	6.080	0.2174	0.9999
	70.00	7.093	0.2470	0.9999
	80.00	8.106	0.2750	0.9999
	90.00	9.119	0.3014	0.9999
	100.00	10.133	0.3271	0.9999
	120.00	12.159	0.3775	0.9999
	140.00	14.186	0.4257	0.9999
	160.00	16.212	0.4720	0.9960
	180.00	18.239	0.5168	0.9870
	200.00	20.265	0.5605	0.9830
	220.00	22.292	0.6035	0.9800
	240.00	24.318	0.6463	0.9752
	260.00	26.345	0.6891	0.9662
	280.00	28.371	0.7324	0.9502
	300.00	30.398	0.7765	0.9270
	310.00	31.411	0.8195	0.8940
	313.00	31.715	0.8600	0.8600
373.15	10.00	1.013	0.0362	0.9740
	20.00	2.027	0.0715	0.9824
	30.00	3.040	0.1058	0.9878
	40.00	4.053	0.1392	0.9908
	50.00	5.066	0.1716	0.9921
	60.00	6.080	0.2030	0.9921
	70.00	7.093	0.2335	0.9915
	80.00	8.106	0.2630	0.9909
	90.00	9.119	0.2915	0.9909
	100.00	10.133	0.3190	0.9921
423.15	10.00	1.013	0.0329	0.9207
	20.00	2.027	0.0702	0.9360
	30.00	3.040	0.1060	0.9485
	40.00	4.053	0.1402	0.9584
	50.00	5.066	0.1727	0.9656
	60.00	6.080	0.2037	0.9701
	70.00	7.093	0.2330	0.9719
	80.00	8.106	0.2608	0.9710
	90.00	9.119	0.2870	0.9674
	100.00	10.133	0.3115	0.9611

COMPONENTS:	ORIGINAL MEASUREMENTS:
1. Methane; CH_4; [74-82-8] 2. Nonane; C_9H_{20}; [111-84-2]	Savvina, Ya. D. *Tr. Vses. Nauchno-Issled. Inst.* *Prirodn. Gazov.,* <u>1962</u>, *17/25,* 185-196.

VARIABLES:
  Temperature, pressure
	PREPARED BY:
	  C. L. Young

EXPERIMENTAL VALUES:

T/K (t/°C)	$P/kgcm^{-3}$	P/Mpa	K-value methane	nonane
313.2	10	0.98	17.43	0.006
(40)	20	1.96	9.69	0.002
	40	3.92	5.42	0.002
	60	5.88	3.82	0.003
	100	9.81	2.61	0.005
	150	14.7	2.00	0.009
	200	19.6	1.66	0.017
	250	24.5	1.42	0.029
	280	27.5	1.31	0.063
	300	29.4	1.25	0.099
	320	31.4	1.17	0.211
	329	32.3	1.06	0.579
333.2	10	0.98	19.03	0.010
(60)	20	1.96	10.35	0.005
	40	3.92	5.65	0.004
	60	5.88	4.02	0.005
	100	9.81	2.73	0.007
	150	14.7	2.06	0.013
	200	19.6	1.69	0.024
	250	24.5	1.44	0.044
	280	27.5	1.33	0.079
	300	29.4	1.26	0.129
	320	31.4	1.18	0.208
	326	32.0	1.08	0.535

AUXILIARY INFORMATION

METHOD/APPARATUS/PROCEDURE:	SOURCE AND PURITY OF MATERIALS:
Values appear to be determined using apparatus described in ref.(1).	No details given.
	ESTIMATED ERROR:

	REFERENCES:
	1. Savvina, Ya. D.;Velikovskii, A. S.
	     Tr. Vses. Nauchno-Issled. Inst.
	     Prirodn. Gazov., <u>1962</u>, *17/25,* 163.

COMPONENTS:	ORIGINAL MEASUREMENTS:
1. Methane; CH_4; [74-82-8] 2. Nonane; C_9H_{20}; [111-84-2]	Savvina, Ya. D. *Tr. Vses. Nauchno-Issled. Inst.* *Prirodn. Gazov.*, <u>1962</u>, *17/25*, 185-196.

T/K (t/°C)	P/kgcm^{-3}	P/Mpa	K-value methane	nonane
353.2 (80)	20	1.96	10.91	0.007
	40	3.92	5.91	0.006
	60	5.88	4.18	0.007
	100	9.81	2.83	0.009
	150	14.7	2.11	0.015
	200	19.6	2.71	0.029
	250	24.5	1.45	0.067
	280	27.5	1.33	0.109
	300	29.4	1.27	0.163
	322	31.6	1.09	0.535
373.2 (100)	20	1.96	11.78	0.011
	40	3.92	6.26	0.010
	60	5.88	4.39	0.010
	100	9.81	2.93	0.013
	150	14.7	2.16	0.020
	200	19.6	1.74	0.038
	250	24.5	1.48	0.076
	280	27.5	1.32	0.135
	300	29.4	1.21	0.244
	310	30.4	1.14	0.385
	313	30.7	1.11	0.485
393.2 (120)	20	1.96	12.31	0.016
	40	3.92	6.54	0.014
	60	5.88	4.59	0.015
	100	9.81	3.03	0.018
	150	14.7	2.15	0.026
	200	19.6	1.72	0.051
	250	24.5	1.46	0.092
	270	26.5	1.33	0.232
	280	27.5	1.27	0.232
	290	28.4	1.20	0.333
	297	29.1	1.10	0.563
423.2 (150)	20	1.96	12.49	0.029
	40	3.92	6.89	0.026
	60	5.88	4.73	0.027
	100	9.81	2.92	0.030
	150	14.7	2.09	0.039
	200	19.6	1.67	0.080
	250	24.5	1.38	0.167
	270	26.5	1.28	0.339
	278	27.3	1.08	0.681

COMPONENTS:	EVALUATOR:
1. Methane; CH_4; [74-82-8]	Colin L. Young, School of Chemistry, University of Melbourne, Parkville, Victoria 3052, Australia.
2. Decane; $C_{10}H_{22}$; [124-18-5]	
	March 1982

EVALUATION:

    This system has been investigated by four groups.   Reamer *et al.* (1) studied this system at temperatures between 311 K and 511 K, Beaudoin and Kohn (2) between 248 K and 423 K, Lin *et al.* (3) between 423 K and 583 K and Koonce and Kobayashi (4) between 244 K and 278 K.   There appears to be fair agreement between all sets of data where they overlap, therefore all sets are classified as tentative.

    There is good agreement (i.e., within 3%) between the data of Reamer *et al.* (1) and Lin *et al.* (4) at 237.8 °C.   The agreement between the data of Beaudoin and Kohn (2) and Lin *et al.* (3) at 150 °C is fair (i.e., within 5%), the mole fraction solubilities of Beaudoin and Kohn (2) being slightly smaller.   Although there is no directly comparable data from the measurements of Koonce and Kobayashi (4) and Beaudoin and Kohn (2) since the isotherm temperatures are different, it appears that the results of Beaudoin and Kohn (2) are slightly low and extrapolation of the data of Koonce and Kobayashi (4) to higher temperatures would give values in good agreement with those of Reamer *et al.* (1).

References

1.   Reamer, H. H.; Olds, R. H.; Sage, B. H.; Lacey, W. N.
      Ind. Eng. Chem., 1942, *34*, 1526.

2.   Beaudoin, J. M.; Kohn, J. P.
      J. Chem. Eng. Data, 1967, *12*, 189.

3.   Koonce, K. T.; Kobayashi, R.
      J. Chem. Eng. Data, 1964, *9*, 490.

4.   Lin, H.-M.; Sebastian, H. M.; Simnick, J. J.; Chao, K.-C.
      J. Chem. Eng. Data, 1979, *24*, 146.

COMPONENTS:	ORIGINAL MEASUREMENTS:
1. Methane; CH_4; [74-82-8] 2. Decane; $C_{10}H_{22}$; [124-18-5]	Reamer, H. H.; Olds, R. H.; Sage, B. H.; Lacey, W. N. *Ind. Eng. Chem.* 1942, *34*, 1526-1531.

VARIABLES:
 	
    Temperature, pressure
	PREPARED BY:
	
	    C. L. Young

EXPERIMENTAL VALUES:

T/K	P/MPa	Wt-fraction of methane in liquid	in gas	Mole fraction of methane in liquid, x_{CH_4}	in gas, y_{CH_4}
310.92	0.14	0.00080	0.9663	0.00705	0.9961
	0.28	0.00161	0.9816	0.01410	0.9979
	0.41	0.00241	0.9868	0.02098	0.9985
	0.55	0.00321	0.9895	0.02777	0.9988
	0.69	0.00401	0.9910	0.03448	0.9990
	1.38	0.00798	0.9939	0.06661	0.9993
	2.76	0.01587	0.9950	0.12515	0.9994
	4.14	0.02372	0.9951	0.17731	0.9994
	5.52	0.03157	0.9948	0.22432	0.9994
	6.89	0.03963	0.9941	0.26797	0.9993
	8.62	0.04990	0.9925	0.31783	0.9991
	10.34	0.06054	0.9905	0.36373	0.9989
	12.07	0.07180	0.9875	0.40695	0.99858
	13.79	0.08350	0.9835	0.44697	0.99811
	15.51	0.09570	0.9775	0.48421	0.99741
	17.24	0.1082	0.9700	0.5184	0.99653
	18.96	0.1214	0.9615	0.5507	0.99551
	20.68	0.1360	0.9510	0.5827	0.99423
	22.41	0.1505	0.9380	0.6111	0.99260
	24.13	0.1652	0.9220	0.6371	0.99055
	25.86	0.1815	0.9008	0.6630	0.98774
	27.58	0.1984	0.8720	0.6871	0.98372
	29.30	0.2183	0.8388	0.7124	0.97880

(cont.)

AUXILIARY INFORMATION

METHOD/APPARATUS/PROCEDURE:	SOURCE AND PURITY OF MATERIALS:
PVT cell charged with mixture of known composition. Bubble point and dew point determined for various compositions. Pressure measured with pressure balance. Temperature measured using platinum resistance thermometer. Coexisting liquid and gas phase properties determined by graphical means. Details of apparatus in ref. (1).	1. Crude sample, treated for removal of higher alkanes, carbon dioxide and water. Final purity 99.97 mole per cent. 2. Eastman Kodak Co. sample. Distilled several times, dried over sodium. n_{20}^D = 1.4100. Mainly decane isomers.
	ESTIMATED ERROR: $\delta T/K = \pm 0.1$; $\delta P/MPa = \pm 0.005$; δ(wt-fraction) = ± 0.003.

	REFERENCES:
	1. Sage, B. H.; Lacey, W. N.
	   Trans. Am. Inst. Mining and Met.
	   Engnrs.
	   1940, *136*, 136.

COMPONENTS:	ORIGINAL MEASUREMENTS:
1. Methane; CH_4; [74-82-8]	Reamer, H. H.; Olds, R. H.; Sage, B. H.; Lacey, W. N.
2. Decane; $C_{10}H_{22}$; [124-18-5]	*Ind. Eng. Chem.*
	<u>1942</u>, *34*, 1526-1531.

EXPERIMENTAL VALUES:

T/K	P/MPa	Wt-fraction of methane in liquid	in gas	Mole fraction of methane in liquid, x_{CH_4}	in gas, y_{CH_4}
310.92	31.03	0.2408	0.7970	0.7378	0.97209
	32.75	0.2720	0.7490	0.7682	0.9636
	34.47	0.3195	0.6880	0.8064	0.95137
	36.20	0.4110	0.5910	0.8609	0.92763
344.26	0.14	0.00070	0.8414	0.00618	0.9792
	0.28	0.00142	0.9110	0.01246	0.9891
	0.41	0.00213	0.9364	0.01858	0.9924
	0.55	0.00285	0.9496	0.02473	0.9941
	0.69	0.00357	0.9577	0.03080	0.9950
	1.38	0.00714	0.9742	0.05997	0.9970
	2.76	0.01429	0.9823	0.11395	0.9980
	4.14	0.02148	0.9840	0.16299	0.9982
	5.52	0.02875	0.9840	0.20798	0.9982
	6.89	0.03615	0.9827	0.24965	0.9980
	8.62	0.04574	0.9800	0.29835	0.9977
	10.34	0.05558	0.9770	0.34300	0.9974
	12.07	0.06585	0.9732	0.38474	0.9969
	13.79	0.07647	0.9682	0.42348	0.9963
	15.51	0.08750	0.9625	0.45965	0.9956
	17.24	0.09926	0.9533	0.49433	0.9945
	18.96	0.1117	0.9426	0.5273	0.9932
	20.68	0.1252	0.9295	0.5594	0.9915
	22.41	0.1400	0.9115	0.5909	0.9892
	24.13	0.1555	0.8900	0.6203	0.9863
	25.86	0.1730	0.8640	0.6498	0.9826
	27.58	0.1930	0.8330	0.6796	0.9779
	29.30	0.2169	0.7970	0.7107	0.9721
	31.03	0.2469	0.7550	0.7441	0.9647
	32.75	0.2850	0.7010	0.7795	0.9541
	34.47	0.3455	0.6240	0.8340	0.9364
377.59	0.14	0.00059	0.5554	0.00521	0.9172
	0.28	0.00124	0.7151	0.01089	0.9570
	0.41	0.00189	0.7864	0.01652	0.9703
	0.55	0.00253	0.8267	0.02201	0.9769
	0.69	0.00318	0.8538	0.02752	0.9811
	1.38	0.00643	0.9103	0.05429	0.9890
	2.76	0.01299	0.9421	0.10454	0.9931
	4.14	0.01966	0.9505	0.15103	0.9942
	5.52	0.02651	0.9536	0.19457	0.9945
	6.89	0.03352	0.9547	0.23528	0.9947
	8.62	0.04251	0.9530	0.28256	0.9945
	10.34	0.05205	0.9500	0.32754	0.9941
	12.07	0.06210	0.9432	0.37002	0.9933
	13.79	0.07241	0.9338	0.40915	0.9938
	15.51	0.08320	0.9230	0.44600	0.9907
	17.24	0.09440	0.9095	0.48044	0.9889
	18.96	0.1065	0.8935	0.5139	0.9867
	20.68	0.1195	0.8750	0.5463	0.8750
	22.41	0.1345	0.8530	0.5796	0.9809
	24.13	0.1512	0.8270	0.6124	0.9770
	25.86	0.1695	0.7941	0.6442	0.9716
	27.58	0.1920	0.7605	0.6782	0.9657
	29.30	0.2205	0.7182	0.7150	0.9576
	31.03	0.2579	0.6668	0.7551	0.9467
	32.75	0.3195	0.5900	0.8064	0.9274
410.93	0.14	0.00044	0.2414	0.00389	0.7384
	0.28	0.00104	0.4202	0.00915	0.8654
	0.41	0.00163	0.5245	0.01428	0.9073

(cont.)

COMPONENTS:		ORIGINAL MEASUREMENTS:

1. Methane; CH_4; [74-82-8]

2. Decane; $C_{10}H_{22}$; [124-18-5]

Reamer, H. H.; Olds, R. H.;
Sage, B. H.; Lacey, W. N.

Ind. Eng. Chem.

<u>1942</u>, *34*, 1526-1531.

EXPERIMENTAL VALUES:

T/K	P/MPa	Wt-fraction of methane in liquid	in gas	Mole fraction of methane in liquid, x_{CH_4}	in gas, y_{CH_4}
410.93	0.55	0.00223	0.5932	0.01944	0.9282
	0.69	0.00283	0.6429	0.02456	0.9411
	1.38	0.00585	0.7647	0.04961	0.9665
	2.76	0.01202	0.8454	0.09741	0.9798
	4.14	0.01839	0.8751	0.14251	0.9842
	5.52	0.02499	0.8893	0.18525	0.9862
	6.89	0.03180	0.8954	0.22562	0.9870
	8.62	0.04077	0.8980	0.27380	0.9874
	10.34	0.05018	0.8955	0.31911	0.9870
	12.07	0.06020	0.8878	0.36234	0.9860
	13.79	0.07066	0.8772	0.40280	0.9845
	15.51	0.08173	0.8596	0.44120	0.9819
	17.24	0.09355	0.8384	0.47795	0.9787
	18.96	0.1063	0.8167	0.5134	0.9753
	20.68	0.1201	0.7933	0.5477	0.9715
	22.41	0.1360	0.7670	0.5827	0.9669
	24.13	0.1536	0.7342	0.6168	0.9608
	25.86	0.1746	0.6960	0.6524	0.9531
	27.58	0.2015	0.6510	0.6912	0.9430
	29.30	0.2400	0.5920	0.7369	0.9279
	31.03	0.3120	0.4900	0.8009	0.8950
444.26	0.14	0.00019	0.0501	0.00168	0.3187
	0.28	0.00076	0.1727	0.00670	0.6493
	0.41	0.00133	0.2625	0.01168	0.7595
	0.55	0.00191	0.3324	0.01669	0.8154
	0.69	0.00249	0.3878	0.02166	0.8489
	1.38	0.00541	0.5483	0.04603	0.9150
	2.76	0.01146	0.6868	0.09325	0.9511
	4.14	0.01773	0.7441	0.13802	0.9627
	5.52	0.02427	0.7693	0.18077	0.9673
	6.89	0.03105	0.7804	0.22135	0.9693
	8.62	0.04000	0.7861	0.26987	0.9702
	10.34	0.04957	0.7843	0.31632	0.9699
	12.07	0.05977	0.7751	0.36058	0.9683
	13.79	0.07059	0.7600	0.40254	0.9656
	15.51	0.08241	0.7416	0.44343	0.9622
	17.24	0.09520	0.7200	0.4828	0.9580
	18.96	0.1098	0.6988	0.5225	0.9532
	20.68	0.1258	0.6712	0.5607	0.9477
	22.41	0.1449	0.6389	0.6005	0.9401
	24.13	0.1691	0.6093	0.6435	0.9326
	25.86	0.1955	0.5405	0.6831	0.9125
	27.58	0.2390	0.4580	0.7359	0.8823
477.59	0.28	0.00026	0.0295	0.00230	0.2124
	0.41	0.00085	0.0891	0.00749	0.4646
	0.55	0.00145	0.1399	0.01272	0.5907
	0.69	0.00205	0.1837	0.01790	0.6663
	1.38	0.00515	0.3333	0.04391	0.8160
	2.76	0.01138	0.4873	0.09265	0.8940
	4.14	0.01788	0.5609	0.13904	0.9189
	5.52	0.02472	0.6019	0.18357	0.9306
	6.89	0.03183	0.6215	0.22579	0.9358
	8.62	0.04117	0.6308	0.27583	0.9381
	10.34	0.05113	0.6340	0.32342	0.9389
	12.07	0.06155	0.6301	0.36782	0.9379
	13.79	0.07320	0.6189	0.41199	0.9351
	15.51	0.08620	0.6000	0.45558	0.9301
	17.24	0.1007	0.5736	0.49833	0.9227

(cont.)

COMPONENTS: ORIGINAL MEASUREMENTS:

  1. Methane; CH₄; [74-82-8]   Reamer, H. H.; Olds, R. H.;
                                     Sage, B. H.; Lacey, W. N.
  2. Decane; C₁₀H₂₂; [124-18-5]   *Ind. Eng. Chem.*

                                     <u>1942</u>, *34*, 1526-1531.

EXPERIMENTAL VALUES:

T/K	P/MPa	Wt-fraction of methane in liquid	in gas	Mole fraction of methane in liquid, x_{CH_4}	in gas, y_{CH_4}
477.59	18.96	0.1180	0.5440	0.54271	0.9137
	20.68	0.1402	0.5122	0.59125	0.9031
	22.41	0.1711	0.4640	0.64678	0.8848
	24.13	0.2260	0.3930	0.72147	0.8517
510.93	0.14	0.00052	0.0246	0.00459	0.1828
	0.28	0.00118	0.0534	0.01037	0.3335
	0.41	0.00458	0.1675	0.03922	0.6409
	0.55	0.01152	0.3015	0.09370	0.7929
	0.69	0.01880	0.3840	0.14528	0.8469
	1.38	0.02630	0.4300	0.19329	0.8700
	2.76	0.03398	0.4532	0.23783	0.8803
	4.14	0.04392	0.4665	0.28953	0.8858
	5.52	0.05430	0.4700	0.33746	0.8872
	6.89	0.06560	0.4672	0.38378	0.8861
	8.62	0.07780	0.4558	0.42804	0.8814
	10.34	0.09300	0.4328	0.47633	0.8713
	12.07	0.1160	0.3960	0.53791	0.8533
	13.79	0.1530	0.3400	0.61574	0.8205

COMPONENTS:	ORIGINAL MEASUREMENTS:
1. Methane; CH_4; [74-82-8] 2. Decane; $C_{10}H_{22}$; [124-18-5]	Koonce, K. T.; Kobayashi, R. *J. Chem. Eng. Data* *1964*, *9*, 490-494.

VARIABLES:	PREPARED BY:
Temperature, pressure	C. L. Young

EXPERIMENTAL VALUES:

T/K	$P/10^5$Pa	Mole fraction of methane in liquid, x_{CH_4}
277.59	17.34	0.0947
	22.66	0.1234
	28.27	0.1498
	37.70	0.1923
	52.19	0.2525
	63.82	0.2901
266.48	15.66	0.0920
	21.45	0.1246
	34.43	0.1928
	46.28	0.2433
	66.87	0.3185
255.37	15.65	0.0982
	22.02	0.1356
	29.93	0.1783
	41.69	0.2402
	63.19	0.3248
244.26	16.00	0.1103
	21.19	0.1453
	31.88	0.2061
	44.20	0.2640
	69.00	0.3641

AUXILIARY INFORMATION

METHOD/APPARATUS/PROCEDURE:	SOURCE AND PURITY OF MATERIALS:
Non-magnetic stainless steel equilibrium vessel, contents stirred with magnetically operated ball bearing. Pressure measured using dead weight piston gauge. Decane metered into a known amount of methane in cell. Pressure measured after equilibrium established. Details in source.	1. Dried, purity 99.7 mole per cent, 0.2 mole per cent nitrogen. 2. Phillips Petroleum sample, purity 99.35 mole per cent.
	ESTIMATED ERROR: $\delta T/K = \pm0.056$; $\delta P = \pm0.1$-0.15%; $\delta x_{CH_4} = \pm2\%$ (estimated by compiler).
	REFERENCES:

COMPONENTS:	ORIGINAL MEASUREMENTS:
1. Methane; CH_4; [74-82-8] 2. Decane; $C_{10}H_{22}$; [124-18-5]	Beaudoin, J.M; Kohn, J.P. *J. Chem. Engng. Data,* <u>1967</u>, *12,* 189-191

VARIABLES:	PREPARED BY:
Temperature, pressure	C.L. Young

EXPERIMENTAL VALUES:

		Mole fraction of methane	
T/K	P/MPa	in liquid, x_{CH_4}	in vapor, y_{CH_4}
423.15	1.01	0.0324	0.926
	2.03	0.0664	0.964
	3.04	0.0990	0.973
	4.05	0.1311	0.978
	5.07	0.1631	0.980
	6.08	0.1935	0.982
	7.09	0.2214	0.983
373.15	1.01	0.0372	0.988
	2.03	0.0735	0.994
	3.04	0.1080	0.996
	4.05	0.1417	0.996
	5.07	0.1730	0.997
	6.08	0.2022	0.996
	7.09	0.2298	0.996
	8.11	0.2542	0.996
	9.12	0.2766	0.996
	10.13	0.2989	0.996
348.15	1.01	0.0412	0.998
	2.03	0.0789	0.998
	3.04	0.1155	0.998
	4.05	0.1498	0.998
	5.07	0.1829	0.998
	6.08	0.2153	0.998
	7.09	0.2430	0.998

AUXILIARY INFORMATION

METHOD/APPARATUS/PROCEDURE:	SOURCE AND PURITY OF MATERIALS:
Borosilicate glass cell. Temperature measured with platinum resistance thermometer. Pressure measured on Bourdon gauge. Details in ref. (2). Samples of methane added to decane, equilibrated, vapor phase composition calculated assuming ideal gas behaviour liquid phase composition estimated from known overall composition and volumes of both phases.	1. Phillips Petroleum Co. sample, purified as in ref. (1).; final purity 99.5 mole per cent. 2. Phillips Petroleum Co. sample purity 99 mole per cent.

ESTIMATED ERROR:
$\delta T/K = \pm 0.07$; $\delta P/MPa = \pm 0.01$; δx_{CH_4}, $\delta y_{CH_4} = \pm 0.0014$.

REFERENCES:
1. Kohn, J.P.;   *J. Am. Inst. Chem. Engrs. J.*   <u>1961</u>, *7*, 514.   2. Kohn, J.P. Kurata, F.;   *Petrol Process.,* <u>1956</u>, *11*, 57.

COMPONENTS: ORIGINAL MEASUREMENTS:

1. Methane; CH_4; [74-82-8]   Beaudoin, J.M.; Kohn, J.P.

2. Decane; $C_{10}H_{22}$; [124-18-5]   *J. Chem. Engng. Data*, <u>1967</u>, *12*,
                                            189-191

EXPERIMENTAL VALUES:

| | | Mole fraction of methane | |
T/K	P/MPa	in liquid x_{CH_4}	in vapor y_{CH_4}
348.15	8.11	0.2679	0.998
	9.12	0.2920	0.998
	10.13	0.3152	0.998
323.15	1.01	0.0450	–
	2.03	0.0867	–
	3.04	0.1259	–
	4.05	0.1622	–
	5.07	0.1968	–
	6.08	0.2291	–
	7.09	0.2569	–
	8.11	0.2822	–
	9.12	0.3082	–
	10.13	0.3344	–
298.15	1.01	0.0486	–
	2.03	0.0951	–
	3.04	0.1379	–
	4.05	0.1767	–
	5.07	0.2120	–
	6.08	0.2443	–
	7.09	0.2748	–
	8.11	0.3040	–
	9.12	0.3330	–
	10.13	0.3610	–
273.15	1.01	0.0560	–
	2.03	0.1086	–
	3.04	0.1553	–
	4.05	0.1991	–
	5.07	0.2388	–
	6.08	0.2763	–
	7.09	0.3120	–
	8.11	0.3443	–
	9.12	0.3741	–
	10.13	0.4040	–
248.15	1.01	0.0702	–
	2.03	0.1330	–
	3.04	0.1901	–
	4.05	0.2408	–
	5.07	0.2850	–
	6.08	0.3256	–
	7.09	0.3635	–
	8.11	0.4000	–
	9.12	0.4350	–
	10.13	0.4708	–

M—M

COMPONENTS:	ORIGINAL MEASUREMENTS:
1. Methane; CH_4; [74-82-8] 2. Decane; $C_{10}H_{22}$; [124-18-5]	Lin, H-M.; Sebastian, H.M.; Simnick, J.J.; Chao, K-C. *J. Chem. Engng. Data*, 1979, *24*, 146-9.
VARIABLES: Temperature, pressure	PREPARED BY: C. L. Young

EXPERIMENTAL VALUES:

T/K	p/atm.	p/MPa	Mole fraction of methane in liquid, x_{CH_4}	in gas, y_{CH_4}
423.2	30	3.04	0.1075	0.9738
	40	4.05	0.1375	0.9780
	50	5.07	0.1722	0.9801
	60	6.08	0.2035	0.9811
	70	7.09	0.2309	0.9819
511.0	27.2	2.76	0.0914	0.8029
	54.4	5.51	0.1866	0.8725
	85.1	8.62	0.2853	0.8912
	119.1	12.07	0.3855	0.8911
	153.1	15.51	0.4840	0.8737
	170.1	17.24	0.5430	0.8563
	184.4	18.68	0.5946	0.8318
542.8	30.10	3.050	0.0946	0.6795
	50.31	5.098	0.1706	0.7638
	100.05	10.138	0.3508	0.8051
	125.02	12.668	0.4440	0.7901
	149.45	15.143	0.6682	0.7116

AUXILIARY INFORMATION

METHOD/APPARATUS/PROCEDURE:	SOURCE AND PURITY OF MATERIALS:
Flow apparatus with both liquid and gas components continually passing into a mixing tube and then into a cell in which phases separated under gravity. Liquid sample removed from bottom of cell and vapor sample from top of cell. Composition determined by gas chromatography. Details in source and ref. (1).	1. Matheson sample with purity better than 99 mole per cent. 2. Aldrich Chemical Co. sample purity better than 99 mole per cent.
	ESTIMATED ERROR: $\delta T/K = \pm 0.2$; $\delta p/MPa \leqslant \pm 0.03$; δx_{CH_4}, $\delta y_{CH_4} = \pm 2\%$.
	REFERENCES: 1. Simnick, J.J.; Lawson, C.C.; Lin, H-M.; Chao, K-C.; *Am. Inst. Chem. Engnrs. J.*, 1977, *23*, 469.

COMPONENTS:	ORIGINAL MEASUREMENTS:
1. Methane; CH_4; [74-82-8] 2. Decane; $C_{10}H_{22}$; [124-18-5]	Lin, H.-M.; Sebastian, H. M.; Simnick, J. J.; Chao, K.-C. *J. Chem. Engng. Data* <u>1979</u>, *24*, 146-9.

T/K	p/atm.	p/MPa	Mole fraction of methane in liquid, x_{CH_4}	in gas, y_{CH_4}
563.3	29.97	3.037	0.0911	0.5528
	50.04	5.070	0.1744	0.6690
	74.88	7.587	0.2744	0.7118
	99.65	10.097	0.3817	0.7055
	109.99	11.145	0.4399	0.6835
	114.62	11.614	0.4652	0.6604
583.1	30.24	3.064	0.0857	0.4133
	50.05	5.071	0.1794	0.5476
	70.25	7.118	0.2834	0.5749
	79.78	8.084	0.3481	0.5646
	85.23	8.636	0.4032	0.5177

COMPONENTS:	EVALUATOR:
1. Methane; CH_4; [74-82-8] 2. Alkanes ($C_{16}-C_{32}$)	Colin L. Young Department of Physical Chemistry, University of Melbourne. Parkville, Victoria, 3052 Australia. February 1986.

CRITICAL EVALUATION:

Hexadecane; $C_{16}H_{34}$; [544-76-3]

There is surprising little experimental data for the systems methane + higher alkanes at elevated pressures. For the system methane + hexadecane the data of Sultanov et al. (1) are substantially in agreement with the data of Lin et al. (2) for the liquid phase compositions in the overlapping range of temperature but there are significant discrepancies between the vapor phase compositions. The vapor compositions for the Lin et al data are probably more precise in the lower pressure region and are also probably more accurate. The data of Glaser et al.(3) cover a different temperature range and detailed comparison between the data and those of ref. (1) and (2) is impossible. Glaser et al. data which is classified as tentative can be used to derive pressure-composition sections as given in the table below.

T/K	290.0	300.0	310.0	320.0	330.0	340.0	350.0	360.0
				p/MPa				
0.977		63.075	61.087	59.352	57.727	56.211	54.804	
0.942		66.714	64.566	62.702	61.051	59.491	58.033	56.685
0.952		67.859	65.680	63.803	62.104	60.556	59.104	57.759
0.927	70.346	67.848	65.709	63.849	62.208	60.681	59.313	
0.887		66.999	64.960	63.204	61.659	60.241	58.939	
0.824	59.082	57.485	56.267	55.081	54.111	53.275	52.374	51.604
0.703		36.769	36.619	36.515	36.448	36.372	36.310	36.198
0.600		23.310	23.608	24.029	24.378	24.593	24.839	25.060
0.497		16.026	16.497	16.691	17.344	17.707		
0.342		8.548	8.863	9.183	9.470	9.736	9.988	
0.295	6.478	6.856	7.130	7.406	7.650	7.888	8.071	
0.184		4.141	4.313	4.490	4.632	4.770	4.905	
0.889		2.151	2.245	2.335	2.418	2.500	2.568	

Eicosane; $C_{20}H_{42}$; [112-95-8]

The system eicosane + methane has been investigated by Puri and Kohn (4). The data when extrapolated to 1 atmosphere pressure gives a solubility which is about 15 per cent lower than that which would be expected from the data of Chappelow and Prausnitz (5) and therefore should be regarded with caution.

Dotriacontane; $C_{32}H_{66}$; [544-85-4]

The system methane + dotriacontane has been studied by Cordeiro et al.(6) The solubility extrapolated to 1 atmosphere pressure is about 5 per cent smaller than might be expected from extrapolation from the data of Lin and Parcher (7) on other similar alkanes. Within the combined uncertainty of the two extrapolations this agreement is satisfactory and the data classified as tentative.

References.
1. Lin, H.-M.; Sebastian, H. M.; Chao, K.-C.
   J. Chem. Eng. Data, 1980, 25, 252-257.
2. Sultanov, R. G.; Skripka, V. G.; Namoit, A. Yu.
   Gazov. Delo. 1972, 10, 43.
3. Glaser, M.; Peters, C. J.; Van der Kooi, H. J.; Lichtenthaler, R. N.;
   J. Chem. Thermodyn., 1985, 17, 803.
4. Puri, S.; Kohn, J. P.;
   J. Chem. Eng. Data, 1970, 15, 372.
5. Chappelow, C. C.; Prausnitz, J. M.;
   Am. Inst. Chem. Engnrs. J., 1974, 20, 1097.
6. Cordeiro, D. J.; Luks, K. D.; Kohn, J. P.;
   Ind. Eng. Chem. Process. Des. Develop. 1973, 12, 47.
7. Lin, P. J.; Parcher, J. F.;
   J. Chromatog. Sci., 1982, 20, 33.

COMPONENTS:	ORIGINAL MEASUREMENTS:
1. Methane; CH_4; [74-82-8] 2. Hexadecane; $C_{16}H_{34}$; [544-76-3]	Sultanov, R. G.; Skripka, V. G.; Namiot, A. Yu. *Gazov. Delo.* <u>1972</u>, *10*, 43-6.
VARIABLES: Temperature, pressure	PREPARED BY: C. L. Young

EXPERIMENTAL VALUES:

T/K	P/MPa	Mole fraction of methane in liquid, x_{CH_4}	in gas, y_{CH_4}	T/K	P/MPa	Mole fraction of methane in liquid, x_{CH_4}	in gas, y_{CH_4}
373.15	4.9	0.181	–	473.15	29.4	0.6370	0.9860
	9.8	0.321	–		39.2	0.7700	0.9725
	19.6	0.508	–		44.1	0.8600	0.9360
	29.4	0.637	–		45.1	0.9020	0.9020
	39.2	0.733	0.9990	523.15	4.9	0.1840	0.8400
	49.0	0.8125	0.9950		9.8	0.3180	0.8880
	53.9	0.8575	0.9835		19.6	0.5240	0.8810
	56.4	0.8980	0.9770		29.4	0.6800	0.9540
423.15	4.9	0.1750	0.9992		36.8	0.8070	0.9160
	9.8	0.3075	0.9980		38.2	0.8620	0.8620
	19.6	0.4940	0.9975	573.15	4.9	0.1880	0.9530
	29.4	0.6250	0.9920		9.8	0.3250	0.9620
	39.2	0.7360	0.9865		19.6	0.0555(+)	0.9550
	49.0	0.8560	0.9700		29.4	0.7550	0.8820
	51.6	0.8950	0.9450		30.4	0.8200	0.8200
	52.4	0.9275	0.9275	623.15	4.9	0.2030	0.9890
473.15	4.9	0.1760	0.9960		9.8	0.3640	0.8900
	9.8	0.3210	0.9960		19.6	0.6640	0.9862
	19.6	0.5000	0.9930		21.2	0.7825	0.7825

+ This composition is obviously a typographical error.

AUXILIARY INFORMATION

METHOD/APPARATUS/PROCEDURE:	SOURCE AND PURITY OF MATERIALS:
Static equilibrium cell fitted with magnetic stirrer, details in ref. (1). Samples of coexisting phases analysed by freezing out hexadecane and estimating methane volumetrically.	1. Purity 99.9 mole per cent. 2. "Pure sample", boiling point 286.3-287.3 °C.
	ESTIMATED ERROR: $\delta T/K = \pm0.3$; $\delta P/MPa = \pm0.1$; δx_{CH_4}, $\delta y_{CH_4} = \pm0.002$ (estimated by compiler).
	REFERENCES: 1. Sultanov, R. G.; Skripka, V. G.; Namiot, A. Yu. *Gazov. Prom.* <u>1971</u>, *16* (4), 6.

COMPONENTS:	ORIGINAL MEASUREMENTS:
1. Methane; CH₄; [74-82-8] 2. Hexadecane; $C_{16}H_{34}$; [544-76-3]	Lin, H.-M.; Sebastian, H.M.; Chao, K.-C. *J. Chem. Engng. Data.* 1980, *25*, 252-257.

VARIABLES:	PREPARED BY:
Temperature, pressure	C.L. Young

EXPERIMENTAL VALUES:

T/K	P/atm	P/MPa	Mole fraction in liquid, x_{CH_4}	in vapor, y_{CH_4}
462.45	20.02	2.029	0.0801	0.99544
	30.71	3.112	0.1187	0.99658
	49.8	5.046	0.1824	0.99722
	100.0	10.13	0.3207	0.99718
	149.9	15.19	0.4326	0.99636
	200.9	20.36	0.5193	0.99467
	249.3	25.26	0.5958	0.99206
542.65	20.50	2.077	0.0831	0.9580
	30.23	3.063	0.1208	0.9687
	50.0	5.07	0.1884	0.9765
	99.5	10.08	0.3322	0.9808
	149.9	15.19	0.4539	0.9798
	200.6	20.33	0.5512	0.9754
	222.5	22.54	0.6229	0.9719
623.15	20.71	2.098	0.0836	0.7930
	31.39	3.181	0.1265	0.8453
	50.0	5.07	0.2032	0.8865
	99.7	10.10	0.3716	0.9132
	150.3	15.23	0.5178	0.9097
	176.1	17.84	0.5968	0.8970
	201.3	20.40	0.7371	0.8733
703.55	20.87	2.115	0.0697	0.3097
	30.77	3.118	0.1363	0.4632
	49.8	5.046	0.2822	0.5099

AUXILIARY INFORMATION

METHOD/APPARATUS/PROCEDURE:	SOURCE AND PURITY OF MATERIALS:
Flow apparatus with both liquid and gas components continually passing into a mixing tube and then into a cell in which phases separated under gravity. Liquid sample removed from bottom of cell and vapor sample from top of cell. Composition of samples found by stripping out gas and estimating amount of solute volumetrically and solvent gravimetrically. Temperature measured with thermocouple and pressure with Bourdon gauge. Details in ref. (1).	1. Matheson sample, purity better than 99 mole per cent. 2. Matheson, Coleman and Bell sample, purity better than 99 mole per cent.

ESTIMATED ERROR:

$\delta T/K = \pm 0.1$; $\delta P/MPa = \pm 0.01$; δx_{CH_4}, $\delta y_{CH_4} > \pm 1.0\%$

(estimated by compiler).

REFERENCES:
1. Simnick, J.J.; Lawson, C.C.; Lin, H.M.; Chao, K.C.

Am. Inst. Chem. Engnrs. J. 1977, *23*, 469.

COMPONENTS:	ORIGINAL MEASUREMENTS:
1. Methane; CH_4 ; [74-82-8] 2. Hexadecane; $C_{16}H_{34}$; [544-76-3]	Glaser, M.; Peters, C. J.; Van der Kooi, H. J.; Lichtenthaler, R. N. *J. Chem. Thermodyn.* 1985, *17*, 803-815.
VARIABLES: Temperature, pressure	PREPARED BY: C. L. Young

EXPERIMENTAL VALUES:

Experimentally determined two-phase boundaries
(x, mole fraction of methane)

T/K	P/MPa	T/K	P/Mpa	T/K	P/Mpa	T/K	P/Mpa
			$x = 0.977$ (l+g)				
293.20	64.56	307.35	61.58	322.08	59.00	336.62	56.70
297.49	63.60	312.30	60.68	326.95	58.20	341.65	55.98
302.62	62.54	317.15	59.84	331.86	57.44	346.49	55.30
						351.30	54.62
			$x = 0.977$ (s+g)				
285.84	70.10	285.83	72.10	285.82	74.10	285.75	80.10
						285.74	84.10
			$x = 0.962$ (l+g)				
293.27	68.36	312.30	64.12	327.01	61.52	346.62	58.44
297.91	67.22	317.26	63.18	331.92	60.74	351.33	57.88
303.42	65.92	322.25	62.32	336.40	59.98	356.28	57.20
				341.62	59.26	361.07	56.54
			$x = 0.962$ (s+g)				
286.28	72.60	286.35	76.10	286.44	80.10	286.54	83.60
						286.61	86.80
							cont.

AUXILIARY INFORMATION

METHOD/APPARATUS/PROCEDURE:	SOURCE AND PURITY OF MATERIALS:
Cailletet apparatus used up to pressures of 20 MPa and an autoclave with two sapphire windows was used at higher pressures. Both systems provided with magnetic stirring facility. Pressure measured with pressure balance and temperatures measured with platinum resistance thermometer. Details of Cailletet apparatus given in ref (1),(2) and source; details of autoclave given in ref (1),(3) and source.	1. Matheson Research Grade sample, purity checked using gas chromatography. 2. Merck sample, purity at least 99.9 mole per cent, checked by gas chromatography.

ESTIMATED ERROR:

REFERENCES:

1. Van der Kooi, H. J.; PhD. thesis, Delft. Univ. Tech., 1981.
2. De Loos, Th. W.; Van der Kooi, H. J.; Poot, W.; Ott, P. L. *Delft. Progr.Rep*, 1983, *8*, 200.
3. Van Wilie, G. S. A.; Diepen, G. A. M., *Rec.Trav.Chim.Pays Bas.*, (1961), *80*, 659.

COMPONENTS:	ORIGINAL MEASUREMENTS:
1. Methane; CH_4 ; [74-82-8] 2. Hexadecane; $C_{16}H_{34}$;[544-76-3]	Glaser, M.; Peters, C. J.; Van der Kooi, H. J.; Lichtenthaler, R. N. *J. Chem. Thermodyn.* <u>1985</u>, *17*, 803-815.

EXPERIMENTAL VALUES:

T/K	P/MPa	T/K	P/MPa	T/K	P/MPa	T/K	P/MPa

$x = 0.952$ (l+g)

T/K	P/MPa	T/K	P/MPa	T/K	P/MPa	T/K	P/MPa
292.16	69.82	312.34	65.22	391.92	61.80	351.41	58.91
297.04	68.57	317.27	64.39	336.91	61.02	356.72	58.20
301.81	67.44	322.19	63.42	341.60	60.32	361.47	57.56
307.45	66.20	327,07	62.58	346.44	59.61		

$x = 0.952$ (s+g)

286.23	72.90	286.33	75.60	286.53	80.10

$x = 0.927$ (l+g)

289.05	70.58	307.98	66.12	327.51	62.60	347.05	59.72
292.82	69.64	312.78	65.16	332.42	61.84	351.84	59.06
297.68	68.40	317.61	64.26	337.33	61.10	356.78	58.44
303.03	67.16	322.63	63.41	341.88	60.40		

$x = 0.927$ (s+g)

286.43	72.60	286.50	73.10	286.56	74.10	286.69	76.40
						286.83	78.10

$x = 0.887$ (l+g)

292.23	68.84	307.59	65.42	322.21	62.84	336.96	60.66
297.01	67.68	312.40	64.52	327.10	62.08	341.72	60.01
302.58	66.44	317.31	63.66	331.95	61.38	346.69	59.36
						351.42	58.76

$x = 0.887$ (s+l)

286.98	76.60	287.00	76.90	287.17	78.20	287.41	80.90
				287.29	79.70	287.64	83.30

$x = 0.824$ (l+g)

287.74	59.48	307.49	56.60	327.13	54.38	346.54	52.68
295.22	58.20	312.34	55.95	331.83	53.94	351.29	52.27
297.47	57.85	317.25	55.38	336.85	53.50	356.11	51.90
302.62	57.15	322.21	54.85	341.99	53.12	361.36	51.56

$x = 0.824$ (s+l)

285.84	62.10	286.11	64.00	286.58	67.50	286.84	71.00

$x = 0.703$ (l+g)

292.72	36.94	312.08	36.59	331.68	36.44	351.95	36.30
297.49	36.82	316.99	36.54	336.58	36.40	356.83	36.24
302.31	36.73	321.82	36.50	341.45	36.36	361.44	36.18
307.21	36.66	326.79	36.46	346.36	36.32		

$x = 0.703$ (s+l)

285.26	40.10	285.62	42.10	285.78	43.00	286.15	45.00

$x = 0.600$ (l+g)

293.19	23.10	312.17	23.76	331.91	24.44	352.39	24.90
297.90	23.22	317.18	23.94	336.77	24.52	354.44	24.98
302.66	23.40	322.11	24.10	341.63	24.64	361.46	25.06
307.58	23.46	326.96	24.26	346.56	24.78		

cont.

COMPONENTS:	ORIGINAL MEASUREMENTS:
1. Methane; CH_4 ; [74-82-8] 2. Hexadecane; $C_{16}H_{34}$; [544-76-3]	Glaser, M.; Peters, C. J.; Van der Kooi, H. J.; Lichtenthaler, R. N. *J. Chem. Thermodyn.* <u>1985</u>, *17*, 803-815.

EXPERIMENTAL VALUES:

T/K	P/MPa	T/K	P/MPa	T/K	P/MPa	T/K	P/MPa

$x = 0.600$ (s+l)

T/K	P/MPa	T/K	P/MPa	T/K	P/MPa	T/K	P/MPa
286.61	26.30	285.92	27.80	286.03	28.10	286.32	29.30

$x = 0.497$ (l+g)

T/K	P/MPa	T/K	P/MPa	T/K	P/MPa	T/K	P/MPa
293.32	15.69	308.05	16.41	322.77	17.08	337.36	17.61
297.93	15.92	312.94	16.63	327.64	17.26	342.36	17.79
303.03	16.18	317.83	16.86	332.38	17.43	347.37	17.94

$x = 0.497$ (s+l)

T/K	P/MPa	T/K	P/MPa	T/K	P/MPa	T/K	P/MPa
285.86	16.24	285.97	16.63	286.14	17.51	286.26	17.90
						286.37	18.39

$x = 0.342$ (l+g)

T/K	P/MPa	T/K	P/MPa	T/K	P/MPa	T/K	P/MPa
290.89	8.19	307.92	8.79	327.70	9.41	347.33	9.92
293.43	8.29	313.02	8.97	332.68	9.54	352.61	10.05
298.09	8.48	317.65	9.11	337.47	9.67	357.38	10.14
303.26	8.65	322.76	9.27	342.53	9.80		

$x = 0.296$ (l+g)

T/K	P/MPa	T/K	P/MPa	T/K	P/MPa	T/K	P/MPa
288.86	6.48	303.24	6.94	322.78	7.48	342.52	7.94
290.56	6.55	308.05	7.07	327.76	7.60	347.53	8.02
293.27	6.63	312.98	7.22	332.68	7.71	352.47	8.13
298.07	6.79	317.90	7.35	337.59	7.83		

$x = 0.296$ (s+l)

T/K	P/MPa	T/K	P/MPa	T/K	P/MPa	T/K	P/MPa
288.21	7.46	288.48	8.58	288.74	9.76	289.37	12.21
				288.86	10.15	289.47	12.56

$x = 0.184$ (l+g)

T/K	P/MPa	T/K	P/MPa	T/K	P/MPa	T/K	P/MPa
298.17	4.106	313.09	4.366	327.74	4.597	342.77	4.805
303.52	4.205	318.41	4.465	332.71	4.675	347.64	4.877
308.13	4.283	323.12	4.533	337.65	4.742	352.56	4.935
						356.33	4.994

$x = 0.184$ (s+l)

T/K	P/MPa	T/K	P/MPa	T/K	P/MPa	T/K	P/MPa
290.21	8.441	290.41	9.317	290.69	10.500	291.31	12.954
				291.01	11.771	291.60	13.984

$x = 0.111$ (l+g)

T/K	P/MPa	T/K	P/MPa	T/K	P/MPa	T/K	P/MPa
290.27	2.061	308.20	2.228	328.23	2.399	343.26	2.522
292.17	2.087	313.21	2.277	333.26	2.453	348.24	2.556
293.22	2.093	318.20	2.320	338.25	2.488	363.23	2.659
303.25	2.184	323.27	2.360				

$x = 0.111$ (s+l)

T/K	P/MPa	T/K	P/MPa	T/K	P/MPa	T/K	P/MPa
290.27	3.189	290.68	4.955	291.52	8.339	292.22	11.086
290.67	4.906	291.07	6.574	291.93	9.958	293.38	11.920

COMPONENTS:	ORIGINAL MEASUREMENTS:
1. Methane; CH_4; [74-82-8] 2. Eicosane; $C_{20}H_{42}$; [112-95-8]	Puri, S.; Kohn, J. P.; *J. Chem. Eng. Data,* 1970, *15,* 372-374.

VARIABLES:
 	
   Pressure
	PREPARED BY:
	
	   C. L. Young

EXPERIMENTAL VALUES:

T/K	P/MPa	Mole fraction of methane, x
313.15	0.51	0.0179
	1.01	0.0510
	1.52	0.0785
	2.03	0.0785
	2.53	0.1268
	3.04	0.1490
	3.55	0.1701
	4.05	0.1910
	4.56	0.2108
	5.07	0.2300
	5.57	0.2489
	6.08	0.2677
solid-liquid-gas equilibrium		
319.15	0.684	0.0348
308.95	0.897	0.0465
308.75	1.177	0.0620
308.55	1.479	0.0775
308.35	1.864	0.0961
308.15	2.260	0.1145
307.95	2.746	0.1360
307.75	3.294	0.1592
307.55	3.921	0.1860
307.35	4.603	0.2142
307.15	5.352	0.2450

AUXILIARY INFORMATION

METHOD/APPARATUS/PROCEDURE:	SOURCE AND PURITY OF MATERIALS:

Borosilicate glass cell. Tempera-ture measured with platinum resist-ance thermometer. Pressure measured with Bourdon gauge. Samples of methane added to eicosane and equilibriated. Liquid phase composition estimated fromn known overall composition and volume of both phases. Details in ref. (1).	1. Matheson pure grade, further
	   purified by distillation and
	   absorption. Final purity better
	   than 99.5 mole per cent.
	2. Humphrey Wilkinson sample,
	   minimum purity 99 mole per cent

	ESTIMATED ERROR:
	        $\partial T/K = \pm0.25$; $\partial P/MPa = \pm0.05$;
	        ∂x = ±0.002

	REFERENCES:
	
	   1. Lee, K. H.; Kohn, J. P.;
	      J. Chem. Eng. Data, 1969, *14*
	      292.

COMPONENTS:	ORIGINAL MEASUREMENTS:
1. Methane; CH_4; [74-82-8] 2. Dotriacontane; $C_{32}H_{66}$; [544-85-4]	Cordeiro, D. J.; Luks, K. D.; Kohn, J. P. *Ind. Eng. Chem. Process. Des. Develop.* <u>1973</u>, *12*, 47-51.
VARIABLES: Temperature, pressure	PREPARED BY: C. L. Young

EXPERIMENTAL VALUES:

T/K	P/MPa	Mole fraction of methane in liquid, x_{CH_4}
343.15	1.585	0.1000
	2.025	0.1250
	2.475	0.1500
	2.960	0.1750
	3.480	0.2000
	4.040	0.2250
	4.630	0.2500
	5.250	0.2750
	5.900	0.3000
	6.550	0.3250

Molar volume data in source.

AUXILIARY INFORMATION

METHOD/APPARATUS/PROCEDURE:	SOURCE AND PURITY OF MATERIALS:
A known amount of gas added to a known amount of solvent in a 10 cm^3 glass equilibrium cell. Liquid phase composition determined from overall composition and volume of both phases. Details in source.	1. Phillips Petroleum pure grade samples, purity better than 99 mole per cent. 2. Humphrey Chemical Co. sample, minimum purity 97 mole per cent.
	ESTIMATED ERROR: $\delta T/K = \pm 0.02$; $\delta P/MPa = \pm 0.007$; $\delta x_{CH_4} = \pm 0.001$ (estimated by compiler).
	REFERENCES:

COMPONENTS:	ORIGINAL MEASUREMENTS:
1. Methane; CH₄; [74-82-8]	Price, A. R.; Kobayashi, R.
2. Ethane; C₂H₆; [74-84-0]	*J. Chem. Engng. Data*
3. Propane; C₃H₈; [74-98-6]	1959, *4*, 40-52.

VARIABLES:	PREPARED BY:
	C. L. Young

EXPERIMENTAL VALUES:

Mole fractions

T/K (T/°F)	P/MPa	P/psi	x_{CH_4}	$x_{C_2H_6}$	$x_{C_3H_8}$	y_{CH_4}	$y_{C_2H_6}$	$y_{C_3H_8}$
			in liquid			in vapor		
283.15 (50)	0.689	100	0.0028		0.998	0.057		0.943
			0.0022	0.0080	0.9897	0.0452	0.0280	0.9268
				0.0236	0.9764		0.0720	0.9280
	1.379	200	0.044		0.960	0.471		0.529
			0.044		0.962	0.471		0.529
			0.0396	0.0655	0.8966	0.412	0.112	0.476
			0.0270	0.160	0.811	0.275	0.284	0.441
			0.0128	0.264	0.7193	0.127	0.476	0.397
				0.357	0.643		0.612	0.388
	2.758	400	0.128		0.872	0.685		0.315
			0.110	0.175	0.715	0.557	0.193	0.250
			0.0697	0.500	0.4303	0.306	0.535	0.159
			0.0620	0.553	0.385	0.282	0.570	0.148
			0.0302	0.7578	0.212	0.115	0.7989	0.0861
				0.9066	0.0934		0.9616	0.0384
				0.9108	0.0892		0.9606	0.0394
	4.137	600	0.216		0.784	0.762		0.238
			0.214	0.151	0.635	0.663	0.133	0.204
			0.204	0.178	0.618	0.654	0.150	0.196
			0.178	0.376	0.446	0.521	0.332	0.147
			0.174	0.457	0.369	0.487	0.382	0.131
			0.137	0.647	0.216	0.371	0.5501	0.0789
	(cont.)		0.113	0.768	0.119	0.286	0.6668	0.0472

AUXILIARY INFORMATION

METHOD/APPARATUS/PROCEDURE:	SOURCE AND PURITY OF MATERIALS:
Recirculating vapor flow apparatus with modified Jerguson sight gauge for equilibrium cell. Vapor re-cycled with magnetic pump. Pressure measured with Bourdon pressure gauge and temperature measured with thermocouple. Samples of liquid and gas analysed by infra-red spectrometry and gas chromatographic analysis.	1. Phillips Petroleum Co. research grade, purity 99.5 mole per cent.
	2. Phillips Petroleum Co. research grade, purity 99.9 mole per cent.
	3. Phillips Petroleum Co. pure grade, purity 99.0 mole per cent.
	ESTIMATED ERROR:
	$\delta T/K = \pm0.06$; $\delta P/MPa = \pm1\%$;
	δx, δy = ±2% (estimated by compiler).
	REFERENCES:

COMPONENTS:	ORIGINAL MEASUREMENTS:
1. Methane; CH_4; [74-82-8]	Price, A. R.; Kobayashi, R.
2. Ethane; C_2H_6; [74-84-0]	*J. Chem. Engng. Data*
3. Propane; C_3H_8; [74-98-6]	<u>1959</u>, *4*, 40-52.

EXPERIMENTAL VALUES:

T/K (T/°F)	P/MPa	P/psi	Mole fractions in liquid x_{CH_4}	$x_{C_2H_6}$	$x_{C_3H_8}$	in vapor y_{CH_4}	$y_{C_2H_6}$	$y_{C_3H_8}$
283.15 (50)	5.516	800	0.300		0.700	0.788		0.212
			0.286	0.121	0.593	0.7224	0.0906	0.187
			0.268	0.359	0.373	0.588	0.282	0.130
			0.255	0.454	0.291	0.534	0.357	0.109
			0.237	0.566	0.197	0.4632	0.455	0.0818
	6.895	1000	0.413		0.587	0.805		0.195
			0.389	0.169	0.442	0.707	0.126	0.167
			0.385	0.225	0.390	0.680	0.164	0.156
			0.373	0.357	0.270	0.600	0.268	0.132
			0.349	0.505	0.146	0.4826	0.431	0.0864
			0.3755	0.5241	0.1004	0.4345	0.490	0.0755
	7.584	1100	0.451		0.549	0.803		0.197
			0.4365	0.0409	0.5226	0.7746	0.0306	0.1948
			0.430	0.110	0.460	0.7341	0.0849	0.181
			0.438	0.192	0.370	0.692	0.143	0.165
			0.438	0.223	0.339	0.665	0.173	0.162
			0.435	0.259	0.306	0.630	0.209	0.161
			0.440	0.376	0.184	0.504	0.351	0.145
	8.274	1200	0.498		0.502	0.784		0.216
			0.4997	0.0375	0.4628	0.7558	0.0302	0.214
			0.504	0.176	0.320	0.673	0.143	0.184
			0.508	0.243	0.249	0.584	0.219	0.197
255.37 (0)	0.689	100	0.0358		0.9642	0.573		0.427
			0.0335	0.0174	0.9491	0.5589	0.0336	0.4075
			0.0279	0.1284	0.8437	0.400	0.246	0.354
			0.0210	0.166	0.813	0.325	0.349	0.326
			0.0107	0.272	0.7173	0.159	0.534	0.307
				0.373	0.627		0.692	0.308
	1.379	200	0.0904		0.9096	0.768		0.232
			0.0917	0.0127	0.8956	0.7717	0.0133	0.215
			0.0702	0.250	0.6798	0.560	0.270	0.170
			0.0457	0.4843	0.470	0.359	0.519	0.122
			0.0127	0.8133	0.174	0.0778	0.8746	0.0476
				0.9013	0.0987		0.9717	0.0283
	2.758	400	0.199		0.801	0.862		0.138
			0.159	0.347	0.494	0.6985	0.217	0.0845
			0.1557	0.5293	0.315	0.6003	0.343	0.0567
			0.1364	0.7243	0.1393	0.4869	0.4852	0.0279
	4.137	600	0.311		0.689	0.891		0.109
			0.293	0.266	0.441	0.781	0.1377	0.0813
			0.288	0.343	0.369	0.7514	0.184	0.0646
			0.272	0.481	0.247	0.6896	0.260	0.0504
			0.271	0.6204	0.1086	0.6138	0.361	0.0252
	5.516	800	0.415		0.585	0.899		0.101
			0.391	0.188	0.421	0.8239	0.0939	0.0822
			0.378	0.354	0.268	0.7727	0.1702	0.0517
			0.380	0.418	0.202	0.7318	0.224	0.0442
			0.380	0.5142	0.1058	0.6795	0.296	0.0245
	6.895	1000	0.522		0.478	0.895		0.105
			0.515	0.0780	0.407	0.8623	0.0443	0.0934
			0.512	0.187	0.301	0.8174	0.1027	0.0799
			0.514	0.239	0.247	0.791	0.140	0.0690
			0.5161	0.3378	0.1461	0.7234	0.226	0.0506
			0.5258	0.4467	0.0275	0.6357	0.349	0.0153

(cont.)

COMPONENTS: ORIGINAL MEASUREMENTS:

1. Methane; CH_4; [74-82-8]   Price, A. R.; Kobayashi, R.

2. Ethane; C_2H_6; [74-84-0]   *J. Chem. Engng. Data*

3. Propane; C_3H_8; [74-98-6]   <u>1959</u>, *4*, 40-52.

EXPERIMENTAL VALUES:

T/K	P/MPa	P/psi	Mole fractions in liquid			Mole fractions in vapor		
			x_{CH_4}	$x_{C_2H_6}$	$x_{C_3H_8}$	y_{CH_4}	$y_{C_2H_6}$	$y_{C_3H_8}$
255.37	7.584	1100	0.582		0.418	0.888		0.112
(0)			0.5759	0.0581	0.366	0.8619	0.0341	0.104
			0.585	0.107	0.308	0.8377	0.0635	0.0988
			0.584	0.157	0.259	0.8237	0.0928	0.0835
			0.587	0.192	0.221	0.7947	0.1231	0.0822
			0.582	0.228	0.190	0.7707	0.154	0.0753
			0.5917	0.2483	0.1600	0.7545	0.1758	0.0697
			0.6043	0.259	0.1367	0.7484	0.1855	0.0661
			0.604	0.273	0.123	0.7142	0.2177	0.0681
	8.274	1200	0.637		0.363	0.873		0.127
			0.6456	0.0874	0.267	0.8218	0.0625	0.1157
			0.653	0.106	0.241	0.8158	0.0751	0.1091
			0.650	0.118	0.232	0.795	0.089	0.116
			0.693	0.161	0.146	0.721	0.144	0.135
	8.963	1300	0.708		0.292	0.845		0.155
			0.7103	0.0297	0.260	0.8294	0.0246	0.146
			0.7315	0.0345	0.234	0.8086	0.0322	0.1592
			0.7364	0.0396	0.224	0.7978	0.0384	0.1638
235.37	0.689	100	0.0769		0.9231	0.840		0.160
(-50)			0.0614	0.1718	0.7668	0.7195	0.1628	0.1177
			0.0486	0.302	0.6494	0.605	0.292	0.103
			0.0479	0.399	0.5531	0.5287	0.380	0.0913
			0.0467	0.448	0.5053	0.4952	0.425	0.0798
			0.0261	0.7259	0.248	0.280	0.682	0.0380
	1.379	200	0.146		0.854	0.9216		0.0784
			0.125	0.268	0.607	0.804	0.140	0.0560
			0.125	0.368	0.507	0.7478	0.206	0.0462
			0.119	0.627	0.254	0.6383	0.333	0.0287
			0.0968	0.9032		0.505	0.495	
	2.758	400	0.296		0.704	0.9493		0.0507
			0.292	0.126	0.582	0.920	0.0383	0.0417
			0.271	0.289	0.440	0.8791	0.0907	0.0302
			0.274	0.448	0.278	0.8245	0.154	0.0215
			0.270	0.512	0.218	0.8084	0.1742	0.0174
			0.277	0.523	0.200	0.7986	0.1880	0.0134
			0.278	0.571	0.151	0.7901	0.194	0.0159
			0.282	0.718		0.724	0.276	
	4.137	600	0.438		0.562	0.9585		0.0415
			0.440	0.145	0.415	0.9179	0.0485	0.0336
			0.424	0.150	0.426	0.919	0.0446	0.0364
			0.443	0.243	0.314	0.8952	0.0790	0.0258
			0.441	0.366	0.193	0.8637	0.119	0.0173
			0.438	0.562		0.800	0.200	
	5.516	800	0.581		0.419	0.959		0.0410
			0.598	0.109	0.293	0.9275	0.362	0.0363
			0.611	0.249	0.140	0.8901	0.0886	0.0213
			0.622	0.378		0.832	0.168	
	6.895	1000	0.736		0.264	0.9458		0.0542
			0.7609	0.0651	0.174	0.9216	0.0331	0.0453
			0.7662	0.0892	0.1446	0.9026	0.0504	0.0470
			0.7749	0.1261	0.0990	0.8838	0.0758	0.0404
			0.7837	0.1491	0.0672	0.8631	0.1001	0.0368
			0.8332	0.1318	0.0350	0.8363	0.1288	0.0349

                                                                (cont.)

COMPONENTS:	ORIGINAL MEASUREMENTS:
1. Methane; CH_4; [74-82-8]	Price, A. R.; Kobayashi, R.
2. Ethane; C_2H_6; [74-84-0]	*J. Chem. Engng. Data*
3. Propane; C_3H_8; [74-98-6]	<u>1959</u>, *4*, 40-52.

EXPERIMENTAL VALUES:

T/K	P/MPa	P/psi	Mole fractions in liquid			in vapor		
			x_{CH_4}	$x_{C_2H_6}$	$x_{C_3H_8}$	y_{CH_4}	$y_{C_2H_6}$	$y_{C_3H_8}$
199.82	0.689	100	0.125		0.875	0.9591		0.0409
(-100)			0.118	0.144	0.738	0.9186	0.0472	0.0342
			0.110	0.320	0.570	0.8651	0.110	0.0249
			0.110	0.580	0.310	0.7916	0.194	0.0144
			0.101	0.899		0.680	0.320	
	1.379	200	0.222		0.778	0.9792		0.0208
			0.235	0.121	0.644	0.9615	0.0228	0.0157
			0.232	0.435	0.333	0.9070	0.0840	0.00897
			0.236	0.483	0.281	0.8926	0.0983	0.00909
			0.250	0.750		0.814	0.186	
	2.758	400	0.477		0.523	0.9855		
			0.492	0.175	0.333	0.963	0.0266	0.0104
			0.496	0.242	0.262	0.9562	0.0364	0.00741
			0.504	0.308	0.188	0.9446	0.0486	0.00685
			0.5191	0.386	0.0949	0.9321	0.0630	0.00486
			0.528	0.472		0.9165	0.0835	
	4.137	600	0.744		0.256	0.9852		0.0148
			0.745	0.0520	0.203	0.9775	0.0119	0.0106
			0.7647	0.1612	0.0741	0.9583	0.0366	0.00512
			0.784	0.216		0.9498	0.0502	
172.04	0.689	100	0.238		0.762	0.9932		0.0068
(-150)			0.242	0.258	0.500	0.9686	0.0282	0.0032
			0.240	0.275	0.485	0.9653	0.0308	0.0039
			0.249	0.446	0.305	0.9538	0.0433	0.0030
			0.252	0.6762	0.0718	0.9299	0.0682	0.0019
			0.251	0.749		0.9312	0.0688	
	1.379	200	0.502		0.498	0.9955		0.0045
			0.514	0.208	0.278	0.9806	0.0167	0.0027
			0.528	0.319	0.153	0.9742	0.0234	0.0024
			0.5532	0.400	0.0468	0.969	0.0294	0.0016
			0.561	0.439		0.970	0.030	
144.26	0.689	100	0.802		0.198	0.9993		0.0007
(-200)			0.8075	0.0735	0.119	0.9962	0.0021	
			0.8187	0.124	0.0573	0.9961	0.0029	
			0.883	0.167		0.9965	0.0035	

COMPONENTS:	ORIGINAL MEASUREMENTS:
1. Methane; CH_4; [74-82-8] 2. Ethane; C_2H_6; [74-84-0] 3. Propane; C_3H_8; [74-98-6]	Parikh, J. S.; Bukacek, R. F.; Graham, L.; Leipziger, S. *J. Chem. Eng. Data* 1984, *29*, 300-303.
VARIABLES:	PREPARED BY: C. L. Young

EXPERIMENTAL VALUES:

Mole fraction of methane = 0.8511 ± 0.0032
Mole fraction of ethane = 0.1007 ± 0.0023
Mole fraction of propane = 0.0480 ± 0.0005

Dew Points		Bubble Points	
T/K	P/MPa	T/K	P/MPa
210.15	0.6893		
223.26	1.379		
223.71	1.440		
231.04	2.068	144.15	0.7155
236.21	2.757	155.26	1.145
239.87	3.446	166.48	1.747
242.59	4.136	174.82	2.302
244.26	4.825	181.48	2.899
244.43	5.514	183.15	3.000
244.26	5.967	191.48	3.730
243.15	6.465	200.93	4.599
241.87	6.760	205.37	5.162
239.82	6.978	210.93	5.789
237.59	7.134	216.37	6.304
234.15	7.235	219.28	6.494
230.37	7.154	221.98	6.716
227.59	7.037	223.87	6.833
225.37	6.914	224.21[a]	6.850
225.04	6.896		
224.54	6.863		
224.21	6.850	[a] critical point	

AUXILIARY INFORMATION

METHOD/APPARATUS/PROCEDURE:	SOURCE AND PURITY OF MATERIALS:
Thick-walled Pyrex equilibrium cells used of approximately 29×10^{-6} m^3 volume. Cell contents stirred by ball bearing which could be moved by an external magnet. Fluid thermostat used. Dew and bubble points of samples of fixed composition determined visually either by varying pressure at fixed temperature or by varying temperature at fixed pressure. Details in source.	1. Purity 99.99 moles per cent. 2. Purity 99.90 moles per cent. 3. Purity 99.99 moles per cent.
	ESTIMATED ERROR: $\delta T/K = \pm 0.05$; $\delta P/P = \pm 0.001$.
	REFERENCES:

COMPONENTS:	ORIGINAL MEASUREMENTS:
1. Methane; CH_4; [74-82-8] 2. Ethane; C_2H_6; [74-84-0] 3. Pentane; C_5H_{12}; [109-66-0]	Billman, G. W.; Sage, B. H.; Lacey, W. N. *Trans. Am. Inst. Mech. Min. Engnrs.* 1948, *174*, 13-24.

VARIABLES:	PREPARED BY:
	C. L. Young

EXPERIMENTAL VALUES:

T/K (T/°F)	P/MPa (P/psi)	Mole fractions					
		in liquid			in vapor		
		x_{CH_4}	$x_{C_2H_6}$	$x_{C_5H_{12}}$	y_{CH_4}	$y_{C_2H_6}$	$y_{C_5H_{12}}$
310.93 (100)	3.45 (500)	0.154	0.0284	0.818	0.904	0.0377	0.0583
		0.115	0.223	0.662	0.652	0.297	0.0519
		0.0947	0.327	0.578	0.517	0.431	0.0511
		0.0550	0.514	0.431	0.275	0.681	0.0440
		0.0000	0.736	0.264	0.000	0.965	0.0349
	6.89 (1000)	0.263	0.211	0.526	0.762	0.188	0.0499
		0.246	0.314	0.440	0.674	0.282	0.0454
		0.237	0.396	0.368	0.596	0.360	0.0443
		0.224	0.444	0.331	0.548	0.405	0.0468
		0.214	0.515	0.271	0.483	0.473	0.0438
		0.198	0.602	0.200	0.394	0.562	0.0441
		0.196	0.759	0.0451	0.196	0.758	0.0450
	10.34 (1500)	0.420	0.152	0.428	0.814	0.125	0.0608
		0.412	0.291	0.297	0.689	0.244	0.0670
		0.413	0.378	0.209	0.588	0.334	0.0779
		0.462	0.443	0.0948	0.461	0.443	0.0953
	13.79 (2000)	0.580	0.0217	0.399	0.899	0.0173	0.0833
		0.587	0.116	0.297	0.801	0.0981	0.101
		0.599	0.140	0.261	0.763	0.121	0.116

(cont.)

AUXILIARY INFORMATION

METHOD/APPARATUS/PROCEDURE:	SOURCE AND PURITY OF MATERIALS:
Static equilibrium cell, agitated mechanically. Samples withdrawn so as not to disturb equilibrium. Pressure measured with pressure balance. Samples analysed by low temperature fractionation.	1. Commercial sample of about 99.65 mole per cent purity; carbon dioxide about 0.3 mole per cent and 0.04 mole per cent heavier hydrocarbon. 2. Carbide and Carbon Chemicals Corp. sample, fractionated; final purity about 99.8 mole per cent. 3. Phillips Petroleum Co. sample, purity 99.5 mole per cent.

ESTIMATED ERROR:

$\delta T/K = \pm 0.12$; $\delta P/MPa = \pm 0.015$;

δx, $\delta y = \pm 0.002$.

REFERENCES:

COMPONENTS:	ORIGINAL MEASUREMENTS:
1. Methane; CH_4; [74-82-8]	Billman, G. W.; Sage, B. H.; Lacey, W. N.
2. Ethane; C_2H_6; [74-84-0]	*Trans. Am. Inst. Mech. Min. Engnrs.*
3. Pentane; C_5H_{12}; [109-66-0]	1948, *174*, 13-24.

EXPERIMENTAL VALUES:

Smoothed Data

T/K (T/°F)	P/MPa (P/psi)	C^a	Mole fractions					
			in liquid			in vapor		
			x_{CH_4}	$x_{C_2H_6}$	$x_{C_5H_{12}}$	y_{CH_4}	$y_{C_2H_6}$	$y_{C_5H_{12}}$
310.93 (100)	3.45 (500)	0.0	0.160	0.000	0.840	0.940	0.000	0.0603
		0.2	0.124	0.1752	0.701	0.712	0.233	0.0547
		0.4	0.0868	0.365	0.548	0.469	0.482	0.0490
		0.6	0.0417	0.575	0.383	0.200	0.759	0.0414
	6.89 (1000)	0.0	0.308	0.000	0.693	0.9471	0.0000	0.0530
		0.2	0.276	0.145	0.579	0.8225	0.128	0.0497
		0.4	0.247	0.301	0.452	0.6834	0.270	0.0461
		0.6	0.220	0.468	0.312	0.5265	0.428	0.0459
		0.8	0.188	0.650	0.163	0.3472	0.612	0.0406
	10.34 (1500)	0.0	0.440	0.000	0.560	0.9412	0.0000	0.0588
		0.2	0.423	0.115	0.462	0.8460	0.0942	0.0601
		0.4	0.415	0.234	0.351	0.7428	0.194	0.0636
		0.6	0.414	0.352	0.235	0.6204	0.306	0.0739
	13.79 (2000)	0.0	0.579	0.000	0.421	0.9205	0.0000	0.0796
		0.2	0.582	0.0836	0.334	0.8381	0.0690	0.0930
		0.4	0.607	0.157	0.236	0.7347	0.1379	0.1272

$$^a\ C = \frac{\text{mole fraction of ethane}}{\text{mole fraction of ethane + mole fraction of pentane}}$$

COMPONENTS:	ORIGINAL MEASUREMENTS:
1. Methane; CH_4; [74-82-8] 2. Ethane; C_2H_6; [74-84-0] 3. Heptane; C_7H_{16}; [142-82-5]	Van Horn, L. D.; Kobayashi, R. *J. Chem. Engng. Data* <u>1967</u>, *12*, 294-303.

VARIABLES:	PREPARED BY:
Temperature, pressure	C. L. Young

EXPERIMENTAL VALUES:

T/K (T/°F)	P/MPa (P/psi)	Mole fractions				
		in liquid			in vapor	
		x_{CH_4}	$x_{C_2H_6}$	$x_{C_7H_{16}}$	y_{CH_4}	$y_{C_2H_6}$
244.26	0.689	0.042	0.132	0.826	0.786	0.214
(-20)	(100)	0.032	0.252	0.716	0.595	0.405
	1.38	0.084	0.235	0.681	0.786	0.214
	(200)	0.067	0.453	0.480	0.595	0.405
	2.76	0.179	0.203	0.618	0.891	0.109
	(400)	0.165	0.390	0.445	0.786	0.214
	4.14	0.254	0.248	0.498	0.891	0.109
	(600)	0.250	0.474	0.276	0.786	0.214
	5.52	0.331	0.138	0.531	0.9456	0.0544
	(800)	0.342	0.264	0.394	0.891	0.109
	6.89	0.396	0.141	0.463	0.9456	0.0544
	(1000)	0.393	0.268	0.339	0.891	0.109
233.15	0.689	0.047	0.172	0.781	0.786	0.214
(-40)	(100)	0.037	0.328	0.635	0.595	0.405
	1.38	0.094	0.316	0.590	0.786	0.214
	(200)	0.082	0.605	0.313	0.595	0.405
	2.76	0.197	0.262	0.541	0.891	0.109
	(400)	0.194	0.513	0.293	0.786	0.214
	4.14	0.298	0.157	0.545	0.9456	0.0544
	(600)	0.296	0.312	0.392	0.891	0.109
	5.52	0.380	0.163	0.457	0.9456	0.0544
	(800)	0.391	0.317	0.292	0.891	0.109

(cont.)

AUXILIARY INFORMATION

METHOD/APPARATUS/PROCEDURE:	SOURCE AND PURITY OF MATERIALS:
The solubilities were determined by measurement of retention volumes using gas chromatography. The method uses methane as a carrier gas, ethane as an injected sample and heptane as the stationary phase. The technique is described in the source and ref. (1).	1 and 2. Major impurity nitrogen. Total impurities less than 0.05 mole per cent. 3. Research grade.

ESTIMATED ERROR:

$\delta T/K = \pm 0.05$; $\delta P/psi = \pm 1$, $P \leq 1,000$ psia, ± 2, $P \geq 1,000$ psia; δx, $\delta y = \pm 1.5\%$.

REFERENCES:

1. Koonce, K. T.

   Ph.D. thesis, Rice University,

   Houston, <u>1963</u>.

COMPONENTS:	ORIGINAL MEASUREMENTS:
1. Methane; CH_4; [74-83-8]	Van Horn, L. D.; Kobayashi, R.
2. Ethane; C_2H_6; [74-84-0]	*J. Chem. Engng. Data*
3. Heptane; C_7H_{16}; [142-82-5]	<u>1967</u>, *12*, 294-303.

EXPERIMENTAL VALUES:

		Mole fractions				
		in liquid			in vapor	
T/K (T/°F)	P/MPa (P/psi)	x_{CH_4}	$x_{C_2H_6}$	$x_{C_7H_{16}}$	y_{CH_4}	$y_{C_2H_6}$
233.15 (-40)	6.89 (1000)	0.453	0.155	0.392	0.9456	0.0544
222.04 (-60)	0.689 (100)	0.055	0.243	0.702	0.786	0.214
		0.044	0.471	0.485	0.595	0.405
	1.38 (200)	0.120	0.222	0.658	0.891	0.109
		0.112	0.437	0.451	0.786	0.214
	2.76 (400)	0.243	0.175	0.582	0.9456	0.0544
		0.240	0.348	0.412	0.891	0.109
	4.14 (600)	0.346	0.200	0.454	0.9456	0.0544
		0.352	0.390	0.258	0.891	0.109
	5.52 (800	0.440	0.197	0.363	0.9456	0.0544

COMPONENTS:	ORIGINAL MEASUREMENTS:
1. Methane; CH_4; [74-82-8] 2. Propane; C_3H_8; [74-98-6] 3. Butane; C_4H_{10}; [106-97-8]	Wiese, H. C.; Jacobs, J.; Sage, B. H. *J. Chem. Eng. Data* <u>1970</u>, *15*, 82-91.

VARIABLES:	PREPARED BY:
Temperature, pressure liquid phase composition	C. L. Young

EXPERIMENTAL VALUES:

T/K (T/°F)	P/MPa (P/psia)	†Compo-sition factor	Mole fraction in liquid x_{CH_4}	$x_{C_3H_8}$	$x_{C_4H_{10}}$	in vapor y_{CH_4}	$y_{C_3H_8}$	$y_{C_4H_{10}}$
277.6 (40)	1.38 (200)	0.0	0.0808	0	0.9192	0.8888	0	0.1112
		0.2	0.0762	0.1848	0.7390	0.8273	0.0819	0.0909
		0.4	0.0710	0.3716	0.5574	0.7636	0.1659	0.0705
		0.6	0.0662	0.5603	0.3735	0.6990	0.2524	0.0486
		0.8	0.0609	0.7513	0.1878	0.6327	0.3422	0.0251
		1.0	0.0549	0.9451	0	0.5627	0.4373	0
	3.45 (500)	0.0	0.1913	0	0.8087	0.9369	0	0.0631
		0.2	0.1996	0.1601	0.6403	0.9102	0.0391	0.0507
		0.4	0.2001	0.3200	0.4799	0.8825	0.0788	0.0387
		0.6	0.1987	0.4808	0.3205	0.8523	0.1215	0.0263
		0.8	0.1962	0.6431	0.1608	0.8180	0.1685	0.0135
		1.0	0.1923	0.8077	0	0.7819	0.2181	0
	6.89 (1000)	0.0	0.3651	0	0.6349	0.9461	0	0.0539
		0.2	0.3949	0.1210	0.4841	0.9275	0.0281	0.0444
		0.4	0.4030	0.2388	0.3582	0.9068	0.0574	0.0358
		0.6	0.4071	0.3557	0.2371	0.8835	0.0904	0.0261
		0.8	0.4119	0.4705	0.1176	0.8567	0.1289	0.0144
		1.0	0.4226	0.4774	0	0.8208	0.1792	0
	8.62 (1250)	0.0	0.4513	0	0.5487	0.9407	0	0.0593
		0.2	0.4918	0.1016	0.4066	0.9203	0.0276	0.0521
		0.4	0.5070	0.1972	0.2958	0.9004	0.0571	0.0426
		0.6	0.5154	0.2908	0.1939	0.8749	0.0923	0.0329
		0.8	0.5278	0.3778	0.0944	0.8470	0.1335	0.0195
		1.0	0.5492	0.4508	0 (cont.)	0.8222	0.1778	0

AUXILIARY INFORMATION

METHOD/APPARATUS/PROCEDURE:	SOURCE AND PURITY OF MATERIALS:
PVT cell charged with mixture of known composition. Pressure measured with pressure balance and temperature measured using a platinum resistance thermometer. Details in ref. (1). Samples of coexisting phases analysed by GC. Details in source.	1. Texaco sample, passed over calcium chloride, activated charcoal. Ascarite and anhydrous calcium sulfate at pressures in excess of 3 MPa, purity 99.99 mole per cent. 2 and 3. Phillips Petroleum Co. samples, degassed, purities 99.99 and 99.95 mole per cent, respectively.

ESTIMATED ERROR:

$\delta T/K = \pm 0.01$; $\delta P/MPa = \pm 0.1\%$;

δx_{CH_4}, $\delta y_{CH_4} = \pm 0.005$ or better.

REFERENCES:

1. Sage, B. H.; Lacey, W. W.
   Trans. Am. Inst. Mining Met.
   Engnrs.
   <u>1940</u>, *136*, 136.

COMPONENTS: ORIGINAL MEASUREMENTS:

1. Methane; CH_4; [74-82-8]   Wiese, H. C.; Jacobs, J.;
2. Propane; C_3H_8; [74-98-6]   Sage, B. H.
3. Butane; C_4H_{10}; [106-97-8]   *J. Chem. Eng. Data*
                                         1970, *15*, 82-91.

EXPERIMENTAL VALUES:

			Mole fraction					
		†Compo-						
T/K	P/MPa	sition	in liquid			in vapor		
(T/°F)	(P/psia)	factor	x_{CH_4}	$x_{C_3H_8}$	$x_{C_4H_{10}}$	y_{CH_4}	$y_{C_3H_8}$	$y_{C_4H_{10}}$
277.6	10.34	0.0	0.5390	0	0.4610	0.9262	0	0.0738
(40)	(1500)	0.2	0.5970	0.0806	0.3224	0.9035	0.0297	0.0668
		0.4	0.6157	0.1537	0.2306	0.8747	0.0645	0.0608
		0.6	0.6391	0.2165	0.1444	0.8360	0.1114	0.0526
	11.72	0.0	0.6194	0	0.3806	0.9044	0	0.0956
	(1700)	0.2	0.7017	0.0597	0.2387	0.8585	0.0348	0.1067
310.9	1.38	0.0	0.0530	0	0.9470	0.7027	0.	0.2974
(100)	(200)	0.2	0.0452	0.1910	0.7638	0.5742	0.1833	0.2425
		0.4	0.0368	0.3853	0.5779	0.4447	0.3689	0.1864
		0.6	0.0272	0.5837	0.3891	0.3130	0.5574	0.1296
		0.8	0.0165	0.7868	0.1967	0.1823	0.7498	0.0679
		1.0	0.0049	0.9951	0	0.0521	0.9479	0
	3.45	0.0	0.1556	0	0.8444	0.8473	0	0.1527
	(500)	0.2	0.1514	0.1697	0.6789	0.7902	0.0831	0.1267
		0.4	0.1459	0.3417	0.5125	0.7316	0.1700	0.0984
		0.6	0.1386	0.5168	0.3446	0.6653	0.2641	0.0706
		0.8	0.1304	0.6957	0.1739	0.5953	0.3669	0.0378
		1.0	0.1235	0.8765	0	0.5209	0.4791	0
	6.89	0.0	0.3171	0	0.6829	0.8809	0	0.1191
	(1000)	0.2	0.3201	0.1360	0.5440	0.8430	0.0545	0.1025
		0.4	0.3209	0.2717	0.4075	0.8036	0.1132	0.0833
		0.6	0.3191	0.4086	0.2724	0.7582	0.1798	0.0621
		0.8	0.3217	0.5426	0.1357	0.7122	0.2525	0.0353
		1.0	0.3271	0.6729	0	0.6635	0.3365	0
	8.62	0.0	0.3974	0	0.6026	0.8786	0	0.1214
	(1250)	0.2	0.4051	0.1190	0.4760	0.8425	0.0506	0.1069
		0.4	0.4111	0.2356	0.3533	0.8038	0.1055	0.0907
		0.6	0.4181	0.3492	0.2328	0.7604	0.1688	0.0708
		0.8	0.4276	0.4579	0.1145	0.7143	0.2427	0.0430
		1.0	0.4511	0.5489	0	0.6766	0.3234	0
	10.34	0.0	0.4799	0	0.5201	0.8665	0	0.1335
	(1500)	0.2	0.4966	0.1007	0.4028	0.8249	0.0505	0.1246
		0.4	0.5152	0.1939	0.2909	0.7783	0.1087	0.1130
		0.6	0.5403	0.2758	0.1839	0.7103	0.1904	0.0993
	11.72	0.0	0.5586	0	0.4414	0.8440	0	0.1560
	(1700)	0.2	0.5910	0.0818	0.3272	0.7801	0.0540	0.1659
344.3	1.38	0.0	0.0256	0	0.9744	0.3517	0	0.6483
(220)	(200)	0.2	0.0091	0.1982	0.7929	0.1171	0.3439	0.5391
	3.45	0.0	0.1201	0	0.8799	0.6796	0	0.3204
	(500)	0.2	0.1086	0.1783	0.7131	0.5771	0.1477	0.2753
		0.4	0.0955	0.3618	0.5427	0.4754	0.3032	0.2214
		0.6	0.0803	0.5518	0.3679	0.3702	0.4702	0.1597
		0.8	0.0634	0.7493	0.1873	0.2650	0.6481	0.0869
		1.0	0.0433	0.9567	0	0.1550	0.8450	0
	6.89	0.0	0.2717	0	0.7283	0.7567	0	0.2433
	(1000)	0.2	0.2669	0.1466	0.5865	0.6944	0.0933	0.2123
		0.4	0.2622	0.2951	0.4427	0.6285	0.1954	0.1761
		0.6	0.2625	0.4463	0.2975	0.5527	0.3131	0.1342
		0.8	0.2556	0.5955	0.1489	0.4640	0.4547	0.0814
		1.0	0.2800	0.7200	0	0.3558	0.6442	0
	8.62	0.0	0.3482	0	0.6518	0.7588	0	0.2412
	(1250)	0.2	0.3549	0.1290	0.5161	0.6992	0.0842	0.2166
		0.4	0.3598	0.2561	0.3842	0.6314	0.1795	0.1891
		0.6	0.3718	0.3769	0.2513	0.5377	0.3006	0.1617
	10.34	0.0	0.4329	0	0.5671	0.7439	0	0.2561
	(1500)	0.2	0.4692	0.1062	0.4246	0.6610	0.0814	0.2576
	11.72	0.0	0.5103	0	0.4897	0.7036	0	0.2964
	(1700)							

(cont.)

COMPONENTS:	ORIGINAL MEASUREMENTS:
1. Methane; CH_4; [74-82-8]	Wiese, H. C.; Jacobs, J.;
2. Propane; C_3H_8; [74-98-6]	Sage, B. H.
3. Butane; C_4H_{10}; [106-97-8]	*J. Chem. Eng. Data*
	<u>1970</u>, *15*, 82-91.

EXPERIMENTAL VALUES:

			Mole fraction					
		[†]Compo-	in liquid			in vapor		
T/K (T/°F)	P/MPa (P/psia)	sition factor	x_{CH_4}	$x_{C_3H_8}$	$x_{C_4H_{10}}$	y_{CH_4}	$y_{C_3H_8}$	$y_{C_4H_{10}}$
377.6 (220)	3.45 (500)	0.0	0.0783	0	0.9217	0.4200	0	0.5800
		0.2	0.0554	0.1889	0.7557	0.2628	0.2233	0.5139
		0.4	0.0328	0.3869	0.5803	0.1342	0.4573	0.4085
		0.6	0.0042	0.5975	0.3983	0.0142	0.7062	0.2796
	6.89 (1000)	0.0	0.2361	0	0.7639	0.5630	0	0.4370
		0.2	0.2311	0.1538	0.6151	0.4615	0.1375	0.4010
		0.4	0.2316	0.3074	0.4610	0.3560	0.2876	0.3564
	8.62 (1250)	0.0	0.3200	0	0.6800	0.5338	0	0.4662

$$[†]\text{Composition factor} = \frac{\text{Moles of propane in liquid}}{\text{Moles of propane in liquid + Moles of butane in liquid}}$$

COMPONENTS:	ORIGINAL MEASUREMENTS:
1. Methane; CH_4; [74-82-8] 2. Propane; C_3H_8; [74-98-6] 3. Pentane; C_5H_{12}; [109-66-0]	Carter, R. T.; Sage, B. H. Lacey, W. N. *Trans. Am. Inst. Met. Min. Engnrs.* *1941, 142,* 170-177.
VARIABLES:	PREPARED BY: C. L. Young

EXPERIMENTAL VALUES:

		Mole fraction					
T/K	P/MPa	in liquid			in vapor		
(T/°F)	(P/psi)	x_{CH_4}	$x_{C_3H_8}$	$x_{C_5H_{12}}$	y_{CH_4}	$y_{C_3H_8}$	$y_{C_5H_{12}}$
310.9	3.45	0.139	0.446	0.415	0.755	0.218	0.027
(100)	(500)	0.136	0.743	0.121	0.607	0.374	0.019
		0.147	0.324	0.529	0.796	0.164	0.040
		0.156	0.105	0.739	0.895	0.054	0.051
	6.89	0.306	0.589	0.105	0.716	0.262	0.023
	(1000)	0.311	0.548	0.141	0.723	0.257	0.020
		0.302	0.220	0.478	0.871	0.088	0.041
		0.311	0.087	0.602	0.922	0.036	0.042
	10.34	0.469	0.289	0.242	0.812	0.145	0.043
	(1500)	0.468	0.299	0.233	0.811	0.145	0.044
		0.470	0.301	0.229	0.808	0.148	0.044
		0.493	0.364	0.143	0.754	0.203	0.042
		0.454	0.187	0.359	0.862	0.085	0.053
		0.446	0.068	0.486	0.918	0.030	0.052
	13.79	0.820	0.103	0.077	0.818	0.104	0.078
	(2000)	0.668	0.151	0.181	0.669	0.151	0.180
		0.609	0.096	0.295	0.856	0.057	0.087
		0.631	0.149	0.220	0.798	0.097	0.105
		0.600	0.066	0.334	0.875	0.042	0.083

(cont.)

AUXILIARY INFORMATION

METHOD/APPARATUS/PROCEDURE:	SOURCE AND PURITY OF MATERIALS:
Static equilibrium cell, agitated mechanically. Samples withdrawn so as not to disturb equilibrium. Pressure measured with pressure balance. Samples analysed by low temperature fractionation.	1. Commercial sample, dried and carbon dioxide removed. Purity better than 99.9 mole per cent. 2. Phillips Petroleum Co. sample, purity better than 99.95 mole per cent. 3. Phillips Petroleum Co. sample, purity about 99.5 mole per cent, major impurity 2-methylbutane.
	ESTIMATED ERROR: $\delta T/K = \pm 0.06$; $\delta P/MPa = \pm 0.015$; $\delta x, \delta y = \pm 0.001$.
	REFERENCES:

COMPONENTS: ORIGINAL MEASUREMENTS:

1. Methane; CH_4; [74-82-8]   Carter, R. T.; Sage, B. H.;
                                      Lacey, W. N.
2. Propane; C_3H_8; [74-98-6]
                                      Trans. Am. Inst. Met. Min. Engnrs.
3. Pentane; C_5H_{12}; [109-66-0]   <u>1941</u>, *142*, 170-177.

EXPERIMENTAL VALUES:

Smoothed Data

| T/K (T/°F) | P/MPa (P/psi) | Composition[a] parameter, C | Mole fraction | | | | | |
| | | | in liquid | | | in vapor | | |
			x_{CH_4}	$x_{C_3H_8}$	$x_{C_5H_{12}}$	y_{CH_4}	$y_{C_3H_8}$	$y_{C_5H_{12}}$
310.9 (100)	3.45 (500)	0	0.160	0.0	0.840	0.940	0.0	0.060
		0.2	0.152	0.170	0.678	0.866	0.085	0.049
		0.4	0.143	0.343	0.514	0.788	0.173	0.039
		0.6	0.135	0.519	0.346	0.708	0.262	0.030
		0.8	0.126	0.699	0.175	0.613	0.365	0.022
		1.0	0.118	0.882	0	0.516	0.484	0.0
	6.89 (1000)	0	0.308	0.0	0.692	0.947	0.00	0.053
		0.2	0.306	0.139	0.555	0.901	0.056	0.043
		0.4	0.304	0.278	0.418	0.853	0.112	0.035
		0.6	0.302	0.419	0.279	0.799	0.174	0.027
		0.8	0.304	0.557	0.139	0.723	0.255	0.022
		1.0	0.321	0.679	0	0.64	0.36	0
	10.34 (1500)	0	0.941	0.0	0.059	0.941	0	0.059
		0.2	0.900	0.049	0.051	0.900	0.049	0.051
		0.4	0.853	0.099	0.048	0.853	0.099	0.048
		0.6	0.794	0.160	0.046	0.794	0.160	0.046
		0.8	0.711	0.246	0.043	0.711	0.246	0.043
	13.79 (2000)	0	0.914	0	0.086	0.914	0	0.086
		0.2	0.871	0.046	0.083	0.871	0.046	0.083
		0.4	0.801	0.096	0.103	0.801	0.096	0.103

[a]
$$C = \frac{\text{Mole fraction of propane}}{\text{Mole fraction of propane + Mole fraction of pentane}}$$

COMPONENTS:	ORIGINAL MEASUREMENTS:
1. Methane; CH_4; [74-82-8] 2. Propane; C_3H_8; [74-98-6] 3. Pentane; C_5H_{12}; [109-66-0]	Dourson, R. H.; Sage, B. H.; Lacey, W. N. *Trans. Am. Inst. Met. Min. Engnrs.* *1943*, *151*, 206-215.
VARIABLES:	PREPARED BY: C. L. Young

EXPERIMENTAL VALUES:

T/K (T/°F)	P/MPa (P/psi)	in liquid x_{CH_4}	$x_{C_3H_8}$	Mole fraction $x_{C_5H_{12}}$	y_{CH_4}	in vapor $y_{C_3H_8}$	$y_{C_5H_{12}}$
344.3 (160)	3.45 (500)	0.119	0.141	0.740	0.768	0.106	0.126
		0.121	0.175	0.704	0.736	0.152	0.112
		0.116	0.259	0.625	0.661	0.238	0.101
		0.107	0.421	0.472	0.563	0.350	0.087
		0.103	0.474	0.423	0.524	0.406	0.070
		0.098	0.522	0.380	0.485	0.439	0.076
		0.075	0.693	0.232	0.352	0.595	0.053
	6.89 (1000)	0.261	0.124	0.615	0.840	0.068	0.093
		0.272	0.287	0.441	0.761	0.157	0.082
		0.247	0.399	0.354	0.678	0.249	0.073
		0.248	0.491	0.261	0.613	0.324	0.063
		0.246	0.583	0.171	0.543	0.406	0.051
	10.34 (1500)	0.401	0.181	0.418	0.798	0.105	0.097
		0.414	0.321	0.265	0.679	0.223	0.098
		0.434	0.357	0.209	0.634	0.272	0.094
	13.79 (2000)	0.538	0.056	0.406	0.820	0.038	0.142
		0.540	0.140	0.320	0.770	0.076	0.154
377.6 (220)	3.45 (500)	0.091	0.160	0.749	0.571	0.190	0.239
		0.062	0.377	0.561	0.345	0.451	0.204
		0.061	0.389	0.550	0.329	0.472	0.199
		0.043	0.570	0.387	0.170	0.666	0.164
		0.010	0.697	0.293	0.059	0.809	0.132

(cont.)

AUXILIARY INFORMATION

METHOD/APPARATUS/PROCEDURE:	SOURCE AND PURITY OF MATERIALS:
Static equilibrium cell, agitated mechanically. Samples withdrawn so as not to disturb equilibrium. Pressure measured with pressure balance. Samples analysed by low temperature fractionation.	1. Commercial sample, dried and carbon dioxide removed. Purity better than 99.9 mole per cent. 2. Phillips Petroleum Co. sample, purity better than 99.5 mole per cent. 3. Phillips Petroleum Co. sample, about 99.5 mole per cent; major impurity 2-methylbutane.
	ESTIMATED ERROR: $\delta T/K = \pm0.06$; $\delta P/MPa = \pm0.015$; δx, $\delta y = \pm0.001$.
	REFERENCES:

COMPONENTS:	ORIGINAL MEASUREMENTS:
1. Methane; CH_4; [74-82-8]	Dourson, R. H.; Sage, B. H.; Lacey, W. N.
2. Propane; C_3H_8; [74-98-6]	
3. Pentane; C_5H_{12}; [109-66-0]	*Trans. Am. Inst. Met. Min. Engnrs.* 1943, *151*, 206-215.

EXPERIMENTAL VALUES:

		Mole fraction					
		in liquid			in vapor		
T/K (T/°F)	P/MPa (P/psi)	x_{CH_4}	$x_{C_3H_8}$	$x_{C_5H_{12}}$	y_{CH_4}	$y_{C_3H_8}$	$y_{C_5H_{12}}$
377.6 (220)	6.89 (1000)	0.228	0.138	0.634	0.714	0.106	0.180
		0.223	0.192	0.585	0.669	0.155	0.176
		0.214	0.306	0.480	0.566	0.264	0.170
		0.208	0.370	0.421	0.521	0.318	0.161
		0.197	0.516	0.287	0.397	0.463	0.140
	10.34 (1500)	0.390	0.133	0.477	0.689	0.108	0.203
		0.388	0.228	0.384	0.598	0.194	0.388
		0.422	0.262	0.316	0.542	0.239	0.219

Smoothed Data

			Mole fraction					
			in liquid			in vapor		
T/K (T/°F)	P/MPa (P/psi)	C^a	x_{CH_4}	$x_{C_3H_8}$	$x_{C_5H_{12}}$	y_{CH_4}	$y_{C_3H_8}$	$y_{C_5H_{12}}$
344.3 (160)	3.45 (500)	0.0	0.136	0.000	0.864	0.867	0.000	0.133
		0.2	0.124	0.175	0.701	0.743	0.145	0.112
		0.4	0.112	0.355	0.533	0.612	0.297	0.091
		0.6	0.095	0.543	0.362	0.472	0.461	0.067
		0.8	0.074	0.741	0.185	0.322	0.639	0.039
		1.0	0.043	0.957	0.000	0.159	0.841	0.000
	6.89 (1000)	0.0	0.274	0.000	0.726	0.893	0.000	0.107
		0.2	0.266	0.147	0.587	0.826	0.080	0.094
		0.4	0.261	0.296	0.443	0.752	0.168	0.080
		0.6	0.255	0.447	0.298	0.657	0.277	0.066
		0.8	0.245	0.604	0.151	0.512	0.439	0.049
		1.0	0.297	0.703	0.000	0.361	0.639	0.000
	10.34 (1500)	0.0	0.398	0.000	0.602	0.885	0.000	0.115
		0.2	0.397	0.121	0.482	0.829	0.068	0.103
		0.4	0.400	0.240	0.360	0.758	0.147	0.095
		0.6	0.412	0.353	0.235	0.645	0.258	0.097
	13.79 (2000)	0.0	0.545	0.000	0.455	0.855	0.000	0.145
		0.2	0.546	0.091	0.363	0.789	0.073	0.138
377.6	3.45 (500)	0.0	0.115	0.000	0.885	0.728	0.000	0.272
		0.2	0.088	0.182	0.730	0.547	0.215	0.238
		0.4	0.065	0.374	0.561	0.359	0.441	0.200
		0.6	0.039	0.577	0.384	0.167	0.680	0.153
	6.89 (1000)	0.0	0.251	0.000	0.749	0.798	0.000	0.202
		0.2	0.230	0.154	0.616	0.698	0.122	0.181
		0.4	0.213	0.315	0.472	0.577	0.260	0.163
		0.6	0.208	0.475	0.317	0.436	0.420	0.144
	10.34 (1500)	0.0	0.382	0.000	0.618	0.794	0.000	0.206
		0.2	0.374	0.125	0.501	0.695	0.099	0.206
	0.4	0.4	0.385	0.246	0.369	0.577	0.207	0.216

$$C = \frac{\text{mole fraction of propane}}{\text{mole fraction of propane + mole fraction of pentane}}$$

a

COMPONENTS:	ORIGINAL MEASUREMENTS:
1. Methane; CH_4; [74-82-8] 2. Propane; C_3H_8; [74-98-6] 3. Heptane; C_7H_{16}; [142-82-5]	Van Horn, L. D.; Kobayashi, R. *J. Chem. Engng. Data* 1967, *12*, 294-303.
VARIABLES: Temperature, pressure	PREPARED BY: C. L. Young

EXPERIMENTAL VALUES:

T/K (T/°F)	P/MPa (P/psi)	Mole fractions				
		in liquid			in vapor	
		x_{CH_4}	$x_{C_3H_8}$	$x_{C_7H_{16}}$	y_{CH_4}	$y_{C_3H_8}$
244.26 (-20)	0.689 (100)	0.048	0.132	0.820	0.9595	0.0405
		0.050	0.255	0.695	0.9233	0.0767
		0.047	0.428	0.525	0.8691	0.1309
	1.38 (200)	0.100	0.127	0.773	0.9784	0.0216
		0.098	0.236	0.666	0.9595	0.0405
		0.102	0.450	0.448	0.9233	0.0767
	2.76 (400)	0.190	0.178	0.632	0.9800	0.0200
		0.197	0.362	0.441	0.9595	0.0405
	4.14 (600)	0.264	0.199	0.537	0.9800	0.0200
		0.268	0.198	0.534	0.9784	0.0216
		0.273	0.404	0.323	0.9595	0.0405
	5.52 (800)	0.361	0.193	0.446	0.9784	0.0216
		0.405	0.353	0.242	0.9595	0.0405
	6.89 (1000)	0.426	0.172	0.402	0.9784	0.0216
233.15 (-40)	0.689 (100)	0.056	0.101	0.843	0.9800	0.0200
		0.057	0.205	0.738	0.9595	0.0405
		0.056	0.397	0.547	0.9233	0.0767
	1.38 (200)	0.112	0.180	0.708	0.9800	0.0200
		0.116	0.360	0.524	0.9595	0.0405

(cont.)

AUXILIARY INFORMATION

METHOD/APPARATUS/PROCEDURE:	SOURCE AND PURITY OF MATERIALS:
The solubilities were determined by measurement of retention volumes using gas chromatography. The method uses methane as a carrier gas, propane as an injected solute and heptane as the stationary phase. The technique is described in the source and in ref. (1).	1 and 2. Major impurities were carbon dioxide and nitrogen amount to about 0.2 mole per cent. 3. Research grade.
	ESTIMATED ERROR: $\delta T/K = \pm 0.05$; $\delta P/psi = \pm 1$, $P \leq$ 1,000 psia, ± 2, $P \geq 1,000$ psia; δx, $\delta y = \pm 1.5\%$.
	REFERENCES: 1. Koonce, K. T. *Ph.D. thesis, Rice University,* Houston, 1963.

COMPONENTS:	ORIGINAL MEASUREMENTS:

COMPONENTS:

1. Methane; CH_4 ; [74-82-8]
2. Propane; C_3H_8 ; [74-98-6]
3. Heptane; C_7H_{16} ; [142-82-5]

ORIGINAL MEASUREMENTS:

Van Horn, L. D.; Kobayashi, R.

J. Chem. Engng. Data

<u>1967</u>, *12*, 294-303.

EXPERIMENTAL VALUES:

Mole fractions

T/K (T/°F)	P/MPa (P/psi)	in liquid			in vapor	
		x_{CH_4}	$x_{C_3H_8}$	$x_{C_7H_{16}}$	y_{CH_4}	$y_{C_3H_8}$
233.15	2.76	0.211	0.128	0.661	0.9902	0.0098
	(400)	0.218	0.259	0.523	0.9800	0.0200
		0.224	0.373	0.403	0.9713	0.0287
		0.233	0.524	0.243	0.9595	0.0405
	4.14	0.307	0.143	0.550	0.9894	0.0106
	(600)	0.333	0.383	0.284	0.9713	0.0287
	5.52	0.385	0.132	0.483	0.9894	0.0106
	(800)	0.414	0.332	0.254	0.9713	0.0287
	6.89	0.454	0.109	0.437	0.9894	0.0106
	(1000)	0.501	0.193	0.306	0.9784	0.0216
222.04	0.689	0.065	0.080	0.855	0.9902	0.0098
(-60)	(100)	0.065	0.158	0.776	0.9800	0.0200
		0.065	0.235	0.700	0.9713	0.0287
		0.067	0.331	0.602	0.9595	0.0405
		0.070	0.651	0.279	0.9233	0.0767
	1.38	0.131	0.275	0.594	0.9800	0.0200
	(200)	0.134	0.402	0.464	0.9713	0.0287
	2.76	0.239	0.184	0.577	0.9902	0.0098
	(400)	0.244	0.204	0.552	0.9894	0.0106
		0.242	0.388	0.370	0.9800	0.0200
	4.14	0.353	0.195	0.452	0.9894	0.0106
	(600)	0.390	0.362	0.248	0.9800	0.0200
	5.52	0.459	0.156	0.385	0.9894	0.0106
	(800)					
210.93	0.689	0.076	0.134	0.790	0.9902	0.0098
(-80)	(100)	0.078	0.282	0.640	0.9800	0.0200
		0.079	0.411	0.510	0.9713	0.0287
	1.38	0.145	0.224	0.631	0.9902	0.0098
	(200)	0.166	0.455	0.379	0.9800	0.0200
	2.76	0.298	0.290	0.412	0.9902	0.0098
	(400)					
	4.14	0.432	0.241	0.327	0.9902	0.0098
	(600)					
199.82	0.689	0.095	0.248	0.657	0.9902	0.0098
(-100)	(100)	0.104	0.498	0.398	0.9800	0.0200
	1.38	0.197	0.386	0.417	0.9902	0.0098
	(200)					
	2.76	0.396	0.398	0.206	0.9902	0.0098
	(400)					

COMPONENTS:	ORIGINAL MEASUREMENTS:
1. Methane; CH_4; [74-82-8] 2. Propane; C_3H_8; [74-98-6] 3. Heptane; C_7H_{16}; [142-82-5]	Koonce, K. T.; Kobayashi, R. *J. Chem. Eng. Data* <u>1964</u>, *9*, 494-501.

VARIABLES:	PREPARED BY:
Temperature, pressure	C. L. Young

EXPERIMENTAL VALUES:

T/K (T/°F)	P/psi	P/MPa	Mole fractions in liquid x_{CH_4}	$x_{C_3H_8}$	$x_{C_7H_{16}}$	in vapor y_{CH_4}	$y_{C_3H_8}$
233.15 (-40)	97	0.67	0.0662	0.0959	0.8379	0.9792	0.0208
	201	1.39	0.118	0.175	0.707	0.9792	0.0208
	400	2.76	0.239	0.249	0.512	0.9792	0.0208
	605	4.17	0.338	0.261	0.401	0.9792	0.0208
	791	5.45	0.453	0.277	0.320	0.9792	0.0208
	991	6.83	0.560	0.176	0.264	0.9792	0.0208
244.26 (-20)	107	0.74	0.0560	0.147	0.797	0.9569	0.0431
	240	1.65	0.111	0.275	0.614	0.9569	0.0431
	422	2.91	0.195	0.375	0.430	0.9569	0.0431
	603	4.16	0.304	0.378	0.318	0.9569	0.0431
	795	5.48	0.414	0.339	0.247	0.9569	0.0431
	993	6.85	0.517	0.275	0.208	0.9569	0.0431

AUXILIARY INFORMATION

METHOD/APPARATUS/PROCEDURE:	SOURCE AND PURITY OF MATERIALS:
The solubilities were determined by measurement of retention volumes using gas chromatography. The method uses methane as a carrier gas, propane as an injected solute and heptane as the stationary phase. The technique is described in the source and ref. (1).	1. Sample dried; purity 99.7 mole per cent, 0.2 mole per cent nitrogen and 0.1 mole per cent ethane. 2. Phillips Petroleum Co. sample, purity 99.5 mole per cent. 3. Phillips Petroleum Co. research grade sample, purity 99.90 mole per cent.

ESTIMATED ERROR:

$\delta T/K = \pm 0.1$; $\delta P/MPa = \pm 2\%$;

δx, $\delta y = \pm 6\%$ (estimated by compiler).

REFERENCES:

1. Koonce, K. T.
   Ph.D. Thesis
   Rice University, Houston, <u>1963</u>.

COMPONENTS:	ORIGINAL MEASUREMENTS:
1. Methane; CH_4; [74-82-8] 2. Propane; C_3H_8; [74-98-6] 3. Octane; C_8H_{18}; [111-65-9]	Hottovy, J. D.; Kohn, J. P.; Luks, K. D. *J. Chem. Eng. Data* 1982, *27*, 298-302.

VARIABLES:	PREPARED BY:
	C. L. Young

EXPERIMENTAL VALUES:

Data for octane-lean liquid phase.

Type of data	T/K	P/atm	P/MPa	Mole fractions $x_{C_3H_8}$	$x_{C_8H_{18}}$	Molar volume /cm^{3}mol^{-1}
K(L_1-L_2=V)	213.07	67.7	6.86	0.078	0.0083	73.7
	210.65	65.1	6.60	0.065	0.0048	75.4
	206.83	60.9	6.17	0.050	0.0039	78.3
	205.72	59.5	6.03	0.044	0.0036	80.3
	204.72	58.7	5.95	0.039	0.0026	83.8
	201.92	56.0	5.67	0.028	0.0016	88.4
Q(S-L_1-L_2-V)	199.70	53.8	5.45	0.027	0.0022	77.8
	199.61	53.5	5.42	0.029	0.0026	73.7
	199.05	52.3	5.30	0.039	0.0051	68.6
	198.90	51.7	5.24	0.043	0.0061	66.5
	198.58	51.1	5.18	0.047	0.0073	64.8
	198.49	50.8	5.15	0.051	0.0091	64.3
	198.03	49.5	5.02	0.061	0.0125	61.1
	197.90	49.8	5.05	0.063	0.0134	61.8
	197.86	49.1	4.98	0.066	0.0146	61.3
	197.61	49.0	4.98	0.069	0.0158	60.8
	197.31	48.3	4.89	0.076	0.0190	59.7
	197.25	47.8	4.84	0.079	0.0212	59.1
	196.81	47.0	4.76	0.107	0.0385	58.9
LCST(L_1=L_2-V)	211.06	64.8	6.57	0.123	0.0316	62.9
	210.53	64.1	6.49	0.120	0.0319	61.4

(cont.)

AUXILIARY INFORMATION

METHOD/APPARATUS/PROCEDURE:	SOURCE AND PURITY OF MATERIALS:

Glass equilibrium cell.
Temperature measured with platinum resistance thermometer and pressure with Bourdon gauge. Stoichiometry and volumetric measurements were used to compute liquid phase compositions and molar volumes.
Details in ref. 1 and source.
	1. Linde Ultrapure grade, purity
	   99.97 mole per cent.
	
	2. Linde Instrument grade, purity
	   99.5 mole per cent.
	
	3. Humphrey-Wilkinson grade, purity
	   99 mole per cent.

ESTIMATED ERROR:
$\delta T/K = \pm 0.03$; $\delta P/MPa = \pm 0.07$;
$\delta x_{C_3H_8} = \pm 3.5\%$; $\delta x_{C_8H_{18}} = \pm 2\%$
(octane rich phase); $x\delta_{C_8H_{18}} = \pm 8\%$
(octane lean phase).

REFERENCES:

1. Kohn, J. P.
   Am. Inst. Chem. Engnrs. J.
   1961, *7*, 514.

COMPONENTS:	ORIGINAL MEASUREMENTS:
1. Methane; CH_4; [74-82-8]	Hottovy, J. D.; Kohn, J. P.;
2. Propane; C_3H_8; [74-98-6]	Luks, K. D.
3. Octane; C_8H_{18}; [111-65-0]	*J. Chem. Eng. Data*
	<u>1982</u>, *27*, 298-302.

EXPERIMENTAL VALUES:

Data for octane-lean liquid phase.

Type of data	T/K	P/atm	P/MPa	Mole fractions $x_{C_3H_8}$	$x_{C_8H_{18}}$	Molar volume /cm^{3}mol^{-1}
LCST(L_1=L_2-V)	205.99	58.4	5.92	0.134	0.0519	61.4
	204.93	57.1	5.79	0.130	0.0503	61.4
	203.96	55.8	5.65	0.128	0.0493	60.6
	201.11	52.3	5.30	0.131	0.0619	60.8
	199.71	50.4	5.11	0.126	0.0594	60.1
	197.82	48.2	4.88	0.121	0.0567	59.9
L_1-L_2-V	209.54	63.5	6.43	0.075	0.0083	59.9
	208.79	61.0	6.18	0.111	0.0283	61.1
	207.79	61.7	6.25	0.069	0.0073	64.5
	207.04	59.0	5.98	0.106	0.0268	61.5
	205.31	56.9	5.77	0.101	0.0257	62.0
	204.00	57.6	5.84	0.050	0.0057	72.5
	204.00	57.3	5.81	0.055	0.0070	68.4
	204.00	57.2	5.80	0.059	0.0071	66.3
	204.00	56.0	5.67	0.087	0.0022	61.4
	203.47	55.3	5.60	0.088	0.0024	60.0
	203.20	56.2	5.69	0.059	0.0090	63.1
	203.20	56.8	5.76	0.049	0.0550	70.8
	202.74	56.5	5.72	0.039	0.0045	76.4
	202.23	55.2	5.59	0.052	0.0072	64.2
	202.00	54.9	5.56	0.054	0.0081	67.2
	202.00	54.7	5.54	0.059	0.0097	65.8
	202.00	54.3	5.50	0.066	0.0111	64.6
	202.00	54.0	5.47	0.073	0.0129	62.9
	202.00	53.3	5.40	0.089	0.0243	60.5
	202.00	55.2	5.59	0.049	0.0070	69.3
	201.57	54.6	5.53	0.041	0.0050	66.7
	201.04	54.2	5.49	0.048	0.0069	68.2
	200.70	51.8	5.25	0.090	0.0247	59.5
	200.09	54.1	5.48	0.028	0.0023	76.7
	200.10	53.6	5.43	0.038	0.0044	71.1
	200.12	53.2	5.39	0.043	0.0054	68.6
	200.11	53.1	5.38	0.046	0.0062	67.5
	200.10	52.9	5.36	0.049	0.0770	67.1
	200.10	52.2	5.29	0.060	0.0114	63.5
	200.10	51.9	5.26	0.066	0.0134	63.1
	200.10	51.4	5.21	0.076	0.0169	61.4
	200.09	51.0	5.17	0.092	0.0255	60.0
	200.03	53.2	5.39	0.064	0.0134	63.2
	199.99	53.1	5.38	0.044	0.0065	65.1
	199.94	53.3	5.40	0.039	0.0052	67.6
	199.03	50.1	5.08	0.084	0.0226	59.3
	198.42	49.3	4.00	0.085	0.0230	59.3
	198.40	49.7	5.04	0.074	0.0209	60.4
	198.00	49.3	5.00	0.065	0.0143	61.5
	198.00	48.9	4.95	0.079	0.0197	60.6
	198.00	48.4	4.90	0.094	0.0285	60.6

(cont.)

COMPONENTS:	ORIGINAL MEASUREMENTS:
1. Methane; CH_4; [74-82-8]	Hottovy, J. K.; Kohn, J. P.;
2. Propane; C_3H_8; [74-98-6]	Luks, K. D.
3. Octane; C_8H_{18}; [111-65-9]	*J. Chem. Eng. Data*
	1982, 27, 298-302.

EXPERIMENTAL VALUES:

Data for octane-rich liquid phase.

Type of data	T/K	P/atm	P/MPa	Mole fractions $x_{C_3H_8}$	$x_{C_8H_{18}}$	Molar volume /cm^{3}mol^{-1}
K(L_1-L_2=V)	212.96	67.6	6.85	0.150	0.052	62.8
	210.40	64.2	6.51	0.165	0.092	64.5
	208.91	63.2	6.40	0.164	0.107	68.1
	205.54	59.5	6.03	0.160	0.153	69.4
	202.86	56.8	5.76	0.144	0.194	71.1
	201.50	55.7	5.64	0.136	0.205	70.1
	200.98	55.2	5.59	0.132	0.217	71.6
Q(S-L_1-L_2-V)	199.17	52.5	5.32	0.129	0.218	71.6
	199.05	52.1	5.28	0.130	0.212	70.5
	198.71	51.5	5.22	0.133	0.198	69.3
	198.69	51.4	5.21	0.135	0.197	69.9
	198.35	50.3	5.10	0.135	0.180	66.5
	197.69	48.7	4.93	0.142	0.146	65.0
	197.06	47.4	4.80	0.141	0.091	61.2
L_1-L_2-V	208.35	61.5	6.23	0.141	0.055	60.8
	207.04	59.9	6.07	0.137	0.053	61.0
	206.52	59.2	6.00	0.135	0.052	61.0
	206.82	59.8	6.06	0.154	0.085	62.5
	205.30	57.6	5.84	0.141	0.068	60.2
	204.47	58.6	5.94	0.147	0.151	63.6
	204.09	58.0	5.88	0.151	0.157	66.6
	204.00	56.8	5.76	0.154	0.132	64.1
	204.00	56.7	5.75	0.152	0.126	63.3
	204.00	56.5	5.72	0.151	0.116	62.3
	205.93	54.5	5.52	0.147	0.091	61.4
	202.88	54.4	5.51	0.135	0.064	60.2
	202.35	55.9	5.66	0.134	0.1690	63.6
	202.00	56.0	5.67	0.141	0.2000	70.4
	202.00	55.9	5.66	0.139	0.1940	68.6
	202.00	55.7	5.64	0.139	0.1890	68.1
	202.00	54.5	5.52	0.152	0.1490	65.5
	202.00	54.0	5.47	0.151	0.1300	63.5
	202.00	53.7	5.44	0.148	0.1140	62.4
	201.65	52.8	5.35	0.144	0.0890	61.4
	201.65	55.8	5.65	0.132	0.1980	68.0
	200.48	53.9	5.46	0.133	0.1980	68.9
	200.48	54.0	5.47	0.133	0.2000	68.9
	200.30	51.0	5.17	0.141	0.0870	61.2
	200.08	53.8	5.45	0.130	0.2180	71.5
	200.08	53.8	5.45	0.130	0.2160	71.2
	200.07	53.7	5.44	0.131	0.2130	70.7
	200.10	53.3	5.40	0.135	0.2020	70.0
	200.10	53.2	5.39	0.136	0.1960	69.0
	200.10	52.8	5.35	0.138	0.1850	68.0
	200.07	52.3	5.30	0.148	0.1620	66.0
	200.07	52.1	5.28	0.149	0.1520	65.2
	200.07	51.9	5.26	0.149	0.1420	64.4
	200.10	51.4	5.21	0.147	0.1260	62.9
	200.10	51.0	5.17	0.141	0.1030	60.3
	199.37	50.9	5.16	0.143	0.1480	65.4
	199.33	49.9	5.06	0.138	0.0850	61.2
	199.25	49.7	5.04	0.129	0.0700	59.8
	198.00	49.0	4.96	0.144	0.145	64.7
	198.00	48.7	4.93	0.146	0.126	63.6
	198.00	48.3	4.89	0.142	0.102	62.5
	198.14	48.1	4.87	0.135	0.083	60.8

COMPONENTS:	ORIGINAL MEASUREMENTS:
1. Methane; CH_4; [74-82-8] 2. Propane; C_3H_8; [74-98-6] 3. Decane; $C_{10}H_{22}$; [124-18-5]	Koonce, K. T.; Kobayashi, R. *J. Chem. Eng. Data* <u>1964</u>, *9*, 494-501.

VARIABLES:	PREPARED BY:
Temperature, pressure	C. L. Young

EXPERIMENTAL VALUES:

T/K (T/°F)	P/MPa	Mole fractions				
		in liquid			in vapor	
		x_{CH_4}	$x_{C_3H_8}$	$x_{C_{10}H_{22}}$	y_{CH_4}	$y_{C_3H_8}$
224.26 (-20)	0.134	0.0100	0.0	0.9900	1.0	0.0
	0.252	0.0185	0.0	0.9815	1.0	0.0
	0.445	0.0315	0.0	0.9685	1.0	0.0
	0.689	0.0481	0.0 -	0.9519	1.0	0.0
	1.04	0.0719	0.0	0.9281	1.0	0.0
	1.64	0.114	0.0	0.886	1.0	0.0
	2.41	0.162	0.0	0.838	1.0	0.0
	3.11	0.203	0.0	0.797	1.0	0.0
	4.42	0.269	0.0	0.731	1.0	0.0
	5.84	0.330	0.0	0.670	1.0	0.0
	6.78	0.364	0.0	0.636	1.0	0.0
	0.310	0.0216	0.0363	0.9421	0.9792	0.0208
	0.710	0.0487	0.0765	0.8748	0.9792	0.0208
	1.39	0.0942	0.128	0.7778	0.9792	0.0208
	2.08	0.139	0.165	0.696	0.9792	0.0208
	2.76	0.182	0.182	0.636	0.9792	0.0208
	4.05	0.251	0.202	0.547	0.9792	0.0208
	5.52	0.322	0.193	0.485	0.9792	0.0208
	6.89	0.383	0.170	0.447	0.9792	0.0208
	0.569	0.0383	0.124	0.838	0.9569	0.0431
	1.28	0.0847	0.241	0.674	0.9569	0.0431
	1.95	0.125	0.315	0.560	0.9569	0.0431

(cont.)

AUXILIARY INFORMATION

METHOD/APPARATUS/PROCEDURE:	SOURCE AND PURITY OF MATERIALS:
The solubilities were determined by measurement of retention volumes using gas chromatography. The method uses methane as a carrier gas, propane as an injected solute and decane as the stationary phase. The technique is described in the source and ref. (1).	1. Sample dried, purity 99.7 mole per cent, 0.2 mole per cent nitrogen and 0.1 mole per cent ethane. 2. Phillips Petroleum sample, purity 99.5 mole per cent. 3. Phillips Petroleum research grade sample, purity 99.35 mole per cent.

	ESTIMATED ERROR:
	$\delta T/K = \pm 0.1$; δP/MPa $= \pm 2\%$; δx, $\delta y = \pm 6\%$ (estimated by compiler).

	REFERENCES:
	1. Koonce, K. T. *Ph.D. Thesis* Rice University, Houston, <u>1963</u>.

COMPONENTS:	ORIGINAL MEASUREMENTS:
1. Methane; CH_4; [74-82-8]	Koonce, K. T.; Kobayashi, R.
2. Propane; C_3H_8; [74-98-6]	*J. Chem. Eng. Data*
3. Decane; $C_{10}H_{22}$; [124-18-5]	<u>1964</u>, *9*, 494-501.

EXPERIMENTAL VALUES:

T/K (T/°F)	P/MPa	x_{CH_4}	$x_{C_3H_8}$	$x_{C_{10}H_{22}}$	y_{CH_4}	$y_{C_3H_8}$

T/K (T/°F)	P/MPa	Mole fractions in liquid x_{CH_4}	$x_{C_3H_8}$	$x_{C_{10}H_{22}}$	in vapor y_{CH_4}	$y_{C_3H_8}$
244.26 (-20)	2.63	0.177	0.359	0.464	0.9569	0.0431
	4.01	0.259	0.395	0.346	0.9569	0.0431
	5.41	0.352	0.365	0.283	0.9569	0.0431
	6.95	0.429	0.306	0.265	0.9569	0.0431
	0.362	0.0239	0.131	0.845	0.9310	0.0690
	0.793	0.0517	0.268	0.680	0.9310	0.0690
	1.36	0.0870	0.379	0.534	0.9310	0.0690
	2.17	0.149	0.519	0.332	0.9310	0.0690
	2.81	0.198	0.585	0.217	0.9310	0.0690
255.37 (0)	0.141	0.0100	0.0	0.9900	1.0	0.0
	0.201	0.0140	0.0	0.9860	1.0	0.0
	0.343	0.0213	0.0	0.9787	1.0	0.0
	0.545	0.0351	0.0	0.9649	1.0	0.0
	0.910	0.0578	0.0	0.9422	1.0	0.0
	1.73	0.108	0.0	0.892	1.0	0.0
	2.63	0.161	0.0	0.839	1.0	0.0
	3.12	0.188	0.0	0.812	1.0	0.0
	3.76	0.219	0.0	0.781	1.0	0.0
	4.80	0.270	0.0	0.730	1.0	0.0
	6.16	0.322	0.0	0.678	1.0	0.0
	0.234	0.0151	0.0184	0.9665	0.9792	0.0208
	0.434	0.0275	0.0339	0.9386	0.9792	0.0208
	0.710	0.0447	0.0516	0.9037	0.9792	0.0208
	1.39	0.0851	0.0897	0.8252	0.9792	0.0208
	2.05	0.126	0.116	0.758	0.9792	0.0208
	2.74	0.163	0.135	0.702	0.9792	0.0208
	4.21	0.245	0.152	0.603	0.9792	0.0208
	5.47	0.297	0.154	0.549	0.9792	0.0208
	6.96	0.346	0.144	0.510	0.9792	0.0208
	0.372	0.0232	0.0588	0.9180	0.9569	0.0431
	0.738	0.0454	0.111	0.844	0.9569	0.0431
	1.40	0.0839	0.184	0.732	0.9569	0.0431
	2.08	0.125	0.238	0.637	0.9569	0.0431
	2.77	0.165	0.275	0.560	0.9569	0.0431
	4.17	0.245	0.304	0.451	0.9569	0.0431
	5.48	0.306	0.304	0.390	0.9569	0.0431
	6.86	0.361	0.276	0.363	0.9569	0.0431
	0.431	0.0261	0.109	0.865	0.9310	0.0690
	0.807	0.0480	0.192	0.760	0.9310	0.0690
	1.39	0.0817	0.289	0.629	0.9310	0.0690
	2.08	0.122	0.373	0.505	0.9310	0.0690
	2.73	0.168	0.421	0.411	0.9310	0.0690
	4.05	0.248	0.466	0.286	0.9310	0.0690
	5.59	0.336	0.448	0.216	0.9310	0.0690
	6.89	0.394	0.404	0.202	0.9310	0.0690
	0.558	0.0327	0.187	0.780	0.9056	0.0994
	0.896	0.0517	0.281	0.667	0.9056	0.0944
	1.46	0.0831	0.400	0.517	0.9056	0.0944
	2.09	0.128	0.502	0.370	0.9056	0.0944
	2.76	0.178	0.569	0.253	0.9056	0.0944
277.59 (40)	0.154	0.0091	0.0	0.9909	1.0	0.0
	0.203	0.0118	0.0	0.9882	1.0	0.0
	0.343	0.0203	0.0	0.9797	1.0	0.0
	0.565	0.0331	0.0	0.9669	1.0	0.0
	0.883	0.0508	0.0	0.9492	1.0	0.0
	1.24	0.0699	0.0	0.9301	1.0	0.0
	1.24	0.0962	0.0	0.9038	1.0	0.0
	2.41	0.129	0.0	0.871	1.0	0.0

(cont.)

COMPONENTS:	ORIGINAL MEASUREMENTS:
1. Methane; CH_4; [74-82-8]	Koonce, K. T.; Kobayashi, R.
2. Propane; C_3H_8; [74-98-6]	*J. Chem. Eng. Data*
3. Decane; $C_{10}H_{22}$; [124-18-5]	1964, *9*, 494-501.

EXPERIMENTAL VALUES:

T/K (T/°F)	P/MPa	Mole fractions in liquid x_{CH_4}	$x_{C_3H_8}$	$x_{C_{10}H_{22}}$	in vapor y_{CH_4}	$y_{C_3H_8}$
277.59	3.44	0.178	0.0	0.822	1.0	0.0
(40)	4.81	0.237	0.0	0.763	1.0	0.0
	5.61	0.265	0.0	0.735	1.0	0.0
	6.12	0.282	0.0	0.718	1.0	0.0
	6.79	0.303	0.0	0.697	1.0	0.0
	0.345	0.0200	0.0137	0.9663	0.9792	0.0208
	0.710	0.0406	0.0276	0.9318	0.9792	0.0208
	1.39	0.0759	0.0484	0.8757	0.9792	0.0208
	2.05	0.110	0.0646	0.825	0.9792	0.0208
	2.73	0.144	0.0773	0.779	0.9792	0.0208
	4.13	0.206	0.0929	0.701	0.9792	0.0208
	5.44	0.265	0.0981	0.637	0.9792	0.0208
	7.05	0.322	0.100	0.578	0.9792	0.0208
	0.317	0.0180	0.0268	0.9552	0.9569	0.0431
	0.703	0.0391	0.0555	0.9054	0.9569	0.0431
	1.37	0.0730	0.0980	0.8290	0.9569	0.0431
	2.07	0.108	0.133	0.759	0.9569	0.0431
	2.74	0.143	0.156	0.701	0.9569	0.0431
	4.06	0.204	0.188	0.608	0.9569	0.0431
	5.42	0.259	0.199	0.542	0.9569	0.0431
	6.91	0.309	0.200	0.491	0.9569	0.0431
	0.341	0.0188	0.0437	0.9375	0.9569	0.0431
	0.779	0.0421	0.0965	0.8614	0.9569	0.0431
	1.39	0.0727	0.156	0.771	0.9310	0.0690
	2.08	0.106	0.207	0.687	0.9310	0.0690
	2.76	0.141	0.246	0.613	0.9310	0.0690
	4.09	0.202	0.289	0.509	0.9310	0.0690
	5.43	0.258	0.307	0.435	0.9310	0.0690
	6.82	0.307	0.300	0.393	0.9310	0.0690
	0.290	0.0156	0.0539	0.9305	0.9056	0.0944
	0.663	0.0348	0.114	0.851	0.9056	0.0944
	1.35	0.0686	0.208	0.723	0.9056	0.0944
	2.03	0.100	0.277	0.623	0.9056	0.0944
	2.72	0.138	0.327	0.535	0.9056	0.0944
	4.05	0.199	0.385	0.416	0.9056	0.0944
	5.47	0.260	0.405	0.335	0.9056	0.0944
	6.19	0.287	0.410	0.303	0.9056	0.0944
	6.91	0.312	0.400	0.288	0.9056	0.0944
	0.303	0.0156	0.0748	0.9096	0.8691	0.1309
	0.512	0.0261	0.121	0.853	0.8691	0.1309
	0.717	0.0362	0.169	0.795	0.8691	0.1309
	1.39	0.0674	0.286	0.647	0.8691	0.1309
	2.05	0.0973	0.375	0.528	0.8691	0.1309
	2.76	0.136	0.442	0.422	0.8691	0.1309
	4.12	0.204	0.513	0.283	0.8691	0.1309
	4.76	0.234	0.530	0.236	0.8691	0.1309
	5.52	0.267	0.530	0.203	0.8691	0.1309
	6.21	0.293	0.528	0.180	0.8691	0.1309
	6.89	0.312	0.498	0.190	0.8691	0.1309
	0.290	0.0144	0.0830	0.9026	0.8373	0.1627
	0.493	0.0242	0.148	0.828	0.8373	0.1627
	0.703	0.0342	0.204	0.762	0.8373	0.1627
	1.38	0.0644	0.355	0.581	0.8373	0.1627
	2.07	0.0949	0.464	0.441	0.8373	0.1627
	2.77	0.135	0.548	0.317	0.8373	0.1627
	4.13	0.209	0.623	0.168	0.8373	0.1627
294.26	0.214	0.0119	0.0	0.9881	1.0	0.0
(70)	0.355	0.0195	0.0	0.9805	1.0	0.0

(cont.)

COMPONENTS:	ORIGINAL MEASUREMENTS:
1. Methane; CH_4; [74-82-8]	Koonce, K. T.; Kobayashi, R.
2. Propane; C_3H_8; [74-98-6]	*J. Chem. Eng. Data*
3. Decane; $C_{10}H_{22}$; [124-18-5]	1964, *9*, 494-501.

EXPERIMENTAL VALUES:

T/K (T/°F)	P/MPa	Mole fractions in liquid x_{CH_4}	$x_{C_3H_8}$	$x_{C_{10}H_{22}}$	in vapor y_{CH_4}	$y_{C_3H_8}$
294.26 (70)	0.372	0.0203	0.0	0.9797	1.0	0.0
	0.634	0.0339	0.0	0.9661	1.0	0.0
	1.71	0.0862	0.0	0.9138	1.0	0.0
	2.63	0.130	0.0	0.870	1.0	0.0
	3.54	0.172	0.0	0.828	1.0	0.0
	4.81	0.217	0.0	0.783	1.0	0.0
	6.28	0.265	0.0	0.735	1.0	0.0
	0.283	0.0153	0.0076	0.9771	0.9792	0.0208
	0.485	0.0258	0.0130	0.9612	0.9792	0.0208
	0.710	0.0370	0.0173	0.9457	0.9792	0.0208
	1.39	0.0704	0.0319	0.8977	0.9792	0.0208
	2.07	0.102	0.0441	0.854	0.9792	0.0208
	2.74	0.132	0.0535	0.815	0.9792	0.0208
	4.12	0.191	0.0658	0.743	0.9792	0.0208
	5.52	0.243	0.0746	0.682	0.9792	0.0208
	6.86	0.280	0.0806	0.743	0.9792	0.0208
	0.290	0.0153	0.0165	0.682	0.9792	0.0208
	0.486	0.0253	0.0263	0.9484	0.9569	0.0431
	0.703	0.0360	0.0362	0.9278	0.9569	0.0431
	1.39	0.0684	0.0664	0.8652	0.9569	0.0431
	2.08	0.0999	0.0902	0.8099	0.9569	0.0431
	2.90	0.137	0.113	0.750	0.9569	0.0431
	4.13	0.191	0.136	0.673	0.9569	0.0431
	5.34	0.236	0.150	0.614	0.9569	0.0431
	6.74	0.275	0.157	0.568	0.9310	0.0690
	0.303	0.0155	0.0260	0.9585	0.9310	0.0690
	0.510	0.0257	0.0426	0.9317	0.9310	0.0690
	0.758	0.0378	0.0627	0.8995	0.9310	0.0690
	1.32	0.0638	0.101	0.835	0.9310	0.0690
	2.01	0.0940	0.140	0.766	0.9310	0.0690
	2.79	0.129	0.174	0.697	0.9310	0.0690
	4.11	0.186	0.209	0.605	0.9310	0.0690
	5.47	0.234	0.233	0.533	0.9310	0.0690
	6.83	0.274	0.242	0.484	0.9310	0.0690
	3.62	0.0179	0.0427	0.9394	0.9056	0.0944
	7.45	0.0359	0.0874	0.8767	0.9056	0.0944
	1.39	0.0647	0.143	0.792	0.9056	0.0944
	2.07	0.0941	0.196	0.710	0.9056	0.0944
	2.77	0.128	0.234	0.638	0.9056	0.0944
	4.10	0.184	0.283	0.533	0.9056	0.0944
	5.52	0.234	0.318	0.448	0.9056	0.0944
	6.87	0.274	0.323	0.403	0.9056	0.0944
	0.331	0.0158	0.0545	0.9297	0.8691	0.1309
	0.689	0.0320	0.109	0.859	0.8691	0.1309
	1.34	0.0599	0.196	0.744	0.8691	0.1309
	2.07	0.0905	0.269	0.641	0.8691	0.1309
	2.73	0.124	0.317	0.559	0.8691	0.1309
	4.09	0.179	0.376	0.445	0.8691	0.1309
	5.54	0.231	0.408	0.361	0.8691	0.1309
	6.85	0.268	0.422	0.310	0.8691	0.1309
	0.414	0.0189	0.0847	0.8969	0.8373	0.1627
	0.752	0.0334	0.148	0.819	0.8373	0.1627
	1.38	0.0598	0.242	0.698	0.8373	0.1627
	2.08	0.0881	0.329	0.583	0.8373	0.1627
	2.81	0.122	0.390	0.488	0.8373	0.1627
	4.19	0.182	0.462	0.356	0.8373	0.1627
	5.47	0.225	0.496	0.279	0.8373	0.1627
	6.88	0.268	0.507	0.225	0.8373	0.1627

COMPONENTS:	ORIGINAL MEASUREMENTS:
1. Methane; CH_4; [74-82-8] 2. Propane; C_3H_8; [74-98-6] 3. Decane; $C_{10}H_{22}$; [124-18-5]	Wiese, H. C.; Reamer, H. H.; Sage, B. H. *J. Chem. Eng. Data* <u>1970</u>, *15*, 75-82.

VARIABLES:	PREPARED BY:
Temperature, pressure, liquid phase composition	C. L. Young

EXPERIMENTAL VALUES:

T/K (T/°F)	P/MPa (P/psia)	†Composition factor	Mole fraction in liquid			Mole fraction in vapor		
			x_{CH_4}	$x_{C_3H_8}$	$x_{C_{10}H_{22}}$	y_{CH_4}	$y_{C_3H_8}$	$y_{C_{10}H_{22}}$
277.6 (40)	2.76 (400)	0.0	0.1350	0	0.8650	0.9997	0	0.0003
		0.2	0.1352	0.1730	0.6919	0.9497	0.0500	0.0003
		0.4	0.1364	0.3454	0.5181	0.8998	0.0999	0.0003
		0.6	0.1380	0.5172	0.3448	0.8485	0.1513	0.0002
		0.8	0.1411	0.6871	0.1718	0.7971	0.2027	0.0002
		1.0	0.1471	0.8529	0	0.7505	0.2495	0
	6.89 (1000)	0.0	0.2845	0	0.7115	0.9994	0	0.0006
		0.2	0.3017	0.1397	0.5586	0.9692	0.0303	0.0005
		0.4	0.3110	0.2756	0.4154	0.9378	0.0617	0.0005
		0.6	0.3245	0.4053	0.2702	0.9022	0.0973	0.0005
		0.8	0.3466	0.5227	0.1307	0.8579	0.1417	0.0004
		1.0	0.4226	0.4774	0	0.8208	0.1792	0
	13.79 (2000)	0.0	0.4851	0	0.5149	0.9985	0	0.0015
		0.2	0.4940	0.1012	0.4048	0.9671	0.0312	0.0017
		0.4	0.5109	0.1956	0.2935	0.9299	0.0682	0.0020
		0.6	0.5473	0.2716	0.1811	0.8853	0.1116	0.0030
		0.8	0.6181	0.3055	0.0764	0.8190	0.1734	0.0076
	20.68 (3000)	0.0	0.6115	0	0.3885	0.9948	0	0.0051
		0.2	0.6182	0.0764	0.3054	0.9599	0.0338	0.0064
		0.4	0.6455	0.1418	0.2127	0.9160	0.0735	0.0105
		0.6	0.7100	0.1740	0.1160	0.8485	0.1270	0.0245
	27.58 (4000)	0.0	0.7015	0	0.2985	0.9863	0	0.0137
		0.2	0.7181	0.0564	0.2255	0.9389	0.0353	0.0239
		0.4	0.7797	0.0881	0.1322	0.8820	0.0694	0.0485

(cont.)

AUXILIARY INFORMATION

METHOD/APPARATUS/PROCEDURE:	SOURCE AND PURITY OF MATERIALS:
PVT cell charged with mixture of known composition. Pressure measured with pressure balance and temperature measured using a platinum resistance thermometer. Details in ref. (1). Samples of coexisting phases analysed by G.C. Details in source.	1. Texaco sample, purified by passage over drying agents charcoal and CO removed, final purity 99.77 mole per cent. 2. Phillips Petroleum Co. sample, purity 99.99 mole per cent. 3. Phillips Petroleum Co. sample, purity 99.49 mole per cent.

ESTIMATED ERROR:

$\delta T/K = \pm 0.01$; $\delta P/MPa = \pm 0.1\%$;

δx_{CH_4}, $\delta y_{CH_4} = \pm 0.005$ or better.

REFERENCES:

1. Sage, B. H.; Lacey, W. W.
 Trans. Am. Inst. Mining Met. Engnrs.
 <u>1940</u>, *136*, 136.

COMPONENTS:	ORIGINAL MEASUREMENTS:
1. Methane; CH_4; [74-82-8]	Wiese, H. C.; Reamer, H. H.;
2. Propane; C_3H_8; [74-98-6]	Sage, B. H.
3. Decane; $C_{10}H_{22}$; [124-18-5]	*J. Chem. Eng. Data*
	<u>1970</u>, *15*, 75-82.

EXPERIMENTAL VALUES:

T/K (T/°F)	P/MPa (P/psia)	†Compo- sition factor	\multicolumn{3}{c}{Mole fraction in liquid}			\multicolumn{3}{c}{in vapor}		
			x_{CH_4}	$x_{C_3H_8}$	$x_{C_{10}H_{22}}$	y_{CH_4}	$y_{C_3H_8}$	$y_{C_{10}H_{22}}$
310.9	2.76	0.0	0.1251	0	0.8749	0.9994	0	0.0006
(100)	(400)	0.2	0.1113	0.1777	0.7110	0.8950	0.1044	0.0006
		0.4	0.0994	0.3603	0.5404	0.7850	0.2144	0.0006
		0.6	0.0918	0.5449	0.3633	0.6724	0.3270	0.0006
		0.8	0.0862	0.7310	0.1828	0.5573	0.4423	0.0004
		1.0	0.0845	0.9155	0	0.4472	0.5528	0
	6.89	0.0	0.2679	0	0.7321	0.9993	0	0.0007
	(1000)	0.2	0.2579	0.1484	0.5937	0.9466	0.0527	0.0008
		0.4	0.2533	0.2987	0.4480	0.8910	0.1081	0.0008
		0.6	0.2558	0.4465	0.2977	0.8200	0.1791	0.0009
		0.8	0.2634	0.5893	0.1473	0.7351	0.2640	0.0009
		1.0	0.3231	0.6769	0	0.6635	0.3365	0
	13.79	0.0	0.4469	0	0.5541	0.9981	0	0.0019
	(2000)	0.2	0.4437	0.1113	0.4450	0.9552	0.0427	0.0021
		0.4	0.4463	0.2215	0.3322	0.9038	0.0936	0.0026
		0.6	0.4619	0.3229	0.2153	0.8395	0.1568	0.0038
		0.8	0.5168	0.3865	0.0966	0.7457	0.2453	0.0090
	20.68	0.0	0.5827	0	0.4173	0.9942	0	0.0058
	(3000)	0.2	0.5824	0.0835	0.3340	0.9533	0.0394	0.0073
		0.4	0.5944	0.1622	0.2434	0.8989	0.0892	0.0119
		0.6	0.6344	0.2193	0.1462	0.8244	0.1528	0.0228
	27.58	0.0	0.6870	0	0.3130	0.9837	0	0.0163
	(4000)	0.2	0.6950	0.0610	0.2440	0.9383	0.0367	0.0250
		0.4	0.7283	0.1087	0.1630	0.8677	0.0819	0.0504
410.9	2.76	0.0	0.0974	0	0.9026	0.9798	0	0.0202
(280)	(400)	0.2	0.0632	0.1874	0.7494	0.5872	0.3925	0.0203
		0.4	0.0251	0.3900	0.5850	0.2081	0.7702	0.0217
	6.89	0.0	0.2256	0	0.7744	0.9870	0	0.0130
	(1000)	0.2	0.1941	0.1612	0.6447	0.7973	0.1881	0.0146
		0.4	0.1591	0.3363	0.5045	0.5992	0.3844	0.0164
		0.6	0.1209	0.5274	0.3516	0.3930	0.5881	0.0189
		0.8	0.0791	0.7367	0.1842	0.1808	0.7883	0.0309
	13.79	0.0	0.4028	0	0.5972	0.9845	0	0.0155
	(2000)	0.2	0.3791	0.1242	0.4967	0.8871	0.0942	0.0187
		0.4	0.3588	0.2565	0.3847	0.7715	0.2056	0.0230
		0.6	0.3472	0.3917	0.2611	0.6311	0.3333	0.0356
	20.68	0.0	0.5476	0	0.4524	0.9715	0	0.0285
	(3000)	0.2	0.5381	0.0924	0.3695	0.8971	0.0676	0.0353
		0.4	0.5397	0.1841	0.2762	0.8032	0.1456	0.0512
		0.6	0.5950	0.2425	0.1625	0.6915	0.2205	0.0880
	27.58	0.0	0.6912	0	0.3088	0.9430	0	0.0570
	(4000)	0.2	0.7116	0.0577	0.2308	0.8690	0.0476	0.0834
477.6	2.76	0.0	0.0926	0	0.9074	0.8851	0	0.1149
(400)	(400)	0.2	0.0364	0.1927	0.7709	0.3090	0.5782	0.1128
	6.89	0.0	0.2258	0	0.7742	0.9358	0	0.0642
	(1000)	0.2	0.1768	0.1646	0.6586	0.6717	0.2580	0.0703
		0.4	0.1228	0.3509	0.5263	0.4010	0.5193	0.0797
		0.6	0.0579	0.5653	0.3767	0.1409	0.7524	0.1067
	13.79	0.0	0.4119	0	0.5811	0.9351	0	0.0649
	(2000)	0.2	0.3815	0.1237	0.4948	0.7963	0.1274	0.0763
		0.4	0.3571	0.2572	0.3857	0.6375	0.2638	0.0987
		0.6	0.3485	0.3909	0.2606	0.4425	0.3968	0.1607
	20.68	0.0	0.5912	0	0.4088	0.9030	0	0.0970
	(3000)	0.2	0.5984	0.0803	0.3213	0.7843	0.0745	0.1413
510.9	2.76	0.0	0.0937	0	0.9063	0.7928	0	0.2072
(460)	(400)	0.2	0.0223	0.1955	0.7822	0.1626	0.6184	0.2190
	6.89	0.0	0.2378	0	0.7622	0.8803	0	0.1197
	(1000)	0.2	0.1759	0.1648	0.6593	0.5935	0.2687	0.1378

(cont.)

COMPONENTS:	ORIGINAL MEASUREMENTS:
1. Methane; CH_4; [74-82-8]	Wiese, H. C.; Reamer, H. H.;
2. Propane; C_3H_8; [74-98-6]	Sage, B. H.
3. Decane; $C_{10}H_{22}$; [124-18-5]	*J. Chem. Eng. Data*
	<u>1970</u>, *15*, 75-82.

EXPERIMENTAL VALUES:

				Mole fraction				
				in liquid			in vapor	
T/K	*P*/MPa (*P*/psia)	[†]Composition factor	x_{CH_4}	$x_{C_3H_8}$	$x_{C_{10}H_{22}}$	y_{CH_4}	$y_{C_3H_8}$	$y_{C_{10}H_{22}}$
510.9 (460)	6.89 (1000)	0.4	0.1080	0.3568	0.5352	0.2947	0.5406	0.1647
		0.6	0.0333	0.5800	0.3867	0.0583	0.7192	0.2225
	13.79 (2000)	0.0	0.4280	0	0.5720	0.8815	0	0.1185
		0.2	0.4062	0.1188	0.4751	0.7230	0.1282	0.1489
		0.4	0.3899	0.2440	0.3661	0.5396	0.2536	0.2068

$$[†]\text{Composition factor} = \frac{\text{Mole fraction of propane in liquid}}{\text{Mole fraction of propane in liquid} + \text{Mole fraction of decane in liquid}}$$

COMPONENTS:	ORIGINAL MEASUREMENTS:
1. Methane; CH_4; [74-82-8] 2. Butane; C_4H_{10}; [106-97-8] 3. Octane; C_8H_{18}; [111-65-9]	Hottovy, J. D.; Kohn, J. P.; Luks, K. D. *J. Chem. Eng. Data* 1982, *27*, 298-302.
VARIABLES:	PREPARED BY: C. L. Young

EXPERIMENTAL VALUES:

Data for octane-lean liquid phase.

Type of data	T/K	P/atm	P/MPa	Mole fractions $x_{C_4H_{10}}$	$x_{C_8H_{18}}$	Molar volume /cm^{3}mol^{-1}
K(L_1-L_2=V)	206.44	63.2	6.40	0.045	0.0046	72.9
	204.28	60.3	6.11	0.029	0.0025	79.5
	203.28	59.3	6.01	0.023	0.0029	82.6
	202.09	57.4	5.82	0.020	0.0019	86.3
	201.77	57.2	5.80	0.022	0.0230	83.5
	197.96	53.0	5.37	0.012	0.0017	90.5
Q(S-L_1-L_2-V)	197.57	52.6	5.33	0.012	0.0017	87.0
	197.45	52.3	5.30	0.017	0.0031	72.8
	197.02	51.5	5.22	0.022	0.0039	70.6
	196.78	50.9	5.16	0.025	0.0046	68.9
	196.25	49.9	5.06	0.031	0.0060	67.6
	196.20	50.0	5.07	0.029	0.0059	65.9
	195.84	49.2	4.99	0.037	0.0075	67.3
	195.65	49.0	4.96	0.036	0.0075	64.4
	195.59	48.9	4.95	0.040	0.0081	64.5
	195.39	48.6	4.92	0.042	0.0089	62.8
	195.07	47.9	4.85	0.047	0.1030	62.4
	194.85	47.5	4.81	0.052	0.0097	62.5
	194.07	46.1	4.67	0.061	0.0170	59.4
	193.83	45.7	4.63	0.068	0.0202	59.1
	193.49	45.1	4.57	0.088	0.0314	59.7

(cont.)

AUXILIARY INFORMATION

METHOD/APPARATUS/PROCEDURE:	SOURCE AND PURITY OF MATERIALS:

Glass equilibrium cell.
Temperature measured with platinum resistance thermometer and pressure with Bourdon gauge. Stoichiometry and volumetric measurements were used to compute liquid phase compositions and molar volumes.
Details in ref. 1 and source.
	1. Linde Ultrapure grade, purity
	   99.97 mole per cent.
	
	2. Linde Instrument grade, purity
	   99.5 mole per cent.
	
	3. Humphrey-Wilkinson grade, purity
	   99 mole per cent.

ESTIMATED ERROR:
δT/K = ±0.03; δP/MPa = ±0.07;
$\delta x_{C_4H_{10}}$ = ±3.5%; $\delta x_{C_8H_{18}}$ = ±2%
(octane rich phase); $\delta x_{C_8H_{18}}$ = ±8%
(octane lean phase).

REFERENCES:

1. Kohn, J. P.
   Am. Inst. Chem. Engnrs. J.
   1961, *7*, 514.

COMPONENTS:	ORIGINAL MEASUREMENTS:

1. Methane; CH_4; [74-82-8]

2. Butane; C_4H_{10}; [106-97-8]

3. Octane; C_8H_{18}; [111-65-9]

Hottovy, J. D.; Kohn, J. P.;

Luks, K. D.

J. Chem. Eng. Data

1982, *27*, 298-302.

EXPERIMENTAL VALUES:

Data for octane-lean liquid phase.

Type of data	T/K	$P/$atm	$P/$MPa	Mole fractions		Molar volume /cm^{3}mol^{-1}
				$x_{C_4H_{10}}$	$x_{C_8H_{18}}$	
LCST(L_1=L_2-V)	204.61	60.2	6.10	0.086	0.0186	60.5
	202.47	57.2	5.80	0.102	0.0278	60.5
	197.16	49.8	5.05	0.106	0.0383	59.5
	195.64	47.8	4.84	0.108	0.0425	60.9
	193.48	45.0	4.56	0.098	0.0384	60.0
L_1-L_2-V	206.00	62.5	6.33	0.053	0.0071	63.9
	204.92	61.0	6.18	0.048	0.0059	69.5
	204.00	59.6	6.04	0.063	0.0077	64.4
	204.00	59.4	6.02	0.070	0.0141	62.9
	206.00	62.5	6.33	0.053	0.0071	63.9
	204.92	61.0	6.18	0.048	0.0059	69.5
	204.00	59.6	6.04	0.063	0.0077	64.4
	204.00	59.4	6.02	0.070	0.0141	62.9
	202.00	57.3	5.81	0.035	0.0048	70.1
	201.95	57.0	5.78	0.026	0.0050	76.3
	202.00	57.0	5.78	0.044	0.0072	64.6
	202.00	56.8	5.76	0.035	0.0082	65.3
	202.00	56.7	5.75	0.073	0.0126	61.6
	202.00	56.5	5.72	0.072	0.0145	62.0
	201.15	56.3	5.70	0.036	0.0097	70.3
	200.00	55.2	5.59	0.022	0.0031	77.3
	200.00	55.0	5.57	0.023	0.0037	73.0
	200.00	54.9	5.56	0.028	0.0041	70.4
	200.00	54.8	5.55	0.030	0.0047	68.3
	200.00	54.6	5.53	0.036	0.0055	67.6
	200.01	54.4	5.51	0.040	0.0064	65.8
	200.00	53.8	5.45	0.087	0.0229	60.0
	198.00	52.6	5.33	0.023	0.0042	71.8
	197.99	52.3	5.30	0.028	0.0054	69.4
	198.00	52.1	5.28	0.033	0.0063	66.8
	198.00	52.0	5.27	0.036	0.0063	65.6
	198.00	51.9	5.26	0.039	0.0072	64.3
	198.00	51.2	5.19	0.069	0.0161	61.0
	196.00	49.3	5.00	0.035	0.0075	63.9
	196.00	49.1	4.98	0.042	0.0091	62.9
	196.00	48.5	4.91	0.075	0.0203	59.2

(cont.)

COMPONENTS:

1. Methane; CH_4; [74-82-8]
2. Butane; C_4H_{10}; [106-97-8]
3. Octane; C_8H_{18}; [111-65-9]

ORIGINAL MEASUREMENTS:

Hottovy, J. D.; Kohn, J. P.;
Luks, K. D.

J. Chem. Eng. Data
<u>1982</u>, *27*, 298-302.

EXPERIMENTAL VALUES:

Data for octane-rich liquid phase.

Type of data	T/K	P/atm	P/MPa	Mole fractions $x_{C_4H_{10}}$	$x_{C_8H_{18}}$	Molar volume /cm^{3}mol^{-1}
K(L_1-L_2=V)	206.39	63.0	6.38	0.113	0.032	62.6
	205.63	62.2	6.30	0.127	0.039	61.1
	203.90	59.9	6.07	0.145	0.063	59.4
	203.62	59.5	6.03	0.149	0.068	60.3
	203.41	59.3	6.01	0.153	0.072	62.5
	201.45	56.9	5.77	0.159	0.107	64.1
	200.20	55.6	5.63	0.156	0.131	66.0
	198.93	54.1	5.48	0.149	0.161	68.4
Q(S-L_1-L_2-V)	196.31	50.2	5.09	0.145	0.162	68.3
	196.09	49.8	5.05	0.147	0.158	68.1
	195.07	48.1	4.87	0.146	0.128	65.6
	194.85	47.6	4.82	0.148	0.123	65.5
	194.10	46.3	4.69	0.142	0.098	63.2
	193.91	46.0	4.66	0.144	0.097	61.7
L_1-L_2-V	204.00	59.6	6.04	0.109	0.030	59.9
	204.00	59.5	6.03	0.114	0.033	60.1
	203.79	59.1	5.99	0.110	0.030	60.2
	203.00	58.6	5.94	0.150	0.068	61.0
	202.00	57.1	5.79	0.151	0.071	62.1
	202.00	57.0	5.78	0.146	0.067	61.6
	202.00	56.6	5.73	0.130	0.049	60.2
	200.00	55.2	5.59	0.152	0.127	64.4
	200.00	55.2	5.59	0.152	0.128	64.7
	200.00	54.6	5.53	0.151	0.104	63.2
	200.00	54.6	5.53	0.153	0.103	63.8
	200.00	54.0	5.47	0.137	0.064	61.3
	200.00	53.9	5.46	0.139	0.063	61.5
	200.00	53.7	5.44	0.121	0.045	60.3
	198.00	52.7	5.34	0.146	0.161	67.7
	198.00	52.6	5.33	0.147	0.160	68.2
	198.00	52.3	5.30	0.150	0.130	65.2
	198.00	52.2	5.29	0.153	0.127	66.0
	198.00	51.7	5.24	0.146	0.105	63.0
	198.00	51.6	5.23	0.150	0.101	63.6
	198.00	51.6	5.23	0.151	0.100	63.8
	198.00	51.2	5.19	0.137	0.067	61.0
	198.00	51.1	5.18	0.131	0.060	60.2
	196.00	49.4	5.01	0.149	0.129	65.4
	196.00	49.3	5.00	0.149	0.123	64.9
	196.00	48.9	4.95	0.148	0.103	63.5
	196.00	48.8	4.94	0.147	0.098	63.2
	196.00	48.5	4.91	0.127	0.060	59.7
	196.00	48.4	4.90	0.122	0.055	60.4
	195.00	47.1	4.77	0.118	0.054	59.4

COMPONENTS:	ORIGINAL MEASUREMENTS:
1. Methane; CH_4; [74-82-8] 2. 2,2-Dimethylpropane (neopentane); C_5H_{12}; [463-82-1] 3. Pentane; C_5H_{12}; [109-66-0]	Prodany, N. W.; Williams, B. *J. Chem. Engng. Data* 1971, *16*, 1-6.

VARIABLES:	PREPARED BY:
	C. L. Young

EXPERIMENTAL VALUES:

T/K = 344.26   (T/°F = 160)

		Mole fractions					
		in liquid			in vapor		
P/MPa	p/psi	Comp.1	Comp.2	Comp.3	Comp.1	Comp.2	Comp.3
3.47	503	0.1407	0.2165	0.6429	0.8451	0.0577	0.0972
5.18	751	0.2058	0.1997	0.5945	0.8714	0.0459	0.0827
6.94	1006	0.2781	0.1760	0.5459	0.8787	0.0406	0.0807
8.63	1251	0.3375	0.1641	0.4985	0.8778	0.0404	0.0818
10.38	1505	0.3998	0.1482	0.4520	0.8699	0.0410	0.0891
12.13	1759	0.4607	0.1314	0.4079	0.8546	0.0429	0.1025
13.88	2013	0.5500	0.1112	0.3388	0.8052	0.0536	0.1413
14.62	2120	0.6019	0.0976	0.3005	0.7753	0.0591	0.1655

AUXILIARY INFORMATION

METHOD/APPARATUS/PROCEDURE:	SOURCE AND PURITY OF MATERIALS:
Stirred equilibrium cell fitted with vapor and liquid sampling valves. Temperature measured with mercury in glass thermometer. Pressure measured with Bourdon gauge. Cell charged with components and contents equilibrated. Vapor and liquid samples withdrawn through pressure lock systems. Analysed using gas chromatography. Details in source.	1. Phillips Petroleum Co. sample, 99.3 mole per cent (0.6 mole per cent nitrogen, 0.1 mole per cent ethane). 2 and 3. Phillips Petroleum Co. samples, purities 99.8 and 99.9 mole per cent, respectively.
	ESTIMATED ERROR: $\delta T/K = \pm 0.3$; $\delta P/MPa = \pm 0.02$; $\delta x_{CH_4} = \pm 0.75\%$.
	REFERENCES:

COMPONENTS:	ORIGINAL MEASUREMENTS:
1. Methane; CH_4; [74-82-8] 2. 2-Methylbutane (*isopentane*); C_5H_{12}; [78-78-4] 3. Pentane; C_5H_{12}; [109-66-0]	Prodany, N. W.; Williams, B. *J. Chem. Engng. Data* <u>1971</u>, *16*, 1-6.

VARIABLES:	PREPARED BY:
	C. L. Young

EXPERIMENTAL VALUES:

T/K (T/°F)	P/MPa	p/psi	Mole fractions in liquid Comp.1	Comp.2	Comp.3	in vapor Comp.1	Comp.2	Comp.3
344.26 (160)	3.47	504	0.1386	0.2227	0.6387	0.8712	0.0395	0.0892
	5.21	755	0.2114	0.2058	0.5828	0.8939	0.0305	0.0755
	6.92	1003	0.2738	0.1888	0.5374	0.8971	0.0296	0.0733
	10.29	1493	0.4057	0.1536	0.4408	0.8890	0.0307	0.0803
	13.62	1975	0.5038	0.1290	0.3672	0.8488	0.0405	0.1107
	13.76	1995	0.5210	0.1229	0.3561	0.8419	0.0422	0.1159
	15.64	2268	0.5934	0.1039	0.3207	0.7585	0.0632	0.1783
377.59 (220)	3.50	507	0.1202	0.2198	0.6599	0.7445	0.0701	0.1854
	5.19	753	0.1867	0.2047	0.6086	0.7878	0.0577	0.1545
	6.86	995	0.2513	0.1876	0.5611	0.8096	0.0510	0.1394
	8.71	1263	0.3191	0.1716	0.5093	0.8100	0.0509	0.1391
	10.47	1519	0.3887	0.1540	0.4573	0.8006	0.0527	0.1466
	12.17	1765	0.4537	0.1391	0.4072	0.7706	0.0613	0.1681
	14.11	2047	0.5552	0.1117	0.3331	0.7470	0.0652	0.1878
410.93 (280)	3.72	539	0.0998	0.2196	0.6806	0.5633	0.1162	0.3205
	3.73	541	0.1030	0.2321	0.6649	0.5682	0.1197	0.3120
	5.22	757	0.1642	0.2143	0.6215	0.6292	0.1020	0.2687
	5.24	760	0.1622	0.2017	0.6361	0.6303	0.0967	0.2730
	6.90	1001	0.2311	0.1866	0.5822	0.6623	0.0866	0.2510
	7.11	1031	0.2417	0.1956	0.5628	0.6646	0.0910	0.2444
	8.64	1253	0.3064	0.1644	0.5291	0.6738	0.0809	0.2452
	8.65	1255	0.3042	0.1797	0.5161	0.6745	0.0870	0.2385
	10.79	1565	0.4533	0.1384	0.4083	0.6156	0.0994	0.2850

AUXILIARY INFORMATION

METHOD/APPARATUS/PROCEDURE:	SOURCE AND PURITY OF MATERIALS:
Stirred equilibrium cell fitted with vapor and liquid sampling valves. Temperature measured with mercury in glass thermometer. Pressure measured with Bourdon gauge. Cell charged with components and contents equilibrated. Vapor and liquid samples withdrawn through pressure lock systems. Analysed using gas chromatography. Details in source.	1. Phillips Petroleum Co. sample, 99.3 mole per cent (0.6 mole per cent nitrogen, 0.1 mole per cent ethane). 2 and 3. Phillips Petroleum Co. samples, purities both 99.9 mole per cent.

ESTIMATED ERROR:
$\delta T/K = \pm 0.3$; $\delta P/MPa = \pm 0.02$; $\delta x_{CH_4} = \pm 0.75\%$.

REFERENCES:

COMPONENTS:	ORIGINAL MEASUREMENTS:
1. Methane; CH_4; [74-82-8] 2. Pentane; C_5H_{12}; [109-66-0] 3. Octane; C_8H_{18}; [111-65-9]	Merrill, R. C.; Luks, K. D.; Kohn, J. P. *J. Chem. Eng. Data* <u>1983</u>, *28*, 210-215.

VARIABLES:	PREPARED BY:
	C. L. Young

EXPERIMENTAL VALUES:

Phases in equilibrium	T/K	P/atm	P/MPa	Mole fractions of pentane $x_{C_5H_{12}}$	octane $x_{C_8H_{18}}$
			Data for octane-lean phase, L_2		
$L_1,L_2 \equiv V$	196.96	52.60	5.330	0.0077	0.0018
	196.97	52.60	5.330	0.0078	0.0019
	197.73	53.08	5.378	0.0090	0.0027
	200.61	57.05	5.781	0.01620	0.0011
	200.71	57.08	5.784	0.01450	0.0019
S,L_1,L_2,V	191.86	44.81	4.540	0.0236	0.0044
	192.15	45.93	4.654	0.0192	0.0027
	192.83	46.37	4.698	0.0179	0.0035
	193.22	47.43	4.806	0.0151	0.0023
L_1,L_2,V	190.00	41.93	4.249	0.0631	0.0091
	192.00	44.93	4.553	0.0310	0.0045
	192.00	45.05	4.565	0.0240	0.0075
	192.00	45.21	4.581	0.0252	0.0048
	194.00	47.84	4.847	0.0191	0.0066
	194.00	47.94	4.858	0.0174	0.0052
	194.00	48.01	4.865	0.0165	0.0050
	194.00	48.51	4.915	0.0132	0.0033
	194.00	48.58	4.922	0.0135	0.0029
	194.00	48.65	4.929	0.0113	0.0026
	194.00	48.72	4.937	0.0122	0.0034
	196.00	50.69	5.136	0.0154 (cont.)	0.0056

AUXILIARY INFORMATION

METHOD/APPARATUS/PROCEDURE:	SOURCE AND PURITY OF MATERIALS:
Glass equilibrium cell. Temperature measured with platinum resistance thermometer and pressure with Bourdon gauge. Stoichiometry and volumetric measurements were used to obtain liquid phase compositions and molar volumes. Gas phase assumed to be pure methane except for $L_1,L_2 = V$ equilibrium. Molar volume data in source. Details of apparatus in ref. (1).	1. Linde, Ultra Pure grade sample, purity 99.97 moles per cent. 2. Phillips Petroleum Co. sample, purity 99 moles per cent. 3. Humphrey Chemical Co. sample, purity 99 moles per cent.

	ESTIMATED ERROR:
	$\delta T/K = \pm 0.03$; $\delta P/MPa = \pm 0.007$; $\delta x/x = \pm 0.02$ in L_1 phase, $\quad\quad\quad \pm 0.08$ in L_2 phase.

	REFERENCES:
	1. Hottovy, J. D.; Kohn, J. P.; Luks, K. D. *J. Chem. Eng. Data* <u>1981</u>, *26*, 135.

COMPONENTS:	ORIGINAL MEASUREMENTS:
1. Methane; CH_4; [74-82-8]	Merrill, R. C.; Luks, K. D.;
2. Pentane; C_5H_{12}; [109-66-0]	Kohn, J. P.
3. Octane; C_8H_{18}; [111-65-9]	*J. Chem. Eng. Data*
	<u>1983</u>, *28*, 210-215.

EXPERIMENTAL VALUES:

Phases in equilibrium	T/K	P/atm	P/MPa	Mole fractions of pentane $x_{C_5H_{12}}$	octane $x_{C_8H_{18}}$
L_1,L_2,V	196.00	50.89	5.156	0.0140	0.0043
	196.00	51.23	5.191	0.0099	0.0028
	196.00	51.37	5.205	0.0084	0.0022
	196.00	51.44	5.212	0.0079	0.0020
	198.00	53.14	5.384	0.0366	0.0020
	198.00	53.28	5.399	0.0248	0.0014
	198.00	53.47	5.418	0.0151	0.0009
	200.00	55.75	5.649	0.0399	0.0020
	200.00	55.81	5.655	0.0339	0.0020
	200.00	55.89	5.663	0.0332	0.0013
	200.00	55.96	5.670	0.0306	0.0016
		Data for octane-rich phase, L_1			
$L_1,L_2 \equiv V$	195.95	51.44	5.212	0.176	0.137
	196.31	51.85	5.254	0.220	0.113
	196.67	52.33	5.302	0.189	0.114
	199.84	49.34	4.999	0.172	0.031
	200.35	56.84	5.759	0.104	0.015
	201.28	58.41	5.918	0.135	0.014
S,L_1,L_2,V	189.32	41.71	4.226	0.163	0.081
	190.75	43.62	4.420	0.169	0.104
	191.79	45.11	4.571	0.185	0.096
$L_1 \equiv L_2,V$	193.58	46.97	4.759	0.100	0.018
	193.75	47.10	4.772	0.104	0.019
	196.09	50.12	5.078	0.090	0.013
	196.25	50.41	5.108	0.093	0.014
	197.90	52.72	5.342	0.091	0.010
	199.22	54.55	5.527	0.079	0.007
	199.54	55.07	5.580	0.086	0.008
L_1,L_2,V	190.00	42.27	4.283	0.153	0.058
	190.00	42.37	4.293	0.163	0.071
	192.00	44.89	4.548	0.162	0.061
	192.00	44.97	4.557	0.170	0.074
	192.00	45.25	4.585	0.180	0.109
	192.00	45.39	4.599	0.193	0.096
	192.00	45.45	4.605	0.199	0.103
	194.00	47.43	4.806	0.175	0.061
	194.00	47.64	4.827	0.169	0.064
	194.00	47.82	4.845	0.174	0.076
	194.00	47.97	4.861	0.176	0.107
	194.00	48.31	4.895	0.180	0.090
	194.00	48.38	4.902	0.202	0.105
	194.00	48.38	4.902	0.229	0.118
	196.00	49.92	5.058	0.130	0.024
	196.00	50.12	5.078	0.130	0.023
	196.00	50.56	5.123	0.176	0.061
	196.00	50.64	5.131	0.176	0.067
	196.00	50.83	5.150	0.181	0.079
	196.00	51.17	5.185	0.183	0.091
	196.00	51.27	5.205	0.225	0.116
	198.00	53.07	5.377	0.147	0.028
	198.00	53.69	5.440	0.181	0.063
	198.00	53.89	5.460	0.179	0.068
	200.00	55.78	5.652	0.104	0.0133
	200.00	55.99	5.673	0.121	0.014
	200.00	56.31	5.706	0.129	0.015

COMPONENTS:	ORIGINAL MEASUREMENTS:
1. Methane; CH_4; [74-82-8] 2. Hexane; C_6H_{14}; [110-54-3] 3. Octane; C_8H_{18}; [111-65-9]	Merrill, R. C.; Luks, K. D.; Kohn, J. P. *J. Chem. Eng. Data* *1983*, *28*, 210-215.

VARIABLES:	PREPARED BY:
	C. L. Young

EXPERIMENTAL VALUES:

Phases in equilibrium	T/K	P/atm	P/MPa	Mole fractions of hexane $x_{C_6H_{14}}$	octane $x_{C_8H_{18}}$
Data for octane-lean phase, L_2					
$L_1, L_2 = V$	193.15	48.24	4.892	0.0054	0.0034
	193.20	48.44	4.908	0.0	0.0156
	193.69	48.96	4.961	0.0017	0.0012
	193.85	48.94	4.959	0.0017	0.0021
	194.14	49.37	5.002	0.0098	0.0018
	194.29	49.63	5.029	0.0051	0.0013
S, L_1, L_2, V	181.38	32.76	3.319	0.0349	0.0053
	181.92	33.38	3.382	0.0331	0.0046
	182.61	34.20	3.465	0.0292	0.0042
	182.20	36.16	3.664	0.0185	0.0067
	187.62	40.59	4.113	0.0134	0.0030
	191.94	46.29	4.690	0.0068	0.0029
	192.00	46.65	4.727	0.0059	0.0021
	192.20	46.56	4.718	0.0075	0.0036
L_1, L_2, V	182.00	33.38	3.382	0.0347	0.0034
	182.00	33.38	3.382	0.0335	0.0033
	182.00	33.44	3.388	0.0397	0.0019
	182.00	33.44	3.388	0.0437	0.0037
	182.00	33.45	3.389	0.0313	0.0039
	182.00	33.51	3.395	0.0330	0.0039
	182.00	33.64	3.409	0.0478 (cont.)	0.0022

AUXILIARY INFORMATION

METHOD/APPARATUS/PROCEDURE:	SOURCE AND PURITY OF MATERIALS:
Glass equilibrium cell. Temperature measured with platinum resistance thermometer and pressure with Bourdon gauge. Stoichiometry and volumetric measurements were used to obtain liquid phase compositions and molar volumes. Gas phase assumed to be pure methane except for L_1, L_2 = V equilibrium. Molar volume data in source. Details of apparatus in ref. (1).	1. Linde Ultra Pure grade sample, purity 99.97 moles per cent. 2. and 3. Humphrey Chemical Co. samples of purity 99 moles per cent.

ESTIMATED ERROR:
$\delta T/K = \pm0.03$; $\delta P/MPa = \pm0.007$; $\delta x/x = \pm0.02$ in L_1 phase ±0.08 in L_2 phase.

REFERENCES:
1. Hottovy, J. D.; Kohn, J. P.; Luks, K. D. *J. Chem. Eng. Data* *1981*, *26*, 135.

COMPONENTS:	ORIGINAL MEASUREMENTS:
1. Methane; CH_4; [74-82-8]	Merrill, R. C.; Luks, K. D.;
2. Hexane; C_6H_{14}; [110-54-3]	Kohn, J. P.
3. Octane; C_8H_{18}; [111-65-9]	*J. Chem. Eng. Data*
	1983, *28*, 210-215.

EXPERIMENTAL VALUES:

Phases in equilibrium	T/K	P/atm	P/MPa	Mole fractions of hexane $x_{C_6H_{14}}$	octane $x_{C_8H_{18}}$
L_1,L_2,V	180.00	31.33	3.175	0.0429	0.0053
	180.00	31.33	3.175	0.0405	0.0050
	180.00	31.40	3.182	0.0505	0.0040
	178.00	29.28	2.967	0.0798	0.0038
	178.00	29.29	2.968	0.0449	0.0056
	184.00	35.55	3.602	0.0479	0.0036
	184.00	35.62	3.609	0.0317	0.0043
	184.00	35.62	3.609	0.0395	0.0035
	184.00	35.62	3.609	0.0396	0.0031
	184.00	35.69	3.616	0.0250	0.0069
	184.00	35.69	3.616	0.0278	0.0090
	184.00	35.69	3.616	0.0359	0.0055
	184.00	35.76	3.623	0.0340	0.0034
	184.00	35.82	3.629	0.0379	0.0030
	184.00	35.96	3.644	0.0261	0.0042
	186.00	37.87	3.837	0.0280	0.0036
	186.00	37.93	3.843	0.0335	0.0022
	186.00	38.00	3.850	0.0343	0.0006
	186.00	38.00	3.850	0.0291	0.0022
	186.00	38.07	3.857	0.0312	0.0021
	186.00	38.14	3.865	0.0219	0.0025
	186.00	38.20	3.871	0.0239	0.0027
	186.00	38.20	3.871	0.0307	0.0011
	186.00	38.21	3.872	0.0334	0.0006
	188.00	40.31	4.084	0.0213	0.0037
	188.00	40.31	4.084	0.0286	0.0022
	188.00	40.52	4.106	0.0259	0.0027
	188.00	40.58	4.112	0.0274	0.0020
	188.00	40.59	4.113	0.0191	0.0062
	188.00	40.59	4.113	0.0287	0.0022
	188.00	40.72	4.126	0.0281	0.0019
	188.00	40.93	4.147	0.0119	0.0037
	190.00	42.90	4.347	0.0255	0.0018
	190.00	42.90	4.347	0.0249	0.0018
	190.00	43.17	4.374	0.0188	0.0016
	190.00	43.17	4.374	0.0234	0.0016
	190.00	43.17	4.374	0.0245	0.0017
	190.00	43.52	4.410	0.0108	0.0021
	192.00	45.96	4.657	0.0199	0.0013
	192.00	46.23	4.684	0.0123	0.0037
	192.00	46.30	4.691	0.0089	0.0025
	192.00	46.36	4.697	0.0046	0.0038
	192.00	46.58	4.720	0.0050	0.0027
	194.00	48.68	4.933	0.0149	0.0001
	194.00	49.09	4.974	0.0074	0.0034
	194.00	49.16	4.981	0.0039	0.0011

Data for octane-rich phase, L_1

$L_1,L_2 \equiv V$	193.06	48.21	4.885	0.2450	0.2008
	193.35	48.49	4.913	0.2554	0.1446
	193.56	49.00	4.965	0.2045	0.1387
	193.76	49.06	4.971	0.2288	0.1297
	194.24	49.53	5.019	0.2711	0.0201
	194.86	50.24	5.091	0.2119	0.0390
	195.02	50.31	5.098	0.2537	0.0103

(cont.)

COMPONENTS:	ORIGINAL MEASUREMENTS:
1. Methane; CH_4; [74-82-8]	Merrill, R. C.; Luks, K. D.;
2. Hexane; C_6H_{14}; [110-54-3]	Kohn, J. P.
3. Octane; C_8H_{18}; [111-65-9]	*J. Chem. Eng. Data*
	<u>1983</u>, *28*, 210-215.

EXPERIMENTAL VALUES:

Phases in equilibrium	T/K	P/atm	P/MPa	Mole fractions of hexane $x_{C_6H_{14}}$	octane $x_{C_8H_{18}}$
S,L_1,L_2,V	174.48	25.88	2.622	0.1819	0.0261
	175.16	26.49	2.684	0.1752	0.0265
	176.62	27.99	2.836	0.1997	0.0371
	184.80	36.36	3.684	0.2467	0.0898
	189.96	43.82	4.440	0.2042	0.1368
	190.51	44.60	4.519	0.2126	0.1427
	191.91	46.81	4.743	0.2206	0.1786
	192.36	47.14	4.776	0.2333	0.1977
$L_1 \equiv L_2,V$	174.69	26.23	2.658	0.1247	0.0097
	174.84	26.17	2.652	0.1507	0.0087
	175.94	27.25	2.761	0.1557	0.0121
	177.14	28.39	2.877	0.0999	0.0056
	177.81	28.88	2.926	0.1505	0.0087
	178.29	29.48	2.987	0.1292	0.0063
	178.43	29.62	3.001	0.1411	0.0068
	179.02	29.16	2.955	0.1547	0.0089
L_1,L_2,V Binary mixture	182.91	34.18	3.463	0.1592	0.0
L_1,L_2,V	176.00	27.19	2.755	0.1587	0.0138
	176.00	27.31	2.767	0.1944	0.0264
	178.00	29.02	2.940	0.1491	0.0083
	178.00	29.16	2.955	0.1757	0.0154
	178.00	29.21	2.960	0.1959	0.0270
	178.00	29.28	2.967	0.2014	0.0379
	178.00	29.28	2.967	0.1555	0.0087
	178.00	29.29	2.968	0.1777	0.0139
	180.00	31.14	3.155	0.1881	0.0165
	180.00	31.26	3.167	0.2040	0.0384
	180.00	31.26	3.167	0.2053	0.0278
	180.00	31.33	3.175	0.1770	0.0088
	180.00	31.33	3.175	0.1902	0.0147
	180.00	31.40	3.182	0.1821	0.0103
	180.00	31.46	3.188	0.1861	0.0098
	182.00	33.23	3.367	0.2277	0.0313
	182.00	33.24	3.368	0.2033	0.0117
	182.00	33.25	3.369	0.2105	0.0184
	182.00	33.37	3.381	0.2276	0.0434
	182.00	33.44	3.388	0.2202	0.0171
	184.00	35.54	3.601	0.2451	0.0333
	184.00	35.68	3.615	0.2488	0.0465
	184.00	35.69	3.616	0.2176	0.0125
	184.00	35.75	3.622	0.2232	0.0113
	184.00	35.76	3.623	0.2276	0.0198
	184.00	35.96	3.644	0.2415	0.0188
	184.00	36.02	3.650	0.2414	0.0188
	186.00	37.77	3.827	0.2232	0.0128
	186.00	37.81	3.831	0.2022	0.0173
	186.00	37.92	3.842	0.2392	0.0326
	186.00	38.06	3.856	0.2375	0.0446
	186.00	38.07	3.857	0.2256	0.0114
	186.00	38.13	3.864	0.2380	0.0881
	186.00	38.20	3.871	0.2379	0.0185
	186.00	38.20	3.871	0.2478	0.0847
	188.00	40.25	4.078	0.2508	0.0184
	188.00	40.32	4.085	0.2320	0.0202
	188.00	40.38	4.092	0.2670	0.0359

(cont.)

COMPONENTS:	ORIGINAL MEASUREMENTS:
1. Methane; CH_4; [74-82-8]	Merrill, R. C.; Luks, K. D.;
2. Hexane; C_6H_{14}; [110-54-3]	Kohn, J. P.
3. Octane; C_8H_{18}; [111-65-9]	*J. Chem. Eng. Data*
	<u>1983</u>, *28*, 210-215.

EXPERIMENTAL VALUES:

Phases in equilibrium	T/K	*P*/atm	*P*/MPa	Mole fractions of	
				hexane $x_{C_6H_{14}}$	octane $x_{C_8H_{18}}$
L_1,L_2,V	188.00	40.44	4.098	0.2349	0.0441
	188.00	40.52	4.106	0.2363	0.0237
	188.00	40.52	4.106	0.2391	0.0186
	188.00	40.65	4.119	0.2402	0.0263
	188.00	40.65	4.119	0.2378	0.0279
	188.00	40.72	4.126	0.2390	0.0220
	188.00	40.75	4.129	0.2301	0.0132
	190.00	42.83	4.340	0.2354	0.0135
	190.00	42.90	4.347	0.2431	0.0224
	190.00	42.91	4.348	0.2384	0.0208
	190.00	42.97	4.354	0.2437	0.0348
	190.00	43.17	4.374	0.2433	0.0244
	190.00	43.17	4.374	0.2445	0.0268
	190.00	43.17	4.374	0.2451	0.0247
	190.00	43.17	4.374	0.2376	0.0444
	190.00	43.23	4.380	0.2271	0.0826
	190.00	43.30	4.387	0.2318	0.0793
	192.00	45.93	4.654	0.2754	0.0236
	192.00	45.96	4.657	0.2605	0.0485
	192.00	45.96	4.657	0.2685	0.0383
	192.00	46.09	4.670	0.2793	0.0260
	192.00	46.16	4.677	0.2898	0.1047
	194.00	48.62	4.926	0.2634	0.0243
	194.00	48.69	4.934	0.2593	0.0370
	194.00	48.81	4.946	0.2586	0.0221
	194.00	48.95	4.960	0.2452	0.0453
	194.00	49.05	4.970	0.2753	0.0224

COMPONENTS:	ORIGINAL MEASUREMENTS:
1. Methane; CH_4; [74-82-8] 2. Decane; $C_{10}H_{22}$; [124-18-5] 3. Dotriacontane; $C_{32}H_{66}$; [544-85-4]	Cordeiro, D. J.; Luks, K. D.; Kohn, J. P. *Ind. Eng. Chem. Process. Des.* *Develop.* <u>1973</u>, *12*, 47-51.
VARIABLES:	PREPARED BY: C. L. Young

EXPERIMENTAL VALUES:

T/K	P/MPa^*	P/atm	Mole fraction of comp. 3 in liquid, $x_{C_{32}H_{66}}$	Mole fraction of comp. 1 in liquid, x_{CH_4}
330	0.0	0	0.3457	0.0
	0.507	5	0.3443	0.0263
	1.01	10	0.3425	0.0515
	2.03	20	0.3383	0.0986
	3.04	30	0.3337	0.1419
	4.05	40	0.3290	0.1817
	5.07	50	0.3241	0.2184
	6.08	60	0.3192	0.2522
	7.09	70	0.3142	0.2835
	8.10	80	0.3093	0.3123
	9.12	90	0.3043	0.3390
	10.13	100	0.2994	0.3638
335	0.0	0	0.5622	0.000
	0.507	5	0.5584	0.0286
	1.01	10	0.5543	0.0559
	2.03	20	0.5455	0.1068
	3.04	30	0.5367	0.1534
	4.05	40	0.5279	0.1961
	5.07	50	0.5192	0.2353
	6.08	60	0.5107	0.2712
	7.09	70	0.5022	0.3042

(cont.)

AUXILIARY INFORMATION

METHOD/APPARATUS/PROCEDURE:	SOURCE AND PURITY OF MATERIALS:

A known amount of gas was added to a known amount of solvent of known composition in a 10 cm³ glass equilibrium cell. Liquid phase composition determined from the overall composition and volume of both phases. Details in source.	1. and 2. Phillips Petroleum Co.
	   pure grade samples, purity better
	   than 99 mole per cent.
	
	 3. Humphrey Chemical Co. sample,
	   minimum purity 97 mole per cent.

	ESTIMATED ERROR: $\delta T/K = \pm 0.02$; $\delta P/MPa = \pm 0.007$; $\delta x_{CH_4} = \pm 0.001$ (estimated by compiler).
	REFERENCES:

COMPONENTS:

1. Methane; CH_4; [74-82-8]

2. Decane; $C_{10}H_{22}$; [124-18-5]

3. Dotriacontane; $C_{32}H_{66}$; [544-85-4]

ORIGINAL MEASUREMENTS:

Cordeiro, D. J.; Luks, K. D.; Kohn, J. P.

Ind. Eng. Chem. Process. Des. Develop.

<u>1973</u>, *12*, 47-51.

EXPERIMENTAL VALUES:

T/K	P/MPa*	P/atm	Mole fraction of comp. 3 in liquid, $x_{C_{32}H_{66}}$	Mole fraction of comp. 1 in liquid, x_{CH_4}
335	8.10	80	0.4940	0.3346
	9.12	90	0.4859	0.3626
	10.13	100	0.4780	0.3886
340	0.0	0	0.8412	0.000
	0.507	5	0.8362	0.0323
	1.01	10	0.8310	0.0631
	1.52	15	0.8258	0.0926
	2.03	20	0.8206	0.1208
	2.53	25	0.8154	0.1477
	3.04	30	0.8103	0.1734

* calculated by compiler.

COMPONENTS:	ORIGINAL MEASUREMENTS:
1. Methane; CH_4; [74-82-8] 2. Nitrogen; N_2; [7727-37-9] 3. Butane; C_4H_{10}; [106-97-8]	Merrill, R. C. Jr.; Luks, K. D.; Kohn, J. P. *Adv. Cryog. Eng.*, <u>1984</u>, *29*, 949-955.
VARIABLES: Temperature, pressure	PREPARED BY: C. L. Young

EXPERIMENTAL VALUES:

Raw data for the Butane rich liquid phase

type	T/K	p/bar	$x_{C_4H_{10}}$	x_{N_2}	molar volume v/cm mol^{-1}
K	141.29	42.30	0.555	0.1002	58.1
	142.79	43.12	0.613	0.1216	67.6
	149.54	47.46	0.485	0.1517	59.1
	174.56	62.02	0.154	0.1961	64.3
	175.38	62.57	0.129	0.1897	57.3
Q	117.15	10.30	0.349	0.0948	51.1
	118.06	11.20	0.393	0.0930	53.5
	119.15	12.58	0.424	0.1005	54.7
	120.22	14.09	0.469	0.1029	55.9
	121.46	16.36	0.542	0.1086	60.0
	124.57	22.10	0.567	0.1050	56.1
	124.57	22.44	0.670	0.1244	67.3
UCST	120.31	11.34	0.117	0.2469	43.5
	127.82	15.96	0.108	0.2658	44.4
	130.74	18.24	0.0941	0.2844	45.1
	139.38	24.91	0.118	0.2668	46.7
	141.58	26.92	0.128	0.2595	47.5
LCST	150.45	35.33	0.129	0.2591	49.2
	157.61	42.71	0.142	0.2452	50.9
	158.98	44.03	0.116	0.2563	51.0
	161.79	47.20	0.122	0.2530	51.5
	168.83	54.78	0.128	0.2386	53.9
	174.50	60.85	0.086	0.2527	54.6

AUXILIARY INFORMATION

METHOD/APPARATUS/PROCEDURE:	SOURCE AND PURITY OF MATERIALS:
Glass equilibrium cell. Temperature measured with platinum resistance thermometer and pressure with Bourdon gauge. Stoichiometry and volumetric measurements were used to obtain liquid phase compositions and molar volumes. Molar volume data in source. Details of apparatus in ref (1). and source.	1. Linde Ultra pure sample, stated purity of 99.97 mole per cent. 2. Linde high purity sample, stated purity 99.99 mole per cent. 3. Linde Instrument grade, stated purity 99.5 mole per cent.
	ESTIMATED ERROR:
	REFERENCES: 1. Hottovy, J. D.; Kohn, J. P.; Luks, K. D. *J. Chem. Eng. Data*, <u>1981</u>, *26*, 135.

COMPONENTS:	ORIGINAL MEASUREMENTS:
1. Methane; CH_4; [74-82-8] 2. Nitrogen; N_2; [7727-37-9] 3. Butane; C_4H_{10}; [106-97-8]	Merrill, R. C. Jr.; Luks, K. D.; Kohn, J. P. *Adv. Cryog. Eng.*, <u>1984</u>, *29*, 949-955

EXPERIMENTAL VALUES:

Raw data for the Butane rich liquid phase

type	T/K	p/bar	$x_{C_4H_{10}}$	x_{N_2}	molar volume v/cm mol^{-1}
L-L-G	128.65	17.12	0.266	0.1607	48.3
	128.65	17.13	0.281	0.1472	51.1
	128.65	18.03	0.331	0.1375	51.6
	128.65	18.29	0.356	0.1344	52.5
	128.65	18.44	0.346	0.1458	50.7
	128.65	19.34	0.399	0.1259	54.5
	128.65	20.30	0.444	0.1247	55.7
	128.65	20.64	0.452	0.1218	55.8
	128.65	22.36	0.527	0.1203	60.2
	128.65	26.23	0.570	0.0910	57.2
	128.65	26.51	0.674	0.1058	68.8
	138.97	25.12	0.245	0.1781	50.0
	138.97	25.47	0.268	0.1740	50.1
	138.97	26.79	0.326	0.1540	52.7
	138.97	27.05	0.341	0.1481	53.0
	138.97	27.05	0.331	0.1612	51.6
	138.97	27.60	0.361	0.1463	53.7
	138.97	28.17	0.377	0.1442	54.8
	138.97	30.02	0.429	0.1378	56.4
	138.97	31.53	0.461	0.1308	57.3
	138.97	33.12	0.512	0.1310	60.7
	138.97	37.74	0.614	0.1232	66.8
	138.97	38.99	0.543	0.1160	56.4
	138.97	39.06	0.622	0.1402	66.4
	149.27	34.37	0.208	0.1949	50.0
	149.27	36.15	0.277	0.1739	51.5
	149.27	37.19	0.306	0.1680	52.5
	149.27	37.34	0.316	0.1621	53.7
	149.27	38.02	0.326	0.1646	52.6
	149.27	38.84	0.352	0.1534	54.6
	149.27	45.67	0.466	0.1431	58.3
	149.27	47.05	0.497	0.1445	60.6
	159.52	49.12	0.297	0.1770	53.7
	159.52	50.36	0.312	0.1629	54.0
	159.52	52.08	0.353	0.1558	56.1
	169.72	56.15	0.197	0.2700	64.8

Raw data for the butane lean liquid phase

type	T/K	p/bar	$x_{C_4H_{10}}$	x_{N_2}	molar volume v/cm mol^{-1}
K	135.36	38.72	0.0110	0.8158	69.0
	139.67	41.67	0.0053	0.7861	85.1
	142.94	43.27	0.0021	0.7599	99.3
	143.08	43.60	0.0052	0.7402	75.2
	145.15	44.85	0.0035	0.7247	82.2
	152.05	47.40	0.0100	0.6666	108.5
	163.25	52.36	0.0095	0.6035	119.2
	167.93	57.88	0.0160	0.4401	77.6
	168.13	56.70	0.0267	0.4873	95.8
	170.51	59.32	0.0309	0.4294	78.6
	171.42	59.40	0.0375	0.4290	84.9
	174.36	61.94	0.0406	0.3627	73.8
Q	121.24	16.15	0.0189	0.5457	43.1
	121.83	16.78	0.0201	0.5615	43.8
	121.84	17.19	0.0221	0.5726	43.8
	127.88	29.96	0.0106	0.8386	54.2

COMPONENTS:
 1. Methane; CH_4 ; [74-82-8]
 2. Nitrogen; N_2 ; [7727-37-9]
 3. Butane; C_4H_{10}; [106-97-8]

ORIGINAL MEASUREMENTS:
 Merrill, R. C. Jr.; Luks, K. D.;
 Kohn, J. P.
 Adv. Cryog. Eng., 1984, *29*, 949-955.

EXPERIMENTAL VALUES:

Raw data for the Butane lean liquid phase

type	T/K	p/bar	$x_{C_4H_{10}}$	x_{N_2}	molar volume v/cm mol^{-1}
L	128.65	16.58	0.1203	0.2657	45.2
	128.65	16.78	0.0965	0.2983	45.1
	128.65	16.79	0.0944	0.2959	44.3
	128.65	16.85	0.0906	0.2901	45.0
	128.65	19.95	0.0294	0.4569	44.7
	128.65	21.27	0.0197	0.5186	45.0
	128.65	22.22	0.0163	0.5485	45.3
	128.65	22.50	0.0186	0.5668	45.9
	128.65	22.98	0.0166	0.5805	45.9
	128.65	30.51	0.0177	0.8232	54.1
	128.65	30.92	0.0172	0.8361	55.1
	138.97	24.64	0.1118	0.2744	46.9
	138.97	24.78	0.1018	0.2779	46.8
	138.97	25.54	0.0698	0.3285	46.3
	138.97	26.10	0.1011	0.2817	46.2
	138.97	28.43	0.0303	0.4370	47.6
	138.97	30.63	0.0222	0.4961	48.4
	138.97	31.20	0.0188	0.5200	49.0
	138.97	31.89	0.0167	0.5411	50.1
	138.97	32.42	0.0165	0.5509	49.9
	149.27	34.37	0.0907	0.2901	49.0
	149.27	34.44	0.0958	0.2918	48.3
	149.27	34.64	0.0821	0.3035	48.9
	149.27	34.78	0.0765	0.3079	48.9
	149.27	37.54	0.0400	0.4025	50.7
	149.27	38.02	0.0375	0.4034	50.8
	149.27	39.46	0.0270	0.4432	51.7
	149.27	42.28	0.0138	0.5118	55.1
	159.52	45.54	0.1071	0.3241	54.8
	159.52	46.22	0.0576	0.3375	52.5
	159.52	52.01	0.0188	0.4785	63.7
	169.72	56.01	0.0659	0.2905	54.8

COMPONENTS:	ORIGINAL MEASUREMENTS:
1. Methane; CH_4; [74-82-8] 2. Nitrogen; N_2; [7727-37-9] 3. Pentane; C_5H_{12}; [109-66-0]	Merrill, R. C.; Luks, K. D.; Kohn, J. P. *J. Chem. Eng. Data* <u>1984</u>, *29*, 272-276.

VARIABLES:	PREPARED BY:
	C. L. Young

EXPERIMENTAL VALUES:

Phases in equilibrium	T/K	P/atm	P/MPa	Mole fractions Pentane, $x_{C_5H_{12}}$	Nitrogen, x_{N_2}
		Composition of L_1, Pentane rich phase			
$L_1,L_2 = V$	149.10	34.73	3.519	0.5679	0.0342
	152.68	38.13	3.864	0.5368	0.0524
	158.96	49.29	4.994	0.5220	0.1088
	163.05	50.79	5.146	0.4407	0.0944
	166.51	50.11	5.077	0.4198	0.0815
	171.98	53.51	5.422	0.3553	0.0924
	176.82	54.60	5.532	0.3134	0.0876
	177.36	54.40	5.512	0.3422	0.0583
	178.81	54.19	5.491	0.2760	0.0794
	180.96	55.83	5.657	0.2664	0.0758
	182.80	56.16	5.690	0.2589	0.0738
	190.61	58.42	5.919	0.1767	0.0646
S,L_1,L_2,V	133.79	16.22	1.643	0.3975	0.0875
	134.32	18.88	1.913	0.4569	0.0838
	134.95	19.96	2.022	0.5178	0.0876
$L_1 = L_2,V$	153.08	25.55	2.589	0.1519	0.1238
	162.19	33.51	3.395	0.1946	0.0965
	169.78	39.29	3.981	0.1344	0.1123
	172.88	41.94	4.250	0.1474	0.1043
	174.09	43.04	4.361	0.1472	0.1005
					(cont.)

AUXILIARY INFORMATION

METHOD/APPARATUS/PROCEDURE:	SOURCE AND PURITY OF MATERIALS:
Glass equilibrium cell. Temperature measured with platinum resistance thermometer and pressure with Bourdon gauge. Stoichiometry and volumetric measurements were used to obtain liquid phase compositions and molar volumes. Molar volume data in source. Details of apparatus in refs. (1) and (2).	1. Linde "Ultra Pure" grade, purity 99.97 mole per cent. 2. Linde "High Purity" grade, purity 99.99 mole per cent. 3. Humphrey Chemical Co. sample, Purity 99 mole per cent.
	ESTIMATED ERROR: $\delta T/K = \pm0.03$; $\delta P/MPa = \pm0.07$; $\delta x/x < 0.036$.
	REFERENCES: 1. Hottovy, J. D.; Kohn, J. P.; Luks, K. D. *J. Chem. Eng. Data* <u>1981</u>, *26*, 135. 2. Hottovy, J. D.; Kohn, J. P.; Luks, K. D. *J. Chem. Eng. Data* <u>1982</u>, *27*, 298.

COMPONENTS:	ORIGINAL MEASUREMENTS:
1. Methane; CH_4; [74-82-8]	Merrill, R. C.; Luks, K. D.;
2. Nitrogen; N_2; [7727-37-9]	Kohn, J. P.
3. Pentane; C_5H_{12}; [109-66-0]	*J. Chem. Eng. Data*
	<u>1984</u>, *29*, 272-276.

EXPERIMENTAL VALUES:

Phases in equilibrium	T/K	P/atm	P/MPa	Mole fractions Pentane, $x_{C_5H_{12}}$	Nitrogen, x_{N_2}
$L_1 = L_2$,V	177.06	45.49	4.609	0.1064	0.1118
	180.04	47.94	4.858	0.1049	0.1081
	180.04	48.01	4.865	0.1096	0.1022
	181.71	49.30	4.995	0.0900	0.1064
	183.11	50.59	5.126	0.0780	0.1043
	184.09	51.21	5.189	0.1160	0.0989
	187.13	53.52	5.422	0.1181	0.0896
	189.09	56.31	5.706	0.1586	0.0705
	140.00	19.56	1.982	0.3532	0.0785
	140.00	20.44	2.071	0.3813	0.0848
	140.00	21.94	2.223	0.4191	0.0784
	140.00	23.43	2.374	0.4442	0.0806
	140.00	24.32	2.464	0.5076	0.0866
	140.00	25.54	2.588	0.5295	0.0907
	150.00	28.20	2.857	0.3550	0.0847
	150.00	28.61	2.899	0.3617	0.0913
	150.00	31.87	3.229	0.4215	0.0817
	150.00	32.69	3.312	0.4304	0.0852
	150.00	35.14	3.561	0.5042	0.0945
	150.00	35.48	3.595	0.5150	0.0940
	150.00	37.72	3.822	0.5510	0.1002
	160.00	32.02	3.244	0.2201	0.0935
	160.00	32.22	3.265	0.2318	0.0964
	160.00	32.41	3.284	0.2321	0.0953
	160.00	32.62	3.305	0.2409	0.0935
	160.00	33.37	3.381	0.2614	0.0926
	160.00	33.50	3.394	0.2662	0.0885
	160.00	35.01	3.547	0.2950	0.1477
	160.00	36.43	3.691	0.3234	0.0928
	160.00	44.32	4.491	0.4931	0.0972
	160.00	46.71	4.733	0.4631	0.0859
	165.00	36.23	3.671	0.2088	0.0928
	165.00	36.44	3.692	0.2203	0.0932
	165.00	37.05	3.754	0.2339	0.0979
	165.00	37.11	3.760	0.2532	0.0892
	165.00	37.72	3.822	0.2553	0.0908
	165.00	38.27	3.878	0.2644	0.0926
	165.00	39.08	3.960	0.2795	0.0879
	165.00	40.44	4.098	0.3097	0.0962
	165.00	41.60	4.215	0.3236	0.0938
	165.00	44.46	4.505	0.3658	0.0843
	165.00	48.34	4.898	0.4100	0.0904
	165.00	48.34	4.898	0.4113	0.0826
	170.00	41.07	4.161	0.2182	0.0900
	170.00	41.75	4.230	0.2389	0.0839
	170.00	42.22	4.278	0.2499	0.0922
	170.00	42.28	4.284	0.2478	0.0911
	170.00	43.23	4.380	0.2649	0.0875
	170.00	43.64	4.422	0.2701	0.0866
	170.00	45.07	4.567	0.2929	0.0830
	170.00	46.16	4.677	0.3131	0.0924
	170.00	47.52	4.815	0.3318	0.0896
	170.00	50.92	5.159	0.3694	0.0945
	170.00	52.01	5.270	0.3817	0.0902
	175.00	43.92	4.450	0.1511	0.1019
	175.00	44.12	4.470	0.1634	0.1049
	175.00	46.44	4.706	0.2333	0.0843
	175.00	47.04	4.766	0.2426	0.0874

(cont.)

COMPONENTS:	ORIGINAL MEASUREMENTS:
1. Methane; CH_4; [74-82-8]	Merrill, R. C.; Luks, K. D.;
2. Nitrogen; N_2; [7727-37-9]	Kohn, J. P.
3. Pentane; C_5H_{12}; [109-66-0]	*J. Chem. Eng. Data*
	<u>1984</u>, *29*, 272-276.

EXPERIMENTAL VALUES:

Phases in equilibrium	T/K	P/atm	P/MPa	Mole fractions	
				Pentane, $x_{C_5H_{12}}$	Nitrogen, x_{N_2}
$L_1 = L_2$,V	175.00	48.07	4.871	0.2611	0.0828
	175.00	48.14	4.878	0.2632	0.0751
	175.00	49.15	4.980	0.2756	0.0839
	175.00	49.97	5.063	0.2865	0.0826
	175.00	50.99	5.167	0.2993	0.0856
	175.00	52.28	5.297	0.3087	0.0942
	180.00	48.27	4.891	0.1584	0.0913
	180.00	48.75	4.940	0.1751	0.0856
	180.00	52.57	5.327	0.2448	0.0802
	180.00	52.62	5.332	0.2550	0.0837
	180.00	54.54	5.526	0.2822	0.0830

Composition of L_2, Pentane lean phase

Phases in equilibrium	T/K	P/atm	P/MPa	Pentane, $x_{C_5H_{12}}$	Nitrogen, x_{N_2}
$L_1,L_2 = V$	153.83	46.84	4.746	0.0090	0.6232
	171.77	53.31	5.402	0.0042	0.3762
	173.07	53.85	5.456	0.0063	0.3436
	174.15	53.72	5.443	0.0045	0.3578
	174.49	54.19	5.491	0.0081	0.3040
	174.85	54.60	5.532	0.0043	0.3414
	175.41	53.79	5.450	0.0117	0.3387
	176.02	54.60	5.532	0.0081	0.2979
	177.44	55.01	5.574	0.0059	0.3000
	179.19	55.29	5.602	0.0164	0.2468
	183.03	56.03	5.677	0.0253	0.2230
	187.19	57.25	5.801	0.0127	0.1779
	187.75	57.59	5.835	0.0167	0.1576
	188.26	57.66	5.842	0.0100	0.1655
	188.83	57.72	5.848	0.0098	0.1576
	188.86	58.00	5.877	0.0165	0.1457
	192.18	58.27	5.904	0.0028	0.1344
S,L_1,L_2,V	132.28	12.01	1.217	0.1206	0.1835
	132.64	12.62	1.279	0.0718	0.1612
	133.26	14.05	1.424	0.0508	0.1943
	135.67	22.76	2.306	0.0052	0.4194
L_1,L_2,V	140.00	16.98	1.720	0.0744	0.1626
	140.00	18.34	1.858	0.0551	0.1966
	140.00	18.88	1.913	0.0488	0.2140
	140.00	24.39	2.471	0.0093	0.3591
	140.00	26.63	2.698	0.0004	0.4265
	150.00	24.53	2.486	0.0614	0.1632
	150.00	26.57	2.692	0.0427	0.2180
	150.00	26.98	2.734	0.0368	0.2211
	150.00	35.48	3.595	0.0101	0.3912
	150.00	36.91	3.740	0.0042	0.4290
	160.00	34.32	3.477	0.0469	0.1987
	160.00	36.57	3.705	0.0322	0.2378
	160.00	37.32	3.781	0.0287	0.2545
	160.00	40.66	4.120	0.0186	0.3066
	165.00	39.77	4.030	0.0373	0.2135
	165.00	43.03	4.360	0.0239	0.2607
	165.00	43.38	4.395	0.0224	0.2677
	165.00	48.61	4.925	0.0104	0.3564
	170.00	42.35	4.291	0.0458	0.1787
	170.00	44.73	4.532	0.0325	0.2124
	170.00	45.35	4.595	0.0297	0.2221
	170.00	48.07	4.871	0.0188	0.2609

(cont.)

COMPONENTS:	ORIGINAL MEASUREMENTS:
1. Methane; CH_4; [74-82-8] 2. Nitrogen; N_2; [7727-37-9] 3. Pentane; C_5H_{12}; [109-66-0]	Merrill, R. C.; Luks, K. D.; Kohn, J. P. *J. Chem. Eng. Data* <u>1984</u>, *29*, 272-276.

EXPERIMENTAL VALUES:

Phases in equilibrium	T/K	P/atm	P/MPa	Mole fractions	
				Pentane, $x_{C_5H_{12}}$	Nitrogen, x_{N_2}
L_1,L_2,V	170.00	48.48	4.912	0.0189	0.2694
	170.00	50.58	5.125	0.0170	0.2791
	170.00	51.27	5.195	0.0117	0.3195
	175.00	47.25	4.788	0.0384	0.1773
	175.00	48.14	4.878	0.0330	0.1909
	175.00	48.48	4.912	0.0320	0.1913
	175.00	50.38	5.105	0.0226	0.2214
	175.00	51.06	5.174	0.0197	0.2355
	175.00	51.68	5.236	0.0182	0.2428
	175.00	51.74	5.243	0.0178	0.2404
	180.00	49.70	5.036	0.0467	0.1414
	180.00	51.00	5.168	0.0364	0.1644
	180.00	51.20	5.188	0.0364	0.1611
	180.00	53.03	5.373	0.0245	0.1929
	180.00	53.04	5.374	0.0264	0.1900
	180.00	53.10	5.380	0.0249	0.1878

COMPONENTS:	ORIGINAL MEASUREMENTS:
1. Methane; CH_4; [74-82-8] 2. Nitrogen; N_2; [7727-37-9] 3. Hexane; C_6H_{14}; [110-54-3]	Poston, R. S.; McKetta, J. J. *Am. Inst. Chem. Engnrs. J.* 1965, *11*, 917-920.

VARIABLES:	PREPARED BY:
	C. L. Young

EXPERIMENTAL VALUES:

T/K (T/°F)	P/MPa (P/psi)	Mole fraction					
		in liquid			in vapor		
		x_{CH_4}	x_{N_2}	$x_{C_6H_{14}}$	y_{CH_4}	y_{N_2}	$y_{C_6H_{14}}$
310.9 (100)	3.45 (500)	0.110	0.020	0.870	0.610	0.374	0.016
	6.89	0.044	0.080	0.876	0.308	0.674	0.018
	(1000)	0.265	0.009	0.726	0.909	0.071	0.020
	10.34	0.083	0.113	0.804	0.202	0.787	0.011
	(1500)	0.132	0.097	0.771	0.302	0.685	0.013
		0.162	0.084	0.754	0.385	0.603	0.012
		0.371	0.016	0.613	0.913	0.066	0.021
	13.79	0.324	0.078	0.598	0.617	0.361	0.022
	(2000)	0.338	0.074	0.588	0.649	0.335	0.016
		0.461	0.020	0.519	0.385	0.066	0.021
		0.461	0.020	0.519	0.913	0.066	0.021
	17.24	0.229	0.105	0.666	0.408	0.578	0.014
	(2500)	0.490	0.061	0.449	0.757	0.223	0.020
		0.390	0.099	0.511	0.608	0.376	0.016
		0.294	0.124	0.582	0.498	0.486	0.016
		0.759	0.145	0.614	0.402	0.580	0.018
	20.68	0.172	0.199	0.629	0.242	0.740	0.018
	(3000)	0.140	0.202	0.658	0.202	0.780	0.018
		0.094	0.216	0.690	0.135	0.852	0.013
		0.587	0.071	0.342	0.821	0.159	0.020
		0.681	0.062	0.257	0.843	0.127	0.030
		0.646	0.044	0.310	0.872	0.109	0.019
		0.806	0.060	0.134	0.886	0.094	0.020
	(cont.)	0.720	0.058	0.222	0.863	0.118	0.019

AUXILIARY INFORMATION

METHOD/APPARATUS/PROCEDURE:	SOURCE AND PURITY OF MATERIALS:
Stainless steel glass windowed cell. Vapor recycled using high pressure magnetic pump. Pressure measured using Bourdon gauge and temperature measured using thermocouples. Samples of both phases withdrawn at constant pressure and analysed by gas chromatography. Details of apparatus in source and ref. (1).	1 and 3. Phillips Petroleum Co. research grade. 2. Research grade.

	ESTIMATED ERROR:
	$\delta T/K = \pm 0.1$; $\delta P/MPa = \pm 0.015$; δx_{CH_4}, $\delta y_{CH_4} = \pm 0.002$.

	REFERENCES:
	1. Roberts, L. R.; McKetta, J. J. *Am. Inst. Chem. Engnrs. J.* 1961, *7*, 173.

COMPONENTS:	ORIGINAL MEASUREMENTS:
1. Methane; CH_4; [74-82-8]	Poston, R. S.; McKetta, J. J.
2. Nitrogen; N_2; [7727-37-9]	*Am. Inst. Chem. Engnrs. J.*
3. Hexane; C_6H_{14}; [110-54-3]	<u>1965</u>, *11*, 917-920.

EXPERIMENTAL VALUES:

		Mole fraction					
T/K	P/MPa	in liquid			in vapor		
(T/°F)	(P/psi)	x_{CH_4}	x_{N_2}	$x_{C_6H_{14}}$	y_{CH_4}	y_{N_2}	$y_{C_6H_{14}}$
310.9	20.68						
(100)	(3000)	Single phase			0.108	0.075	0.033
	24.13	0.398	0.160	0.442	0.524	0.450	0.026
	(3500)	0.378	0.165	0.457	0.536	0.434	0.030
		0.460	0.151	0.389	0.616	0.340	0.044
		0.507	0.133	0.360	0.679	0.283	0.038
		0.603	0.132	0.265	0.715	0.246	0.039
		0.591	0.118	0.291	0.726	0.248	0.026
		0.664	0.104	0.232	0.782	0.160	0.058
		0.675	0.104	0.221	0.783	0.142	0.075
		Single phase			0.787	0.110	0.103
	27.58	0.552	0.168	0.280	0.652	0.293	0.055
	(4000)	0.571	0.158	0.271	0.672	0.272	0.056
		0.507	0.184	0.309	0.609	0.349	0.042
		0.591	0.160	0.249	0.690	0.256	0.054
		0.222	0.239	0.539	0.312	0.670	0.018
		0.364	0.218	0.418	0.443	0.522	0.035
		Single phase			0.715	0.203	0.082
	31.03	0.499	0.238	0.263	0.576	0.372	0.052
	(4500)	0.519	0.235	0.246	0.401	0.334	0.067
		0.304	0.257	0.439	0.388	0.579	0.033
		0.230	0.270	0.500	0.293	0.689	0.018
		Single phase			0.597	0.308	0.095
	34.47	0.458	0.296	0.246	0.516	0.400	0.084
	(5000)	0.383	0.268	0.349	0.459	0.495	0.046
		0.330	0.293	0.377	0.395	0.568	0.037
		0.254	0.304	0.442	0.308	0.671	0.021
		0.474	0.295	0.231	0.501	0.400	0.099
		0.141	0.331	0.528	0.169	0.811	0.020
		Single phase			0.515	0.372	0.113
344.3	3.45	0.096	0.013	0.891	0.698	0.241	0.061
(160)	(500)	0.034	0.041	0.925	0.186	0.757	0.057
	6.89	0.157	0.042	0.801	0.725	0.228	0.047
	(1000)	0.050	0.080	0.870	0.162	0.796	0.042
	10.34	0.209	0.070	0.721	0.555	0.406	0.039
	(1500)	0.084	0.114	0.802	0.248	0.719	0.033
	13.79	0.371	0.047	0.582	0.768	0.182	0.050
	(2000)	0.221	0.103	0.676	0.374	0.586	0.040
		0.102	0.146	0.752	0.216	0.753	0.031
	17.24	0.418	0.072	0.510	0.727	0.216	0.057
	(2500)	0.280	0.114	0.606	0.507	0.449	0.044
		0.170	0.158	0.672	0.319	0.640	0.041
		0.112	0.175	0.713	0.200	0.763	0.037
		0.067	0.194	0.739	0.152	0.804	0.044
	20.68	0.675	0.039	0.286	0.818	0.056	0.126
	(3000)	0.629	0.052	0.319	0.830	0.087	0.083
		0.640	0.041	0.319	0.181	0.072	0.109
		0.597	0.064	0.339	0.818	0.100	0.082
		0.563	0.071	0.366	0.790	0.127	0.083
310.9	20.68	0.422	0.112	0.466	0.656	0.277	0.067
(100)	(3000)	0.237	0.175	0.588	0.507	0.438	0.055
		0.041	0.232	0.727	0.061	0.896	0.043
		Single phase			0.806	0.050	0.144

(cont.)

COMPONENTS:	ORIGINAL MEASUREMENTS:
1. Methane; CH_4; [74-82-8]	Poston, R. S.; McKetta, J. J.
2. Nitrogen; N_2; [7727-37-9]	*Am. Inst. Chem. Engnrs. J.*
3. Hexane; C_6H_{14}; [110-54-3]	1965, *11*, 917-920.

EXPERIMENTAL VALUES:

T/K (T/°F)	P/MPa (P/psi)	x_{CH_4}	x_{N_2}	$x_{C_6H_{14}}$	y_{CH_4}	y_{N_2}	$y_{C_6H_{14}}$
310.9 (100)	24.13 (3500)	0.105	0.249	0.646	0.177	0.777	0.046
		0.301	0.190	0.509	0.388	0.556	0.056
		0.401	0.166	0.433	0.570	0.362	0.068
		0.460	0.149	0.391	0.606	0.343	0.051
		0.521	0.151	0.328	0.656	0.278	0.066
		0.563	0.138	0.299	0.691	0.220	0.089
		Single phase			0.689	0.174	0.137
	27.58 (4000)	0.154	0.273	0.573	0.213	0.740	0.047
		0.283	0.236	0.481	0.396	0.540	0.064
		0.369	0.217	0.414	0.492	0.413	0.095
		0.477	0.222	0.301	0.551	0.297	0.152
		Single phase			0.549	0.255	0.196
	31.03 (4500)	0.148	0.313	0.539	0.198	0.740	0.062
		0.347	0.289	0.364	0.409	0.477	0.114
		0.366	0.293	0.341	0.420	0.454	0.126
		Single phase			0.443	0.360	0.197
	34.47 (5000)	0.025	0.375	0.600	0.043	0.908	0.049
		0.118	0.373	0.509	0.143	0.794	0.060
		0.258	0.350	0.392	0.319	0.576	0.105
		0.304	0.363	0.333	0.346	0.517	0.137
		0.304	0.364	0.332	0.347	0.516	0.137
		Single phase			0.340	0.429	0.221
377.6 (220)	3.45 (500)	0.049	0.025	0.926	0.384	0.491	0.125
	6.89 (1000)	0.161	0.034	0.805	0.646	0.264	0.090
		0.090	0.056	0.854	0.383	0.528	0.089
	10.34 (1500)	0.041	0.118	0.841	0.116	0.811	0.073
		0.121	0.086	0.793	0.377	0.544	0.079
		0.224	0.049	0.727	0.639	0.269	0.092
	13.79 (2000)	0.372	0.045	0.583	0.744	0.160	0.096
		0.252	0.086	0.662	0.537	0.378	0.085
		0.146	0.121	0.733	0.332	0.592	0.076
		0.079	0.162	0.759	0.178	0.748	0.074
	17.24 (2500)	0.454	0.058	0.488	0.741	0.158	0.101
		0.331	0.106	0.563	0.579	0.319	0.102
		0.287	0.129	0.584	0.511	0.393	0.096
		0.189	0.159	0.652	0.337	0.573	0.090
		0.101	0.187	0.712	0.204	0.719	0.077
		0.072	0.208	0.720	0.140	0.786	0.074
	20.68 (3000)	0.070	0.260	0.670	0.124	0.796	0.080
		0.439	0.198	0.561	0.393	0.509	0.098
		0.369	0.150	0.481	0.605	0.288	0.107
		0.462	0.124	0.414	0.640	0.219	0.141
		0.514	0.107	0.379	0.669	0.185	0.146
		0.540	0.103	0.357	0.684	0.173	0.143
		Single phase			0.681	0.143	0.176
	24.13 (3500)	0.300	0.224	0.476	0.429	0.437	0.134
		0.304	0.220	0.436	0.441	0.412	0.147
		0.375	0.226	0.399	0.489	0.358	0.153
		0.482	0.319	0.199	0.437	0.233	0.330
		0.230	0.255	0.515	0.330	0.556	0.114
		0.015	0.338	0.647	0.061	0.854	0.085
		Single phase			0.485	0.300	0.215
	27.58 (4000)	0.190	0.337	0.473	0.259	0.618	0.123
		0.222	0.326	0.452	0.290	0.577	0.133

(cont.)

COMPONENTS:		ORIGINAL MEASUREMENTS:

COMPONENTS:

1. Methane; CH₄; [74-82-8]

2. Nitrogen; N₂; [7727-37-9]

3. Hexane; C₆H₁₄; [110-54-3]

ORIGINAL MEASUREMENTS:

Poston, R. S.; McKetta, J. J.

Am. Inst. Chem. Engnrs. J.

1965, *11*, 917-920.

EXPERIMENTAL VALUES:

T/K (T/°F)	P/MPa (P/psi)	Mole fraction					
		in liquid			in vapor		
		x_{CH_4}	x_{N_2}	$x_{C_6H_{14}}$	y_{CH_4}	y_{N_2}	$y_{C_6H_{14}}$
377.6 (220)	27.58 (4000)	0.263	0.324	0.413	0.324	0.503	0.173
		0.286	0.327	0.387	0.342	0.485	0.173
		0.328	0.336	0.336	0.367	0.427	0.206
		0.231	0.338	0.431	0.297	0.563	0.140
		Single phase			0.371	0.367	0.262
	31.03 (4500)	0.130	0.413	0.457	0.166	0.699	0.135
		0.186	0.423	0.391	0.228	0.611	0.161
		0.202	0.419	0.379	0.228	0.599	0.173
		0.163	0.408	0.429	0.210	0.639	0.151
		0.201	0.414	0.385	0.227	0.568	0.205
		Single phase			0.207	0.549	0.244
	34.47 (5000)	0.088	0.488	0.424	0.097	0.760	0.143
		0.127	0.516	0.357	0.144	0.695	0.161
		0.135	0.524	0.341	0.142	0.670	0.188
		Single phase			0.166	0.640	0.194
410.9 (280)	3.45 (500)	0.044	0.027	0.929	0.335	0.417	0.248
	6.89 (1000)	0.131	0.039	0.830	0.539	0.293	0.168
		0.080	0.061	0.859	0.332	0.500	0.168
	10.34 (1500)	0.055	0.147	0.798	0.140	0.715	0.145
		0.134	0.088	0.778	0.379	0.474	0.147
		0.199	0.064	0.737	0.551	0.298	0.151
	13.79 (2000)	0.393	0.038	0.569	0.721	0.098	0.181
		0.157	0.129	0.714	0.324	0.520	0.156
		0.086	0.173	0.741	0.184	0.670	0.146
		0.066	0.181	0.753	0.141	0.720	0.139
	17.24 (2500)	0.076	0.239	0.685	0.148	0.696	0.156
		0.152	0.215	0.633	0.267	0.570	0.163
		0.326	0.143	0.531	0.511	0.295	0.194
		0.372	0.118	0.510	0.565	0.218	0.217
		0.410	0.102	0.488	0.596	0.173	0.231
		0.429	0.098	0.473	0.598	0.164	0.238
		0.462	0.095	0.443	0.613	0.138	0.249
		Single phase			0.599	0.116	0.285
	20.68 (3000)	0.072	0.317	0.611	0.106	0.741	0.153
		0.122	0.289	0.589	0.196	0.636	0.168
		0.210	0.271	0.519	0.309	0.492	0.199
		0.239	0.246	0.515	0.338	0.450	0.212
		0.279	0.246	0.475	0.369	0.400	0.231
		0.326	0.252	0.422	0.375	0.370	0.255
		Single phase			0.387	0.334	0.279
	24.13 (3500)	0.165	0.370	0.465	0.213	0.550	0.237
		0.189	0.387	0.424	0.221	0.535	0.240
		0.191	0.380	0.429	0.232	0.524	0.244
		0.147	0.373	0.480	0.190	0.584	0.226
		0.079	0.385	0.536	0.106	0.705	0.189
		0.223	0.418	0.359	0.233	0.460	0.307
	27.58 (4000)	0.010	0.489	0.501	0.013	0.775	0.212
		0.043	0.495	0.462	0.047	0.711	0.242
		0.043	0.484	0.473	0.052	0.716	0.232
		0.064	0.507	0.429	0.076	0.650	0.274
		Single phase			0.095	0.589	0.319

(cont.)

COMPONENTS: ORIGINAL MEASUREMENTS:

1. Methane; CH$_4$; [74-82-8]   Poston, R. S.; McKetta, J. J.

2. Nitrogen; N$_2$; [7727-37-9]   *Am. Inst. Chem. Engnrs. J.*

3. Hexane; C$_6$H$_{14}$; [110-54-3]   <u>1965</u>, *11*, 917-920.

EXPERIMENTAL VALUES:

T/K (T/°F)	P/MPa (P/psi)	Mole fraction in liquid			in vapor		
		x_{CH_4}	x_{N_2}	$x_{C_6H_{14}}$	y_{CH_4}	y_{N_2}	$y_{C_6H_{14}}$
444.3 (340)	3.45 (500)	0.089	0.001	0.910	0.471	0.061	0.468
		0.052	0.008	0.940	0.406	0.136	0.458
		0.046	0.021	0.933	0.310	0.256	0.434
	6.89 (1000)	0.194	0.005	0.801	0.642	0.032	0.326
		0.107	0.053	0.840	0.402	0.280	0.318
		0.054	0.080	0.866	0.220	0.483	0.297
		0.052	0.085	0.863	0.160	0.550	0.290
	10.34 (1500)	0.353	0.008	0.639	0.652	0.015	0.333
		0.312	0.028	0.660	0.618	0.074	0.308
		0.278	0.049	0.673	0.564	0.137	0.299
		0.217	0.080	0.703	0.468	0.257	0.275
		0.029	0.171	0.800	0.076	0.681	0.243
		0.026	0.188	0.786	0.062	0.698	0.240
	13.79 (2000)	0.126	0.193	0.681	0.219	0.477	0.304
		0.272	0.139	0.589	0.409	0.227	0.364
		0.335	0.121	0.544	0.418	0.174	0.408
		Single phase			0.390	0.140	0.470
	17.24 (2500)	0.039	0.328	0.633	0.055	0.628	0.317
		0.105	0.333	0.562	0.136	0.488	0.376
		Single phase			0.169	0.409	0.422

COMPONENTS:	ORIGINAL MEASUREMENTS:
1. Methane; CH_4; [74-82-8] 2. Nitrogen; N_2; [7727-37-9] 3. Hexane; C_6H_{14}; [110-54-3]	Merrill, R. C.; Luks, K. D.; Kohn, J. P. *J. Chem. Eng. Data* 1984, *29*, 272-276.

VARIABLES:	PREPARED BY:
	C. L. Young

EXPERIMENTAL VALUES:

Composition of Hexane rich phase

Phases in equilibrium	T/K	P/atm	P/MPa	Mole fractions	
				Hexane, $x_{C_6H_{14}}$	Nitrogen, x_{N_2}
$L_1, L_2 = V$	176.11	51.40	5.208	0.4020	0.0599
	181.01	50.92	5.159	0.3106	0.2222
	185.23	51.81	5.250	0.2988	0.0362
	186.63	51.62	5.230	0.3211	0.0444
	189.65	51.62	5.230	0.2972	0.0260
	189.89	51.06	5.174	0.3091	0.0262
	190.61	51.30	5.198	0.3040	0.0153
	190.70	51.06	5.174	0.2983	0.0165
	191.95	50.05	5.071	0.2858	0.0093
	194.41	51.19	5.187	0.2589	0.0048
S, L_1, L_2, V	164.07	22.11	2.240	0.2005	0.0222
	164.20	23.16	2.347	0.2224	0.0256
	164.23	23.51	2.382	0.2259	0.0275
	164.46	24.39	2.471	0.2521	0.0305
	164.51	24.67	2.500	0.2557	0.0329
	164.52	25.34	2.568	0.2643	0.0299
	164.65	25.95	2.629	0.2818	0.0319
	165.13	27.94	2.831	0.3019	0.0379
	166.11	34.18	3.463	0.2917	0.2368
	166.76	38.15	3.866	0.3995	0.0175
	167.64	45.68	4.629	0.4726	0.0635

(cont.)

AUXILIARY INFORMATION

METHOD APPARATUS/PROCEDURE:	SOURCE AND PURITY OF MATERIALS:
Glass equilibrium cell. Temperature measured with platinum resistance thermometer and pressure with Bourdon gauge. Stoichiometry and volumetric measurements were used to obtain liquid phase compositions and molar volumes. Molar volume data in source. Details of apparatus in refs. (1) and (2).	1. Linde "Ultra Pure" grade, purity 99.97 mole per cent. 2. Linde "High Purity" grade, purity 99.99 mole per cent. 3. Phillips Petroleum Co. sample, purity 99 mole per cent.

ESTIMATED ERROR:

$\delta T/K = \pm 0.03$; δP/MPa = ± 0.07;

$\delta x/x < 0.09$ in L_1 phase; < 0.03

in L_2 phase.

REFERENCES:
1. Hottovy, J. D.; Kohn, J. P.;
   Luks, K. D.
   J. Chem. Eng. Data 1981, *26*, 135.

2. Hottovy, J. D.; Kohn, J. P.;
   Luks, K. D.
   J. Chem. Eng. Data 1982, *27*, 298.

COMPONENTS:	ORIGINAL MEASUREMENTS:
1. Methane; CH_4; [74-82-8]	Merrill, R. C.; Luks, K. D.;
2. Nitrogen; N_2; [7727-37-9]	Kohn, J. P.
3. Hexane; C_6H_{14}; [110-54-3]	*J. Chem. Eng. Data*
	1984, *29*, 272-276.

EXPERIMENTAL VALUES:

Phases in equilibrium	T/K	P/atm	P/MPa	Mole fractions Hexane, $x_{C_6H_{14}}$	Nitrogen, x_{N_2}
$L_1 = L_2,V$	163.75	21.60	2.189	0.1001	0.0319
	165.12	23.40	2.371	0.2153	0.0246
	165.76	22.76	2.306	0.0930	0.0318
	167.19	22.34	2.264	0.1084	0.0354
	168.86	25.41	2.575	0.1975	0.0207
	170.64	25.95	2.629	0.1097	0.0278
	173.05	27.45	2.781	0.1160	0.0248
	177.27	31.06	3.147	0.1858	0.0143
	182.02	33.37	3.381	0.1195	0.0072
L_1,L_2,V	165.00[a]	22.79	2.309	0.1955	0.0226
	165.00	22.99	2.329	0.2076	0.0234
	165.00	23.84	2.416	0.2271	0.0290
	165.00	24.19	2.451	0.2341	0.0312
	165.00	24.80	2.513	0.2513	0.0314
	165.00	25.08	2.541	0.2544	0.0339
	165.00	25.68	2.602	0.2694	0.0301
	165.00	26.36	2.671	0.2791	0.0320
	170.00	26.43	2.678	0.2049	0.0197
	170.00	27.69	2.806	0.2361	0.0241
	170.00	28.41	2.879	0.2488	0.0293
	170.00	28.48	2.886	0.2330	0.0290
	170.00	29.08	2.947	0.2582	0.0301
	170.00	30.37	3.077	0.2787	0.0284
	170.00	30.91	3.132	0.2827	0.0312
	170.00	32.43	3.286	0.3044	0.0639
	170.00	41.62	4.217	0.4162	0.0573
	170.00	48.54	4.918	0.4770	0.0484
	175.00	31.05	3.146	0.2224	0.0200
	175.00	32.38	3.281	0.2499	0.0226
	175.00	32.52	3.295	0.2526	0.0240
	175.00	32.97	3.341	0.2570	0.0287
	175.00	33.52	3.396	0.2636	0.0284
	175.00	33.71	3.416	0.2658	0.0300
	175.00	34.79	3.525	0.2809	0.0291
	175.00	35.47	3.594	0.2907	0.0282
	175.00	43.57	4.415	0.2930	0.0366
	175.00	44.54	4.513	0.3736	0.0539
	175.00	50.85	5.152	0.4213	0.0591
	180.00	33.78	3.423	0.2047	0.0125
	180.00	35.75	3.622	0.2419	0.0183
	180.00	37.76	3.826	0.2663	0.0227
	180.00	37.93	3.843	0.2675	0.0260
	180.00	38.81	3.932	0.2790	0.0290
	180.00	38.82	3.933	0.2777	0.0277
	180.00	39.83	4.036	0.2878	0.0289
	180.00	41.19	4.174	0.3040	0.0278
	180.00	43.05	4.362	0.3215	0.0359
	180.00	48.90	4.955	0.3559	0.0444
	180.00	49.56	5.022	0.3038	0.0416
	185.00	38.47	3.898	0.2245	0.0070
	185.00	41.75	4.230	0.2637	0.0225
	185.00	43.44	4.402	0.2816	0.0265
	185.00	43.68	4.426	0.2849	0.0219
	185.00	44.39	4.498	0.2882	0.0255
	185.00	44.80	4.539	0.2928	0.0252
	185.00	45.08	4.568	0.2910	0.0254
	185.00	47.45	4.808	0.3132	0.0257
	185.00	49.58	5.024	0.3373	0.0309

(cont.)

COMPONENTS:	ORIGINAL MEASUREMENTS:
1. Methane; CH_4; [74-82-8]	Merrill, R. C.; Luks, K. D.;
2. Nitrogen; N_2; [7727-37-9]	Kohn, J. P.
3. Hexane; C_6H_{14}; [110-54-3]	*J. Chem. Eng. Data*
	<u>1984</u>, *29*, 272-276.

EXPERIMENTAL VALUES:

Phases in equilibrium	T/K	P/atm	P/MPa	Mole fractions	
				Hexane, $x_{C_6H_{14}}$	Nitrogen, x_{N_2}
L_1,L_2,V	190.00	44.80	4.539	0.2530	0.0071
	190.00	44.94	4.554	0.2503	0.0045
	190.00	45.41	4.601	0.2678	0.0176
	190.00	47.26	4.789	0.2745	0.0192
	190.00	49.50	5.016	0.2903	0.0369
	190.00	50.10	5.076	0.2925	0.0230
	190.00	50.42	5.109	0.2940	0.0316

[a] Compositions at 165 K estimated from cross plots of data for L_1, L_2 = V and L_1, L_2, V in equilibrium.

Compositions of Hexane lean phase

L_1,L_2 = V	168.48	50.25	5.092	0.0047	0.4119
	169.90	50.80	5.147	0.0037	0.3748
	176.96	51.46	5.214	0.0115	0.2490
	179.67	51.73	5.242	0.0138	0.2203
	181.95	51.80	5.249	0.0148	0.1973
	183.61	51.81	5.250	0.0043	0.1784
	184.25	49.62	5.028	0.0052	0.1707
	185.04	51.73	5.242	0.0187	0.1617
	186.90	51.75	5.244	0.0063	0.1270
	189.01	51.61	5.229	0.0079	0.1005
	190.19	51.61	5.229	0.0212	0.0420
	191.10	51.54	5.222	0.0125	0.0732
S,L_1,L_2,V	164.25	22.75	2.305	0.0804	0.0446
	164.52	25.34	2.568	0.0528	0.0692
	165.28	29.96	3.036	0.0305	0.1142
	165.35	29.56	2.995	0.0281	0.1079
	165.53	30.37	3.077	0.0319	0.1147
	165.53	30.71	3.112	0.0298	0.1225
	165.65	29.43	2.982	0.0368	0.1054
	165.94	33.50	3.394	0.0193	0.1469
	166.18	35.55	3.602	0.0204	0.1706
L_1,L_2,V	170.00	31.80	3.222	0.0329	0.0867
	170.00	32.00	3.242	0.0311	0.0923
	170.00	34.05	3.450	0.0258	0.1099
	170.00	34.86	3.532	0.0236	0.1188
	170.00	38.26	3.877	0.0177	0.1495
	170.00	39.43	3.995	0.0157	0.1750
	170.00	40.04	4.057	0.0193	0.1805
	175.00	36.56	3.704	0.0278	0.0891
	175.00	37.04	3.753	0.0264	0.0946
	175.00	39.08	3.960	0.0195	0.1047
	175.00	39.63	4.016	0.0193	0.1180
	175.00	39.76	4.029	0.0193	0.1233
	175.00	40.37	4.090	0.0177	0.1272
	175.00	41.81	4.236	0.0155	0.1412
	175.00	44.05	4.463	0.0124	0.1652
	175.00	45.55	4.615	0.0078	0.1875
	180.00	37.17	3.766	0.0359	0.0459
	180.00	37.24	3.773	0.0356	0.0489
	180.00	45.48	4.608	0.0135	0.1218
	180.00	45.68	4.629	0.0136	0.1272
	180.00	45.76	4.637	0.0154	0.1187
	180.00	45.89	4.650	0.0154	0.1287

(cont.)

COMPONENTS:	ORIGINAL MEASUREMENTS:
1. Methane; CH_4; [74-82-8] .	Merrill, R. C.; Luks, K. D.;
2. Nitrogen; N_2; [7727-37-9]	Kohn, J. P.
3. Hexane; C_6H_{14}; [110-54-3]	*J. Chem. Eng. Data*
	<u>1984</u>, *29*, 272-276.

EXPERIMENTAL VALUES:

Phases in equilibrium	T/K	*P*/atm	*P*/MPa	Mole fractions	
				Hexane, $x_{C_6H_{14}}$	Nitrogen, x_{N_2}
L_1, L_2, V	180.00	48.82	4.947	0.0109	0.1686
	180.00	49.29	4.994	0.0074	0.1674
	180.00	51.05	5.173	0.0078	0.1963
	185.00	40.51	4.105	0.0352	0.0286
	185.00	42.89	4.346	0.0392	0.0517
	190.00	44.47	4.506	0.0261	0.0176
	190.00	45.83	4.644	0.0248	0.0352
	190.00	49.57	5.023	0.0146	0.0610

COMPONENTS:	ORIGINAL MEASUREMENTS:
1. Methane; CH_4; [74-82-8]	Wang, R. H.; McKetta, J. J.
2. Carbon dioxide; CO_2; [124-38-9]	*J. Chem. Engng. Data*
3. Butane; C_4H_{10}; [106-97-8]	<u>1964</u>, *9*, 30-35.

VARIABLES:	PREPARED BY:
Temperature, pressure	C. L. Young

EXPERIMENTAL VALUES:

Mole fractions

T/K (T/°F)	P/MPa (P/psi)	in liquid x_{CH_4}	in liquid $x_{C_4H_{10}}$	x_{CO_2}	in vapor y_{CH_4}	in vapor $y_{C_4H_{10}}$	y_{CO_2}
310.93 (100)	2.76 (400)	0.122	0.878	–	0.824	0.176	–
		0.0936	0.8453	0.0611	0.6570	0.1787	0.1643
		0.0858	0.8163	0.0979	0.5612	0.1784	0.2604
		0.0665	0.8031	0.1304	0.4687	0.1727	0.3586
		0.0495	0.7803	0.1702	0.3609	0.1780	0.4611
		0.0332	0.7347	0.2321	0.1926	0.1782	0.6292
		0.0221	0.7233	0.2546	0.1247	0.1800	0.6953
		–	0.6996	0.3004	–	0.1814	0.8186
	5.52 (800)	0.254	0.746	–	0.877	0.123	–
		0.2336	0.6873	0.0791	0.7798	0.1202	0.1000
		0.2045	0.6508	0.1447	0.7156	0.1100	0.1744
		0.1861	0.5718	0.2421	0.5873	0.1062	0.3065
		0.1517	0.5227	0.3256	0.4851	0.1053	0.4096
		0.1002	0.4396	0.4602	0.3271	0.0977	0.5752
		0.0433	0.3263	0.6304	0.1308	0.0892	0.7800
		–	0.2594	0.7406	–	0.0803	0.9197
	8.27 (1200)	0.880	0.120	–	0.381	0.619	–
		0.7887	0.1172	0.0941	0.3554	0.5572	0.0874
		0.7008	0.1112	0.1880	0.3206	0.5102	0.1692
		0.5498	0.1150	0.3352	0.2752	0.4246	0.3002
		0.4603	0.1173	0.4224	0.2556	0.3700	0.3744
		0.4383	0.1100	0.4517	0.2430	0.3549	0.4021
		0.3817	0.1087	0.5096	0.2152	0.3287	0.4561

(cont.)

AUXILIARY INFORMATION

METHOD/APPARATUS/PROCEDURE:	SOURCE AND PURITY OF MATERIALS:
Stainless steel windowed equilibrium cell with magnetic pump for re-circulating vapor. Samples analysed by gas chromatography and mass spectrometry. Some details given in source and in ref. (1).	1. Phillips Petroleum Co. research grade sample, purity at least 99.9 mole per cent.
	2. Matheson Co. bone dry grade, purity at least 99.8 mole per cent.
	3. Phillips Petroleum Co. research grade, purity at least 99.9 mole per cent.

ESTIMATED ERROR:

$\delta T/K = \pm 0.03$ at 310 K; ± 0.3 at 178 K

$\delta P/MPa = \pm 0.2\%$; δx, $\delta y = \pm 0.001$.

REFERENCES:

1. Wang, R. H.
   Ph.D. thesis, University of Texas,
   Austin, <u>1963</u>.

COMPONENTS: ORIGINAL MEASUREMENTS:

1. Methane; CH_4; [74-82-8]   Wang, R. H.; McKetta, J. J.

2. Carbon dioxide; CO_2; [124-38-9]   *J. Chem. Engng. Data*

3. Butane; C_4H_{10}; [106-97-8]   1964, *9*, 30-35.

EXPERIMENTAL VALUES:

Mole fractions

T/K (T/°F)	P/MPa (P/psi)	in liquid			in vapor		
		x_{CH_4}	$x_{C_4H_{10}}$	x_{CO_2}	y_{CH_4}	$y_{C_4H_{10}}$	y_{CO_2}
310.93	8.27	0.3282	0.1118	0.5600	0.1928	0.3052	0.5020
(100)	(1200)	0.2191	0.1087	0.6722	0.1496	0.2496	0.6008
		0.1478	0.1291	0.7231	0.1332	0.2003	0.6665
	11.72	0.599	0.441	–	0.844	0.156	–
	(1700)	0.5562	0.4050	0.0388	0.7972	0.1566	0.0462
		0.5426	0.3636	0.0938	0.7344	0.1589	0.1067
		0.5308	0.3344	0.1348	0.6798	0.1714	0.1488
		0.5337	0.2984	0.1679	0.6086	0.2014	0.1900
277.59	2.76	0.159	0.841	–	0.928	0.072	–
(40)	(400)	0.1484	0.8078	0.0438	0.8567	0.0635	0.0798
		0.1310	0.7821	0.0869	0.7963	0.0635	0.1402
		0.1002	0.7146	0.1852	0.6394	0.0706	0.2900
		0.0904	0.6942	0.2154	0.5837	0.0739	0.3424
		0.0698	0.6514	0.2788	0.4812	0.0688	0.4500
		0.0509	0.5993	0.3498	0.3705	0.0778	0.5517
		0.0289	0.5248	0.4463	0.2060	0.0748	0.7192
		0.0157	0.4704	0.5139	0.0912	0.0790	0.8298
		–	0.4279	0.5721	–	0.0782	0.9218
	5.52	0.296	0.704	–	0.945	0.055	–
	(800)	0.2801	0.6456	0.0743	0.8420	0.0598	0.0982
		0.2607	0.5932	0.1461	0.7444	0.0560	0.1996
		0.2480	0.5338	0.2182	0.6648	0.0604	0.2748
		0.2292	0.4624	0.3084	0.5749	0.0504	0.3747
		0.2070	0.3932	0.3998	0.5040	0.0638	0.4322
		0.1886	0.3389	0.4725	0.4674	0.0474	0.4852
		0.1740	0.2802	0.5458	0.3806	0.0623	0.5571
		0.1723	0.2009	0.6268	0.2864	0.0736	0.6400
		0.1845	0.1488	0.6667	0.2297	0.0954	0.6749
	8.27	0.426	0.574	–	0.942	0.058	–
	(1200)	0.4383	0.4851	0.0762	0.8715	0.0503	0.0782
		0.4273	0.4430	0.1297	0.8186	0.0502	0.1312
		0.4309	0.3862	0.1829	0.7611	0.0489	0.1900
		0.4288	0.3152	0.2560	0.6756	0.0506	0.2738
		0.4208	0.2700	0.3092	0.6197	0.0526	0.3277
		0.4152	0.2212	0.3636	0.5333	0.0767	0.3900
		0.4145	0.1833	0.4022	0.4680	0.1042	0.4278
	11.72	0.612	0.388	–	0.899	0.101	–
	(1700)	0.6321	0.3376	0.0303	0.8543	0.1094	0.0364
		0.6468	0.2921	0.0611	0.8100	0.1202	0.0698
		0.6588	0.2459	0.0953	0.7562	0.1401	0.1037
244.26	2.76	0.194	0.806	–	0.973	0.027	–
(-20)	(400)	0.1777	0.7137	0.1086	0.9276	0.0247	0.0477
		0.1511	0.5887	0.2602	0.8676	0.0226	0.1098
		0.1483	0.4755	0.3762	0.8109	0.0212	0.1679
		0.1204	0.3902	0.4894	0.6956	0.0198	0.2846
		0.1100	0.3234	0.5666	0.6128	0.0208	0.3664
		0.0755	0.2208	0.7037	0.5343	0.0136	0.4521
		0.0692	0.0966	0.8342	0.4550	0.0113	0.5337
		0.0992	0.0546	0.8462	0.2920	0.0102	0.6978
	5.52	0.370	0.630	–	0.975	0.025	–
	(800)	0.3509	0.5779	0.0712	0.9495	0.0201	0.0304
		0.3349	0.5289	0.1362	0.9223	0.0199	0.0578
		0.3084	0.4321	0.2595	0.8765	0.0198	0.1037
		0.2887	0.3411	0.3702	0.8288	0.0174	0.1538
		0.2995	0.2312	0.4693	0.7920	0.0191	0.1889
		0.2885	0.2012	0.5103	0.7498	0.0217	0.2285
(cont.)		0.2906	0.1439	0.5655	0.7238	0.0281	0.2481

COMPONENTS:	ORIGINAL MEASUREMENTS:
1. Methane; CH_4; [74-82-8]	Wang, R. H.; McKetta, J. J.
2. Carbon dioxide; CO_2; [124-38-9]	*J. Chem. Engng. Data*
3. Butane; C_4H_{10}; [106-97-8]	<u>1964</u>, *9*, 30-35.

EXPERIMENTAL VALUES:

		Mole fractions					
		in liquid				in vapor	
T/K (T/°F)	P/MPa (P/psi)	x_{CH_4}	$x_{C_4H_{10}}$	x_{CO_2}	y_{CH_4}	$y_{C_4H_{10}}$	y_{CO_2}
244.26 (-20)	5.52 (800)	0.3183	0.0913	0.5904	0.7005	0.0128	0.2867
	8.27 (1200)	0.534	0.466	–	0.971	0.029	–
		0.5269	0.3897	0.0834	0.9056	0.0305	0.0639
		0.5251	0.3241	0.1508	0.8486	0.0292	0.1222
		0.5271	0.2595	0.2134	0.7980	0.0257	0.1763
		0.5488	0.1717	0.2795	0.7374	0.0240	0.2386
		0.5836	0.0966	0.3198	0.6762	0.0296	0.2932
	11.72 (1700)	0.831	0.169	–	0.902	0.098	–
		0.8504	0.1282	0.0214	0.8868	0.1000	0.0132
210.93 (-80)	2.76 (400)	0.335	0.665	–	0.990	0.010	–
		0.3253	0.6357	0.0393	0.9649	0.0097	0.0254
		0.3217	0.6188	0.0595	0.9503	0.0096	0.0401
		0.3181	0.6019	0.0800	0.9400	0.0037	0.0563
		0.3068	0.5777	0.1155	0.9112	0.0085	0.0803
		0.2802	0.4606	0.2592	0.8104	0.0084	0.1812
		0.2700	0.3600	0.3700	0.7277	0.0123	0.2600
		0.2801	0.1797	0.5402	0.5799	0.0198	0.4003
		0.3011	0.0856	0.6133	0.5160	0.0204	0.4636
	5.52 (800)	0.6700	0.3300	–	0.990	0.010	–
		0.6753	0.3055	0.0192	0.9698	0.0146	0.0156
		0.6758	0.2835	0.0407	0.9570	0.0098	0.0332
		0.6848	0.1855	0.1297	0.8760	0.0099	0.1141
		0.6995	0.1209	0.1796	0.8275	0.0102	0.1623
		0.7167	0.0811	0.2022	0.7988	0.0100	0.1912
	6.89 (1000)	0.848	0.152	–	0.970	0.030	–
		0.8674	0.1103	0.0223	0.9582	0.0251	0.0167
		0.8896	0.0742	0.0362	0.9433	0.0270	0.0297
177.59 (-140)	2.76 (400)	0.850	0.150	–	0.996	0.004	–
		0.8389	0.1168	0.0443	0.9802	0.0031	0.0167
		0.8455	0.0701	0.0844	0.9659	0.0042	0.0299
		0.8737	0.0207	0.1056	0.9482	0.0056	0.0462

COMPONENTS:	ORIGINAL MEASUREMENTS:
1. Methane; CH_4; [74-82-8] 2. Carbon dioxide; CO_2; [124-38-9] 3. Butane; C_4H_{10}; [106-97-8]	Saxena, A. C.; Robinson, D. B. *Can. J. Chem. Engng.* 1969, *47*, 69-75.
VARIABLES:	PREPARED BY: C. L. Young

EXPERIMENTAL VALUES:

		Mole fraction					
			in liquid			in vapor	
T/K	P/MPa	x_{CH_4}	x_{CO_2}	$x_{C_4H_{10}}$	y_{CH_4}	y_{CO_2}	$y_{C_4H_{10}}$
(T/°F)	(P/psia)						
310.93	2.758	0.070	0.132	0.798	0.472	0.349	0.179
(100)	(400)	0.084	0.099	0.807	0.541	0.281	0.178
		0.030	0.225	0.745	0.231	0.587	0.182
		0.122	-	0.878	0.824	-	0.176
		-	0.300	0.700	-	0.820	0.180
	5.516	0.169	0.275	0.556	0.466	0.412	0.122
	(800)	0.224	0.095	0.682	0.713	0.158	0.129
		0.073	0.549	0.378	0.199	0.694	0.107
		0.254	-	0.746	0.877	-	0.123
		-	0.735	0.265	-	0.920	0.080
	8.27	-	-	-	0.464	0.429	0.107
	(1200)	0.212	0.445	0.343	0.399	0.487	0.114
		0.260	0.313	0.427	0.520	0.369	0.111
		0.152	0.595	0.253	0.237	0.656	0.107
		0.336	0.122	0.542	0.743	0.140	0.117
		0.203	0.458	0.339	-	-	-
		0.381	-	0.619	0.880	-	0.120
277.59	2.758	0.155	0.090	0.755	0.732	0.183	0.085
(40)	(400)	0.117	0.178	0.705	0.571	0.334	0.095
		0.094	0.297	0.608	0.414	0.487	0.099
		0.029	0.479	0.492	0.175	0.735	0.090

(cont.)

AUXILIARY INFORMATION

METHOD/APPARATUS/PROCEDURE:	SOURCE AND PURITY OF MATERIALS:
Variable volume cell with sample confined between two moveable pistons. Volume varied by mercury injection. Details of design given in source. Cell fitted with windows. Samples of gas and liquid analysed by gas chromatography. Temperature measured with iron-constantan thermocouple. Pressure measured with Bourdon gauge.	1. Matheson sample, purity varied between 99.3 and 99.7 per cent as determined by gas chromatography. 2. Canadian Liquid Air Co., purity at least 99.5 per cent. 3. Matheson instrument grade, purity 99.5 mole per cent.
	ESTIMATED ERROR: $\delta T/K = \pm 0.03$; $\delta P/MPa = \pm 0.003$ up to 6 MPa, ± 0.02 above 6 MPa; δx, $\delta y = \pm 1\%$.
	REFERENCES:

COMPONENTS: ORIGINAL MEASUREMENTS:

1. Methane; CH_4; [74-82-8]   Saxena, A. C.; Robinson, D. B.
2. Carbon dioxide; CO_2; [124-38-9]   *Can. J. Chem. Engng.*
3. Butane; C_4H_{10}; [106-97-8]   <u>1969</u>, *47*, 69-75.

EXPERIMENTAL VALUES:

		Mole fraction					
T/K	P/MPa	in liquid				in vapor	
(T/°F)	(P/psia)	x_{CH_4}	x_{CO_2}	$x_{C_4H_{10}}$	y_{CH_4}	y_{CO_2}	$y_{C_4H_{10}}$
277.59	2.758						
(40)	(400)	0.198	–	0.802	0.928	–	0.072
		–	0.568	0.432	–	0.930	0.070
	5.516	0.271	0.250	0.479	0.663	0.284	0.053
	(800)	0.331	0.141	0.528	0.777	0.169	0.054
		0.093	0.827	0.080	0.287	0.699	0.014
		0.210	0.417	0.373	0.504	0.447	0.049
		0.407	–	0.593	0.956	–	0.044
		0.062	0.938	–	0.214	0.786	–
	8.27	0.476	0.166	0.358	0.789	0.157	0.054
	(1200)	0.209	0.765	0.026	0.331	0.660	0.009
		0.328	0.467	0.205	0.553	0.403	0.044
		0.567	–	0.433	0.950	–	0.050
		0.194	0.806	–	0.308	0.692	–
244.26	2.758	0.136	0.590	0.274	–	–	–
(-20)	(400)	0.043	0.947	0.010	–	–	–
		0.122	0.638	0.240	0.518	0.473	0.009
		0.159	0.491	0.350	0.578	0.396	0.026
		0.266	0.094	0.640	0.861	0.114	0.025
		0.212	0.278	0.510	0.699	0.277	0.024
		0.0395	0.9605	–	0.419	0.581	–
		0.294	–	0.706	0.957	–	0.043
	5.516	0.335	0.529	0.136	0.682	0.310	0.008
	(800)	0.386	0.402	0.232	0.746	0.240	0.014
		0.410	0.328	0.262	0.780	0.207	0.013
		0.467	0.112	0.421	0.895	0.091	0.014
		0.410	0.327	0.263	0.775	0.208	0.017
		0.188	0.812	–	0.625	0.375	–
		0.499	–	0.501	0.978	–	0.022
	8.27	0.678	0.059	0.263	0.927	0.047	0.026
		0.647	0.136	0.217	0.879	0.095	0.026
		0.624	0.211	0.165	0.837	0.146	0.017
		0.708	–	0.292	0.972	–	0.028

COMPONENTS:	ORIGINAL MEASUREMENTS:
1. Methane; CH_4; [74-82-8] 2. Carbon dioxide; CO_2; [124-38-9] 3. Hexane; C_6H_{14}; [110-54-3]	Merrill, R. C.; Luks, K. D.; Kohn, J. P. *J. Chem. Eng. Data* 1983, *28*, 210-215.

VARIABLES:	PREPARED BY:
	C. L. Young

EXPERIMENTAL VALUES:

Phases in equilibrium	T/K	P/atm	P/MPa	Mole fractions of hexane $x_{C_6H_{14}}$	carbon dioxide x_{CO_2}

Data for hexane-lean phase, L_2

Phases in equilibrium	T/K	P/atm	P/MPa	$x_{C_6H_{14}}$	x_{CO_2}
L_1, L_2, V	204.00	57.46	5.822	0.0331	0.0841
	204.00	57.96	5.873	0.0304	0.0850
	202.00	55.96	5.670	0.0214	0.0083
	200.00	53.45	5.416	0.0310	0.0783
	200.00	53.65	5.436	0.0299	0.0751
	198.00	50.93	5.160	0.0371	0.0791
	198.00	52.02	5.271	0.0233	0.0404
	198.00	52.49	5.319	0.0209	0.0267
	196.00	49.09	4.974	0.0323	0.0436
	196.00	49.30	4.995	0.0284	0.0394
	196.00	49.36	5.001	0.0285	0.0259
	196.00	49.36	5.001	0.0303	0.0263
	196.00	49.43	5.008	0.0256	0.0267
	196.00	50.38	5.105	0.0204	0.0181
	196.00	50.44	5.111	0.0182	0.0251
	196.00	50.59	5.126	0.0206	0.0188
	196.00	50.66	5.133	0.0159	0.0029
	196.00	51.00	5.168	0.0178	0.0022
	194.00	46.17	4.678	0.0437	0.0283
	194.00	46.44	4.706	0.0397	0.0400
	194.00	47.52	4.815	0.0292	0.0238

(cont.)

AUXILIARY INFORMATION

METHOD/APPARATUS/PROCEDURE:	SOURCE AND PURITY OF MATERIALS:
Glass equilibrium cell. Temperature measured with platinum resistance thermometer and pressure with Bourdon gauge. Stoichiometry and volumetric measurements were used to obtain liquid phase compositions and molar volumes. Gas phase composition assumed to be the same as in binary (ref. (2)). Molar volume data in source. Details of apparatus in ref. (1).	1. Linde Ultra Pure grade sample, purity 99.97 moles per cent. 2. Matheson "Coleman Grade" sample, purity 99.99 moles per cent. 3. Humphrey Chemical Co. sample, purity 99 moles per cent.

ESTIMATED ERROR:
$\delta T/K = \pm 0.03$; $\delta P/MPa = \pm 0.007$; $\delta x/x = \pm 0.02$ in L_1 phase, ± 0.08 in L_2 phase.

REFERENCES:
1. Hottovy, J.D.; Kohn, J.P.; Luks, K.D. *J. Chem. Eng. Data* 1981, *26*, 135.

2. Mraw, S.C.; Hwang, S.-C.; Kobayashi, R. *J. Chem. Eng. Data* 1978, *23*, 135.

COMPONENTS:	ORIGINAL MEASUREMENTS:
1. Methane; CH_4; [74-82-8]	Merrill, R. C.; Luks, K. D.;
2. Carbon dioxide; CO_2; [124-38-9]	Kohn, J. P.
3. Hexane; C_6H_{14}; [110-54-3]	*J. Chem. Eng. Data*
	<u>1983</u>, *28*, 210-215.

EXPERIMENTAL VALUES:

Phases in equilibrium	T/K	P/atm	P/MPa	Mole fractions of hexane $x_{C_6H_{14}}$	carbon dioxide x_{CO_2}
L_1,L_2,V	194.00	47.73	4.836	0.0210	0.0046
	194.00	47.80	4.843	0.0246	0.0012
	194.00	47.87	4.850	0.0274	0.0000
	192.00	43.85	4.443	0.0551	0.0305
	192.00	44.67	4.526	0.0374	0.0124
	192.00	44.87	4.546	0.0266	0.0047
	192.00	45.15	4.575	0.0361	0.0134
	192.00	45.49	4.609	0.0291	0.0031
	190.00	42.22	4.278	0.0466	0.0139
	190.00	42.43	4.299	0.0387	0.0032
	190.00	42.56	4.312	0.0411	0.0101
	190.00	42.70	4.327	0.0361	0.0035
	188.00	39.37	3.989	0.0623	0.0143
	188.00	39.77	4.030	0.0468	0.0042
	188.00	39.91	4.044	0.0473	0.0039
	188.00	40.11	4.064	0.0455	0.0035
	186.00	37.60	3.810	0.0592	0.0035
$L_1,L_2 = V$	197.15	52.09	5.278	0.0055	0.0014
	199.80	54.26	5.426	0.0090	0.0255
	200.05	55.35	5.608	0.0078	0.0392
	200.18	54.53	5.525	0.0073	0.0234
	200.19	54.33	5.505	0.0044	0.0194
	201.83	55.96	5.670	0.0080	0.0364
	202.33	56.37	5.712	0.0104	0.0522
	202.91	56.85	5.760	0.0149	0.0628
	203.98	57.86	5.863	0.0137	0.0747

Data for hexane-rich phase, L_1

$L_1 \equiv L_2,V$	184.14	35.41	3.588	0.1305	0.0157
	184.76	35.96	3.644	0.1495	0.0176
	189.61	41.13	4.167	0.1566	0.0310
	189.85	41.39	4.194	0.1698	0.0260
	190.69	42.36	4.292	0.1701	0.0252
	191.13	42.90	4.347	0.1707	0.0338
	193.74	45.36	4.596	0.1306	0.0454
	199.18	51.82	5.251	0.1408	0.0743
	199.34	51.75	5.244	0.1251	0.0674
	202.63	55.70	5.644	0.1161	0.0901
	188.41	39.50	4.002	0.1141	0.0310
	191.17	42.49	4.305	0.0850	0.0323
$L_1,L_2 = V$	196.89	52.22	5.291	0.2365	0.0123
	197.76	52.28	5.297	0.2304	0.0344
	197.92	52.42	5.311	0.2232	0.0428
	198.10	52.89	5.359	0.2264	0.0334
	198.13	52.90	5.360	0.2334	0.0079
	199.87	54.27	5.499	0.2035	0.0883
	203.24	57.06	5.782	0.1423	0.1380
	203.66	57.67	5.843	0.1482	0.0848
L_1,L_2,V	204.00	58.07	5.884	0.1152	0.1078
	204.00	58.21	5.898	0.1184	0.1018
	202.00	55.16	5.589	0.1302	0.0807
	202.00	55.36	5.609	0.1261	0.1226
	202.00	55.43	5.616	0.1364	0.0938
	202.00	55.63	5.637	0.1548	0.0664

(cont.)

COMPONENTS:	ORIGINAL MEASUREMENTS:
1. Methane; CH_4; [74-82-8]	Merrill, R. C.; Luks, K. D.;
2. Carbon dioxide; CO_2; [124-38-9]	Kohn, J. P.
3. Hexane; C_6H_{14}; [110-54-3]	*J. Chem. Eng. Data*
	<u>1983</u>, *28*, 210-215.

EXPERIMENTAL VALUES:

Phases in equilibrium	T/K	P/atm	P/MPa	Mole fractions of hexane $x_{C_6H_{14}}$	carbon dioxide x_{CO_2}
L_1,L_2,V	200.00	52.70	5.340	0.1345	0.0774
	198.00	51.21	5.189	0.1778	0.0609
	198.00	51.34	5.202	0.1740	0.0599
	198.00	52.08	5.277	0.2135	0.0000
	198.00	52.42	5.311	0.2079	0.0249
	198.00	52.68	5.338	0.2332	0.0214
	196.00	48.49	4.913	0.1626	0.0659
	196.00	49.49	5.015	0.1980	0.0419
	196.00	49.96	5.062	0.2179	0.0278
	196.00	50.04	5.070	0.2074	0.0353
	196.00	50.17	5.083	0.2171	0.0279
	196.00	50.38	5.105	0.2285	0.0312
	196.00	50.65	5.132	0.2313	0.0017
	196.00	50.73	5.140	0.2322	0.0171
	194.00	46.30	4.691	0.1820	0.0398
	194.00	46.83	4.745	0.2063	0.0293
	194.00	46.91	4.753	0.2004	0.0377
	194.00	46.97	4.759	0.2098	0.0322
	194.00	46.99	4.761	0.2030	0.0286
	194.00	47.04	4.766	0.2120	0.0317
	194.00	47.46	4.809	0.2151	0.0240
	194.00	47.74	4.837	0.2218	0.0172
	192.00	44.12	4.470	0.1595	0.0330
	192.00	44.20	4.479	0.1845	0.0250
	192.00	44.25	4.484	0.1905	0.0277
	192.00	44.25	4.484	0.1951	0.0291
	192.00	44.27	4.486	0.1847	0.0256
	192.00	44.32	4.491	0.1918	0.0294
	192.00	44.80	4.539	0.2136	0.0075
	192.00	44.95	4.555	0.2082	0.0187
	192.00	45.19	4.579	0.2166	0.0185
	190.00	41.61	4.216	0.1623	0.0244
	190.00	42.23	4.279	0.1936	0.0184
	190.00	42.56	4.312	0.2044	0.0204
	188.00	39.83	4.036	0.1927	0.0228
	188.00	39.83	4.036	0.1934	0.0245
	188.00	39.90	4.043	0.1983	0.0092
	186.00	37.32	3.781	0.1645	0.0059
	186.00	37.45	3.795	0.1534	0.0188
	186.00	37.52	3.802	0.1690	0.0203

COMPONENTS:	ORIGINAL MEASUREMENTS:
1. Methane; CH_4; [74-82-8] 2. Carbon dioxide; CO_2; [124-38-9] 3. Octane; C_8H_{18}; [111-65-9]	Hottovy, J. D.; Kohn, J. P.; Luks, K. D. *J. Chem. Engng. Data* <u>1982</u>, *27*, 298-302.

VARIABLES:	PREPARED BY:
	C. L. Young

EXPERIMENTAL VALUES: Data for octane-lean liquid phase.

Type of data	T/K	P/atm	P/MPa	Mole fractions x_{CO_2}	$x_{C_8H_{18}}$	Molar volume /cm³mol⁻¹
K(L_1-L_2=V)	219.73	69.7	7.06	0.251	0.0146	67.5
	217.32	67.0	6.79	0.221	0.0094	73.1
	214.71	64.3	6.52	0.200	0.0075	74.9
	212.20	62.0	6.28	0.179	0.0055	78.3
	211.03	60.9	6.17	0.163	0.0040	82.7
	207.61	57.9	5.87	0.146	0.0044	79.9
	204.83	55.7	5.64	0.119	0.0029	84.7
	204.04	55.0	5.57	0.107	0.0039	89.8
Q(S-L_1-L_2-V)	202.40	53.7	5.44	0.110	0.0045	77.0
	202.27	53.2	5.39	0.114	0.0056	73.3
	202.16	52.6	5.33	0.123	0.0070	69.4
	202.14	52.5	5.32	0.131	0.0072	68.0
	201.98	51.5	5.22	0.146	0.0097	64.3
	201.43	48.9	4.95	0.205	0.0223	56.8
	201.41	48.7	4.93	0.199	0.0297	58.0
	201.38	48.0	4.86	0.227	0.0302	55.9
	201.08	47.1	4.77	0.253	0.0515	54.5
LCST(L_1=L_2-V)	213.54	61.5	6.22	0.298	0.0800	58.1
	218.21	67.3	6.82	0.291	0.0704	58.2
	215.33	63.7	6.45	0.301	0.0646	57.5
	211.73	59.2	6.00	0.290	0.0719	58.5
	210.14	57.5	5.83	0.289	0.0750	57.0

(cont.)

AUXILIARY INFORMATION

METHOD/APPARATUS/PROCEDURE:	SOURCE AND PURITY OF MATERIALS:
Glass equilibrium cell. Temperature measured with platinum resistance thermometer and pressure with Bourdon gauge. Stoichiometry and volumetric measurements were used to compute liquid phase compositions and molar volumes. Details in ref. 1 and source.	1. Linde Ultrapure grade, purity 99.97 mole per cent. 2. Matheson "Coleman Grade", purity 99.99 mole per cent. 3. Humphrey-Wilkinson grade, purity 99 mole per cent.
	ESTIMATED ERROR: δT/K = ±0.03; δP/MPa = ±0.07; $\delta x_{C_3H_8}$ = ±3.5%; $\delta x_{C_8H_{18}}$ = ±2% (octane rich phase); $\delta x_{C_8H_{18}}$ = ±8% (octane lean phase).
	REFERENCES: 1. Kohn, J. P. *Am. Inst. Chem. Engnrs. J.* <u>1961</u>, *7*, 514.

COMPONENTS: ORIGINAL MEASUREMENTS:

1. Methane; CH_4; [74-82-8]   Hottovy, J. D.; Kohn, J. P.;
2. Carbon dioxide; CO_2; [124-38-9]   Luks, K. D.
3. Octane; C_8H_{18}; [111-65-9]   *J. Chem. Eng. Data*
                                           1982, *27*, 298-302.

EXPERIMENTAL VALUES: Data for octane-lean liquid phase.

Type of data	T/K	P/atm	P/MPa	Mole fractions x_{CO_2}	$x_{C_8H_{18}}$	Molar volume /cm^{3}mol^{-1}
LCST(L_1=L_2-V)	206.97	53.5	5.42	0.278	0.0712	56.1
	206.29	52.8	5.35	0.280	0.0848	56.5
	205.69	51.9	5.26	0.275	0.0833	56.3
	203.84	50.1	5.08	0.273	0.0817	55.8
L_1-L_2-V	216.00	65.5	6.64	0.237	0.0133	66.3
	216.00	65.3	6.62	0.247	0.0198	62.3
	214.00	63.2	6.40	0.234	0.0157	64.1
	214.00	62.8	6.36	0.246	0.0218	61.6
	214.00	62.5	6.33	0.256	0.0274	60.6
	212.00	60.8	6.16	0.230	0.0173	62.3
	212.00	60.5	6.13	0.241	0.0209	61.2
	212.00	60.1	6.09	0.255	0.0293	60.9
	212.00	59.9	6.07	0.264	0.0335	59.5
	210.00	58.7	5.95	0.228	0.0183	61.0
	210.00	58.5	5.93	0.231	0.0201	59.3
	210.00	57.5	5.83	0.259	0.0355	59.0
	208.00	56.7	5.75	0.208	0.0151	61.6
	208.00	56.1	5.68	0.223	0.0199	59.6
	208.00	55.9	5.66	0.231	0.0232	59.2
	208.00	55.4	5.61	0.244	0.0294	57.8
	208.00	55.3	5.60	0.249	0.0309	57.6
	208.00	55.2	5.59	0.259	0.0363	56.5
	208.00	54.9	5.56	0.272	0.0517	56.0
	206.00	55.9	5.66	0.154	0.0093	69.3
	206.00	54.5	5.52	0.205	0.0167	60.2
	206.00	54.2	5.49	0.212	0.0182	59.4
	206.00	53.6	5.43	0.226	0.0239	57.9
	206.00	53.4	5.41	0.234	0.0265	57.5
	206.00	52.8	5.35	0.252	0.0374	56.7
	206.00	52.7	5.34	0.261	0.0399	55.3
	204.00	54.9	5.56	0.115	0.0037	81.7
	204.00	54.6	5.53	0.130	0.0063	75.4
	204.00	54.5	5.52	0.134	0.0062	71.6
	204.00	54.2	5.49	0.142	0.0080	69.6
	204.00	53.3	5.40	0.160	0.0111	65.2
	204.00	52.0	5.27	0.203	0.0185	58.8
	204.00	51.6	5.23	0.214	0.0214	57.7
	204.00	51.4	5.21	0.216	0.0231	57.4
	204.00	51.2	5.19	0.224	0.0253	57.0
	204.00	51.0	5.17	0.235	0.0307	56.3
	204.00	50.6	5.13	0.251	0.0395	55.1
	204.00	50.4	5.11	0.255	0.0445	55.8
	204.00	50.2	5.09	0.265	0.0555	54.9
	202.00	51.6	5.23	0.145	0.0099	65.0
	202.00	49.6	5.03	0.205	0.0212	57.2
	202.00	49.3	5.00	0.212	0.0239	56.4
	202.00	48.8	4.94	0.226	0.0301	56.3
	202.00	48.1	4.87	0.253	0.0472	54.7
	202.00	47.9	4.85	0.255	0.0521	55.0

(cont.)

COMPONENTS:	ORIGINAL MEASUREMENTS:
1. Methane; CH_4; [74-82-8]	Hottovy, J. D.; Kohn, J. P.;
2. Carbon dioxide; CO_2; [124-38-9]	Luks, K. D.
3. Octane; C_8H_{18}; [111-65-9]	*J. Chem. Eng. Data*
	1982, *27*, 298-302.

EXPERIMENTAL VALUES:

Data for octane-rich liquid phase.

Type of data	T/K	P/atm	P/MPa	Mole fractions x_{CO_2}	$x_{C_8H_{18}}$	Molar volume /cm^{3}mol^{-1}
K(L_1-L_2=V)	220.78	70.7	7.16	0.322	0.0718	58.5
	220.08	70.7	7.16	0.323	0.0828	59.0
	218.99	68.8	6.97	0.316	0.0987	59.5
	218.28	67.9	6.88	0.312	0.1104	61.0
	217.43	67.2	6.81	0.209	0.1199	61.0
	216.97	66.6	6.75	0.302	0.1269	61.0
	212.93	62.7	6.35	0.268	0.1777	65.2
	207.18	57.6	5.84	0.208	0.2410	69.8
	203.01	54.3	5.50	0.158	0.2823	72.7
Q(S-L_1-L_2-V)	202.32	53.3	5.40	0.156	0.2846	73.4
	201.89	51.2	5.19	0.183	0.2528	69.9
	201.87	50.9	5.16	0.192	0.2423	68.6
	201.82	50.3	5.10	0.202	0.2303	67.0
	201.45	48.4	4.90	0.239	0.1830	62.9
	201.36	47.9	4.85	0.254	0.1717	61.6
	201.23	47.5	4.81	0.246	0.1473	60.0
L_1-L_2-V	216.00	65.2	6.61	0.289	0.1213	59.8
	216.00	65.0	6.59	0.297	0.1087	59.1
	216.00	64.7	6.56	0.306	0.0883	58.3
	214.00	62.5	6.33	0.292	0.1185	59.7
	214.00	62.2	6.30	0.297	0.1006	58.9
	214.00	62.1	6.29	0.302	0.0894	57.8
	212.00	61.3	6.21	0.260	0.1729	63.6
	212.00	60.9	6.17	0.268	0.1619	63.7
	212.00	60.1	6.09	0.297	0.1198	58.6
	212.00	59.7	6.05	0.288	0.1094	59.7
	212.00	59.7	6.05	0.292	0.099	57.7
	210.00	58.5	5.93	0.259	0.166	63.2
	210.00	58.1	5.89	0.267	0.157	63.1
	210.00	58.0	5.88	0.270	0.150	61.6
	210.00	57.4	5.82	0.286	0.109	58.4
	208.00	57.0	5.78	0.240	0.197	65.2
	208.00	55.9	5.66	0.263	0.163	62.1
	208.00	55.4	5.61	0.272	0.144	60.6
	208.00	55.5	5.62	0.267	0.151	61.8
	206.00	55.9	5.66	0.202	0.239	69.4
	206.00	55.2	5.59	0.214	0.225	68.7
	206.00	54.3	5.50	0.237	0.193	65.0
	206.00	53.5	5.42	0.255	0.171	62.5
	206.00	53.0	5.37	0.264	0.144	60.4
	206.00	52.9	5.36	0.271	0.136	59.6
	204.00	53.7	5.44	0.192	0.247	69.5
	204.00	53.6	5.43	0.237	0.188	63.9
	204.00	53.2	5.39	0.200	0.236	68.6
	204.00	50.5	5.12	0.263	0.137	59.7
	202.00	51.3	5.20	0.183	0.252	69.9
	303.00	51.0	5.17	0.191	0.244	69.2
	202.00	50.5	5.12	0.198	0.233	68.2
	202.00	49.1	4.98	0.236	0.184	63.6
	202.00	48.4	4.90	0.263	0.155	60.8

COMPONENTS:	ORIGINAL MEASUREMENTS:
1. Methane; CH_4; [74-82-8] 2. Hydrogen sulfide; H_2S; [7783-06-4] 3. Butane; C_4H_{10}; [106-97-8]	Saxena, A. C.; Robinson, D. B. *Can. J. Chem. Engng.* 1969, *47*, 69-75.

VARIABLES:	PREPARED BY:
	C. L. Young

EXPERIMENTAL VALUES: Mole fraction

T/K (T/°F)	P/MPa (P/psia)	in liquid			in vapor		
		x_{CH_4}	x_{H_2S}	$x_{C_4H_{10}}$	y_{CH_4}	y_{H_2S}	$y_{C_4H_{10}}$
310.93 (100)	2.758 (400)	0.1220	–	0.8780	0.8240	–	0.1760
		0.0010	0.9990	–	0.0120	0.988	–
		0.0738	0.5212	0.4050	0.3992	0.4907	0.1101
		0.0690	0.6307	0.3003	0.3523	0.5479	0.0998
		0.1189	0.1411	0.7400	–	–	–
		0.0745	0.5726	0.3529	–	–	–
		0.0392	0.7808	0.1800	0.2141	0.7280	0.0579
		–	–	–	0.4940	0.3809	0.1251
		–	–	–	0.6751	0.1487	0.1762
		0.1236	0.0885	0.7878	0.7465	0.0840	0.1695
		0.1086	0.2692	0.6222	0.5955	0.2543	0.1502
		0.1401	0.1055	0.7544	–	–	–
	5.516 (800)	0.2540	–	0.7460	0.8770	–	0.1230
		0.0510	0.9490	–	0.4050	0.5950	–
		0.1053	0.7413	0.1534	–	–	–
		0.1247	0.6392	0.2361	–	–	–
		–	–	–	0.5700	0.3400	0.0900
		0.2155	0.2545	0.5300	–	–	–
		0.1867	0.4073	0.4060	0.6061	0.3016	0.0923
		0.2503	0.1569	0.5928	0.7240	0.1530	0.1230
		0.2442	0.1184	0.6374	0.7926	0.0877	0.1196
		0.2093	0.3584	0.4323	0.6309	0.2655	0.1035

(cont.)

AUXILIARY INFORMATION

METHOD/APPARATUS/PROCEDURE:	SOURCE AND PURITY OF MATERIALS:
Variable volume cell with sample confined between two moveable pistons. Volume varied by mercury injection. Details of design given in source. Cell fitted with windows. Samples of gas and liquid analysed by gas chromatography. Temperature measured with iron-constantan thermocouple. Pressure measured with Bourdon gauge.	1. Matheson sample, purity varied between 99.3 and 99.7 per cent as determined by gas chromatography. 2. Matheson C.P. grade, purity 99.5 mole per cent. 3. Matheson instrument grade, purity 99.5 mole per cent.

ESTIMATED ERROR:

$\delta T/K = \pm 0.03$; $\delta P/MPa = \pm 0.003$ up to 6 MPa, ± 0.02 above 6 MPa;
δx, $\delta y = \pm 1\%$.

REFERENCES:

COMPONENTS:	ORIGINAL MEASUREMENTS:
1. Methane; CH_4; [74-82-8]	Saxena, A. C.; Robinson, D. B.
2. Hydrogen sulfide; H_2S; [7783-06-4]	*Can. J. Chem. Engng.*
3. Butane; C_4H_{10}; [106-97-8]	<u>1969</u>, *47*, 69-75.

EXPERIMENTAL VALUES:

Mole fraction

T/K (T/°F)	P/MPa (P/psia)	in liquid			in vapor		
		x_{CH_4}	x_{H_2S}	$x_{C_4H_{10}}$	y_{CH_4}	y_{H_2S}	$y_{C_4H_{10}}$
310.93 (100)	8.27 (1200)	0.3810	–	0.6190	0.8800	–	0.1200
		0.1100	0.8900	–	0.5080	0.4910	–
		0.3471	0.2599	0.3930	0.7084	0.1834	0.1082
		0.3022	0.3887	0.3091	0.6450	0.2550	0.1000
		0.1909	0.7136	0.0955	0.5160	0.4559	0.0281
		0.2068	0.6826	0.1106	–	–	–
		0.2810	0.4780	0.2401	0.5720	0.3508	0.0772
		0.2290	0.6203	0.1502	–	–	–
		0.3810	0.1463	0.4727	0.7640	0.1170	0.1191
277.59 (40)	2.758 (400)	0.1977	–	0.8023	0.9278	–	0.0722
		0.0350	0.9650	–	0.5130	0.4870	–
		0.1286	0.4574	0.4140	0.5979	0.3049	0.0972
		0.1663	0.3153	0.5184	0.6932	0.2087	0.0982
		0.1685	0.1900	0.6415	0.7461	0.1436	0.1103
		0.1131	0.6153	0.2716	0.5620	0.3678	0.0702
		0.1688	0.2717	0.5596	0.6962	0.1887	0.1151
		0.1726	0.2653	0.5621	–	–	–
	5.516 (800)	0.4070	–	0.5930	0.9560	–	0.0440
		0.0900	0.9100	–	0.6900	0.3100	–
		0.3002	0.3826	0.3172	0.7610	0.1861	0.0529
		–	–	–	0.7260	0.2240	0.0500
		0.2694	0.4781	0.2525	0.7168	0.2321	0.0511
		0.2576	0.4849	0.2576	–	–	–
		0.3449	0.2460	0.4092	0.8708	0.0930	0.0362
		0.3549	0.2368	0.4083	0.8558	0.0933	0.0509
		0.3688	0.1639	0.4673	0.8870	0.0683	0.0447
		0.3790	0.1878	0.4332	0.8986	0.0635	0.0379
		0.3756	0.1616	0.4629	–	–	–
		0.3845	0.1402	0.4753	0.9015	0.0543	0.0442
	8.27 (1200)	0.5670	–	0.4330	0.9496	–	0.0504
		0.1620	0.8380	–	0.7270	0.2730	–
		0.4350	0.3588	0.2062	0.7555	0.1793	0.0452
		0.5145	0.1797	0.3058	0.8706	0.0747	0.0547
		0.5210	0.1634	0.3156	0.8810	0.0638	0.0550
		0.5479	0.1482	0.3039	0.9074	0.0526	0.0400
		–	–	–	0.9110	0.0412	0.0477
		0.5406	0.1419	0.3175	0.9217	0.0455	0.0328
244.26 (-20)	2.758 (400)	0.2935	–	0.7065	0.9566	–	0.0434
		0.0535	0.9465	–	0.7970	0.2030	–
		0.2305	0.3202	0.4493	0.8808	0.0959	0.0233
		0.2592	0.1814	0.5594	0.9090	0.0620	0.0290
		0.1925	0.4635	0.3440	0.8490	0.1255	0.0255
		0.1589	0.5677	0.2733	0.8365	0.1404	0.0232
		0.2297	0.3186	0.4516	0.8879	0.0853	0.0268
	5.516 (800)	0.4991	–	0.5009	0.9783	–	0.0217
		0.1371	0.8629	–	0.8786	0.1214	–
		0.4673	0.1512	0.3815	0.9392	0.0390	0.0218
		0.4740	0.1464	0.3796	0.9367	0.0325	0.0308
		0.4215	0.3150	0.2635	0.9070	0.0777	0.0153
		0.3623	0.4739	0.1638	0.8888	0.0963	0.0149
		0.3816	0.4199	0.1985	0.8968	0.0896	0.0135

(cont.)

COMPONENTS:	ORIGINAL MEASUREMENTS:
1. Methane; CH_4; [74-82-8]	Saxena, A. C.; Robinson, D. B.
2. Hydrogen sulfide; H_2S; [7783-06-4]	*Can. J. Chem. Engng.*
3. Butane; C_4H_{10}; [106-97-8]	1969, *47*, 69-75.

EXPERIMENTAL VALUES:

		Mole fraction					
T/K	P/MPa	in liquid			in vapor		
(T/°F)	(P/psia)	x_{CH_4}	x_{H_2S}	$x_{C_4H_{10}}$	y_{CH_4}	y_{H_2S}	$y_{C_4H_{10}}$
244.26	8.27	0.7078	–	0.2922	0.9721	–	0.0279
(-20)	(1200)	0.2110	0.7890	–	0.8580	0.1420	–
		0.6029	0.2494	0.1476	–	–	–
		0.6146	0.2232	0.1622	0.9019	0.0749	0.0232
		0.5580	0.3232	0.1188	0.8914	0.0944	0.0142

COMPONENTS:	EVALUATOR:
(1) Methane; CH_4; [74-82-8]	H. Lawrence Clever Chemistry Department

(2) Cycloalkanes
  Cyclohexane
  Methylcyclohexane
  Dimethylcyclohexanes
  Cyclooctane   Bicyclohexyl
	Emory University
	Atlanta, GA 30322 USA
	
	1984, January

CRITICAL EVALUATION:

The Solubility of Methane in Cycloalkanes at Partial

Pressures up to 200 kPa (*ca*. 2 atm).

Values of the solubility of methane in cycloalkanes are reported in seven papers by various volumetric methods used at a total pressure of about one atmosphere. With the exception of the methane + cyclohexane system, there are not enough measurements on any one system to recommend solubility values. Most of the data are classed as tentative.

Methane + Cyclohexane; C_6H_{12}; [110-82-7]

Guerry (ref. 1), Lannung and Gjaldbaek (ref. 2), and Ben-Naim and Yaacobi (ref. 3) report solubility data on the system. Guerry's data are about 15 percent smaller than the data of the others and his data are classed as doubtful. The smoothed data of Lannung and Gjaldbaek and of Ben-Naim and Yaacobi agree within 0.30 percent between 288 and 303 K.

The combined sets of data were fitted by the method of least squares to obtain the equation

$$\ln x_1 = -6.74545 + 3.06826/(T/100 \text{ K})$$

with a standard error about the regression line of 2.0×10^{-5}. The temperature independent thermodynamic changes from the equation are

$$\Delta \overline{H}_1^\circ/\text{kJ mol}^{-1} = -2.55 \qquad \text{and} \qquad \Delta \overline{S}_1^\circ/\text{J K}^{-1} \text{ mol}^{-1} = -56.1$$

The smoothed solubility data and partial molal Gibbs energy of solution are in Table 1.

Table 1. Solubility of methane in cyclohexane. Recommended mole fraction solubility at 101.325 kPa (1 atm) partial pressure of methane and the partial molal Gibbs energy of solution as a function of temperature.

T/K	Mol Fraction $10^3 x_1$	$\Delta \overline{G}_1^\circ/\text{kJ mol}^{-1}$
283.15	3.47	13.329
293.15	3.35	13.890
298.15	3.29	14.170
303.15	3.24	14.451
313.15	3.13	15.012

Methane + Methylcyclohexane; C_2H_{14}; [108-87-2]

Only Field, Wilhelm and Battino (ref. 4) report solubility data on this system. These solubility values at three temperatures were treated by a linear regression to obtain the equation

$$\ln x_1 = -7.54994 + 6.01428/(T/100 \text{ K})$$

with a standard error about the regression line of 2.6×10^{-6}.

The temperature independent thermodynamic changes from the equation are

$$\Delta \bar{H}_1^\circ / \text{kJ mol}^{-1} = -5.00 \quad \text{and} \quad \Delta \bar{S}_1^\circ / \text{J K}^{-1} \text{mol}^{-1} = -62.8$$

The smoothed solubility and partial molal Gibbs energy of solution values are in Table 2.

Table 2. Solubility of methane in methylcyclohexane. Tentative values of the mole fraction solubility at 101.325 kPa (1 atm) methane partial pressure and partial molal Gibbs energy of solution as a function of temperature.

T/K	Mol Fraction $10^3 x_1$	$\Delta \bar{G}_1^\circ / \text{kJ mol}^{-1}$
283.15	4.40	12.774
293.15	4.09	13.401
298.15	3.96	13.715
303.15	3.83	14.029
313.15	3.59	14.657

Methane + trans-1,2-Dimethylcyclohexane; C_8H_{16}; [6876-23-9]

Methane + cis-1,2-Dimethylcyclohexane; C_8H_{16}; [2207-01-4]

Methane + trans-1,3-Dimethylcyclohexane; C_8H_{16}; [2207-03-6]

 + cis-1,3-Dimethylcyclohexane; C_8H_{16}; [638-04-0]

Methane + trans-1,4-Dimethylcyclohexane; C_8H_{16}; [2207-04-7]

 + cis-1,4-Dimethylcyclohexane; C_8H_{16}; [624-24-3]

The solubility data on the four systems were reported by Geller, Battino, and Wilhelm (ref. 5). Measurements were reported for only two temperatures, thus the partial molal enthalpy and entropy of solution are possibly less reliable than for systems with measurements at additional temperatures. The values of the thermodynamic changes on solution are

Thermo changes	trans-1,2-DMC	cis-1,2-DMC	trans-1,3-DMC/ cis-1,3-DMC 41/59 mol%	trans-1,4-DMC/ cis-1,4-DMC 30/70 mol%
$\Delta \bar{H}_1^\circ / \text{kJ mol}^{-1}$	-4.11	-4.54	-4.50	-5.86
$\Delta \bar{S}_1^\circ / \text{J K}^{-1} \text{mol}^{-1}$	-59.1	-61.2	-60.5	-64.9

The data on each system were fitted by a linear regression to a two constant equation

$$\ln x_1 = A_1 + A_2 / (T/100 \text{ K})$$

Values for A_1 and A_2 for each system are given below.

COMPONENTS:	EVALUATOR:

(1) Methane; CH_4; [74-82-8]
 	
(2) Cycloalkanes
    Cyclohexane
    Methylcyclohexane
    Dimethylcyclohexanes
    Cyclooctane   Bicyclohexyl
	H. Lawrence Clever
	Chemistry Department
	Emory University
	Atlanta, GA 30322 USA
	
	1984, January

CRITICAL EVALUATION:

Constants	*trans*-1,2-DMC	*cis*-1,2-DMC	*trans*-1,3-DMC/ *cis*-1,3-DMC 41/59 mol%	*trans*-1,4-DMC/ *cis*-1,4-DMC 30/70 mol%
A_1	-7.11257	-7.36300	-7.27244	-7.81023
A_2	4.94415	5.45917	5.41233	7.04831

Smoothed values of the solubility are in Table 3.

Table 3.   Solubility of methane in dimethylcyclohexanes. Tentative values
of the mole fraction solubility at 101.325 kPa (1 atm) partial
methane pressure as a function of temperature.

	methane Mol Fraction, $10^3 x_1$			
T/K	*trans*-1,2-DMC	*cis*-1,2-DMC	*trans*-1,3-DMC/ *cis*-1,3-DMC 41/59 mol%	*trans*-1,4-DMC/ *cis*-1,4-DMC 30/70 mol%
298.15	4.28	3.96	4.27	4.31
303.15	4.16	3.84	4.14	4.15
313.15	3.95	3.63	3.91	3.85

Methane + Cyclooctane; C_8H_{16}; [296-64-8]

Wilcock, Battino and Wilhelm (ref. 6) report the solubility of methane
in cyclooctane at three temperatures between 288.89 and 313.45 K. A linear
regression of the data gives the equation

$$\ln x_1 = -7.43325 + 4.68345/(T/100 \text{ K})$$

with a standard error about the regression line of 9.8×10^{-5}.

The temperature independent thermodynamic changes on solution from
the equation are

$$\Delta \bar{H}_1^\circ/\text{kJ mol}^{-1} = -3.89 \quad \text{and} \quad \Delta \bar{S}_1^\circ/\text{J K}^{-1} \text{ mol}^{-1} = -61.8$$

Smoothed values of the solubility and partial molal Gibbs energy of solution are in Table 4.

Table 4.   Solubility of methane in cyclooctane. Tentative values of the
           mole fraction solubility at 101.325 kPa (1 atm) partial methane
           pressure and partial molal Gibbs energy of solution as a func-
           tion of temperature.

T/K	Mol Fraction $10^3 x_1$	$\Delta \overline{G}_1^\circ / kJ\ mol^{-1}$
293.15	2.92	14.223
298.15	2.84	14.532
303.15	2.77	14.841
313.15	2.64	15.459

Methane + Bicyclohexyl; $C_{12}H_{22}$; [92-51-3]

Cukor and Prausnitz (ref. 7) report eight values of the solubility of
methane in bicyclohexyl at 25 degrees intervals between 300 and 475 K.
The Henry's constants reported by the authors have been converted to mole
fraction values at 101.325 kPa (1 atm) methane partial pressure and fitted
by a linear regression to obtain the equation

$$\ln x_1 = -20.76150 + 25.20566/(T/100\ K) + 6.07641\ \ln(T/100\ K)$$

with a standard error about the regression line of 3.9×10^{-5}. The three
constant equation gives thermodynamic changes in enthalpy and entropy that
change with temperature. Values at several temperatures are below:

T/K	$\Delta \overline{H}_1^\circ / kJ\ mol^{-1}$	$\Delta \overline{S}_1^\circ / J\ K^{-1}\ mol^{-1}$	$\Delta \overline{C}_{p_p}^\circ / J\ K^{-1}\ mol^{-1}$
298.15	−5.89	−66.9	50.5
323.15	−4.63	−62.8	50.5
373.15	−2.10	−55.6	50.5
423.15	+0.42	−49.2	50.5
473.15	+2.95	−43.6	50.5

Smoothed values of the solubility and partial molal Gibbs energy are in
Table 5.   The minimum solubility occurs at 415 K.

Table 5.   Solubility of methane in bicyclohexyl. Tentative values of the
           mole fraction solubility at 101.325 kPa (1 atm) methane partial
           pressure and partial molal Gibbs energy of solution as a func-
           tion of temperature.

T/K	Mol Fraction $10^3 x_1$	$\Delta \overline{G}_1^\circ / kJ\ mol^{-1}$
298.15	3.45	14.054
303.15	3.32	14.387
313.15	3.10	15.039
323.15	2.93	15.675
373.15	2.47	18.631
423.15	2.38	21.247
473.15	2.50	23.565

References

1. Guerry, D. Jr. Ph.D. thesis, 1944, Vanderbilt University,
    Nashville, TN.

2. Lannung, A.; Gjaldbaek, J. C.   *Acta Chem. Scand.* 1960, *14*, 1124.

3. Ben-Naim, A.; Yaacobi, M.   *J. Phys. Chem.* 1974, *14*, 1124.

4. Field, L. R.; Wilhelm, E.; Battino, R. *J. Chem. Thermodyn.* 1974,
    6, 237.

5. Geller, E. B.; Battino, R.; Wilhelm, E. *J. Chem. Thermodyn.* 1976,
    8, 197.

6. Wilcock, R. J.; Battino, R.; Wilhelm, E. *J. Chem. Thermodyn.* 1977,
    9, 111.

7. Cukor, P. M.; Prausnitz, J. M. *J. Phys. Chem.* 1972, *76*, 598.

COMPONENTS:	ORIGINAL MEASUREMENTS:
(1) Methane; CH_4; [74-82-8] (2) Cyclohexane; C_6H_{12}; [110-82-7]	Lannung, A.; Gjaldbaek, J. C. *Acta Chem. Scand.* **1960**, *14*, 1124 - 1128.

VARIABLES:	PREPARED BY:
T/K = 291.15 - 310.15 p_1/kPa = 101.325 (1 atm)	J. Chr. Gjaldbaek

EXPERIMENTAL VALUES:

T/K	Mol Fraction $10^3 x_1$	Bunsen Coefficient α/cm^3(STP)cm^{-3}atm^{-1}	Ostwald Coefficient L/cm^{3}cm^{-3}
291.15	3.38	0.702	0.748
291.15	3.37	0.699	0.745
298.15	3.28	0.677	0.739
298.15	3.26	0.673	0.735
310.15	3.18	0.645	0.732
310.15	3.16	0.641	0.728

Smoothed Data: For use between 291.15 and 310.15 K.

$$\ln x_1 = -6.6984 + 2.9232/(T/100 \text{ K})$$

The standard error about the regression line is 1.78 x 10^{-5}.

T/K	Mol Fraction $10^3 x_1$
298.15	3.29
308.15	3.18

AUXILIARY INFORMATION

METHOD/APPARATUS/PROCEDURE:
A calibrated all-glass combined manometer and bulb containing degassed solvent and the gas was placed in an air thermostat and shaken until equilibrium (1).

The absorbed volume of gas is calculated from the initial and final amounts, both saturated with solvent vapor. The amount of solvent is determined by the weight of displaced mercury.

The values are at 101.325 kPa (1 atm) pressure assuming Henry's law is obeyed.

SOURCE AND PURITY OF MATERIALS:
(1) Methane. Generated from magnesium methyl iodide. Purified by fractional distillation. Specific gravity corresponds with mol wt 16.08.

(2) Cyclohexane. Poulenc Frères. Shaken with fuming sulfuric acid and washed with water. Dried and distilled over phosphorus pentoxide. M.p./°C = 6.3.

ESTIMATED ERROR:
$$\delta T/K = \pm 0.05$$
$$\delta x_1/x_1 = \pm 0.015$$

REFERENCES:
1. Lannung, A.
J. Am. Chem. Soc. **1930**, *52*, 68.

COMPONENTS:	ORIGINAL MEASUREMENTS:
1. Methane; CH_4; [74-82-8] 2. Cyclohexane; C_6H_{12}; [110-82-7]	Ben-Naim, A.; Yaacobi, M. *J. Phys. Chem.*, 1974,*78*,175-8
VARIABLES: Temperature	PREPARED BY: C.L. Young

EXPERIMENTAL VALUES:

T/K	Ostwald coefficient[*], L	Mole fraction[+] at partial pressure of 101.3 kPa, x_{CH_4}
283.15	0.7603	0.00348
288.15	0.7520	0.00341
293.15	0.7450	0.00334
298.15	0.7395	0.00333
303.15	0.7353	0.00322

[*] Smoothed values obtained from the equation.

$$kT \ln L = 1,822.9 - 12,053 \, (T/K) + 0.01791 \, (T/K)^2 \; cal \; mol^{-1}$$
where k is in units of cal mol^{-1} K^{-1}

[+] calculated by compiler assuming the ideal gas law for methane.

AUXILIARY INFORMATION

METHOD/APPARATUS/PROCEDURE:	SOURCE AND PURITY OF MATERIALS:
The apparatus was similar to that described by Ben-Naim and Baer (1) and Wen and Hung (2). It consists of three main parts, a dissolution cell of 300 to 600 cm^3 capacity, a gas volume measuring column, and a manometer. The solvent is degassed in the dissolution cell, the gas is introduced and dissolved while the liquid is kept stirred by a magnetic stirrer immersed in the water bath. Dissolution of the gas results in the change in the height of a column of mercury which is measured by a cathetometer.	1. Matheson sample, purity 99.97 mol per cent. 2. AR grade.

ESTIMATED ERROR:

$$\delta T/K = \pm 0.1; \quad \delta x_{CH_4} = \pm 2\%$$

REFERENCES:

1. Ben-Naim, A.; Baer, S. *Trans. Faraday Soc.* 1963,*59*, 2735.

2. Wen, W.-Y.; Hung, J.H. *J. Phys. Chem.* 1970,*74*,170.

COMPONENTS:	ORIGINAL MEASUREMENTS:
(1) Methane; CH_4; [74-82-8] (2) Methylcyclohexane; C_7H_{14}; [108-87-2]	Field, L. R.; Wilhelm, E.; Battino, R. *J. Chem. Thermodyn.* <u>1974</u>, *6*, 237 - 243.
VARIABLES: T/K: 284.28 - 313.28 P/kPa: 101.325 (1 atm)	PREPARED BY: H. L. Clever

EXPERIMENTAL VALUES:

T/K	Mol Fraction $10^3 x_1$	Bunsen Coefficient $\alpha/cm^3 (STP) cm^{-3} atm^{-1}$	Ostwald Coefficient $L/cm^3 cm^{-3}$
284.28	4.363	0.778	0.8095
298.16	3.957	0.694	0.7570
313.28	3.587	0.618	0.7086

The gas solubility values were adjusted to an oxygen partial pressure of 101.325 kPa (1 atm) by Henry's law.

The Bunsen coefficients were calculated by the compiler.

Smoothed Data: For use between 283.15 and 313.28 K.

$$\ln x_1 = -7.5499 + 6.0143/(T/100 \text{ K})$$

The standard error about the regression line is 2.64×10^{-6}.

T/K	Mol Fraction $10^3 x_1$
283.15	4.401
293.15	4.094
298.15	3.955
303.15	3.826
313.15	3.591

AUXILIARY INFORMATION

METHOD/APPARATUS/PROCEDURE:	SOURCE AND PURITY OF MATERIALS:
The solubility apparatus is based on the design of Morrison and Billett (1) and the version used is described by Battino, Evans, and Danforth (2). The degassing apparatus is that described by Battino, Banzhof, Bogan, and Wilhelm (3). Degassing. Up to 500 cm³ of solvent is placed in a flask of such size that the liquid is about 4 cm deep. The liquid is rapidly stirred, and vacuum is intermittently applied through a liquid N_2 trap until the permanent gas residual pressure drops to 5 microns. Solubility Determination. The degassed solvent is passed in a thin film down a glass spiral tube containing solute gas plus the solvent vapor at a total pressure of one atm. The volume of gas absorbed is found by difference between the initial and final volumes in the buret system. The solvent is collected in a tared flask and weighed.	(1) Methane. Either Matheson Co., Inc. or Air Products and Chemicals, Inc. Purest grade available, minimum purity greater than 99 mole per cent. (2) Methylcyclohexane. Phillips Petroleum Co. Pure Grade. Distilled.
	ESTIMATED ERROR: $\delta T/K = 0.03$ $\delta P/mmHg = 0.5$ $\delta x_1/x_1 = 0.005$
	REFERENCES: 1. Morrison, T. J.; Billett, F. *J. Chem. Soc.* <u>1948</u>, 2033. 2. Battino, R.; Evans, F. D.; Danforth, W. F. *J. Am. Oil Chem. Soc.* <u>1968</u>, *45*, 830. 3. Battino, R.; Banzhof, M.; Bogan, M.; Wilhelm, E. *Anal. Chem.* <u>1971</u>, *43*, 806.

COMPONENTS:	ORIGINAL MEASUREMENTS:
(1) Methane; CH_4; [74-82-8] (2) cis-1,2-Dimethylcyclohexane; C_8H_{16}; [2207-01-4]	Geller, E. B.; Battino, R. Wilhelm, E. J. Chem. Thermodyn. <u>1976</u>, 8, 197-202.

VARIABLES:	PREPARED BY:
T/K: 297.95, 312.99 p/kPa: 101.325 (1 atm)	H. L. Clever

EXPERIMENTAL VALUES:

T/K	Mol Fraction $10^3 x_1$	Bunsen Coefficient α/cm^3 (STP) cm^{-3} atm^{-1}	Ostwald Coefficient L/cm^3 cm^{-3}
297.95	3.963	0.6297	0.6869
312.99	3.629	0.5678	0.6506

The Bunsen coefficients were calculated by the compiler assuming ideal
gas behavior.

The solubility values were adjusted to a methane partial pressure of
101.325 kPa by Henry's law.

Smoothed Data: The equation is based on only two pair of experimental
                 points and should be used with caution.

                 For use between 297.95 and 312.99 K

$$\ln x_1 = -7.3720 + 5.4868/(T/100K)$$

T/K	Mol Fraction $10^3 x_1$
298.15	3.959
308.15	3.730

AUXILIARY INFORMATION

METHOD/APPARATUS/PROCEDURE:	SOURCE AND PURITY OF MATERIALS:

The solubility apparatus is based on the design of Morrison and Billett (1) and the version used is described by Battino, Evans, and Danforth (2). The degassing apparatus is that described by Battino, Banzhof, Bogan, and Wilhelm (3).	(1) Methane. Matheson Co., Inc.
	    Stated to be 99.97 mole percent.
	
	(2) cis-1,2-Dimethylcyclohexane.
	    Chemical Samples Co. Fractionally distilled and stored in dark. Refractive index (NaD, 298.15 K) 1.4337.

Degassing. Up to 500 cm^3 of solvent
is placed in a flask of such size
that the liquid is about 4 cm deep.
The liquid is rapidly stirred, and
vacuum is intermittently applied
through a liquid N_2 trap until the
permanent gas residual pressure drops
to 5 microns.

ESTIMATED ERROR:
δT/K = 0.03 δP/mmHg = 0.5 $\delta x_1/x_1$ = 0.005

Solubility Determination. The degassed solvent is passed in a thin
film down a glass helical tube containing solute gas plus the solvent
vapor at a total pressure of one atm.
The volume of gas absorbed is found
by differences between the initial
and final volumes in the buret
system. The solvent is collected
in a tared flask and weighed.

REFERENCES:
1. Morrison, T. J.; Billett, F.
   J. Chem. Soc. <u>1948</u>, 2033.
2. Battino, R.; Evans, F. D.;
   Danforth, W. F.
   Chem. Soc. <u>1968</u>, 45, 830.
3. Battino, R.; Banzhof, M.;
   Bogan, M.; Wilhelm, E.
   Anal. Chem. <u>1971</u>, 43, 806.

COMPONENTS:	ORIGINAL MEASUREMENTS:
(1) Methane; CH_4; [74-82-8] (2) *trans*-1,2-Dimethylcyclohexane; C_8H_{16}; [6876-23-9]	Geller, E. B.; Battino, R. Wilhelm, E. *J. Chem. Thermodyn.* <u>1976</u>, *8*, 197-202.

VARIABLES:	PREPARED BY:
T/K: 297.93 - 313.00 p/kPa: 101.325 (1 atm)	H. L. Clever

EXPERIMENTAL VALUES:

T/K	Mol Fraction $10^3 x_1$	Bunsen Coefficient α/cm^3(STP)cm^{-3}atm^{-1}	Ostwald Coefficient L/cm^{3}cm^{-3}
297.93	4.275	0.6622	0.7223
298.06	4.264	0.6604	0.7207
298.08	4.274	0.6610	0.7224
298.08	4.274	0.6619	0.7224
298.13	4.314	0.6681	0.7292
313.00	3.954	0.6031	0.6911

The Bunsen coefficients were calculated by the compiler assuming ideal gas behavior.

The solubility values were adjusted to a methane partial pressure of 101.325 kPa by Henry's law.

Smoothed Data: For use between 297.93 and 313.00 K

$$\ln x_1 = -7.1244 + 4.9808/(T/100\text{K})$$

T/K	Mol Fraction $10^3 x_1$
298.15	4.280
308.15	4.054

AUXILIARY INFORMATION

METHOD/APPARATUS/PROCEDURE:	SOURCE AND PURITY OF MATERIALS:

The solubility apparatus is based on the design of Morrison and Billett (1) and the version used is described by Battino, Evans, and Danforth (2). The degassing apparatus is that described by Battino, Banzhof, Bogan, and Wilhelm (3).
 	
 Degassing. Up to 500 cm³ of solvent is placed in a flask of such size that the liquid is about 4 cm deep. The liquid is rapidly stirred, and vacuum is intermittently applied through a liquid N₂ trap until the permanent gas residual pressure drops to 5 microns.
 	
 Solubility Determination. The degassed solvent is passed in a thin film down a glass helical tube containing solute gas plus the solvent vapor at a total pressure of one atm. The volume of gas absorbed is found by differences between the initial and final volumes in the buret system. The solvent is collected in a tared flask and weighed.
	(1) Methane. Matheson Co., Inc.
	    Stated to be 99.97 mole percent.
	
	 (2) *trans*-1,2-Dimethylcyclohexane.
	    Chemical Samples Co. Frac-
	    tionally distilled and stored in
	    dark. Refractive index (NaD,
	    298.15 K) 1.4248.

ESTIMATED ERROR:
δT/K = 0.03 δP/mmHg = 0.5 $\delta x_1/x_1$ = 0.005

REFERENCES:
1. Morrison, T. J.; Billett, F.
   J. Chem. Soc. <u>1948</u>, 2033.
2. Battino, R.; Evans, F. D.;
   Danforth, W. F. *J. Am. Oil
   Chem. Soc.* <u>1968</u>, *45*, 830.
3. Battino, R.; Banzhof, M.;
   Bogan, M.; Wilhelm, E.
   Anal. Chem. <u>1971</u>, *43*, 806.

COMPONENTS:	ORIGINAL MEASUREMENTS:
(1) Methane; CH_4; [74-82-8]	Geller, E. B.; Battino, R. Wilhelm, E.
(2) *trans*-1,4-Dimethylcyclohexane, 30 mol %; C_8H_{16}; [2207-04-7]	*J. Chem. Thermodyn.* <u>1976</u>, *8*, 197-202.
(3) *cis*-1,4-Dimethylcyclohexane, 70 mol %; C_8H_{16}; [624-24-3]	

VARIABLES:	PREPARED BY:
T/K: 298.08, 313.11 p/kPa: 101.325 (1 atm)	H. L. Clever

EXPERIMENTAL VALUES:

T/K	Mol Fraction $10^3 x_1$	Bunsen Coefficient $\alpha/cm^3(STP)cm^{-3}atm^{-1}$	Ostwald Coefficient $L/cm^3 cm^{-3}$
298.08	4.315	0.6686	0.7296
313.11	3.852	0.5873	0.6732

The Bunsen coefficients were calculated by the compiler assuming ideal gas behavior.

The solubility values were adjusted to a methane partial pressure of 101.325 kPa (1 atm) by Henry's law.

AUXILIARY INFORMATION

METHOD/APPARATUS/PROCEDURE:	SOURCE AND PURITY OF MATERIALS:
The solubility apparatus is based on the design of Morrison and Billett (1) and the version used is described by Battino, Evans, and Danforth (2). The degassing apparatus is that described by Battino, Banzhof, Bogan, and Wilhelm (3). Degassing. Up to 500 cm³ of solvent is placed in a flask of such size that the liquid is about 4 cm deep. The liquid is rapidly stirred, and vacuum is intermittently applied through a liquid N_2 trap until the permanent gas residual pressure drops to 5 microns. Solubility Determination. The degassed solvent is passed in a thin film down a glass helical tube containing solute gas plus the solvent vapor at a total pressure of one atm. The volume of gas absorbed is found by differences between the initial and final volumes in the buret system. The solvent is collected in a tared flask and weighed.	(1) Methane. Matheson Co., Inc. Stated to be 99.97 mole percent. (2) *trans*-1,4-Dimethylcyclohexane. (3) *cis*-1,4-Dimethylcyclohexane. Chemical Samples Co. The binary mixture used as received. Composition determined by refractive index by authors.
	ESTIMATED ERROR: δT/K = 0.03 δP/mmHg = 0.5 $\delta x_1/x_1$ = 0.005
	REFERENCES: 1. Morrison, T. J.; Billett, F. *J. Chem. Soc.* <u>1948</u>, 2033. 2. Battino, R.; Evans, F. D.; Danforth, W. F. *J. Am. Oil Chem. Soc.* <u>1968</u>, *45*, 830. 3. Battino, R.; Banzhof, M.; Bogan, M.; Wilhelm, E. *Anal. Chem.* <u>1971</u>, *43*, 806.

COMPONENTS:	ORIGINAL MEASUREMENTS:

COMPONENTS:

(1) Methane; CH_4; [74-82-8]

(2) trans-1,3-Dimethylcyclohexane,
    41 mol %; C_8H_{16}; [2207-03-6]

(3) cis-1,3-Dimethylcyclohexane,
    59 mol %; C_8H_{16}; [638-04-0]

ORIGINAL MEASUREMENTS:

Geller, E. B.; Battino, R.
Wilhelm, E.

J. Chem. Thermodyn. 1976, 8,
197-202.

VARIABLES:
    T/K: 298.41, 313.08
    p/kPa: 101.325 (1 atm)

PREPARED BY:

H. L. Clever

EXPERIMENTAL VALUES:

T/K	Mol Fraction $10^3 x_1$	Bunsen Coefficient α/cm^3(STP)cm^{-3}atm^{-1}	Ostwald Coefficient L/cm^{3}cm^{-3}
298.41	4.259	0.6572	0.7180
313.08	3.912	0.5945	0.6814

The Bunsen coefficients were calculated by the compiler assuming ideal
gas behavior.

The solubility values were adjusted to a methane partial pressure of
101.325 kPa (1 atm) by Henry's law.

AUXILIARY INFORMATION

METHOD/APPARATUS/PROCEDURE:
The solubility apparatus is based on
the design of Morrison and Billett
(1) and the version used is described
by Battino, Evans, and Danforth (2).
The degassing apparatus is that
described by Battino, Banzhof, Bogan,
and Wilhelm (3).

Degassing. Up to 500 cm^3 of solvent
is placed in a flask of such size
that the liquid is about 4 cm deep.
The liquid is rapidly stirred, and
vacuum is intermittently applied
through a liquid N_2 trap until the
permanent gas residual pressure drops
to 5 microns.

Solubility Determination. The degassed solvent is passed in a thin
film down a glass helical tube containing solute gas plus the solvent
vapor at a total pressure of one atm.
The volume of gas absorbed is found
by differences between the initial
and final volumes in the buret
system. The solvent is collected
in a tared flask and weighed.

SOURCE AND PURITY OF MATERIALS:
(1) Methane. Matheson Co., Inc.
    Stated to be 99.97 mole percent.

(2) trans-1,3-Dimethylcyclohexane.

(3) cis-1,3-Dimethylcyclohexane.
    Chemical Samples Co. Binary
    mixture used as received.
    Authors analyzed mixture by
    refractive index.

ESTIMATED ERROR:
    δT/K = 0.03
    δP/mmHg = 0.5
    $\delta x_1 / x_1$ = 0.005

REFERENCES:
1. Morrison, T. J.; Billett, F.
   J. Chem. Soc. 1948, 2033.
2. Battino, R.; Evans, F. D.;
   Danforth, W. F. J. Am. Oil
   Chem. Soc. 1968, 45, 830.
3. Battino, R.; Banzhof, M.;
   Bogan, M.; Wilhelm, E.
   Anal. Chem. 1971, 43, 806.

COMPONENTS:	ORIGINAL MEASUREMENTS:
(1) Methane; CH_4; [74-82-8] (2) Cyclooctane; C_8H_{16}; [292-64-8]	Wilcock, R. J.; Battino, R.; Wilhelm, E. *J. Chem. Thermodyn.* <u>1977</u>, *9*, 111 - 115.

VARIABLES:	PREPARED BY:
T/K: 288.89 - 313.45 P/kPa: 101.325 (1 atm)	H. L. Clever

EXPERIMENTAL VALUES:

T/K	Mol Fraction $10^3 x_1$	Bunsen Coefficient α	Ostwald Coefficient L
288.89	3.042	0.5106	0.5400
298.21	2.765	0.4599	0.5021
313.45	2.664	0.4365	0.5009

The Bunsen coefficients were calculated by the compiler.

The solubility values were adjusted to a methane partial pressure of 101.325 kPa by Henry's law.

Smoothed Data: For 288.15 to 313.15 K

$$\ln x_1 = -7.4333 + 4.6835/(T/100K)$$

The standard error about the regression line is 9.80×10^{-5}.

T/K	Mol Fraction $10^3 x_1$
288.15	3.00
298.15	2.84
308.15	2.70

AUXILIARY INFORMATION

METHOD/APPARATUS/PROCEDURE:

The solubility apparatus is based on the design of Morrison and Billett (1) and the version used is described by Battino, Evans, and Danforth (2). The degassing apparatus is that described by Battino, Banzhof, Bogan, and Wilhelm (3).

Degassing. Up to 500 cm³ of solvent is placed in a flask of such size that the liquid is about 4 cm deep. The liquid is rapidly stirred, and vacuum is intermittently applied through a liquid N_2 trap until the permanent gas residual pressure drops to 5 microns.

Solubility Determination. The degassed solvent is passed in a thin film down a glass spiral tube containing solute gas plus the solvent vapor at a total pressure of one atm. The volume of gas absorbed is found by difference between the initial and final volumes in the buret system. The solvent is collected in a tared flask and weighed.

SOURCE AND PURITY OF MATERIALS:

(1) Methane. Matheson Co., Inc. Minimum mole per cent purity is 99.97.

(2) Cyclooctane. Chemical Samples Co. 99 mole per cent, distilled, refractive index (NaD, 298.15 K) 1.4562.

ESTIMATED ERROR:
$$\delta T/K = 0.03$$
$$\delta P/mmHg = 0.5$$
$$\delta x_1/x_1 = 0.005$$

REFERENCES:

1. Morrison, T. J.; Billett, F. *J. Chem. Soc.* <u>1948</u>, 2033.

2. Battino, R.; Evans, F. D.; Danforth, W. F. *J. Am. Oil Chem. Soc.* <u>1968</u>, *45*, 830.

3. Battino, R.; Banzhof, M.; Bogan, M.; Wilhelm, E. *Anal. Chem.* <u>1971</u>, *43*, 806.

COMPONENTS:	ORIGINAL MEASUREMENTS:
(1) Methane; CH_4; [74-82-8] (2) Cyclic hydrocarbons; C_6H_{10} and C_6H_{12}	Guerry, D. Jr. Ph.D. thesis, <u>1944</u> Vanderbilt University Nashville, TN Thesis Director: L. J. Bircher

VARIABLES:	PREPARED BY:
T/K: 293.15, 298.15 P/kPa: 101.325 (1 atm)	H. L. Clever

EXPERIMENTAL VALUES:

T/K	Mol Fraction x_1 x 10^4	Bunsen Coefficient α	Ostwald Coefficient L
Cyclohexene; C_6H_{10}; [110-83-8]			
293.15	24.8	0.551	0.591
298.15	24.6	0.543	0.593
Cyclohexane; C_6H_{12}; [110-82-7]			
293.15	29.2	0.607	0.651
298.15	28.3	0.585	0.639

The Ostwald coefficients were calculated by the
compiler.

AUXILIARY INFORMATION

METHOD/APPARATUS/PROCEDURE:

A Van Slyke-Neill Manometric Apparatus manufactured by the Eimer
and Amend Co. was used.

The procedure of Van Slyke (1) for
pure liquids was modified (2) so
that small solvent samples (2 cm^3)
could be used with almost complete
recovery of the sample.

An improved temperature control
system was used.

SOURCE AND PURITY OF MATERIALS:

(1) Methane. Prepared by hydrolysis
    of crystaline methyl Grignard
    reagent. Passed through conc.
    H_2SO_4, solid KOH, and Dririte.

(2) Hydrocarbons. Both were
    Eastman Kodak Co. products.
    They were purified by standard
    methods, and distilled from Na
    in a nitrogen atm.

SOURCE AND PURITY OF MATERIALS:
Cyclohexene. B.p. (756.6 mmHg)
t/°C 82.35 - 82.50 (corr.).

Cyclohexane. B.p. (760.7 mmHg)
t/°C 80.90 (corr.).

Data on density, refractive index
and vapor pressure are in the
thesis.

ESTIMATED ERROR:

$$\delta T/K = 0.05$$

REFERENCES:
1. Van Slyke, D. D.
   J. Biol. Chem. <u>1939</u>, *130*, 545.

2. Ijams, C. C.
   Ph.D. thesis, <u>1941</u>
   Vanderbilt University

COMPONENTS:	ORIGINAL MEASUREMENTS:
1. Methane; CH_4; [74-82-8] 2. 1,1'-Bicyclohexyl; $C_{12}H_{22}$; [92-51-3]	Cukor, P.M.; Prausnitz, J.M.; *J. Phys. Chem.* 1972, *76*, 598-601
VARIABLES: Temperature	PREPARED BY: C.L. Young

EXPERIMENTAL VALUES:

T/K	Henry's Constant[a] /atm	Mole fraction of methane[b] in liquid, x_{CH_4}
300	298	0.00336
325	341	0.00293
350	378	0.00265
375	406	0.00246
400	422	0.00237
425	425	0.00235
450	415	0.00241
475	392	0.00255

a.   Quoted in supplementary material for original paper

b.   Calculated by compiler for a partial pressure of 1 atmosphere.

AUXILIARY INFORMATION

METHOD/APPARATUS/PROCEDURE:	SOURCE AND PURITY OF MATERIALS:
Volumetric apparatus similar to that described by Dymond and Hildebrand (1). Pressure measured with a null detector and precision gauge. Details in ref. (2).	No details given
	ESTIMATED ERROR: $\delta T/K = \pm 0.05$; $\delta x_{CH_4} = \pm 2\%$

	REFERENCES:
	1. Dymond, J.; Hildebrand, J.H.
	     Ind. Eng. Chem. Fundam. 1967, *6*,
	     130.
	
	2. Cukor, P.M.; Prausnitz, J.M.
	     Ind. Eng. Chem. Fundam. 1971, *10*,
	     638.

COMPONENTS:	EVALUATOR:
1. Methane; CH_4; [74-82-8] 2. Cyclohexane; C_6H_{12}; [110-82-7]	Colin L. Young Department of Physical Chemistry, University of Melbourne. Parkville, Victoria, 3052 Australia. February 1986.

CRITICAL EVALUATION:

This system has been fairly extensively investigated by Russian workers but there are serious doubts as to the reliability of some of the early work, ref (1-3). Legret, Richon and Renon (4) classified the data of Stepanov and Vybornova (5) as having methane mole fractions of better than 2 per cent but the original article was unavailable to us. The most extensive study is that of Reamer et al. (6). Their data are thought to be fairly reliable and are classified as tentative. Since these workers did not, however, provide raw experimental data it is difficult to establish the reliability of the smoothed data with certainty. The earlier data of Sage et al. (7) are very limited in extent and are superseded by this groups later measurements (6).

The recent data of Brunner et al. (8) are in reasonanble agreement with the more extensive data of Sage et al. (6). The data of Schoch et al. (9) are only of moderate precision but are in reasonable agreement with the data of Reamer et al. (6). Therefore the data given in ref (8) and (9) support the classification of tentative for the data of Reamer et al. (6). However, in view of the fact that ref. (6) only reports smoothed data the data cannot be unreservedly classified as recommended.

The data of Frolich et al. (10) were presented in small graphical form and are thought to be of low accuracy are and classified as doubtful.

References.

1. Savvina, Ya. D.; Velikovskii, A. S.;
   Zh. Fiz. Khim., 1956, 30, 1596.
2. Savvina, Ya. D.;
   Tr. Vses. Nauch. Isseled. Inst. Pridod. Gazov., 1962, 17-25, 185.
3. Stepanov, G. S.;
   Gazov. Delo., 1970, 1, 26.
4. Legret, D.; Richon, D.; Renon, H.;
   Fluid Phase Equilib., 1984, 17, 323.
5. Stepanov, G. S.; Vybornova, Ya. I.;
   Gazov. Delo. Nauch. Tekhn. Sb., 1964, 10, 9.
6. Reamer, H. H.; Sage, B. H.; Lacey, W. N.;
   Chem. Eng. Data Ser.3. 1958, 3, 240.
7. Sage, B. H.; Webster, D. C.; Lacey, W. N.;
   Ind. Eng. Chem., 1936, 38, 1045.
8. Brunner, E.; Maier, S.; Windhaber, K.;
   J. Phys. E., 1984, 17, 44.
9. Schoch, E. P.; Hoffmann, A. E.; Mayfield, F. D.;
   Ind. Eng. Chem., 1940, 32, 1351.
10. Frolich, K.; Tauch, E. J.; Hogan, J. J.; Peer, A. A.;
    Ind. Eng. Chem., 1931, 23, 548.

COMPONENTS:	ORIGINAL MEASUREMENTS:
1. Methane; CH₄; [74-82-8] 2. Cyclohexane; C₆H₁₂; [110-82-7]	Frolich, P.K.; Tauch, E.J.; Hogan, J.J.; Peer, A.A. *Ind. Eng. Chem.* <u>1931</u>, *23*, 548-550.

VARIABLES:	PREPARED BY:
Pressure	C.L. Young

EXPERIMENTAL VALUES:

T/K	P/MPa	Solubility*	Mole fraction of methane in liquid,[+] x_{CH_4}
298.15	1.0	6	0.026
	2.0	14	0.059
	3.0	23	0.093
	4.0	34	0.131
	5.0	44	0.164
	6.0	56	0.199
	7.0	68	0.232

* Data taken from graph in original article. Volume of gas measured at 101.325 kPa and 298.15 K dissolved by unit volume of liquid measured under the same conditions.

+ Calculated by compiler.

AUXILIARY INFORMATION

METHOD/APPARATUS/PROCEDURE:	SOURCE AND PURITY OF MATERIALS:
Static equilibrium cell. Liquid saturated with gas and after equilibrium established samples removed and analysed by volumetric method. Allowance was made for the vapor pressure of the liquid and the solubility of the gas at atmospheric pressure. Details in source.	Stated that the materials were the highest purity available. Purity 98 to 99 mole per cent.
	ESTIMATED ERROR: $\delta T/K = \pm 0.1$; $\delta x_{CH_4} = \pm 5\%$
	REFERENCES:

COMPONENTS:	ORIGINAL MEASUREMENTS:
1. Methane; CH_4; [74-82-8] 2. Cyclohexane; C_6H_{12}; [110-82-7]	Sage, B. H.; Webster, D. C.; Lacey, W. N. *Ind. Eng. Chem.* <u>1936</u>, *28*, 1045-1047.

VARIABLES:	PREPARED BY:
	C. L. Young

EXPERIMENTAL VALUES:

T/K (T/°F)	p/psi	P/MPa[†]	Mass fraction of methane	Mole fraction[†] of methane, x_{CH_4}
310.9 (100)	2045 2554	14.10 17.61	0.1001 0.1344	0.3683 0.4487
344.3 (160)	2196 2698	15.14 18.60	0.1001 0.1344	0.3683 0.4487
377.6 (220)	2240 2734	15.44 18.85	0.1001 0.1344	0.3683 0.4487

[†] calculated by compiler.

AUXILIARY INFORMATION

METHOD/APPARATUS/PROCEDURE:	SOURCE AND PURITY OF MATERIALS:
PVT cell charged with mixture of known composition. Pressure measured with pressure balance. Bubble point determined from the discontinuity in the pressure, volume isotherm. Details of apparatus in ref. (1).	1. Prepared from natural gas, treated for removal of higher alkanes, carbon dioxide and water vapor. Final purity 99.9 mole per cent. 2. Eastman Kodak Co. sample, used without further purification.

	ESTIMATED ERROR: $\delta T/K = \pm 0.1$; $\delta P/MPa = \pm 0.02$; $\delta x_{CH_4} = \pm 0.002$ (estimated by compiler).

REFERENCES:

1. Sage, B. H.; Lacey, W. N.
   Ind. Eng. Chem.
   <u>1934</u>, *26*, 103.

COMPONENTS:	ORIGINAL MEASUREMENTS:
1. Methane; CH_4 ; [74-82-8] 2. Cyclohexane; C_6H_{12} ; [110-82-7]	Schoch, E. P.; Hoffmann, A. E.; Mayfield, F. D. *Ind. Eng. Chem.* 1940, *32*, 1351-3.

VARIABLES:	PREPARED BY:
Temperature, pressure	C. L. Young

EXPERIMENTAL VALUES:

T/K	P/MPa	Mole fraction of methane in liquid, x_{CH_4}	T/K	P/MPa	Mole fraction of methane in liquid, x_{CH_4}
311.08	4.15	0.1185	344.26	17.35	0.4166
	7.708	0.2160		20.24	0.4812
	11.17	0.2996		23.81	0.5677
	14.81	0.3833		25.72	0.6358
	18.06	0.4561		26.83	0.6916
	20.49	0.5117		27.08	0.7385
	21.90	0.5447		26.92	0.7838
	23.44	0.5870	377.59	4.71	0.1189
	25.07	0.6330		8.756	0.2173
	25.83	0.6939		12.42	0.2998
	26.23	0.7414		16.20	0.3855
	26.32	0.7810		20.84	0.5000
344.26	4.47	0.1175		23.48	0.5784
	8.467	0.2165		25.06	0.6613
	10.49	0.2599		25.33	0.7175
	14.00	0.3423		25.28	0.7754

AUXILIARY INFORMATION

METHOD/APPARATUS/PROCEDURE:	SOURCE AND PURITY OF MATERIALS:
Rocking equilibrium cell fitted with stirring paddles. Temperature measured with Beckmann thermometer calibrated against standard platinum resistance thermometer. Pressure measured with Bourdon gauge. Samples injected into cell using mercury displacement. Equilibrium pressure measured. Bubble point determined from change in slope of pressure-volume isotherms. Details in ref. (1).	1. Crude sample treated for removal of oxygen, carbon dioxide, water vapor and liquids condensible at 200 K; distilled. 2. Eastman Kodak Co. sample distilled.

	ESTIMATED ERROR: $\delta T/K = \pm 0.01$ at 311.08 K; ± 0.03 at higher temperatures; $\delta P/MPa = \pm 0.01$; $\delta x_{CH_4} = \pm 0.001$ (estimated by compiler).

	REFERENCES:
	1. Schoch, E. P.; Hoffmann, A. E.;
	   Kasperik, A. S.; Lightfoot, J. H.;
	   Mayfield, F. D.
	   Ind. Eng. Chem. 1940, *32*, 788.

COMPONENTS:	ORIGINAL MEASUREMENTS:
1. Methane; CH_4; [74-82-8] 2. Cyclohexane; C_6H_{12}; [110-82-7]	Reamer, H. H.; Sage, B. H.; Lacey, W. N. *Ind. Eng. Chem.* 1958, *3*, 240-245.

VARIABLES:	PREPARED BY:
	C. L. Young

EXPERIMENTAL VALUES:

T/K (T/°F)	P/MPa	p/psi	Mole fraction of methane in liquid, x_{CH_4}	in vapor, y_{CH_4}
294.3	1.38	200	0.0440	0.9891
(70)	2.76	400	0.0870	0.9924
	4.14	600	0.1288	0.9934
	5.52	800	0.1693	0.9938
	6.89	1000	0.2086	0.9938
	8.62	1250	0.2560	0.9931
	10.34	1500	0.3022	0.9920
	12.07	1750	0.3468	0.9901
	13.79	2000	0.3901	0.9873
	15.51	2250	0.4331	0.9844
	17.24	2500	0.4750	0.9805
	18.96	2750	0.5170	0.9740
	20.68	3000	0.5581	0.9661
	24.13	3500	0.6392	0.9390
	27.58	4000	0.7350	0.8489
	28.20	4090	0.765	0.765
310.9	1.38	200	0.0414	0.9793
(100)	2.76	400	0.0920	0.9860
	4.14	600	0.1217	0.9876
	5.52	800	0.1601	0.9883
	6.89	1000	0.1977	0.9885
	8.62	1250	0.2430	0.9876

(cont.)

AUXILIARY INFORMATION

METHOD/APPARATUS/PROCEDURE:	SOURCE AND PURITY OF MATERIALS:
PVT cell charged with mixture of known composition. Pressure measured with pressure balance. Temperature measured using platinum resistance thermometer. Details in ref. (1). Gas samples analysed by condensing cyclohexane out in cold trap. Bubble point determined from discontinuity in pressure-volume isotherm for fixed total composition.	1. Sample treated for removal of carbon dioxide and water vapor. Purity about 99.9 mole per cent. 2. Phillips Petroleum Co. research grade sample, purity 99.98 mole per cent.

ESTIMATED ERROR:

$\delta T/K = \pm0.05$; $\delta P/MPa = \pm0.01$;

δx_{CH_4}, $\delta y_{CH_4} = \pm0.002$.

REFERENCES:

1. Sage, B. H.; Lacey, W. N.
Trans. Am. Inst. Mining Met. Engnrs.
1940, *136*, 136.

COMPONENTS:	ORIGINAL MEASUREMENTS:
1. Methane; CH_4; [74-82-8]	Reamer, H. H.; Sage, B. H.; Lacey, W. N.
2. Cyclohexane; C_6H_{12}; [110-82-7]	*Ind. Eng. Chem.*
	<u>1958</u>, *3*, 240-245.

EXPERIMENTAL VALUES:

T/K (T/°F)	P/MPa	p/psi	Mole fraction of methane in liquid, x_{CH_4}	in vapor, y_{CH_4}
310.9	10.34	1500	0.2870	0.9860
(100)	12.07	1750	0.3300	0.9840
	13.79	2000	0.3720	0.9810
	15.51	2250	0.4129	0.9770
	17.24	2500	0.4540	0.9710
	18.96	2750	0.4959	0.9640
	20.68	3000	0.5365	0.9539
	24.13	3500	0.6201	0.9270
	27.58	4000	0.7274	0.8263
	27.85	4040	0.758	0.758
344.3	1.38	200	0.0365	0.9380
(160)	2.76	400	0.0740	0.9616
	4.14	600	0.1103	0.9671
	5.52	800	0.1462	0.9700
	6.89	1000	0.1812	0.9709
	8.62	1250	0.2244	0.9712
	10.34	1500	0.2670	0.9700
	12.07	1750	0.3086	0.9678
	13.79	2000	0.3505	0.9649
	15.51	2250	0.3911	0.9598
	17.24	2500	0.4323	0.9540
	18.96	2750	0.4746	0.9459
	20.68	3000	0.5180	0.9370
	24.13	3500	0.6070	0.9002
	26.75	3880	0.737	0.737
377.6	1.38	200	0.0318	0.8437
(220)	2.76	400	0.0677	0.9065
	4.14	600	0.1028	0.9249
	5.52	800	0.1373	0.9334
	6.89	1000	0.1714	0.9381
	8.62	1250	0.2134	0.9417
	10.34	1500	0.2548	0.9410
	12.07	1750	0.2963	0.9399
	13.79	2000	0.3374	0.9370
	15.51	2250	0.3780	0.9310
	17.24	2500	0.4191	0.9220
	18.96	2750	0.4610	0.9109
	20.68	3000	0.5079	0.8960
	24.13	3500	0.6090	0.8270
	25.44	3690	0.711	0.711
410.9	1.38	200	0.0248	0.6520
(280)	2.76	400	0.0603	0.7990
	4.14	600	0.0951	0.8464
	5.52	800	0.1295	0.8709
	6.89	1000	0.1634	0.8853
	8.62	1250	0.2054	0.8939
	10.34	1500	0.2471	0.8967
	12.07	1750	0.2886	0.8961
	13.79	2000	0.3297	0.8918
	15.51	2250	0.3708	0.8829
	17.24	2500	0.4134	0.8690
	18.96	2750	0.4615	0.8501
	20.68	3000	0.5141	0.8210
	23.10	3350	0.667	0.667

(cont.)

COMPONENTS:	ORIGINAL MEASUREMENTS:
1. Methane; CH_4; [74-82-8]	Reamer, H. H.; Sage, B. H.; Lacey, W. N.
2. Cyclohexane; C_6H_{12}; [110-82-7]	*Ind. Eng. Chem.*
	<u>1958</u>, *3*, 240-245.

EXPERIMENTAL VALUES:

T/K (T/°F)	P/MPa	p/psi	Mole fraction of methane in liquid, x_{CH_4}	in vapor, y_{CH_4}
444.3 (340)	1.38	200	0.0148	0.3653
	2.76	400	0.0512	0.6354
	4.14	600	0.0870	0.7236
	5.52	800	0.1224	0.7673
	6.89	1000	0.1566	0.7891
	8.62	1250	0.1984	0.8019
	10.34	1500	0.2392	0.8059
	12.07	1750	0.2820	0.8079
	13.79	2000	0.3250	0.8031
	15.51	2250	0.3697	0.7886
	17.24	2500	0.4193	0.7644
	18.96	2750	0.4781	0.7000
	20.06	2910	0.608	0.608

COMPONENTS:	ORIGINAL MEASUREMENTS:
1. Methane; CH₄; [74-82-8] 2. Cyclohexane; C₆H₁₂; [110-82-7]	Savvina, Ya. D. *Tr. Vses. Nauchno-Issled. Inst.* *Prirodn. Gazov.,*1962, *17/25*, 185-196.
VARIABLES:	PREPARED BY:
Temperature, pressure	C. L. Young

EXPERIMENTAL VALUES:

T/K (t/°C)	P/kgcm⁻³	P/Mpa	K-value methane	cyclohexane
313.2	20	1.96	15.85	0.052
(40)	50	4.90	6.59	0.021
	100	9.81	3.33	0.028
	150	14.7	2.36	0.040
	200	19.6	1.83	0.065
	220	21.6	1.69	0.081
	250	24.5	1.42	0.180
	265	26.0	1.22	0.413
	269	26.4	1.10	0.651
333.2	20	1.96	16.10	0.070
(60)	50	4.90	6.90	0.030
	100	9.81	3.45	0.037
	150	14.7	2.43	0.053
	200	19.6	1.85	0.084
	230	22.6	1.61	0.134
	250	24.5	1.43	0.214
	263	25.8	1.20	0.452
	266	26.1	1.10	0.662
353.2	50	4.90	7.11	0.038
(80)	100	9.81	3.61	0.043
	150	14.7	2.45	0.064
	200	19.6	1.87	0.110
	220	21.6	1.70	0.156
	240	23.5	1.47	0.227
	250	24.5	1.34	0.327

AUXILIARY INFORMATION

METHOD/APPARATUS/PROCEDURE:	SOURCE AND PURITY OF MATERIALS:
Values appear to be determined using apparatus described in ref.1.	No Details given.
	ESTIMATED ERROR:

	REFERENCES:
	1. Savvina, Ya. D.;Velikovskii, A. S.
	
	   Tr. Vses. Nauchno-Issled. Inst.
	   Prirodn. Gazov., 1962,*17/25*, 163.

COMPONENTS:	ORIGINAL MEASUREMENTS:
1. Methane; CH_4; [74-82-8] 2. Cyclohexane; C_6H_{12}; [110-82-7]	Savvina, Ya. D. *Tr. Vses. Nauchno-Issled. Inst.* *Prirodn. Gazov.*, 1962, *17/25*, 185-196.

Experimental Values:

T/K (t/°C)	P/kgcm^{-3}	P/Mpa	K-value methane	cyclohexane
353.2(80)	262	26.0	1.07	0.765
373.2	30	2.94	11.50	0.086
(100)	50	4.90	7.61	0.048
	100	9.81	3.69	0.052
	150	14.7	2.51	0.081
	200	19.6	1.88	0.134
	220	21.6	1.63	0.177
	240	23.5	1.39	0.306
	250	24.5	1.21	0.498
	254	24.9	1.03	0.875
393.2	20	1.96	17.87	0.167
(120)	50	4.90	7.61	0.073
	100	9.81	3.78	0.077
	150	14.7	2.44	0.108
	200	19.6	1.81	0.186
	220	21.6	1.57	0.254
	240	23.5	1.25	0.473
	246	24.1	1.04	0.863
423.2	30	2.94	12.93	0.170
(150)	50	4.90	7.20	0.097
	100	9.81	3.56	0.107
	150	14.7	2.31	0.143
	200	19.6	1.65	0.279
	220	21.6	1.33	0.458
	227	22.3	1.09	0.788

COMPONENTS:	ORIGINAL MEASUREMENTS:
(1) Methane; CH_4; [74-82-8] (2) Cyclohexane; C_6H_{12}; [110-82-7]	Brunner, E.; Maier, S.; Windhaber, K. *J. Phys. E:* <u>1984</u>. *17*, 44-8.
VARIABLES: T/K = 311.0, 344.3 p_t/MPa = 3.05 -18.32	PREPARED BY: H. L. Clever

EXPERIMENTAL VALUES:

Temperature		Total Pressure	Mol Fraction	Molar Volume
$t/^0C$	T/K	p_t/MPa	x_1	v /cm^3 mol^{-1}
37.8	311.0	0.0214	0	110.5
		5.04	0.1465	102.5
		10.95	0.3041	93.1
		13.48	0.3651	89.8
		16.15	0.4289	87.0
71.1	344.3	0.0737	0	115.4
		3.05	0.0814	111.4
		5.94	0.1597	106.9
		8.95	0.2357	102.5
		11.77	0.3023	97.8
		15.57	0.3929	93.5
		18.32	0.4555	89.9

The Kelvin temperatures were added by the compiler.

The first line at each temperature gives the vapor pressure and molar volume of pure cyclohexane.

AUXILIARY INFORMATION

METHOD/APPARATUS/PROCEDURE:

The measuring method consists in metering known masses of components 1 and 2 into the measuring cell with continuous thorough stirring until a transition from the homogeneous to the heterogeneous state, or *vice versa*, is observed.

The measuring cell is one of three specially constructed cells described in the paper.

From the masses metered in and the temperature-corrected cell volume, the boiling point or the dew point as well as the densities are obtained.

The *pvT* data of Angus *et al.* (ref 1) was used. The average deviation of the experimental bubble points and molar volumes from the smoothed values obtained by Reamer *et al.* (ref 2) is less than 0.5 percent.

SOURCE AND PURITY OF MATERIALS:

(1) Methane. Messer-Griesheim. Purity stated to be 99.9 percent.

(2) Cyclohexane. BASF. Stated to be 99.99 percent purity.

ESTIMATED ERROR:
$$\delta T/K = \pm\ 0.1$$
$$\delta p/p = \pm\ 0.002$$
$$\delta x_1/x_1 = \pm\ 0.02$$
$$\delta v/v = \pm\ 0.02$$

REFERENCES:

1. Angus, S.; Armstrong, B.; de Reuck. *Methane. Int. thermo tables of the fluid state-5* <u>1978</u>, Pergamon.

2. Reamer,H.H.; Sage,B.H.;Lacey,W.N. *J. Chem. Eng. Data* <u>1958</u>, *3*, 240.

COMPONENTS:	ORIGINAL MEASUREMENTS:
1. Methane; CH_4 ; [74-82-8] 2. Hexane; C_6H_{14} ; [110-54-3] 3. Cyclohexane; C_6H_{12}; [110-82-7]	Velikovskii, V. S. ; Stepanova, G. S. Vybornova, Ya. I. *Gazov. Prom.*, <u>1965</u>, *10*(6), 45-49.

VARIABLES:
  Temperature, pressure
	PREPARED BY:
	  C. L. Young

EXPERIMENTAL VALUES:

T/K	p /kg cm	p /MPa	Mole fractions in liquid			Mole fractions in vapor		
273.15	50	4.9	0.230	0.560	0.210	0.9969	0.0025	0.0006
	50	4.9	0.220	0.347	0.433	0.9976	0.0014	0.0010
	50	4.9	0.205	0.165	0.630	0.9982	0.0006	0.0012
	100	9.8	0.415	0.430	0.155	0.9935	0.0050	0.0015
	100	9.8	0.395	0.270	0.335	0.9945	0.0027	0.0028
	100	9.8	0.360	0.130	0.510	0.9955	0.0010	0.0035
	150	14.7	0.595	0.290	0.115	0.9815	0.0140	0.0045
	150	14.7	0.550	0.200	0.250	0.9830	0.0080	0.0090
	150	14.7	0.500	0.100	0.400	0.9840	0.0035	0.0125
	200	19.6	0.775	0.155	0.070	0.9260	0.0540	0.0200
	200	19.6	0.709	0.125	0.166	0.9460	0.0260	0.0280
	200	19.6	0.632	0.068	0.300	0.9600	0.0080	0.0320
	210	20.6	0.860	0.100	0.040	0.8600	0.1000	0.0400
	230	22.6	0.855	0.065	0.080	0.8550	0.0650	0.0800
	250	24.5	0.800	0.045	0.155	0.8950	0.0250	0.0830
	255	25.0	0.849	0.033	0.118	0.8490	0.0330	0.1180
293.15	50	4.9	0.203	0.573	0.324	0.9870	0.0080	0.0022
	50	4.9	0.190	0.355	0.455	0.9900	0.0055	0.0045
	50	4.9	0.170	0.175	0.655	0.9920	0.0027	0.0053
	100	9.8	0.382	0.444	0.174	0.9850	0.0120	0.0030
	100	9.8	0.358	0.280	0.362	0.9870	0.0070	0.0060
	100	9.8	0.325	0.130	0.545	0.9890	0.0030	0.0080
	150	14.7	0.545	0.310	0.145	0.9750	0.0190	0.0060
	150	14.7	0.510	0.213	0.277	0.9780	0.0110	0.0110

AUXILIARY INFORMATION

METHOD APPARATUS/PROCEDURE:	SOURCE AND PURITY OF MATERIALS:

Details of method given in
ref. (1).
	1. Purity 98.5 mole per cent,
	   1.5 mole per cent nitrogen.
	2. and 3. Purity checked by
	   refractive index, density and
	   boiling point.

	ESTIMATED ERROR:

	REFERENCES:
	  1. Velikovski, A. S.; Pokrovskii,
	     V. K.; Stepanova, G. S.;
	     Rasamot, M. S.
	     Gazov. Prom.,<u>1958</u> no. 10.

1. Methane; CH_4 ; [74-82-8] 2. Hexane; C_6H_{14} ; [110-54-3] 3. Cyclohexane; C_6H_{12}; [110-82-7]	Velikovskii, V. S. ; Stepanova, G. S. Vybornova, Ya. I. *Gazov. Prom.*, 1965, *10*(6), 45-49.

T/K	p /kg cm	p /MPa	Mole fractions in liquid			Mole fractions in vapor		
293.15	150	14.7	0.460	0.105	0.435	0.9800	0.0050	0.0150
	200	19.6	0.720	0.200	0.080	0.9380	0.0460	0.0160
	200	19.6	0.662	0.142	0.196	0.9540	0.0220	0.0240
	200	19.6	0.590	0.080	0.330	0.9570	0.0100	0.0330
	221	21.7	0.845	0.115	0.040	0.8450	0.1150	0.0400
	238	23.3	0.837	0.072	0.091	0.8370	0.0720	0.0910
	250	24.5	0.740	0.050	0.210	0.9030	0.0190	0.0780
	262	25.7	0.828	0.038	0.134	0.828	0.0380	0.1340
313.15	50	4.9	0.192	0.585	0.223	0.9800	0.0160	0.0040
	50	4.9	0.180	0.360	0.460	0.9830	0.0095	0.0075
	50	4.9	0.165	0.180	0.655	0.9855	0.0045	0.0100
	100	9.8	0.360	0.460	0.180	0.9770	0.0180	0.0050
	100	9.8	0.332	0.290	0.378	0.9790	0.0110	0.0100
	100	9.8	0.300	0.150	0.540	0.9810	0.0050	0.0140
	150	14.7	0.510	0.348	0.142	0.9645	0.0270	0.0085
	150	14.7	0.475	0.225	0.300	0.9660	0.0170	0.0170
	150	14.7	0.435	0.125	0.440	0.9710	0.0080	0.0210
	200	19.6	0.665	0.240	0.095	0.9300	0.0520	0.0180
	200	19.6	0.620	0.160	0.220	0.9450	0.0260	0.0290
	200	19.6	0.567	0.090	0.343	0.9520	0.0120	0.0360
	225	22.1	0.827	0.126	0.047	0.8270	0.1260	0.047
	230	22.6	0.650	0.072	0.278	0.9270	0.0160	0.0570
	241	23.6	0.815	0.082	0.103	0.8150	0.0820	0.1030
	262	25.7	0.809	0.042	0.1490	0.8090	0.0420	0.1490
333.15	50	4.9	0.179	0.592	0.228	0.9660	0.0265	0.0075
	50	4.9	0.160	0.366	0.474	0.9690	0.0160	0.0150
	50	4.9	0.141	0.190	0.669	0.9720	0.0080	0.0200
	100	9.8	0.338	0.476	0.191	0.9635	0.0280	0.0085
	100	9.8	0.306	0.302	0.392	0.9660	0.0170	0.0170
	100	9.8	0.277	0.165	0.558	0.9695	0.0085	0.0220
	150	14.7	0.480	0.365	0.155	0.9520	0.0360	0.0120
	150	14.7	0.449	0.235	0.316	0.9560	0.0220	0.0220
	150	14.7	0.415	0.133	0.452	0.9600	0.0110	0.0290
	200	19.6	0.645	0.250	0.110	0.9160	0.0610	0.0230
	200	19.6	0.591	0.173	0.236	0.9260	0.0360	0.0380
	200	19.6	0.545	0.100	0.355	0.9360	0.160	0.0480
	210	20.6	0.685	0.220	0.095	0.8970	0.0750	0.0280
	210	20.6	0.626	0.157	0.217	0.9170	0.0400	0.0430
	210	20.6	0.572	0.095	0.333	0.9250	0.0200	0.0550
	223	21.9	0.805	0.140	0.055	0.8050	0.1400	0.0550
	239	23.4	0.800	0.089	0.111	0.8000	0.0890	0.1110
	259	25.4	0.795	0.045	0.160	0.7950	0.0450	0.1600

COMPONENTS:	EVALUATOR:

COMPONENTS:

1. Methane; CH_4; [74-82-8]

2. Ethene; C_2H_4; [74-85-1]

EVALUATOR:

Colin L. Young,
School of Chemistry,
University of Melbourne,
Parkville, Victoria 3052,
Australia.

March 1982

EVALUATION:

This system has been investigated by several workers. The data of Miller, Kidnay and Hiza (1) are classified as recommended and cover the temperature range 150 to 190 K. These data are in good agreement with the less extensive data of Hsi and Lu (2), the latter data are classified as tentative. The data of Volova (3) are in fair agreement with those of Miller *et al.* (1) but the isotherms at 143 K and 127 K are of questionable accuracy. The limited data of Sagara *et al.* (4) in the temperature range 198 to 248 K are not directly comparable with those of Miller *et al.* (1) but are classified as tentative.

The data of Guter *et al.* (5) are not considered here as the experimental data were presented in small graphical form and are rejected.

References

1. Miller, R. C.; Kidnay, A. J.; Hiza, M. J.
   J. Chem. Thermodyn., <u>1977</u>, *9*, 167.

2. Hsi, C.; Lu, B. C.-Y.
   Can. J. Chem. Eng., <u>1971</u>, *49*, 140.

3. Volova, L. M.
   Zh. Fiz. Khim., <u>1940</u>, *14*, 268.

4. Sagara, H.; Arai, Y.; Saito, S.
   J. Chem. Eng. Japan, <u>1972</u>, *5*, 339.

5. Guter, M.; Newitt, D. M.; Ruhemann, M.
   Proc. Roy. Soc. London, <u>1940</u>, *A176*, 140.

COMPONENTS:	ORIGINAL MEASUREMENTS:
1. Methane; CH_4; [74-82-8] 2. Ethene; C_2H_4; [74-85-1]	Hsi, C.; Lu, B. C.-Y. *Can. J. Chem. Eng.* <u>1971</u>, *49*, 140-143. (Supplementary data)

VARIABLES:	PREPARED BY:
	C. L. Young

EXPERIMENTAL VALUES:

T/K (T/°F)	P/psi	P/MPa	Mole fraction of methane in liquid, x_{CH_4}	in vapor, y_{CH_4}
148.1 (-193.1)	30.50	0.2103	0.1523	0.8969
	36.00	0.2482	0.1798	0.9070
	42.10	0.2903	0.2262	0.9240
	45.80	0.3158	0.2419	0.9294
	45.55	0.3141	0.2442	0.9374
	59.90	0.4130	0.3423	0.9510
	60.20	0.4151	0.3448	0.9503
	60.00	0.4137	0.3483	0.9510
	65.35	0.4506	0.4004	0.9553
	72.00	0.4964	0.4329	0.9619
	71.85	0.4954	0.4359	0.9593
	78.00	0.5378	0.4938	0.9668
	79.25	0.5464	0.5034	0.9690
	85.00	0.5861	0.5522	0.9716
	87.00	0.5998	0.5759	0.9730
	89.00	0.6136	0.5812	0.9694
	89.30	0.6157	0.5911	0.9733
	92.50	0.6378	0.6197	0.9765
	99.05	0.6829	0.6814	0.9810
	105.00	0.7239	0.7483	0.9832
	114.80	0.7915	0.8194	0.9894
	123.13	0.8490	0.8921	0.9937

(cont.)

AUXILIARY INFORMATION

METHOD/APPARATUS/PROCEDURE:	SOURCE AND PURITY OF MATERIALS:
Recirculating vapor flow apparatus constructed of 100 ml Jerguson gauge. Temperature measured using copper-constantan thermocouples. Pressure measured with Bourdon gauge. Cell charged and vapor recirculated with magnetic pump for 2 or more hours. Samples of vapor and liquid removed at constant pressure and analysed using gas chromatography. Helium was used as a carrier gas.	1. Matheson research grade, purity 99.99 mole per cent. 2. Matheson research grade, purity 99.98 mole per cent.

ESTIMATED ERROR: $\delta T/K = \pm 0.02$; $\delta P/MPa \sim \pm 0.005$; δx_{CH_4}, $\delta y_{CH_4} = \pm 1\%$ (estimated by compiler).

REFERENCES:

COMPONENTS:	ORIGINAL MEASUREMENTS:
1. Methane; CH_4; [74-82-8]	Hsi, C.; Lu, B. C.-Y.
2. Ethene; C_2H_4; [74-85-1]	*Can. J. Chem. Eng.*
	1971, *49*, 140-143.
	(Supplementary data)

| EXPERIMENTAL VALUES: | | | Mole fraction of methane | |
T/K (T/°F)	P/psi	P/MPa	in liquid, x_{CH_4}	in vapor, y_{CH_4}
148.1 (-193.1)	136.30	0.9398	0.9851	0.9975
159.2	42.25	0.2913	0.1284	0.8263
(-173.1)	47.00	0.3241	0.1469	0.8402
	50.50	0.3482	0.1629	0.8572
	64.24	0.4429	0.2153	0.8930
	67.05	0.4623	0.2335	0.8934
	76.80	0.5295	0.2726	0.9150
	86.76	0.5982	0.3163	0.9202
	96.00	0.6619	0.3617	0.9341
	104.50	0.7205	0.4073	0.9377
	115.25	0.7946	0.4634	0.9490
	132.00	0.9102	0.5458	0.9564
	145.00	0.9997	0.6210	0.9666
	168.05	1.1587	0.7385	0.9758
	180.80	1.2466	0.8125	0.9830
	185.00	1.2755	0.8325	0.9850
	188.70	1.3010	0.8653	0.9891
168.7	28.41	0.1959	0.0408	0.5197
(-156.1)	35.00	0.2413	0.0557	0.6010
	44.25	0.3051	0.0840	0.6902
	48.00	0.3309	0.0919	0.7215
	47.55	0.3278	0.0938	0.7194
	60.05	0.4140	0.1265	0.7761
	61.30	0.4226	0.1352	0.7887
	75.42	0.5200	0.1719	0.8165
	77.02	0.5310	0.1772	0.8230
	93.01	0.6413	0.2302	0.8631
	97.00	0.6688	0.2398	0.8575
	110.00	0.7584	0.2811	0.8784
	122.50	0.8446	0.3251	0.8972
	132.00	0.9101	0.3553	0.9070
	140.00	0.9653	0.3875	0.9110
	145.69	1.0045	0.4080	0.9237
	156.00	1.0756	0.4403	0.9256
	160.27	1.1050	0.4570	0.9300
	162.00	1.1170	0.4617	0.9294
	170.80	1.1776	0.5040	0.9358
	176.50	1.2169	0.5132	0.9380
	185.00	1.2755	0.5450	0.9430
	191.00	1.3169	0.5711	0.9449
	195.00	1.3445	0.5843	0.9479
	196.50	1.3548	0.5848	0.9483
	206.0	1.4203	0.6192	0.9518
	210.0	1.4479	0.6380	0.9540
	209.5	1.4445	0.6430	0.9621
	228.3	1.5741	0.6890	0.9615
	227.8	1.5706	0.7042	0.9608
	230.9	1.5920	0.7225	0.9654
	235.0	1.6203	0.7320	0.9670
	235.0	1.6203	0.7327	0.9700
	247.5	1.7065	0.7794	0.9733
	250.0	1.7237	0.7860	0.9750
	249.2	1.7182	0.7893	0.9776
	256.5	1.7685	0.8037	0.9767
	260.1	1.7933	0.8330	0.9850
	264.5	1.8237	0.8424	0.9800
	271.5	1.8719	0.8648	0.9856
	275.0	1.8961	0.8780	0.9860
	280.3	1.9326	0.8816	0.9841
	284.0	1.9581	0.8962	0.9920
	293.0	2.0202	0.9250	0.9914
	306.6	2.1139	0.9690	0.9967

COMPONENTS:	ORIGINAL MEASUREMENTS:
1. Methane; CH_4; [74-82-8] 2. Ethene (Ethylene); C_2H_4; [74-85-1]	Sagara, H.; Arai, Y.; Saito, S. *J. Chem. Engng. Japan* <u>1972</u>, *5*, 339-348.

VARIABLES:
 	
        Temperature, pressure
	PREPARED BY:
	
	      C. L. Young

EXPERIMENTAL VALUES:

T/K	P/MPa	Mole fraction of methane in liquid, x_{CH_4}	in gas, y_{CH_4}
198.15	1.02	0.107	0.623
	2.03	0.311	0.789
	2.03	0.312	0.799
	4.03	0.738	0.912
223.15	2.03	0.123	0.443
	4.05	0.398	0.708
	5.57	0.616	0.764
248.15	3.04	0.0809	0.242
	4.05	0.172	0.385
	4.56	0.242	0.453
	5.07	0.284	0.478
	6.08	0.395	0.513

AUXILIARY INFORMATION

METHOD/APPARATUS/PROCEDURE:	SOURCE AND PURITY OF MATERIALS:
Static stainless steel cell of capacity 5×10^5 mm^3 fitted with magnetic stirrer and sampling valves. Cell enclosed in cryostat. Temperature measured with thermocouple. Pressure measured using Bourdon gauge. Gases added to cell and equilibrated. Samples of liquid and gas withdrawn and analysed using a gas chromatograph with thermal conductivity detector. Details in source.	1. Takachiho Chemical Industry Co. sample, purity 99.9 mole per cent. 2. Takachiho Chemical Industry Co. Ltd., sample purity 99.5 mole per cent.
	ESTIMATED ERROR: $\delta T/K = \pm 0.1$; $\delta P/MPa = \pm 0.01$; δx_{CH_4}, $\delta y_{CH_4} = \pm 1\%$.
	REFERENCES:

COMPONENTS:	ORIGINAL MEASUREMENTS:
1. Methane; CH_4; [74-82-8] 2. Ethene; C_2H_4; [74-85-1]	Miller, R.C.; Kidnay, A.J.; Hiza, M.J. *J. Chem. Thermodynamics*, <u>1977</u>, *9*, 167-178.
VARIABLES: Temperature, pressure	PREPARED BY: C.L. Young

EXPERIMENTAL VALUES:

T/K	$P/10^5$Pa	Mole fraction of methane in liquid, x_{CH_4}	in vapor, y_{CH_4}
150.00	1.547	0.0848	0.8338
	2.039	0.1237	0.8783
	3.035	0.2090	0.9244
	3.950	0.2896	0.9455
	5.022	0.4043	0.9627
	5.991	0.5118	0.9720
	7.02	0.6327	0.9790
	8.03	0.7580	0.9859
	9.03	0.8724	0.9927
160.00	1.798	0.0559	0.6952
	3.055	0.1260	0.8279
	5.024	0.2391	0.9032
	6.94	0.3653	0.9352
	8.51	0.4755	0.9537
	9.97	0.5913	0.9655
	11.45	0.7116	0.9754
	12.92	0.8156	0.9837
	13.91	0.8866	0.9887
170.00	2.028	0.0328	0.4815
	4.038	0.1105	0.7463
	6.034	0.1951	0.8372
	8.00	0.2811	0.8842

AUXILIARY INFORMATION

METHOD/APPARATUS/PROCEDURE:	SOURCE AND PURITY OF MATERIALS:
Vapor-recirculation system similar to that in ref. 1 and 2. Pressure measured with Bourdon gauge, temperature measured with platinum resistance thermometer. Samples of liquid and vapor analysed by gas chromatography. Details in source.	1. Purity 99.99 mole per cent. 2. Purity 99.98 mole per cent.

ESTIMATED ERROR:

δT/K = ±0.02; $\delta P/10^5$Pa = ±0.01 up to 0.6 MPa ±0.05 above 0.6 MPa; δx_{CH_4}, δy_{CH_4} = ±0.001.

REFERENCES:
1. Duncan, A.G. and Hiza, M.J.; *Adv. Cryogen. Engng.* <u>1970</u>, *15*, 42

2. Hiza, M.J.; Duncan, A.G. *Rev. Sci. Inst.* <u>1969</u>, *40*, 513.

COMPONENTS:	ORIGINAL MEASUREMENTS:
1. Methane; CH_4; [74-82-8]	Miller, R.C.; Kidnay, A.J.; Hiza, M.J.
2. Ethene; C_2H_4; [74-85-1]	J. Chem. Thermodynamics, 1977, 9, 167-178.

EXPERIMENTAL VALUES:

		Mole fraction of methane	
T/K	$P/10^5$Pa	in liquid, x_{CH_4}	in vapor, y_{CH_4}
170.00	10.01	0.3778	0.9128
	12.17	0.4872	0.9339
	14.07	0.5859	0.9489
	16.01	0.6897	0.9612
	18.00	0.7891	0.9728
	18.03	0.7918	0.9740
	19.96	0.8797	0.9840
180.00	3.227	0.0400	0.4340
	4.306	0.0706	0.5794
	6.101	0.1251	0.7065
	8.11	0.1879	0.7816
	11.99	0.3193	0.8589
	16.17	0.4703	0.9059
	20.30	0.6155	0.9319
	24.03	0.7586	0.9552
	27.26	0.8628	0.9716
190.00	4.323	0.0297	0.3123
	6.182	0.0723	0.5207
	7.98	0.1162	0.6290
	10.11	0.1695	0.7110
	15.12	0.3025	0.8146
	19.91	0.4411	0.8671
	25.16	0.5826	0.9028
	29.99	0.7050	0.9289
	37.20	0.8783	0.9623

COMPONENTS:	ORIGINAL MEASUREMENTS:
1. Methane; CH_4; [74-82-8] 2. Ethene; C_2H_4; [74-85-1] 3. Ethane; C_2H_6; [74-84-0]	Hsi, C.; Lu, B. C.-Y. *Can. J. Chem. Eng.* <u>1971</u>, *49*, 140-143. (Supplementary data)

VARIABLES:	PREPARED BY:
	C. L. Young

EXPERIMENTAL VALUES:

$T/K = 159.2$ (T/°F = -173.1)

Mole fraction

P/psi	P/MPa	in liquid			in vapor		
		x_{CH_4}	$x_{C_2H_4}$	$x_{C_2H_6}$	y_{CH_4}	$y_{C_2H_4}$	$y_{C_2H_6}$
38.14	0.2630	0.1240	0.6812	0.1948	0.8203	0.1572	0.0225
60.43	0.4167	0.2150	0.6116	0.1734	0.8905	0.0957	0.0448
84.31	0.5813	0.3147	0.5185	0.1668	0.9251	0.0647	0.0102
102.00	0.7033	0.4058	0.4578	0.1364	0.9401	0.0528	0.0071
110.00	0.7584	0.4517	0.4161	0.1322	0.9504	0.0430	0.0066
127.97	0.8823	0.5367	0.3572	0.1061	0.9596	0.0356	0.0048
147.03	1.0137	0.6378	0.2840	0.0782	0.9722	0.0246	0.0032
162.89	1.1231	0.7230	0.2163	0.0607	0.9777	0.0201	0.0022
38.00	0.2620	0.1241	0.5775	0.2984	0.8268	0.1437	0.0295
59.90	0.4130	0.2200	0.5189	0.2611	0.8966	0.0830	0.0204
85.33	0.5883	0.3313	0.4422	0.2265	0.9306	0.0565	0.0129
104.95	0.7236	0.4241	0.3803	0.1956	0.9505	0.0406	0.0089
127.25	0.8774	0.5399	0.2986	0.1615	0.9623	0.0302	0.0075
136.29	0.9397	0.5878	0.2706	0.1416	0.9660	0.0280	0.0060
38.70	0.2668	0.1297	0.4356	0.4347	0.8516	0.1086	0.0398
64.42	0.4442	0.2419	0.3812	0.3769	0.9122	0.0611	0.0267
86.70	0.5978	0.3382	0.3280	0.3338	0.9389	0.0420	0.0191
103.00	0.7102	0.4136	0.2981	0.2883	0.9520	0.0341	0.0139
123.00	0.8481	0.5138	0.2499	0.2363	0.9638	0.0260	0.0102
143.43	0.9889	0.6255	0.1911	0.1834	0.9748	0.0180	0.0072
36.00	0.2482	0.1196	0.1844	0.6960	0.8787	0.0529	0.0684
60.19	0.4150	0.2179	0.1693	0.6128	0.9282	0.0314	0.0404

(cont.)

AUXILIARY INFORMATION

METHOD/APPARATUS/PROCEDURE:	SOURCE AND PURITY OF MATERIALS:
Recirculating vapor flow apparatus constructed of 100 ml Jerguson gauge. Temperature measured using copper-constantan thermocouples. Pressure measured with Bourdon gauge. Cell charged and vapor recirculated with magnetic pump for 2 or more hours. Samples of vapor and liquid removed at constant pressure and analysed using gas chromatography. Helium was used as a carrier gas.	1, 2, 3. Matheson research grade samples, purities 99.99 mole per cent, 99.98 mole per cent and 99.9 mole per cent, respectively.
	ESTIMATED ERROR: $\delta T/K = \pm0.02$; $\delta P/MPa \sim \pm0.005$; δx, $\delta y = \pm1\%$ (estimated by compiler).
	REFERENCES:

COMPONENTS:	ORIGINAL MEASUREMENTS:
1. Methane; CH_4; [74-82-8]	Hsi, C.; Lu, B. C.-Y.
2. Ethene; C_2H_4; [74-85-1]	*Can. J. Chem. Eng.*
3. Ethane; C_2H_6; [74-84-0]	<u>1971</u>, *49*, 140-143.
	(Supplementary data)

EXPERIMENTAL VALUES:

$$T/K = 159.2 \quad (T/°F = -173.1)$$

		Mole fraction					
		in liquid			in vapor		
P/psi	P/MPa	x_{CH_4}	$x_{C_2H_4}$	$x_{C_2H_6}$	y_{CH_4}	$y_{C_2H_4}$	$y_{C_2H_6}$
80.00	0.5516	0.3040	0.1453	0.5507	0.9506	0.0215	0.0279
100.86	0.6954	0.3990	0.1262	0.4748	0.9637	0.0161	0.0202
121.08	0.8348	0.5027	0.1066	0.3907	0.9725	0.0117	0.0158
141.22	0.9737	0.6051	0.0814	0.3135	0.9810	0.0077	0.9887
154.16	1.0629	0.6761	0.0661	0.2578	0.9839	0.0061	0.0100
58.56	0.4038	0.2127	0.2388	0.5485	0.9151	0.0445	0.0404
73.50	0.5068	0.2782	0.2101	0.5117	0.9398	0.0319	0.0341
90.62	0.6248	0.3574	0.1910	0.4516	0.9540	0.0243	0.0217
105.00	0.7239	0.4228	0.1732	0.4040	0.9629	0.0204	0.0167
117.21	0.8081	0.4867	0.1575	0.3558	0.9687	0.0174	0.0139
130.00	0.8963	0.5535	0.1371	0.3094	0.9747	0.0140	0.0113
150.35	1.0366	0.6587	0.1056	0.2357	0.9804	0.0108	0.0088
33.95	0.2341	0.1143	0.2456	0.6401	0.8551	0.0754	0.0695
54.22	0.3738	0.1973	0.2218	0.5809	0.9139	0.0451	0.0410
73.78	0.5087	0.2770	0.2060	0.5170	0.9383	0.0333	0.0284
89.75	0.6188	0.3526	0.1827	0.4647	0.9563	0.0233	0.0204
111.95	0.7719	0.4588	0.1571	0.3841	0.9658	0.0184	0.0158
130.22	0.8978	0.5518	0.1281	0.3201	0.9720	0.0149	0.0131

COMPONENTS:	ORIGINAL MEASUREMENTS:
1. Methane; CH_4 ; [74-82-8] 2. Ethene; C_2H_4 ; [74-85-1] 3. 2-Methylpropane; C_4H_{10} ; [75-28-5]	Benedict, M.; Solomon, E.; Rubin, L. C. *Ind. Eng. Chem.* 1945, *37*, 55-59.
VARIABLES:	PREPARED BY: C. L. Young

EXPERIMENTAL VALUES:

T/K (T/°C)	P/MPa (P/atm)	Mole fraction in liquid			Mole fraction in vapor		
		x_{CH_4}	$x_{C_2H_4}$	$x_{C_4H_{10}}$	y_{CH_4}	$y_{C_2H_4}$	$y_{C_4H_{10}}$
310.93 (37.78)	3.447 (34.02)	0.125_3	0.124_5	0.752	0.576_5	0.224	0.199_5
		0.069	0.279	0.652	0.335_5	0.481_5	0.183
		0.039	0.372	0.589	0.187_5	0.640	0.172_5
		0.000	0.497	0.503	0.000	0.840_5	0.159_5
344.26 (71.11)	3.447 (34.02)	0.083	0.078	0.839	0.418	0.163_5	0.418_5
		0.056_5	0.150	0.793_5	0.269	0.325_5	0.405_5
		0.030	0.215	0.755	0.148	0.458_5	0.393_5
		0.000	0.296	0.704	0.000	0.624	0.376
310.93 (37.78)	6.895 (68.05)	0.284	0.170_5	0.545_5	0.657	0.187	0.156
		0.252	0.270	0.383_5	0.552	0.297	0.151
		0.215_5	0.383_5	0.401	0.434	0.420_5	0.145_5
		0.189	0.462	0.349	0.357_5	0.500	0.142_5
		0.157_5	0.563	0.279_5	0.257	0.603_5	0.139_5
344.26 (71.11)	6.895 (68.05)	0.217_5	0.181	0.6015	0.459	0.229_5	0.311_5
		0.159	0.334	0.507	0.283	0.408	0.309
		0.113_5	0.458_5	0.428	0.162_5	0.5175	0.320

AUXILIARY INFORMATION

METHOD/APPARATUS/PROCEDURE:	SOURCE AND PURITY OF MATERIALS:
Static equilibrium cell. Complete gas and liquid phases removed by mercury injection. Gas and liquid samples analysed by determination of gas density and mole per cent of olefins present.	1. Crude sample dried and carbon dioxide removed by passage over Ascarite and Drierite. Mass spectrometry revealed 0.35 mole per cent ethane. 2. Anesthesia grade, purity about 99.5 mole per cent. 3. Phillips Petroleum sample, purity better than 99 mole per cent.
	ESTIMATED ERROR:
	REFERENCES:

COMPONENTS:	ORIGINAL MEASUREMENTS:
1. Methane; CH_4; [74-82-8] 2. Ethene; C_2H_4; [74-85-1] 3. 2-Methylpropane; C_4H_{10}; [75-28-5] 4. Hexadecane; $C_{16}H_{34}$; [544-76-3]	Solomon, E. *Chem. Eng. Progr. Symp. Ser. No. 3* <u>1952</u>, *48*, 93-97.

VARIABLES:	PREPARED BY:
Temperature, pressure	C. L. Young

EXPERIMENTAL VALUES:

1 MPa = 145.04 psi

		Mole fractions						
		in liquid				in gas		
T/K (T/°F)	P/psi	x_{CH_4}	$x_{C_2H_4}$	$x_{C_4H_{10}}$	$x_{C_{16}H_{34}}$	y_{CH_4}	$y_{C_2H_4}$	$y_{C_4H_{10}}$
310.93 (100)	500	0.035	0.328	0.373	0.264	0.2305	0.667	0.1025
	500	0.0495	0.3065	0.232	0.412	0.3225	0.617	0.060
377.59 (220)	500	0.095	0.0745	0.019	0.8115	0.7595	0.2135	0.009
	500	0.0245	0.1975	0.1805	0.5975	0.2195	0.654	0.1265
	500	0.029	0.2225	0.1035	0.645	0.218	0.708	0.074
	1000	0.132	0.2125	0.1525	0.503	0.5235	0.400	0.0765

AUXILIARY INFORMATION

METHOD/APPARATUS/PROCEDURE:	SOURCE AND PURITY OF MATERIALS:
Static equilibrium cell. Complete gas and liquid phases removed by mercury injection. Gas sample analysed by determination of gas density and mole per cent of olefins present. Light hydrocarbons stripped from liquid phase in debutanization still leaving heavy hydrocarbon. Liquid hydrocarbons analysed by same method as gas samples. Details in source and ref. (1).	1. Crude sample dried and carbon dioxide removed by passage over Ascarite and Drierite. Mass spectroscopy revealed 0.35 mole per cent ethane. 2. Anesthesia grade, purity about 99.5 mole per cent. 3. Phillips Petroleum sample, purity better than 99 mole per cent. 4. No details given.
	ESTIMATED ERROR:
	$\delta T/K = \pm 0.02$; $\delta P/P = \pm 0.1\%$; δx, $\delta y = \pm 0.0015$.
	REFERENCES:
	1. Benedict, M.; Solomon, E.; Rubin, L. C. *Ind. Eng. Chem.* <u>1945</u>, *37*, 55.

COMPONENTS:	ORIGINAL MEASUREMENTS:

1. Methane; CH_4; [74-82-8]
 	
2. Ethene; C_2H_4; [74-85-1]
 	
3. 2-Methylpropane; C_4H_{10};
  [75-28-5]
 	
4. 1,1'-Bicyclohexyl; $C_{12}H_{22}$;
  [92-51-3]
	Solomon, E.
	
	Chem. Eng. Progr. Symp. Ser. No. 3
	
	1952, *48*, 93-97.

VARIABLES:	PREPARED BY:
Temperature, pressure	C. L. Young

EXPERIMENTAL VALUES:

1 MPa = 145.04 psi

Mole fractions

T/K (T/°F)	P/psi	in liquid				in gas		
		x_{CH_4}	$x_{C_2H_4}$	$x_{C_4H_{10}}$	$x_{C_{12}H_{22}}$	y_{CH_4}	$y_{C_2H_4}$	$y_{C_4H_{10}}$
310.93	500	0.0495	0.1985	0.152	0.600	0.434	0.518	0.048
(100)	500	0.055	0.224	0.425	0.296	0.3725	0.5045	0.123
377.59	500	0.0235	0.1385	0.1265	0.7115	0.2635	0.6195	0.117
(220)	500	0.035	0.1025	0.159	0.7035	0.4005	0.4585	0.141

AUXILIARY INFORMATION

METHOD/APPARATUS/PROCEDURE:	SOURCE AND PURITY OF MATERIALS:
Static equilibrium cell. Complete gas and liquid phases removed by mercury injection. Gas sample analysed by determination of gas density and mole per cent of olefins present. Light hydrocarbons stripped from liquid phase in debutanization still leaving heavy hydrocargon. Liquid hydrocarbons analysed by same method as gas samples. Details in source and ref. (1).	1. Crude sample dried and carbon dioxide removed by passage over Ascarite and Drierite. Mass spectroscopy revealed 0.35 mole per cent ethane. 2. Anesthesia grade, purity about 99.5 mole per cent. 3. Phillips Petroleum sample, purity better than 99 mole per cent. 4. No details given.

ESTIMATED ERROR:

$\delta T/K = \pm 0.02$; $\delta P/P = \pm 0.1\%$;

δx, $\delta y = \pm 0.0015$.

REFERENCES:

1. Benedict, M.; Solomon, E.;
   Rubin, L. C.

   Ind. Eng. Chem.

   1945, *37*, 55.

COMPONENTS:	ORIGINAL MEASUREMENTS:

1. Methane; CH_4; [74-82-8]
 	
2. Ethene; C_2H_4; [74-85-1]
 	
3. 2-Methylpropane; C_4H_{10};
   [75-28-5]
 	
4. Methylnaphthalene; $C_{11}H_{10}$;
   [1321-94-4]
	Solomon, E.
	
	Chem. Eng. Progr. Symp. Ser. No. 3
	
	<u>1952</u>, *48*, 93-97.

VARIABLES:	PREPARED BY:
Temperature, pressure	C. L. Young

EXPERIMENTAL VALUES:

1 MPa = 145.04 psi

T/K (T/°F)	P/psi	Mole fractions						
		in liquid				in gas		
		x_{CH_4}	$x_{C_2H_4}$	$x_{C_4H_{10}}$	$x_{C_{11}H_{10}}$	y_{CH_4}	$y_{C_2H_4}$	$y_{C_4H_{10}}$
310.93 (100)	500	0.0405	0.079	0.203	0.6775	0.632	0.265	0.103
	500	0.0395	0.141	0.287	0.5325	0.457	0.4155	0.1275
377.59 (220)	500	0.013	0.1065	0.068	0.8125	0.2455	0.649	0.1055
	500	0.0215	0.082	0.060	0.8365	0.3945	0.5085	0.097
	500	0.023	0.075	0.1085	0.7935	0.3905	0.441	0.1685
	1000	0.0635	0.0965	0.198	0.642	0.514	0.2975	0.1885

AUXILIARY INFORMATION

METHOD/APPARATUS/PROCEDURE:	SOURCE AND PURITY OF MATERIALS:
Static equilibrium cell. Complete gas and liquid phases removed by mercury injection. Gas sample analysed by determination of gas density and mole per cent of olefins present. Light hydrocarbons stripped from liquid phase in debutanization still leaving heavy hydrocarbon. Liquid hydrocarbons analysed by same method as gas samples. Details in source and ref. (1).	1. Crude sample dried and carbon dioxide removed by passage over Ascarite and Drierite. Mass spectroscopy revealed 0.35 mole per cent ethane. 2. Anesthesia grade, purity about 99.5 mole per cent. 3. Phillips Petroleum sample, purity better than 99 mole per cent. 4. No details given.

ESTIMATED ERROR:

$\delta T/K = \pm 0.02$; $\delta P/P = \pm 0.1\%$;

δx, $\delta y = \pm 0.0015$.

REFERENCES:

1. Benedict, M.; Solomon, E.;
   Rubin, L. C.
   Ind. Eng. Chem.
   <u>1945</u>, *37*, 55.

COMPONENTS:	ORIGINAL MEASUREMENTS:
1. Methane; CH_4; [74-82-8] 2. Ethene; C_2H_4; [74-85-1] 3. 2-Methylpropane; C_4H_{10}; [75-28-5] 4. Gas oil	Solomon, E. *Chem. Eng. Progr. Symp. Ser. No. 3* <u>1952</u>, *48*, 93-97.

VARIABLES:	PREPARED BY:
Temperature, pressure	C. L. Young

EXPERIMENTAL VALUES:

1 MPa = 145.04 psi

Mole fractions

T/K (T/°F)	P/psi	x_{CH_4}	$x_{C_2H_4}$	$x_{C_4H_{10}}$	$x_{Gas oil}$	y_{CH_4}	$y_{C_2H_4}$	$y_{C_4H_{10}}$
			in liquid				in gas	
310.93	500	0.0785	0.1955	0.1775	0.5485	0.5515	0.402	0.0465
(100)	500	0.0945	0.125	0.6965	0.084	0.521	0.302	0.177
377.59	500	0.060	0.1225	0.1425	0.675	0.494	0.4015	0.1045
(220)	500	0.058	0.106	0.2475	0.5885	0.457	0.352	0.191
	1000	0.1055	0.211	0.192	0.4915	0.473	0.422	0.105

Properties of Gas oil: Distillation properties   °F

Initial boiling point	447	
5%	579	Percentage
10	582	recovery 98.0
20	590	
30	596	Percentage
40	604	residue 1.8
50	612	
60	623	Percentage
70	642	loss 0.2
80	664	
90	700	Molecular
95	734	weight 275
End point	746	

AUXILIARY INFORMATION

METHOD/APPARATUS/PROCEDURE:	SOURCE AND PURITY OF MATERIALS:
Static equilibrium cell. Complete gas and liquid phases removed by mercury injection. Gas sample analysed by determination of gas density and mole per cent of olefins present. Light hydrocarbons stripped from liquid phase in debutanization still leaving heavy hydrocarbon. Liquid hydrocarbons analysed by same method as gas samples. Details in source and ref. (1).	1. Crude sample dried and carbon dioxide removed by passage over Ascarite and Drierite. Mass spectroscopy revealed 0.35 mole per cent ethane. 2. Anesthesia grade, purity about 99.5 mole per cent. 3. Phillips Petroleum sample, purity better than 99 mole per cent. 4. Details given above.

ESTIMATED ERROR:
$\delta T/K = \pm 0.02$; $\delta P/P = \pm 0.1\%$;
δx, $\delta y = \pm 0.0015$.

REFERENCES:

1. Benedict, M.; Solomon, E.; Rubin, L. C.

   Ind. Eng. Chem.

   <u>1945</u>, *37*, 55.

COMPONENTS:	ORIGINAL MEASUREMENTS:
1. Methane; CH_4; [74-82-8]	Solomon, E.
2. Ethene; C_2H_4; [74-85-1]	*Chem. Eng. Progr. Symp. Ser. No. 3*
3. 2-Methylpropane; C_4H_{10}; [75-28-5]	*1952*, *48*, 93-97.
4. Hydroformer Still Bottoms	

VARIABLES:	PREPARED BY:
Temperature, pressure	C. L. Young

EXPERIMENTAL VALUES:

1MPa = 145.04 psi

				Mole fractions				
		in liquid				in gas		
T/K (T/°F)	P/psi	x_{CH_4}	$x_{C_2H_4}$	$x_{C_4H_{10}}$	x_{HSB}	y_{CH_4}	$y_{C_2H_4}$	$y_{C_4H_{10}}$
377.59	500	0.0255	0.089	0.052	0.8335	0.4075	0.5115	0.081
(220)	500	0.0285	0.063	0.1385	0.770	0.440	0.352	0.208
	1000	0.071	0.070	0.174	0.685	0.598	0.224	0.178

HSB - Hydroformer Still Bottoms

Properties of Hydroformer Still Bottoms:

Distillation properties	°F		
Initial boiling point	553		
5%	584	Percentage recovery	96.5
10	593		
20	603		
30	611	Percentage residue	3.5
40	620		
50	629		
60	643	Percentage loss	0.0
70	659		
80	683		
90	727	Molecular weight	209
95	756		
End point	760		

AUXILIARY INFORMATION

METHOD/APPARATUS/PROCEDURE:	SOURCE AND PURITY OF MATERIALS:
Static equilibrium cell. Complete gas and liquid phases removed by mercury injection. Gas sample analysed by determination of gas density and mole per cent of olefins present. Light hydrocarbons stripped from liquid phase in debutanization still leaving heavy hydrocarbon. Liquid hydrocarbons analysed by same method as gas samples. Details in source and ref. (1).	1. Crude sample dried and carbon dioxide removed by passage over Ascarite and Drierite. Mass spectroscopy revealed 0.35 mole per cent ethane. 2. Anesthesia grade, purity about 99.5 mole per cent. 3. Phillips Petroleum sample, purity better than 99 mole per cent. 4. Details given above.
	ESTIMATED ERROR:
	$\delta T/K = \pm 0.02$; $\delta P/P = \pm 0.1\%$; δx, $\delta y = \pm 0.0015$.
	REFERENCES:
	1. Benedict, M.; Solomon, E.; Rubin, L. C. *Ind. Eng. Chem.* *1945*, *37*, 55.

COMPONENTS:	EVALUATOR:
(1) Methane; CH_4; [74-82-8]	H. Lawrence Clever
	Chemistry Department
(2) Cycloalkenes	Emory University
Cyclohexene	Atlanta, GA 30322 USA
Pinene	
	1984, January

CRITICAL EVALUATION:

Methane + Cyclohexene; C_6H_{10}; [110-83-8]

Methane + 2,6,6-Trimethylbicyclo[3.1.1]hept-2-ene or pinene; $C_{10}H_{16}$; [80-56-8]

Guerry (ref. 2) reported the solubility of methane in cyclohexene at 293.15 and 298.15 K and McDaniel (ref. 1) reported the solubility of methane in pinene at five temperatures between 293.15 and 328.35 K. Other methane solubility values reported by these authors have proved unreliable, often being too small by 20 to 50 percent. Thus these data are classed as doubtful.

Guerry's data leads to a partial molal enthalpy of solution of -1.18 kJ mol^{-1} of methane in cyclohexene and McDaniels data leads to a value of -8.13 kJ mol^{-1} of methane in pinene. McDaniel's value appears to be too large and Guerry's value too small when the enthalpies are compared to more reliable values in other hydrocarbon solvents.

The smoothed solubility data which should be used with caution because they are probably both too small and of incorrect temperature coefficient are in Table 1.

Table 1. Solubility of methane in cyclohexene and pinene. Mole fraction solubility at 101.325 kPa partial pressure methane.

T/K	Mol Fraction, $10^3 x_1$	
	Cyclohexene	Pinene
293.15	2.48	3.29
298.15	2.46	3.11
303.15	–	2.95
313.15	–	2.66
323.15	–	2.41

References

1. McDaniel, A. S. *J. Phys. Chem.* 1911, *15*, 587-610.

2. Guerry, D. Ph.D. thesis, Vanderbilt University, 1944.

COMPONENTS:	ORIGINAL MEASUREMENTS:
(1) Methane; CH_4; [74-82-8]	McDaniel, A. S.
(2) 2,6,6-Trimethylbicyclo[3.1.1] hept-2-ene or pinene; $C_{10}H_{16}$; [80-56-8]	*J. Phys. Chem.* <u>1911</u>, *15*, 587-610.

VARIABLES:	PREPARED BY:
$T/K = 293.15 - 328.35$ $p_1/kPa = 101.3$ (1 atm)	H. L. Clever

EXPERIMENTAL VALUES:

Temperature		Mol Fraction	Bunsen Coefficient[a]	Ostwald Coefficient[b]
$t/°C$	T/K	$10^3 x_1$	α	L/cm^3 cm^{-3}
20.0	293.15	3.21	0.4565	0.4888
25.0	298.15	2.98	0.4235	0.4623[c]
30.1	303.25	2.93	0.4163	0.4620
39.1	312.25	2.76	0.3914	0.4472
45.0	318.15	2.69	0.3811	0.4440
55.2	328.35	2.17	0.3076	0.3694

[a] Bunsen coefficient, α/cm^3 (STP) cm^{-3} atm^{-1}.

[b] Listed as absorption coefficient in the original paper. Interpreted to be equivalent to Ostwald coefficient by compiler.

[c] Ostwald coefficient (absorption coefficient) estimated as 298.15 K value by author.

[d] Mole fraction and Bunsen coefficient values calculated by compiler assuming ideal gas behavior.

EVALUATOR'S COMMENT: McDaniel's data should be used with caution. His values are often 20 percent or more too small when compared with more reliable data.

AUXILIARY INFORMATION

METHOD/APPARATUS/PROCEDURE:	SOURCE AND PURITY OF MATERIALS:
The apparatus is all glass. It consists of a gas buret connected to a contacting vessel. The solvent is degassed by boiling under reduced pressure. Gas pressure or volume is adjusted using mercury displacement. Equilibration is achieved at atm pressure by hand shaking, and incrementally adding gas to the contacting chamber. Solubility measured by obtaining total uptake of gas by known volume of the solvent.	(1) Methane. Prepared by reaction of methyl iodide with zinc-copper. Passed through water and sulfuric acid. (2) Pinene.
	ESTIMATED ERROR: $\delta L/L \geq -0.20$
	REFERENCES:

COMPONENTS:	EVALUATOR:

(1) Methane; CH_4; [74-82-8]
 	
(2) Arenes
    Benzene   Methylbenzene
    Dimethylbenzenes
    1,1'-Methylenebisbenzene
    1-Methylnahthalene
    Decahydronaphthalene
	H. Lawrence Clever
	Chemistry Department
	Emory University
	Atlanta, GA 30322 USA
	
	1984, January

CRITICAL EVALUATION:

The Solubility of Methane in Arenes at Partial

Pressures up to 200 kPa (*ca.* 2 atm).

The solubility of methane in arenes at methane partial pressures up to 200 kPa is reported in nine papers. With one exception all of the solubilities were measured by volumetric methods at a total pressure near 101 kPa (*ca.* 1 atm). The exception (ref. 5) used a GLC system to analyze the saturated solution.

The partial molal enthalpy change on solution is less exothermic for the arenes than for the alkanes. For the methane + benzene it is -1.23 kJ mol^{-1} while for the other methyl substituted benzenes it ranges from -2.6 to -4.3 kJ mol^{-1}. Most of the methyl substituted enthalpy change values are intermediate between the benzene and alkane values. Enthalpy changes based on the temperature coefficient of solubility are probably not reliable enough to base any far reaching conclusions on when they show as little difference as do these values.

Methane + Benzene; C_6H_6; [71-43-2]

McDaniel (ref. 1) reports four solubility values between 295.25 and 323.05 K; Horiuti (ref. 2) reports four values between 286.25 and 333.15 K, Lannung and Gjaldbaek (ref. 3) report six values between 291.15 and 310.15 K and Hayduk and Buckley (ref. 4) report one value at 298.15 K.

The smoothed McDaniel data are smaller by 9 percent at 293 K and 35 percent at 323 K than the data in (ref. 2,3). McDaniel's data are rejected.

The smoothed data of Horiuti and of Lannung and Gjaldbaek agree within 0.5 to 1.5 percent between 288 and 308 K with the Lannung and Gjaldbaek solubility values the larger. Hayduk and Buckley's single value at 298.15 K is between 1 and 2 percent smaller than the values from the other two papers.

The data of Horiuti, Lannung and Gjaldbaek, and Hayduk and Buckley were combined in a linear regression to obtain the equation

$$\ln x_1 = -6.66791 + 1.47786/(T/100 \text{ K})$$

with a standard error about the regression line of 1.7×10^{-5}.

The temperature independent thermodynamic changes calculated from the constants of the equation are

$$\Delta \bar{H}_1^\circ/\text{kJ mol}^{-1} = -1.23 \quad \text{and} \quad \Delta \bar{S}_1^\circ/\text{J K}^{-1} \text{ mol}^{-1} = -55.4$$

Smoothed values of the mole fraction solubility and partial molal Gibbs energy of solution are given in Table 1.

Table 1.   Solubility of methane in benzene. Tentative mole fraction
           solubility at 101.325 (1 atm) methane partial pressure and
           partial Gibbs energy of solution as a function of temperature.

T/K	Mol Fraction $10^3 x_1$	$\Delta \bar{G}_1^\circ/kJ\ mol^{-1}$
283.15	2.14	14.469
293.15	2.10	15.023
298.15	2.05	15.300
303.15	2.07	15.578
313.15	2.04	16.132
323.15	2.01	16.686
333.15	1.98	17.241

Methane + Methylbenzene; C_7H_8; [108-88-3]

   McDaniel (ref. 1) reports four values of the solubility of methane in
methylbenzene between 303.15 and 333.15 K and Field, Wilhelm and Battino
(ref. 7) report three values of temperatures between 284.28 and 313.17 K.

   The McDaniel data are 10 percent smaller than the Field et al. data
over the temperature interval of common measurement. Although both data
sets are classed as tentative the data of Field, Wilhelm and Battino are
preferred and the tentative values below are based on their data.

   A linear regression of the data of Field et al. gives the equation

$$\ln x_1 = -7.79695 + 5.22144/(T/100\ K)$$

with a standard error about the regression line of 1.66×10^{-4}.

   The thermodynamic changes for the transfer of one mole of gas from
the gas phase to the infinitely dilute solution are

$$\Delta H_1^\circ/kJ\ mol^{-1} = -4.34 \qquad \text{and} \qquad \Delta S_1^\circ/J\ K^{-1}\ mol^{-1} = -64.8$$

We are concerned that this is larger by about 25 percent than any of the
other enthalpies of solution in benzene and methyl substituted benzenes.

   Smoothed values of the mole fraction solubility and partial molal
Gibbs energy of solution are in Table 2.

Table 2.   Solubility of methane in methylbenzene. Tentative values of the
           mole fraction solubility at 101.325 kPa methane partial pressure
           and partial molal Gibbs energy of solution as a function of
           temperature.

T/K	Mol Fraction $10^3 x_1$	$\Delta \bar{G}_1^\circ/kJ\ mol^{-1}$
283.15	2.60	14.014
293.15	2.44	14.663
298.15	2.37	14.987
303.15	2.30	15.311
313.15	2.18	15.959

COMPONENTS:	EVALUATOR:

(1) Methane; CH_4; [74-82-8]
 	
(2) Arenes
    Benzene   Methylbenzene
    Dimethylbenzenes
    1,1'-Methylenebisbenzene
    1-Methylnahthalene
    Decahydronaphthalene
	H. Lawrence Clever
	Chemistry Department
	Emory University
	Atlanta, GA 30322 USA
	
	1984, January

CRITICAL EVALUATION:

Methane + 1,2-Dimethylbenzene; C_8H_{10}; [95-47-6]

Methane + 1,3-Dimethylbenzene; C_8H_{10}; [108-38-3]

Methane + 1,4-Dimethylbenzene; C_8H_{10}; [106-42-3]

Byrne, Battino and Wilhelm (ref. 9) report between 5 and 7 solubility values each for these systems over the 283 to 313 K temperature interval. McDaniel reports four solubility values for the methane + 1,3-dimethylbenzene system at temperatures between 294.25 and 333.15 K.

All of the data are classed as tentative. For the methane + 1,3-dimethylbenzene system McDaniel's smoothed data are 5.2 to 5.4 percent smaller than the smoothed results of Byrne, Battino and Wilhelm. The agreement is within the experimental error. However, the Byrne *et al.* data are preferred and their data are the basis of all the smoothed data presented here. The data were fitted to the following equations by a linear regression.

Methane + 1,2-dimethylbenzene

$$\ln x_1 = -7.03873 + 3.13979/(T/100 \text{ K})$$

Methane + 1,3-dimethylbenzene

$$\ln x_1 = -7.06540 + 3.44764/(T/100 \text{ K})$$

Methane + 1,4-dimethylbenzene

$$\ln x_1 = -7.20727 + 4.01581/(T/100 \text{ K})$$

with standard errors about the regression lines of 7.9×10^{-6}, 9.7×10^{-6}, and 18.2×10^{-6}, respectively.

The temperature independent thermodynamic changes for the transfer of one mole of methane from the gas phase to the infinitely dilute solution are given in Table 3.

Table 3. Thermodynamic values for the transfer of methane from the gas to the infinitely dilute solution in dimethylbenzenes.

Solvent	$\Delta \overline{H}_1^{\circ}/\text{kJ mol}^{-1}$	$\Delta \overline{S}_1^{\circ}/\text{J K}^{-1} \text{ mol}^{-1}$	$\Delta \overline{G}_1^{\circ}/\text{kJ mol}^{-1}$
1,2-Dimethylbenzene	-2.61	-58.5	14.838[a]
1,3-Dimethylbenzene	-2.87	-58.7	14.648[a]
1,4-Dimethylbenzene	-3.34	-59.9	14.527[a]

[a] The partial molar Gibbs energy values are for a temperature of 298.15 K.

The smoothed solubility values are given in Table 4.

Table 4. Tentative values of the solubility of methane in the dimethyl-
          benzenes. Mole fraction solubility at 101.325 kPa partial
          pressure methane as a function of temperature.

| | Mole Fraction, $10^3 x_1$ | | |
T/K	1,2-Dimethylbenzene	1,3-Dimethylbenzene	1,4-Dimethylbenzene
283.15	2.66	2.89	3.06
293.15	2.56	2.77	2.91
298.15	2.51	2.71	2.85
303.15	2.47	2.66	2.79
313.15	2.39	2.57	2.67

Methane + 1,1'-Methylenebisbenzene; $C_{13}H_{12}$; [101-81-5]

Cukor and Prausnitz (ref. 6) report Henry's constant values at
25 degree intervals from 300 to 475 K. The solubility values were converted to mole fraction values at a 101.325 kPa methane partial pressure
assuming ideal gas behavior and Henry's law to be obeyed. The results
are classed as tentative.

The solubility shows a minimum in the temperature range. The values
were fitted to a three constant equation by a linear regression to obtain

$$\ln x_1 = -12.53866 + 10.12809/(T/100 \text{ K}) + 2.59305 \ln(T/100 \text{ K})$$

with a standard error about the regression line of 1.26×10^{-5}.

Values of the thermodynamic changes for the transfer of methane from
the gas phase at 101.325 kPa to the infinitely dilute solution at several
temperatures are as follows:

T/K	$\Delta \bar{H}^\circ_1$/kJ mol^{-1}	$\Delta \bar{S}^\circ_1$/J K^{-1} mol^{-1}	$\Delta \bar{C}^\circ_{p_1}$/J K^{-1} mol^{-1}
298.15	-1.99	-59.1	21.6
323.15	-1.45	-57.4	21.6
373.15	-0.38	-54.3	21.6
423.15	+0.70	-51.6	21.6
473.15	1.78	-49.2	21.6

Smoothed values of the solubility of methane and the partial molal Gibbs
energy as a function of temperature are given in Table 5. The temperature
of minimum solubility is 391 K.

Table 5. Tentative values of the solubility of methane in
          1-methylnaphthalene. Mole fraction solubility of 101.325 kPa
          (1 atm) methane partial pressure and partial molal Gibbs energy
          as a function of temperature.

T/K	Mol Fraction $10^3 x_1$	$\Delta \bar{G}^\circ_1$/kJ mol^{-1}
298.15	1.58	15.988
323.15	1.47	17.531
348.15	1.40	19.018
373.15	1.37	20.453
391	1.36	21.621
398.15	1.36	21.839
423.15	1.38	23.181
448.15	1.40	24.479
473.15	1.44	25.737

COMPONENTS:	EVALUATOR:
(1) Methane; CH_4; [74-82-8] (2) Arenes Benzene Methylbenzene Dimethylbenzenes 1,1'-Methylenebisbenzene 1-Methylnahthalene Decahydronaphthalene	H. Lawrence Clever Chemistry Department Emory University Atlanta, GA 30322 USA 1984, January

CRITICAL EVALUATION:

Methane + Decahydronaphthalene; $C_{10}H_{18}$; [91-17-8]

Lenoir, Renault, and Renon (ref. 5) report solubility values at two temperatures of 298.2 and 323.2 K. The values are classed as tentative, but there is some concern regarding the results because of the large magnitude of the enthalpy of solution relative to similar solvents.

The solubility data are reproduced by the equation

$$\ln x_1 = -8.7721 + 8.7034/(T/100\ K)$$

which gives temperature independent thermodynamic changes for the transfer of methane from the gas at 101.325 kPa to the infinitely dilute solution of

$$\bar{H}_1^\circ/kJ\ mol^{-1} = -7.24 \qquad \text{and} \qquad \bar{S}_1^\circ/J\ K^{-1}\ mol^{-1} = -72.9$$

The values are of larger magnitude than normally observed for either alkane or arene solvents.

Smoothed values of solubility and partial molal Gibbs energy of solution are given in Table 6.

Table 6.   Tentative values of the solubility of methane in decahydro-
           naphthalene or decalin. Mole fraction solubility at 101.325 kPa
           methane partial pressure and partial molal Gibbs energy of
           solution as a function of temperature.

T/K	Mol Fraction $10^3 x_1$	$\Delta \bar{G}_1^\circ/kJ\ mol^{-1}$
298.15	2.87	14.513
303.15	2.74	14.877
313.15	2.50	15.607
323.15	2.29	16.336

References

1. McDaniel, A. S. *J. Phys. Chem.* 1911, *15*, 587.

2. Horiuti, J. *Sci. Pap. Inst. Phys. Chem. Res. (JPN)*, 1931/32, *17*, 125.

3. Lannung, A.; Gjaldbaek, J. C. *Acta Chem. Scand.* 1960, *14*, 1124.

4. Hayduk, W.; Buckley, W. D. *Can. J. Chem. Eng.* 1971, *49*, 667.

5. Lenoir, J-Y.; Renault, P.; Renon, H. *J. Chem. Eng. Data*, 1971, *16*, 340.

6. Cukor, P. M.; Prausnitz, J. M. *J. Phys. Chem.* 1972, *76*, 598.

7. Field, L. R.; Wilhelm, E.; Battino, R. *J. Chem. Thermodyn.* 1974,
    6, 237.

8. Chappelow, C. C.; Prausnitz, J. M. *Am. Inst. Chem. Engnrs. J.* 1974,
    20, 1097.

9. Byrne, J. E.; Battino, R.; Wilhelm, E. *J. Chem. Thermodyn.* 1975,
    7, 515.

COMPONENTS:	ORIGINAL MEASUREMENTS:
(1) Methane; CH_4; [74-82-8] (2) Benzene; C_6H_6; [71-43-2]	McDaniel, A. S. *J. Phys. Chem.* <u>1911</u>, *15*, 587-610.

VARIABLES:	PREPARED BY:
T/K = 295.25 - 323.05 p_1/kPa = 101.3 (1 atm)	H. L. Clever

EXPERIMENTAL VALUES:

Temperature		Mol Fraction	Bunsen Coefficient[a]	Ostwald Coefficient[b]
$t/°C$	T/K	$10^3 x_1$	α	$L/cm^3 \ cm^{-3}$
22.1	295.25	1.82	0.4600	0.4954
25.0	298.15	1.77	0.4438	0.4844[c]
35.0	308.15	1.60	0.3976	0.4484
40.1	313.25	1.49	0.3661	0.4198
49.9	323.05	1.27	0.3081	0.3645

[a] Bunsen coefficient, α/cm^3 (STP) $cm^{-3} \ atm^{-1}$.

[b] Listed as absorption coefficient in the original paper. Interpreted to be equivalent to Ostwald coefficient by compiler.

[c] Ostwald coefficient (absorption coefficient) estimated as 298.15 K value by author.

[d] Mole fraction and Bunsen coefficient values calculated by compiler assuming ideal gas behavior.

EVALUATOR'S COMMENT: McDaniel's data should be used with caution. His values are often 20 percent or more too small when compared with more reliable data.

AUXILIARY INFORMATION

METHOD/APPARATUS/PROCEDURE:	SOURCE AND PURITY OF MATERIALS:
The apparatus is all glass. It consists of a gas buret connected to a contacting vessel. The solvent is degassed by boiling under reduced pressure. Gas pressure or volume is adjusted using mercury displacement. Equilibration is achieved at atm pressure by hand shaking, and incrementally adding gas to the contacting chamber. Solubility measured by obtaining total uptake of gas by known volume of the solvent.	(1) Methane. Prepared by reaction of methyl iodide with zinc-copper. Passed through water and sulfuric acid. (2) Benzene. Source not given.

	ESTIMATED ERROR:
	$\delta L/L \geq -0.20$

	REFERENCES:

COMPONENTS:	ORIGINAL MEASUREMENTS:
(1) Methane; CH_4; [74-82-8] (2) Benzene; C_6H_6; [71-43-2]	Horiuti, J. *Sci. Pap. Inst. Phys. Chem. Res.* *(Jpn)* 1931/32, *17*, 125 - 256.
VARIABLES: T/K: 286.25 - 333.15 p_1/kPa: 101.325 (1 atm)	PREPARED BY: M. E. Derrick H. L. Clever

EXPERIMENTAL VALUES:

T/K	Mol Fraction $10^3 x_1$	Bunsen Coefficient α/cm^3(STP)cm^{-3}atm^{-1}	Ostwald Coefficient L/cm^{3}cm^{-3}
286.25	2.130	0.5427	0.5687
293.15	2.094	0.5292	0.5680
303.15	2.056	0.5133	0.5697
333.15	1.972	0.4745	0.5787

The mole fraction and Bunsen coefficient values were calculated by the compiler with the assumption the gas is ideal and that Henry's law is obeyed.

Smoothed Data: For use between 286.25 and 333.15 K.

$$\ln x_1 = -6.6908 + 1.5351/(T/100K)$$

The standard error about the regression line is 6.44 x 10^{-6}.

T/K	Mol Fraction $10^3 x_1$
288.15	2.116
298.15	2.079
308.15	2.044
318.15	2.013
328.15	1.983

AUXILIARY INFORMATION

METHOD/APPARATUS/PROCEDURE:

The apparatus consists of a gas buret, a solvent reservoir, and an absorption pipet. The volume of the pipet is determined at various meniscus heights by weighing a quantity of water. The meniscus height is read with a cathetometer.

The dry gas is introduced into the degassed solvent. The gas and solvent are mixed with a magnetic stirrer until saturation. Care is taken to prevent solvent vapor from mixing with the solute gas in the gas buret. The volume of gas is determined from the gas buret readings, the volume of solvent is determined from the meniscus height in the absorption pipet.

SOURCE AND PURITY OF MATERIALS:

(1) Methane. Aluminum carbide was prepared from aluminum and soot carbon. The aluminum carbide was treated with hot water. The gas evolved was scrubbed to remove impurities, dried and fractionated. Final product had a density, ρ/g dm^{-3} = 0.7168±0.0003 at normal conditions.

(2) Benzene. Merck. Extra pure and free of sulfur. Refluxed with sodium amalgam, distilled. Boiling point(760 mmHg) 80.18°C.

ESTIMATED ERROR:
$$\delta T/K = 0.05$$
$$\delta x_1/x_1 = 0.01$$

REFERENCES:

COMPONENTS:	ORIGINAL MEASUREMENTS:
(1) Methane; CH_4; [74-82-8] (2) Benzene; C_6H_6; [71-43-2]	Lannung, A.; Gjaldbaek, J. C. *Acta Chem. Scand.* 1960, *14*, 1124 - 1128.

VARIABLES:	PREPARED BY:
T/K = 291.15 - 310.15 p_1/kPa = 101.325 (1 atm)	J. Chr. Gjaldbaek

EXPERIMENTAL VALUES:

T/K	Mol Fraction $10^3 x_1$	Bunsen Coefficient α/cm^3(STP)cm^{-3}atm^{-1}	Ostwald Coefficient L/cm^{3}cm^{-3}
291.15	2.13	0.537	0.572
291.15	2.11	0.533	0.568
298.15	2.09	0.523	0.571
298.15	2.09	0.523	0.571
310.15	2.05	0.506	0.574
310.15	2.08	0.514	0.584

Smoothed Data: For use between 291.15 and 310.15 K.

$$\ln x_1 = -6.5770 + 1.2198/(T/100 \text{ K})$$

The standard error about the regression line is 1.37×10^{-5}.

T/K	Mol Fraction $10^3 x_1$
298.15	2.095
308.15	2.07

AUXILIARY INFORMATION

METHOD/APPARATUS/PROCEDURE:	SOURCE AND PURITY OF MATERIALS:
A calibrated all-glass combined manometer and bulb containing degassed solvent and the gas was placed in an air thermostat and shaken until equilibrium (1). The absorbed volume of gas is calculated from the initial and final amounts, both saturated with solvent vapor. The amount of solvent is determined by the weight of displaced mercury. The values are at 101.325 kPa (1 atm) pressure assuming Henry's law is obeyed.	(1) Methane. Generated from magnesium methyl iodide. Purified by fractional distillation. Specific gravity corresponds with mol wt 16.08. (2) Benzene. Kahlbaum. "Zur molekulargewichsbestimmung." M.p./°C = 5.48.

ESTIMATED ERROR:
$\delta T/K$ = ± 0.05 $\delta x_1/x_1$ = ± 0.015

REFERENCES:
1. Lannung, A.   *J. Am. Chem. Soc.* 1930, *52*, 68.

COMPONENTS:	ORIGINAL MEASUREMENTS:
(1) Methane; CH_4; [74-82-8] (2) Benzene; C_6H_6; [71-43-2]	Hayduk, W.; Buckley, W.D. *Can. J. Chem. Eng.* <u>1971</u>, *49*, 667-671.

VARIABLES:	PREPARED BY:
T/K: 298.15 P/kPa: 101.325	W. Hayduk

EXPERIMENTAL VALUES:

T/K	Ostwald Coefficient[1] L/cm^3 cm^{-3}	Bunsen Coefficient[2] α/cm^3 (STP)$cm^{-3}atm^{-1}$	Mole Fraction 10^4 x_1	$10^{-4}\Delta G°$[3] /J mol^{-1}
298.15	0.565	0.518	20.6	1.533

[1]Original data.

[2]Calculated by compiler.

[3]Calculated by compiler from the following equation:

$$\Delta G°/\text{J mol}^{-1} = -RT \ln x_1$$

AUXILIARY INFORMATION

METHOD/APPARATUS/PROCEDURE:	SOURCE AND PURITY OF MATERIALS:
A volumetric method using a glass apparatus was employed. Degassed solvent contacted the gas while flowing as a thin film, at a constant rate, through an absorption spiral into a solution buret. A constant solvent flow was obtained by means of a calibrated syringe pump. The solution at the end of the spiral was considered saturated. Dry gas was maintained at atmospheric pressure in a gas buret by mechanically raising the mercury level in the buret at an adjustable rate. The solubility was calculated from the constant slope of volume of gas dissolved and volume of solvent injected.	1. Matheson Co. Specified as ultra high purity grade of 99.97 per cent. 2. Canlab. Chromatoquality grade of specified minimum purity of 99.0 per cent.

METHOD/APPARATUS/PROCEDURE (cont.)	ESTIMATED ERROR:
Degassing was accomplished using a two stage vacuum process described by Clever et al. (1).	δT/K = 0.1 $\delta x_1/x_1$ = 0.01

REFERENCES:

1.   Clever, H.L.; Battino, R.; Saylor, J.H.; Gross, P.M.
J. Phys. Chem. <u>1971</u>, *61*, 1078.

COMPONENTS:	ORIGINAL MEASUREMENTS:
(1) Methane; CH_4; [74-82-8] (2) Methylbenzene or toluene; C_7H_8; [108-88-3]	McDaniel, A. S. *J. Phys. Chem.* <u>1911</u>, *15* , 587-610.

VARIABLES:	PREPARED BY:
T /K = 298.15 - 333.15 p_1 /kPa = 101.3 (1 atm)	H. L. Clever

EXPERIMENTAL VALUES:

Temperature		Mol Fraction	Bunsen Coefficient[a]	Ostwald Coefficient[b]
t/°C	T/K	10^3 x_1	α	L/cm^3 cm^{-3}
25.0	298.15	2.11	0.4450	0.4852[c]
30.0	303.15	2.06	0.4300	0.4778
40.1	313.25	1.97	0.4080	0.4675
50.2	323.35	1.88	0.4013	0.4545
60.0	333.15	1.83	0.3690	0.4502

[a] Bunsen coefficient, α/cm^3 (STP) cm^{-3} atm^{-1}.

[b] Listed as absorption coefficient in the original paper.
Interpreted to be equivalent to Ostwald coefficient by compiler.

[c] Ostwald coefficient (absorption coefficient) estimated as
298.15 K value by author.

[d] Mole fraction and Bunsen coefficient values calculated by
compiler assuming ideal gas behavior.

EVALUATOR'S COMMENT: McDaniel's data should be used with caution.
His values are often 20 percent or more too small when compared
with more reliable data.

AUXILIARY INFORMATION

METHOD/APPARATUS/PROCEDURE:	SOURCE AND PURITY OF MATERIALS:
The apparatus is all glass. It consists of a gas buret connected to a contacting vessel. The solvent is degassed by boiling under reduced pressure. Gas pressure or volume is adjusted using mercury displacement. Equilibration is achieved at atm pressure by hand shaking, and incrementally adding gas to the contacting chamber. Solubility measured by obtaining total uptake of gas by known volume of the solvent.	(1) Methane. Prepared by reaction of methyl iodide with zinc-copper. Passed through water and sulfuric acid. (2) Methylbenzene.
	ESTIMATED ERROR: $\delta L/L \geq -0.20$
	REFERENCES:

COMPONENTS:	ORIGINAL MEASUREMENTS:
(1) Methane; CH_4; [74-82-8] (2) Methylbenzene or toluene; C_7H_8; [108-88-3]	Field, L. R.; Wilhelm, E.; Battino, R. *J. Chem. Thermodyn.* 1974, 6, 237 - 243.

VARIABLES:	PREPARED BY:
T/K: 284.28 - 313.17 P/kPa: 101.325 (1 atm)	H. L. Clever

EXPERIMENTAL VALUES:

T/K	Mol Fraction $10^3 x_1$	Bunsen Coefficient $\alpha/cm^3 (STP) cm^{-3} atm^{-1}$	Ostwald Coefficient $L/cm^3 cm^{-3}$
284.28	2.656	0.567	0.5899
297.80	2.240	0.507	0.5534
313.17	2.240	0.464	0.5315

The gas solubility values were adjusted to an oxygen partial pressure of 101.325 kPa (1 atm) by Henry's law.

The Bunsen coefficients were calculated by the compiler.

Smoothed Data: For use between 284.15 and 313.17 K.

$$\ln x_1 = -7.7969 + 5.2214/(T/100 \text{ K})$$

The standard error about the regression line is 1.66×10^{-4}.

T/K	Mol Fraction $10^3 x_1$
283.15	2.60
293.15	2.44
298.15	2.37
303.15	2.30
313.15	2.18

AUXILIARY INFORMATION

METHOD/APPARATUS/PROCEDURE:	SOURCE AND PURITY OF MATERIALS:
The solubility apparatus is based on the design of Morrison and Billett (1) and the version used is described by Battino, Evans, and Danforth (2). The degassing apparatus is that described by Battino, Banzhof, Bogan, and Wilhelm (3). Degassing. Up to 500 cm^3 of solvent is placed in a flask of such size that the liquid is about 4 cm deep. The liquid is rapidly stirred, and vacuum is applied intermittently through a liquid N_2 trap until the permanent gas residual pressure drops to 5 microns. Solubility Determination. The degassed solvent is passed in a thin film down a glass spiral tube containing solute gas plus the solvent vapor at a total pressure of one atm. The volume of gas absorbed is found by difference between the initial and final volumes in the buret system. The solvent is collected in a tared flask and weighed.	(1) Methane. Either Matheson Co., Inc. or Air Products and Chemicals, Inc. Purest grade available, minimum purity greater than 99 mole per cent. (2) Methylbenzene. Phillips Petroleum. Pure Grade. Distilled.

	ESTIMATED ERROR:
	$\delta T/K = 0.03$ $\delta P/mmHg = 0.5$ $\delta x_1/x_1 = 0.005$

REFERENCES:
1. Morrison, T. J.; Billett, F. *J. Chem. Soc.* 1948, 2033.
2. Battino, R.; Evans, F. D.; Danforth, W. F. *J. Am. Oil Chem. Soc.* 1968, 45, 830.
3. Battino, R.; Banzhof, M.; Bogan, M.; Wilhelm, E. *Anal. Chem.* 1971, 43, 806.

COMPONENTS:	ORIGINAL MEASUREMENTS:
(1) Methane; CH_4; [74-82-8] (2) 1,2-Dimethylbenzene or o-xylene; C_8H_{10}; [95-47-6]	Byrne, J. E.; Battino, R.; Wilhelm, E. J. Chem. Thermodyn. <u>1975</u>, 7, 515-522.

VARIABLES:	PREPARED BY:
T/K: 283.24 - 313.17 p_1/kPa: 101.325 (1 atm)	H. L. Clever A. L. Cramer

EXPERIMENTAL VALUES:

T/K	Mol Fraction $10^3 x_1$	Bunsen Coefficient α/cm^3 (STP) cm^{-3} atm^{-1}	Ostwald Coefficient L/cm^3 cm^{-3}
283.24	2.667	0.5016	0.5201
283.40	2.651	0.4985	0.5172
298.16	2.507	0.4647	0.5073
313.08	2.396	0.4378	0.5018
313.17	2.390	0.4367	0.5007

The Bunsen coefficients were calculated by the compiler. The solubility values were adjusted to a methane partial pressure of 101.325 kPa (1 atm) by Henry's law.

Smoothed Data: For use between 283.15 and 313.17 K.

$$\ln x_1 = -7.0387 + 3.1398/(T/100 \text{ K})$$

The standard error about the regression line is 7.91 x 10^{-6}.

T/K	Mol Fraction $10^3 x_1$
283.15	2.659
293.15	2.560
298.15	2.515
303.15	2.471
313.15	2.391

AUXILIARY INFORMATION

METHOD/APPARATUS/PROCEDURE:
The solubility apparatus is based on the design of Morrison and Billett (1) and the version used is described by Battino, Evans, and Danforth (2). The degassing apparatus is that described by Battino, Banzhof, Bogan, and Wilhelm (3).

Degassing. Up to 500 cm^3 of solvent is placed in a flask of such size that the liquid is about 4 cm deep. The liquid is rapidly stirred, and vacuum is intermittently applied through a liquid N_2 trap until the permanent gas residual pressure drops to 5 microns.

Solubility Determination. The de-gassed solvent is passed in a thin film down a glass helical tube con-taining solute gas plus the solvent vapor at a total pressure of one atm. The volume of gas absorbed is found by difference between the initial and final volumes in the buret system. The solvent is col-lected in a tared flask and weighed.

SOURCE AND PURITY OF MATERIALS:
(1) Methane. Either Air Products & Chemicals, Inc., or Matheson Co., Inc. 99 mol per cent or better.

(2) 1,2-Dimethylbenzene. Phillips Petroleum Co. Pure grade.

ESTIMATED ERROR:
$$\delta T/K = 0.03$$
$$\delta P/\text{mmHg} = 0.5$$
$$\delta x_1/x_1 = 0.005$$

REFERENCES:
1. Morrison, T. J.; Billett, F. J. Chem. Soc. <u>1948</u>, 2033.
2. Battino, R.; Evans, F. D.; Danforth, W. F. J. Am. Oil Chem. Soc. <u>1968</u>, 45, 830.
3. Battino, R.; Banzhof, M.; Bogan, M.; Wilhelm, E. Anal. Chem. <u>1971</u>, 43, 806.

COMPONENTS:	ORIGINAL MEASUREMENTS:
(1) Methane; CH_4; [74-82-8] (2) 1,3-Dimethylbenzene or m-xylene; C_8H_{10}; [108-38-3]	McDaniel, A. S. *J. Phys. Chem.* <u>1911</u>, *15*, 587-610.

VARIABLES:	PREPARED BY:
T/K = 294.25 - 333.15 p_1/kPa = 101.3 (1 atm)	H. L. Clever

EXPERIMENTAL VALUES:

Temperature		Mol Fraction	Bunsen Coefficiant[a]	Ostwald Coefficient[b]
$t/°C$	T/K	$10^3 x_1$	α	L/cm^3 cm^{-3}
21.1	294.25	2.61	0.4778	0.5146
25.0	298.15	2.57	0.4669	0.5096[c]
30.5	303.65	2.50	0.4529	0.5028
50.0	323.15	2.37	0.4203	0.4972
60.0	333.15	2.27	0.3992	0.4870

[a] Bunsen coefficient, α/cm^3(STP) cm^{-3} atm^{-1}.

[b] Listed as absorption coefficient in the original paper. Interpreted to be equivalent to Ostwald coefficient by compiler.

[c] Ostwald coefficient (absorption coefficient) estimated as 298.15 K value by author.

[d] Mole fraction and Bunsen coefficient values calculated by compiler assuming ideal gas behavior.

EVALUATOR'S COMMENT: McDaniel's data should be used with caution. His values are often 20 percent or more too small when compared with more reliable data.

AUXILIARY INFORMATION

METHOD/APPARATUS/PROCEDURE:	SOURCE AND PURITY OF MATERIALS:
The apparatus is all glass. It consists of a gas buret connected to a contacting vessel. The solvent is degassed by boiling under reduced pressure. Gas pressure or volume is adjusted using mercury displacement. Equilibration is achieved at atm pressure by hand shaking, and incrementally adding gas to the contacting chamber. Solubility measured by obtaining total uptake of gas by known volume of the solvent.	(1) Methane. Prepared by reaction of methyl iodide with zinc-copper. Passed through water and sulfuric acid. (2) 1,3-Dimethylbenzene.

	ESTIMATED ERROR: $\delta L/L \geq -0.20$
	REFERENCES:

COMPONENTS:	ORIGINAL MEASUREMENTS:
(1) Methane; CH_4; [74-82-8] (2) 1,3-Dimethylbenzene or *m*-xylene; C_8H_{10}; [108-38-3]	Byrne, J. E.; Battino, R.; Wilhelm, E. *J. Chem. Thermodyn.* <u>1975</u>, *7*, 515-522.

VARIABLES:	PREPARED BY:
T/K: 283.13 - 313.21 p_1/kPa: 101.325 (1 atm)	H. L. Clever A. L. Cramer

EXPERIMENTAL VALUES:

T/K	Mol Fraction $10^3 x_1$	Bunsen Coefficient α/cm^3(STP)cm^{-3}atm^{-1}	Ostwald Coefficient L/cm^{3}cm^{-3}
283.13	2.885	0.5330	0.5525
283.21	2.891	0.5341	0.5538
298.09	2.709	0.4928	0.5378
298.17	2.713	0.4935	0.5387
313.14	2.556	0.4581	0.5252
313.14	2.584	0.4631	0.5309
313.21	2.569	0.4603	0.5278

The Bunsen coefficients were calculated by the compiler. The solubility values were adjusted to a methane partial pressure of 101.325 kPa (1 atm) by Henry's law.

Smoothed Data: For use between 283.15 and 313.21 K.

$$\ln x_1 = -7.0654 + 3.4476/(T/100 \text{ K})$$

The standard error about the regression line is 9.72 x 10^{-6}.

T/K	Mol Fraction $10^3 x_1$
283.15	2.886
293.15	2.769
298.15	2.715
303.15	2.663
313.15	2.568

AUXILIARY INFORMATION

METHOD/APPARATUS/PROCEDURE:	SOURCE AND PURITY OF MATERIALS:
The solubility apparatus is based on the design of Morrison and Billett (1) and the version used is described by Battino, Evans, and Danforth (2). The degassing apparatus is that described by Battino, Banzhof, Bogan, and Wilhelm (3). Degassing. Up to 500 cm^3 of solvent is placed in a flask of such size that the liquid is about 4 cm deep. The liquid is rapidly stirred, and vacuum is applied intermittently through a liquid N$_2$ trap until the permanent gas residual pressure drops to 5 microns. Solubility Determination. The degassed solvent is passed in a thin film down a glass helical tube containing solute gas plus the solvent vapor at a total pressure of one atm. The volume of gas absorbed is found by difference between the initial and final volumes in the buret system. The solvent is collected in a tared flask and weighed.	(1) Methane. Either Air Products & Chemicals, Inc., or Matheson Co., Inc. 99 mol per cent or better. (2) 1,3-Dimethylbenzene. Phillips Petroleum Co. Pure grade. ESTIMATED ERROR: δT/K = 0.03 δP/mmHg = 0.5 $\delta x_1/x_1$ = 0.005 REFERENCES: 1. Morrison, T. J.; Billett, F. *J. Chem. Soc.* <u>1948</u>, 2033. 2. Battino, R.; Evans, F. D.; Danforth, W. F. *J. Am. Oil Chem. Soc.* <u>1968</u>, *45*, 830. 3. Battino, R.; Banzhof, M.; Bogan, M.; Wilhelm, E. *Anal. Chem.* <u>1971</u>, *43*, 806.

COMPONENTS:	ORIGINAL MEASUREMENTS:
(1) Methane; CH_4; [74-82-8] (2) 1,4-Dimethylbenzene or *p*-xylene; C_8H_{10}; [106-42-3]	Byrne, J. E.; Battino, R.; Wilhelm, E. *J. Chem. Thermodyn.* 1975, *7*, 515-522.

VARIABLES:	PREPARED BY:
T/K: 287.94 - 313.18 p_1/kPa: 101.325 (1 atm)	H. L. Clever

EXPERIMENTAL VALUES:

T/K	Mol Fraction $10^3 x_1$	Bunsen Coefficient $\alpha/cm^3 (STP) cm^{-3} atm^{-1}$	Ostwald Coefficient $L/cm^3 cm^{-3}$
287.94	2.995	0.5490	0.5787
288.18	2.976	0.5454	0.5754
298.12	2.831	0.5134	0.5603
298.14	2.878	0.5219	0.5697
313.14	2.675	0.4778	0.5477
313.18	2.666	0.4761	0.5459

The Bunsen coefficients were calculated by the compiler. The solubility values were adjusted to a methane partial pressure of 101.325 kPa (1 atm) by Henry's law.

Smoothed Data: For use between 287.94 and 313.18 K.

$$\ln x_1 = -7.2073 + 4.0158/(T/100 \text{ K})$$

The standard error about the regression line is 1.82×10^{-5}.

T/K	Mol Fraction $10^3 x_1$
293.15	2.916
298.15	2.850
303.15	2.788
313.15	2.672

AUXILIARY INFORMATION

METHOD/APPARATUS/PROCEDURE:	SOURCE AND PURITY OF MATERIALS:
The solubility apparatus is based on the design of Morrison and Billett (1) and the version used is described by Battino, Evans, and Danforth (2). The degassing apparatus is that described by Battino, Banzhof, Bogan, and Wilhelm (3).	(1) Methane. Either Air Products & Chemicals, Inc., or Matheson Co., Inc. 99 mol per cent or better. (2) 1,4-Dimethylbenzene. Phillips Petroleum Co. Pure grade. Used as received.

Degassing. Up to 500 cm^3 of solvent is placed in a flask of such size that the liquid is about 4 cm deep. The liquid is rapidly stirred, and vacuum is intermittently applied through a liquid N_2 trap until the permanent gas residual pressure drops to 5 microns.

Solubility Determination. The degassed solvent is passed in a thin film down a glass helical tube containing solute gas plus the solvent vapor at a total pressure of one atm. The volume of gas absorbed is found by difference between the initial and final volumes in the buret system. The solvent is collected in a tared flask and weighed.

ESTIMATED ERROR:
$$\delta T/K = 0.01$$
$$\delta P/mmHg = 0.5$$
$$\delta x_1/x_1 = 0.005$$

REFERENCES:

1. Morrison, T. J.; Billett, F. *J. Chem. Soc.* 1948, 2033.
2. Battino, R.; Evans, F. D.; Danforth, W. F. *J. Am. Oil Chem. Soc.* 1968, *45*, 830.
3. Battino, R.; Banzhof, M.; Bogan, M.; Wilhelm, E. *Anal. Chem.* 1971, *43*, 806.

COMPONENTS:	ORIGINAL MEASUREMENTS:
1. Methane; CH_4; [74-82-8] 2. 1-Methylnaphthalene; $C_{11}H_{10}$; [1321-94-4]	Chappelow, C.C.; Prausnitz, J.M. *Am. Inst. Chem. Engnrs. J.* <u>1974</u>, *20*, 1097-1104.

VARIABLES:
 	
  Temperature
	PREPARED BY:
	
	  C.L. Young

EXPERIMENTAL VALUES:

T/K	Henry's Constant[a] /atm	Mole fraction[b] of methane at 1 atm. partial pressure, x_{CH_4}
300	644	0.00155
325	676	0.00148
350	705	0.00142
375	728	0.00137
400	740	0.00135
425	738	0.00136
450	719	0.00139
475	680	0.00147

a. Authors stated measurements were made at several pressures and values of solubility used were all within the Henry's Law region.

b. Calculated by compiler assuming linear relationship between mole fraction and pressure.

AUXILIARY INFORMATION

METHOD/APPARATUS/PROCEDURE:	SOURCE AND PURITY OF MATERIALS:
Volumetric apparatus similar to that described by Dymond and Hildebrand (1). Pressure measured with a null detector and precision gauge. Details in ref. (2).	Solvent degassed, no other details given.

ESTIMATED ERROR:

$$\delta T/K = \pm 0.1; \quad \delta x_{CH_4} = \pm 1\%$$

REFERENCES:
1. Dumond, J.; Hildebrand, J.H. *Ind.Eng.Chem.Fundam.* <u>1967</u>,*6*,130.
2. Cukor, P.M.; Prausnitz, J.M. *Ind.Eng.Chem.Fundam.* <u>1971</u>,*10*,638.

COMPONENTS:	ORIGINAL MEASUREMENTS:
1. Methane; CH_4; [74-82-8] 2. 1,1´-Methylenebisbenzene, (Diphenylmethane); $C_{13}H_{12}$; [101-81-5]	Cukor, P.M.; Prausnitz, J.M. *J. Phys. Chem.* 1972, *76*, 598-601
VARIABLES: Temperature	PREPARED BY: C.L. Young

EXPERIMENTAL VALUES:

T/K	Henry's Constant[a] /atm	Mole fraction of methane[b] in liquid, x_{CH_4}
300	555	0.00180
325	579	0.00173
350	597	0.00168
375	608	0.00164
400	611	0.00164
425	608	0.00164
450	597	0.00168
475	579	0.00173

a.   Quoted in supplementary material for original paper

b.   Calculated by compiler for a partial pressure of 1 atmosphere

---

AUXILIARY INFORMATION

METHOD/APPARATUS/PROCEDURE:	SOURCE AND PURITY OF MATERIALS:
Volumetric apparatus similar to that described by Dymond and Hildebrand (1). Pressure measured with a null detector and precision gauge. Details in ref. (2).	No details given
	ESTIMATED ERROR: $\delta T/K = \pm0.05$; $\delta x_{CH_4} = \pm2\%$
	REFERENCES: 1. Dymond, J.; Hildebrand, J.H. *Ind. Eng. Chem. Fundam.* 1967, *6*, 130. 2. Cukor, P.M., Prausnitz, J.M. *Ind. Eng. Chem. Fundam.* 1971, *10*, 638.

COMPONENTS:	ORIGINAL MEASUREMENTS:
1. Methane; CH_4; [74-82-8] 2. Decahydronaphthalene,(Decalin); $C_{10}H_{18}$; [91-17-8]	Lenoir, J-Y.; Renault, P.; Renon, H. *J. Chem. Eng. Data*, <u>1971</u>, *16*, 340-2.

VARIABLES:	PREPARED BY:
Temperature	C. L. Young

EXPERIMENTAL VALUES:

T/K	Henry's constant H_{CH_4}/atm	Mole fraction at 1 atm* x_{CH_4}
298.2	348	0.00287
323.2	437	0.00229

* Calculated by compiler assuming a linear function of P_{CH_4} vs x_{CH_4}, i.e., x_{CH_4} (1 atm) = $1/H_{CH_4}$.

AUXILIARY INFORMATION

METHOD/APPARATUS/PROCEDURE:	SOURCE AND PURITY OF MATERIALS:
A conventional gas-liquid chromato-graphic unit fitted with a thermal conductivity detector was used. The carrier gas was helium. The value of Henry's law constant was calculated from the retention time. The value applies to very low partial pressures of gas and there may be a substantial difference from that measured at 1 atm. pressure. There is also considerable uncertainty in the value of Henry's constant since surface adsorption was not allowed for although its possible existence was noted.	(1) L'Air Liquide sample, minimum purity 99.9 mole per cent. (2) Touzart and Matignon or Serlabo sample, purity 99 mole per cent.
	ESTIMATED ERROR: $\delta T/K = \pm 0.1$; $\delta H/atm = \pm 6\%$ (estimated by compiler).
	REFERENCES:

COMPONENTS:	EVALUATOR:
1. Methane; CH_4; [74-82-8] 2. Benzene; C_6H_6; [71-43-2]	Colin L. Young Department of Physical Chemistry, University of Melbourne. Parkville, Victoria, 3052 Australia. February 1986.

CRITICAL EVALUATION:

   The system benzene + methane has been fairly studied extensively but there
is still a need for a definitive study. The temperature and pressure ranges
studied to date are as follows: Frolich et al.(1), 298 K up to 11 MPa,
Schoch et al.(2), 311 K up to 36 MPa, Elbishlawi and Spencer (3) 338.7 K up
to 33 MPa, Lin et al.(4) 421 K to 501 K up to 24 MPa, Savvina (5), 313 K to
423 K up to 38 Mpa, Ipatieff and Monroe (6), 373 K to 523 K up to 16 MPa,
Sage et al. (7), 311 K to 377 K up to 21 MPa, Stepanova et al. (8) at 273 to
333 K up to 36 Mpa and Legret, Richon and Renon, 313.2 K up to 37 MPa.
   The data of Frolich et al.(1) were presented in graphical form and are not
of high accuracy. These data are classified as doubtful. It is possible to
compare the data of Schoch et al., Savvina, Stepanova et al., Sage et al.
and Legret et al. at approximately the same temperature (313.2 K). There are
significant disagreements between the measurements (i.e. up to 0.05 for
liquid mole fractions and 0.035 for the vapor mole fractions. Legret et
al.´s data are broadly consistent with the K-values of Savvina. However, the
data of Schoch et al., Stepanova et al. and Sage et al. are all broadly
consistent with each other but deviate from Legret et al.´s data in the
opposite direction to the data of Savvina. The data of Legret et al. show
less scatter than the data of other workers in this temperature region and
are probably the most accurate.
      It is very difficult to establish which sets of data are the more
reliable at high temperatures. The data of Savvina are in disagreement with
the data of Lin et al. particularly as regards the gas phase composition. At
high temperatures the data of Lin et al. are probably the most accurate. The
older data of Ipatieff and Monroe are considerably less precise than the
more recent data of Lin et al. and while detailed comparison is not possible
it appears that the liquid mole fraction data of Ipatieff and Monroe are
slightly too large.
      Therefore, although there is a need for a definitive study of
this system, the data of Lin et al. (4) and of Legret et al. (7) are
classified as tentative at present.

References.

1. Frolich, P. K.; Tauch, E. J.; Hogan, J. J.; Peer, A. A.
     Ind. Eng. Chem., <u>1931</u>, *23*, 548.
2. Schoch, E. P.; Hoffmann, A. E.; Kasperik, A. S.; Lightfoot, J. H.;
     Mayfield, F. D.
     Ind. Eng. Chem., <u>1940</u>, *32*, 788.
3. Elbishlawi, M.; Spencer. J. R.,
     Ind. Eng. Chem., <u>1951</u>, *43*, 1811.
4. Lin, H.-M.; Sebastian, H. M.; Simnick, J. J.; Chao, K.-C.;
     J. Chem. Eng. Data, <u>1979</u>, *24*, 146.
5. Savvina, Ya. D.;
     Tr. Vses. Nauchno-Issled. Inst. Prir. Gazov., <u>1962</u>, *17*, 189.
6. Ipatieff, V. N.; Monroe, G. S.;
     Ind. Eng. Chem., <u>1942</u>, *14*, 166.
7. Sage, B. H.; Webster, D. C.; Lacey, W. N.;
     Ind. Eng. Chem., <u>1936</u>, *28*, 1045.
8. Stepanova, G. S.; Vybornova, Ya.I.; Velikovskii, A. S.;
     Gaz. Delo. Nauchno-Teckniche, Sbornik., <u>1965</u>, *9*, 3.
9. Legret, D.; Richon, D.; Renon, H.;
     Am. Inst. Chem. Engnrs. J., <u>1981</u>, *27*, 203.

COMPONENTS:	ORIGINAL MEASUREMENTS:
1. Methane; CH_4; [74-82-8] 2. Benzene; C_6H_6; [71-43-2]	Frolich, P.K.; Tauch, E.J.; Hogan, J.J.; Peer, A.A. *Ind. Eng. Chem.* 1931, *23*, 548-550

VARIABLES:
 	
    Pressure,
	PREPARED BY:
	
	    C.L. Young

EXPERIMENTAL VALUES:

T/K	P/MPa	Solubility[*]	Mole fraction of methane in liquid,[+] x_{CH_2}
298.15	1.0	5	0.018
	2.0	11	0.039
	3.0	18	0.062
	4.0	26	0.087
	5.0	34	0.111
	6.0	42	0.133
	7.0	51	0.157
	8.0	60	0.180
	9.0	70	0.204
	10.0	80	0.226
	11.0	90	0.248

*   Data taken from graph in original article. Volume of
    gas measured at 101.325 kPa pressure and 298.15 K
    dissolved by unit volume of liquid measured under the
    same conditions.

+   Calculated by compiler.

                        AUXILIARY INFORMATION

METHOD/APPARATUS/PROCEDURE:	SOURCE AND PURITY OF MATERIALS:
Static equilibrium cell. Liquid saturated with gas and after equilibrium established samples removed and analysed by volumetric method. Allowance was made for the vapor pressure of the liquid and the solubility of the gas at atmospheric pressure. Details in source.	Stated that the materials were the highest purity available. Purity 98 to 99 mole per cent.
	ESTIMATED ERROR: $\delta T/K = \pm 0.1$; $\delta x_{CH_4} = \pm 5\%$
	REFERENCES:

COMPONENTS:	ORIGINAL MEASUREMENTS:
1. Methane; CH_4; [74-82-8 2. Benzene; C_6H_6; [71-43-2]	Sage, B. H.; Webster, D. C.; Lacey, W. N. *Ind. Eng. Chem.* <u>1936</u>, *28*, 1045-1047.

VARIABLES:	PREPARED BY:
	C. L. Young

EXPERIMENTAL VALUES:

T/K (T/°F)	p/psi	P/MPa [†]	Mass fraction of methane	Mole fraction [†] of methane, x_{CH_4}
310.9 (100)	1448	9.983	0.0431	0.1797
	2390	16.48	0.0757	0.2849
344.3 (160)	1456	10.04	0.0431	0.1797
	2354	16.23	0.0757	0.2849
377.6 (220)	1432	9.873	0.0431	0.1797
	2310	15.93	0.0757	0.2849

[†] calculated by compiler.

AUXILIARY INFORMATION

METHOD/APPARATUS/PROCEDURE:	SOURCE AND PURITY OF MATERIALS:
PVT cell charged with mixture of known composition. Pressure measured with pressure balance. Bubble point determined from the discontinuity in the pressure, volume isotherm. Details of apparatus in ref. (1).	1. Prepared from natural gas, treated for removal of higher alkanes, carbon dioxide and water vapor. Final purity 99.9 mole per cent. 2. Thiophene-free sample.

ESTIMATED ERROR:

$\delta T/K = \pm 0.1$; $\delta P/MPa = \pm 0.02$;

$\delta x_{CH} = \pm 0.002$ (estimated by compiler).

REFERENCES:

1. Sage, B. H.; Lacey, W. N.
   Ind. Eng. Chem.
   <u>1934</u>, *26*, 103.

COMPONENTS:	ORIGINAL MEASUREMENTS:
1. Methane; CH_4; [74-82-8] 2. Benzene, C_6H_6; [71-43-2]	Schoch, E. P.; Hoffmann, A. E.; Kasperik, A. S.; Lightfoot, J. H.; Mayfield, F. D. *Ind. Eng. Chem.* 1940, *32*, 788-791.

VARIABLES:	PREPARED BY:
Pressure	C. L. Young

EXPERIMENTAL VALUES:

T/K	P/MPa	Mole fraction of methane in liquid, x_{CH_4}
311.08	10.51	0.1946
	13.71	0.2453
	17.33	0.3025
	20.77	0.3550
	21.46	0.3642
	24.24	0.4075
	28.13	0.4685
	31.48	0.5340
	35.08	0.6285
	35.56	0.6671
	35.84	0.7601

AUXILIARY INFORMATION

METHOD/APPARATUS/PROCEDURE:	SOURCE AND PURITY OF MATERIALS:
Rocking equilibrium cell fitted with stirring paddles. Temperature measured with Beckmann thermometer calibrated against standard platinum resistance thermometer. Pressure measured with Bourdon gauge. Details in source. Samples injected into cell using mercury displacement, equilibrium pressure measured. Bubble point determined from change in slope of pressure-volume isotherms. Details in source.	1. Crude sample treated for removal of oxygen, carbon dioxide, water vapor and liquids condensible at 200 K. Distilled. 2. Sample distilled, details in source.

ESTIMATED ERROR:
$\delta T/K = \pm 0.01$; δP/MPa = ± 0.5%; $\delta x_{CH_4} = \pm 0.001$ (estimated by compiler).

REFERENCES:

COMPONENTS:	ORIGINAL MEASUREMENTS:
1. Methane; CH_4; [74-82-8] 2. Benzene; C_6H_6; [71-43-2]	Ipatieff, V. N.; Monroe, G. S. *Ind. Eng. Chem. Anal. Edn.* <u>1942</u>, *14*, 166-171.

VARIABLES:	PREPARED BY:
	C. L. Young

EXPERIMENTAL VALUES:

T/K	T/°C	P/atm	P/MPa	Solubility[a]	Mole fraction of methane x_{CH_4}
373	100	31	3.1	1.3	0.060
		71	7.2	3.4	0.142
		99	10.0	4.8	0.189
		103	10.4	5.1	0.199
398	125	34	3.4	1.35	0.062
		73	7.4	3.4	0.142
		103	10.4	5.0	0.196
		113	11.4	6.0	0.188
423	150	39	4.0	1.4	0.064
		80	8.1	3.65	0.151
		109	11.0	5.7	0.217
		122	12.4	7.0	0.254
448	175	43	4.4	1.55	0.070
		87	8.8	4.0	0.163
		119	12.1	6.7	0.246
		132	13.4	8.2	0.285
473	200	50	5.1	1.75	0.077
		96	9.7	4.6	0.183
		134	13.6	8.5	0.293
		142	14.4	9.6	0.319

[a] g of methane per 100 g benzene.

(cont.)

AUXILIARY INFORMATION

METHOD/APPARATUS/PROCEDURE:	SOURCE AND PURITY OF MATERIALS:
Rotating bomb of 3.5 dm³ capacity. Pressure measured with a Bourdon gauge and temperature measured with thermocouple. Methane in both liquid and gaseous samples determined by stripping out benzene at low temperature and estimating methane volumetrically. Benzene estimated gravimetrically.	1. Carbide and Carbon Chemicals Corp. sample, containing about 96.0 mole per cent methane and 4.0 mole per cent nitrogen. 2. Baker C.P. thiophene-free sample n_D^{20} 1.5012.
	ESTIMATED ERROR:
	$\delta T/K = \pm 0.5$; $\delta x_{CH_4} = \pm 5\%$ (estimated by compiler).
	REFERENCES:

COMPONENTS:	ORIGINAL MEASUREMENTS:
1. Methane; CH_4; [74-82-8]	Ipatieff, V. N.; Monroe, G. S.
	Ind. Eng. Chem. Anal. Edn.
2. Benzene; C_6H_6; [71-43-2]	<u>1942</u>, *14*, 166-171.

EXPERIMENTAL VALUES:

T/K	T/°C	P/atm	P/MPa	Solubility[a]	Mole fraction of methane x_{CH_4}
498	225	58	5.9	2.1	0.093
		104	10.5	5.55	0.213
		153	15.5	13.2	0.391
		153	15.5	11.9	0.367
523	250	71	7.2	2.7	0.116
		117	11.9	7.9	0.278
458	185	91	9.2	4.2	0.170
338	65	91	9.2	4.0	0.163
488	215	101	10.2	5.1	0.199
393	120	101	10.2	5.0	0.196
		101	10.2	4.9	0.193
513	240	111	11.2	6.5	0.240
428	155	111	11.2	5.9	0.223
395	122	111	11.2	5.9	0.223
525	252	121	12.3	9.0	0.305
450	177	121	12.3	6.9	0.252
420	147	121	12.3	6.9	0.252
468	195	131	13.3	8.1	0.283
445	172	131	13.3	8.1	0.283
481	208	141	14.3	9.6	0.319
471	198	141	14.3	9.5	0.316

[a] g of methane per 100 g benzene.

COMPONENTS:	ORIGINAL MEASUREMENTS:
1. Methane; CH₄; [74-82-8]	Elbishlawi, M.; Spencer, J.R.
2. Benzene; C₆H₆; [71-43-2]	*Ind. Eng. Chem.* <u>1951</u>, *43*, 1811-5

VARIABLES:	PREPARED BY:
Pressure	C.L. Young

EXPERIMENTAL VALUES:

T/K	P/MPa	Mole fraction of methane in liquid, x_{CH_4}	in vapor y_{CH_4}
338.71	0.689	0.014	0.925
	1.034	0.022	0.947
	1.379	0.030	0.957
	2.758	0.060	0.977
	4.137	0.090	0.980
	5.516	0.118	0.980
	6.895	0.146	0.977
	10.34	0.213	0.974
	13.79	0.278	0.964
	17.24	0.340	0.963
	20.68	0.400	0.956
	24.13	0.455	0.950
	27.58	0.514	0.935
	28.96	0.538	0.923
	30.34	0.565	0.900
	31.72	0.603	0.865
	33.09	0.695	0.775

AUXILIARY INFORMATION

METHOD/APPARATUS/PROCEDURE:	SOURCE AND PURITY OF MATERIALS:
Equilibrium cell fitted with vapor sampling port. Calibrated mercury injection pump. Details in source. Components charged into cell, pressure raised by injection of mercury. Cell rocked to establish equilibrium. Portions of mercury withdrawn and curve relating change in volume to pressure obtained. Bubble point established from change in slope. Vapor phase sample analysed. Details in source.	1. Phillips Petroleum Co. pure sample purity 99 mole per cent impurities ethane (~0.5 mole per cent) and nitrogen (~0.3 mole per cent) trace of carbon dioxide. 2. Commercial sample purified by distillation.

ESTIMATED ERROR:

$\delta T/K = \pm 0.7$; $\delta P/MPa = \pm 0.01$;
δx_{CH_4}, $\delta y_{CH_4} = \pm 0.001$.
(estimated by compiler).

REFERENCES:

COMPONENTS:	ORIGINAL MEASUREMENTS:
1. Methane; CH_4; [74-82-8] 2. Benzene; C_6H_6; [71-43-2]	Savvina, Ya. D. *Tr. Vses. Nauchno-Issled. Inst.* *Prirodn. Gazov.,* <u>1962</u>, *17/25,* 185-196.
VARIABLES:	PREPARED BY:
Temperature, pressure	C. L. Young

EXPERIMENTAL VALUES:

T/K (t/ C)	P/kgcm^{-3}	P/Mpa	K-value methane	K-value benzene
313.2 (40)	20	1.96	17.94	0.013
	50	4.90	7.80	0.010
	100	9.81	4.38	0.012
	150	14.7	3.18	0.017
	200	19.6	2.59	0.024
	250	24.5	2.21	0.035
	285	27.9	2.02	0.053
	300	29.4	1.95	0.065
	320	31.4	1.83	0.105
	350	34.3	1.61	0.193
	370	36.3	1.37	0.359
	381	37.4	1.04	0.854
333.2 (60)	20	1.96	18.15	0.021
	50	4.90	8.08	0.015
	100	9.81	4.47	0.020
	150	14.7	3.32	0.028
	200	19.2	2.65	0.044
	250	24.5	2.24	0.059
	280	27.5	2.06	0.082
	300	29.4	1.94	0.101
	320	31.4	1.83	0.130
	340	33.3	1.61	0.224
	350	34.3	1.44	0.324
	362	35.5	1.13	0.703

AUXILIARY INFORMATION

METHOD APPARATUS/PROCEDURE:	SOURCE AND PURITY OF MATERIALS:
Values appear to be determined using apparatus described in ref.(1).	No details given
	ESTIMATED ERROR:
	REFERENCES: 1. Savvina, Ya. D.;Velikovskii, A. S.; *Tr. Vses. Nauchno-Issled. Inst.* *Prirodn. Gazov.,* 1962, *17/25,* 163.

COMPONENTS:	EVALUATOR:
1. Methane; CH_4; [74-82-8] 2. Benzene; C_6H_6; [71-43-2]	Savvina, Ya. D. *Tr. Vses. Nauchno-Issled. Inst.* *Prirodn. Gazov.*,<u>1962</u>, *17/25*, 185-196.

CRITICAL EVALUATION:

T/K (t/ C)	P/kgcm^{-3}	P/Mpa	K-value	
			methane	benzene
353.2	20	1.96	18.63	0.032
(80)	50	4.90	8.57	0.024
	100	9.81	4.58	0.030
	150	14.7	3.39	0.042
	200	19.6	2.72	0.057
	230	22.6	2.42	0.082
	250	24.5	2.27	0.087
	270	26.5	2.10	0.115
	300	29.4	1.88	0.147
	320	31.4	1.67	0.202
	340	33.3	1.31	0.473
	343	33.6	1.05	0.862
373.2	20	1.96	19.77	0.054
(100)	50	4.90	8.80	0.036
	100	9.81	4.67	0.040
	150	14.7	3.43	0.054
	200	19.6	2.69	0.074
	230	22.6	2.35	0.100
	250	24.5	2.19	0.116
	270	26.5	2.00	0.147
	300	29.4	1.72	0.214
	315	30.9	1.48	0.345
	323	31.7	1.17	0.658
393.2	20	1.96	20.46	0.082
(120)	50	4.90	8.82	0.052
	100	9.81	4.85	0.059
	150	14.7	3.37	0.072
	200	19.6	2.56	0.099
	230	22.6	2.25	0.135
	250	24.5	2.08	0.161
	270	26.5	1.90	0.200
	290	28.4	1.59	0.320
	300	29.4	1.36	0.480
	305	29.9	1.07	0.850
423.2	20	1.96	21.00	0.123
(150)	50	4.90	8.33	0.084
	100	9.81	4.36	0.092
	150	14.7	3.26	0.114
	200	19.6	2.42	0.145
	230	22.6	2.08	0.199
	250	24.5	1.80	0.286
	270	26.5	1.15	0.757

COMPONENTS:	ORIGINAL MEASUREMENTS:
1. Methane; CH_4; [74-82-8] 2. Benzene; C_6H_6; [71-43-2]	Lin, H-M.; Sebastian, H.M.; Simnick, J.J.; Chao, K-C. *J. Chem. Engng. Data*, 1979, *24*, 146-9.
VARIABLES: Temperature, pressure	PREPARED BY: C. L. Young

EXPERIMENTAL VALUES:

T/K	p/atm	p/MPa	Mole fraction of methane in liquid, x_{CH_4}	in gas, y_{CH_4}
421.1	19.61	1.987	0.0252	0.6516
	33.01	3.345	0.0502	0.7630
	46.69	4.731	0.0823	0.7994
	99.50	10.082	0.1739	0.8208
	148.29	15.025	0.2651	0.8140
	200.68	20.334	0.3754	0.7809
	239.46	24.263	0.4947	0.7222
461.9	29.82	3.022	0.0346	0.5285
	50.50	5.117	0.0762	0.6439
	99.63	10.095	0.1801	0.7120
	136.79	13.860	0.2609	0.7250
	160.87	16.300	0.3179	0.7163
501.2	50.57	5.124	0.0648	0.4485
	72.01	7.296	0.1145	0.5326
	100.24	10.157	0.1806	0.5715
	125.36	12.702	0.2464	0.5564
	134.89	13.668	0.2842	0.5418
	143.67	14.557	0.3243	0.5153

AUXILIARY INFORMATION

METHOD/APPARATUS/PROCEDURE:	SOURCE AND PURITY OF MATERIALS:
Flow apparatus with both liquid and gas components continually passing into a mixing tube and then into a cell in which phases separated under gravity. Liquid sample removed from bottom of cell and vapor sample from top of cell. Composition determined by gas chromatography. Details in source and ref. (1).	1. Matheson sample with purity better than 99 mole per cent. 2. Mallinckrodt Co. sample, Analytical reagent with 0.5°C boiling point range.
	ESTIMATED ERROR: $\delta T/K = \pm 0.2$; δp/MPa $\leqslant \pm 0.03$; δx_{CH_4}, δy_{CH_4} = $\pm 2\%$
	REFERENCES: 1. Simnick, J.J.; Lawson, C.C.; Lin, H-M.; Chao, K-C. *Am. Inst. Chem. Engnrs. J.*, 1977, *23*, 469.

COMPONENTS:	ORIGINAL MEASUREMENTS:
1. Methane; CH_4; [74-82-8] 2. Benzene; C_6H_6; [71-43-2]	Legret, D.; Richon, D.; Renon, H. *J. Chem. Engng. Data* 1982, *27*, 165-169.
VARIABLES:	PREPARED BY: C. L. Young

EXPERIMENTAL VALUES:

$T/K = 313.2$

$10^{-5}p/Pa$	Mole fraction of methane in liquid, x_{CH_4}	in vapor, y_{CH_4}
36.6	0.0986	0.987
101.0	0.210	0.987[a]
	0.212	−
148.7	0.287	0.986
	0.288	−
199.8	0.369	0.984[a]
	0.367	−
250.3	0.442	0.977
	0.439	−
310.4	0.521	0.951
351.1	0.593	0.883
	0.586	−
368.7	0.661	0.836
	−	0.840
374.2	0.688	0.801
	−	0.805

[a] interpolated values.

AUXILIARY INFORMATION

METHOD/APPARATUS/PROCEDURE:	SOURCE AND PURITY OF MATERIALS:
High pressure static cell fitted with magnetic stirrer. Pressure measured with transducer calibrated by comparison with Heise gauges which were checked periodically calibrated against a dead weight tester. Temperature measured with K type iron-constantan thermo-couples. Sampling microcell used and samples analysed using gas chromatography. Details in ref. (1).	1. Air-Gas sample, purity at least 99.95 volume per cent. 2. Merck sample, stated purity by GC of 99.7 per cent.

ESTIMATED ERROR:

$\delta T/K = \pm 0.25$; $\delta p/MPa = \pm 0.1$;

$\delta x_{CH_4} = \pm 0.01$; $\delta y_{CH_4} = \pm 0.005$.

REFERENCES:

1. Legret, D.; Richon, D.; Renon, H.

   Am. Inst. Chem. Eng. J.

   1981, *27*, 203.

COMPONENTS:	EVALUATOR:
1. Methane; CH_4; [74-82-8] 2. Methylbenzene; C_7H_8; [108-88-3]	Colin L. Young Department of Physical Chemistry, University of Melbourne. Parkville, Victoria, 3052 Australia. February 1986.

CRITICAL EVALUATION:

This system has been studied by several workers and data are available for the temperature range 188.7 K to 543.2 K. At the higher temperatures the data of Chao and coworkers (1) are thought to be reliable. These workers have recently made extensive studies of gas solubilities of hydrogen, methane and carbon dioxide in numerous solvents and, in general, their data are reliable. The data of Savvina and Velikovski (2,3) are in significant disagreement with those of Chao and coworkers (1). As pointed out by the latter workers the mole fraction of toluene in the vapour phase in Savvina data at 423.2 K are probably about one order of magnitude too small.

The low temperature range has been studied by Kobayashi and coworkers (4,5). The data of Chang and Kobayashi (4) are thought to be unreliable and are classified as doubtful. More recent work by Kobayashi and coworkers (5) has indicated considerable error in the earlier work particularly at pressures above 10 MPa which has been attributed to sampling and/or analysis errors. The data of Kobayashi and coworkers (5) for the temperature range 188.7 K to 277.6 K are classified as tentative. Legret, Richon and Renon (6) have made careful measurements at 313.2 K on this system. Comparison of their data with that of Savvina (2) and Elbishlawi and Spencer (7) suggests that the data of Elbishlawi and Spencer is inaccurate and the low temperature data of Savvina is, at least, consistent with the data of Legret, Richon and Renon (6).

Therefore the data of Chao and coworkers (1), Legret, Richon and Renon (6) and Kobayashi and coworkers are classified as tentative. The data of Savvina (2) is thought to be of reasonable accuracy at lower temperatures but inaccurate at the higher temperatures. The data of Chang and Kobayashi (4) and Elbishlawi and Spencer (7) are classified as doubtful.

References.

1. Lin, H.-M.; Sebastian, H. M.; Simnick, J. J.; Chao, K.-C.;
   J. Chem. Eng. Data, <u>1979</u>, *24*, 146.
2. Savvina, Ya. D.
   Tr. Vses. Nauchno-Issled. Inst. Prir. Gazov., <u>1962</u>, *17*, 189.
3. Savvina, Ya. D.; Velikovski, A. S.;
   Zh. Fiz. Khim., <u>1956</u>, *30*, 1596.
4. Chang, H. L.; Kobayashi, R.
   J. Chem. Eng. Data, <u>1967</u>, *12*, 517.
5. Lin, Y. N.; Hwang, S. C.; Kobayashi, R.;
   J. Chem. Eng. Data, <u>1978</u>, *23*, 231.
6. Elbishlawi, M.; Spencer. J. R.,
   Ind. Eng. Chem., <u>1951</u>, *43*, 1811.
7. Legret, D.; Richon, D.; Renon, H.;
   Am. Inst. Chem. Engnrs. J., <u>1981</u>, *27*, 203.

COMPONENTS:	ORIGINAL MEASUREMENTS:
1. Methane; CH_4; [74-82-8] 2. Methylbenzene; C_7H_8; [108-88-3]	Elbishlawi, M.; Spencer, J. R.; *Ind. Eng. Chem.*, 1951, *43*, 1811-5.
VARIABLES: Pressure	PREPARED BY: C. L. Young

EXPERIMENTAL VALUES:

T/K	P/MPa	Mole fraction of methane in liquid, x	in vapor, y
338.71	6.89	0.017	0.973
	20.69	0.052	0.987
	34.48	0.085	0.990
	48.28	0.120	0.990
	68.95	0.252	0.987
	103.43	0.325	0.985
	137.90	0.393	0.980
	172.38	0.452	0.976
	206.85	0.505	0.971
	241.3	0.554	0.962
	275.8	0.604	0.945
	310.3	0.664	0.919
	344.8	0.680	0.910
	351.7	0.700	0.895
	365.4	0.729	0.870

AUXILIARY INFORMATION

METHOD/APPARATUS/PROCEDURE:	SOURCE AND PURITY OF MATERIALS:
Equilibrium cell fitted with vapor sampling port. Calibrated mercury injection pump. Details in source. Components charged into cell, pressure raised by injection of mercury. Cell rocked to establish equilibrium. Portions of mercury withdrawn and curve relating change in volume to pressure obtained. Bubble point established from change in slope. Vapor phase sample analysed. Details in source.	1. Phillips Petroleum Co. pure sample, purity 99 mole per cent: impurities ethane (˜0.5 mole per cent) and nitrogen (˜0.3 mole per cent) and a trace of carbon dioxide. 2. Commercial sample purified by distillation.

ESTIMATED ERROR:
$\partial T/K = \pm 0.7$; $\partial P/MPa = \pm 0.01$;
$\partial x/x, \partial y/y = \pm 0.001$
(estimated by compiler).

REFERENCES:

COMPONENTS:	ORIGINAL MEASUREMENTS:
1. Methane; CH_4; [74-82-8] 2. Methylbenzene, (Toluene); C_7H_8; [108-88-3]	Chang, H.O.; Kobayashi, R. *J. Chem. Engng. Data.* <u>1967</u>, *12*, 517-520.

VARIABLES:
 	
     Temperature, pressure
	PREPARED BY:
	
	     C.L. Young

EXPERIMENTAL VALUES:

		Mole fraction of methane	
T/K	P/MPa	in liquid, x_{CH_4}	in vapor, y_{CH_4}
255.4	0.6895	0.0193	0.9979
	1.379	0.0390	0.9988
	2.758	0.0740	0.9991
	4.137	0.1120	0.9991
	5.516	0.1495	0.9991
	6.895	0.1861	0.9990
	8.618	0.2230	0.9990
	10.34	0.2660	0.9989
	12.07	0.3040	0.9986
	13.79	0.3500	0.9982
	17.24	0.4330	0.9973
	20.68	0.4950	0.9959
	24.13	0.5400	0.9937
244.3	0.6895	0.0209	0.9988
	1.379	0.0410	0.9993
	2.758	0.0815	0.9995
	4.137	0.1211	0.9996
	5.516	0.1609	0.9996
	6.895	0.1989	0.9995
	8.618	0.2465	0.9994
	10.34	0.2900	0.9993
	12.07	0.3300	0.9991
	13.79	0.3650	0.9989
	17.24	0.4450	0.9983

AUXILIARY INFORMATION

METHOD/APPARATUS/PROCEDURE:	SOURCE AND PURITY OF MATERIALS:
Recirculating vapor flow apparatus with magnetic vapor pump. Pressure measured with Bourdon gauge and temperature with thermopile. Samples of both phases analysed using gas chromatography with flame ionization detector. Details in ref. (1) and (2).	1. Matheson Co.sample, purity 99.99 mole per cent. 2. Phillips Petroleum sample, purity 99.96 mole per cent.

	ESTIMATED ERROR:
	
	     $\delta T/K = \pm 0.01$; $\delta P/MPa = \pm 0.015$;
	     $\delta(1-x_{CH_4})$, $\delta(1-y_{CH_4}) = \pm 5\%$ or less.

	REFERENCES: 1. Chang, H.L.; *Ph.D. Thesis.* Rice University, Houston, Texas, <u>1966</u>, 2. Chang, H.L.; Hunt, L.J. and Kobayashi, R. *Am. Inst. Chem. Engnrs. J.*<u>1965</u>, *12*, 1212.

COMPONENTS:	ORIGINAL MEASUREMENTS:
1. Methane; CH₄; [74-82-8]	Chang, H.O.; Kobayashi, R.
2. Methylbenzene, (Toluene); C₇H₈; [108-88-3]	*J. Chem. Engng. Data.* 1967, 12, 517-520.

EXPERIMENTAL VALUES:

T/K	P/MPa	Mole fraction of methane in liquid, x_{CH_4}	in vapor, y_{CH_4}
244.3	20.68	0.5100	0.9974
	24.13	0.5650	0.9959
233.2	0.6895	0.0230	0.9994
	1.379	0.0452	0.9997
	2.758	0.0867	0.9998
	4.137	0.1296	0.9998
	5.516	0.1729	0.9998
	6.895	0.2150	0.9998
	8.618	0.2628	0.9997
	10.34	0.3099	0.9996
	12.07	0.3540	0.9995
	13.79	0.3951	0.9994
	17.24	0.4650	0.9990
	20.68	0.5315	0.9984
	24.13	0.5820	0.9974
222.0	0.6895	0.0251	0.99970
	1.379	0.0499	0.99983
	2.758	0.0972	0.99989
	4.137	0.1425	0.99990
	5.516	0.1920	0.99990
	6.895	0.2328	0.99988
	8.618	0.2840	0.99985
	10.34	0.3300	0.99981
	12.07	0.3760	0.99975
	13.79	0.4198	0.99968
	17.24	0.4979	0.99948
	20.68	0.5670	0.99911
	24.13	0.6251	0.99840
210.9	0.6895	0.0287	0.99988
	1.379	0.0551	0.99993
	2.758	0.1061	0.99995
	4.137	0.1605	0.99995
	5.516	0.2051	0.99995
	6.895	0.2510	0.99994
	8.618	0.3054	0.99993
	10.34	0.3535	0.99991
	12.07	0.4015	0.99989
	13.79	0.4500	0.99985
	17.24	0.5300	0.99975
	20.68	0.6012	0.99955
	24.13	0.6619	0.99913
199.8	0.6895	0.0322	0.99996
	1.379	0.0621	0.99997
	2.758	0.1184	0.99998
	4.137	0.1751	0.99998
	5.516	0.2279	0.99998
	6.895	0.2745	0.99998
	8.618	0.3301	0.99997
	10.34	0.3832	0.99996
	12.07	0.4350	0.99995
	13.79	0.4820	0.99994
	17.24	0.5710	0.99989
	20.68	0.6449	0.99979
	24.13	0.7076	0.99955

COMPONENTS:	ORIGINAL MEASUREMENTS:
1. Methane; CH_4; [74-82-8] 2. Methylbenzene; C_7H_8; [108-88-3]	Savvina, Ya. D. *Tr. Vses. Nauchno-Issled. Inst.* *Prirodn. Gazov.*, <u>1962</u>, *17/25*, 185-196.

VARIABLES:	PREPARED BY:
Temperature, pressure	C. L. Young

EXPERIMENTAL VALUES:

T/K (t/°C)	$P/kgcm^{-3}$	P/Mpa	K-value methane	methylbenzene
313.2	20	1.96	18.75	0.006
(40)	50	4.90	8.36	0.005
	100	9.81	4.64	0.006
	150	14.7	3.29	0.011
	200	19.6	2.60	0.014
	250	24.5	2.17	0.027
	300	29.4	1.86	0.058
	350	34.3	1.62	0.108
	380	37.3	1.49	0.182
	410	40.2	1.35	0.289
	430	42.2	1.20	0.485
	435	42.7	1.10	0.685
333.2	20	1.96	19.43	0.009
(60)	50	4.90	8.86	0.008
	100	9.81	4.83	0.010
	150	14.7	4.43	0.014
	200	19.6	2.65	0.024
	250	24.5	2.20	0.041
	300	29.4	1.86	0.069
	350	34.3	1.58	0.136
	370	36.3	1.48	0.193
	390	38.2	1.39	0.268
	410	40.2	1.25	0.418
	419	41.1	1.09	0.706
353.2	20	1.96	20.54	0.014
(80)	50	4.90	9.43	0.011

AUXILIARY INFORMATION

METHOD/APPARATUS/PROCEDURE:	SOURCE AND PURITY OF MATERIALS:
Values appear to be determined using apparatus described in ref.(1)	No details given.
	ESTIMATED ERROR:
	REFERENCES: 1.Savvina, Ya. D.;Velikovskii, A. S. *Tr.Vses. Nauchno-Issled. Inst.* *Prirodn. Gazov.*, 1962, *17/25*, 163.

COMPONENTS:	ORIGINAL MEASUREMENTS:
1. Methane; CH_4; [74-82-8] 2. Methylbenzene; C_7H_8; [108-88-3]	Savvina, Ya. D. *Tr. Vses. Nauchno-Issled. Inst.* *Prirodn. Gazov.*,1962, *17/25*, 185-196.

EXPERIMENTAL VALUES:

T/K (t/°C)	P/kgcm^{-3}	P/Mpa	K-value methane	methylbenzene
353.2 (80)	100	9.81	5.02	0.014
	150	14.7	3.46	0.023
	200	19.6	2.59	0.035
	250	24.5	2.10	0.055
	300	29.4	1.80	0.089
	350	34.3	1.54	0.162
	380	37.3	1.37	0.289
	390	38.2	1.31	0.349
	400	39.2	1.20	0.500
	404	39.6	1.12	0.659
373.2 (100)	20	1.96	21.75	0.021
	50	4.90	9.75	0.016
	100	9.81	4.96	0.021
	150	14.7	3.32	0.031
	200	19.6	2.50	0.045
	250	24.5	2.06	0.069
	300	29.4	1.75	0.109
	330	32.4	1.58	0.165
	360	35.3	1.41	0.258
	380	37.3	1.24	0.447
	386	37.9	1.12	0.642
393.2 (120)	20	1.96	24.25	0.031
	50	4.90	9.98	0.024
	100	9.81	4.87	0.031
	150	14.7	3.16	0.041
	200	19.6	2.44	0.062
	250	24.5	2.02	0.081
	300	29.4	1.69	0.147
	310	30.4	1.63	0.172
	330	32.4	1.52	0.209
	355	34.8	1.35	0.325
	370	36.3	1.16	0.577
423.2 (150)	20	1.96	19.24	0.041
	50	4.90	9.49	0.035
	100	9.81	4.56	0.041
	150	14.7	3.09	0.055
	200	19.6	2.40	0.073
	250	24.5	1.96	0.105
	300	29.4	1.64	0.178
	330	32.4	1.44	0.290
	345	33.8	1.25	0.472
	350	34.3	1.13	0.676

COMPONENTS:	ORIGINAL MEASUREMENTS:
1. Methane; CH_4; [74-82-8] 2. Methylbenzene; C_7H_8; [108-88-3]	Lin, Y.-N.; Hwang, S.-C.; Kobayashi, R. *J. Chem. Engng. Data* *1978*, *23*, 231-4. (Same data in *Gas Proc. Assoc. Proc.* *57*, *Ann. Conv.*, p.12-17.)

VARIABLES:
 	
        Temperature, pressure
	PREPARED BY:
	
	        C. L. Young

EXPERIMENTAL VALUES:

T/K	T/°F	p/psi	P/MPa	Mole fraction of methane in liquid, x_{CH_4}	in vapor, y_{CH_4}
277.59	40.00	50.0	0.345	0.009669	0.99613
		100.0	0.689	0.01924	0.99785
		200.0	1.379	0.03647	0.99687
		400.0	2.758	0.06838	0.999105
		600.0	4.137	0.1004	0.999199
		800.0	5.516	0.1369	0.999045
		1000	6.895	0.1585	0.99875
		1250	8.618	0.1980	0.99830
		1500	10.34	0.2307	0.99773
		1750	12.07	0.2691	0.99692
		2000	13.79	0.2881	0.99581
		2500	17.24	0.3352	0.99266
		3000	20.68	0.3733	-
		3500	24.13	0.4015	-
		4000	27.58	0.4381	-
		4500	31.03	0.4736	-
		5000	34.47	0.5130	-
		5500	37.92	0.5530	-
		6000	41.37	0.5903	-
		7070†	48.75	0.8259	-
255.37	0.00	50.0	0.345	0.01158	0.999086
		100.0	0.689	0.02084	0.999565

            † critical pressure   (cont.)

AUXILIARY INFORMATION

METHOD/APPARATUS/PROCEDURE:	SOURCE AND PURITY OF MATERIALS:

Liquid phase compositions deter-
mined using a recirculating vapor
flow apparatus fitted with magnetic
pump as described in ref. (1).
Equilibrium cell fitted with glass
windows. Pressure measured with
Bourdon pressure gauges and
temperature measured with a
platinum resistance thermometer.
Liquid samples analysed by GC.
Gas phase concentration determined
using elution technique as given
in ref. (2).
	1. Matheson ultra-high purity sample,
	   at least 99.97 mole per cent
	   methane.
	
	2. Phillips Petroleum Company
	   research grade sample, purity
	   99.94 mole per cent.

	ESTIMATED ERROR: $\delta T/K = \pm0.01$; δp/psi = ±1% of full scale for gauges of range 0-1000, 0-3000, 0-6000 and 1-10000; $\delta x = \pm2\%$ or 0.005; $\delta y = \pm2\%$ or 0.00005 which- ever is the largest.

	REFERENCES:
	1. Mraw, S. C.; Hwang; S.-C.;
	   Kobayashi, R.
	   J. Chem. Engng. Data <u>*1978*</u>, *23*, 135.
	
	2. Hwang, S.-C.; Lin, H.-M.;
	   Chappelear, P. S.; Kobayashi, R.
	   J. Chem. Engng. Data <u>*1976*</u>, *21*, 493.

COMPONENTS:

1. Methane; CH₄; [74-82-8]

2. Methylbenzene; C₇H₈;
   [108-88-3]

ORIGINAL MEASUREMENTS:

Lin, Y.-N.; Hwang, S.-C.;
Kobayashi, R.
J. Chem. Engng. Data
1978, 23, 231-4.
(Same data in Gas Proc. Assoc. Proc.
57, Ann. Conv., p.12-17.)

EXPERIMENTAL VALUES:

T/K	T/°F	p/psi	P/MPa	Mole fraction of methane in liquid, x_{CH_4}	in vapor, y_{CH_4}
255.37	0.00	200.0	1.379	0.04147	0.999710
		400.0	2.758	0.08100	0.999819
		600.0	4.137	0.1225	0.999823
		800.0	5.516	0.1622	0.999767
		1000	6.895	0.1984	0.999589
		1250	8.618	0.2445	0.999250
		1500	10.34	0.2763	0.99877
		1750	12.07	0.3196	0.99792
		2000	13.79	0.3414	0.99667
		2500	17.24	0.3782	0.99351
		3000	20.68	0.4106	-
		3500	24.13	0.4352	-
		4000	27.58	0.4532	-
		4500	31.03	0.4673	-
		5000	34.47	0.4763	-
		5500	37.92	0.4960	-
		6000	41.37	0.5204	-
233.15	-40.00	100.0	0.689	0.02531	
		200.0	1.379	0.04471	
		400.0	2.758	0.08924	
		600.0	4.137	0.1326	
		800.0	5.516	0.1833	
		1000	6.895	0.1958	
		1250	8.618	0.2300	
		1500	10.34	0.2542	
		1750	12.07	0.2744	
		2000	13.79	0.3003	
		2500	17.24	0.3184	
		3000	20.68	0.3495	
		3500	24.13	0.3613	
		4000	27.58	0.3789	
		4500	31.03	0.4033	
		5000	34.47	0.4170	
		5500	37.92	0.4417	
		6000	41.37	0.4474	
188.71	-120.00	100.0	0.689	0.04179	
		200.0	1.379	0.08919	
		400.0	2.758	0.1861	
		600.0	4.137	0.2595	
		630.0	4.344	0.2652	
		630.0	4.344	0.9898*	
		1000	6.895	0.2541	
		1000	6.895	0.9880*	
		3000	20.68	0.2459	
		3000	20.68	0.9869*	

† critical pressure

* methane-rich liquid phase

COMPONENTS:	ORIGINAL MEASUREMENTS:
1. Methane; CH_4; [74-82-8] 2. Methylbenzene; C_7H_8; [108-88-3]	Lin, H-M.; Sebastian, H.M.; Simnick, J.J.; Chao, K-C. *J. Chem. Engng. Data*, <u>1979</u>, *24*, 146-9.

VARIABLES:	PREPARED BY:
Temperature, pressure	C. L. Young

EXPERIMENTAL VALUES:

T/K	p/atm	p/MPa	Mole fraction of methane in liquid, x_{CH_4}	in gas, y_{CH_4}
422.5	19.95	2.021	0.0353	0.8426
	29.88	3.028	0.0545	0.8808
	50.77	5.144	0.0954	0.9100
	99.08	10.039	0.1949	0.9231
	150.66	15.266	0.2879	0.9148
	200.00	20.265	0.3858	0.8981
	246.95	25.022	0.4897	0.8493
462.1	20.03	2.030	0.0280	0.6724
	30.10	3.050	0.0486	0.7499
	49.70	5.036	0.0884	0.8144
	98.83	10.014	0.1897	0.8606
	150.00	15.199	0.2850	0.8593
	199.61	20.225	0.4106	0.8257
	227.37	23.038	0.4925	0.7848
	249.40	25.270	0.6332	0.6780

AUXILIARY INFORMATION

METHOD/APPARATUS/PROCEDURE:	SOURCE AND PURITY OF MATERIALS:
Flow apparatus with both liquid and gas components continually passing into a mixing tube and then into a cell in which phases separated under gravity. Liquid sample removed from bottom of cell and vapor sample from top of cell. Composition determined by gas chromatography. Details in source and ref. (1).	1. Matheson sample with purity better than 99 mole per cent. 2. Mallinckrodt Co. sample. Analytical reagent with $1.0^{0}C$ boiling point range.

ESTIMATED ERROR:
$\delta T/K = \pm 0.2$; δp/MPa $\leqslant \pm 0.03$; δx_{CH_4}, $\delta y_{CH_4} = \pm 2\%$

REFERENCES:
1. Simnick, J.J.; Lawson, C.C.; Lin, H-M.; Chao, K-C.; *Am. Inst. Chem. Engnrs. J.*, <u>1977</u>, *23*, 469.

COMPONENTS:	ORIGINAL MEASUREMENTS:
1. Methane; CH_4; [74-82-8] 2. Methylbenzene; C_7H_8; [108-88-3]	Lin, H.-M.; Sebastian, H. M.; Simnick, J. J.; Chao, K.-C. *J. Chem. Engng. Data* <u>1979</u>, *24*, 146-9.

T/K	p/atm	p/MPa	Mole fraction of methane in liquid, x_{CH_4}	in gas, y_{CH_4}
500.8	19.90	2.016	0.0179	0.3668
	29.56	2.995	0.0379	0.5283
	49.91	5.057	0.0841	0.6712
	99.79	10.111	0.1964	0.7479
	147.28	14.923	0.3098	0.7439
	166.47	16.868	0.3807	0.7120
543.2	30.37	3.077	0.0219	0.2476
	49.90	5.056	0.0718	0.4222
	69.57	7.049	0.1246	0.5039
	99.44	10.076	0.2155	0.5493
	113.60	11.511	0.2736	0.5416

COMPONENTS:	ORIGINAL MEASUREMENTS:
1. Methane; CH_4; [74-82-8] 2. Methylbenzene (toluene); C_7H_8; [108-88-3]	Legret, D.; Richon, D.; Renon, H. J. Chem. Engng. Data 1982, 27, 165-169.

VARIABLES:	PREPARED BY:
	C. L. Young

EXPERIMENTAL VALUES:

$T/K = 313.2$

$10^{-5}p/Pa$	Mole fraction of methane in liquid, x_{CH_4}	in vapor, y_{CH_4}
101.0	0.237	0.991
	0.234	-
152.1	0.316	0.991[a]
	0.321	-
196.4	0.383	0.990
250.3	0.447	0.983[a]
300.1	0.524	0.971
	0.520	-
349.2	0.592	0.954
380.0	-	0.919
387.5	0.653	0.911
	0.652	-
399.5	-	0.894
405.9	-	0.875
408.0	0.679	-
409.3	-	0.864
414.9	0.701	0.837
416.9	-	0.835
420.1	0.725	0.818
	-	0.822
424.2	0.733	0.789
424.5	0.744	0.784

[a] interpolated values.

AUXILIARY INFORMATION

METHOD/APPARATUS/PROCEDURE:	SOURCE AND PURITY OF MATERIALS:
High pressure static cell fitted with magnetic stirrer. Pressure measured with transducer calibrated by comparison with Heise gauges which were checked periodically calibrated against a dead weight tester. Temperature measured with K type iron-constantan thermocouples. Sampling microcell used and samples analysed using gas chromatography. Details in ref. (1).	1. Air-Gas sample, purity at least 99.95 volume per cent. 2. Merck sample, stated purity by GC of 99.5 per cent.

ESTIMATED ERROR:

$\delta T/K = \pm 0.25$; $\delta p/MPa = \pm 0.1$;

$\delta x_{CH_4} = \pm 0.01$; $\delta y_{CH_4} = \pm 0.005$.

REFERENCES:

1. Legret, D.; Richon, D.; Renon, H.

   Am. Inst. Chem. Eng. J.

   1981, 27, 203.

COMPONENTS:	EVALUATOR:
1. Methane; CH_4; [74-82-8]	C. L. Young
2. 1,3-Dimethylbenzene; C_8H_{10}; [108-38-3]	Department of Physical Chemistry, Univ. of Melbourne, Parkville, Victoria, 3052 Australia. Jan. 86

CRITICAL EVALUATION:

This system has been investigated by four groups (1-4). The data of Stepanova and Velikovskii (1) at 293.15 K and 333.15 K were not available to us but have been presented graphically by Legret et al.(2). The data from ref. (1) and (2) are in moderate agreement, the differences above 30 MPa may be due to the different temperature of the measurements. The data of Ng et al. (3) are in fair agreement in the limited range of overlap of the temperature and pressure range. The data of Simnick et al. (4) at high temperatures is consistent with that of Ng et al. (3) but it is impossible to make a very detailed comparison because of the limited overlap of the two sets of data. For a given partial pressure the mole fraction solubility of methane in the liquid is rather insensitive to the temperature.

References.

1. Stepanova, G. S.; Velikovskii, A. S.; *Gazov. Delo.* 1969, *12*, 10.
2. Legret, D.; Richon, D.; Renon, H.; *J. Chem. Eng. Data,* 1982, *27*, 165.
3. Ng, H.-J.; Huang, S. S.-S.; Robinson, D. B.; *J. Chem. Eng. Data* 1982, *27*, 119.
4. Simnick, J. J.; Sebastian, H. M.; Lin, H. M.; Chao, K. C.; *Fluid Phase Equil.* 1979, *3*, 145.

COMPONENTS:	ORIGINAL MEASUREMENTS:
1. Methane; CH_4; [74-82-8] 2. 1,3-Dimethylbenzene; C_8H_{10}; [108-38-3]	Simnick, J. J.; Sebastian, H. M.; Lin, H. M.; Chao, K. C. *Fluid Phase Equilibria*, <u>1979</u>, *3*, 145-154.

VARIABLES:	PREPARED BY:
Temperature, pressure	C. L. Young

EXPERIMENTAL VALUES:

T/K	P/MPa	P/atm	Mole fraction of methane in liquid, x_{CH_4}	in gas, y_{CH_4}
460.75	2.07	20.4	0.0386	0.7970
	2.94	29.0	0.0568	0.8478
	5.05	49.8	0.1019	0.8931
	10.04	99.1	0.2037	0.9144
	15.00	148.0	0.2910	0.9141
	20.19	199.3	0.3935	0.8960
501.55	2.02	19.9	0.0317	0.6094
	3.08	30.4	0.0542	0.7107
	5.08	50.1	0.0999	0.7959
	10.09	99.6	0.2088	0.8480
	15.17	149.7	0.3199	0.8444
	20.19	199.3	0.4553	0.8013
541.85	2.12	20.9	0.0219	0.3225
	3.06	30.2	0.0449	0.4730
	5.11	50.4	0.0953	0.6260
	10.08	99.5	0.2195	0.7216
	15.37	151.7	0.3796	0.6894
	16.18	159.7	0.4140	0.6707
582.35	3.05	30.1	0.0238	0.1845
	5.10	50.3	0.0859	0.3700
	10.03	99.1	0.2635	0.4824
	11.46	113.1	0.4346	0.4782

AUXILIARY INFORMATION

METHOD/APPARATUS/PROCEDURE:	SOURCE AND PURITY OF MATERIALS:
Flow apparatus with both liquid and gaseous components continually passing into a mixing tube and then into a cell in which phases separated under gravity. Liquid sample removed from bottom of cell and vapor sample from top of cell. Composition of samples found by stripping out gas and estimating amount of solvent gravimetrically. Temperature measured with thermocouple and pressure with Bourdon gauge. Details in ref. (1).	1. Matheson sample, purity better than 99 mole per cent. 2. Aldrich Chemical Co. minimum purity 99 mole per cent. Distilled.
	ESTIMATED ERROR: δT/K = ±0.4; δP/MPa = ±0.02; δx_{CH_4}, δy_{CH_4} = ±2%.
	REFERENCES: 1. Simnick, J. J.; Lawson, C. C.; Lin, H. M.; Chao, K. C. *Am. Inst. Chem. Engnrs. J.* <u>1977</u>, *23*, 469.

COMPONENTS:	ORIGINAL MEASUREMENTS:
1. Methane; CH_4; [74-82-8] 2. 1,3-Dimethylbenzene; C_8H_{10}; [108-38-3]	Legret, D.; Richon, D.; Renon, H. *J. Chem. Engng. Data* 1982, *27*, 165-169.

VARIABLES:	PREPARED BY:
	C. L. Young

EXPERIMENTAL VALUES:

$$T/K = 313.2$$

$10^{-5}p$/Pa	Mole fraction of methane in liquid, x_{CH_4}	in vapor, y_{CH_4}
50.6	–	0.998
100.3	0.246	0.998
	0.253	–
150.1	0.345	0.997
	0.350	–
	0.348	–
199.8	0.413	0.995
250.3	0.475	0.990
	0.477	–
299.3	0.535	0.983
349.8	0.595	0.975
400.6	0.645	0.956
440.3	0.710	0.919
	0.712	–
441	0.712	–
459.3	0.745	0.870
459.6	0.749	0.862
465.2	0.783	0.840

AUXILIARY INFORMATION

METHOD/APPARATUS/PROCEDURE:	SOURCE AND PURITY OF MATERIALS:
High pressure static cell fitted **with magnetic stirrer.** Pressure measured with transducer calibrated by comparison with Heise gauges which were checked periodically calibrated against a dead weight tester. Temperature measured with K type iron-constantan thermocouples. Sampling microcell used and samples analysed using gas chromatography. Details in ref. (1).	1. Air-Gas sample, purity at least 99.95 volume per cent. 2. Merck sample, stated purity by GC of 98.5 per cent. Major impurities *o*-xylene (0.5%), *p*-xylene (0.5%) and ethylbenzene (0.5%).

ESTIMATED ERROR:

$\delta T/K = \pm 0.25$; δp/MPa $= \pm 0.1$;

$\delta x_{CH_4} = \pm 0.01$; $\delta y_{CH_4} = \pm 0.005$.

REFERENCES:

1. Legret, D.; Richon, D.; Renon, H.
Am. Inst. Chem. Eng. J.
1981, *27*, 203.

COMPONENTS:	ORIGINAL MEASUREMENTS:
1. Methane; CH₄; [74-82-8] 2. 1,3-Dimethylbenzene; C₈H₁₀; [108-38-3]	Ng, H.-J.; Huang, S. S.-S.; Robinson, D. B. *J. Chem. Engng. Data* __1982__, *27*, 119-122.

VARIABLES:	PREPARED BY:
	C. L. Young

EXPERIMENTAL VALUES:

T/K	p/MPa	Mole fraction of methane in liquid, x_{CH_4}	in vapor, y_{CH_4}
310.9	0.407	0.0136	0.9864
	2.13	0.0583	0.9953
	4.69	0.1299	0.9960
	6.92	0.1699	0.9966
	9.12	0.2214	0.9960
	11.58	0.2514	0.9952
	13.74	0.2954	0.9946
394.3	0.517	0.0118	0.8816
	1.83	0.0407	0.9581
	3.86	0.0866	0.9740
	5.98	0.1431	0.9778
	8.36	0.1774	0.9789
	11.2	0.2301	0.9772
	14.48	0.2951	0.9750
477.6	1.06	0.0156	0.526
	2.50	0.0482	0.768
	4.48	0.0930	0.848
	7.02	0.148	0.871
	9.44	0.204	0.886
	11.78	0.252	0.889
	13.91	0.295	0.879

AUXILIARY INFORMATION

METHOD/APPARATUS/PROCEDURE:	SOURCE AND PURITY OF MATERIALS:
Static equilibrium cell fitted with windows and magnetic stirrer. Temperature of thermostatic liquid measured with platinum resistance thermometer. Pressure measured using dead weight gauge and differential pressure transducer. Samples of vapor and liquid analysed by gas chromatography. Details in refs. (1) and (2).	1. Matheson Co. Ultrahigh-purity sample containing 99.97+ mole per cent methane. 2. Matheson, Coleman and Bell Chromatoquality sample with purity of greater than 99 mole per cent.

ESTIMATED ERROR:

$\delta T/K = \pm 0.06$; $\delta p/MPa = \pm 0.02$;

δx_{CH_4}, $\delta y_{CH_4} = \pm 0.002$.

REFERENCES:

1. Ng, H.-J.; Robinson, D. B.
   J. Chem. Engng. Data
   __1978__, *23*, 325.

2. Ohgaki, K.; Katayama, T.
   J. Chem. Engng. Data
   __1975__, *20*, 264.

COMPONENTS:	ORIGINAL MEASUREMENTS:
1. Methane; CH_4; [74-82-8] 2. 1,3,5-Trimethylbenzene (Mesitylene); C_9H_{12}; [108-67-8]	Legret, D.; Richon, D.; Renon, H. *J. Chem. Engng. Data* <u>1982</u>, *27*, 165-169.
VARIABLES:	PREPARED BY: C. L. Young

EXPERIMENTAL VALUES:

T/K = 313.2

$10^{-5}p$/Pa	Mole fraction of methane in liquid, x_{CH_4}	in vapor, y_{CH_4}
101.7	0.278	0.9995
	0.280	-
144.5	0.355	0.9993
	0.361	-
199.5	0.428	0.998
	0.434	-
249.6	0.495	0.995
	0.501	-
298.4	-	0.989
299.1	0.540	0.992
350.2	0.589	0.984
399.3	0.649	-
399.7	0.650	0.973
449.8	0.704	0.958
500.7	-	0.919
501.1	0.776	0.922
510.4	-	0.901
510.8	0.784	0.903
518.7	-	0.858
519.1	0.808	0.860

AUXILIARY INFORMATION

METHOD/APPARATUS/PROCEDURE:	SOURCE AND PURITY OF MATERIALS:
High pressure static cell fitted with magnetic stirrer. Pressure measured with transducer calibrated by comparison with Heise gauges which were checked periodically calibrated against a dead weight tester. Temperature measured with K type iron-constantan thermo-couples. Sampling microcell used and samples analysed using gas chromatography. Details in ref. (1).	1. Air-Gas sample, purity at least 99.95 volume per cent. 2. Fluka sample, stated purity not less than 99 mole per cent.

ESTIMATED ERROR:

$\delta T/K$ = ±0.25; δp/MPa = ±0.1;

δx_{CH_4} = ±0.01; δy_{CH_4} = ±0.005.

REFERENCES:

1. Legret, D.; Richon, D.; Renon, H.

   Am. Inst. Chem. Eng. J.

   <u>1981</u>, *27*, 203.

COMPONENTS:	ORIGINAL MEASUREMENTS:
1. Methane; CH₄; [74-82-8] 2. 1,3,5-Trimethylbenzene; C₉H₁₂; [108-67-8]	Huang, S. S.-S.; Robinson, D. B. *Can. J. Chem. Eng.* 1985, *63*, 126-130.

VARIABLES:	PREPARED BY:
	C. L. Young

EXPERIMENTAL VALUES:

T/K	P/MPa	Mole fraction of methane in liquid, x_{CH_4}	in vapor, y_{CH_4}	Equilibrium constant k_{CH_4}	$k_{C_9H_{12}}$
310.9	0.345	0.0108	0.9951	92.2	0.00495
	1.36	0.0423	0.9979	23.6	0.00219
	2.77	0.0837	0.9983	11.9	0.00185
	5.48	0.1504	0.9987	6.64	0.00159
	8.07	0.2119	0.9985	4.71	0.00190
	11.82	0.2828	0.9981	3.53	0.00265
	14.27	0.3246	0.9973	3.07	0.00400
394.3	0.479	0.0108	0.9389	86.9	0.0618
	1.46	0.0346	0.9759	28.3	0.0250
	3.01	0.0705	0.9850	14.0	0.0161
	5.61	0.1313	0.9887	7.53	0.0130
	8.92	0.1982	0.9888	4.99	0.0140
	11.93	0.2507	0.9882	3.94	0.0158
	14.59	0.2986	0.9864	3.30	0.0194
477.6	0.655	--	0.5648	57.6	0.440
	1.03	0.0193	0.7204	37.3	0.285
	2.52	0.0543	0.8648	15.9	0.143
	4.32	0.0962	0.9089	9.45	0.101
	6.96	0.1553	0.9280	5.98	0.0852
	9.41	0.2086	0.9340	4.48	0.0834
	11.80	0.2588	0.9345	3.61	0.0884
	14.13	0.3058	0.9318	3.05	0.0982

AUXILIARY INFORMATION

METHOD APPARATUS/PROCEDURE:	SOURCE AND PURITY OF MATERIALS:
Stirred static cell fitted with glass window. Temperature measured with Bourdon gauge. After equilibrium established gas and liquid phases sampled and analysed using gas chromatography with a flame ionisation detector. Details in ref. (1) and source.	1. Ultrahigh purity sample obtained from Matheson, purity at least 99.97 mole per cent. 2. Aldrich Chemical Co. sample, purity better than 99 mole per cent.

	ESTIMATED ERROR:
	δT/K = ±0.06; δP/MPa = ±0.007 (up to 6.9 MPa; ±0.02 (above 6.9 MPa).

	REFERENCES:
	1. Ng, H.-J.; Robinson, D. B. *J. Chem. Eng. Data* 1978, *23*, 325-327.

COMPONENTS:	ORIGINAL MEASUREMENTS:
1. Methane; CH_4; [74-82-8] 2. Butylbenzene; $C_{10}H_{14}$; [104-51-8]	O'Reilly, W. F.; Blumer, T. E.; Luks, K. D.; Kohn, J. P. *J. Chem. Engng. Data* 1976, *21*, 220-222.

VARIABLES:	PREPARED BY:
Temperature, pressure	C. L. Young

EXPERIMENTAL VALUES:

T/K	p/atm	p/kPa	Mole fraction of methane, x_{CH_4}	Molar volume of liquid, v /cm^3 mol^{-1}
343.2	10	1.0	0.0268	160.08
	20	2.0	0.0529	156.90
	30	3.0	0.0782	154.10
	40	4.1	0.1029	151.51
	50	5.1	0.1262	149.02
	60	6.1	0.1476	146.61
373.2	10	1.0	0.0245	166.06
	20	2.0	0.0491	163.00
	30	3.0	0.0732	160.09
	40	4.1	0.0968	157.48
	50	5.1	0.1196	155.00
	60	6.1	0.1411	152.59

AUXILIARY INFORMATION

METHOD/APPARATUS/PROCEDURE:	SOURCE AND PURITY OF MATERIALS:
A known amount of gas added to a known amount of solvent in a 10 cm^3 glass equilibrium cell. Liquid phase composition determined from overall composition and volume of both phases. Details in ref. (1).	1. Phillips Petroleum Co. sample, pure grade, minimum purity 99 mole per cent. 2. Aldrich Chemical Co. sample, purity better than 99 mole per cent.

ESTIMATED ERROR:

$\delta T/K = \pm 0.2$; $\delta p/kPa = \pm 7.0$;

$\delta x_{CH_4} = \pm 0.003$.

REFERENCES:

1. Cordeiro, D. J.; Luks, K. D.; Kohn, J. P.

Ind. Eng. Chem. Proc. Des. Develop.

1973, *12*, 47.

COMPONENTS:	ORIGINAL MEASUREMENTS:
1. Methane; CH_4; [74-82-8]	O'Reilly, W. F.; Blumer, T. E.; Luks, K. D.; Kohn, J. P.
2. Butylbenzene; $C_{10}H_{14}$; [104-51-8]	*J. Chem. Engng. Data*
3. Dotriacontane; $C_{32}H_{66}$; [544-85-4]	<u>1976</u>, *21*, 220-222.

VARIABLES:	PREPARED BY:
Composition, pressure	C. L. Young

EXPERIMENTAL VALUES:

T/K	Mole ratio Butylbenzene/ Dotriacontane	p/atm	p/kPa	Mole fraction of methane, x_{CH_4}	Molar volume of liquid, v /cm³ mol⁻¹
343.2	0.3053	10	1.0	0.0520	457.3
		20	2.0	0.1012	436.3
		30	3.0	0.1480	416.2
		40	4.1	0.1914	398.0
		50	5.1	0.2288	382.2
		60	6.1	0.2591	369.5
	0.7757	10	1.0	0.0505	364.5
		20	2.0	0.0989	353.6
		30	3.0	0.1441	342.7
		40	4.1	0.1840	331.8
		50	5.1	0.2167	320.9
		60	6.1	0.2419	310.0

AUXILIARY INFORMATION

METHOD/APPARATUS/PROCEDURE:	SOURCE AND PURITY OF MATERIALS:
A known amount of gas added to a known amount of solvent in a 10 cm³ glass equilibrium cell. Liquid phase composition determined from overall composition and volume of both phases. Details in ref. (1).	1. Phillips Petroleum Co. sample, pure grade, minimum purity 99 mole per cent. 2. Aldrich Chemical Co. sample, purity better than 99 mole per cent. 3. Humphrey Chemical Co. sample, purity at least 97 mole per cent.

ESTIMATED ERROR:

$\delta T/K = \pm0.2$; $\delta p/kPa = \pm7.0$;

$\delta x_{CH_4} = \pm0.008$.

REFERENCES:

1. Cordeiro, D. J.; Luks, K. D.; Kohn, J. P.

   Ind. Eng. Chem. Proc. Des. Develop.

   <u>1973</u>, *12*, 47.

COMPONENTS:	ORIGINAL MEASUREMENTS:
1. Methane; CH_4; [74-82-8]	O'Reilly, W. F.; Blumer, T. E.; Luks, K. D.; Kohn, J. P.
2. *Trans*-decahydronaphthalene (*Trans*-decalin); $C_{10}H_{18}$; [493-02-7]	*J. Chem. Engng. Data*
3. Phenanthrene; $C_{12}H_{10}$; [85-01-8]	1976, *21*, 220-222.

VARIABLES:	PREPARED BY:
Composition, pressure	C. L. Young

EXPERIMENTAL VALUES:

T/K	Mole ratio Decalin/ Phenanthrene	p/atm	p/kPa	Mole fraction of methane, x_{CH_4}	Molar volume of liquid, v /cm³ mol⁻¹
373.2	0.2885	10	1.0	0.0118	165.30
		20	2.0	0.0423	163.97
		30	3.0	0.0345	162.63
		40	4.1	0.0454	161.31
		50	5.1	0.0562	159.99
		60	6.1	0.0666	158.67
	0.9123	10	1.0	0.0165	165.96
		20	2.0	0.0317	164.18
		30	3.0	0.0468	162.39
		40	4.1	0.0616	160.60
		50	5.1	0.0762	158.81
		60	6.1	0.0904	157.01
	1.2619	10	1.0	0.0138	166.58
		20	2.0	0.0310	164.55
		30	3.0	0.0477	162.50
		40	4.1	0.0642	160.46
		50	5.1	0.0803	158.43
		60	6.1	0.0963	156.40

AUXILIARY INFORMATION

METHOD/APPARATUS/PROCEDURE:	SOURCE AND PURITY OF MATERIALS:
A known amount of gas added to a known amount of solvent in a 10 cm³ glass equilibrium cell. Liquid phase composition determined from overall composition and volume of both phases. Details in ref. (1).	1. Phillips Petroleum Co. sample, pure grade, minimum purity 99 mole per cent. 2. No details given. 3. Aldrich Chemical Co. sample, purity better than 98 mole per cent.

ESTIMATED ERROR:

$\delta T/K = \pm 0.2$; δp/kPa = ±7.0; δx_{CH_4} =±0.001.

REFERENCES:

1. Cordeiro, D. J.; Luks, K. D.; Kohn, J. P.
Ind. Eng. Chem. Proc. Des. Develop.
1973, *12*, 47.

COMPONENTS:	ORIGINAL MEASUREMENTS:
1. Methane; CH_4; [74-82-8] 2. Butylbenzene; $C_{10}H_{14}$; [104-51-8] 3. Phenanthrene; $C_{12}H_{10}$; [85-01-8]	O'Reilly, W. F.; Blumer; T. E.; Luks, K. D.; Kohn, J. P. *J. Chem. Engng. Data* 1976, *21*, 220-222.
VARIABLES: Composition, pressure	PREPARED BY: C. L. Young

EXPERIMENTAL VALUES:

T/K	Mole ratio Butylbenzene/ Phenanthrene	p/atm	p/kPa	Mole fraction of methane, x_{CH_4}	Molar volume of liquid, v /cm^3 mol^{-1}
373.2	0.3082	10	1.0	0.0160	164.33
		20	2.0	0.0279	162.84
		30	3.0	0.0410	161.39
		40	4.1	0.0512	159.99
		50	5.1	0.0617	158.70
		60	6.1	0.0720	157.52
	0.9252	10	1.0	0.0177	164.58
		20	2.0	0.0322	162.82
		30	3.0	0.0463	161.13
		40	4.1	0.0600	159.49
		50	5.1	0.0734	157.86
		60	6.1	0.0864	156.26

AUXILIARY INFORMATION

METHOD/APPARATUS/PROCEDURE:	SOURCE AND PURITY OF MATERIALS:
A known amount of gas added to a known amount of solvent in a 10 cm^3 glass equilibrium cell. Liquid phase composition determined from overall composition and volume of both phases. Details in ref. (1).	1. Phillips Petroleum Co. sample, pure grade, minimum purity 99 mole per cent. 2. Aldrich Chemical Co. sample, purity better than 99 mole per cent. 3. No details given.
	ESTIMATED ERROR: $\delta T/K = \pm0.2$; $\delta p/kPa = \pm7.0$; $\delta x_{CH_4} = \pm0.003$.
	REFERENCES: 1. Cordeiro, D. J.; Luks, K. D.; Kohn, J. P. *Ind. Eng. Chem. Proc. Des. Develop.* 1973, *12*, 47.

COMPONENTS:	ORIGINAL MEASUREMENTS:
1. Methane; CH_4; [74-82-8] 2. 1,2,3,4-Tetrahydronaphthalene (Tetralin); $C_{10}H_{12}$; [119-64-2]	Sebastian, H.M.; Simnick, J.J.; Lin, H-M.; Chao, K-C. *J. Chem. Engng. Data*, 1979, *24*, 149-152.

VARIABLES:	PREPARED BY:
Temperature, pressure	C. L. Young

EXPERIMENTAL VALUES:

T/K	p/atm	p/MPa	Mole fraction of methane in liquid, x_{CH_4}	in gas, y_{CH_4}
461.9	20.71	2.098	0.0310	0.9634
	30.58	3.099	0.0483	0.9735
	49.63	5.029	0.0772	0.9809
	99.10	10.041	0.1494	0.9841
	149.6	15.16	0.2174	0.9835
	199.3	20.19	0.2822	0.9801
	247.6	25.09	0.3397	0.9761
542.8	20.03	2.030	0.0293	0.8112
	30.03	3.043	0.0464	0.8628
	49.96	5.062	0.0787	0.9064
	100.1	10.14	0.1613	0.9303
	149.0	15.10	0.2379	0.9335
	193.5	20.11	0.3151	0.9304
	250.0	25.33	0.3918	0.9183
623.2	30.56	3.096	0.0383	0.5477
	50.09	5.075	0.0806	0.6917
	99.56	10.088	0.1805	0.7757
	149.8	15.18	0.2793	0.7995
	198.5	20.12	0.3915	0.7883
	223.1	22.61	0.4732	0.7415

AUXILIARY INFORMATION

METHOD/APPARATUS/PROCEDURE:	SOURCE AND PURITY OF MATERIALS:
Flow apparatus with both liquid and gas components continually passing into a mixing tube and then into a cell in which phases separated under gravity. Liquid sample removed from bottom of cell and vapor sample from top of cell. Composition determined by gas chromatography. Details in source and ref. (1).	1. Matheson sample with purity better than 99 mole per cent. 2. Aldrich Chemical Co. sample purity 99 mole per cent.

<table>
<tr><td rowspan="3"></td><td>ESTIMATED ERROR:</td></tr>
<tr><td>$\delta T/K = \pm 0.2$; δp/MPa $\leqslant \pm 0.03$;
δx_{CH_4}, δy_{CH_4} = $\pm 2\%$.</td></tr>
<tr><td>REFERENCES:

1. Simnick, J.J.; Lawson, C.C.; Lin, H-M.; Chao, K-C. *Am. Inst. Chem. Engnrs. J.*, 1977, *23*, 469.</td></tr>
</table>

COMPONENTS:	ORIGINAL MEASUREMENTS:
1. Methane; CH_4; [74-82-8] 2. 1,2,3,4-Tetrahydronaphthalene (Tetralin); $C_{10}H_{12}$; [119-64-2]	Sebastian, H. M.; Simnick, J. J.; Lin, H.-M.; Chao, K.-C. *J. Chem. Engng. Data* 1979, *24*, 149-152.

T/K	p/atm	p/MPa	Mole fraction of methane	
			in liquid, x_{CH_4}	in gas, y_{CH_4}
664.6	49.84	5.050	0.0741	0.4950
	99.17	10.045	0.1934	0.6128
	148.8	15.08	0.3514	0.5593
	155.8	15.79	0.3615	0.4191

COMPONENTS:	ORIGINAL MEASUREMENTS:
1. Methane; CH_4; [74-82-8] 2. 1-Methylnaphthalene; $C_{11}H_{10}$; [90-12-0]	Sebastian, H.M.; Simnick, J.J.; Lin, H-M.; Chao, K-C. *J. Chem. Engng. Data*, <u>1979</u>, *24*, 149-152.
VARIABLES: Temperature, pressure	PREPARED BY: C. L. Young

EXPERIMENTAL VALUES:

T/K	p/atm.	p/MPa	Mole fraction of methane in liquid, x_{CH_4}	in gas, y_{CH_4}
464.2	20.74	2.101	0.0281	0.9858
	31.02	3.143	0.0411	0.9891
	50.24	5.091	0.0651	0.9915
	99.16	10.047	0.1254	0.9928
	150.9	15.29	0.1803	0.9922
	199.4	20.20	0.2322	0.9906
	247.6	25.09	0.2787	0.9884
543.6	20.38	2.065	0.0275	0.9071
	30.55	3.095	0.0421	0.9317
	50.32	5.099	0.0697	0.9526
	99.99	10.131	0.1360	0.9648
	149.3	15.13	0.1992	0.9664
	200.0	20.27	0.2598	0.9643
	248.0	25.13	0.3184	0.9603
624.5	20.23	2.050	0.0246	0.6463
	30.55	3.095	0.0408	0.7476
	50.66	5.133	0.0746	0.8252
	100.40	10.173	0.1512	0.8784
	149.0	15.10	0.2275	0.8892
	199.2	20.18	0.3045	0.8917
	247.7	25.10	0.3798	0.8777

AUXILIARY INFORMATION

METHOD/APPARATUS/PROCEDURE:	SOURCE AND PURITY OF MATERIALS:
Flow apparatus with both liquid and gas components continually passing into a mixing tube and then into a cell in which phases separated under gravity. Liquid sample removed from bottom of cell and vapor sample from top of cell. Composition determined by gas chromatography. Details in source and ref. (1).	1. Matheson sample with purity better than 99 mole per cent. 2. Aldrich Chemical Co. sample purity 97 mole per cent. Fractionally distilled under vacuum.
	ESTIMATED ERROR: $\delta T/K = \pm 0.2$; $\delta p/MPa \leqslant \pm 0.03$; δx_{CH_4}, $\delta y_{CH_4} = \pm 2\%$
	REFERENCES: 1. Simnick, J.J.; Lawson, C.C.; Lin, H-M.; Chao, K-C., *Am. Inst. Chem. Engnrs. J.*, <u>1977</u>, *23*, 469.

COMPONENTS:	ORIGINAL MEASUREMENTS:
1. Methane; CH_4; [74-82-8]	Sebastian, H. M.; Simnick, J. J.; Lin, H.-M.; Chao, K.-C.
2. 1-Methylnaphthalene; $C_{11}H_{10}$; [90-12-0]	*J. Chem. Engng. Data* <u>1979</u>, *24*, 149-152.

T/K	p/atm.	p/MPa	Mole fraction of methane in liquid, x_{CH_4}	in gas, y_{CH_4}
704.0	30.07	3.047	0.0286	0.3523
	50.24	5.091	0.0716	0.5215
	100.25	10.158	0.1797	0.6603
	148.8	15.08	0.2951	0.6750
	172.9	17.52	0.3724	0.6481
	181.3	18.37	0.4016	0.6429
	185.7	18.82	0.4544	0.6341

COMPONENTS:	ORIGINAL MEASUREMENTS:
1. Methane; CH_4; [74-82-8] 2. 1-Methylnaphthalene; $C_{11}H_{10}$; [90-12-0]	Henson, B. J.; Tarrer, A. R.; Curtis, C. W.; Guln, J. A. *Ind. Eng. Chem. Process Des. Dev.* *1982*, *21*, 575-579.

VARIABLES:	PREPARED BY:
	C. L. Young

EXPERIMENTAL VALUES:

t/°C	T/K	P/MPa	Mole fraction of methane, x_{CH_4}
102	375	4.79	0.0685
		8.95	0.1185
		11.45	0.1543
202	475	4.95	0.0645
		6.95	0.0890
		11.21	0.1411

AUXILIARY INFORMATION

METHOD/APPARATUS/PROCEDURE:	SOURCE AND PURITY OF MATERIALS:
One gallon static equilibrium cell fitted with magnetic agigator. Samples taken from small volume sample loops through which equilibrium liquid was circulated. Gas in liquid sample as estimated by volumetric technique using a Toffel pump.	1. Matheson sample, purity 99 mole per cent. 2. Aldrich Chemical Co. sample, purity 97 mole per cent.

ESTIMATED ERROR:

$\delta T/K = \pm 1$; $\delta x_{CH_4} = \pm 4\%$
(estimated by compiler).

REFERENCES:

COMPONENTS:	ORIGINAL MEASUREMENTS:

1. Methane; CH$_4$; [74-82-8]
 	
2. 1,1´-Methylenebisbenzene,
   (Diphenylmethane); C$_{13}$H$_{12}$;
   [101-81-5]
	Sebastian, H.M.; Simnick, J.J.;
	Lin, H-M.; Chao, K-C.
	
	J. Chem. Engng. Data, <u>1979</u>, *24*,
	149-152.

VARIABLES:	PREPARED BY:
Temperature, pressure	C. L. Young

EXPERIMENTAL VALUES:

T/K	p/atm	p/MPa	Mole fraction of methane	
			in liquid, x_{CH_4}	in gas, y_{CH_4}
462.5	19.89	2.015	0.0335	0.9893
	30.10	3.050	0.0493	0.9933
	49.83	5.049	0.0792	0.9949
	99.23	10.05	0.1508	0.9953
	149.3	15.13	0.2139	0.9947
	200.0	20.27	0.2731	0.9938
	249.7	25.30	0.3235	0.9918
542.2	19.83	2.009	0.0339	0.9350
	30.24	3.064	0.0516	0.9533
	49.97	5.063	0.0848	0.9661
	98.83	10.01	0.1636	0.9753
	149.9	15.19	0.2339	0.9759
	196.8	19.94	0.2992	0.9738
	248.7	25.20	0.3618	0.9700
623.7	19.83	2.009	0.0320	0.7290
	30.04	3.044	0.0519	0.8029
	49.84	5.050	0.0912	0.8673
	99.65	10.097	0.1845	0.9059
	149.2	15.12	0.2679	0.9156
	199.5	20.21	0.3535	0.9127
	249.7	25.30	0.4388	0.9001

AUXILIARY INFORMATION

METHOD/APPARATUS/PROCEDURE:	SOURCE AND PURITY OF MATERIALS:
Flow apparatus with both liquid and gas components continually passing into a mixing tube and then into a cell in which phases separated under gravity. Liquid sample removed from bottom of cell and vapor sample from top of cell. Composition determined by gas chromatography. Details in source and ref. (1).	1. Matheson sample with purity better than 99 mole per cent. 2. Aldrich Chemical Co. sample purity 99 mole per cent.

ESTIMATED ERROR:
δT/K = ±0.2; δp/MPa $\leqslant$ ±0.03; δx_{CH_4}, δy_{CH_4} = ±2%.

REFERENCES:

1. Simnick, J.J.; Lawson, C.C.;
   Lin, H-M; Chao, K-C. *Am. Inst.*
   Chem. Engnrs. J., <u>1977</u>, *23*, 469.

COMPONENTS:	ORIGINAL MEASUREMENTS:
1. Methane; CH_4; [74-82-8]	Sebastian, H. M.; Simnick, J. J.; Lin, H.-M.; Chao, K.-C.
2. 1,1'-Methylenebisbenzene (Diphenylmethane); $C_{13}H_{12}$; [101-81-5]	*J. Chem. Engng. Data* <u>1979</u>, *24*, 149-152.

T/K	*p*/atm	*p*/MPa	Mole fraction of methane in liquid, x_{CH_4}	in gas, y_{CH_4}
702.9	30.51	3.091	0.0448	0.4407
	49.83	5.049	0.0961	0.5982
	100.3	10.16	0.2228	0.7132
	150.4	15.24	0.3576	0.7230

COMPONENTS:	ORIGINAL MEASUREMENTS:
1. Methane; CH_4; [74-82-8] 2. 9,10-Dihydrophenanthrene $C_{14}H_{12}$; [776-35-2]	Sebastian, H.M.; Lin, H-M.; Chao, K-C. *J. Chem. Engng. Data.* <u>1980</u>,*25*, 379-381.

VARIABLES:
   Temperature, pressure
	PREPARED BY:
	
	        C.L. Young

EXPERIMENTAL VALUES:

T/K	p/atm	p/MPa	Mole fraction of methane in liquid, x_{CH_4}	in gas x_{CH_4}
546.3	20.09	2.036	0.0255	0.99760
	30.22	3.062	0.0371	0.99834
	50.5	5.12	0.0600	0.99884
	99.6	10.09	0.1132	0.99871
	150.3	15.23	0.1628	0.99849
	200.3	20.30	0.2058	0.99806
	250.0	25.33	0.2445	0.99757
542.85	20.04	2.031	0.0268	0.9781
	30.18	3.058	0.0396	0.9481
	50.2	5.087	0.0642	0.9886
	99.8	10.11	0.1227	0.9912
	150.1	15.21	0.1768	0.9914
	199.9	20.25	0.2278	0.9908
	249.7	25.30	0.2744	0.9893
622.5	20.45	2.072	0.0284	0.8925
	30.19	3.059	0.0421	0.9224
	49.9	5.06	0.0693	0.9461
	100.0	10.13	0.1353	0.9629
	150.7	15.27	0.1972	0.9665
	200.1	20.28	0.2576	0.9666
	249.2	25.25	0.3115	0.9643
703.15	20.04	2.031	0.0252	0.6623
	30.01	3.041	0.0429	0.7477
	50.5	5.117	0.0763	0.8301
	99.7	10.10	0.1537	0.8889
	150.7	15.27	0.2246	0.9105

AUXILIARY INFORMATION

METHOD/APPARATUS/PROCEDURE:	SOURCE AND PURITY OF MATERIALS:
Flow apparatus with both liquid and gas components continually passing into a mixing tube and then into a cell in which phases separated under gravity. Liquid sample removed from bottom of cell and vapor sample from top of cell. Composition determined by gas chromatography. Details in source and ref. (1). Some decomposition to phenanthrene occurred at the highest temperature (up to ~5% at the highest pressure)	1. Matheson sample, minimum purity 99 mole per cent. 2. Aldrich Chemical Co. sample purified by zone refining, final purity better than 99 mole per cent as determined using GC.

	ESTIMATED ERROR:
	
	   $\delta T/K = \pm 0.2$; $\delta p/MPa < \pm 0.03$;
	   δx_{CH_4}, $\delta y_{CH_4} = \pm 2\%$

	REFERENCES:
	
	1. Simnick, J.J.; Lawson, C.C.; Lin, H-M. Chao, K-C.
	   Am. Inst. Chem. Engnrs. J. <u>1977</u>
	   23, 469.

COMPONENTS:	ORIGINAL MEASUREMENTS:
1. Methane; CH_4; [74-82-8] 2. Nonane; C_9H_{20}; [111-84-2] 3. 2,2,3-Trimethylbutane; C_7H_{16}; [464-06-2]	Savvina, Ya. D.; Velikovskii, A. S. *Tr. Vses. Nauchno-Issled. Inst.* *Prirod. Gaz.*, <u>1962</u>, *17* 197-202.
VARIABLES: Pressure, solvent composition	PREPARED BY: C. L. Young

EXPERIMENTAL VALUES:

$$T/K = 333.2$$

$P/kgcm^{-2}$	solvent comp^n ^{a}	Mole fraction in liquid			Mole fraction in vapour		
		C_7H_{16}	C_9H_{20}	CH_4	C_7H_{16}	C_9H_{20}	CH_4
200	100		0.417	0.583		0.010	0.990
	25	0.318	0.080	0.602	0.033	0.015	0.952
	0	0.388		0.612	0.073		0.927
220	50	0.194	0.153	0.653	0.025	0.014	0.961
	0	0.286		0.714	0.145		0.855
	100		0.318	0.682		0.014	0.986
250	75	0.083	0.209	0.708	0.018	0.017	0.965
	50	0.138	0.106	0.756	0.030	0.019	0.951
	25	0.125	0.033	0.842	0.092	0.032	0.876
280	75	0.061	0.152	0.787	0.025	0.025	0.950
	50	0.066	0.059	0.875	0.064	0.047	0.889
	100		0.232	0.768		0.030	0.970
300	75	0.045	0.105	0.850	0.048	0.069	0.883

^{a} volume fraction of nonane in original cell charge

AUXILIARY INFORMATION

METHOD/APPARATUS/PROCEDURE:	SOURCE AND PURITY OF MATERIALS:
Values appear to be determined using apparatus described in ref. (1). Composition of liquid phase determined from refractive index measurements.	No details given except purity of methane 99 mole per cent.
	ESTIMATED ERROR:

	REFERENCES:
	
	1. Savvina, Ya. D.;
	   Velikovskii, A. S.;
	   Tr. Vses. Nauchno-Issed.
	   Inst. Prirodn. Gazov.,
	   <u>1962</u>, *17/25,* 163.

COMPONENTS:	ORIGINAL MEASUREMENTS:
1. Methane; CH_4; [74-82-8] 2. Benzene; C_6H_6; [71-43-2] 3. 2,2,3-Trimethylbutane; C_7H_{16}; [464-06-2]	Savvina, Ya. D.; Velikovskii, A. S. *Tr. Vses. Nauchno-Issled. Inst.* *Prirod. Gaz.,* <u>1962</u>, *17*, 197-202.

VARIABLES:	PREPARED BY:
Pressure, solvent composition	C. L. Young

EXPERIMENTAL VALUES:

$$T/K = 333.2$$

$P/kgcm^{-2}$	solvent comp^a	Mole fraction in liquid			Mole fraction in vapour		
		C_7H_{16}	C_6H_6	CH_4	C_7H_{16}	C_6H_6	CH_4
200	100		0.634	0.366		0.028	0.972
	75	0.149	0.426	0.425	0.012	0.029	0.951
	50	0.305	0.214	0.481	0.025	0.033	0.942
	25	0.348	0.105	0.547	0.033	0.035	0.932
	0	0.388		0.612	0.073		0.927
210	100		0.621	0.379		0.029	0.971
	75	0.132	0.422	0.446	0.018	0.032	0.950
	50	0.266	0.236	0.498	0.031	0.034	0.935
	25	0.332	0.094	0.575	0.043	0.038	0.919
	0	0.342		0.658	0.093		0.907
220	100		0.609	0.391		0.031	0.969
	75	0.152	0.391	0.457	0.025	0.034	0.941
	50	0.272	0.205	0.523	0.042	0.036	0.922
	25	0.292	0.109	0.599	0.053	0.042	0.905
	0	0.281		0.719	0.145		0.855
230	100		0.596	0.404		0.032	0.968
	75	0.137	0.393	0.470	0.033	0.040	0.927
	50	0.245	0.393	0.546	0.053	0.042	0.905
	25	0.273	0.091	0.636	0.061	0.044	0.855

^{a} volume fraction of benzene in original cell charge

AUXILIARY INFORMATION

METHOD/APPARATUS/PROCEDURE:	SOURCE AND PURITY OF MATERIALS:
Values appear to be determined using apparatus described in ref. (1). Composition of liquid phase determined from refractive index measurements.	No details given except purity of methane 99 mole per cent.
	ESTIMATED ERROR:
	REFERENCES: 1. Savvina, Ya. D.; Velikovskii, A. S.; *Tr. Vses. Nauchno-Issed.* *Inst. Prirodn. Gazov.,* <u>1962</u>, *17/25*, 163.

COMPONENTS:	ORIGINAL MEASUREMENTS:
1. Methane; CH_4; [74-82-8] 2. Ethane; C_2H_6; [74-84-0] 3. Propane; C_3H_8; [74-98-6] 4. Methylbenzene; C_7H_8; [108-88-3] 5. 1-Methylnaphthalene; $C_{11}H_{10}$; [90-12-0]	Li, Y.-H.; Dillard, K. H.; Robinson, R. L. *J. Chem. Eng. Data* 1981, *26*, 200-204.
VARIABLES:	PREPARED BY:
Temperature	C. L. Young

EXPERIMENTAL VALUES:

T/K (T/°F)	P/MPa (p/psia)	Phase	Mole fractions				
			x_{CH_4}	$x_{C_2H_6}$	$x_{C_3H_8}$	$x_{C_7H_8}$	$x_{C_{11}H_{10}}$
377.6 (220)	1.38 (200)	gas liquid	0.6625 0.01380	0.1805 0.01455	0.1150 0.02195	0.04155 0.5575	0.0004620 0.3920
	2.76 (400)	gas liquid	0.7100 0.03055	0.1670 0.02765	0.09505 0.03595	0.02470 0.5365	0.0003350 0.3695
	4.14 (600)	gas liquid	0.7415 0.05030	0.1590 0.03945	0.08045 0.04460	0.01905 0.5155	0.0003170 0.3500
	5.52 (800)	gas liquid	0.7575 0.06940	0.1520 0.04810	0.07325 0.05090	0.01695 0.4910	0.0003645 0.3405
	6.89 (1000)	gas liquid	0.7695 0.08745	0.1450 0.05390	0.06875 0.05270	0.01650 0.4820	0.0003555 0.3240
	8.62 (1250)	gas liquid	0.7845 0.1115	0.1360 0.06030	0.06260 0.05565	0.01615 0.4635	0.0003990 0.3090
	10.34 (1500)	gas liquid	0.7940 0.1355	0.1295 0.06535	0.05885 0.05735	0.01680 0.4460	0.0004290 0.2960
	12.07 (1750)	gas liquid	0.8030 0.1560	0.1230 0.06855	0.05535 0.05770	0.01780 0.4275	0.0005140 0.2900
	13.79 (2000)	gas liquid	0.8085 0.1755	0.1190 0.07190	0.05285 0.05855	0.01910 0.4120	0.0005610 0.2820

(cont.)

AUXILIARY INFORMATION

METHOD/APPARATUS/PROCEDURE:	SOURCE AND PURITY OF MATERIALS:

Variable volume, windowed phase equilibrium cell was used in which the mixture was confined by a floating piston. Pressure was measured with a Bourdon pressure gauge. Temperature was measured with a platinum resistance thermometer. Samples of vapor and liquid phases analysed by GC using a thermal conductivity detector. Details in ref. (1).	1, 2, 3. Linde samples, purities
	     99.97, 99.0 and 99.5 mole
	     per cent, respectively.
	4. Phillips Petroleum Co. sample,
	     purity better than 97.8 mole
	     per cent.
	5. Aldrich Chemical Co. sample,
	     purity better than 97 mole per
	     cent.

	ESTIMATED ERROR: $\delta T/K = \pm 0.05$; $\delta P/MPa = \pm 2\%$; $\delta x = \pm 2\%$ or 0.0001 (whichever is greater) (estimated by compiler).

	REFERENCES:
	1. Li, Y.-H.; Dillard, K. H.;
	     Robinson, R. L.
	     J. Chem. Eng. Data
	     1981, *26*, 53.

COMPONENTS:	ORIGINAL MEASUREMENTS:
1. Methane; CH_4; [74-82-8] 2. Ethane; C_2H_6; [74-84-0] 3. Propane; C_3H_8; [74-98-6] 4. Methylbenzene; C_7H_8; [108-88-3] 5. 1-Methylnaphthalene; $C_{11}H_{10}$; [90-12-0]	Li, Y.-H.; Dillard, K. H.; Robinson, R. L. *J. Chem. Eng. Data* 1981, *26*, 200-204.

EXPERIMENTAL VALUES:

T/K (T/°F)	P/MPa p/psia	Phase	Mole fractions				
			x_{CH_4}	$x_{C_2H_6}$	$x_{C_3H_8}$	$x_{C_7H_8}$	$x_{C_{11}H_{10}}$
410.9	1.65 (239)	gas liquid	0.6350 0.01615	0.1670 0.01410	0.1110 0.01985	0.08185 0.5730	0.001885 0.3770
	3.03 (439)	gas liquid	0.6795 0.03330	0.1665 0.02510	0.09930 0.03105	0.05335 0.5580	0.001360 0.3525
	4.41 (640)	gas liquid	0.7050 0.05150	0.1610 0.03420	0.09015 0.03835	0.04265 0.5425	0.001025 0.3335
	5.78 (839)	gas liquid	0.7240 0.06905	0.1545 0.04150	0.08255 0.04330	0.03810 0.5240	0.001030 0.3225
	7.14 (1035)	gas liquid	0.7380 0.08810	0.1480 0.04770	0.07655 0.04680	0.03640 0.5050	0.001170 0.3125
	8.91 (1292)	gas liquid	0.7510 0.1165	0.1415 0.05655	0.07155 0.05100	0.03470 0.4285	0.001200 0.2975
	10.63 (1542)	gas liquid	0.7650 0.1415	0.1340 0.06360	0.06620 0.05470	0.03335 0.4560	0.001310 0.2845
	12.36 (1792)	gas liquid	0.7745 0.1610	0.1280 0.06665	0.06185 0.05530	0.03410 0.4405	0.001515 0.2765
	14.09 (2043)	gas liquid	0.7810 0.1855	0.1235 0.07040	0.05980 0.05620	0.03455 0.4250	0.001565 0.2630
444.3	1.41 (204)	gas liquid	0.5565 0.01230	0.1480 0.008945	0.1035 0.01200	0.1855 0.5235	0.006925 0.4435
	2.63 (381)	gas liquid	0.6280 0.02675	0.1550 0.01770	0.09995 0.02160	0.1125 0.5355	0.004590 0.3985
	4.07 (591)	gas liquid	0.6640 0.04575	0.1545 0.02720	0.09460 0.03095	0.08280 0.5235	0.003730 0.3725
	5.47 (793)	gas liquid	0.6870 0.06125	0.1510 0.03375	0.08840 0.03660	0.07035 0.5140	0.003280 0.3545
	6.87 (996)	gas liquid	0.7080 0.08120	0.1455 0.04075	0.08215 0.04070	0.06145 0.4980	0.003050 0.3395
	8.58 (1244)	gas liquid	0.7210 0.1050	0.1405 0.04775	0.07715 0.04495	0.05830 0.4775	0.003205 0.3250
	10.35 (1501)	gas liquid	0.7355 0.1335	0.1350 0.05475	0.07220 0.04890	0.05455 0.4595	0.003025 0.3035
	12.01 (1742)	gas liquid	0.7445 0.1535	0.1300 0.05960	0.06840 0.05075	0.05380 0.4440	0.003190 0.2915
	13.64 (1978)	gas liquid	0.7495 0.1695	0.1260 0.06220	0.06550 0.05155	0.05535 0.4335	0.003725 0.2830
477.6 (400)	1.43 (207)	gas liquid	0.4925 0.009990	0.1305 0.006505	0.08890 0.008065	0.2700 0.4305	0.01810 0.5450
	2.75 (399)	gas liquid	0.5715 0.02390	0.1455 0.01445	0.09450 0.01625	0.1780 0.5070	0.01045 0.4385
	4.21 (611)	gas liquid	0.6150 0.04250	0.1490 0.02350	0.09255 0.02495	0.1350 0.5105	0.008365 0.3985
	5.56 (806)	gas liquid	0.6390 0.05910	0.1480 0.03030	0.08900 0.03050	0.1165 0.5080	0.007335 0.3720

(cont.)

COMPONENTS: ORIGINAL MEASUREMENTS:

1. Methane; CH_4; [74-82-8]   Li, Y.-H.; Dillard, K. H.;
2. Ethane; C_2H_6; [74-84-0]   Robinson, R. L.
3. Propane; C_3H_8; [74-98-6]
4. Methylbenzene; C_7H_8; [108-88-3]   *J. Chem. Eng. Data*
5. 1-Methylnaphthalene; $C_{11}H_{10}$;   1981, *26*, 200-204.
   [90-12-0]

EXPERIMENTAL VALUES:

T/K (T/°F)	P/MPa p/psia	Phase	Mole fractions				
			x_{CH_4}	$x_{C_2H_6}$	$x_{C_3H_8}$	$x_{C_7H_8}$	$x_{C_{11}H_{10}}$
477.6 (400)	7.03 (1019)	gas	0.6605	0.1455	0.08440	0.1030	0.006690
		liquid	0.07470	0.03685	0.03505	0.5015	0.3520
	8.69 (1261)	gas	0.6785	0.1405	0.07900	0.09510	0.006885
		liquid	0.09955	0.04400	0.03955	0.4820	0.3345
	10.29 (1493)	gas	0.6890	0.1370	0.07535	0.09155	0.006990
		liquid	0.1245	0.05060	0.04325	0.4670	0.3150
	12.18 (1766)	gas	0.6700	0.1325	0.07130	0.08915	0.007310
		liquid	0.1480	0.5585	0.04535	0.4515	0.2995
	13.65 (1980)	gas	0.7030	0.1290	0.06865	0.09115	0.008305
		liquid	0.1720	0.06100	0.04825	0.4330	0.2860

COMPONENTS:	ORIGINAL MEASUREMENTS:
1. Methane; CH_4; [74-82-8] 2. Propane; C_3H_8; [74-98-6] 3. Methylbenzene; C_7H_8; [108-88-3]	Van Horn, L. D.; Kobayashi, R. *J. Chem. Engng. Data* <u>1967</u>, *12*, 294-303.

VARIABLES:	PREPARED BY:
Temperature, pressure	C. L. Young

EXPERIMENTAL VALUES:

		Mole fractions				
		in liquid			in vapor	
T/K (T/°F)	P/MPa (P/psi)	x_{CH_4}	$x_{C_3H_8}$	$x_{C_7H_8}$	y_{CH_4}	$y_{C_3H_8}$
(233.15) -40	0.689 (100)	0.030 0.044	0.245 0.558	0.725 0.398	0.9233 0.869	0.0767 0.131
	1.38 (200)	0.052 0.056 0.062 0.109	0.101 0.142 0.222 0.661	0.847 0.802 0.716 0.230	0.9784 0.9713 0.9595 0.9233	0.0216 0.0287 0.0405 0.0767
	2.76 (400)	0.117 0.135 0.168	0.167 0.247 0.444	0.716 0.618 0.388	0.9784 0.9713 0.9595	0.0216 0.0287 0.0405
	4.14 (600)	0.145 0.175 0.216	0.080 0.191 0.290	0.775 0.634 0.494	0.9894 0.9784 0.9713	0.0106 0.0216 0.0287
	5.52 (800)	0.182 0.234 0.278	0.076 0.176 0.260	0.742 0.590 0.462	0.9894 0.9784 0.9713	0.0106 0.0216 0.0287
	6.89 (1000)	0.250	0.143	0.607	0.9784	0.0216

AUXILIARY INFORMATION

METHOD/APPARATUS/PROCEDURE:	SOURCE AND PURITY OF MATERIALS:
The solubilities were determined by measurement of retention volumes using gas chromatography. The method uses methane as carrier gas, propane as an injected solute and toluene as the stationary phase. The technique is described in the source and in ref. (1).	1 and 2. Major impurities were carbon dioxide and nitrogen amounting to about 0.2 mole per cent. 3. Research grade.

ESTIMATED ERROR:
$\delta T/K = \pm0.05$; $\delta P/\text{psi} = \pm1$, $P \leq$ 1,000 psia; ±2, $P \geq 1,000$ psia; δx, $\delta y = \pm1.5\%$.

REFERENCES:
1. Koonce, K. T.
     Ph.D. thesis, Rice University,
   Houston, <u>1963</u>.

COMPONENTS:	ORIGINAL MEASUREMENTS:
1. Methane; CH₄; [74-82-8] 2. Hydrocarbon oil	Sage, B. H.; Backus, H. S.; Lacey, W. N. *Ind. Eng. Chem.* <u>1935</u>, *27*, 686-690.

VARIABLES:	PREPARED BY:
Temperature, pressure	C. L. Young

EXPERIMENTAL VALUES:

T/°F	T/K	P/psia	P/MPa	Solubility, S /wt-%
70.0	294.3	200	1.38	0.36
		400	2.76	0.83
		600	4.14	1.10
		800	5.52	1.48
		1000	6.89	1.88
		1250	8.62	2.40
		1500	10.34	2.93
		1750	12.07	3.50
		2000	13.79	4.14
		2250	15.51	4.86
100.0	311.0	200	1.38	0.35
		400	2.76	0.79
		600	4.14	1.06
		800	5.52	1.42
		1000	6.89	1.79
		1250	8.62	2.27
		1500	10.34	2.78
		1750	12.07	3.32
		2000	13.79	3.92
		2250	15.51	4.54
		2500	17.24	5.22

(cont.)

AUXILIARY INFORMATION

METHOD/APPARATUS/PROCEDURE:	SOURCE AND PURITY OF MATERIALS:
Contents of variable volume cell brought to equilibrium at desired temperature and pressure and volume determined. Volume varied by admission or removal of mercury. Bubble point determined from change in slope of pressure-volume curve.	1. Natural gas sample which was treated for removal of nitrogen, ethane and higher hydrocarbons. Final purity about 99.8 mole per cent. 2. Non-waxy asphalt crude oil with molecular weight of between 335 & 340 (by freezing point depression).

ESTIMATED ERROR:

$\delta T/K = \pm 0.13$; $\delta P/psia = \pm 1$;
$\delta S/S = \pm 0.001$.

REFERENCES:

COMPONENTS:	ORIGINAL MEASUREMENTS:
1. Methane; CH₄; [74-82-8]	Sage, B. H.; Backus, H. S.; Lacey, W. N.
2. Hydrocarbon oil	*Ind. Eng. Chem.* 1935, *27*, 686-690.

EXPERIMENTAL VALUES:

T/°F	T/K	P/psia	P/MPa	Solubility, /wt-%
130.0	327.6	200	1.38	0.34
		400	2.76	0.75
		600	4.14	1.02
		800	5.52	1.36
		1000	6.89	1.72
		1250	8.62	2.17
		1500	10.34	2.65
		1750	12.07	3.15
		2000	13.79	3.69
		2250	15.51	4.26
		2500	17.24	4.88
		2750	18.96	5.4 *
160.0	344.3	200	1.38	0.33
		400	2.76	0.73
		600	4.14	0.98
		800	5.52	1.31
		1000	6.89	1.66
		1250	8.62	2.09
		1500	10.34	2.54
		1750	12.07	3.01
		2000	13.79	3.51
		2250	15.51	4.03
		2500	17.24	4.59
		2750	18.96	5.20
		3000	20.68	5.8 *
190.0	361.0	200	1.38	0.31
		400	2.76	0.70
		600	4.14	0.94
		800	5.52	1.27
		1000	6.89	1.62
		1250	8.62	2.01
		1500	10.34	2.44
		1750	12.07	2.88
		2000	13.79	3.35
		2250	15.51	3.83
		2500	17.24	4.35
		2750	18.96	4.90
		3000	20.68	5.5 *
220.0	377.6	200	1.38	0.30
		400	2.76	0.67
		600	4.14	0.91
		800	5.52	1.23
		1000	6.89	1.05
		1250	8.62	1.94
		1500	10.34	2.35
		1750	12.07	2.77
		2000	13.79	3.21
		2250	15.51	3.66
		2500	17.24	4.31
		2750	18.96	4.64
		3000	20.68	5.16

* extrapolated values.

COMPONENTS:	ORIGINAL MEASUREMENTS:
1. Methane; CH$_4$; [74-82-8] 2. Hydrocarbon Blend (Heavy Naphtha)	Frolich, P.K.; Tauch, E.J.; Hogan, J.J.; Peer, A.A. *Ind. Eng. Chem.* <u>1931</u>, *23*, 548-550

VARIABLES:	PREPARED BY:
Pressure	C.L. Young

EXPERIMENTAL VALUES:

T/K	P/atm	P/MPa	Solubility, S*
298.15	10	1.0	6
	20	2.0	12
	30	3.0	18
	40	4.1	23
	50	5.1	28
	60	6.1	33
	70	7.1	39
	80	8.1	45

* Volume of gas measured at 101.325 kPa pressure and 298.15 K
dissolved by unit volume of liquid measured under the same
conditions.

AUXILIARY INFORMATION

METHOD/APPARATUS/PROCEDURE:	SOURCE AND PURITY OF MATERIALS:
Static equilibrium cell. Liquid saturated with gas and after equilibrium established samples removed and analysed by volumetric method. Allowance was made for the vapor pressure of the liquid and the solubility of the gas at atmospheric pressure. Details in source.	1. Methane was of the highest purity available. 2. Density of 0.8003g cm^{-3} and vapor pressure of 80 mmHg at 298.15 K.

ESTIMATED ERROR:

$\delta T/K = \pm 0.1$; $\delta S = \pm 5\%$

REFERENCES:

COMPONENTS:	ORIGINAL MEASUREMENTS:
1. Methane; CH ; [74-82-8] 2. Hydrocarbon Blend (Gas Oil)	Frolich, P.K.; Tauch, E.J.; Hogan, J.J.; Peer, A.A. *Ind. Eng. Chem.* 1931, *23*, 548-550.

VARIABLES:
 	
        Pressure
	PREPARED BY:
	
	        C.L. Young

EXPERIMENTAL VALUES:

T/K	P/atm	P/MPa	Solubility, S*
298.15	10	1.0	4
	20	2.0	8
	30	3.0	12
	40	4.1	16
	50	5.1	20
	60	6.1	24
	70	7.1	29
	80	8.1	34
	90	9.1	39
	100	10.1	44
	110	11.1	49
	120	12.2	54
	130	13.2	59
	140	14.2	64

* Volume of gas measured at 101.325 kPa pressure and 298.15 K
  dissolved by unit volume of liquid measured under the same
  conditions.

AUXILIARY INFORMATION

METHOD/APPARATUS/PROCEDURE:	SOURCE AND PURITY OF MATERIALS:

Static equilibrium cell. Liquid saturated with gas and after equilibrium established samples removed and analysed by volumetric method. Allowance was made for the vapor pressure of the liquid and the solubility of the gas at atmospheric pressure. Details in source.	1. Methane was of the highest
	    purity available.
	
	2. Density of 0.8319 g cm^{-3} and
	    vapor pressure of 2 mmHg at
	    298.15 K.

	ESTIMATED ERROR: $\delta T/K = \pm 0.1$; $\delta S = \pm 5\%$
	REFERENCES:

COMPONENTS:	ORIGINAL MEASUREMENTS:
1. Methane; CH_4; [74-82-8] 2. Coal Liquids - Distillate from Exxon Donor Solvent Process	Lin, H.-M.; Sebastian, H. M.; Simnick, J. J.; Chao, K.-C. *Ind. Eng. Chem. Process. Des. Dev.* <u>1981</u>, *20*, 253-256.
VARIABLES: Temperature, pressure	PREPARED BY: C. L. Young

EXPERIMENTAL VALUES:

T/K	P/atm	P/MPa	Mole fraction of methane in liquid, x_{CH}	Solubility[#], S
			Sample 1	
188.9	49.6	5.03	0.0926	106.06
189.0	99.5	10.08	0.1828	232.50
188.5	149.2	15.12	0.2646	373.96
189.0	198.3	20.09	0.3375	529.62
188.7	245.9	24.92	0.4049	707.30
268.1	48.9	4.95	0.1001	115.63
268.0	99.5	10.08	0.2010	261.52
268.0	146.7	14.86	0.2867	417.87
268.1	196.9	19.95	0.3617	589.08
267.9	243.9	24.71	0.4438	829.41
			Sample 2	
189.1	50.1	5.08	0.0958	93.190
189.0	100.1	10.14	0.1708	181.31
189.2	150.4	15.24	0.2333	267.78
189.1	198.3	20.09	0.2932	365.04
189.3	251.7	25.50	0.3542	482.68
271.0	49.1	4.98	0.0882	85.12
270.9	99.5	10.08	0.1718	182.61
271.0	149.3	15.13	0.2484	290.80
271.0	199.3	20.19	0.3178	409.99

[#]10^4 × g of methane/g of methane-free oil.   (cont.)

AUXILIARY INFORMATION

METHOD/APPARATUS/PROCEDURE:	SOURCE AND PURITY OF MATERIALS:
Flow apparatus with both liquid and gaseous components continually passing into a mixing tube and then into a cell in which phases separated under gravity. Liquid sample removed from bottom of cell. Volume of vapor kept extremely small so that liquid composition did not change significantly. Composition of liquid sample found by stripping out gas. Details in source and ref. (1).	1. Matheson sample, purity better than 99 mole per cent. 2. See experimental values.

ESTIMATED ERROR:
$\delta T/K = \pm 0.05$; $\delta P/MPa = \pm 0.1\%$ or 0.03 (whichever is greater); $\delta S = \pm 2\%$.

REFERENCES:
1. Simnick, J. J.; Lawson, C. C.; Lin, H.-M.; Chao, K.-C.
Am. Inst. Chem. Eng. J.
<u>1977</u>, *23*, 469.

COMPONENTS: ORIGINAL MEASUREMENTS:

1. Methane; CH₄; [74-82-8]   Lin, H.-M.; Sebastian, H. M.;

2. Coal Liquids -- Distillate from   Simnick. J. J.; Chao, K.-C.
   Exxon Donor Solvent Process   *Ind. Eng. Chem. Process. Des. Dev.*
                                               1981, *20*, 253-256.

EXPERIMENTAL VALUES:

Details of samples

	Sample 1	Sample 2
Fraction boiling range	400-450 °F	500-600°F
Elemental analyses, wt-%		
C	89.09	89.57
H	9.65	10.35
N	0.06	0.13
O	0.65	0.57
S	0.05	0.19
sp gr at 60 °F	0.9320	0.9844

GC distillation

wt-% distilled at °F

wt-%	Sample 1	Sample 2
1	356.7	465.6
5	376.0	485.5
10	388.2	497.6
20	395.2	513.0
30	399.8	523.3
40	404.2	534.6
50	409.5	545.6
60	416.8	553.0
70	426.7	563.3
80	436.5	579.8
90	446.8	595.7
95	453.1	607.4
99	467.8	632.3
100	494.7	666.6

compound type analyses

(wt-% by MS)

	Sample 1	Sample 2
total saturates	28.22	26.91
paraffins	1.88	3.20
total aromatics	71.78	73.09
approximate molecular weight	154.34	182.30
Saybolt viscosity at 100 °F s	27.5	556.9
Saybolt viscosity at 210 °F s	12.6	9.6

COMPONENTS:	ORIGINAL MEASUREMENTS:
1. Methane; CH_4; [74-82-8] 2. Coal Liquid - Distillate from Solvent Refined Coal Process II	Lin, H.-M.; Sebastian, H. M.; Simnick, J. J.; Chao, K.-C. *Ind. Eng. Chem. Process. Des. Dev.* <u>1981</u>, *20*, 253-256.

VARIABLES:
 	
       Pressure
	PREPARED BY:
	
	       C. L. Young

EXPERIMENTAL VALUES:

T/K	P/atm	P/MPa	Mole fraction of methane in liquid, x_{CH_4}	Solubility[#], S
		Sample 1		
269.7	50.1	5.08	0.0933	90.673
269.7	51.0	5.17	0.0949	92.380
269.7	100.7	10.20	0.2050	227.26
269.9	147.6	14.96	0.2933	365.89
269.8	247.3	25.06	0.3974	581.37
		Sample 2		
270.3	51.2	5.19	0.0884	73.402
270.4	99.6	10.09	0.1672	151.93
270.1	150.7	15.27	0.2418	241.29
270.7	239.8	24.30	0.3668	438.35

[#] $10^4 \times$ g of methane/g of methane-free oil.

(cont.)

AUXILIARY INFORMATION

METHOD/APPARATUS/PROCEDURE:	SOURCE AND PURITY OF MATERIALS:
Flow apparatus with both liquid and gaseous components continually passing into a mixing tube and then into a cell in which phases separated under gravity. Liquid sample removed from bottom of cell. Volume of vapor kept extremely small so that liquid composition did not change significantly. Composition of liquid sample found by stripping out gas. Details in source and ref. (1).	1. Matheson sample, purity better than 99 mole per cent. 2. See experimental values.

	ESTIMATED ERROR:
	$\delta T/K = \pm 0.05$; $\delta P/MPa = \pm 0.1\%$ or 0.03 (whichever is greater); $\delta S = \pm 2\%$.

REFERENCES:

1. Simnick, J. J.; Lawson, C. C.; Lin, H.-M.; Chao, K.-C. *Am. Inst. Chem. Eng. J.* <u>1977</u>, *23*, 469.

COMPONENTS:

1. Methane; CH₄; [74-82-8]

2. Coal Liquid - Distillate from
   Solvent Refined Coal Process II

ORIGINAL MEASUREMENTS:

Lin, H.-M.; Sebastian, H. M.;
Simnick, J. J.; Chao, K.-C.
Ind. Eng. Chem. Process. Des. Dev.
<u>1981</u>, *20*, 253-256.

EXPERIMENTAL VALUES:

Details of samples

	Sample 1	Sample 2
Boiling range/°F	500-528	600-632
specific gravity, 60/60°F	0.9826	1.0306
molecular weight, ASTM D 2503	182	212
viscosity, SUS,		
cSt at 100 °F	41.8 (4.82)	74.3 (14.20)
210 °F	-(1.27)	33.5 (2.19)
250 °F	-(0.96)	-(1.53)
distillation, ASTM D 86		
over point, °F	436	566
end point, °F	580	672
5% cond. at °F	452	576
10	456	578
20	462	580
30	470	586
40	476	592
50	484	598
60	492	606
70	502	612
80	516	622
90	536	638
95	558	660
recovery, %	98.0	98.0
residue, %	1.0	1.0
loss, %	1.0	1.0

COMPONENTS:	ORIGINAL MEASUREMENTS:
1. Methane; CH_4; [74-82-8] 2. Creosote oil	Henson, B.J.; Tarrer, A. R.; Curtis, C. W.; Guln, J. A. *Ind. Eng. Chem. Process Des. Dev.* 1982, *21*, 575-579.
VARIABLES:	PREPARED BY: C. L. Young

EXPERIMENTAL VALUES:

t/°C	T/K	P/MPa	Solubility, S g CH_4/g creosote oil
30	303	5.6	0.0062
		8.8	0.0099
100	373	6.6	0.0065
		7.0	0.0063
		13.7	0.0129
		14.0	0.0135
		20.4	0.0191
		20.8	0.0192
200	473	7.5	0.0088
		8.0	0.0091
		15.2	0.0167
		15.3	0.0177
		20.9	0.0233
		21.6	0.0235
300	573	8.6	0.0110
		9.3	0.0120
		13.3	0.0171
		13.7	0.0171
		20.8	0.0264
		22.0	0.0268
400	673	7.5	0.0101
		7.8	0.0110
		14.2	0.0204
		14.6	0.0204
		21.3	0.0303
		22.1	0.0305

AUXILIARY INFORMATION

METHOD/APPARATUS/PROCEDURE:	SOURCE AND PURITY OF MATERIALS:
One gallon static equilibrium cell fitted with magnetic agigator. Samples taken from small volume sample loops through which equilibrium liquid was circulated. Gas in liquid sample as estimated by volumetric technique using a Toffel pump.	1. Matheson sample, purity 99 mole per cent. 2. Produced from Kentucky No. 9 coal. Elemental analysis % C 91.5 ± 0.7; H 6.4 ± 0.05; N 1.05 ± 0.31; S 0.53 ± 0.02.
	ESTIMATED ERROR: $\delta T/K = \pm 1$; $\delta S = \pm 4\%$ (estimated by compiler).
	REFERENCES:

COMPONENTS:	ORIGINAL MEASUREMENTS:
1. Methane; CH₄; [74-82-8]	Henson, B. J.; Tarrer, A. R.;
	Curtis., C. W.; Guln, J. A.
2. SRC recycle solvent	*Ind. Eng. Chem. Process Des. Dev.*
	<u>1982</u>, *21*, 575-579.

VARIABLES:	PREPARED BY:
	C. L. Young

EXPERIMENTAL VALUES:

t/°C	T/K	P/MPa	Solubility, S g CH₄/g recycle oil
100	373	5.4	0.0077
		5.6	0.0079
		11.4	0.0164
		12.0	0.0171
		16.2	0.0227
		16.2	0.0229
		18.8	0.0264
		19.4	0.0267
200	473	6.6	0.0095
		6.7	0.0101
		13.3	0.0194
		13.7	0.0209
		19.4	0.0282
		19.9	0.0300
		20.4	0.0308
300	573	6.0	0.0114
		6.7	0.0104
		14.0	0.0222
		14.5	0.0233
		20.0	0.0326
		20.3	0.0329
400	673	6.7	0.0112
		7.2	0.0119
		14.2	0.0238
		14.6	0.0247
		19.7	0.0335
		20.5	0.0349

AUXILIARY INFORMATION

METHOD/APPARATUS/PROCEDURE:	SOURCE AND PURITY OF MATERIALS:
One gallon static equilibrium cell fitted with magnetic agigator. Samples taken from small volume sample loops through which equilibrium liquid was circulated. Gas in liquid sample as estimated by volumetric technique using a Toffel pump.	1. Matheson sample, purity 99 mole per cent. 2. Produced from Kentucky No. 9 coal. Elemental analysys % C 88.2 ± 0.2; H 8.57 ± 0.12; N 0.49 ± 0.15; S 0.33 ± 0.03.

ESTIMATED ERROR:

$\delta T/K = \pm 1$; $\delta S = \pm 4\%$

(estimated by compiler).

REFERENCES:

COMPONENTS:	ORIGINAL MEASUREMENTS:
1. Methane; CH_4; [74-82-8] 2. Santowax R	Grove, N. H.; Whitley, F. J.; Woolmer, R. N. *J. Appl. Chem.* 1960, *10*, 101-109.
VARIABLES: Temperature, pressure	PREPARED BY: C. L. Young

EXPERIMENTAL VALUES:

T/K	$P/10^5$Pa	Solubility*	Ostwald coefficient, L
510	1.59	10.7	0.269
510	2.33	13.0	0.222
514	3.68	24.3	0.265
595	4.29	30.0	0.302
604	1.92	14.7	0.333
608	2.73	20.0	0.319
680	3.13	23.0	0.332
680	4.89	39.3	0.363
684	2.17	18.3	0.382

* moles of methane per Mg of Santowax R

AUXILIARY INFORMATION

METHOD/APPARATUS/PROCEDURE:	SOURCE AND PURITY OF MATERIALS:
Static cell with null pressure transducer. Pressure measured with Bourdon gauge. Temperature measured with thermocouple. Sample placed in cell and gas added at room temperature. Pressures on both sides of transducer kept approximately equal. Details in source.	1. No details given. 2. Analysis by infra-red method showed sample to be 11.8% o-terphenyl, 56.3% m-terphenyl, 29.3% p-terphenyl, 2.6% diphenyl and higher polyphenyls. Obtained from Monsanto Chemicals Ltd.
	ESTIMATED ERROR: $\delta T/K = \pm 1$; $\delta P/10^5$Pa $= \pm 0.01$; $\delta L_{CH_4} = \pm 10$%.
	REFERENCES:

COMPONENTS:	ORIGINAL MEASUREMENTS:
(1) Methane; CH_4; [74-82-8] (2) Petroleum	Gniewosz, S.; Walfisz, A. Z. Phys. Chem. <u>1887</u>, 1, 70 - 72.

VARIABLES:	PREPARED BY:
T/K = 283.15, 293.15 p/kPa = 101 ("atmospheric")	M. E. Derrick H. L. Clever

EXPERIMENTAL VALUES:

Temperature		Bunsen Coefficient	Ostwald Coefficient
$t/^0$C	T/K	α/cm^3(STP)cm^{-3}atm^{-1}	L/cm^3 cm^{-3}
10	283.15	0.143	
		0.142	
		0.146	
		0.144 Av.	0.149
20	293.15	0.129	
		0.134	
		0.131	
		0.131 Av.	0.141

The Ostwald coefficients were calculated by the compiler.

AUXILIARY INFORMATION

METHOD/APPARATUS/PROCEDURE:	SOURCE AND PURITY OF MATERIALS:
The apparatus consisted of an absorption flask connected to a gas buret by a flexible lead capillary. The system was thermostated in a large water bath. The volume of gas absorbed in a known volume of degassed petroleum was measured directly using the gas buret.	(1) Methane. No information. (2) Petroleum. Russian petroleum. Cleaned by boiling in a large copper flask.
	ESTIMATED ERROR: $\delta\alpha/\alpha$ = ± 0.05 (compiler)
	REFERENCES:

COMPONENTS:	ORIGINAL MEASUREMENTS:
(1) Methane; CH_4; [74-82-8] (2) Mineral oil	Rodman, C. J.; Maude, A. H. *Trans. Am. Electrochem. Soc.* <u>1925</u>, *47*, 71 - 92.

VARIABLES:	PREPARED BY:
T/K = 298.15, 353.15 p_1/kPa = 101.3 (760 mmHg)	H. L. Clever

EXPERIMENTAL VALUES:

Temperature		Bunsen Coefficient	Ostwald Coefficient	Solubility
t/°C	T/K	α/cm^3(STP)cm^{-3}atm^{-1}	L/cm^{3}cm^{-3}	g kg^{-1}
25	298.15	0.381	0.416	0.317
80	353.15	0.164	0.212	0.147

These values appear in the International Critical Tables,
McGraw-Hill Book Co., New York and London, Vol. III,
pp. 261 - 270 where they are credited to an industrial
report edited by A. H. Maude.

AUXILIARY INFORMATION

METHOD/APPARATUS/PROCEDURE:	SOURCE AND PURITY OF MATERIALS:
The apparatus consists of an 180 cm^3 absorption bottle connected to a 100 cm^3 gas buret. The absorption bottle sets in a thermostat, which is attached to a shaking machine. A weighed sample of oil is introduced into the absorption vessel. The sample is degassed by vacuum taking care to avoid excessive foaming. The gas is brought into the system. An initial buret reading taken, and the shaker is started and reading taken every 5 minutes until 2 or 3 constant readings are obtained.	(1) Methane. No information. (2) Mineral oil. A Pennsylvania base oil, 96 per cent saturated hydrocarbons, and distilling between 300 and 400°C. Density at 25°C = 0.840 and at 80°C = 0.800 g cm^{-3}. As a commercial product the oil is known as "Wemco A".
	ESTIMATED ERROR:
	REFERENCES:

COMPONENTS:	ORIGINAL MEASUREMENTS:
(1) Methane; CH_4; [74-82-8] (2) Paraffin Wax	Ridenour, W. P.; Weatherford, W. D.; Capell, R. G. *Ind. Eng. Chem.* <u>1954</u>, *46*, 2376-81.

VARIABLES:	PREPARED BY:
T/K = 345.35 p_1/kPa = 29.00 - 103.48	H. L. Clever

EXPERIMENTAL VALUES:

Temperature		Methane Pressure	Mol Fraction	Bunsen[a] Coefficient	Solubility Coefficient
$t/^0C$	T/K	$p_1/mmHg$	$10^3 x_1$	$\alpha/$	$/cm^3$ (STP) g^{-1}
72.2	345.35	217.5	1.77	0.305	0.113
		339.5	2.72	0.301	0.174
		479.3	3.88	0.303	0.248
		616.5	5.00	0.303	0.320
		776.2	6.32	0.304	0.404

[a] Bunsen coefficient, α/cm^3 (STP) cm^{-3} atm^{-1}.

AUXILIARY INFORMATION

METHOD/APPARATUS/PROCEDURE:

  The apparatus was similar to the equilibrium adsorption apparatus described by Brunaur, Emmett, and Teller (ref 1) for the measurement of the surface area of a solid catalyst.

  A weighed amount of wax was placed in the apparatus. The gas and solvent were equilibrated for 20 to 60 minutes. The gas volume absorbed from the buret system was calculated by the ideal gas law.

  The results of the absorption measurement were checked by a desorption measurement. The results of the two measurments agreed well.

SOURCE AND PURITY OF MATERIALS:
(1) Methane. Ohio Chemical Co. 97.8 % methane, 2.2 % heavier hydro-carbons.

(2) Paraffin wax. Described as 122 ^0F English melting wax. Molecular weight 350, actual melting point 123.2 ^0F (323.8 K), density 0.7716 g cm^{-3} at 293.3 K and 0.7662 g cm^{-3} at 298.0 K.

ESTIMATED ERROR:
$\delta T/K$ = ± 2
$\delta p/mmHg$ = ± 0.2
$\delta \alpha/cm^3$ = ± 0.004 (low pressure) to 0.001 (high press.)

REFERENCES:

1. Brunaur, S.; Emmett, P. H.; Teller, E.
   J. Am. Chem. Soc. <u>1938</u>, *60*, 309.

COMPONENTS:	ORIGINAL MEASUREMENTS:
(1) Methane; CH_4; [74-82-8] (2) Gasoline	Pomeroy, R. D.; Lacey, W. N.; Scudder, N. F.; Stapp, F. P. *Ind. Eng. Chem.* 1933, *25*, 1014-1019.

VARIABLES:	PREPARED BY:
$T/K = 303.15$ $p_1/MPa = 0.990, 1.982$ (9.77, 19.56 atm)	H. L. Clever

EXPERIMENTAL VALUES:

Temperature		Pressure		Solubility[1]
$t/°C$	T/K	p_1/atm	p_1/MPa	$c_s/cm^3\ cm^{-3}$
30	303.15	9.77	0.990	6.00
		19.56	1.982	11.77

[1] Gas volumes measured at 303.15 K (30°C) and 101.325 kPa (1 atm).

AUXILIARY INFORMATION

METHOD/APPARATUS/PROCEDURE:	SOURCE AND PURITY OF MATERIALS:
Measurements were carried out in a brass absorption cell designed for diffusion measurements.	(1) Methane. Gas obtained from a natural gas sample which was treated with activated carbon at pressures up to 70 atm. The methane contained up to 2 per cent ethane and a small amount of nitrogen. (2) Gasoline. Sample after treatment consisted largely of naphthalenes. B.p.(38 mmHg) $t/°C = 79.4 - 88.5$, density $\rho^{30}/g\ cm^{-3} = 0.7894$.

	ESTIMATED ERROR:
	$\delta T/K = \pm\ 0.05$ $\delta C_s/C_s = \pm\ 0.05$ (compiler)

COMPONENTS:	ORIGINAL MEASUREMENT:
(1) Methane; CH_4; [74-82-8]	Treshchina, N. I.
(2) Petroleum Mineralized water	*Trudy Vses. Neft. Nauch.-Issled.* *Geol.-Razvedoch* <u>1955</u>, No. 83, 566-71. *Chem. Abstr.* <u>1958</u>, *52*, 6771c.

EXPERIMENTAL VALUES:

Petroleum Sample		Temperature		Solubility Coefficient[a]
Location	Specific Gravity d_4^{20}	$t/^0C$	T/K	
Koschagyl, Emba oilfield	0.917	20 40 60	293 313 333	0.320 0.300 0.292
Buguruslan, Volga-Ural oilfield	0.913	20 40 60	293 313 333	0.334 0.315 0.308
Koschagyl, Emba oilfield	0.906	20 40 60	293 313 333	0.345 0.300 0.296
Kulsary, Emba oilfield	0.886	20 40 60	293 313 333	0.358 0.325 0.314
Kulsary, Emba oilfield	0.887	20 40 60	293 313 333	0.358 0.324 0.310
Kulsary, Emba oilfield	0.862	20 40 60	293 313 333	0.405 0.350 0.316
Grozny Grozny oilfield	0.835	20 40 60	293 313 333	0.458 0.407 0.328
Kulsary, Emba oilfield	0.813	20 40 60	293 313 333	0.502 0.463 0.428
Kulsary, Emba oilfield	0.782	20 40 60	293 313 333	0.566 0.507 0.478
Kerosene	0.819	20 40 60	293 313 333	0.505 0.436 0.415
Gasoline	0.746	20 40 60	293 313 333	0.745 0.665 0.599

[a] Solubility coefficient appears to be the Bunsen coefficient, $\alpha/cm^3 (STP) cm^{-3} atm^{-1}$.

The petroleum viscosities are 47.8, -, 38.0, 11.4, 11.4, 6.5, -, 3.1, - centistoke at 323 K as one comes down the table above.

Some information on the petroleum compositions are given in the paper.

The solubility of methane (natural gas) in water and mineralized water was given. See next page.

COMPONENTS:	ORIGINAL MEASUREMENTS:
(1) Methane; CH_4; [74-82-8] (2) Petroleum Mineralized water	Treshchina, N. I. *Trudy Vses. Neft. Nauch.-Issled. Geol.-Razvedoch* <u>1955</u>, No. 83, 566-71. *Chem. Abstr.* <u>1958</u>, *52*, 6771c.
VARIABLES: T/K = 293, 313, 333 p_1/kPa = 101.3	PREPARED BY: H. L. Clever

EXPERIMENTAL VALUES:

Temperature		Water Mineral[b] Content	Solubility Coefficient[a]
$t/^\circ C$	T/K	$m/g\ dm^{-3}$	
20	293	0	0.0331
		10	0.0315
		15	0.0305
		25	0.0290
40	313	0	0.0237
		10	0.0226
		15	0.0224
		25	0.0210
60	333	0	0.0200
		10	0.0190
		15	0.0187
		25	0.0180

[a] Appears to be the Bunsen coefficient.

[b] The solid in the mineralized water does not appear to be identified.

AUXILIARY INFORMATION

METHOD/APPARATUS/PROCEDURE:	SOURCE AND PURITY OF MATERIALS:
A detailed diagram of the apparatus is given in the paper.	(1) Methane. A natural gas from the western Ukraine, containing 99 % methane and less than 1 % nitrogen. (2) Petroleum, kerosene, and gasoline. Petroleum from wells in three oil fields. Specific gravity, viscosity, and some information on composition and various fractions was given. See data sheet.
	ESTIMATED ERROR: $\delta\alpha/\alpha = \pm 0.03$ (compiler)
	REFERENCES:

COMPONENTS:	ORIGINAL MEASUREMENTS:
(1) Methane; CH_4; 74-82-8 (2) Petroleum, crude oils	Safronova, T. P.; Zhuze, T. P. *Khim. i Tekhnol. Topliva i Masel* *1958*, *3 (2)*, 41-46. *Chem. Abstr.* *1958*, *52*, 8518d.

| VARIABLES:
    T/K = 293 - 373
  p_1/MPa = up to 34.5 | PREPARED BY:

     H. L. Clever |

EXPERIMENTAL VALUES:

Temperature		Pressure	Solubility Coefficient
$t/^0 C$	T/K	p_1/atm	/cm^3 cm^{-3} atm^{-1}

1. Nebit-Dag (Akchagylian layer) crude oil, Specific gravity, d_4^{20} = 0.8713, kinematic viscosity = 8.70 centistoke.

100	373	50	0.238
		100	0.265
		200	0.2738
		300	0.2673

2. Romashkino oilfield crude oil, Specific gravity, d_4^{20}= 0.8530, kinematic viscosity = 6.54 centistokes.

100	373	50	0.322
		100	0.320
		200	0.332
		300	0.337

3. Surakhany oil field crude oil, Specific gravity, d_4^{20}= 0.8494, kinematic viscosity = 5.19 centistokes.

100	373	50	0.266
		100	0.270
		200	0.281
		300	0.279

AUXILIARY INFORMATION

METHOD/APPARATUS/PROCEDURE:	SOURCE AND PURITY OF MATERIALS:
A detailed diagram of the high pressure apparatus was given in the paper. Many of the data are presented in figures of Solubility/cm^3 cm^{-3} *vs.* p_1/atm. A summary of the graphical data follows:	(1) Methane. Contained 5.1 % nitrogen, 0.05 % carbon dioxide, and 0.10 % carbon monoxide. (2) Petroleum crude oils. Four crude oils. Descriptions given above. Additional information on composition in the paper.

System	Temperatures	Maximum
		Pressure
	$t/^0C$	p_1/atm
1	25, 50, 100	340
2	20, 50, 100	300
3	20, 50, 100	300
4	50	300

ESTIMATED ERROR:
The compiler estimates the data have an uncertainty of 3 to 5 percent.

The fourth system is 4. Tuimazy oil field crude oil, Specific gravity, d_4^{20} = 0.8510, kinematic viscosity = 4.46 centistokes.

REFERENCES:

COMPONENTS:	ORIGINAL MEASUREMENTS:
(1) Methane; CH_4; [74-82-8] (2) Kerosene A-1	Hannaert, H.; Haccuria, M.; Mathieu, M. P. *Ind. Chim. Belge* 1967, *32*, 156-164.

VARIABLES:	PREPARED BY:
$T/K = 233.15 - 293.15$	E. L. Boozer H. L. Clever

EXPERIMENTAL VALUES:

Temperature Interval of Measurements T/K	Methane Mol % Range $10^2 x_1$/mol %	$K\pi\nu$/atm[1] at 293.15 K	Enthalpy of Dissolution ΔH/kcal mol^{-1}	Constant A
233.15-293.15	0.5	187	1.165	3.145

[1] $\log (K\pi\nu/\text{atm}) = A - (\Delta H/\text{cal mol}^-)/(2.3R(T/K))$

The author's definitions are:

$K = y_1/x_1 = \dfrac{\text{mole fraction gas in gas phase}}{\text{mole fraction gas in liquid phase,}}$

π /atm = total pressure,

ν = coefficient of fugacity.

The function, $K\pi\nu$/atm, is equivalent to a Henry's constant in the form

$H_{1,2}$/atm $= (f_1/\text{atm})/x_1$ where f_1 is the fugacity.

AUXILIARY INFORMATION

METHOD/APPARATUS/PROCEDURE:	SOURCE AND PURITY OF MATERIALS:
The authors describe three methods: 1.A. [Saturat. n° 1]. A measure of the static pressure of saturation in an apparatus which gave a precision of 10 - 15 %. 1.B. [Saturat. n° 2]. A measure of the static pressure of saturation in an apparatus which gave a precision of 2 - 5 %. 2. [Chromato]. A Gas liquid chromatographic method estimated to have a precision of 2 - 5 %. 3. [Anal. directe]. Direct analysis of the gaseous and liquid phases. Method 1.B. was used for this system.	(1) Methane. Air Liquide. Purity 99.95 per cent. (2) Kerosene A-1

Distillation range, °C	Density gcm^{-3},20°C	mol wt
A-1 150-280	0.7805	170

ESTIMATED ERROR:

REFERENCES:

COMPONENTS:	ORIGINAL MEASUREMENTS:
(1) Methane; CH_4; [74-82-8] (2) Athabasca bitumen	Svrcek, W.Y.; Mehrotra, A.K. *JCPT, J. Can. Pet. Technol.* 1982, 21, 31-8.

EXPERIMENTAL VALUES:

Temperature		Pressure	Viscosity	Density	Solubility	
$t/°C$	T/K	p_1/MPa	η/Pa s	ρ/g cm^{-3}	/cm^3 cm^{-3}	wt %
26.2	299.4	9.77	14.1	0.992	18.34	1.320
26.4	299.6	8.25	16.0	1.009	14.96	1.058
26.8	300.0	7.04	19.2	1.002	14.10	1.004
27.5	300.7	5.79	21.2	1.000	11.10	0.793
27.9	301.1	4.46	>23.5	1.011	9.18	0.648
27.5	300.7	3.32	>23.5	1.017	6.41	0.450
28.3	301.5	2.32	>23.5	1.025	4.51	0.314
28.2	301.4	1.59	>23.5	1.016	3.29	0.231
44.8	318.0	2.15	6.15	1.018	2.25	0.158
44.9	318.1	4.28	3.72	0.998	6.79	0.486
45.7	318.9	6.39	2.66	0.997	10.68	0.764
44.6	317.8	8.18	2.05	1.014	13.64	0.961
44.0	317.2	9.63	1.72	0.992	15.47	1.113
43.4	316.6	5.09	3.41	0.994	8.81	0.632
45.8	319.0	3.18	4.46	1.000	5.40	0.386
45.7	318.9	1.08	6.42	1.002	1.53	0.109
69.0	342.2	9.65	0.330	0.983	14.18	1.030
67.9	341.1	8.60	0.400	0.981	12.74	0.928
67.5	340.7	7.47	0.470	0.994	11.66	0.838
67.0	340.2	6.29	0.515	0.990	9.96	0.718
67.2	340.4	5.10	0.610	0.995	7.88	0.565
67.6	340.8	3.64	0.725	1.002	5.21	0.371
67.2	340.4	2.45	0.830	0.993	3.06	0.220
67.5	340.7	0.88	0.990	1.010	1.25	0.088
99.8	373.0	9.44	0.086	0.951	12.73	0.945
100.2	373.4	7.82	0.091	0.957	10.99	0.820
100.2	373.4	5.79	0.106	0.964	9.01	0.667
100.7	373.9	3.82	0.118	0.965	4.28	0.317
99.6	372.8	2.34	0.139	0.966	2.66	0.917
99.4	372.6	0.95	0.158	0.976	0.98	0.072

The volume/volume solubility is cm^3 (STP) cm^{-3}.

The density and viscosity values are for the gas saturated bitumen at the temperature and pressure of the solubility measurement. The density is considered reliable to 0.003 g cm^{-3}.

COMPONENTS:	ORIGINAL MEASUREMENTS:
(1) Methane; CH_4; [74-82-8] (2) Athabasca bitumen	Svrcek, W.Y.; Mehrotra, A.K. *JCPT, J. Can. Pet. Technol.* <u>1982</u>, *21*, 31-8.

| VARIABLES:
T/K = 299.4 - 373.9
p_1/MPa = 0.88 - 9.77 | PREPARED BY:

H. L. Clever |

EXPERIMENTAL VALUES:

See preceding page.

The solubility data are
repeated in a second pub-
lication (ref. 2).

AUXILIARY INFORMATION

METHOD/APPARATUS/PROCEDURE:

The design of the gas-solubility
experiment is based on the principle
that the gas that is dissolved in
bitumen will evolve when the pressure
is released and the temperature is
slightly increased. The volume of
the gas so released was measured at
a selected temperature of 100 °C and
atm pressure. The volumetric measurements were performed with a mercury-
filled Ruska pump. The pressure was
monitored with a precision Heise gage.
The sample and expansion chamber were
contained in a temperature controlled
oven. It was assumed that the system
was at equilibrium after the viscosity remained constant for at least
four hours.

SOURCE AND PURITY OF MATERIALS:

(1) Methane. No information.

(2) Athabasca bitumen. Obtained by
     toluene extraction of tar-sands
     of the Athabasca region (ref 1).
     Maltene distillables
       b.p. 600<°C   42.4%
       b.p. 600>°C   36.9%
       Asphaltenes   20.7%
     Above values not significantly
     changed by the experiment.

ESTIMATED ERROR:

REFERENCES:
1. Vorndran, L.D.L.; Serres, A.;
    Donnelly, J.K.; Moore, R.G.;
    Bennion, D.W.
    Can. J. Chem. Eng. <u>1980</u>, *58*, 580.
2. Mehrotra, A.K.; Svrcek, W.Y.
    JCPT, J. Can. Pet. Technol. <u>1982</u>,
    21, 95.

COMPONENTS:	ORIGINAL MEASUREMENTS:
(1) Methane; CH_4; [74-82-8]	Ramanujam, S.; Leipziger, S.; Weil, S. A.
(2) Simulated Light Aromatic Oil	*Ind. Eng. Chem. Process Des. Dev.* 1985, *24*, 107-11.

EXPERIMENTAL VALUES:

T/K	Total Pressure		Equilibrium Liquid-Vapor Mole Fractions							
			Methane		Benzene		Toluene		Decane	
	p_t/atm	p_t/MPa	x_1	y_1	x_2	y_2	x_3	y_3	x_7	y_7
403.3	19.5	1.98	0.035	0.873	0.410	0.092	0.210	0.024	0.119	0.003
	31.0	3.14	0.056	0.907	0.424	0.068	0.201	0.015	0.113	0.003
	51.0	5.17	0.086	0.935	0.431	0.049	0.193	0.011	0.104	0.002
	61.6	6.24	0.104	0.931	0.435	0.051	0.188	0.011	0.099	0.0018
	69.5	7.04	0.118	0.950	0.426	0.038	0.187	0.009	0.099	0.001
	82.7	8.38	0.148	0.957	0.403	0.033	0.183	0.007	0.095	0.001
	106.5	10.79	0.186	0.964	0.377	0.028	0.179	0.005	0.096	0.001
450.0	22.2	2.25	0.031	0.702	0.409	0.209	0.212	0.058	0.128	0.0096
	33.1	3.35	0.046	0.772	0.425	0.160	0.201	0.045	0.118	0.007
	45.6	4.62	0.070	0.817	0.426	0.131	0.195	0.033	0.112	0.007
	53.7	5.44	0.083	0.843	0.427	0.112	0.188	0.028	0.108	0.006
	61.9	6.27	0.101	0.872	0.419	0.091	0.189	0.022	0.105	0.005
	82.0	8.31	0.135	0.898	0.400	0.072	0.182	0.017	0.103	0.004
	96.5	9.78	0.160	0.905	0.389	0.067	0.179	0.016	0.101	0.004
494.4	23.5	2.38	0.023	0.517	0.400	0.304	0.179	0.096	0.133	0.029
	35.9	3.64	0.042	0.593	0.411	0.258	0.187	0.082	0.119	0.022
	40.1	4.06	0.054	0.631	0.411	0.233	0.185	0.075	0.116	0.021
	56.9	5.77	0.082	0.706	0.403	0.195	0.183	0.054	0.109	0.015
	64.5	6.55	0.098	0.725	0.404	0.183	0.182	0.050	0.107	0.014
	74.0	7.50	0.115	0.749	0.389	0.162	0.177	0.044	0.106	0.017
	100.1	10.14	0.151	0.760	0.378	0.159	0.171	0.042	0.102	0.014
550.0	33.1	3.35	0.038	0.368	0.311	0.336	0.159	0.125	0.138	0.065
	40.1	4.06	0.050	0.418	0.319	0.313	0.159	0.112	0.139	0.058
	51.4	5.21	0.074	0.475	0.330	0.274	0.166	0.103	0.130	0.052
	73.8	7.48	0.104	0.570	0.333	0.232	0.165	0.085	0.114	0.039
	89.8	9.10	0.134	0.587	0.337	0.223	0.166	0.081	0.108	0.041
	97.6	9.89	0.143	0.595	0.346	0.228	0.162	0.078	0.103	0.039

Mole fraction in liquid x.

Mole fraction in vapor y.

 The simulated light aromatic oil has 16 components. Only the compositions with respect to benzene, toluene and decane are given above. The three liquids made up 0.901 mole fraction of the oil. The liquid-vapor composition of all 16 components in given in the origianl paper.

 See the next page for the mass fraction and mole fraction composition of the oil.

COMPONENTS:	ORIGINAL MEASUREMENTS:
(1) Methane; CH_4; [74-82-8] (2) Simulated Light Aromatic Oil	Ramanujam, S.; Leipziger, S.; Weil, S. A. *Ind. Eng. Chem. Process Des. Dev.* 1985, *24*, 107-11.

| VARIABLES:
$T/K = 403.0 - 550.0$
$p_t/MPa = 1.98 - 10.79$ | PREPARED BY:

H. L. Clever |

SOURCE AND PURITY OF MATERIALS:

(1) Methane. Matheson Gas Co. Stated to be 99.9 percent purity.

(2) Simulated Light Aromatic Oil. Composition:

No.	Component	Formula	Registry Number	Mass Fraction	Mole Fraction
1.	Benzene	C_6H_6	[71-43-2]	0.458	0.548
2.	Toluene or methylbenzene	C_7H_8	[108-88-3]	0.183	0.186
3.	Octane	C_8H_{18}	[111-65-9]	0.0037	0.003
4.	p-Xylene or 1,4-dimethyl benzene	C_8H_{10}	[106-42-3]	0.092	0.081
5.	o-Xylene or 1,2-dimethyl benzene	C_8H_{10}	[95-47-6]	0.027	0.024
6.	Mesitylene or 1,3,5-trimethyl benzene	C_9H_{12}	[108-67-8]	0.011	0.009
7.	Decane	$C_{10}H_{22}$	[124-18-5]	0.131	0.086
8.	Naphthalene	$C_{10}H_8$	[91-20-3]	0.055	0.040
9.	1-Methylnaphthalene	$C_{11}H_{10}$	[90-12-0]	0.014	0.009
10.	1,1'-Biphenyl	$C_{12}H_{10}$	[92-52-4]	0.0055	0.0034
11.	Acenaphthene or 1,2-Dihydro-naphthylene	$C_{12}H_{10}$	[83-32-9]	0.0046	0.0027
12.	Fluorene or 9-*H*-fluorene	$C_{13}H_{10}$	[86-73-7]	0.0046	0.0026
13.	1-Phenylnaphthalene	$C_{16}H_{12}$	[605-02-7]	0.0055	0.0025
14.	Phenanthrene	$C_{14}H_{10}$	[85-01-8]	0.0046	0.0024
15.	Fluoranthene	$C_{16}H_{10}$	[206-44-0]	0.0018	0.0008
16.	Chrysene	$C_{18}H_{12}$	[218-32-9]	0.0014	0.0006

AUXILIARY INFORMATION

METHOD/APPARATUS/PROCEDURE:

The apparatus is a modification of that of Simnick *et al.* (ref 1). The details are given by Srinivasan (ref 2).

The apparatus is a recirculation type equilibration system. The equilibration cell is a 300 cm³ autoclave. The liquid was recycled by a metering pump.

Each phase was sampled in its respective sampling loop. The samples were analyzed by GLC with a programmable integrator.

All chemicals were used as received. No decomposition of the components, or reaction of the components was observed in the study.

Values are the average of two determinations. Agreement for the major components range from 1 to 5 %.

SOURCE AND PURITY OF MATERIALS:

(2) Simulated oil (continued)
Fisher Scientific Co. Benzene (crystallizable, low thiophene), Toluene (99 %), Octane (Reagent), p-Xylene (certified), o-Xylene (Reagent), naphthalene (scintanalyzed), 1-methylnaphthalene (purified), Biphenyl (Reagent).
Phillips Petroleum Co. Decane (99 %).
Aldrich Chemical Co. Numbers 11 - 16 on list above.

ESTIMATED ERROR:

$$\delta T/K = \pm 0.5$$
$$\delta p_1/p_1 = \pm 0.0025$$

REFERENCES:

1. Simnick, J. J.; Lawson, C. C. Lin, H. M.; Chao, K. C.
   A.I.Ch.E.J. 1877, *23*, 469.
2. Srinivasan, R.
   Ph. D. Thesis, Department Gas Engineering, IIT, Chicago, IL 1981.

COMPONENTS:	ORIGINAL MEASUREMENTS:
(1) Methane; CH_4; [74-82-8]	Fischer, F.; Zerbe, C.
(2) Various pure solvents and petroleum products (see table below)	*Brennstoff-Chem.* <u>1923</u>, *4*, 17-9.

EXPERIMENTAL VALUES:

Temperature		Pressure	Solvent	Gas Volume Evolved	Gas in 1 g of Solvent at 1 atm
$t/^0C$	T/K	p_1/atm	/g	/cm^3	/cm^3

Water; H_2O; [7732-18-5]

20	293	18	71.9	112	0.09

Benzene; C_6H_6; [71-43-2]

23	296	17.5	10.8	97	0.51

Benzene, technical; C_6H_6; [71-43-2]

23	296	17.5	11.5	94	0.47

Dimethylbenzene; C_8H_{10}; [1330-20-7]

23	296	18	10.6	102	0.53

Methanol; CH_4O; [67-56-1]

20	293	17	13.3	104	0.46

Ethanol; C_2H_6O; [64-17-5]

21	294	17.5	9.3	98	0.60

3-Methyl-1-butanol or isoamyl alcohol; $C_5H_{12}O$; [123-51-3]

20	293	17	11.3	84	0.44

Methylphenol or tricresol; C_7H_8O; [1319-77-3]

21	294	17	20.5	92	0.26

1,1'-Oxybisethane or diethylether; $C_4H_{10}O$; [60-29-7]

20	293	18	2.7	44	0.91

2-Propanone or acetone; C_2H_6O; [67-64-1]

20	293	18	7.8	81	0.61

Acetic acid; $C_2H_4O_2$; [64-19-7]

20	293	18	8.7	71	0.45

Trichloromethane; or chloroform; $CHCl_3$; [67-66-3]

20	293	18	15.0	88	0.82

Carbon disulfide; CS_2; [75-15-0]

20	293	18	9.9	64	0.36

Aniline; C_6H_7N; [62-53-3]

20	293	18	33.6	90	0.16

Nitrobenzene; $C_6H_5NO_2$; [98-95-3]

20	293	18	32.0	88	0.16

Petroleum

20	293	14	12.0	94	0.56
20	293	15	12.3	101	0.55

Paraffin oil

20	293	15	10.4	68	0.44

COMPONENTS:	ORIGINAL MEASUREMENTS:
(1) Methane; CH_4; [74-82-8] (2) Various pure solvents and petroleum products (see table below)	Fischer, F.; Zerbe, C. *Brennstoff-Chem.* 1923, *4*, 17-9.

VARIABLES:	PREPARED BY:
T/K = 293, 294, 295 p_1/kPa = 1419 - 1824 Gas volumes and solubilities 101.3 kPa	H. L. Clever

EXPERIMENTAL VALUES:

Temperature		Pressure	Solvent	Gas Volume Evolved	Gas in 1 g of Solvent at 1 atm
$t/°C$	T/K	p_1/atm	w/g	v/cm^3	$/cm^3$
Petroleum ether, boiling point up to 65 $°C$					
22	295	17	2.6	62	1.34
Petroleum ether, boiling point 65 - 100 $°C$					
20	293	18	5.9	89	0.84
Petroleum ether, boiling point 100 - 150 $°C$					
20	293	18	8.6	102	0.66
Urteerkohlenwasserstoffe (low temperature tar hydrocarbons)					
20	293	18	12.3	89	0.40
Urteerphenole (low temperature tar phenols), 250 - 300 $°C$					
20	293	18	14.6	71	0.27
Urteerfraktion (low temperature tar fraction), 250-300 $°C$					
21	294	17	15.3	94	0.36
Braunkohlentreiböl (lignite coal motor oil)					
21	294	18	13.1	92	0.39
Braunkohlenkreosot (lignite coal creosote)					
21	294	17	24.8	94	0.22
Steinkohlenkarbolöl (coal tar oil)					
22	295	18	15.5	89	0.32

AUXILIARY INFORMATION

METHOD/APPARATUS/PROCEDURE:

The gas was pumped into an evacuated steel cylinder which contained the degassed solvent. The cylinder was shaken until a constant pressure indicated equilibrium was attained. Part of the saturated liquid was transferred into a buret where the dissolved gas was extracted at one atm pressure. The saturation pressure was taken as the mean of the cylinder pressure before and after sampling. The change in pressure before and after sampling was two atm. See the earlier paper on oxygen solubility (ref 1) for more information.

EVALUATORS'S COMMENT: These data are of marginal accuracy. They should be used only if more modern values are not available for a system.

Gas volumes measured at atmospheric pressure and the temperature of the measurement.

SOURCE AND PURITY OF MATERIALS:

(1) Methane. Gas sample contained 79.4 % methane, 17.1 % nitrogen, 2.8 % oxygen, and 0.7 % carbon dioxide.

(2) Solvents. Sources not given. Density and sometimes vapor pressure on pure compounds. Boiling points of the petroleum ethers. Data are in (ref 1).

ESTIMATED ERROR:

10 - 25 per cent (compiler).

REFERENCES:

1. Fischer, F.; Pfleiderer, G.
   Z. Anorg. Chem. 1922, *124*, 61.

COMPONENTS:	EVALUATOR:
(1) Methane; CH_4; [74-82-8] (2) Alkanols (alcohols)	H. Lawrence Clever Department of Chemistry Emory University Atlanta, GA 30322 USA 1985, April

CRITICAL EVALUATION:

THE SOLUBILITY OF METHANE IN ALKANOLS AT METHANE PARTIAL PRESSURES UP TO 0.200 MPa (*ca.* 2 ATM).

Nine papers report solubility data on methane + alkanol systems over the C_1 to C_{12} range of alcohols. All of the workers have used volumetric methods. Except for 1-propanol all of the measurements were made in the 283-318 K temperature interval at pressures near 100 kPa. Komarenko and Manzhelii (ref 6) measured the solubility of methane in 1-propanol over the 173-243 K range at a methane partial pressure of 26.7 kPa (200 mmHg).

The results reported by Winkler (ref 3) and Friedel and Gorgeu (ref 1) are qualitative. The Winkler value is rejected, but the value of Friedel and Gorgeu appears useful. The work of McDaniel (ref 2) is poor. His results are 10 to 20 percent smaller than more modern results, and his temperature coefficients of solubility are sometimes of much larger magnitude than the more recent measurements.

Figure 1 shows the mole fraction solubility at 298.15 K and 0.1013 MPa methane partial pressure in the normal alcohols. The line was drawn to follow the results of Lannung and Gjaldbaek (ref 5), Ben-Naim and Yaacobi (ref 7), and Wilcock, Battino, Danforth and Wilhelm (ref 8); the three papers we judge to contain the most reliable data. The results reported by Makranczy, Rusz, and Balog-Megyery (ref 9) are larger than all other results and do not show the decrease in solubility as the alcohol becomes more polar that the other workers show. Although we do not have proof, we suspect the Makranczy *et al.* results are too large. The results of Boyer and Bircher (ref 4) show the effect of the increased polarity of the low molecular weight alcohols on the solubility, but their results also appear to be too large in the high carbon number alcohols. Their temperature coefficients of solubility show larger variations from alcohol to alcohol than do the results of others.

In evaluating ethane, propane, butane and 2-methylpropane solubility in alcohols Hayduk (ref 10) fitted the data to equations of the type

$$\ln x_1 = b_1 + b_2 \ln C_n$$

COMPONENTS:	EVALUATOR:
(1) Methane; CH_4; [74-82-8] (2) Alkanols (alcohols)	H. Lawrence Clever Department of Chemistry Emory University Atlanta, GA 30322 USA 1985, April

CRITICAL EVALUATION:

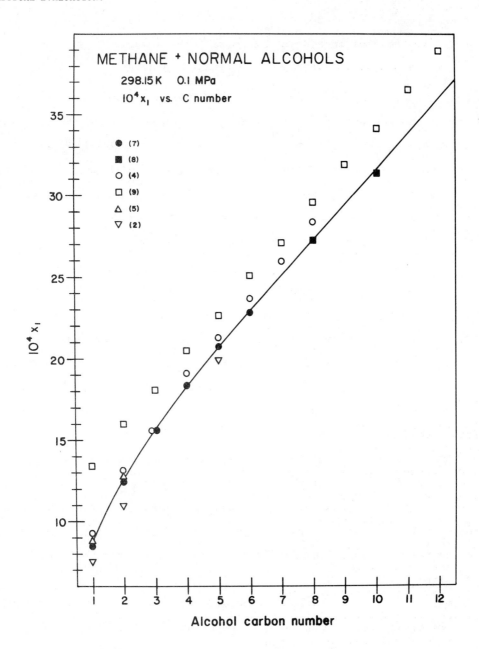

Figure 1. Methane + n-Alcohols. The methane mole fraction solubility
        at 298.15 K and 0.101 MPa vs. alcohol carbon number.

The equation $\ln x_1 = -7.0388 + 0.54046 \ln C_n$ or

$x_1 = (8.7718 \times 10^{-4}) \, C_n^{0.54046}$ reproduces the line of the above

graph within one percent from C_n = 1 to 8. As the carbon number

increases above 8 the equation values are smaller than the line

values.

at 298.15 K where x_1 is the mole fraction solubility and C_n is the alcohol carbon number.

We have done the same for the methane solubility values of Lannung and Gjaldbaek, Ben-Naim and Yaacobi, and Wilcock *et al*. to obtain the equation

$$\ln x_1 = -7.0388 + 0.54046 \ln C_n \quad \text{with } r = 0.9987$$

and for all of the data on Figure 1 to obtain the equation

$$\ln x_1 = -7.0155 + 0.55673 \ln C_n \quad \text{with } r = 0.9728$$

The first equation reproduces the line of Figure 1 with an average deviation of one percent for $C_n = 1$ through 8, but by $C_n = 12$ the calculated value is 6.7 percent low. Thus, equations of this type empirically reproduce the changing solubility with carbon number quite well up to carbon number 8, but are unreliable and give results that are progressively too small as the carbon number increases beyond $C_n = 8$.

The individual systems are discussed below in more detail.

Methane + Methanol; CH_3OH; [67-56-11]

McDaniel (ref 2), Boyer and Bircher (ref 4), Lannung and Gjaldbaek (ref 5), Ben-Naim and Yaacobi (ref 7) and Makranczy, Rusz and Balog-Megyery (ref 9) report values of the solubility of methane in methanol. At 298.15 K the results of Lannung and Gjaldbaek and of Ben-Naim and Yaacobi accord within one percent. Their enthalpies and entropies of solution from the temperature coefficient of solubility agree within 8 and 2 percent, respectively. At 298.15 K the McDaniel value is 14 percent smaller, the Boyer and Bircher value 6 percent larger, and the Makranczy *et al*. value 55 percent larger than the average of the Lannung and Gjaldbaek and the Ben-Naim and Yaacobi values. The temperature coefficient of solubility of McDaniel gives an enthalpy of solution that is four times the magnitude of the other results.

The values of Lannung and Gjaldbaek and of Ben-Naim and Yaacobi were weighted twice, and the single value of Boyer and Bircher was weighted once in a linear regression to obtain the tentative equation for the mole fraction solubility of methane in methanol over the 283.15 to 308.15 K interval

$$\ln x_1 = -8.52458 + 4.40002/(T/100 \text{ K})$$

with a standard error about the regression line of 1.35×10^{-5}. From the constants of the equation the temperature independent thermodynamic

COMPONENTS:	EVALUATOR:
(1) Methane; CH_4; [74-82-8] (2) Alkanols (alcohols)	H. Lawrence Clever Department of Chemistry Emory University Atlanta, GA 30322 USA 1985, April

CRITICAL EVALUATION:

changes for the transfer of one mole of methane from the gas at 0.1 MPa to the infinitely dilute solution are

$$\Delta \overline{H}_1^0 / \text{ kJ mol}^{-1} = -3.66 \text{ and } \Delta \overline{S}_1^0 / \text{ J K}^{-1} \text{ mol}^{-1} = -70.9$$

Smoothed values of the solubility are in Table 1.

Table 1. Solubility of Methane in Methanol. Tentative mole fraction solubility and partial molal Gibbs energy of solution as a function of temperature at a methane partial pressure of 0.1013 MPa.

T/K	$10^4 x_1$	$\Delta \overline{G}_1^0 / \text{kJ mol}^{-1}$
283.15	9.39	16.410
288.15	9.14	16.765
293.15	8.91	17.119
298.15	8.68	17.473
303.15	8.48	17.828
308.15	8.28	18.182

Methane + Ethanol; CH_3CH_2OH; [64-17-5]

The same five papers (ref 2,4,5,7,9) report the solubility of methane in ethanol. Again McDaniel's values are the smallest by about 14 percent, but his enthalpy of solution value agrees with the other workers. His values are doubtful. The single value of Makranczy $et\ al$. at 298.15 K is 26 percent the largest. It is classed as doubtful.

The values of Lannung and Gjaldbaek, Ben-Naim and Yaacobi, and Boyer and Bircher agree within about 3 percent of 298.15 K. Their values are classed as tentative. Their results were combined in a linear regression to obtain the tentative equation for the mole fraction solubility of methane in ethanol of a partial pressure of 0.1013 MPa over the 283.15 to 308.15 K interval

$$\ln x_1 = -8.11131 + 4.31444 / (T/100 \text{ K})$$

with a standard error about the regression line of 1.44×10^{-5}.

From the constants of the equation the temperature independent thermodynamic changes for the transfer of one mole of methane from the gas at

0.1013 MPa to the infinitely dilute solution are

$$\Delta \overline{H}_1^0 / kJ \ mol^{-1} = -3.59 \quad \text{and} \quad \Delta \overline{S}_1^0 / J \ K^{-1} \ mol^{-1} = -67.4$$

Smoothed values of the solubility are in Table 2.

Table 2.   The solubility of Methane in Ethanol. Tentative values of the
           mole fraction solubility and partial molar Gibbs energy of
           solution as a function of temperature at a methane partial
           pressure of 0.1013 MPa.

T/K	$10^4 x_1$	$\Delta \overline{G}_1^0 / kJ \ mol^{-1}$
283.15	13.8	15.508
288.15	13.4	15.846
293.15	13.1	16.183
298.15	12.8	16.520
303.15	12.5	16.857
308.15	12.2	17.194

Methane + 1-Propanol; CH_3CH_2OH; [71-23-8]

Boyer and Bircher (ref 4), Ben-Naim and Yaacobi (ref 7), and Makranczy
et al. (ref 9) report the solubility of methane in 1-propanol in the room
temperature region. Komarenko and Manzhelii (ref 6) measured the solubility of methane at a partial pressure of 26.7 kPa (200 mmHg) over the 173.15
to 243.15 K interval.

Boyer and Bircher, and Ben-Naim and Yaacobi report the same mole
fraction solubility at 298.15 K. Makranczy et al. report a value that is
16 percent larger. The value was not used.

The data are fit by linear regressions to three equations. First,
the data of Komarenko and Manzhelii were calculated for a methane partial
pressure of 0.1013 MPa assuming Henry's law is obeyed, and the equation
for the temperature interval of 173.15 to 243.15 K obtained

$$\ln x_1 = -5.67059 + 3.23393/(T/100 \ K) - 2.047 \ln(T/100K)$$

with a standard error about the regression line of 5.43×10^{-5}.

Second, the data of Boyer and Bircher, and Ben-Naim and Yaacobi were
combined to obtain an equation for the mole fraction solubility over the
283.15 to 308.15 K interval

$$\ln x_1 = -8.05738 + 4.77195 /(T/100 \ K)$$

with a standard error about the regression line of 1.26×10^{-5}.

Third, the three data sets were combined in a linear regression to

COMPONENTS:	EVALUATOR:
(1) Methane; CH_4; [74-82-8] (2) Alkanols (alcohols)	H. Lawrence Clever Department of Chemistry Emory University Atlanta, GA 30322 USA 1985, April

CRITICAL EVALUATION:

obtain the equation for the mole fraction solubility over the 173.15 to

308.15 K interval at a methane partial pressure of 0.1013 MPa.

$$\ln x_1 = -15.5019 + 14.8315/(T/100K) + 3.7230 \ln (T/100 \text{ K})$$

with a standard error about the regression line of 1.01×10^{-4}.

Smoothed data from the three equations are in Table 3.

Table 3.   The Solubility of Methane in 1-Propanol. Tentative values of
the mole fraction solubility as a function of temperature at a
methane partial pressure of 0.1013 MPa from three equations.

	$10^4 x_1$ from equations for temperature interval		
T/K	173-243 K	283-308 K	173-308 K
173.15	72.5		75.0
183.15	58.4		57.9
193.15	47.8		46.4
203.15	39.7		38.4
213.15	33.4		32.6
223.15	28.4		28.3
233.15	24.4		25.1
243.15	21.1		22.6
253.15			20.6
263.15			19.0
273.15			17.8
283.15		17.1	16.8
288.15		16.6	16.4
293.15		16.1	16.0
298.15		15.7	15.6
303.15		15.3	15.3
308.15		14.9	15.0

The thermodynamic changes for the transfer of one mole of methane

from the gas at 0.1013 MPa to the infinitely dilute solution calculated

from the three equations follow Table 3. The equation for the data of
Komarenko and Mazhelii gives enthalpy changes that become more negative as
the temperature increases. The more usual trend is for ΔH to become less
negative as T increases as is seen in the third equation that combines the
three data sets.

The thermodynamic changes calculated from the constants of the fitted
equations are

	173-243 K			283-308 K		173-308 K		
T/K	ΔH	ΔS	ΔC_p	ΔH	ΔS	ΔH	ΔS	ΔC_p
183	-5.81	-74.5	-17.0	-	-	-6.66	-79.2	31.0
233	-6.66	-78.6	-17.0	-	-	-5.11	-71.7	31.0
288	-	-	-	-3.97	-67.0	-3.41	-65.2	31.0
298	-	-	-	-3.97	-67.0	-3.10	-64.1	31.0

Units: kJ mol^{-1} and J K^{-1} mol^{-1}.

Methane + 2-Propanol; $CH_3CHOHCH_3$; [67-63-0]

Only McDaniel (ref 2) reports the solubility of methane in 2-propanol.
At 298.15 K the mole fraction solubility in 2-propanol is about 8 percent
smaller than in 1-propanol. This is contrary to most of our experience
that gases of this molecular weight are more soluble in the branched than
the unbranched carbon solvent. Since McDaniel's solubility values are
usually too small it is reasonable to assume these values are too small.
The enthalpy of solution compares well with those of other workers, indi-
cating the temperature coefficient of solubility may be correct.

The data are classed as tentative, but it is suspected they are at
least 10 percent too small. A linear regression gives the equation for
the 293.15 to 313.15 K interval at a methane partial pressure of 0.1013
MPa

$$\ln x_1 = -7.99145 + 4.31546/(T/100 \text{ K})$$

with a standard error about the regression line of 4.55×10^{-6}.

From the constants of the equation the temperature independent thermodynamic changes for the transfer of one mole of methane from the gas at
0.1013 MPa to the infinitely dilute solution are

$$\Delta \overline{H}_1^0/\text{k J mol}^{-1} = -3.59 \text{ and } \Delta \overline{S}_1^0/\text{J K}^{-1} \text{ mol}^{-1} = -66.4$$

Smoothed solubility values are in Table 4.

COMPONENTS:	EVALUATOR:
(1) Methane; CH_4; [74-82-8] (2) Alkanols (alcohols)	H. Lawrence Clever Department of Chemistry Emory University Atlanta, GA 30322 USA 1985, April

CRITICAL EVALUATION:

Table 4.   The Solubility of Methane in 2-Propanol.   Tentative values of
the mole fraction solubility and partial molal Gibbs energy as
a function of temperature at a methane partial pressure of
0.1013 MPa.

T/K	$10^4 x_1$	$\Delta \bar{G}_1^0/kJ\ mol^{-1}$
293.15	14.7	15.889
298.15	14.4	16.222
303.15	14.0	16.554
308.15	13.7	16.887
313.15	13.4	17.219

Methane + 1-Butanol; C_4H_9OH; [71-36-3]

Boyer and Bircher (ref 4), and Ben-Naim and Yaacobi (ref 7) report
solubility values that agree within 4 percent at 298.15 K and within 5 percent at 303.15 K.   Makranczy *et al.* report a single value which is 10 percent larger than the average of the other two at 298.15 K.

The solubilities of Boyer and Bircher and of Ben-Naim and Yaacobi
were combined in a linear regression to obtain the equation for the mole
fraction solubility over the 283.15 to 308.15 K interval at 0.1013 MPa

$$\ln x_1 = -7.76446 + 4.41862/(T/100\ K)$$

with a standard error about the regression line of 3.75×10^{-5}.

From the constants of the equation the temperature independent thermodynamic changes for the transfer of one mole of methane from the gas at
0.1013 MPa pressure to the infinitely dilute solution are

$$\Delta \bar{H}_1^0/kJ\ mol^{-1} = -3.67 \text{ and } \Delta \bar{S}_1^0/J\ K^{-1}\ mol^{-1} = -64.6$$

Smoothed values of the solubility are in Table 5.

Table 5. Solubility of Methane in 1-Butanol. Tentative values of the
          mole fraction solubility and the partial molar Gibbs energy of
          solution as a function of temperature at a methane partial
          pressure of 0.1013 MPa.

T/K	$10^4 x_1$	$\Delta \bar{G}_1^0$/kJ mol^{-1}
283.15	20.2	14.605
288.15	19.7	14.928
293.15	19.2	15.251
298.15	18.7	15.574
303.15	18.2	15.896
308.15	17.8	16.219

Methane + 2-Methyl-1-propanol; $CH_3CH(CH_3)CH_2OH$: [78-83-1]

Only Winkler's (ref 3) qualitative measurement, which corresponds to
a mole fraction solubility of 13 x 10^{-4}, is available for the system.
The value appears to be much too small and is rejected.

Methane + 1-Pentanol; $CH_3(CH_2)_3CH_2OH$; [71-41-0]

McDaniel (ref 2), Boyer and Bircher (ref 4), Ben-Naim and Yaacobi
(ref 7), and Makranczy et al. (ref 9) report solubility values for the
methane + 1-pentanol system. At 298.15 K the four solubility values
show a range of about 13 percent.

All values are classed as tentative and all values were combined in
a linear regression to obtain the equation for the mole fraction solubility over the 283.15 to 303.15 K interval at 0.1013 MPa pressure.

$\ln x_1 = -7.42305 + 3.7397/(T/100 \text{ K})$

with a standard error about the regression line of 9.30 x 10^{-5}.

The thermodynamic changes for the transfer of one mole of methane
from the gas at 0.1013 MPa to the infinitely dilute solution are

$\Delta \bar{H}_1^0$/kJ mol^{-1} = -3.11 and $\Delta \bar{S}_1^0$/J K^{-1} mol^{-1} = -61.7

Smoothed solubility data are in Table 6.

COMPONENTS:	EVALUATOR:
(1) Methane; CH_4; [74-82-8] (2) Alkanols (alcohols)	H. Lawrence Clever Department of Chemistry Emory University Atlanta, GA USA 1985, April

CRITICAL EVALUATION:

Table 6.   Solubility of Methane in 1-Pentanol. Tentative values of the
           mole fraction solubility and partial molar Gibbs energy of solu-
           tion as a function of temperature at a methane partial pressure
           of 0.1013 MPa.

T/K	$10^4 x_1$	$\Delta \overline{G}_1^0 / kJ \ mol^{-1}$
283.15	22.4	14.366
288.15	21.9	14.675
293.15	21.4	14.983
298.15	20.9	15.292
303.15	20.5	15.600
308.15	20.1	15.909

Methane + 3-Methyl-1-butanol; $CH_3CH(CH_3)CH_2CH_2OH$; [123-51-3]

Friedel and Gorgeu (ref 1) report an absorption experiment at 285.7 K
and 0.1013 MPa. The mole fraction solubility calculated from their measurement is 23×10^{-4} which is of similar magnitude to the solubility of methane in 1-propanol at that temperature. The value is classed as tentative.

Methane + 1-Hexanol; $CH_3(CH_2)_4CH_2OH$; [111-27-3]

Ben-Naim and Yaacobi (ref 7) report the solubility of methane in
1-hexanol at five degree intervals from 283.15 to 303.15 K. Boyer and
Bircher (ref 4) and Makranczy et al. report single solubility values at
298.15 K which are 4 and 10 percent larger, respectively, than the Ben-Naim
and Yaacobi value. All values are classed as tentative, but only the
Ben-Naim and Yaacobi values were used in the linear regression to obtain
the equation for the mole fraction solubility over the 283.15 to 303.15 K
interval at 0.1013 MPa.

$$\ln x_1 = -7.84121 + 5.24282/(T/100 \ K)$$

with a standard error about the regression line of 5.00×10^{-6}.

The thermodynamic changes for the transfer of one mole of methane
from the gas at 0.1013 MPa to the infinitely dilute solution are

$\Delta \overline{H}_1^0/k$ J mol^{-1} = -4.36 and $\Delta \overline{S}_1^0$ J K^{-1} mol^{-1} = -65.2

The smoothed solubility data are in Table 7.

Table 7.   The Solubility of Methane in 1-Hexanol. Tentative values of the mole fraction solubility and the partial molar Gibbs energy of solution as a function of temperature at a methane partial pressure of 0.1013 MPa.

T/K	$10^4 x_1$	$\Delta \overline{G}_1^0/k$ J mol^{-1}
283.15	25.0	14.101
288.15	24.3	14.427
293.15	23.5	14.753
298.15	22.8	15.079
303.15	22.2	15.405

Methane + 1-Heptanol; $CH_3(CH_2)_5CH_2OH$; [111-70-6]

Two solubility values are reported for this system. Boyer and Bircher (ref 4) report a mole fraction solubility of 26.0 x 10^{-4} at 298.15 K and Makranczy *et al.* (ref 9) report a value of 27.1 x 10^{-4} at 298.15 K. The values are classed as tentative.

Methane + 1-Octanol; $CH_3(CH_2)_6CH_2OH$: [111-87-5]

Wilcock, Battino, Danforth, and Wilhelm (ref 8) report solubilities at three temperatures, Boyer and Bircher (ref 4) at two temperatures, and Makranczy *et al.* (ref 9) at one temperature for the system. At 298.15 K Boyer and Bircher's value is 4 percent larger, and the value of Makranczy *et al.* is 8.5 percent larger than the value of Wilcock *et al.* All of the data are classed as tentative, but only the solubilities reported by Wilcock *et al.* and Boyer and Bircher were used in the linear regression to obtain the equation for the mole fraction solubility over the 283.15 to 313.15 interval at 0.1013 MPa.

ln x_1 = -7.43754 + 4.60763/(T/100 K)

with a standard error about the regression line of 6.38 x 10^{-5}.

The thermodynamic changes for the transfer of one mole of methane from the gas at 0.1013 MPa to the infinitely dilute solution are

$\Delta \overline{H}_1^0/kJ$ mol^{-1} = -3.83 and $\Delta \overline{S}_1^0/j$ K^{-1} mol^{-1} = -61.8

The smoothed solubility values are in Table 8.

COMPONENTS:	EVALUATOR:
(1) Methane; CH_4; [74-82-8] (2) Alkanols (alcohols)	H. Lawrence Clever Department of Chemistry Emory University Atlanta, GA USA 1985, April

CRITICAL EVALUATION:

Table 8.   The Solubility of Methane in 1-Octanol. Tentative values of the mole fraction solubility and the partial molar Gibbs energy of solution as a function of temperature at a methane partial pressure of 0.1013 MPa.

T/K	$10^4 x_1$	$\Delta \overline{G}_1^0$/kJ mol^{-1}
283.15	29.97	13.679
288.15	29.13	13.988
293.15	28.35	14.297
298.15	27.61	14.606
303.15	26.92	14.915
308.15	26.26	15.224
313.15	25.64	15.534

Methane + 1-Nonanol; $CH_3(CH_2)_7CH_2OH$; [143-08-8]

Makranczy, Rusz, and Balog-Megyery (ref 9) report a solubility which gives a mole fraction of 31.9 x 10^{-4} at 298.15 K and 0.1013 MPa. The value is classed as tentative.

Methane + 1-Decanol; $CH_3(CH_2)_8CH_2OH$; [112-30-1]

Wilcock *et al*. (ref 8) and Makranczy *et al*. (ref 9) report solubility values at three and one temperatures, respectively. At 298.15 K the Makranczy value is the larger by 9 percent. All data are classed as tentative, but we prefer the data of Wilcock *et al*. which are used in the linear regression to obtain the equation for the mole fraction solubility over the 283.15 to 313.15 K interval at a pressure of 0.1013 MPa.

$$\ln x_1 = -7.2511 + 4.4321/(T/100 \text{ K})$$

with a standard error about the regression line of 3.46 x 10^{-5}.

The thermodynamic changes for the transfer of one mole of methane from the gas at 0.1013 MPa to the infinitely dilute solution are

$$\Delta \overline{H}_1^0/\text{kJ mol}^{-1} = -3.68 \text{ and } \Delta \overline{S}_1^0/\text{J K}^{-1} \text{ mol}^{-1} = -60.3$$

Smoothed solubility values are in Table 9.

Table 9.   The Solubility of Methane in 1-Decanol. Tentative values of the
           mole fraction solubility and partial molar Gibbs energy of solu-
           tion as a function of temperature at a methane partial pressure
           of 0.1013 MPa.

T/K	$10^4 x_1$	$\Delta \bar{G}_1^0 /kJ\ mol^{-1}$
283.15	33.94	13.386
288.15	33.03	13.687
293.15	32.17	13.989
298.15	31.37	14.290
303.15	30.61	14.591
308.15	29.89	14.893
313.15	29.21	15.194

Methane + 1-Undecanol; $CH_3(CH_2)_9CH_2OH$; [112-42-5]

Methane + 1-Dodecanol; $CH_3(CH_2)_{10}CH_2OH$; [112-53-8]

Makranczy, Rusz and Balog-Megyery (ref 9) report solubility values
that correspond to a mole fraction of 36.5×10^{-4} and 38.9×10^{-4}, respectively, for the two systems at 298.15 K and a methane partial pressure at
0.1013 MPa. Both values are classed as tentative; however, it is the
judgement of the evaluator the values may be 8-10 percent too large.

Alcohols and alkanes as solvents for methane show interesting trends
when the enthalpy and entropy changes for the transfer of one mole of
methane from the gas at 0.1013 MPa to the infinitely dilute solution are
compared. The average enthalpy change is -3.72 ± 0.33 for all or -3.71
± 0.14 kJ mol^{-1} when the values for the C_5 and C_6 alcohols are omitted.
The values compare with average enthalpy changes of -4.06 ± 0.52 for all
or -4.05 ± 0.14 kJ mol^{-1} when values for the C_7 and C_{16} alkanes are omitted.
Thus, there is no significant change in $\Delta \bar{H}_1^0$ with carbon number for either
alcohols or alkanes. The 0.34 kJ mol^{-1} more exothermic average $\Delta \bar{H}_1^0$ for alkanes than alcohols falls just at the limit of the uncertainty of the two
averages, and may not be significant.

The entropy change is nearly constant for the alkanes at -57.0 ± 2.3
J K^{-1} mol^{-1} while the entropy change varies from -70.9 for methanol to
-61.8 and -60.3 for 1-octanol and 1-decanol, respectively. As the alcohol
carbon number increases the entropy change approaches to within about 3
J K^{-1} mol^{-1} of the hydrocarbon value.

COMPONENTS:	EVALUATOR:
(1) Methane; CH_4; [74-82-8] (2) Alkanols (alcohols)	H. Lawrence Clever Department of Chemistry Emory University Atlanta, GA 30322 USA 1985, April

CRITICAL EVALUATION:

For the solubility of methane in water at 298.15 K the enthalpy and entropy changes are much more negative being -13.19 kJ mol^{-1} and -132.2 J K^{-1} mol^{-1}, respectively. Comparison of the water enthalpy values with the alkane and alkanol values suggests the methane molecule is located primarily in a hydrocarbon-like environment in both the hydrocarbon and alcohol solvents including even methanol and other small carbon number alcohols. Comparison of the entropy values suggests the methanol is intermediate between water and hydrocarbon but nearer the hydrocarbon as an ordered solution. By about carbon number eight the alcohol and hydrocarbon entropy difference is only about 3 J K^{-1} mol^{-1} with the methane about twice as soluble in the hydrocarbon as the alcohol.

References

1. Friedel, C.; Gorgeu, A. *Compt. rendu* <u>1908</u>, *127*, 590-4.

2. McDaniel, A. S. *J. Phys. Chem.* <u>1911</u>, *15*, 587-610.

3. Winkler, L. W. *Z. Angew. Chem.* <u>1916</u>, *29*, *I*, 218-20.

4. Boyer, F. L.; Bircher, L. J. *J. Phys. Chem.* <u>1960</u>, *64*, 1330-1.

5. Lannung, A.; Gjaldbaek, J. C. *Acta Chem. Scand.* <u>1960</u>, *14*, 1124-8.

6. Komarenki, V. G.; Manzhelii, V. G. *Ukr. Fiz. Zh. (Ukr. Ed.)* <u>1968</u>, *13*, 387-91.

7. Ben-Naim, A.; Yaacobi, M. *J. Phys. Chem.* <u>1974</u>, *78*, 175-8.

8. Wilcock, R. J.; Battino, R.; Danforth, W. F.; Wilhelm, E. *J. Chem. Thermodyn.* <u>1978</u>, *10*, 817-22.

9. Makranczy, J.; Rusz, L.; Balog-Megyery, K. *Hung. J. Ind. Chem.* <u>1979</u>, *7*, 41-6.

10. Hayduk   W. *ETHANE, Solubility Series* <u>1982</u>, *9*, 166-7; *PROPANE, BUTANE, 2-METHYLPROPANE, Solubility Series* <u>1985</u>, *24*, 331-4.

COMPONENTS:	ORIGINAL MEASUREMENTS:
(1) Methane; CH_4; [74-82-8] (2) Methanol; CH_3OH; [67-56-1]	McDaniel, A. S. *J. Phys. Chem.* <u>1911</u>, *15*, 587-610.

VARIABLES:	PREPARED BY:
$T/K = 295.25 - 322.95$ $p_1/kPa = 101.3$ (1 atm)	H. L. Clever

EXPERIMENTAL VALUES:

Temperature		Mol Fraction	Bunsen Coefficient[a]	Ostwald Coefficient[b]
$t/°C$	T/K	$10^3 x_1$	α	$L/cm^3\ cm^{-3}$
22.1	295.25	0.746	0.4102	0.4436
25.0	298.15	0.737	0.4059	0.4431[c]
30.2	303.35	0.704	0.3883	0.4278
40.0	313.15	0.635	0.3436	0.3938
49.8	322.95	0.426	0.2278	0.2695

[a] Bunsen coefficient, α/cm^3 (STP) $cm^{-3}\ atm^{-1}$.

[b] Listed as absorption coefficient in the original paper. Interpreted to be equivalent to Ostwald coefficient by compiler.

[c] Ostwald coefficient (absorption coefficient) estimated as 298.15 K value by author.

[d] Mole fraction and Bunsen coefficient values calculated by compiler assuming ideal gas behavior.

EVALUATOR'S COMMENT: McDaniel's data should be used with caution. His values are often 20 percent or more too small when compared with more reliable data.

AUXILIARY INFORMATION

METHOD/APPARATUS/PROCEDURE:	SOURCE AND PURITY OF MATERIALS:
The apparatus is all glass. It consists of a gas buret connected to a contacting vessel. The solvent is degassed by boiling under reduced pressure. Gas pressure or volume is adjusted using mercury displacement. Equilibration is achieved at atm pressure by hand shaking, and incrementally adding gas to the contacting chamber. Solubility measured by obtaining total uptake of gas by known volume of the solvent.	(1) Methane. Prepared by reaction of methyl iodide with zinc-copper. Passed through water and sulfuric acid. (2) Methanol. Source not given, purity stated to be 99 per cent.
	ESTIMATED ERROR: $\delta L/L \geq -0.20$ (compiler)
	REFERENCES:

COMPONENTS:	ORIGINAL MEASUREMENTS:
(1) Methane; CH_4; [74-82-8] (2) Methanol; CH_3OH; [67-56-1]	Boyer, F. L.; Bircher, L. J. *J. Phys. Chem.* <u>1960</u>, *64*, 1330 - 1331.

VARIABLES:	PREPARED BY:
T/K: 298.15 P/kPa: 101.325 (1 atm)	M. E. Derrick H. L. Clever

EXPERIMENTAL VALUES:

T/K	Mol Fraction $10^4 x_1$	Bunsen Coefficient[1] α	Ostwald Coefficient $L/cm^3 \ cm^{-3}$
298.15	9.19	0.506	0.552 ± 0.004

[1] α/cm^3 (STP) cm^{-3} atm^{-1}

The Bunsen coefficient was calculated by the compiler.

The mole fraction solubility was taken from Boyer's thesis (1).

Boyer's thesis gives the equations:

$$\log x_1 = -3.062 + 0.565 \log C \quad \text{for 298.15 K}$$
$$\log x_1 = -3.091 + 0.579 \log C \quad \text{for 308.15 K}$$

where C is the number of normal alcohol carbon atoms. Most of the mole fraction solubility values given in the paper were calculated from the equation at 298.15 K.

AUXILIARY INFORMATION

METHOD/APPARATUS/PROCEDURE:	SOURCE AND PURITY OF MATERIALS:
A commercial Van Slyke blood gas apparatus (E. H. Sargent Co.) was modified by the authors. The total pressure of the gas and the solvent vapor in the solution chamber was adjusted to a pressure of one atm. The pressure was maintained at one atm during the solution process. The saturated solution was transferred to a bulb below the lower stopcock of the extraction vessel and sealed off. The gas and solvent vapor were then brought to volume over mercury. See (2) for details of the extraction procedure.	1. Methane. Phillips Petroleum Co. Stated to be 99.9 mol per cent. 2. Methanol. Source not given. Treated by standard methods to remove aldehydes and ketones, then dried and distilled.

	ESTIMATED ERROR: $\delta T/K = \pm 0.01$ $\delta L/cm^3 = \pm 0.003$

	REFERENCES: 1. Boyer, F. L., Ph.D. thesis, <u>1959</u> Vanderbilt Univ., Nashville, TN 2. Peters, J. P.; Van Slyke, D. D. *Quantitative Clinical Chemistry* Baltimore, MD, 1932, Volume II.

COMPONENTS:	ORIGINAL MEASUREMENTS:
(1) Methane; CH_4; [74-82-8] (2) Methanol; CH_4O; [67-56-1]	Lannung, A.; Gjaldabek, J. C. *Acta Chem. Scand.* <u>1960</u>, *14*, 1124 - 1128.

VARIABLES:
     T/K = 291.15 - 310.15
    p_1/kPa = 101.325 (1 atm)
	PREPARED BY:
	     J. Chr. Gjaldbaek

EXPERIMENTAL VALUES:

T/K	Mol Fraction $10^4 x_1$	Bunsen Coefficient α/cm^3 (STP) $cm^{-3}atm^{-1}$	Ostwald Coefficient $L/cm^3 cm^{-3}$
291.15	9.01	0.500	0.533
291.15	9.01	0.500	0.533
298.15	8.71	0.479	0.523
310.15	8.10	0.440	0.500
310.15	8.30	0.451	0.512

Smoothed Data: For use between 291.15 and 310.15 K.

$$\ln x_1 = -8.5538 + 4.4905/(T/100\ K)$$

The standard error about the regression line is 8.23×10^{-6}.

T/K	Mol Fraction $10^4 x_1$
298.15	8.69
308.15	8.28

AUXILIARY INFORMATION

METHOD/APPARATUS/PROCEDURE:
A calibrated all-glass combined
manometer and bulb containing
degassed solvent and the gas was
placed in an air thermostat and
shaken until equilibrium (1).

The absorbed volume of gas is
calculated from the initial and
final amounts, both saturated with
solvent vapor. The amount of
solvent is determined by the
weight of displaced mercury.

The values are at 101.325 kPa
(1 atm) pressure assuming Henry's
law is obeyed.

SOURCE AND PURITY OF MATERIALS:
(1) Methane. Generated from
    magnesium methyl iodide.
    Purified by fractional distilla-
    tion. Specific gravity cor-
    responds with mol wt 16.08.

(2) Methanol. B.A.S.F. Distilled
    over magnesium.

ESTIMATED ERROR:
$$\delta T/K = \pm\ 0.05$$
$$\delta x_1/x_1 = \pm\ 0.015$$

REFERENCES:
1. Lannung, A.
   J. Am. Chem. Soc. <u>1930</u>, *52*, 68.

COMPONENTS:	ORIGINAL MEASUREMENTS:
1. Methane; CH_4; [74-82-8] 2. Methanol; CH_4O; [67-56-1]	Ben-Naim, A.; Yaacobi, M. *J. Phys. Chem.* <u>1974</u>,*78*,175-8

VARIABLES:	PREPARED BY:
Temperature	C.L. Young

EXPERIMENTAL VALUES:

T/K	Ostwald coefficient,[*] L	Mole fraction[+] at partial pressure of 101.3 kPa, x_{CH_4}
283.15	0.5437	0.000935
288.15	0.5364	0.000912
293.15	0.5278	0.000888
298.15	0.5180	0.000862
303.15	0.5070	0.000834

[*] Smoothed values obtained from the equation.

$$kT \ln L = -2,604.5 + 18.546 \ (T/K) - 0.03729 \ (T/K)^2 \ \text{cal mol}^{-1}$$
where k is in units of cal $\text{mol}^{-1}\text{K}^{-1}$

[+] calculated by compiler assuming the ideal gas law for methane.

AUXILIARY INFORMATION

METHOD/APPARATUS/PROCEDURE:	SOURCE AND PURITY OF MATERIALS:
The apparatus was similar to that described by Ben-Naim and Baer (1) and Wen and Hung (2). It consists of three main parts, a dissolution cell of 300 to 600 cm^3 capacity, a gas volume measuring column, and a manometer. The solvent is degassed in the dissolution cell, the gas is introduced and dissolved while the liquid is kept stirred by a magnetic stirrer immersed in the water bath. Dissolution of the gas results in the change in the height of a column of mercury which is measured by a cathetometer.	1. Matheson sample, purity 99.97 mol per cent. 2. AR grade.

	ESTIMATED ERROR:
	$\delta T/K = \pm 0.1$; $\delta x_{CH_4} = \pm 2\%$ (estimated by compiler).

	REFERENCES:
	1. Ben-Naim, A.; Baer, S. *Trans. Faraday. Soc.* <u>1963</u>,*59*, 2735. 2. Wen, W.-Y. Hung J.H. *J. Phys. Chem.* <u>1970</u>,*74*, 170.

M—T*

COMPONENTS:	ORIGINAL MEASUREMENTS:
1. Methane; CH_4; [74-82-8] 3. Methanol; CH_4O; [67-56-1]	Makranczy, J.; Rusz, L.; Balog-Megyery, K. *Hung. J. Ind. Chem.* <u>1979</u>, *7*, 41-6.

VARIABLES:	PREPARED BY:
	C.L. Young

EXPERIMENTAL VALUES:

T/K	P^+/kPa	Ostwald coefficient	Mole fraction of methane*, x_{CH_4}
298.15	101.3	0.808	0.001343

*   calculated by compiler

AUXILIARY INFORMATION

METHOD/APPARATUS/PROCEDURE:	SOURCE AND PURITY OF MATERIALS:
Apparently the volumetric apparatus described in ref. (1) was modified for use at temperatures above 0°C. The apparatus was designed to be operated at a partial pressure of sulfur dioxide of 760 torr.	No details given.
	ESTIMATED ERROR: $\delta x_{CH_4} = \pm 3\%$
	REFERENCES: 1. Bodor, E.; Bor, Gy.; Mohai, B.; Sipos, G. *Veszpremi Vegyip. Egy. Kozl.* <u>1957</u>, *1*, 55. *Chem. Abstr.* <u>1961</u>, *55*, 3175h

COMPONENTS:	ORIGINAL MEASUREMENTS:
(1) Methane; CH_4; [74-82-8] (2) Ethanol; C_2H_5OH; [64-17-5]	McDaniel, A. S. *J. Phys. Chem.* 1911, *15*, 587-610.

VARIABLES:	PREPARED BY:
T/K = 295.35 - 313.15 p_1/kPa = 101.3	H. L. Clever

EXPERIMENTAL VALUES:

Temperature		Mol Fraction	Bunsen Coefficient[a]	Ostwald Coefficient[b]
t/°C	T/K	$10^3 x_1$	α	L/cm^3 cm^{-3}
22.2	295.35	1.116	0.4282	0.4628
25.0	298.15	1.096	0.4197	0.4581[c]
30.1	303.25	1.064	0.4051	0.4503
40.0	313.15	0.998	0.3771	0.4323

[a] Bunsen coefficient, α/cm^3 (STP) cm^{-3} atm^{-1}.

[b] Listed as absorption coefficient in the original paper. Interpreted to be equivalent to Ostwald coefficient by compiler.

[c] Ostwald coefficient (absorption coefficient) estimated as 298.15 K value by author.

[d] Mole fraction and Bunsen coefficient values calculated by compiler assuming ideal gas behavior.

EVALUATOR'S COMMENT: McDaniel's data should be used with caution. His values are often 20 percent or more too small when compared with more reliable data.

AUXILIARY INFORMATION

METHOD/APPARATUS/PROCEDURE:	SOURCE AND PURITY OF MATERIALS:
The apparatus is all glass. It consists of a gas buret connected to a contacting vessel. The solvent is degassed by boiling under reduced pressure. Gas pressure or volume is adjusted using mercury displacement. Equilibration is achieved at atm pressure by hand shaking, and incrementally adding gas to the contacting chamber. Solubility measured by obtaining total uptake of gas by known volume of the solvent.	(1) Methane. Prepared by reaction of methyl iodide with zinc-copper. Passed through water and sulfuric acid. (2) Ethanol. Source not given. Purity stated to be 99.8 per cent.
	ESTIMATED ERROR: $\delta L/L \geq -0.20$
	REFERENCES:

COMPONENTS:	ORIGINAL MEASUREMENTS:
(1) Methane; CH_4; [74-82-8] (2) Ethanol; C_2H_5OH; [64-17-5]	Boyer, F. L.; Bircher, L. J. *J. Phys. Chem.* <u>1960</u>, *64*, 1330-1331.

VARIABLES:	PREPARED BY:
T/K: 298.15 P/kPa: 101.325 (1 atm)	M. E. Derrick H. L. Clever

EXPERIMENTAL VALUES:

T/K	Mol Fraction $10^4 x_1$	Bunsen Coefficient[1] α	Ostwald Coefficient $L/cm^3 \ cm^{-3}$
298.15	13.0	0.494	0.539 ± 0.003

[1] α/cm^3 (STP) $cm^{-3} \ atm^{-1}$

The Bunsen coefficient was calculated by the compiler.

The mole fraction solubility was taken from Boyer's thesis (1).

See the methanol data sheet for the equations relating the mole fraction solubility and the number of normal alcohol carbon numbers.

AUXILIARY INFORMATION

METHOD/APPARATUS/PROCEDURE:	SOURCE AND PURITY OF MATERIALS:
A commercial Van Slyke blood gas apparatus (E. H. Sargent Co.) was modified by the authors. The total pressure of the gas and the solvent vapor in the solution chamber was adjusted to a pressure of one atm. The pressure was maintained at one atm during the solution process. The saturated solution was transferred to a bulb below the lower stopcock of the extraction vessel and sealed off. The gas and solvent vapor were then brought to volume over mercury. See (2) for details of the extraction procedure.	(1) Methane. Phillips Petroleum Co. Stated to be 99.9 mol per cent. (2) Ethanol. Source not given. Treated by standard methods to remove aldehydes and ketones, then dried and distilled.

	ESTIMATED ERROR: $\delta T/K = \pm 0.01$ $\delta L/cm^3 = \pm 0.003$

	REFERENCES: 1. Boyer, F. L., Ph.D. thesis, <u>1959</u> Vanderbilt Univ., Nashville, TN 2. Peters, J. P.; Van Slyke, D. D. *Quantitative Clinical Chemistry* Baltimore, MD, 1932, Volume II.

COMPONENTS:	ORIGINAL MEASUREMENTS:
(1) Methane; CH_4; [74-82-8] (2) Ethanol; C_2H_6O; [64-17-5]	Lannung, A.; Gjaldbaek, J. C. *Acta Chem. Scand.* <u>1960</u>, *14*, 1124 - 1128.

VARIABLES:	PREPARED BY:
T/K = 291.15 - 310.15 p_1/kPa = 101.325 (1 atm)	J. Chr. Gjaldbaek

EXPERIMENTAL VALUES:

T/K	Mol Fraction $10^3 x_1$	Bunsen Coefficient α/cm^3(STP)cm^{-3}atm^{-1}	Ostwald Coefficient L/cm^{3}cm^{-3}
291.15	1.33	0.511	0.545
291.15	1.33	0.512	0.546
298.15	1.28	0.487	0.532
298.15	1.28	0.490	0.535
310.15	1.21	0.454	0.515
310.15	1.21	0.456	0.518

Smoothed Data: For use between 291.15 and 310.15 K.

$$\ln x_1 = -8.1618 + 4.4789/(T/100 \text{ K})$$

The standard error about the regression line is 1.53 x 10^{-6}.

T/K	Mol Fraction $10^3 x_1$
298.15	1.28
308.15	1.22

AUXILIARY INFORMATION

METHOD/APPARATUS/PROCEDURE:

A calibrated all-glass combined manometer and bulb containing degassed solvent and the gas was placed in an air thermostat and shaken until equilibrium (1).

The absorbed volume of gas is calculated from the initial and final amounts, both saturated with solvent vapor. The amount of solvent is determined by the weight of displaced mercury.

The values are at 101.325 kPa (1 atm) pressure assuming Henry's law is obeyed.

SOURCE AND PURITY OF MATERIALS:

(1) Methane. Generated from magnesium methyl iodide. Purified by fractional distillation. Specific gravity corresponds with mol wt 16.08.

(2) Ethanol. Alcohol absolutus Ph. Dan. Distilled twice over quick lime.

ESTIMATED ERROR:

$$\delta T/K = \pm 0.05$$
$$\delta x_1/x_1 = \pm 0.015$$

REFERENCES:
1. Lannung, A.
   J. Am. Chem. Soc. <u>1930</u>, *52*, 68.

COMPONENTS:	ORIGINAL MEASUREMENTS:
1. Methane; CH_4; [74-82-8] 2. Ethanol; C_2H_6O; [64-17-5]	Ben-Naim, A.; Yaacobi, M. J. Phys. Chem. 1974, 78, 175-8

VARIABLES:	PREPARED BY:
Temperature,	C.L. Young

EXPERIMENTAL VALUES:

T/K	Ostwald coefficient,[*] L	Mole fraction[+] at partial pressure of 101.3 kPa, x_{CH_4}
283.15	0.5567	0.00138
288.15	0.5468	0.00134
293.15	0.5370	0.00128
298.15	0.5272	0.00126
303.15	0.5175	0.00123

* Smoothed values obtained from the equation.

$$kT \ln L = 255.3 + 2.636 \ (T/K) - 0.01024 \ (T/K)^2 \ cal \ mol^{-1}$$
where k is in units of $cal \ mol^{-1} \ K^{-1}$

+ calculated by compiler assuming the ideal gas law for methane.

AUXILIARY INFORMATION

METHOD/APPARATUS/PROCEDURE:	SOURCE AND PURITY OF MATERIALS:
The apparatus was similar to that described by Ben-Naim and Baer (1) and Wen and Hung (2). It consists of three main parts, a dissolution cell of 300 to 600 cm^3 capacity, a gas volume measuring column, and a manometer. The solvent is degassed in the dissolution cell, the gas is introduced and dissolved while the liquid is kept stirred by a magnetic stirrer immersed in the water bath. Dissolution of the gas results in the change in the height of a column of mercury which is measured by a cathetometer.	1. Matheson sample, purity 99.97 mol per cent. 2. AR grade.

ESTIMATED ERROR:
$\delta T/K = \pm 0.1$; $\delta x_{CH_4} = \pm 2\%$
(estimated by compiler)

REFERENCES:

1. Ben-Naim, A.; Baer, S. Trans. Faraday Soc. 1963, 59, 2735.

2. Wen, W.-Y.; Hung, J. H. J. Phys. Chem. 1970, 74, 170

COMPONENTS:	ORIGINAL MEASUREMENTS:
1. Methane; CH_4; [74-82-8] 2. Ethanol; C_2H_6O; [64-17-5]	Makranczy, J.; Rusz, L.; Balog-Megyery, K. *Hung. J. Ind. Chem.* <u>1979</u>, *7*, 41-6

VARIABLES:	PREPARED BY:
	C.L. Young

EXPERIMENTAL VALUES:

T/K	P/kPa	Ostwald coefficient	Mole fraction of methane*, x_{CH_4}
298.15	101.3	0.666	0.00160

* calculated by compiler

AUXILIARY INFORMATION

METHOD/APPARATUS/PROCEDURE:	SOURCE AND PURITY OF MATERIALS:
Apparently the volumetric apparatus described in ref. (1) was modified for use at temperatures above 0°C. The apparatus was designed to be operated at a partial pressure of sulfur dioxide of 760 torr.	No details given.

ESTIMATED ERROR:

$$\delta x_{CH_4} = \pm 3\%$$

REFERENCES:

1. Bodor, E.; Bor, Gy.; Mohai, B.; Sipos. G.
   Veszpremi Vegyip. Egy. Kozl.
   <u>1957</u>, *1*, 55.
   Chem. Abstr. <u>1961</u>, *55*, 3175h

COMPONENTS:	ORIGINAL MEASUREMENTS:
(1) Methane; CH_4; [74-82-8] (2) 1-Propanol; C_3H_7OH; [71-23-8]	Boyer, F. L.; Bircher, L. J. *J. Phys. Chem.* <u>1960</u>, *64*, 1330 - 1331.

VARIABLES:	PREPARED BY:
T/K: 298.15, 308.15 P/kPa: 101.325 (1 atm)	M. E. Derrick H. L. Clever

EXPERIMENTAL VALUES:

T/K	Mol Fraction $10^4 x_1$	Bunsen Coefficient[1] α	Ostwald Coefficient $L/cm^3\ cm^{-3}$
298.15	15.6	0.467	0.510 ± 0.004
308.15	15.1	0.449	0.506 ± 0.002

[1] $\alpha/cm^3 (STP)\ cm^{-3}\ atm^{-1}$

The Bunsen coefficients were calculated by the compiler.

The mole fraction solubilities were taken from Boyer's thesis (1).

See the methanol data sheet for the equations relating the mole fraction solubility and the number of normal alcohol carbon numbers.

AUXILIARY INFORMATION

METHOD/APPARATUS/PROCEDURE:	SOURCE AND PURITY OF MATERIALS:
A commercial Van Slyke blood gas apparatus (E. H. Sargent Co.) was modified by the authors. The total pressure of the gas and the solvent vapor in the solution chamber was adjusted to a pressure of one atm. The pressure was maintained at one atm during the solution process. The saturated solution was transferred to a bulb below the lower stopcock of the extraction vessel and sealed off. The gas and solvent vapor were then brought to volume over mercury. See (2) for details of the extraction procedure.	(1) Methane. Phillips Petroleum Co. Stated to be 99.9 mol per cent. (2) 1-Propanol. Source not given. Treated by standard methods to remove aldehydes and ketones, then dried and distilled.

ESTIMATED ERROR:
$$\delta T/K = \pm\ 0.01$$
$$\delta L/cm^3 = \pm\ 0.004\ \text{(at 298.15)}$$
$$\pm\ 0.002\ \text{(at 208.15)}$$

REFERENCES:
1. Boyer, F. L., Ph.D. thesis, <u>1959</u> Vanderbilt Univ., Nashville, TN

2. Peters, J. P.; Van Slyke, D. D. *Quantitative Clinical Chemistry* Baltimore, MD, 1932, Volume II.

COMPONENTS:	ORIGINAL MEASUREMENTS:
(1) Methane; CH_4; [74-82-8] (2) 1-Propanol; C_3H_8O; [71-23-8]	Komarenko, V. G.; Manzhelii, V. G. *Ukr. Fiz. Zh. (Ukr. Ed.)* <u>1968</u>, *13*, 387-391. *Ukr. Phys. J. (Engl. Tranl.)* <u>1968</u>, *13*, 273-276.

VARIABLES:	PREPARED BY:
T/K = 173.15 - 243.15 p_1/kPa = 26.664 (200 mmHg)	H. L. Clever

EXPERIMENTAL VALUES:

Temperature		Mol Fraction p_1/mmHg = 200 $10^3 x_1$	Mol Fraction p_1/mmHg = 760 $10^3 x_1$
t/^{0}C	T/K		
-100	173.15	1.924	7.31
- 90	183.15	1.511	5.74
- 80	193.15	1.261	4.79
- 70	203.15	1.048	3.98
- 60	213.15	0.887	3.37
- 50	223.15	0.744	2.83
- 40	233.15	0.637	2.42
- 30	243.15	0.558	2.12

The compiler added the Kelvin temperatures.

The compiler calculated the mole fraction solubility values at 760 mmHg assuming Henry's law is obeyed.

AUXILIARY INFORMATION

METHOD/APPARATUS/PROCEDURE:	SOURCE AND PURITY OF MATERIALS:
The solvent was degassed by vacuum. A thin layer of alcohol, cooled to 125-175 K, was kept for 20 h in a vacuum maintained at 10^{-3} mmHg. The degassed liquid was sealed under vacuum in an ampule which was placed in the apparatus. The apparatus consisted of a manostat, a mercury compensator, and a solubility cell divided by a mercury seal. A gas pressure of 200 mmHg and the temperature were established. The foil ends of the ampule were pierced. The gas dissolved as the liquid flowed through a series of small cups. The amount of gas dissolved was measured by the rise in mercury level in the compensator. Some measurements were made at 400 mmHg gas pressure. The results confirmed that Henry's law was obeyed.	(1) Methane. Source not given. Purity by chromatographic method was 99.78 percent. (2) 1-Propanol. Purified and analyzed in the All-Union Sci. Res. Inst. for Single Crystals and High-Purity Substances. Purity 99.97 weight percent.

ESTIMATED ERROR:
δT/K = ± 0.05 δp_1/mmHg = ± 0.01 $\delta x_1/x_1$ = ± 0.005

REFERENCES:

COMPONENTS:	ORIGINAL MEASUREMENTS:
1. Methane; CH_4; [74-82-8] 2. 1-Propanol; C_3H_8O; [71-23-8]	Ben-Naim, A.; Yaacobi, M. *J. Phys. Chem.* <u>1974</u>,*78*,175-8

VARIABLES:	PREPARED BY:
Temperature	C.L. Young

EXPERIMENTAL VALUES:

T/K	Ostwald coefficient,[*] L	Mole fraction[+] at partial pressure of 101.3 kPa, x_{CH_4}
283.15	0.5417	0.00172
288.15	0.5282	0.00166
293.15	0.5174	0.00161
298.15	0.5090	0.00156
303.15	0.5029	0.00152

[*] Smoothed values obtained from the equation.

$$kT \ln L = 4{,}378.1 - 29.028 \, (T/K) + 0.04361 \, (T/K)^2 \text{ cal mol}^{-1}$$
where k is in units of cal mol^{-1} K^{-1}

[+] calculated by compiler assuming the ideal gas law for methane.

AUXILIARY INFORMATION

METHOD/APPARATUS/PROCEDURE:	SOURCE AND PURITY OF MATERIALS:
The apparatus was similar to that described by Ben-Naim and Baer (1) and Wen and Hung (2). It consists of three main parts, a dissolution cell of 300 to 600 cm^3 capacity, a gas volume measuring column, and a manometer. The solvent is degassed in the dissolution cell, the gas is introduced and dissolved while the liquid is kept stirred by a magnetic stirrer immersed in the water bath. Dissolution of the gas results in the change in the height of a column of mercury which is measured by a cathetometer.	1. Matheson sample, purity 99.97 mol per cent. 2. CP grade.

	ESTIMATED ERROR:
	$\delta T/K = \pm 0.1$; $\delta x_{CH_4} = \pm 2\%$ (estimated by compiler)

	REFERENCES:
	1. Ben-Naim, A.; Baer, S. *Trans. Faraday Soc.* <u>1963</u>,*59*, 2735. 2. Wen, W.-Y.; Hung, J.H. *J. Phys. Chem.*<u>1970</u>,*74*, 170.

COMPONENTS:	ORIGINAL MEASUREMENTS:
1. Methane; CH_4; [74-82-8] 2. 1-Propanol; C_3H_8O; [71-23-8]	Makranczy, J.; Rusz, L.; Balog-Megyery, K. *Hung. J. Ind. Chem.* <u>1979</u>, *7*, 41-6.

VARIABLES:	PREPARED BY:
	C.L. Young

EXPERIMENTAL VALUES:

T/K	P/kPa	Ostwald coefficient	Mole fraction of methane*, x_{CH_4}
298.15	101.3	0.589	0.00181

* calculated by compiler

AUXILIARY INFORMATION

METHOD/APPARATUS/PROCEDURE:	SOURCE AND PURITY OF MATERIALS:
Apparently the volumetric apparatus described in ref. (1) was modified for use at temperatures above 0°C. The apparatus was designed to be operated at a partial pressure of sulfur dioxide of 760 torr.	No details given.

ESTIMATED ERROR:

$$\delta x_{CH_4} = \pm 3\%$$

REFERENCES:

1. Bodor, E.; Bor, Gy.; Mohai, B.; Sipos, G.
Veszpremi Vegyip. Egy. Kozl.
<u>1957</u>, *1*, 55.
Chem. Abstr. <u>1961</u>, *55*, 3175h

COMPONENTS:	ORIGINAL MEASUREMENTS:
(1) Methane; CH_4; [74-82-8] (2) 2-Propanol or isopropyl alcohol; C_3H_7OH; [67-63-0]	McDaniel, A. S. J. Phys. Chem. 1911, 15, 587-610.

VARIABLES:	PREPARED BY:
T/K = 294.65 - 313.15 p_1/kPa = 101.3 (1 atm)	H. L. Clever

EXPERIMENTAL VALUES:

Temperature		Mol Fraction	Bunsen Coefficient[a]	Ostwald Coefficient[b]
$t/°C$	T/K	$10^3 x_1$	α	$L/cm^3\ cm^{-3}$
21.5	294.65	1.46	0.4275	0.4620
25.0	298.15	1.44	0.4200	0.4585[c]
29.9	303.05	1.41	0.4081	0.4532
40.0	313.15	1.34	0.3837	0.4400

[a] Bunsen coefficient, α/cm^3 (STP) $cm^{-3}\ atm^{-1}$.

[b] Listed as absorption coefficient in the original paper. Interpreted to be equivalent to Ostwald coefficient by compiler.

[c] Ostwald coefficient (absorption coefficient) estimated as 298.15 K value by author.

[d] Mole fraction and Bunsen coefficient values calculated by compiler assuming ideal gas behavior.

EVALUATOR'S COMMENT: McDeniel's data should be used with caution. His values are often 20 percent or more too small when compared with more reliable data.

AUXILIARY INFORMATION

METHOD/APPARATUS/PROCEDURE:	SOURCE AND PURITY OF MATERIALS:
The apparatus is all glass. It consists of a gas buret connected to a contacting vessel. The solvent is degassed by boiling under reduced pressure. Gas pressure or volume is adjusted using mercury displacement. Equilibration is achieved at atm pressure by hand shaking, and incrementally adding gas to the contacting chamber. Solubility measured by obtaining total uptake of gas by known volume of the solvent.	(1) Methane. Prepared by reaction of methyl iodide with zinc-copper. Passed through water and sulfuric acid. (2) 2-Propanol. Source not given.
	ESTIMATED ERROR: $\delta L/L \geq -0.20$
	REFERENCES:

COMPONENTS:	ORIGINAL MEASUREMENTS:
(1) Methane; CH_4; [74-82-8] (2) 1-Butanol; C_4H_9OH; [71-36-3]	Boyer, F. L.; Bircher, L. J. *J. Phys. Chem.* <u>1960</u>, *64*, 1330 - 1331.

VARIABLES:	PREPARED BY:
T/K: 298.15, 308.15 P/kPa: 101.325 (1 atm)	M. E. Derrick H. L. Clever

EXPERIMENTAL VALUES:

T/K	Mol Fraction $10^4 x_1$	Bunsen Coefficient[1] α	Ostwald Coefficient $L/cm^3\ cm^{-3}$
298.15	19.1	0.466	0.509 ± 0.002
308.15	18.2	0.443	0.500 ± 0.005

[1] α/cm^3 (STP) cm^{-3} atm^{-1}

The Bunsen coefficients were calculated by the compiler.

The mole fraction solubilities were taken from Boyer's thesis (1).

See the methanol data sheet for the equations relating the mole fraction solubility and the number of normal alcohol carbon numbers.

AUXILIARY INFORMATION

METHOD/APPARATUS/PROCEDURE:	SOURCE AND PURITY OF MATERIALS:
A commercial Van Slyke blood gas apparatus (E. H. Sargent Co.) was modified by the authors. The total pressure of the gas and the solvent vapor in the solution chamber was adjusted to a pressure of one atm. The pressure was maintained at one atm during the solution process. The saturated solution was transferred to a bulb below the lower stopcock of the extraction vessel and sealed off. The gas and solvent vapor were then brought to volume over mercury. See (2) for details of the extraction procedure.	(1) Methane. Phillips Petroleum Co. Stated to be 99.9 mol per cent. (2) 1-Butanol. Source not given. Treated by standard methods to remove aldehydes and ketones, then dried and distilled.

ESTIMATED ERROR:
$$\delta T/K = \pm\ 0.01$$
$$\delta L/cm^3 = \pm\ 0.002\ \text{(at 298.15)}$$
$$\pm\ 0.005\ \text{(at 308.15)}$$

REFERENCES:
1. Boyer, F. L., Ph.D. thesis, <u>1959</u> Vanderbilt Univ., Nashville, <u>TN</u>

2. Peters, J. P.; Van Slyke, D. D. *Quantitative Clinical Chemistry* Baltimore, MD, 1932, Volume II.

COMPONENTS:	ORIGINAL MEASUREMENTS:
1. Methane; CH_4; [74-82-8] 2. 1-Butanol; $C_4H_{10}O$; [71-36-3]	Ben-Naim, A.; Yaacobi, M. *J. Phys. Chem.* <u>1974</u>,*78*,175-8

VARIABLES:
      Temperature,
	PREPARED BY:
	      C.L. Young

EXPERIMENTAL VALUES:

T/K	Ostwald coefficient,[*] L	Mole fraction[+] at partial pressure of 101.3 kPa, x_{CH_4}
283.15	0.5194	0.00203
288.15	0.5115	0.00197
293.15	0.5016	0.00191
298.15	0.4898	0.00184
303.15	0.4765	0.00177

* Smoothed values obtained from the equation.

$$kT \ln L = -4{,}090.4 + 29.065 \ (T/K) - 0.05623 \ (T/K)^2 \ \text{cal mol}^{-1}$$
where k is in units of cal mol^{-1} K^{-1}

+ calculated by compiler assuming the ideal gas law for methane.

AUXILIARY INFORMATION

METHOD/APPARATUS/PROCEDURE:	SOURCE AND PURITY OF MATERIALS:
The apparatus was similar to that described by Ben-Naim and Baer (1) and Wen and Hung (2). It consists of three main parts, a dissolution cell of 300 to 600 cm^3 capacity, a gas volume measuring column, and a manometer. The solvent is degassed in the dissolution cell, the gas is introduced and dissolved while the liquid is kept stirred by a magnetic stirrer immersed in the water bath. Dissolution of the gas results in the change in the height of a column of mercury which is measured by a cathetometer.	1. Matheson sample, purity 99.97 mol per cent. 2. AR grade.

	ESTIMATED ERROR:
	      $\delta T/K = \pm 0.1$; $\delta x_{CH_4} = \pm 2\%$
	      (estimated by compiler).

	REFERENCES: 1. Ben-Naim, A.; Baer, S. *Trans. Faraday Soc.* <u>1963</u>,*59*, 2735. 2. Wen, W.-Y.; Hung, J.H. *J. Phys. Chem.* <u>1970</u>,*74*, 170

COMPONENTS:	ORIGINAL MEASUREMENTS:
1. Methane; CH_4; [74-82-8] 2. 1-Butanol; $C_4H_{10}O$; [71-36-3]	Makranczy, J.; Rusz, L.; Balog-Megyery, K. *Hung. J. Ind. Chem.* <u>1979</u>, *7*, 41-6

VARIABLES:	PREPARED BY:
	C.L. Young

EXPERIMENTAL VALUES:

T/K	P/kPa	Ostwald coefficient	Mole fraction of methane *, x_{CH_4}
298.15	101.3	0.546	0.00205

* calculated by compiler

AUXILIARY INFORMATION

METHOD/APPARATUS/PROCEDURE:	SOURCE AND PURITY OF MATERIALS:
Apparently the volumetric apparatus described in ref. (1) was modified for use at temperatures above 0°C. The apparatus was designed to be operated at a partial pressure of sulfur dioxide of 760 torr.	No details given

ESTIMATED ERROR:

$$\delta x_{CH_4} = \pm 3\%$$

REFERENCES:
1. Bodor, E.; Bor, Gy.; Mohai, B.; Sipos, G.
Veszpremi Vegyip. Egy. Kozl. <u>1957</u>, *1*, 55.
Chem. Abstr. <u>1961</u>, *55*, 3175h

COMPONENTS:	ORIGINAL MEASUREMENTS:
(1) Methane; CH_4; [74-82-8] (2) 2-Methyl-1-propanol or isobutyl alcohol; $C_4H_{10}O$; [78-83-1]	Winkler, L. W. *Z. Angew. Chem.* <u>1916</u>, *29, I,* 218-20.
VARIABLES:	PREPARED BY: H. L. Clever

EXPERIMENTAL VALUES:

     The author states that the absorption coefficient
of methane in isobutyl alcohol at room temperature
in near 1/3. Compared to methane solubility values
in other alcohols the value appears to be too small
and is classed as doubtful. The small value may be
due to water in the alcohol.

AUXILIARY INFORMATION

METHOD/APPARATUS/PROCEDURE:	SOURCE AND PURITY OF MATERIALS:
No information.	No information.
	ESTIMATED ERROR:
	REFERENCES:

COMPONENTS:	ORIGINAL MEASUREMENTS:
(1) Methane; CH_4; [74-82-8] (2) 1-Pentanol or amyl alcohol; $C_5H_{11}OH$; [71-41-0]	McDaniel, A. S. *J. Phys. Chem.* <u>1911</u>, *15*, 587-610.
VARIABLES: T/K = 295.15 - 303.25 p_1/kPa = 101.3 (1 atm)	PREPARED BY: H. L. Clever

EXPERIMENTAL VALUES:

Temperature		Mol Fraction	Bunsen Coefficient[a]	Ostwald Coefficient[b]
t/°C	T/K	$10^3 x_1$	α	L/cm^3 cm^{-3}
22.0	295.15	2.02	0.4196	0.4532
25.0	298.15	1.99	0.4123	0.4500[c]
30.1	303.25	1.95	0.4002	0.4444

[a] Bunsen coefficient, α/cm^3(STP) cm^{-3} atm^{-1}.

[b] Listed as absorption coefficient in the original paper.
Interpreted to be equivalent to Ostwald coefficient by compiler.

[c] Ostwald coefficient (absorption coefficient) estimated as
298.15 K value by author.

[d] Mole fraction and Bunsen coefficient values calculated by
compiler assuming ideal gas behavior.

EVALUATOR'S COMMENT: McDaniel's data should be used with caution.
His values are often 20 percent or more too small when compared
with more reliable data.

AUXILIARY INFORMATION

METHOD/APPARATUS/PROCEDURE:	SOURCE AND PURITY OF MATERIALS:
The apparatus is all glass. It consists of a gas buret connected to a contacting vessel. The solvent is degassed by boiling under reduced pressure. Gas pressure or volume is adjusted using mercury displacement. Equilibration is achieved at atm pressure by hand shaking, and incrementally adding gas to the contacting chamber. Solubility measured by obtaining total uptake of gas by known volume of the solvent.	(1) Methane. Prepared by reaction of methyl iodide with zinc-copper. Passed through water and sulfuric acid. (2) 1-Pentanol. Source not given.
	ESTIMATED ERROR: $\delta L/L \geq -0.20$
	REFERENCES:

COMPONENTS:	ORIGINAL MEASUREMENTS:
(1) Methane; CH_4; [74-82-8] (2) 1-Pentanol; $C_5H_{11}OH$; [71-41-0]	Boyer, F. L.; Bircher, L. J. *J. Phys. Chem.* <u>1960</u>, *64*, 1330 - 1331.

VARIABLES:
    T/K: 298.15, 308.15
    P/kPa: 101.325 (1 atm)
	PREPARED BY:
	        M. E. Derrick
	        H. L. Clever

EXPERIMENTAL VALUES:

T/K	Mol Fraction $10^4 x_1$	Bunsen Coefficient[1] α	Ostwald Coefficient $L/cm^3\ cm^{-3}$
298.15	21.5	0.442	0.483 ± 0.005
308.15	21.1	0.429	0.484 ± 0.010

[1] α/cm^3 (STP) $cm^{-3}\ atm^{-1}$

The Bunsen coefficients were calculated by the compiler.

The mole fraction solubilities were taken from Boyer's thesis (1).

See the methanol data sheet for the equations relating the mole fraction solubility and the number of normal alcohol carbon numbers.

AUXILIARY INFORMATION

METHOD/APPARATUS/PROCEDURE:	SOURCE AND PURITY OF MATERIALS:
A commercial Van Slyke blood gas apparatus (E. H. Sargent Co.) was modified by the authors. The total pressure of the gas and the solvent vapor in the solution chamber was adjusted to a pressure of one atm. The pressure was maintained at one atm during the solution process. The saturated solution was transferred to a bulb below the lower stopcock of the extraction vessel and sealed off. The gas and solvent vapor were then brought to volume over mercury. See (2) for details of the extraction procedure.	(1) Methane. Phillips Petroleum Co. Stated to be 99.9 mol per cent. (2) 1-Pentanol. Source not given. Treated by standard methods to remove aldehydes and ketones, then dried and distilled.

	ESTIMATED ERROR:
	        $\delta T/K$ = ± 0.01
	    $\delta L/cm^3$ = ± 0.005 (at 298.15)
	        ± 0.010 (at 308.15)

	REFERENCES: 1. Boyer, F. L., Ph.D. thesis, <u>1959</u> Vanderbilt Univ., Nashville, <u>TN</u> 2. Peters, J. P.; Van Slyke, D. D. *Quantitative Clinical Chemistry* Baltimore, MD, 1932, Volume II.

COMPONENTS:	ORIGINAL MEASUREMENTS:
1. Methane; CH_4; [74-82-8] 2. 1-Pentanol; $C_5H_{12}O$; [71-41-0]	Ben-Naim, A.; Yaacobi, M. *J. Phys. Chem.* <u>1974</u>,*78*, 175-8.

VARIABLES:
  Temperature
	PREPARED BY:
	  C.L. Young

EXPERIMENTAL VALUES:

T/K	Ostwald coefficient,[*] *L*	Mole fraction[+] at partial pressure of 101.3 kPa, x_{CH_4}
283.15	0.4925	0.00227
288.15	0.4843	0.00220
293.15	0.4760	0.00214
298.15	0.4676	0.00207
303.15	0.4592	0.00202

[*] Smoothed values obtained from the equation

$$kT \ln L = -390.8 + 3.230 \, (T/K) - 0.01150 \, (T/K)^2 \text{ cal mol}^{-1}$$
where k is in units of cal mol^{-1} K^{-1}

[+] calculated by compiler assuming the ideal gas law for methane.

AUXILIARY INFORMATION

METHOD/APPARATUS/PROCEDURE:	SOURCE AND PURITY OF MATERIALS:
The apparatus was similar to that described by Ben-Naim and Baer (1) and Wen and Hung (2). It consists of three main parts, a dissolution cell of 300 to 600 cm^3 capacity, a gas volume measuring column, and a manometer. The solvent is degassed in the dissolution cell, the gas is introduced and dissolved whicle the liquid is kept stirred by a magnetic stirrer immersed in the water bath. Dissolution of the gas results in the change in the height of a column of mercury which is measured by a cathetometer.	1. Matheson sample, purity 99.9 mol per cent. 2. AR grade.

	ESTIMATED ERROR:
	
	  $\delta T/K = \pm 0.1$; $\delta x_{CH_4} = \pm 2\%$
	  (estimated by compiler).

	REFERENCES:
	
	1. Ben-Naim, A.; Baer, S.
	  Trans. Faraday Soc. <u>1963</u>,*59*, 2735.
	
	2. Wen, W.-Y.; Hung, J.H.
	  J. Phys. Chem. <u>1970</u>,*74*,170

COMPONENTS:	ORIGINAL MEASUREMENTS:

1. Methane; CH_4; [74-82-8]
 	
2. 1-Pentanol; $C_5H_{12}O$; [71-41-0]
       or
    1-Hexanol; $C_6H_{14}O$; [111-27-3]
	Makranczy, J.; Rusz, L.;
	Balog-Megyery, K.
	
	Hung. J. Ind. Chem. <u>1979</u>, *7*, 41-6

VARIABLES:	PREPARED BY:
	C.L. Young

EXPERIMENTAL VALUES:

T/K	P/kPa	Ostwald coefficient	Mole fraction of methane *, x_{CH_4}
		1-Pentanol	
298.15	101.3	0.513	0.00227
		1-Hexanol	
298.15	101.3	0.491	0.00251

* calculated by compiler.

AUXILIARY INFORMATION

METHOD/APPARATUS/PROCEDURE:	SOURCE AND PURITY OF MATERIALS:
Apparently the volumetric apparatus described in ref. (1) was modified for use at temperatures above 0°C. The apparatus was designed to be operated at a partial pressure of sulfur dioxide of 760 torr.	No details given

ESTIMATED ERROR:

$$\delta x_{CH_4} = \pm 3\%$$

REFERENCES:

1. Bodor, E.; Bor, Gy.; Mohai, B.;
    Sipos, G.
    Veszpremi Vegyip. Egy. Kozl.
    <u>1957</u>, *1*, 55.
    Chem. Abstr. <u>1961</u>, *55*, 3175h

COMPONENTS:	ORIGINAL MEASUREMENTS:
(1) Methane; CH_4; [74-82-8] (2) 3-Methyl-1-butanol; $C_5H_{12}O$; [123-51-3]	Friedel, C.; Gorgeu, A. *Compt. rendu* 1908, *127*, 590-4.

VARIABLES:	PREPARED BY:
T/K = 285.7 p/kPa = 101.3	H. L. Clever

EXPERIMENTAL VALUES:

Temperature		Pressure	Solubility Volume Methane/
$t/^0C$	T/K	p/m	Volume Alcohol
12.5	285.7	0.760	0.5

AUXILIARY INFORMATION

METHOD/APPARATUS/PROCEDURE:	SOURCE AND PURITY OF MATERIALS:
In the original paper the alcohol was named simply amyl alcohol. However, the boiling point corresponds to the alcohol later named primary isoamyl alcohol or 3-methyl-1-butanol.	(1) Methane. Prepared by authors by the decomposition of dimethyl mercury. (2) 3-Methyl-1-butanol. Prepared by the authors. Boiling point 130-132 ^0C.
	ESTIMATED ERROR:
	REFERENCES:

COMPONENTS:	ORIGINAL MEASUREMENTS:
(1) Methane; CH_4; [74-82-8] (2) 1-Hexanol; $C_6H_{13}OH$; [111-27-3]	Boyer, F. L.; Bircher, L. J. *J. Phys. Chem.* <u>1960</u>, *64*, 1330 - 1331.

VARIABLES:	PREPARED BY:
T/K: 298.15 P/kPa: 101.325 (1 atm)	M. E. Derrick H. L. Clever

EXPERIMENTAL VALUES:

T/K	Mol Fraction $10^4 x_1$	Bunsen Coefficient[1] α	Ostwald Coefficient $L/cm^3\ cm^{-3}$
298.15	23.7	0.425	0.464 ± 0.002

[1] α/cm^3 (STP) cm^{-3} atm^{-1}

The Bunsen coefficient was calculated by the compiler.

The mole fraction solubility was taken from Boyer's thesis (1).

See the methanol data sheet for the equations relating the mole fraction solubility and the number of normal alcohol carbon numbers.

AUXILIARY INFORMATION

METHOD/APPARATUS/PROCEDURE:	SOURCE AND PURITY OF MATERIALS:
A commercial Van Slyke blood gas apparatus (E. H. Sargent Co.) was modified by the authors. The total pressure of the gas and the solvent vapor in the solution chamber was adjusted to a pressure of one atm. The pressure was maintained at one atm during the solution process. The saturated solution was transferred to a bulb below the lower stopcock of the extraction vessel and sealed off. The gas and solvent vapor were then brought to volume over mercury. See (2) for details of the extraction procedure.	(1) Methane. Phillips Petroleum Co. Stated to be 99.9 mol per cent. (2) 1-Hexanol. Source not given. Treated by standard methods to remove aldehydes and ketones, then dried and distilled.

ESTIMATED ERROR:

$$\delta T/K = \pm\ 0.01$$
$$\delta L/cm^3 = \pm\ 0.002$$

REFERENCES:
1. Boyer, F. L., Ph.D. thesis, <u>1959</u> Vanderbilt Univ., Nashville, TN

2. Peters, J. P.; Van Slyke, D. D. *Quantitative Clinical Chemistry* Baltimore, MD, 1932, Volume II.

COMPONENTS:	ORIGINAL MEASUREMENTS:
1. Methane; CH_4; [74-82-8] 2. 1-Hexanol; $C_6H_{14}O$; [111-27-3]	Ben-Naim, A.; Yaacobi, M. *J. Phys. Chem.* 1974,*78*, 175-8

VARIABLES:
    Temperature
	PREPARED BY:
	    C.L. Young

EXPERIMENTAL VALUES:

T/K	Ostwald coefficient,[*] L	Mole fraction[+] at partial pressure of 101.3 kPa, x_{CH_4}
283.15	0.4727	0.00251
288.15	0.4622	0.00242
293.15	0.4535	0.00235
298.15	0.4663	0.00228
303.15	0.4404	0.00222

[*] Smoothed values obtained from the equation.

$kT \ln L = 3.087.6 - 20.591 \, (T/K) + 0.02895 \, (T/K)^2$ cal mol^{-1}
where k is in units of cal mol^{-1} K^{-1}

[+] calculated by compiler assuming the ideal gas law for methane.

AUXILIARY INFORMATION

METHOD/APPARATUS/PROCEDURE:	SOURCE AND PURITY OF MATERIALS:
The apparatus was similar to that described by Ben-Naim and Baer (1) and Wen and Hung (2). It consists of three main parts, a dissolution cell of 300 to 600 cm^3 capacity, a gas volume measuring column, and a manometer. The solvent is degassed in the dissolution cell, the gas is introduced and dissolved while the liquid is kept stirred by a magnetic stirrer immersed in the water bath. Dissolution of the gas results in the change in the height of a column of mercury which is measured by a cathetometer.	1. Matheson sample, purity 99.97 mol per cent. 2. AR grade.

ESTIMATED ERROR:

$\delta T/K = \pm 0.1$; $\delta x_{CH_4} = \pm 2\%$
(estimated by compiler)

REFERENCES:
1. Ben-Naim, A.; Baer, S.
   Trans. Faraday Soc. 1963,*59*, 2735.

2. Wen, W.-Y.; Hung, J.H.
   J. Phys. Chem. 1970,*74*,170.

COMPONENTS:	ORIGINAL MEASUREMENTS:
(1) Methane; CH_4; [74-82-8] (2) 1-Heptanol; $C_7H_{15}OH$; [111-70-6]	Boyer, F. L.; Bircher, L. J. *J. Phys. Chem.* <u>1960</u>, *64*, 1330 - 1331.

VARIABLES:	PREPARED BY:
T/K: 298.15 P/kPa: 101.325 (1 atm)	M. E. Derrick H. L. Clever

EXPERIMENTAL VALUES:

T/K	Mol Fraction $10^4 x_1$	Bunsen Coefficient[1] α	Ostwald Coefficient $L/cm^3 \ cm^{-3}$
298.15	26.0	0.410	0.448 ± 0.004

[1] α/cm^3 (STP) $cm^{-3} \ atm^{-1}$

The Bunsen coefficient was calculated by the compiler.

The mole fraction solubility was taken from Boyer's thesis (1).

See the methanol data sheet for the equations relating the mole fraction solubility and the number of normal alcohol carbon numbers.

AUXILIARY INFORMATION

METHOD/APPARATUS/PROCEDURE:	SOURCE AND PURITY OF MATERIALS:
A commercial Van Slyke blood gas apparatus (E. H. Sargent Co.) was modified by the authors. The total pressure of the gas and the solvent vapor in the solution chamber was adjusted to a pressure of one atm. The pressure was maintained at one atm during the solution process. The saturated solution was transferred to a bulb below the lower stopcock of the extraction vessel and sealed off. The gas and solvent vapor were then brought to volume over mercury. See (2) for details of the extraction procedure.	(1) Methane. Phillips Petroleum Co. Stated to be 99.9 mol per cent. (2) 1-Heptanol. Source not given. Treated by standard methods to remove aldehydes and ketones, then dried and distilled.
	ESTIMATED ERROR: $\delta T/K = \pm 0.01$ $\delta L/cm^3 = \pm 0.004$
	REFERENCES: 1. Boyer, F. L., Ph.D. thesis, <u>1959</u> Vanderbilt Univ., Nashville, <u>TN</u> 2. Peters, J. P.; Van Slyke, D. D. *Quantitative Clinical Chemistry* Baltimore, MD, 1932, Volume II.

COMPONENTS:	ORIGINAL MEASUREMENTS:

1. Methane; CH_4; [74-82-8]
 	
2. 1-Heptanol; $C_7H_{16}O$; [111-70-6]
      or
  1-Octanol; $C_8H_{18}O$; [111-87-5]
	Makranczy, J.; Rusz, L.;
	Balog-Megyery, K.
	
	Hung. J. Ind. Chem. <u>1979</u>, *7*, 41-6.

VARIABLES:	PREPARED BY:
	C.L. Young

EXPERIMENTAL VALUES:

T/K	P/kPa	Ostwald coefficient	Mole fraction of methane*, x_{CH_4}
		1-Heptanol	
298.15	101.3	0.469	0.00271
		1-Octanol	
298.15	101.3	0.633[+] (0.458)	0.00408[+] (0.00296)

+ appears to be an error in original table, probable value given in parentheses.

* calculated by compiler.

AUXILIARY INFORMATION

METHOD/APPARATUS/PROCEDURE:	SOURCE AND PURITY OF MATERIALS:
Apparently the volumetric apparatus described in ref. (1) was modified for use at temperatures above 0°C. The apparatus was designed to be operated at a partial pressure of sulfur dioxide of 760 torr.	No details given.

ESTIMATED ERROR:

$$\delta x_{CH_4} = \pm 3\%$$

REFERENCES:

1. Bodor, E.; Bor, Gy.; Mohai, B.
Sipos, G.
Veszpremi Vegyip. Egy. Kozl.
<u>1957</u>, *1*, 55.
Chem. Abstr. <u>1961</u>, *55*, 3175h

COMPONENTS:	ORIGINAL MEASUREMENTS:
(1) Methane; CH_4; [74-82-8] (2) 1-Octanol; $C_8H_{17}OH$; [111-87-5]	Boyer, F. L.; Bircher, L. J. *J. Phys. Chem.* 1960, *64*, 1330 - 1331.

VARIABLES:	PREPARED BY:
T/K: 298.15, 308.15 P/kPa: 101.325 (1 atm)	M. E. Derrick H. L. Clever

EXPERIMENTAL VALUES:

T/K	Mol Fraction $10^4 x_1$	Bunsen Coefficient[1] α	Ostwald Coefficient $L/cm^3\ cm^{-3}$
298.15	28.4	0.399	0.436 ± 0.004
308.15	26.4	0.372	0.420 ± 0.005

[1] α/cm^3 (STP) $cm^{-3}\ atm^{-1}$

The Bunsen coefficients were calculated by the compiler.

The mole fraction solubilities were taken from Boyer's thesis (1).

See the methanol data sheet for the equations relating the mole fraction solubility and the number of normal alcohol carbon numbers.

AUXILIARY INFORMATION

METHOD/APPARATUS/PROCEDURE:

A commercial Van Slyke blood gas apparatus (E. H. Sargent Co.) was modified by the authors.

The total pressure of the gas and the solvent vapor in the solution chamber was adjusted to a pressure of one atm. The pressure was maintained at one atm during the solution process. The saturated solution was transferred to a bulb below the lower stopcock of the extraction vessel and sealed off. The gas and solvent vapor were then brought to volume over mercury. See (2) for details of the extraction procedure.

SOURCE AND PURITY OF MATERIALS:

(1) Methane. Phillips Petroleum Co. Stated to be 99.9 mol per cent.

(2) 1-Octanol. Source not given. Treated by standard methods to remove aldehydes and ketones, then dried and distilled.

ESTIMATED ERROR:
$$\delta T/K = \pm\ 0.01$$
$$\delta L/cm^3 = \pm\ 0.004\ (at\ 298.15)$$
$$\pm\ 0.005\ (at\ 308.15)$$

REFERENCES:

1. Boyer, F. L., Ph.D. thesis, 1959 Vanderbilt Univ., Nashville, TN

2. Peters, J. P.; Van Slyke, D. D. *Quantitative Clinical Chemistry* Baltimore, MD, 1932, Volume II.

COMPONENTS:	ORIGINAL MEASUREMENTS:
(1) Methane; CH_4; [74-82-8] (2) 1-Octanol; $C_8H_{17}OH$; [111-87-5]	Wilcock, R. J.; Battino, R.; Danforth, W. F.; Wilhelm, E. J. Chem. Thermodyn. <u>1978</u>, 10, 817 - 822.
VARIABLES: T/K: 283.23 - 313.46 p/kPa: 101.325 (1 atm)	PREPARED BY: H. L. Clever

EXPERIMENTAL VALUES:

T/K	Mol Fraction $10^3 x_1$	Bunsen Coefficient α/cm^3 (STP) cm^{-3} atm^{-1}	Ostwald Coefficient L/cm^3 cm^{-3}
283.23	2.992	0.4292	0.4450
298.08	2.687	0.3807	0.4154
313.46	2.548	0.3562	0.4088

The Bunsen coefficients were calculated by the compiler.

It is assumed that the gas is ideal and that Henry's law is obeyed.

Smoothed Data: For use between 283.15 to 313.15 K

$$\ln x_1 = -7.4910 + 4.7335/(T/100K)$$

The standard error about the regression line is 5.46×10^{-5}.

T/K	Mol Fraction $10^3 x_1$
283.15	2.970
293.15	2.805
298.15	2.730
303.15	2.660
313.15	2.530

AUXILIARY INFORMATION

METHOD/APPARATUS/PROCEDURE:	SOURCE AND PURITY OF MATERIALS:
The solubility apparatus is based on the design of Morrison and Billett (1) and the version used is described by Battino, Evans, and Danforth (2). The degassing apparatus is that described by Battino, Banzhof, Bogan, and Wilhelm (3). Degassing. Up to 500 cm^3 of solvent is placed in a flask of such size that the liquid is about 4 cm deep. The liquid is rapidly stirred, and vacuum is intermittently applied through a liquid N_2 trap until the permanent gas residual pressure drops to 5 microns. Solubility Determination. The degassed solvent is passed in a thin film down a glass helical tube containing solute gas plus the solvent vapor at a total pressure of one atm. The volume of gas absorbed is found by difference between the initial and final volumes in the buret system. The solvent is collected in a tared flask and weighed.	(1) Methane. Matheson Co., Inc. Minimum mole per cent purity stated to be 99.97. (2) 1-Octanol. Eastman Organic Chemicals. Distilled, density at 298.15 K, ρ/g cm^{-3} 0.8247.
	ESTIMATED ERROR: $\delta T/K = 0.02$ $\delta P/mmHg = 0.5$ $\delta x_1/x_1 = 0.01$
	REFERENCES: 1. Morrison, T. J.; Billett, F. J. Chem. Soc. <u>1948</u>, 2033. 2. Battino, R.; Evans, F. D.; Danforth, W. F. J. Am. Oil Chem. Soc. <u>1968</u>, 45, 830. 3. Battino, R.; Banzhof, M.; Bogan, M.; Wilhelm, E. Anal. Chem. <u>1971</u>, 43, 806.

COMPONENTS:	ORIGINAL MEASUREMENTS:

1. Methane; CH_4; [74-82-8]
2. 1-Nonanol; $C_9H_{20}O$; [143-08-8]
 	
    or
    1-Decanol; $C_{10}H_{22}O$; [112-30-1]
	Makranczy, J.; Rusz, L.;
	Balog-Megyery, K.
	
	Hung. J. Ind. Chem. 1979, *7*, 41-6.

VARIABLES:	PREPARED BY:
	C.L. Young

EXPERIMENTAL VALUES:

T/K	P/kPa	Ostwald coefficient	Mole fraction of methane *, x_{CH_4}
		1-Nonanol	
298.15	101.3	0.448	0.00319
		1-Decanol	
298.15	101.3	0.437	0.00341

* calculated by compiler

AUXILIARY INFORMATION

METHOD/APPARATUS/PROCEDURE:	SOURCE AND PURITY OF MATERIALS:
Apparently the volumetric apparatus described in ref. (1) was modified for use at temperatures above 0°C. The apparatus was designed to be operated at a partial pressure of sulfur dioxide of 760 torr.	No details given

ESTIMATED ERROR:

$$\delta x_{CH_4} = \pm 3\%$$

REFERENCES:
1. Bodor, E.; Bor, Gy.; Mohai, B.; Sipos. G.
Veszpremi Vegyip Egy. Kozl. 1957, *1*, 55.
Chem. Abstr. 1961, *55*, 3175h

COMPONENTS:	ORIGINAL MEASUREMENTS:
(1) Methane; CH_4; [74-82-8] (2) 1-Decanol; $C_{10}H_{21}OH$; [112-30-1]	Wilcock, R. J.; Battino, R.; Danforth, W. F.; Wilhelm, E. *J. Chem. Thermodyn.* 1978, *10*, 817 - 822.

VARIABLES:	PREPARED BY:
T/K: 284.04 - 313.37 p/kPa: 101.325 (1 atm)	H. L. Clever

EXPERIMENTAL VALUES:

T/K	Mol Fraction $10^3 x_1$	Bunsen Coefficient α/cm^3 (STP) cm^{-3} atm^{-1}	Ostwald Coefficient L/cm^3 cm^{-3}
284.04	3.362	0.3984	0.4143
298.08	3.166	0.3709	0.4048
313.37	2.905	0.3361	0.3856

The Bunsen coefficients were calculated by the compiler.

It is assumed that the gas is ideal and that Henry's law is obeyed.

Smoothed Data: For use between 283.15 to 313.15 K

$$\ln x_1 = -7.2511 + 4.4321/(T/100K)$$

The standard error about the regression line is 3.46×10^{-5}.

T/K	Mol Fraction $10^3 x_1$
283.15	3.394
293.15	3.217
298.15	3.137
303.15	3.061
313.15	2.921

AUXILIARY INFORMATION

METHOD/APPARATUS/PROCEDURE:

The solubility apparatus is based on the design of Morrison and Billett (1) and the version used is described by Battino, Evans, and Danforth (2). The degassing apparatus is that described by Battino, Banzhof, Bogan, and Wilhelm (3).

Degassing. Up to 500 cm^3 of solvent is placed in a flask of such size that the liquid is about 4 cm deep. The liquid is rapidly stirred, and vacuum is intermittently applied through a liquid N_2 trap until the permanent gas residual pressure drops to 5 microns.

Solubility Determination. The degassed solvent is passed in a thin film down a glass helical tube containing solute gas plus the solvent vapor at a total pressure of one atm. The volume of gas absorbed is found by difference between the initial and final volumes in the buret system. The solvent is collected in a tared flask and weighed.

SOURCE AND PURITY OF MATERIALS:

(1) Methane. Matheson Co., Inc. Minimum mole percent purity stated to be 99.97.

(2) 1-Decanol. Eastman Organic Chemicals. Distilled, density at 298.15 K, ρ/g cm^{-3} 0.8206.

ESTIMATED ERROR:
$$\delta T/K = 0.02$$
$$\delta P/mmHg = 0.5$$
$$\delta x_1/x_1 = 0.01$$

REFERENCES:

1. Morrison, T. J.; Billett, F. *J. Chem. Soc.* 1948, 2033.

2. Battino, R.; Evans, F. D.; Danforth, W. F. *J. Am. Oil Chem. Soc.* 1968, *45*, 830.

3. Battino, R.; Banzhof, M.; Bogan, M.; Wilhelm, E. *Anal. Chem.* 1971, *43*, 806.

COMPONENTS:	ORIGINAL MEASUREMENTS:

1. Methane; CH_4; [64-82-8]
2. 1-Undecanol; $C_{11}H_{24}O$; [112-42-5]
    or
    1-Dodecanol; $C_{12}H_{26}O$; [112-53-8]
	Makranczy, J.; Rusz, L.;
	Balog-Megyery. K.
	
	Hung. J. Ind. Chem. <u>1979</u>, *7*, 41-6

VARIABLES:	PREPARED BY:
	C.L. Young

EXPERIMENTAL VALUES:

T/K	P/kPa	Ostwald coefficient	Mole fraction of methane *, x_{CH_4}
		1-Undecanol	
298.15	101.3	0.431	0.00365
		1-Dodecanol	
298.15	101.3	0.426	0.00389

* calculated by compiler

AUXILIARY INFORMATION

METHOD/APPARATUS/PROCEDURE:	SOURCE AND PURITY OF MATERIALS:
Apparently the volumetric apparatus described in ref. (1) was modified for use at temperatures above 0°C. The apparatus was designed to be operated at a partial pressure of sulfur dioxide of 760 torr.	No details given.

ESTIMATED ERROR:

$$\delta x_{CH_4} = \pm 3\%$$

REFERENCES:

1. Bodor, E.; Bor, Gy.; Mohai, B.;
   Sipos. G.
   Veszpremi. Vegyip. Egy. Kozl.
   <u>1957</u>, *1*, 55.
   Chem. Abstr. <u>1961</u>, *55*, 3175h

COMPONENTS:	EVALUATOR:
1. Methane; CH_4; [74-82-8] 2. Methanol; CH_4O; [67-56-1]	Colin L. Young Department of Physical Chemistry, University of Melbourne. Parkville, Victoria, 3052 Australia. February 1986.

EVALUATION:

       This system has been investigated at elevated pressures by five groups (1-5). The most detailed study has been undertaken by Francesconi et al.(4) and their data are classified as tentative over the temperature range 303 K to 513 K for pressures up to 300 MPa. There are large differences between the early work of Kritchevsky and Koroleva (1) and the data reported in ref (4). At 70 MPa deviations up to 20 per cent in the gas phase mole fraction of methanol are evident. The data Shenderei et al. (2) were determined at high pressure but these workers only reported Henry's law constants. These data which cover the temperature range 213 K to 250 K are classified as tentative.

    Yarym-Agaev et al. (5) investigated this system in the temperature range 298 K to 338 K at pressures up to 12.5 MPa. However there is virtually no overlap with the pressure range of the data reported by Francesconi et al. (4). The data of Yarym-Agaev et al (5) is classied as tentative. Hemmaplardh and King (3) have investigated the solubility of methanol in compressed methane up to 6 MPa. but detailed consideration of this data falls outside the scope of the present work.

References

1. Kritchevsky, I. and Koroleva, M.
   Acta Physicochim. USSR, <u>1941</u>, *15*, 327.
2. Shenderei, E. P.; Zelvinski, Y. D.; Ivanovsky, F. P.
   Gazov. Prom., <u>1961</u>, *6(3)*, 42.
3. Hemmaplardh, B.; King, A. D.;
   J. Phys. Chem., <u>1972</u>, *76*, 2170.
4. Francesconi, A. Z.; Lentz, H.; Franck, E. U.
   J. Phys. Chem., <u>1981</u>, *85*, 3303.
5. Yarym-Agaev, N. L.; Sinyavskaya, R. P.; Koliushko, I. L.; Levinton, L. Ya.
   Zh. Prikl. Khim., <u>1985</u>, *58(1)*, 165.

COMPONENTS:	ORIGINAL MEASUREMENTS:
1. Methane; CH_4; [74-82-8] 2. Methanol; CH_4O; [67-56-1]	Shenderei, E. P.; Zelvinski, Y. D.; Ivanovsky, F. P. *Gazov. Prom.*, 1961, *6(3)*, 42-45.

VARIABLES:	PREPARED BY:
Temperature	C. L. Ycung

EXPERIMENTAL VALUES:

T/K	Henry´s constant atm	Mole fraction of methane[a]
248	800.0	0.00125
238	756.0	0.00132
223	662.0	0.00151
213	610.0	0.00164

[a] extrapolated to 1 atm pressure, calculated by compiler

AUXILIARY INFORMATION

METHOD/APPARATUS/PROCEDURE:	SOURCE AND PURITY OF MATERIALS:
Static equilibrium cell fitted with a magnetic stirrer. Temperature measured with a copper-constantan thermopile. Pressure measured with Bourdon gauge. Sample analysed by separating components by freezing out methanol. Details in ref. (1).	1. Obtained from fermentation, free from higher hydrocarbons carbon dioxide removed by passing through potassium hydroxide solution. 2. No details given

ESTIMATED ERROR:
$\partial T/K = \pm 0.1$;
$\partial x/x = \pm 0.04$ (compiler)

REFERENCES:

1. Shenderei, E. R.; Zelvenski, Ya. D.; Ivanovskii, F. P.; *Khim. Prom.*, 1959, 50.

COMPONENTS:	ORIGINAL MEASUREMENTS:
1. Methane; CH_4; [74-82-8] 2. Methanol; CH_4O; [67-56-1]	Yarym-Agaev, N. L.; Sinyavskaya, R. P.; Koliushko, I.L.; Levinton, L. Ya.; *Zh. Prikl. Khim.*, 1985, *58(1)*, 165-8.

VARIABLES:
  Temperature, pressure
	PREPARED BY:
	  C. L. Young

EXPERIMENTAL VALUES:

T/K	P/MPa	Mole fraction of methane in liquid	in vapour
298.2	2.5	0.02277	0.99192
	5.0	0.04404	0.99487
	7.5	0.06615	0.99574
	10.0	0.08080	0.99608
	12.5	0.09437	0.99564
313.2	2.5	0.02279	0.98477
	5.0	0.04164	0.98830
	7.5	0.06043	0.99011
	10.0	0.07793	0.99134
	12.5	0.09200	0.99276
338.2	2.5	0.02211	0.95433
	5.0	0.03934	0.97068
	7.5	0.05771	0.97567
	10.0	0.07239	0.97759
	12.5	0.08650	0.97956

AUXILIARY INFORMATION

METHOD/APPARATUS/PROCEDURE:	SOURCE AND PURITY OF MATERIALS:
Flow method: dry methane passed through a series of six saturators containing methanol, each fitted with a diffuser. gas then passed through a demister fitted with packed gauze. Flow rate of methane was about 200 cm^3hr. Gas then passed through a heated needle valve to near atmospheric pressure. Samples of gas analysed either GC or by freezing out methanol and estimating gravimetrically and estimating methane volumetrically.	1. Purity 99.95 mass per cent. 2. Contained 0.05 mass per cent water.

	ESTIMATED ERROR:
	  $\partial T/K = \pm 0.1$; $\partial p/MPa = \pm 0.05$;
	  $\partial x/x = \pm 0.003$ (estimated by compiler)

	REFERENCES:

COMPONENTS:	ORIGINAL MEASUREMENTS:
1. Methane; CH_4; [74-82-8] 2. Methanol; CH_4O; [67-56-1]	Francesconi, A. Z.; Lentz, H.; Franck, E. U. *J. Phys. Chem.* <u>1981</u>, *85*, 3303-7.

VARIABLES:	PREPARED BY:
	C. L. Young

EXPERIMENTAL VALUES:

T/K	t/°C	P/MPa	Mole fraction of methane, x_{CH_4}
502.2	229.0	11.7	0.997
500.2	227.0	11.9	
495.2	222.0	12.2	
482.2	209.0	13.2	
468.2	195.0	13.9	
435.2	162.0	14.4	
317.2	44.0	18.4	
485.2	212.0	14.0	0.110
471.2	198.0	14.6	
426.7	153.5	15.4	
338.7	65.5	18.0	
318.2	45.0	21.9	
468.2	195.0	22.2	0.290
463.2	190.0	23.5	
457.7	184.5	25.4	
438.2	165.0	28.3	
409.2	136.0	32.6	
379.7	106.5	37.7	
373.2	100.0	40.6	
351.7	78.5	44.8	
328.2	55.0	49.0	
322.2	49.0	61.5	

(cont.)

AUXILIARY INFORMATION

METHOD/APPARATUS/PROCEDURE:	SOURCE AND PURITY OF MATERIALS:
Static equilibrium cell fitted with a magnetic stirrer and movable piston and single colorless sapphire indow. PVT data obtained for mixtures of fixed and known compo-sitions. Details of apparatus and procedure in source and ref. (1).	1. Messer-Griesheim sample, purity 99.95 mole per cent or better. 2. Research grade Merck sample, contained less than 0.01 mole per cent water.

ESTIMATED ERROR:

$\delta T/K = \pm 1.0$; $\delta x_{CH_4} = \pm 0.002$.

REFERENCES:

1. Lentz, H.
   Rev. Sci. Instrum.
   <u>1969</u>, *40*, 341.

COMPONENTS:	ORIGINAL MEASUREMENTS:
1. Methane; CH_4; [74-82-8]	Francesconi, A. Z.; Lentz, H.; Franck, E. U.
2. Methanol; CH_4O; [67-56-1]	*J. Phys. Chem.*
	1981, *85*, 3303-7.

EXPERIMENTAL VALUES:

T/K	t/°C	P/MPa	Mole fraction of methane, x_{CH_4}
457.2	184.0	9.2	0.359
471.2	198.0	10.7	
481.2	208.0	13.4	
483.2	210.0	16.6	
479.2	206.0	18.1	
476.2	203.0	19.8	
467.2	194.0	23.7	
458.2	185.0	25.1	
453.2	180.0	26.0	
450.2	177.0	28.3	
443.7	170.5	30.1	
436.2	163.0	32.5	
427.2	154.0	35.7	
430.2	157.0	9.3	0.464
445.7	172.5	11.7	
455.2	182.0	15.5	
464.2	191.0	19.4	
465.2	192.0	22.9	
458.2	185.0	26.0	
442.8	169.6	29.8	
435.2	162.0	32.8	
425.8	152.6	35.8	
417.2	144.0	39.6	0.464
406.2	133.0	44.4	
392.8	119.6	51.0	
381.7	108.5	58.0	
373.7	100.5	64.3	
364.2	91.0	71.3	
345.4	72.2	80.6	
336.0	62.8	88.9	
324.0	50.8	101.1	
418.2	145.0	12.2	0.549
433.2	160.0	15.2	
443.2	170.0	18.0	
451.7	178.5	22.5	
451.2	178.0	26.6	
442.2	169.0	30.3	
433.7	160.5	33.4	
419.9	146.7	40.0	
395.2	122.0	50.8	
376.7	103.5	63.7	
356.8	83.6	76.8	
341.2	68.0	94.2	
322.2	49.0	126.8	
314.8	41.6	141.9	
305.2	32.0	168.7	
297.2	24.0	198.6	
433.2	160.0	21.5	0.700
437.2	164.0	23.9	
437.2	164.0	24.7	
432.2	159.0	28.9	
422.2	149.0	34.0	
402.2	129.0	44.5	
391.7	118.5	51.4	

(cont.)

COMPONENTS:	ORIGINAL MEASUREMENTS:
1. Methane; CH₄; [74-82-8] 2. Methanol; CH₄O; [67-56-1]	Francesconi, A. Z.; Lentz, H.; Franck, E. U. *J. Phys. Chem.* <u>1981</u>, *85*, 3303-7.

EXPERIMENTAL VALUES:

T/K	t/°C	P/MPa	Mole fraction of methane, x_{CH_4}
381.7	108.5	58.6	0.700
366.8	93.6	71.5	
354.5	81.3	86.5	
344.0	70.8	102.1	
336.4	63.2	115.0	
325.6	52.4	145.0	
321.6	48.4	164.9	
317.6	44.4	190.4	
313.8	40.6	208.0	
311.7	38.5	227.8	
309.4	36.2	245.8	
308.2	35.0	270.5	
305.8	32.6	292.3	
410.2	137.0	22.8	0.816
413.2	140.0	25.2	
415.2	142.0	30.0	
410.2	137.0	32.6	
398.7	125.5	37.0	
391.2	118.0	42.3	
378.2	105.0	50.0	
364.2	91.0	61.0	
349.2	76.0	78.3	
337.2	64.0	101.3	
328.7	55.5	121.5	
320.2	47.0	159.0	
313.5	40.3	186.1	
312.3	39.1	220.6	
309.2	36.0	247.5	
308.7	35.5	275.0	
307.0	33.8	293.5	
380.2	107.0	34.0	0.9158
373.2	100.0	40.5	
363.2	90.0	53.0	
349.2	76.0	65.3	
342.2	69.0	88.5	
327.2	54.0	99.5	
322.2	49.0	121.6	
318.2	45.0	141.5	
303.2	30.0	248.0	

COMPONENTS:	ORIGINAL MEASUREMENTS:
1. Methane; CH_4; [74-82-8] 2. Ethanol; C_2H_6O; [64-17-5]	Frolich, P.K.; Tauch, E.J.; Hogan, J.J.; Peer, A.A. *Ind. Eng. Chem.* <u>1931</u>,23,548-550

VARIABLES:	PREPARED BY:
Pressure	C.L. Young

EXPERIMENTAL VALUES:

T/K	P/MPa	Solubility*,S	Mole fraction of methane in liquid, x_{CH_4}
298.15	1.0	4	0.0095
	2.0	8	0.019
	3.0	12	0.028
	4.0	16	0.037
	5.0	20	0.046
	6.0	24	0.054
	7.0	27.5	0.062
	8.0	31.2	0.070
	9.0	36.5	0.081
	10.0	40.5	0.089
	11.0	45	0.097
	12.0	49	0.105

* Data taken from graph in original article. Volume of gas measured at 101.325 kPa and 298.15 K dissolved by unit volume of liquid measured under the same conditions.

+ calculated by compiler.

AUXILIARY INFORMATION

METHOD/APPARATUS/PROCEDURE:	SOURCE AND PURITY OF MATERIALS:
Static equilibrium cell. Liquid saturated with gas and after equilibrium established samples removed and analysed by volumetric method. Allowance was made for vapor pressure of liquid and the solubility of the gas at atmospheric pressure. Details in source.	Stated that the materials were the highest purity available.
	ESTIMATED ERROR: $\delta T/K = \pm 0.1$; $\delta x_{CH_4} = \pm 5\%$
	REFERENCES:

COMPONENTS:	ORIGINAL MEASUREMENTS:
1. Methane; CH_4; [74-82-8] 2. 2-Propanol; C_3H_8O; [67-63-0]	Frolich, P.K.; Tauch, E.J.; Hogan, J.J.; Peer, A.A. *Ind. Eng. Chem.* <u>1931</u>,*23*, 548-550

VARIABLES:
 	
          Pressure
	PREPARED BY:
	
	          C.L. Young

EXPERIMENTAL VALUES:

T/K	P/MPa	Solubility[*],S	Mole fraction of methane in liquid,[+] x_{CH_4}
298.15	1.0	4	0.012
	2.0	9	0.028
	3.0	14	0.042
	4.0	19	0.056
	5.0	25	0.073
	6.0	30.5	0.088
	7.0	36.5	0.103
	8.0	42.5	0.118
	9.0	48	0.131

*   Data taken from graph in original article. Volume of gas measured at 101.325 kPa pressure and 298.15K dissolved by unit volume of liquid measured under the same conditions.

+   calculated by compiler.

AUXILIARY INFORMATION

METHOD/APPARATUS/PROCEDURE:	SOURCE AND PURITY OF MATERIALS:
Static equilibrium cell. Liquid saturated with gas and after equilibrium established samples removed and analysed by volumetric method. Allowance was made for vapor pressure of liquid and the solubility of the gas at atmospheric pressure. Details in source.	Stated that the materials were the highest purity available.
	ESTIMATED ERROR: $\delta T/K = \pm 0.1$; $\delta x_{CH_4} = \pm 5\%$
	REFERENCES:

COMPONENTS:	EVALUATOR:
(1) Methane; CH_4; [74-82-8] (2) Cyclohexanol; $C_6H_{12}O$; [108-93-0]	H. Lawrence Clever Chemistry Department Emory University Atlanta, GA 30322 USA 1985, April

CRITICAL EVALUATION:

Cauquil (ref 1) and Lannung and Gjaldbaek (ref 2) have measured the solubility of methane in cyclohexanol. The single solubility value of Cauquil is rejected because it is only about one-half the magnitude of the values reported by Lannung and Gjaldbaek, workers whose solubility data are usually reliable.

The mole fraction solubility values of Lannung and Gjaldbaek were treated by a linear regression to obtain the equation

$$\ln x_1 = -7.8296 + 3.4376/(T/100 \text{ K})$$

with a standard error about the regression line of 1.12×10^{-5}.

The equation was treated to obtain the thermodynamic changes for the transfer of one mole of methane from the gas at a pressure of 0.101325 MPa to the infinitely dilute solution of

$$\Delta H_1^0/\text{kJ mol}^{-1} = -2.86 \text{ and } \Delta S_1^0/\text{J K}^{-1} \text{ mol}^{-1} = -65.1.$$

Smoothed values of the mole fraction solubility are in Table 1.

Table 1. The solubility of methane in cyclohexanol. Tentative values of the mole fraction solubility and partial molal Gibbs energy of solution as a function of temperature at a methane partial pressure of 0.101325 MPa.

T/K	$10^3 x_1$	$\Delta G_1^0/\text{kJ mol}^{-1}$
298.15	1.26	16.551
303.15	1.24	16.876
308.15	1.21	17.202

1. Cauquil, G. *J. Chim. Phys.* 1927, *24*, 53-5.

2. Lannung, A.; Gjaldbaek, J. C. *Acta Chem. Scand.* 1960, *14*, 1124-8.

COMPONENTS:	ORIGINAL MEASUREMENTS:
(1) Methane; CH_4; [74-82-8] (2) Cyclohexanol; $C_6H_{12}O$; [108-93-0]	Cauquil, G. *J. Chim. Phys.* <u>1927</u>, *24*, 53-55.

| VARIABLES:
$T/K = 299$
$p_1/kPa = 102$ | PREPARED BY:

          H. L. Clever |

EXPERIMENTAL VALUES:

   The author states that one liter of cyclohexanol absorbs
133 cm^3 methane at 26 ^{0}C and 765 mmHg.

   The compiler calculates an Ostwald coefficient of
$L/cm^3cm^{-3} = 0.133$ and a mole fraction solubilty at 101.325
kPa of $x_1 = 5.6 \times 10^{-4}$ at 299 K.

AUXILIARY INFORMATION

METHOD/APPARATUS/PROCEDURE:	SOURCE AND PURITY OF MATERIALS:
The apparatus appears to be of the Bunsen type. The initial and final volumes of gas in contact with the liquid were measured at 26 ^{0}C and a known pressure. The liquid vapor pressure was ignored.	(1) Methane. No information. (2) Cyclohexanol. Distilled, boiling point 160.9 ^{0}C at 766 mmHg. Degassed and tested to be air free.
	ESTIMATED ERROR: $\delta L/L = \pm 0.05$ (compiler)
	REFERENCES:

COMPONENTS:	ORIGINAL MEASUREMENTS:
(1) Methane; CH_4; [74-82-8] (2) Cyclohexanol; $C_6H_{12}O$; [108-93-0]	Lannung, A.; Gjaldbaek, J. C. *Acta Chem. Scand.* 1960, *14*, 1124 - 1128.
VARIABLES: $T/K = 298.15 - 310.15$ $p_1/kPa = 101.325$ (1 atm)	PREPARED BY: J. Chr. Gjaldbaek

EXPERIMENTAL VALUES:

T/K	Mol Fraction $10^3 x_1$	Bunsen Coefficient $\alpha/cm^3(STP)cm^{-3}atm^{-1}$	Ostwald Coefficient L/cm^3cm^{-3}
298.15	1.25	0.263	0.287
298.15	1.27	0.268	0.292
310.15	1.20	0.251	0.285
310.15	1.21	0.253	0.287

Smoothed Data: For use between 298.15 and 310.15 K.

$$\ln x_1 = -7.8296 + 3.4376/(T/100 \text{ K})$$

The standard error about the regression line is 1.12×10^{-5}.

T/K	Mol Fraction $10^3 x_1$
298.15	1.26
308.15	1.22

AUXILIARY INFORMATION

METHOD/APPARATUS/PROCEDURE:	SOURCE AND PURITY OF MATERIALS:
A calibrated all-glass combined manometer and bulb containing degassed solvent and the gas was placed in an air thermostat and shaken until equilibrium (1). The absorbed volume of gas is calculated from the initial and final amounts, both saturated with solvent vapor. The amount of solvent is determined by the weight of displaced mercury. The values are at 101.325 kPa (1 atm) pressure assuming Henry's law is obeyed.	(1) Methane. Generated from magnesium methyl iodide. Purified by fractional distillation. Specific gravity corresponds with mol wt 16.08. (2) Cyclohexanol. Poulenc Frères. "pur." Fractionated twice *in vacuo*. M.p./°C = 23.6 - 23.9.

	ESTIMATED ERROR: $\delta T/K = \pm 0.05$ $\delta x_1/x_1 = \pm 0.015$

REFERENCES:
1. Lannung, A.
   J. Am. Chem. Soc. 1930, *52*, 68.

COMPONENTS:	EVALUATOR:
(1) Methane; CH_4; [74-82-8] (2) Ethers	H. Lawrence Clever Department of Chemistry Emory University Atlanta, GA 30322 USA 1985, April

CRITICAL EVALUATION:

AN EVALUATION OF THE SOLUBILITY OF METHANE IN ETHERS AT PRESSURES UP TO 0.2 MPa ($\approx$2 ATM).

Seven papers report values of the methane solubility in three dialkyl ethers, four cyclic ethers, and one aromatic alkyl ether. All of the laboratories but one used conventional volumetric procedures. One used a GLC method (ref 4) at low methane partial pressure without correction for surface adsorption. The results are discussed in the three sections below.

I.   Methane + 1,1'-oxybisalkanes

Methane + 1,1'-oxybisethane or diethyl ether [60-29-7]

Christoff (ref 1) reported methane solubility values at 273.15 and 283.15 K. Horiuti (ref 2) reported seven solubility values at temperatures between 192.75 and 293.15 K. Horiuti's work is usually reliable. His values are 5 to 7 percent larger than the Christoff values in the 273 to 283 K temperature interval. The Horiuti values are classed as tentative and the Christoff values as doubtful.

A linear regression was applied to the Horiuti values which omitted the solubility value of 211.55 K to obtain the equation

$$\ln x_1 = -11.6184 + 11.0406/(T/100 \text{ K}) + 2.3007 \ln (T/100 \text{ K})$$

with a standard error about the regression line of 2.96×10^{-5}.

The thermodynamic changes for the transfer of one mole of methane from the gas phase at 0.101325 MPa to the infinitely dilute solution were calculated from the equation to be

T/K	ΔH_1^0/kJ mol^{-1}	ΔS_1^0/J K^{-1} mol^{-1}	$\Delta C_{p_1}^0$/J K^{-1} mol^{-1}
213.15	−5.10	−63.0	19.1
243.15	−4.53	−60.5	19.1
273.15	−3.95	−58.2	19.1
293.15	−3.57	−56.9	19.1

Smoothed values of the mole fraction solubility follow in Table 1.

Table 1. Solubility of methane in diethyl ether. Tentative values of the mole fraction solubility as a function of temperature at a methane partial pressure of 0.101325 MPa.

T/K	$10^3 x_1$	T/K	$10^3 x_1$
193.15	12.43		
203.15	10.54	273.15	5.17
213.15	9.12	278.15	5.01
223.15	8.03	283.15	4.87
233.15	7.19	288.15	4.74
243.15	6.52	293.15	4.62
253.15	5.97		
263.15	5.53		

Methane + 1,1'-oxybispropane or dipropyl ether [111-43-3]

Guerry (ref 3) measured the methane solubility at 293.15 and 298.15 K. The value reported at 293.15 K appears to be in error and it is rejected. The value of 298.15 K corresponds to a mole fraction solubility of 4.38 x 10^{-3} at a methane partial pressure of 0.101325 MPa and it is classed as tentative.

Methane + 1,1'-oxybis-2-propanol or dipropylene glycol [110-98-5]

Lenoir, Renault and Renon (ref 4) report the methane solubility at temperatures of 298.2, 323.2 and 343.2 K by a GLC retention time method. The maximum methane partial pressure was 2 mmHg. The three solubility values were converted to mole fraction solubilities at atmospheric pressure and treated by a linear regression to obtain the equation

$$\ln x_1 = -10.52708 + 9.7177/(T/100\ K)$$

with a standard error about the regression line 1.73 x 10^{-5}.

The thermodynamic changes for the transfer of one mole of methane from the gas at 0.101325 MPa to the infinitely dilute solution are

$$\Delta H_1^0/k\ J\ mol^{-1} = -8.08 \quad and \quad \Delta S_1^0/J\ K^{-1}\ mol^{-1} = -87.5$$

These values are of larger than average magnitude for many organic solvents. They may reflect either a more water-like solvent when several O and OH groups are present or problems in converting the reported Henry's constant at very low pressure to reliable values at atmospheric pressure. Smoothed mole fraction solubility values are in Table 2.

COMPONENTS:	EVALUATOR:
(1) Methane; CH_4; [74-82-8] (2) Ethers	H. Lawrence Clever Department of Chemistry Emory University Atlanta, GA 30322 USA 1985, April

CRITICAL EVALUATION:

Table 2.   The solubility of methane in 1,1'-oxybis-2-propanol. Tentative
values of the mole fraction solubility and partial molal Gibbs
energy as a function of temperature at a methane partial pressure
of 0.101325 MPa.

T/K	$10^4 x_1$	ΔG_1^0/kJ mol^{-1}
298.2	6.97	18.021
303.2	6.61	18.458
313.2	5.97	19.333
323.2	5.42	20.209
333.2	4.95	21.084
343.2	4.55	21.959

II.   Methane + Cyclic Ethers

Methane + 1,4-Dioxane [123-91-1]

Guerry (ref 3), Ben-Naim and Yaacobi (ref 5), and Gallardo, Urieta and
Gutierrez Losa (ref 7) report methane solubilities in 1,4-dioxane. Guerry's two values at 293.15 and 298.15 K are 3 and 6.5 percent, respectively,
smaller than the values of the other workers. They are classed as doubtful. The results of Ben-Naim and Yaacobi and of Gallardo et $al.$ agree
with 0.5 percent at all temperatures between 283.15 and 303.15 K. Both
laboratories used a similar apparatus and experimental technique. The
Ben-Naim and Yaacobi and the Gallardo et $al.$ values were combined in a
linear regression to obtain the equation for the 283.15 to 303.15 K temperature interval

$\ln x_1 = -6.91003 + 0.95526/(T/100 \text{ K})$

with a standard error about the regression line of 5.03×10^{-6}.

The thermodynamic changes for the transfer of one mole of methane
from the gas at 0.101325 MPa to the infinitely dilute solution are found
from the equation to be

ΔH_1^0/kJ mol^{-1} = -0.79_4 and ΔS_1^0/J K^{-1} mol^{-1} = -57.5.

This is a much smaller enthalpy of solution than normally found for organic solvents.   The smoothed mole fraction solubility values follow in Table 3.

Table 3.   The solubility of methane in 1,4-dioxane.   Tentative values of
the mole fraction solubility and partial molal Gibbs energy of
solution as a function of temperature of a methane partial pressure at 0.101325 MPa.

T/K	$10^4 x_1$	$\Delta G_1^0/kJ\ mol^{-1}$
283.15	13.9_8	15.473
288.15	13.9_0	15.761
293.15	13.8_2	16.048
298.15	13.7_5	16.335
303.15	13.6_7	16.622

Methane + Tetrahydrofuran [109-99-9]

Methane + 2,3-Dihydropyran [25512-65-0]

Methane + Tetrahydro-2H-pyran [142-68-7]

Guerry (ref 3) measured the solubility of methane at 293.15 and
298.15 K in each of the three solvents.   His values show a negligible
temperature coefficient of the mole fraction solubility, which imply a
zero to slightly positive partial molal enthalpy of solution.

Guerry's mole fraction solubility values, which are classed as tentative, are:

Solvent and Formula	Mole Fraction Solubility, $10^3 x_1$, at temperatures of	
	293.15 K	298.15 K
Tetrahydrofuran, C_4H_8O	1.94	1.94
2,3-Dihydropyran, C_5H_8O	1.94	1.94
Tetrahydro-2H-pyran, $C_5H_{10}O$	2.32	2.33

III. Methane + aryl alkyl ethers

Methane + methoxybenzene or anisole [100-66-3]

Gibanel, Urieta, and Gutierrez Losa (ref 6) measured the methane solubility at five degree intervals between 283.15 and 303.15 K.   Their values
are classed as tentative.

The data were fit by a linear regression to obtain the equation for
the 283.15 to 303.15 K interval

COMPONENTS:	EVALUATOR:
(1) Methane; CH_4; [74-82-8] (2) Ethers	H. Lawrence Clever Department of Chemistry Emory University Atlanta, GA 30322 USA 1985, April

CRITICAL EVALUATION:

$$\ln x_1 = -7.0524 + 1.8282/(T/100 \text{ K})$$

with a standard error about the regression line of 8.87×10^{-7}.

From the equation the thermodynamic changes for the transfer of one mole of methane from the gas at 0.101325 MPa to the infinitely dilute solution are

$$\Delta H_1^0/\text{kJ mol}^{-1} = -1.52 \quad \text{and} \quad \Delta S_1^0/\text{J K}^{-1} \text{ mol}^{-1} = -58.6$$

Smoothed values of the mole fraction solubility are in Table 4.

Table 4. Solubility of methane in methoxybenzene. Tentative values of the mole fraction solubility and partial molal Gibbs energy of solution as a function of temperature at a methane partial pressure of 0.101325 MPa.

T/K	$10^4 x_1$	$\Delta G_1^0/\text{kJ mol}^{-1}$
283.15	16.50	15.083
288.15	16.32	15.376
293.15	16.14	15.669
298.15	15.98	15.962
303.15	15.82	16.255

References

1. Christoff, A. *Z. Phys. Chem.* <u>1912</u>, *79*, 456-60.

2. Horiuti, J. *Sci. Pap. Inst. Phys. Chem. Res. (Jpn)* <u>1931/32</u>, *17*, 125-256.

3. Guerry, D. Jr., Ph.D. thesis, <u>1944</u>, Vanderbilt University, Nashville, TN USA.

4. Lenoir, J.-Y.; Renault, P.; Renon, H. *J. Chem. Eng. Data* <u>1971</u>, *16*, 340-342.

5. Ben-Naim, A.; Yaacobi, M. *J. Phys. Chem.* <u>1974</u>, *78*, 175-178.

6. Gibanel, F.; Urieta, J. S.; Gutierrez Losa, C. *J. Chim. Phys. Phys.-Chim. Biol.* <u>1981</u>, *78*, 171-174.

7. Gallardo, M.A.; Urieta, J. S.; Gutierrez Losa, C. *J. Chim. Phys. Phys.-Chim. Biol.* <u>1983</u>, *80*, 621-625.

COMPONENTS:	ORIGINAL MEASUREMENTS:
(1) Methane; CH_4; [74-82-8] (2) 1,1'-Oxybisethane or diethyl ether; $C_4H_{10}O$; [60-29-7]	Christoff, A. Z. Phys. Chem. <u>1912</u>, 79, 456-60.

VARIABLES:	PREPARED BY:
T/K = 273.15, 283.15 p_1/kPa = atmospheric	H. L. Clever

EXPERIMENTAL VALUES:

Temperature		Mol Fraction	Bunsen Coefficient	Ostwald Coefficient
$t/^0C$	T/K	$10^3 x_1$	$\alpha/cm^3 (STP) cm^{-3} atm^{-1}$	$L/cm^3 cm^{-3}$
0	273.15	4.77	1.066	1.066
10	283.15	4.50	0.992	1.028

The mole fraction and Bunsen coefficient values were calculated by the compiler assuming ideal gas behavior.

AUXILIARY INFORMATION

METHOD/APPARATUS/PROCEDURE:	SOURCE AND PURITY OF MATERIALS:
The apparatus is an Ostwald type as described by Just (ref 1), and modified by Skirrow (ref 2). The apparatus consists of a thermostated gas buret and an absorption flask. The modification involves the use of vapor free gas in the gas buret. A correction is made for the vapor pressure of the solvent. A steel capillary tube with a stopcock, which prevents the the gas and the solvent vapor from mixing in the buret, is used to connect the absorption flask and the buret.	(1) Methane. Prepared by the author. Treated with sulfuric acid and palladium to remove related gases and hydrogen. (2) Diethyl ether. Merck . Stated to be pure and anhydrous.

	ESTIMATED ERROR:
	$\delta L/L$ = ± 0.03

	REFERENCES:

	1. Just, G.
	   Z. Phys. Chem. <u>1901</u>, 37, 342.
	
	2. Skirrow, F. W.
	   Z. Phys. Chem. <u>1902</u>, 41, 139.

COMPONENTS:	ORIGINAL MEASUREMENTS:
(1) Methane; CH_4: [74-82-8] (2) 1,1'-Oxybisethane or diethyl ether; $C_4H_{10}O$; [60-29-7]	Horiuti, J. *Sci. Pap. Inst. Phys. Chem. Res. (Jpn)* 1931/32, *17*, 125 - 256.
VARIABLES: T/K: 192.75 - 293.15 p_1/kPa: 101.325 (1 atm)	PREPARED BY: M. E. Derrick H. L. Clever

EXPERIMENTAL VALUES:

T/K	Mol Fraction $10^3 x_1$	Bunsen Coefficient α/cm^3 (STP) cm^{-3} atm^{-1}	Ostwald Coefficient L/cm^3 cm^{-3}
192.75	12.49	3.146	2.220
196.45	11.76	2.946	2.119
211.55	9.247	2.265	1.754
231.85	7.314	1.739	1.476
252.55	5.978	1.379	1.275
273.15	5.171	1.157	1.157
293.15	4.624	1.004	1.078

The mole fraction and Bunsen coefficient values were calculated by the compiler with the assumption the gas is ideal and that Henry's law is obeyed.

Smoothed Data: For use between 192.75 and 273.15 K.

The 211.55 K value was omitted from the linear regression.

$$\ln x_1 = -11.6184 + 11.0406/(T/100K) + 2.3007 \ln (T/100K)$$

The standard error about the regression line is 2.96×10^{-5}.

T/K	Mol Fraction $10^3 x_1$
198.15	11.41
213.15	9.12
228.15	7.59
243.15	6.52
258.15	5.74
273.15	5.17
288.15	4.74

AUXILIARY INFORMATION

METHOD/APPARATUS/PROCEDURE:	SOURCE AND PURITY OF MATERIALS:
The apparatus consists of a gas buret, a solvent reservoir, and an absorption pipet. The volume of the pipet is determined at various meniscus heights by weighing a quantity of water. The meniscus height is read with a cathetometer. The dry gas is introduced into the degassed solvent. The gas and solvent are mixed with a magnetic stirrer until saturation. Care is taken to prevent solvent vapor from mixing with the solute gas in the gas buret. The volume of gas is determined from the gas buret readings, the volume of solvent is determined from the meniscus height in the absorption pipet.	(1) Methane. Aluminum carbide was prepared from aluminum and soot carbon. The aluminum carbide was treated with hot water. The gas evolved was scrubbed to remove impurities, dried and fractionated. Final product had a density, ρ/g dm^{-3} = 0.7168±0.0003 at normal conditions. (2) Diethyl ether. Merck. "For analysis grade". Stored over sodium amalgam until evolution of gas ceased. Distilled, boiling point constant within 0.01°C.
	ESTIMATED ERROR: δT/K = 0.05 $\delta x_1/x_1$ = 0.01
	REFERENCES:

COMPONENTS:	ORIGINAL MEASUREMENTS:
(1) Methane; CH_4; [74-82-8] (2) 1,1'-Oxybispropane or dipropyl ether; $C_6H_{14}O$; [111-43-3]	Guerry, D. Jr. Ph.D. thesis, <u>1944</u> Vanderbilt University Nashville, TN Thesis Director: L. J. Bircher

VARIABLES:	PREPARED BY:
T/K: 293.15, 298.15 P/kPa: 101.325 (1 atm)	H. L. Clever

EXPERIMENTAL VALUES:

T/K	Mol Fraction x_1 x 10^4	Bunsen Coefficient α	Ostwald Coefficient L
293.15		0.198 (sic)	0.212 (?)
298.15		0.713	0.778

The Ostwald coefficients were calculated by the compiler.

AUXILIARY INFORMATION

METHOD/APPARATUS/PROCEDURE:	SOURCE AND PURITY OF MATERIALS:
A Van Slyke-Neill Manometric Apparatus manufactured by the Eimer and Amend Co. was used. The procedure of Van Slyke (1) for pure liquids was modified (2) so that small solvent samples (2 cm^3) could be used with almost complete recovery of the sample. An improved temperature control system was used.	(2) Dipropyl ether. Eastman Kodak Co. Refluxed four hours over Na, then distilled from Na in a N_2 atm. B.p. (746.2 mmHg) t/°C 89.03 - 89.28 (corr.). Refractive index, density, and vapor pressure data are in the thesis.

SOURCE AND PURITY OF MATERIALS:	ESTIMATED ERROR:
(1) Methane. Prepared by hydrolysis of crystaline methyl Grignard reagent. Passed through conc. H_2SO_4, solid KOH, and Dririte.	$\delta T/K = 0.05$

REFERENCES:
 1. Van Slyke, D. D.
    J. Biol. Chem. <u>1939</u>, *130*, 545.

 2. Ijams, C. C.
    Ph.D. thesis, <u>1941</u>
    Vanderbilt University

COMPONENTS:	ORIGINAL MEASUREMENTS:
1. Methane; CH_4; [74-82-8] 2. 1,2-Epoxyethane (Ethylene oxide); C_2H_4O; [75-21-8]	Hess, L. G.; Tilton, V. V. *Ind. Eng. Chem.* <u>1950</u>, *42*, 1251-2158.

VARIABLES:	PREPARED BY:
	C. L. Young

EXPERIMENTAL VALUES:

t^a/C	T^b/K	Total pressure,[a] pounds per square inch /psig	Mass percent[a] in solution	Mole fraction[b] /x_1
30	303.2	30	0.02	0.0005
30	303.2	40	0.06	0.0016
30	303.2	50	0.11	0.0030
45	318.2	50	0.06	0.0016
45	318.2	60	0.11	0.0030
45	318.2	70	0.15	0.0041

[a] Original data.

[b] Calculated by compiler.

AUXILIARY INFORMATION

METHOD/APPARATUS/PROCEDURE:	SOURCE AND PURITY OF MATERIALS:
High pressure, steel flow apparatus consisting of two presaturators for the gas and an equilibrium vessel containing a stirrer operated by a solenoid. The gas is supersaturated in the first saturator at a temperature 10 K above the equilibration temperature. A steady flow of gas is made for at least 2 h after which liquid and vapor samples are withdrawn for analysis at 1-h intervals. Equilibrium indicated by constant consecutive compositions of both phases. Details in ref. (1).	Source and purities not available.

	ESTIMATED ERROR:
	$\delta T/K$ = 0.1; $\delta x_1/x$ = $\delta H/H$ = 0.10 (estimated by compiler).

	REFERENCES:
	1. Wan, S.-W.; Dodge, B. F. *Ind. Eng. Chem.* <u>1940</u>, *32*, 95.

COMPONENTS:	ORIGINAL MEASUREMENTS:
1. Methane; CH₄; [74-82-8] 2. 1,2-Epoxyethane (Ethylene oxide); C_2H_2O; [75-21-8]	Olson, J. D. *J. Chem. Engng. Data* <u>1977</u>, *22*, 326-329.

VARIABLES:	PREPARED BY:
	C. L. Young

EXPERIMENTAL VALUES:

t/°C	T/K	Henry's constant[a] /atm	Mole fraction[b] x_{CH_4}
0	273.15	613	0.001631
25	298.15	614	0.001629
50	323.15	595	0.001681

[a] Original data; Henry's constant extrapolated to zero gas partial pressure.

[b] Mole fraction calculated by compiler assuming constant H and gas partial pressure of 101.325 kPa.   Note the normal boiling point of the solvent is 286.7 K.

AUXILIARY INFORMATION

METHOD/APPARATUS/PROCEDURE:	SOURCE AND PURITY OF MATERIALS:
Accurate gravimetric method for determining masses of solvent and gas charged into stainless steel bomb of predetermined volume. Gas introduced at pressures of up to 840 kPa measured by Bourdon gauge. Equilibration by shaking for 2 to 4 h aided by several loose balls in bomb. Pressure measurements along with known volumes and masses of gas and solvent permitted calculation of Henry's constant. Detailed volume change corrections made for both phases.	1. Matheson research grade sample, purity 99.99 mole per cent. 2. UCC commercial grade. GC analysis indicated volatile impurities of less than 100 ppm.

ESTIMATED ERROR:

$\delta T/K = \pm 0.10$; $\delta H/H$(RMS) = 0.03.

REFERENCES:

COMPONENTS:	ORIGINAL MEASUREMENTS:
1. Methane; CH_4; [74-82-8] 2. 1,4-Dioxane; $C_4H_8O_2$; [123-91-1]	Ben-Naim, A.; Yaacobi, M. *J. Phys. Chem.* <u>1974</u>, *78*, 175-8.

VARIABLES:
 	
    Temperature
	PREPARED BY:
	
	    C.L. Young

EXPERIMENTAL VALUES:

T/K	Ostwald coefficient,[*] L	Mole fraction[+] at partial pressure of 101.3 kPa, x_{CH_4}
283.15	0.3841	0.00139
288.15	0.3876	0.00139
293.15	0.3907	0.00138
298.15	0.3933	0.00138
303.15	0.3956	0.00137

*   Smoothed values obtained from the equation.

$$kT \ln L = 1,099.8 + 4.781 \, (T/K) - 0.00988 \, (T/K)^2 \text{ cal mol}^{-1}$$
where k is in units of cal mol^{-1} K^{-1}

+   calculated by compiler assuming the ideal gas law for methane.

AUXILIARY INFORMATION

METHOD/APPARATUS/PROCEDURE:	SOURCE AND PURITY OF MATERIALS:
The apparatus was similar to that described by Ben-Naim and Baer (1) and Wen and Hung (2). It consists of three main parts, a dissolution cell of 300 to 600 cm^3 capacity, a gas volume measuring column, and a manometer. The solvent is degassed in the dissolution cell, the gas is introduced and dissolved while the liquid is kept stirred by a magnetic stirrer immersed in the water bath. Dissolution of the gas results in the change in the height of a column of mercury which is measured by a cathetometer.	1. Matheson sample, purity 99.9 mol per cent. 2. AR grade.

ESTIMATED ERROR:
$$\delta T/K = \pm 0.1; \quad \delta x_{CH_4} = \pm 2\%$$
(estimated by compiler)

REFERENCES:

1. Ben-Naim, A.; Baer, S. *Trans. Faraday Soc.* <u>1963</u>, *59*, 2735.

2. Wen, W.-Y.; Hung, J.H. *J. Phys. Chem.* <u>1970</u>, *74*, 170

COMPONENTS:	ORIGINAL MEASUREMENTS:
(1) Methane; CH_4; [74-82-8] (2) 1,4-Dioxane; $C_4H_8O_2$; [123-91-1]	Gallardo, M. A.; Urieta, J. S.; Gutierrez Losa, C. *J. Chim. Phys. Phys. Chim. Biol.* 1983, *80*, 621-5.

VARIABLES:
       $T/K = 285.15 - 303.15$
    $p_1/kPa = 101.3$
	PREPARED BY:
	
	         H. L. Clever

EXPERIMENTAL VALUES:

T/K	Mol Fraction $10^3 x_1$
285.15	1.400
289.15	1.393
293.15	1.383
298.15	1.372
303.15	1.361

The authors fitted the data to the equation

$$-\ln x_1 = 0.461 \ln (T/K) + 3.965$$

from which they obtained thermodynamic changes for the transfer of one mole methane from the gas at one atm to the hypothetical $x_1 = 1$ solution of

$$\Delta H^0/kJ\ mol^{-1} = -1.14 \quad \text{and}$$

$$\Delta S^0/J\ K^{-1}\ mol^{-1} = -59.$$

AUXILIARY INFORMATION

METHOD/APPARATUS/PROCEDURE:	SOURCE AND PURITY OF MATERIALS:

The apparatus is similar to that used by Ben-Naim and Baer (ref 1). It is described in detail in an earlier paper (ref 2).
 	
   Literature 1,4-dioxane vapor pressure data were fitted to the equation
 	
    $\ln (p/kPa) = -4591.3/(T/K)$
                 $+ 16.98$
	(1) Methane. Sociedad Espanola del Oxigeno. 99.95 mol percent.
	
	 (2) 1,4-Dioxane. Merck and Co. Purity checked by GLC to be $\geq$ 99 percent.

	ESTIMATED ERROR:
	       $\delta T/K = \pm\ 0.1$
	    $\delta p_1/mmHg = \pm\ 0.04$
	     $\delta x_1/\ x_1 = \pm\ 0.007$ (authors ref 2)

REFERENCES:

1. Ben Naim, A.; Baer, S.
   Trans. Faraday Soc. .1963, *59*,2735

2. Carnicer, J.; Gibanel, F.;
   Urieta,J.S.; Gutierrez Losa, C.
   Rev. Acad. Ciencias Zaragoza
   1979, *34*, 115-22.

COMPONENTS:	ORIGINAL MEASUREMENTS:
(1) Methane; CH_4; [74-82-8] (2) 1,4-Dioxane; $C_4H_8O_2$; [123-91-1]	Gallardo, M. A.; Urieta, J. S.; Gutierrez Losa, C. *J. Chim. Phys. Phys.-Chim. Biol.* 1983, *80*, 621-5.

VARIABLES:	PREPARED BY:
$T/K = 285.15 - 303.15$ $P_1/kPa = 101.325$	H. L. Clever

EXPERIMENTAL VALUES:

T/K	Mol Fraction $10^4 x_1$
285.15	14.0_0
289.15	13.9_3
293.15	13.8_3
298.15	13.7_2
303.15	13.6_1

The authors fit their data to the equation

$$- \ln x_1 = - 0.461 \ln (T/K) + 3.965$$

from which they obtained

$$\Delta H_1^0/kJ\ mol^{-1} = - 9.04 \quad and$$

$$\Delta S_1^0/J\ K^{-1}\ mol^{-1} = -69.$$

AUXILIARY INFORMATION

METHOD/APPARATUS/PROCEDURE:	SOURCE AND PURITY OF MATERIALS:
The solubility apparatus was similar to that used by Ben-Naim and Baer (ref 1). It consisted of a gas buret, mercury manometer, and solution vessel. The solvent was degassed in the solution vessel. Measurements were carried out on the vapor saturated gas. The vapor pressure of the solvent was taken from the literature.	(1) Methane. Sociedad Espanol del Oxigeno. Stated to be 99.95 percent pure. (2) 1,4-Dioxane. Merck. Purity equal or better than 99 percent. Checked by GLC.

ESTIMATED ERROR:
$\delta T/K = \pm 0.1$ $\delta x_1/x_1 = \pm 0.01$

REFERENCES:
1. Ben-Naim, A.; Baer, S.   *Trans. Faraday Soc.* 1963, *59*, 2735.

COMPONENTS:	ORIGINAL MEASUREMENTS:
(1) Methane; CH_4; [74-82-8] (2) Cyclic ethers: C_4H_8O, $C_4H_8O_2$, C_5H_8O, and $C_5H_{10}O$	Guerry, D. Jr. Ph.D. thesis, <u>1944</u> Vanderbilt University Nashville, TN Thesis Director: L. J. Bircher

VARIABLES:	PREPARED BY:
T/K: 293.15, 298.15 P/kPa: 101.325 (1 atm)	H. L. Clever

EXPERIMENTAL VALUES:

T/K	Mol Fraction $x_1 \times 10^4$	Bunsen Coefficient α	Ostwald Coefficient L
Tetrahydrofuran; C_4H_8O; [109-99-9]			
293.15	19.4	0.537	0.576
298.15	19.4	0.532	0.581
1,4-Dioxane; $C_4H_8O_2$; [123-91-1]			
293.15	13.4	0.352	0.378
298.15	12.9	0.338	0.369
2,3-Dihydropyran; C_5H_8O; [25512-65-6]			
293.15	19.4	0.480	0.515
298.15	19.4	0.478	0.522
Tetrahydro-2H-pyran; $C_5H_{10}O$; [142-68-7]			
293.15	23.2	0.536	0.575
298.15	23.3	0.536	0.585

The Ostwald coefficients were calculated by the compiler.

AUXILIARY INFORMATION

METHOD/APPARATUS/PROCEDURE:	SOURCE AND PURITY OF MATERIALS:
The apparatus was a modified Van Slyke-Neill Manometric Apparatus manufactured by the Eimer and Amend Co. The procedure of Van Slyke (1) for pure liquids was modified (2) so that small solvent samples (2 cm^3) could be used with almost 100 per cent recovery of the sample. An improved temperature control system was used.	Tetrahydrofuran. Eastman Kodak Co. B.p.(752.7 mmHg) t/°C 65.50 - 65.54. 1,4-Dioxane. Eastman Kodak Co. B.p.(743.7 mmHg) t/°C 100.81-100.82. Dihydro-2H-pyran. Prepared from tetrahydrofurfuryl alcohol. B.p. (743.6 mmHg) t/°C 84.81 - 84.89. Tetrahydro-2H-pyran. Prepared by catalytic reduction of dihydro-2H-pyran. B.p.(750.6 mmHg) t/°C 87.51 - 87.52. All b.p. are corrected.

SOURCE AND PURITY OF MATERIALS:	ESTIMATED ERROR:
(1) Methane. Prepared by hydrolysis of crystaline methyl Grignard reagent. Passed through conc. H_2SO_4, solid KOH, and Dririte. (2) Cyclic ethers. The ethers were fractionally distilled from over Na in a nitrogen atmosphere. In addition to the solubility data the thesis contains measured values of refractive index, density, vapor pressure and b.p.	$\delta T/K = 0.05$
	REFERENCES: 1. Van Slyke, D. D. *J. Biol. Chem.* <u>1939</u>, *130*, 545. 2. Ijams, C. C. Ph.D. thesis, <u>1941</u> Vanderbilt University

COMPONENTS:	ORIGINAL MEASUREMENTS:
1. Methane; CH_4; [74-82-8] 2. Methoxybenzene (*Anisole*); C_7H_8O; [100-66-3]	Gibanel, F.; Urieta, J. S.; Gutierrez Losz, C. *J. Chim. Phys.* __1981__, *78*, 171-174.

VARIABLES:	PREPARED BY:
	C. L. Young

EXPERIMENTAL VALUES:

T/K	10^4 Mole fraction of methane at 1 atm partial pressure $10^4 x_{CH_4}$
283.15	16.51_2
288.15	16.30_8
293.15	16.14_9
298.15	15.97_6
303.15	15.81_8

Smoothing equation given in source

$$\ln x_{CH_4} = 0.625 \ln(T/K) + 2.876.$$

AUXILIARY INFORMATION

METHOD/APPARATUS/PROCEDURE:	SOURCE AND PURITY OF MATERIALS:
Solubility apparatus was similar to that used by Ben-Naim and Baer (1), consisting essentially of a gas buret, mercury manometer and solution vessel. The solvent was degassed in the solution vessel. Measurements were carried out on the saturated gas. It appears that the mole fraction at a partial pressure of 1 atmosphere was estimated from the raw experimental data by assuming that Henry's law is obeyed and that the partial pressure of solvent in the gas phase is given by Raoult's law.	1. Sociedad Espanola del Oxigeno sample, purity 99.95 mole per cent. 2. Fluka product, purity equal to or better than 99 mole per cent, checked by GC.

ESTIMATED ERROR:
$\delta T/K = \pm 0.1$; $\delta x_{CH_4} = \pm 4\%$ (estimated by compiler).

REFERENCES:
1. Ben-Naim, A.; Baer, S. *Trans. Faraday Soc.* __1963__, *59*, 2735.

COMPONENTS:	EVALUATOR:
(1) Methane; CH_4; [74-82-8] (2) 2-Propanone or acetone; C_3H_6O; [67-64-1]	H. Lawrence Clever Department of Chemistry Emory University Atlanta, GA 30322 USA 1985, April

CRITICAL EVALUATION:

Horiuti (ref 1) measured seven values of the solubility of methane in 2-propanone between 196.55 and 303.15 K. Lannung and Gjaldbaek (ref 2) measured six values between 291.15 and 310.15 K. Both laboratories have the reputation of carrying out reliable measurements. The Lannung and Gjaldbaek mole fraction solubilities run about two percent larger than the Horiuti values over the common 291 -310 K temperature interval.

Both sets of data are classed as tentative. All data in both papers were combined in a linear regression to obtain the equation for the 193 - 313 K temperature interval of

$$\ln x_1 = - 13.6388 + 11.385/(T/100 \text{ K}) + 3.2398 \ln (T/100 \text{ K})$$

with a standard error about the regression line of 2.32×10^{-5}.

The equation gives temperature dependent values of the enthalpy and entropy changes for the transfer of one mole of methane from the gas at 0.101325 MPa to the infinitely dilute solution of:

T/K	ΔH_1^0/kJ mol^{-1}	ΔS_1^0/J K^{-1} mol^{-1}	ΔC_{p1}^0/J K^{-1} mol^{-1}
213.15	-3.72	-66.1	26.9
243.15	-2.92	-62.5	26.9
273.15	-2.11	-59.4	26.9
298.15	-1.43	-57.0	26.9

Smoothed values of the mole fraction solubility are in Table 1.

Table 1. The solubility of methane in 2-propanone. Tentative values of the mole fraction solubility as a function of temperature at a methane partial pressure of 0.101325 MPa.

T/K	$10^3 x_1$	T/K	$10^3 x_1$
193.15	3.655	283.15	1.938
203.15	3.221	288.15	1.913
213.15	2.893	293.15	1.891
223.15	2.642	298.15	1.872
233.15	2.447	303.15	1.855
243.15	2.293	308.15	1.840
253.15	2.172	313.15	1.827
263.15	2.075		
273.15	1.999		

A three constant smoothing equation for the Horiuti data only appears on the Horiuti data sheet. The two equations give negligably different solubility values between 193 and 253 K. From 283 to 313 K the equation above gives mole fraction solubility values that range from 0.94 to 1.73 percent larger than the equation based on only the Horiuti data.

REFERENCES:

1. Horiuti, J. *Sci. Pap. Inst. Phys. Chem. Res. (Jpn)* <u>1931/32</u>, *17*, 125 - 256.

2. Lannung, A.; Gjaldbaek, J. C. *Acta Chem. Scand.* <u>1960</u>, *14*, 1124 - 8.

COMPONENTS:	ORIGINAL MEASUREMENTS:
(1) Methane; CH_4; [74-82-8] (2) 2-Propanone or acetone; C_3H_6O; [67-64-1]	Horiuti, J. *Sci. Pap. Inst. Phys. Chem. Res. (Jpn)* <u>1931/32</u>, *17*, 125 - 256.

VARIABLES:	PREPARED BY:
T/K: 196.55 - 303.15 p_1/kPa: 101.325 (1 atm)	M. E. Derrick H. L. Clever

EXPERIMENTAL VALUES:

T/K	Mol Fraction $10^3 x_1$	Bunsen Coefficient α/cm^3(STP)cm^{-3}atm^{-1}	Ostwald Coefficient L/cm^{3}cm^{-3}
196.55	3.496	1.213	0.8726
212.55	2.909	0.9894	0.7699
232.15	2.463	0.8169	0.6943
251.35	2.187	0.7078	0.6513
273.15	1.982	0.6232	0.6232
293.15	1.877	0.5744	0.6165
303.15	1.822	0.5497	0.6101

The mole fraction and Bunsen coefficient values were calculated by the compiler with the assumption the gas is ideal and that Henry's law is obeyed.

Smoothed Data: For use between 196.55 and 303.15 K.

$$\ln x_1 = -13.1623 + 10.8092/(T/100K) + 2.9683 \ln (T/100K)$$

The standard error about the regression line is 5.85×10^{-6}.

T/K	Mol Fraction $10^3 x_1$	T/K	Mol Fraction $10^3 x_1$
198.15	3.423	273.15	1.985
213.15	2.895	288.15	1.893
228.15	2.538	298.15	1.847
243.15	2.289	308.15	1.811
258.15	2.112		

AUXILIARY INFORMATION

METHOD/APPARATUS/PROCEDURE:	SOURCE AND PURITY OF MATERIALS:
The apparatus consists of a gas buret, a solvent reservoir, and an absorption pipet. The volume of the pipet is determined at various meniscus heights by weighing a quantity of water. The meniscus height is read with a cathetometer. The dry gas is introduced into the degassed solvent. The gas and solvent are mixed with a magnetic stirrer until saturation. Care is taken to prevent solvent vapor from mixing with the solute gas in the gas buret. The volume of gas is determined from the gas buret readings, the volume of solvent is determined from the meniscus height in the absorption pipet.	(1) Methane. Aluminum carbide was prepared from aluminum and soot carbon. The alumnium carbide was treated with hot water. The gas evolved was scrubbed to remove impurities, dried and fractionated. Final product had a density, ρ/g dm^{-3} = 0.7168±0.0003 at normal conditions. (2) Acetone. Nippon Pure Chemical Co. or Merck. Extra pure grade. Recrystallized with sodium sulfite and stored over calcium chloride. Fractionated, boiling point (760 mmHg) 56.09°C.
	ESTIMATED ERROR: δT/K = 0.05 $\delta x_1/x_1$ = 0.01
	REFERENCES:

COMPONENTS:	ORIGINAL MEASUREMENTS:
(1) Methane; CH_4; [74-82-8] (2) 2-Propanone or acetone; C_3H_6O;[67-64-1]	Lannung, A.; Gjaldbaek, J. C. *Acta Chem. Scand.* <u>1960</u>, *14*, 1124 - 1128.

VARIABLES:	PREPARED BY:
T/K = 291.15 - 310.15 p_1/kPa = 101.325 (1 atm)	J. Chr. Gjaldbaek

EXPERIMENTAL VALUES:

T/K	Mol Fraction $10^3 x_1$	Bunsen Coefficient α/cm^3(STP)cm^{-3}atm^{-1}	Ostwald Coefficient L/cm^{3}cm^{-3}
291.15	1.94	0.593	0.632
291.15	1.94	0.594	0.633
298.15	1.87	0.566	0.618
298.15	1.85	0.562	0.613
310.15	1.84	0.548	0.622
310.15	1.84	0.546	0.620

Smoothed Data: For use between 291.15 and 310.15 K.

$$\ln x_1 = -7.0633 + 2.3565/(T/100 \text{ K})$$

The standard error about the regression line is 2.46 x 10^{-5}.

T/K	Mol Fraction $10^3 x_1$
298.15	1.89
308.15	1.84

AUXILIARY INFORMATION

METHOD/APPARATUS/PROCEDURE:

A calibrated all-glass combined manometer and bulb containing degassed solvent and the gas was placed in an air thermostat and shaken until equilibrium (1).

The absorbed volume of gas is calculated from the initial and final amounts, both saturated with solvent vapor. The amount of solvent is determined by the weight of displaced mercury.

The values are at 101.325 kPa (1 atm) pressure assuming Henry's law is obeyed.

SOURCE AND PURITY OF MATERIALS:

(1) Methane. Generated from magnesium methyl iodidie. Purified by fractional distillation. Specific gravity corresponds with mol wt 16.08.

(2) 2-Propanone. Kahlbaum. "Zur analyse". Contained no water, aldehyde or acid.

ESTIMATED ERROR:

$$\delta T/K = \pm 0.05$$
$$\delta x_1/x_1 = \pm 0.015$$

REFERENCES:

1. Lannung, A.
   J. Am. Chem. Soc. <u>1930</u>, *52*, 68.

COMPONENTS:	ORIGINAL MEASUREMENTS:
(1) Methane; CH_4; [74-82-8] (2) 2-Propanone or acetone; C_3H_6O; [67-64-1]	Hronec, M.; Hagara, A.; Ilavský, J. *Petrochemia* <u>1983</u>, *23* (2/3), 111-5.

VARIABLES:	PREPARED BY:
T/K = 238 - 278 p_t/kPa = 199 - 401	H. L. Clever

EXPERIMENTAL VALUES:

Temperature		Total Pressure	Mol Fraction	Kuenen Coefficient
$t/^0C$	T/K	p_t/kPa	$10^3 x_1$	S/cm^3 (STP) g^{-1}
-35	238	199	1.6 2.1	0.6 0.8
		300	3.4	1.3
		401	8.0 8.8	3.1 **3.4**
-15	258	199	0.21	0.08
		300	1.8	0.7
		401	5.2 3.6	2.0 1.4
5	278	401	2.6	1.0

The mole fraction solubility values were calculated by the compiler assuming ideal behavior.

Acetone vapor pressures given in the paper are:

$t/^0C$	-35	-15	5
p_2/kPa	1.075	3.823	12.107

AUXILIARY INFORMATION

METHOD/APPARATUS/PROCEDURE:	SOURCE AND PURITY OF MATERIALS:
A volumetric method described in the paper.	(1) Methane. Analyzed by GLC. Contained 2 % impurity which was mostly nitrogen. (2) Acetone. Contained 0.5 % water.

ESTIMATED ERROR:

δp_t/kPa = ± 5
$\delta x_1/x_1$ = ± 0.20 (compiler)

REFERENCES:

COMPONENTS:	ORIGINAL MEASUREMENTS:

(1) Methane; CH_4; [74-82-8]
 	
(2) 2-Propanone or acetone;
   C_3H_6O; [67-64-1]
	Yokoyama, C.; Masuoka, H.; Aral, K.;
	   Saito, S.
	
	J. Chem. Eng. Data 1985, *30*,
	177-9.

VARIABLES:
   $T/K = 298.2, 323,2$
   $p_t/MPa = 1.06 - 11.75$
	PREPARED BY:
	
	   H. L. Clever

EXPERIMENTAL VALUES:

Temperature		Total Pressure	Mol Fraction	
$t/^0C$	T/K	p_t/MPa	Liquid	Vapor
			x_1	y_1
25.0	298.2	1.71	0.0367	0.9753
		2.28	0.0434	0.9789
		3.55	0.0670	0.9866
		4.51	0.0911	0.9871
		5.49	0.1116	0.9880
		7.08	0.1443	0.9873
		8.19	0.1598	0.9872
		9.16	0.1866	0.9870
		10.10	0.1997	0.9868
		11.68	0.2287	0.9853
50.0	323.2	1.06	0.0153	0.9124
		1.50	0.0223	0.9363
		2.07	0.0341	0.9523
		3.07	0.0509	0.9627
		4.28	0.0725	0.9698
		5.00	0.0822	0.9707
		5.98	0.0994	0.9713
		7.05	0.1200	0.9718
		8.25	0.1360	0.9729
		9.59	0.1647	0.9729
		10.73	0.1782	0.9715
		11.75	0.1950	0.9711

AUXILIARY INFORMATION

METHOD/APPARATUS/PROCEDURE:	SOURCE AND PURITY OF MATERIALS:
The equipment consists of an equilibration system and an analysis system. The procedures are essentially the same as those used by King *et al.* (ref 1) and Kubota *et al.* (ref 2). The equilibration system is in a thermostated water bath. The analysis system is in an air bath at 100 ^0C to avoid condensation problems. Details of degassing, equilibration and sampling procedures were not given. The composition analysis was made by gas chromatograph and digital integrator. Calibration curves were obtained from mixtures of known composition.	(1) Methane. Takachiho Kagaku Co., Ltd. Used as received. (2) 2-Propanone. Dojin Yakugaku Ltd. Used as received. A trace analysis of the components found no measurable impurities. The samples were used without further purification.

ESTIMATED ERROR:
$$\delta T/K = \pm 0.05$$
$$\delta p_t/MPa = \pm 0.01$$
$$\delta x_1/x_1 = \pm 0.015$$

REFERENCES:
1. King,M.B.;Alderson,D.A.;Fallah,F.;
   Kassim,D.M.;Sheldon,J.R.;Mahmud,R.
   Chemical Engineering at Super-
   critical Conditions;Paulatis,M.E.
   et al.,Editors, Ann Arbor Science,
   1983, p. 31.
2. Kubota,H.;Inatome,H.;Tanaka,Y.;
Makita,T. *J.Chem.Eng.Jpn.* 1983,*16*,99.

COMPONENTS:	ORIGINAL MEASUREMENTS:

(1) Methane; CH_4; [74-82-8]
 	
(2) Cyclopentanone; C_5H_8O;
    [120-92-3]
	Gallardo, M. A.; López, M. C.
	    Urieta, J. S.; Gutierrez Losa, C.
	
	IUPAC Conference of Chemical Thermo-
	dynamics, 1984, Paper No. 47.

VARIABLES:	PREPARED BY:
T/K = 273.15 - 303.15 p_1/kPa = 101.3	H. L. Clever

EXPERIMENTAL VALUES:

T/K	Mol Fraction $10^4 x_1$
273.15	15.7
283.15	15.2
293.15	14.7
298.15	14.5
303.15	14.3

The authors fit their data to the equation

$$- \ln x_1 = 0.887 \ln (T/K) + 1.481$$

from which they obtained the thermodynamic ch ges

ΔH_1^0/kJ mol^{-1} = - 2.20   and

ΔS_1^0/J K^{-1} mol^{-1} = -62.

AUXILIARY INFORMATION

METHOD/APPARATUS/PROCEDURE:	SOURCE AND PURITY OF MATERIALS:
The solubility apparatus was similar to that used by Ben-Naim and Baer (ref 1). It consisted of a gas buret, mercury manometer, and solution vessel. The solvent was degassed in the solution vessel. Measurements were carried out on the vapor saturated gas.	(1) Methane. Sociedad Espanol del Oxigeno. Stated to be 99.95 percent pure. (2) Cyclopentanone.

ESTIMATED ERROR:
$\delta T/K$ = ± 0.1 $\delta x_1/x_1$ = ± 0.01

REFERENCES:
1. Ben-Naim, A.; Baer, S.   *Trans. Faraday Soc.* 1963, *59*, 2735.

COMPONENTS:	ORIGINAL MEASUREMENTS:
(1) Methane; CH_4; [74-82-8]	Guerry, D. Jr.
(2) Cyclohexanone; $C_6H_{10}O$; [108-94-1]	Ph.D. thesis, 1944 Vanderbilt University Nashville, TN Thesis Director: L. J. Bircher

VARIABLES:	PREPARED BY:
T/K: 293.15, 298.15 P/kPa: 101.325 (1 atm)	H. L. Clever

EXPERIMENTAL VALUES:

T/K	Mol Fraction x_1 x 10^4	Bunsen Coefficient α	Ostwald Coefficient L
293.15	16.1	0.349	0.375
298.15	16.1	0.347	0.379

The Ostwald coefficients were calculated by the compiler.

AUXILIARY INFORMATION

METHOD/APPARATUS/PROCEDURE:

A Van Slyke-Neill Manometric Apparatus manufactured by the Eimer and Amend Co. was used.

The procedure of Van Slyke (1) for pure liquids was modified (2) so that small solvent samples (2 cm^3) could be used with almost complete recovery of the sample.

An improved temperature control system was used.

SOURCE AND PURITY OF MATERIALS:

(1) Methane. Prepared by hydrolysis of crystaline methyl Grignard reagent. Passed through conc. H_2SO_4, solid KOH, and Dririte.

(2) Cyclohexanone. Eastman Kodak Co. Purified, distilled, b.p. (754.5 mmHg) t/°C 155.19. Refractive index, density, and vapor pressure data are in the thesis.

ESTIMATED ERROR:

$$\delta T/K = 0.05$$

REFERENCES:
1. Van Slyke, D. D.
   J. Biol. Chem. 1939, 130, 545.

2. Ijams, C. C.
   Ph.D. thesis, 1941
   Vanderbilt University

COMPONENTS:	ORIGINAL MEASUREMENTS:
(1) Methane; CH_4; [74-82-8] (2) Acetic acid, methyl ester or methyl acetate; $C_3H_6O_2$; [79-20-9]	Horiuti, J. *Sci. Pap. Inst. Phys. Chem. Res.* *(Jpn)* 1931/32, *17*, 125 - 256.

VARIABLES:	PREPARED BY:
T/K: 196.55 - 303.15 p_1/kPa: 101.325 (1 atm)	M. E. Derrick H. L. Clever

EXPERIMENTAL VALUES:

T/K	Mol Fraction $10^3 x_1$	Bunsen Coefficient α/cm^3(STP)cm^{-3}atm^{-1}	Ostwald Coefficient L/cm^{3}cm^{-3}
196.55	3.284	1.052	0.7571
212.55	2.834	0.8901	0.6926
231.55	2.482	0.7614	0.6454
252.75	2.245	0.6704	0.6203
273.15	2.087	0.6068	0.6068
293.15	1.985	0.5620	0.6032
303.15	1.932	0.5395	0.5987

The mole fraction and Bunsen coefficient values were calculated by the compiler with the assumption the gas is ideal and that Henry's law is obeyed.

Smoothed Data: For use between 196.55 and 303.15 K.

The 293.15 K value was omitted from the linear regression.

$$\ln x_1 = -11.2954 + 8.1182/(T/100K) + 2.1375 \ln (T/100K)$$

The standard error about the regression line is 9.46×10^{-6}.

T/K	Mol Fraction $10^3 x_1$	T/K	Mol Fraction $10^3 x_1$
198.15	3.226	273.15	2.080
213.15	2.826	288.15	1.997
228.15	2.544	298.15	1.955
243.15	2.340	303.15	1.937
258.15	2.191		

AUXILIARY INFORMATION

METHOD/APPARATUS/PROCEDURE:	SOURCE AND PURITY OF MATERIALS:
The apparatus consists of a gas buret, a solvent reservoir, and an absorption pipet. The volume of the pipet is determined at various meniscus heights by weighing a quantity of water. The meniscus height is read with a cathetometer. The dry gas is introduced into the degassed solvent. The gas and solvent are mixed with a magnetic stirrer until saturation. Care is taken to prevent solvent vapor from mixing with the solute gas in the gas buret. The volume of gas is determined from the gas buret readings, the volume of solvent is determined from the meniscus height in the absorption pipet.	(1) Methane. Aluminum carbide was prepared from aluminum and soot carbon. The aluminum carbide was treated with hot water. The gas evolved was scrubbed to remove impurities, dried and fractionated. Final product had a density, ρ/g dm^{-3} = 0.7168±0.0003 at normal conditions. (2) Methyl acetate. Merck. Extra pure grade. Dried with P_2O_5. Distilled several times. Boiling point (760 mmHg) 57.12°C.

ESTIMATED ERROR:
$$\delta T/K = 0.05$$
$$\delta x_1/x_1 = 0.01$$

REFERENCES:

COMPONENTS:	ORIGINAL MEASUREMENTS:

1. Methane; CH$_4$; [74-82-8]
2. Carbon dioxide; CO$_2$; [124-38-9]
3. Hydrogen; H$_2$; [1333-74-0]
4. Nitrogen; N$_2$; [7727-37-9]
5. 4-Methyl-1,3-dioxolan-2-one,
   (Propylene carbonate); C$_4$H$_6$O$_3$;
   [108-32-7]
	Rusz, L.
	Veszpremi. Vegyip. Egy. Kozl.
	<u>1968</u>, *11*, 169-180.

VARIABLES:	PREPARED BY:
Temperature, pressure	C. L. Young

EXPERIMENTAL VALUES:

T/K	Total pressure		Gas	Partial pressure		α	Mole fraction
	p/atm	p/Mpa		p/atm	p/MPa		in liquid
283.2	10.6	1.07	CO$_2$	2.3	0.23	8.7	0.038
			N$_2$	2.1	0.21	0.22	0.0010
			H$_2$	6.0	0.61	1.68	0.0076
			CH$_4$	0.2	0.02	0.13	0.0006
	18.6	1.88	CO$_2$	4.1	0.42	14.5	0.0620
			N$_2$	3.5	0.35	0.73	0.0033
			H$_2$	10.6	1.07	2.10	0.0095
			CH$_4$	0.4	0.04	0.24	0.0011
	25.4	2.57	CO$_2$	6.2	0.63	19.8	0.0828
			N$_2$	4.7	0.48	1.05	0.0048
			H$_2$	13.9	1.41	3.85	0.0172
			CH$_4$	0.6	0.06	0.10	0.0005
	31.3	3.17	CO$_2$	11.3	1.14	41.2	0.158
			N$_2$	6.2	0.63	1.61	0.0073
			H$_2$	12.9	1.31	2.74	0.0123
			CH$_4$	0.9	0.09	0.61	0.0028
	38.6	3.91	CO$_2$	14.2	1.44	56.5	0.205
			N$_2$	13.4	1.36	2.52	0.0114
			H$_2$	10.1	1.02	2.83	0.0127
			CH$_4$	0.9	0.09	0.32	0.0015
293.2	9.6	9.7	CO$_2$	1.4	0.14	4.9	0.022
			N$_2$	2.8	0.28	0.45	0.0020
			H$_2$	5.1	0.52	1.16	0.0053
			CH$_4$	0.3	0.03	0.11	0.0005

AUXILIARY INFORMATION

METHOD APPARATUS/PROCEDURE:	SOURCE AND PURITY OF MATERIALS:
Volumetric method. Pressure measured when known amounts of gas were added ,in increments, to a known amount of liquid in a vessel of known dimensions. Exact procedure for calculating solubility not clear.	
	ESTIMATED ERROR:
	REFERENCES:

COMPONENTS:	ORIGINAL MEASUREMENTS:

1. Methane; CH_4; [74-82-8]
2. Carbon dioxide; CO_2; [124-38-9]
3. Hydrogen; H_2; [1333-74-0]
4. Nitrogen; N_2; [7727-37-9]
5. 4-Methyl-1,3-dioxolan-2-one,
   (Propylene carbonate); $C_4H_6O_3$;
   [108-32-7]
	Rusz, L.
	Veszpremi. Vegyip. Egy. Kozl.
	<u>1968</u>, *11*, 169-180.

EXPERIMENTAL VALUES:

T/K	Total pressure p/atm	p/Mpa	Gas	Partial pressure p/atm	p/MPa	α	Mole fract. in liquid
293.2	15.5	15.7	CO_2	1.3	0.13	4.7	0.0210
			N_2	4.9	0.50	0.96	0.0044
			H_2	9.0	0.91	1.94	0.0088
			CH_4	0.3	0.03	0.18	0.0008
	26.2	2.65	CO_2	3.3	0.33	7.7	0.0339
			N_2	5.3	0.54	1.25	0.0057
			H_2	17.1	1.73	4.45	0.0199
			CH_4	0.5	0.05	0.24	0.0011
	34.9	3.54	CO_2	6.8	0.69	16.6	0.0703
			N_2	11.2	1.13	1.75	0.0079
			H_2	16.2	1.64	3.05	0.0137
			CH_4	0.7	0.07	0.32	0.0015
	42.1	4.27	CO_2	13.2	1.34	36.3	0.142
			N_2	15.4	1.56	3.32	0.0149
			H_2	12.5	1.27	3.15	0.0142
			CH_4	1.0	0.10	0.12	0.0005
303.2	8.8	0.88	CO_2	1.2	0.12	3.4	0.0153
			N_2	3.1	0.31	0.81	0.0037
			H_2	4.4	0.45	1.22	0.0055
			CH_4	0.1	0.01	0.16	0.0007
	21.2	2.15	CO_2	2.8	0.28	5.6	0.0249
			N_2	4.2	0.43	0.72	0.0033
			H_2	15.0	1.52	3.55	0.0159
			CH_4	0.2	0.02	0.19	0.00087
	30.4	3.08	CO_2	7.4	0.75	14.8	0.0632
			N_2	7.1	0.72	2.20	0.0099
			H_2	15.6	1.58	4.70	0.0210
			CH_4	0.3	0.03	0.20	0.0009
	39.6	4.01	CO_2	8.9	0.90	16.9	0.0715
			N_2	9.3	0.94	1.92	0.0087
			H_2	20.9	2.12	5.16	0.0230
			CH_4	0.5	0.05	0.14	0.0006
	43.7	4.43	CO_2	14.6	1.48	30.8	0.1231
			N_2	14.2	1.44	3.15	0.0142
			H_2	14.1	1.43	3.05	0.0137
			CH_4	0.8	0.08	0.25	0.0011
313.2	10.7	1.08	CO_2	1.5	0.15	2.5	0.0113
			N_2	2.6	0.26	0.28	0.0013
			H_2	6.4	0.65	1.55	0.0070
			CH_4	0.2	0.02	0.09	0.0004
	21.2	2.15	CO_2	5.4	0.55	6.9	0.0305
			N_2	4.8	0.49	0.84	0.0038
			H_2	10.6	1.07	2.23	0.0101
			CH_4	0.4	0.04	0.31	0.0014
	29.1	2.95	CO_2	7.4	0.75	9.6	0.0419
			N_2	7.3	0.74	1.13	0.0051
			H_2	13.9	1.41	3.65	0.0164
			CH_4	0.5	0.05	0.23	0.0010

cont.

COMPONENTS:	ORIGINAL MEASUREMENTS:
1. Methane; CH_4; [74-82-8] 2. Carbon dioxide; CO_2; [124-38-9] 3. Hydrogen; H_2; [1333-74-0] 4. Nitrogen; N_2; [7727-37-9] 5. 4-Methyl-1,3-dioxolan-2-one, (Propylene carbonate); $C_4H_6O_3$; [108-32-7]	Rusz, L. *Veszpremi. Vegyip. Egy. Kozl.*, <u>1968</u>, *11*, 169-180.

EXPERIMENTAL VALUES:

T/K	Total pressure p/atm	p/Mpa	Gas	Partial pressure p/atm	p/MPa	α	Mole fraction in liquid
313.2	3.43	3.48	CO_2	9.3	0.94	15.4	0.0656
			N_2	10.2	1.03	15.1	0.0068
			H_2	14.1	1.43	2.74	0.0123
			CH_4	0.7	0.07	0.12	0.0005
	43.4	4.40	CO_2	15.2	1.54	23.4	0.0964
			N_2	14.8	1.50	2.74	0.0123
			H_2	12.5	1.27	3.38	0.0152
			CH_4	0.9	0.09	0.15	0.0007

a
mL of gas absorbed (reduced to 0 °C and 1 atmosphere) per g of solvent.

COMPONENTS:	ORIGINAL MEASUREMENTS:

1. Methane; CH_4; [74-82-8]
2. 4-Methyl-1,3-dioxolan-2-one,
   (propylene carbonate); $C_4H_6O_3$;
   [108-32-7]
	Rusz, L.
	Veszpremi. Vegyip. Egy. Kozl.,
	1968, *11*, 169-180.

VARIABLES:	PREPARED BY:
Temperature, pressure	C. L. Young

EXPERIMENTAL VALUES:

T/K	Total pressure p/atm	p/Mpa	a α	Mole fraction of methane,
283.2	11.8	1.20	3.9	0.017
	15.1	1.78	4.5	0.020
	20.6	2.43	5.9	0.026
	25.0	2.53	5.7	0.025
293.2	10.8	1.09	2.9	0.013
	15.6	1.58	3.7	0.017
	21.6	2.19	6.7	0.030
	25.6	2.59	7.5	0.033
303.2	11.4	1.16	3.3	0.015
	15.8	1.60	4.7	0.021
	21.8	2.21	5.1	0.023
	25.9	2.62	5.4	0.024
313.2	13.7	1.39	2.8	0.013
	20.0	2.03	4.5	0.020
	24.8	2.51	6.4	0.028
	28.0	2.84	6.3	0.028

a
   mL of gas absorbed (reduced to 0^0C and 1 atmosphere) per g of solvent

AUXILIARY INFORMATION

METHOD/APPARATUS/PROCEDURE:	SOURCE AND PURITY OF MATERIALS:
Volumetric method. Pressure measured when known amounts of gas were added ,in increments, to a known amount of liquid in a vessel of known dimensions. Exact procedure for calculating solubility not clear.	
	ESTIMATED ERROR:
	REFERENCES:

COMPONENTS:	ORIGINAL MEASUREMENTS:

1. Methane; CH_4 ; [74-82-8]
2. Carbon dioxide; CO_2 ; [124-38-9]
3. Nitrogen; N_2 ; [7727-37-9]
4. 1,2,3-Propanetriol,triacetate,
   (glycerol triacetate); $C_9H_{14}O_6$;
   [102-76-1]
	Makranczy, J.; Maleczkine, S. M.;
	Rusz, L.
	Veszpremi. Vegyip. Egy. Kozl.
	<u>1965</u>, *9*, 95-105.

VARIABLES:	PREPARED BY:
Temperature, pressure	C. L. Young

EXPERIMENTAL VALUES:

$$T/K = 293.2$$

Total pressure /Mpa	Partial pressure			α_{CO_2}	α_{CH_4}	Mole fractions	
	p_{CO_2}/MPa	p_{CH_4}/Mpa	p_{N_2}/Mpa			x_{CO_2}	x_{CH_4}
10.6	6.3	3.1	1.0	24.7	1.3	0.194	0.012
16.4	10.2	4.6	1.6	34.6	3.1	0.252	0.029
24.2	12.7	8.2	3.3	51.0	12.6	0.332	0.109
29.0	17.4	7.5	4.1	76.0	9.1	0.425	0.081
41.8	19.8	16.0	6.0	92.5	14.6	0.474	0.124

α = mL of gas absorbed (reduced to 0^0C and 1 atmosphere) per g of solvent

---

AUXILIARY INFORMATION

METHOD/APPARATUS/PROCEDURE:	SOURCE AND PURITY OF MATERIALS:
Volumetric method. Pressure measured when known amounts of gas were added ,in increments, to a known amount of liquid in a vessel of known dimensions. Exact procedure for calculating solubility not clear.	
	ESTIMATED ERROR:
	REFERENCES:

COMPONENTS:	ORIGINAL MEASUREMENTS:
1. Methane; CH_4 ; [74-82-8] 2. Carbon dioxide; CO_2 ; [124-38-9] 3. Nitrogen; N_2 ; [7727-37-9] 4. 2-[2-(methoxyethoxy)-ethoxy]- ethanol acetate, (methoxytri- ethylene gylcol acetate); $C_9H_{18}O_5$; [3610-27-3]	Makranczy, J.; Maleczkine, S. M.; Rusz, L. *Veszpremi. Vegyip. Egy. Kozl.* 1965, *9*, 95-105.

VARIABLES:	PREPARED BY:
Temperature, pressure	C. L. Young

EXPERIMENTAL VALUES:

$$T/K = 293.2$$

Total pressure /MPa	Partial pressure			α_{CO_2}	α_{CH_4}	Mole fractions	
	p_{CO_2}/MPa	p_{CH_4}/Mpa	p_{N_2}/Mpa			x_{CO_2}	x_{CH_4}
10.6	6.5	2.8	1.3	26.8	2.4	0.198	0.022
16.3	9.7	4.8	1.8	38.5	4.8	0.261	0.042
23.9	13.4	7.7	2.8	61.5	6.2	0.361	0.054
31.1	18.9	8.3	3.9	95.5	9.0	0.467	0.064
41.8	22.5	15.2	4.1	116.0	13.1	0.516	0.107

α= mL of gas absorbed (reduced to 0 °C and 1 atmosphere) per g of solvent

AUXILIARY INFORMATION

METHOD/APPARATUS/PROCEDURE:	SOURCE AND PURITY OF MATERIALS:
Volumetric method. Pressure measured when known amounts of gas were added ,in increments, to a known amount of liquid in a vessel of known dimensions. Exact procedure for calculating solubility not clear.	
	ESTIMATED ERROR:
	REFERENCES:

COMPONENTS:	ORIGINAL MEASUREMENTS:
1. Methane; CH_4; [74-82-8] 2. Carbon dioxide; CO_2; [124-38-9] 3. Nitrogen; N_2; [7727-37-9] 4. 2,2´-{1,2-ethanediylbis-(oxy)]bis- ethanol, (triethylene glycol); $C_6H_{14}O_4$; [112-27-6]	Makranczy, J.; Maleczkine, S. M.; Rusz, L. *Veszpremi. Vegyip. Egy. Kozl.* 1965, *9*, 95-105.

VARIABLES:	PREPARED BY:
Temperature, pressure	C. L. Young

EXPERIMENTAL VALUES:

$$T/K = 293.2$$

Total pressure /MPa	Partial pressure			Mole fractions			
	p_{CO_2}/MPa	p_{CH_4}/Mpa	p_{N_2}/Mpa	α_{CO_2}	α_{CH_4}	x_{CO_2}	x_{CH_4}
12.2	7.8	3.2	1.2	11.8	2.7	0.0732	0.0177
18.4	11.8	4.4	2.2	18.5	2.9	0.1102	0.0190
27.2	17.3	7.6	2.3	26.2	3.4	0.1492	0.0222
36.2	24.1	9.1	2.9	37.6	6.9	0.2010	0.0441
41.7	26.7	11.4	3.6	44.2	8.1	0.2283	0.0514

α=mL of gas absorbed (reduced to 0^0C and 1 atmosphere) per g of solvent

---

AUXILIARY INFORMATION

METHOD/APPARATUS/PROCEDURE:	SOURCE AND PURITY OF MATERIALS:
Volumetric method. Pressure measured when known amounts of gas were added ,in increments, to a known amount of liquid in a vessel of known dimensions. Exact procedure for calculating solubility not clear.	
	ESTIMATED ERROR:
	REFERENCES:

COMPONENTS:	ORIGINAL MEASUREMENTS:

1. Methane; CH_4; [74-82-8]
2. Carbon dioxide; CO_2; [124-38-9]
3. Nitrogen; N_2; [7727-37-9]
4. 4-Methyl-1,3-dioxolan-2-one,
   (propylene carbonate); $C_4H_6O_3$;
   [108-32-7]
	Makranczy, J.; Maleczkine, S. M.;
	Rusz, L.
	Veszpremi. Vegyip. Egy. Kozl.
	1965, *9*, 95-105.

VARIABLES:	PREPARED BY:
Temperature, pressure	C. L. Young

EXPERIMENTAL VALUES:

$$T/K = 293.2$$

Total pressure /MPa	Partial pressure			α_{CO_2}	α_{CH_4}	Mole fractions	
	P_{CO_2}/MPa	P_{CH_4}/Mpa	P_{N_2}/Mpa			x_{CO_2}	x_{CH_4}
8.7	4.4	3.6	0.7	14.3	1.3	0.060	0.0059
16.6	8.2	7.1	1.3	27.6	5.1	0.116	0.023
25.4	12.3	11.1	2.0	41.9	9.2	0.160	0.040
30.9	14.4	13.7	2.8	53.9	10.4	0.197	0.045
35.7	15.7	14.8	5.2	64.2	12.3	0.226	0.053

α= mL of gas absorbed (reduced to 0^0C and 1 atmosphere) per g of solvent

AUXILIARY INFORMATION

METHOD/APPARATUS/PROCEDURE:	SOURCE AND PURITY OF MATERIALS:
Volumetric method. Pressure measured when known amounts of gas were added ,in increments, to a known amount of liquid in a vessel of known dimensions. Exact procedure for calculating solubility not clear.	
	ESTIMATED ERROR:
	REFERENCES:

COMPONENTS:	ORIGINAL MEASUREMENTS:

1. Methane; CH_4; [74-82-8]
2. Carbon dioxide; CO_2; [124-38-9]
3. Nitrogen; N_2; [7727-37-9]
4. 2-[2-(Butoxyethoxy)-ethoxy]-
   ethanol acetate, (butoxytri-
   ethylene gylcol acetate);
   $C_{12}H_{24}O_5$;
	Makranczy, J.; Maleczkine, S. M.;
	Rusz, L.
	Veszpremi. Vegyip. Egy. Kozl.
	<u>1965</u>, *9*, 95-105.

VARIABLES:	PREPARED BY:
Temperature, pressure	C. L. Young

EXPERIMENTAL VALUES:

$$T/K = 293.2$$

Total pressure /MPa	Partial pressure					Mole fractions	
	p_{CO_2}/MPa	p_{CH_4}/Mpa	p_{N_2}/Mpa	α_{CO_2}	α_{CH_4}	x_{CO_2}	x_{CH_4}
10.8	6.7	3.1	1.0	26.3	2.9	0.225	0.031
16.7	10.5	4.3	1.9	38.2	5.6	0.297	0.058
24.6	15.0	6.6	3.0	60.7	8.1	0.402	0.082
29.8	18.9	8.4	2.5	82.4	10.4	0.477	0.103
40.8	22.8	12.1	5.9	100.0	14.3	0.525	0.137

α = mL of gas absorbed (reduced to 0^0C and 1 atmosphere) per g of solvent

AUXILIARY INFORMATION

METHOD/APPARATUS/PROCEDURE:	SOURCE AND PURITY OF MATERIALS:
Volumetric method. Pressure measured when known amounts of gas were added ,in increments, to a known amount of liquid in a vessel of known dimensions. Exact procedure for calculating solubility not clear.	
	ESTIMATED ERROR:
	REFERENCES:

COMPONENTS:	ORIGINAL MEASUREMENTS:

1. Methane; CH_4;   [74-82-8]
 	
2. 4-Methyl-1,3-dioxolan-2-one;
    (Propylene carbonate); $C_4H_6O_3$;
    [108-32-7]
	Lenoir, J-Y.; Renault, P.; Renon, H.
	
	J. Chem. Eng. Data, <u>1971</u>, *16*, 340-2.

VARIABLES:	PREPARED BY:
	C. L. Young

EXPERIMENTAL VALUES:

T/K	Henry's constant H_{CH_4}/atm	Mole fraction at 1 atm* x_{CH_4}
298.2	1140	0.000877
323.2	1340	0.000746
343.2	1430	0.000699

* Calculated by compiler assuming a linear function of P_{CH_4} vs x_{CH_4}, i.e., x_{CH_4} (1 atm) = $1/H_{CH_4}$.

AUXILIARY INFORMATION

METHOD/APPARATUS/PROCEDURE:	SOURCE AND PURITY OF MATERIALS:
A conventional gas-liquid chromato-graphic unit fitted with a thermal conductivity detector was used. The carrier gas was helium. The value of Henry's law constant was calculated from the retention time. The value applies to very low partial pressures of gas and there may be a substantial difference from that measured at 1 atm. pressure. There is also considerable uncertainty in the value of Henry's constant since surface adsorption was not allowed for although its possible existence was noted.	(1) L'Air Liquide sample, minimum purity 99.9 moler per cent. (2) Touzart and Matignon or Serlabo sample, purity 99 mole per cent.
	ESTIMATED ERROR: $\delta T/K = \pm 0.1$; δH/atm = $\pm 6\%$ (estimated by compiler).
	REFERENCES:

COMPONENTS:	ORIGINAL MEASUREMENTS:

1. Methane; CH₄; [74-82-8]
 	
2. 4-Methyl-1,3-dioxolan-2-one,
   (Propylene carbonate); C₄H₆O₃;
   [108-32-7]
	Parcher, J. F.; Bell, M. L.;
	Lin, P. J.
	Adv. Chromat.
	<u>1984</u>, *24*, 227-246.

VARIABLES:	PREPARED BY:
	C. L. Young

EXPERIMENTAL VALUES:

T/K (t/°C)	Henry's law constant, H/atm	Mole fraction [a] extrapolated to 1 atm, x_{CH_4}
283.2 (10)	1500	0.00067
293.2 (20)	1400	0.00071
303.2 (30)	1300	0.00077
313.2 (40)	1400	0.00071

$$R \ d \ln H/d(1/T) = +0.5 \text{ kcal mol}^{-1} = 2 \text{ kJ mol}^{-1}.$$

[a] Calculated by compiler assuming $x = 1/H$.

AUXILIARY INFORMATION

METHOD/APPARATUS/PROCEDURE:	SOURCE AND PURITY OF MATERIALS:
Henry's law constant determined from retention volume of gas on a chromatographic column. Helium was used as a carrier gas and a mass spectrometer used as a detector. The measured Henry's law constants were independent of sample size, flow rate and composition of injected sample. Details given in ref. 1.	No details given.
	ESTIMATED ERROR: δH/atm = ±200.
	REFERENCES: 1. Lin, P. J. and Parcher, J. F. *J. Chromat. Sci.* <u>1982</u>, *20*, 33.

COMPONENTS:	ORIGINAL MEASUREMENTS:

1. Methane; CH_4; [74-82-8]
 	
2. Hexadecane; $C_{16}H_{34}$; [544-76-3]
 	
3. 4-Methyl-1,3-dioxolan-2-one,
    (Propylene carbonate); $C_4H_6O_3$;
    [108-32-7]
	Parcher, J. F.; Bell, M. L.;
	Lin, P. J.
	Adv. Chromat.
	<u>1984</u>, *24*, 227-246.

VARIABLES:	PREPARED BY:
	C. L. Young

EXPERIMENTAL VALUES:

T/K (t/°C)	Mole fraction of Component 2	Henry's law constant, H/atm	Mole fraction [a] extrapolated to 1 atm, x_{CH_4}
293.2 (20)	0.30	460	0.00217
303.2 (30)		480	0.00208
313.2 (40)		470	0.00213
293.2 (20)	0.62	250	0.00400
303.2 (30)		257	0.00389
313.2 (40)		270	0.00370

[a] Calculated by compiler assuming $x = 1/H$.

AUXILIARY INFORMATION

METHOD/APPARATUS/PROCEDURE:	SOURCE AND PURITY OF MATERIALS:
Henry's law constant determined from retention volume of gas on a chromatographic column. Helium was used as a carrier gas and a mass spectrometer used as a detector. The measured Henry's law constants were independent of sample size, flow rate and composition of injected sample. Details given in ref. 1.	No details given.
	ESTIMATED ERROR: $\delta x_{CH_4} = \pm 10\%$ (estimated by compiler).
	REFERENCES: 1. Lin, P. J. and Parcher, J. F. *J. Chromat. Sci.* <u>1982</u>, *20*, 33.

COMPONENTS:	ORIGINAL MEASUREMENTS:
1. Methane; CH_4; [74-82-8] 2. 4-Methyl-1,3-dioxolan-2-one, (Propylene carbonate); $C_4H_6O_3$; [108-32-7]	Shakhova, S.F.; Zubchenko, Yu.P. *Khim. Prom.* <u>1973</u>, *49*, 595-6.

VARIABLES:	PREPARED BY:
Temperature, pressure	C.L. Young

EXPERIMENTAL VALUES:

T/K	$P/10^5$Pa	Mole fraction of methane in liquid, x_{CH_4}	α^+ vol/vol
298.15	43.67	0.0256	6.9
	63.22	0.0363	9.9
	84.81	0.0482	13.3
	113.28	0.0584	16.3
323.15	47.93	0.0270	7.3
	73.56	0.0405	11.1
	79.44	0.0430	11.8
	88.76	0.0457	12.6
	112.77	0.0567	15.8
	146.11	0.0691	19.5
	147.63	0.0691	19.5

+ quoted in original, appears to be volume of gas at
   T/K = 273.15 and P = 1 atmosphere absorbed by unit
   volume of liquid at room temperature.

AUXILIARY INFORMATION

METHOD/APPARATUS/PROCEDURE:	SOURCE AND PURITY OF MATERIALS:
Rocking autoclave. Mixture stirred by ball in rocking autoclave. Samples of liquid analysed by a volumetric method. Details in ref. (1).	1. Purity 97.8 mole per cent. 2. No details given.

ESTIMATED ERROR:
δT/K = ±0.1; δP/10^5Pa = ±0.1;
δx_{CH_4} = ±5%.

(estimated by compiler).

REFERENCES:
1. Shakhova, S.F.; Zubchenko, Yu.P.; Kaplan, L.K.

Khim. Prom. <u>1973</u>, *5*, 108.

COMPONENTS:	ORIGINAL MEASUREMENTS:
1. Methane; CH_4; [74-82-8] 2. Oxybispropanol, (Dipropylene glycol); $C_6H_{14}O_3$; [25265-71-8]	Lenoir, J-Y.; Renault, P.; Renon, H. *J. Chem. Eng. Data*, <u>1971</u>, *16*, 340-2.

VARIABLES:	PREPARED BY:
Temperature	C. L. Young

EXPERIMENTAL VALUES:

T/K	Henry's constant H_{CH_4}/atm	Mole fraction at 1 atm* x_{CH_4}
298.2	1450	0.000690
323.2	1800	0.000556
343.2	2230	0.000448

* Calculated by compiler assuming a linear function of P_{CH_4} vs x_{CH_4}, i.e., x_{CH_4} (1 atm) = $1/H_{CH_4}$.

AUXILIARY INFORMATION

METHOD/APPARATUS/PROCEDURE:	SOURCE AND PURITY OF MATERIALS:
A conventional gas-liquid chromato-graphic unit fitted with a thermal conductivity detector was used. The carrier gas was helium. The value of Henry's law constant was calculated from the retention time. The value applies to very low partial pressures of gas and there may be a substantial difference from that measured at 1 atm. pressure. There is also considerable uncertainty in the value of Henry's constant since surface adsorption was not allowed for although its possible existence was noted.	(1) L'Air Liquide sample, minimum purity 99.9 mole per cent. (2) Touzart and Matignon or Serlabo sample, purity 99 mole per cent.
	ESTIMATED ERROR: $\delta T/K = \pm 0.1$; δH/atm = ± 6% (estimated by compiler).
	REFERENCES:

COMPONENTS:	ORIGINAL MEASUREMENTS:
1. Methane; CH_4; [74-82-8] 2. Benzenemethanol; (Benzyl alcohol); C_7H_8O; [100-51-6]	Lenoir, J-Y.; Renault, P.; Renon, H. *J. Chem. Eng. Data*, <u>1971</u>, *16*, 340-2.

VARIABLES:	PREPARED BY:
	C. L. Young

EXPERIMENTAL VALUES:

T/K	Henry's constant H_{CH_4}/atm	Mole fraction at 1 atm* x_{CH_4}
298.2	1030	0.000971

* Calculated by compiler assuming a linear function of P_{CH_4} vs x_{CH_4}, i.e., x_{CH_4} (1 atm) = $1/H_{CH_4}$.

AUXILIARY INFORMATION

METHOD/APPARATUS/PROCEDURE:	SOURCE AND PURITY OF MATERIALS:
A conventional gas-liquid chromato-graphic unit fitted with a thermal conductivity detector was used. The carrier gas was helium. The value of Henry's law constant was calculated from the retention time. The value applies to very low partial pressures of gas and there may be a substantial difference from that measured at 1 atm. pressure. There is also considerable uncertainty in the value of Henry's constant since surface adsorption was not allowed for although its possible existence was noted.	(1) L'Air Liquide sample, minimum purity 99.9 mole per cent. (2) Touzart and Matignon or Serlabo sample, purity 99 mole per cent.
	ESTIMATED ERROR: $\delta T/K = \pm 0.1$; $\delta H/atm = \pm 6\%$ (estimated by compiler).
	REFERENCES:

COMPONENTS:	ORIGINAL MEASUREMENTS:
1. Methane; CH_4; [74-82-8] 2. Phenol; C_6H_6O; [108-95-2]	Lenoir, J-Y.; Renault, P.; Renon, H. *J. Chem. Eng. Data,* <u>1971</u>, *16*, 340-3.

VARIABLES:	PREPARED BY:
	C. L. Young

EXPERIMENTAL VALUES:

T/K	Henry's constant H_{CH_4}/atm	Mole fraction at 1 atm* x_{CH_4}
323.2	1990	0.000503

* Calculated by compiler assuming a linear function of P_{CH_4} vs x_{CH_4}, i.e., x_{CH_4} (1 atm) = $1/H_{CH_4}$.

AUXILIARY INFORMATION

METHOD/APPARATUS/PROCEDURE:	SOURCE AND PURITY OF MATERIALS:
A conventional gas-liquid chromatographic unit fitted with a thermal conductivity detector was used. The carrier gas was helium. The value of Henry's law constant was calculated from the retention time. The value applies to very low partial pressures of gas and there may be a substantial difference from that measured at 1 atm. pressure. There is also considerable uncertainty in the value of Henry's constant since surface adsorption was not allowed for although its possible existence was noted.	(1) L'Air Liquide sample, minimum purity 99.9 mole per cent. (2) Touzart and Matignon or Serlabo sample, purity 99 mole per cent.
	ESTIMATED ERROR: $\delta T/K = \pm 0.1$; δH/atm = ± 6% (estimated by compiler).
	REFERENCES:

COMPONENTS:	ORIGINAL MEASUREMENTS:
1. Methane; CH_4; [74-82-8] 2. 1,2,3-Propanetriol, triacetate, (Glycerol triacetate); $C_9H_{14}O_6$; [102-76-1]	Shakhova, S.F.; and Zubchenko, Yu.P. *Khim. Prom.* <u>1973</u>, *49*, 595-6.

VARIABLES:	PREPARED BY:
Temperature, pressure	C.L. Young

EXPERIMENTAL VALUES:

T/K	$P/10^5 Pa$	Mole fraction of methane in liquid, x_{CH_4}	α^+ vol/vol
298.15	35.87	0.0640	8.1
	40.73	0.0706	9.0
	50.97	0.0828	10.7
	70.42	0.1070	14.2
	75.59	0.1144	15.3
	90.08	0.1306	17.8
	104.56	0.1512	21.1
343.15	62.31	0.0941	12.3
	85.32	0.3970	78.0
	95.14	0.4220	86.0
	113.28	0.4614	101.5
	114.80	0.4639	102.5
	127.47	0.4882	113.0

+ quoted in original, appears to be volume of gas at $T/K = 273.15$ and $P = 1$ atmosphere absorbed by unit volume of liquid at room temperature.

AUXILIARY INFORMATION

METHOD/APPARATUS/PROCEDURE:	SOURCE AND PURITY OF MATERIALS:
Rocking autoclave. Mixture stirred by ball in rocking autoclave. Samples of liquid analysed by a volumetric method. Details in ref. (1).	1. Purity 97.8 mole per cent. 2. No details given.

ESTIMATED ERROR:
$\delta T/K = \pm 0.1$; $\delta P/10^5 Pa = \pm 0.1$;
$\delta x_{CH_4} = \pm 5\%$.
(estimated by compiler)

REFERENCES:
1. Shakhova, S.F.; Zubchenko, Yu.P. Kaplan, L.K.

Khim. Prom. <u>1973</u>, *5*, 108.

COMPONENTS:	ORIGINAL MEASUREMENTS:
1. Methane; CH₄; [74-82-8] 2. 3-Methylphenol, (*m*-cresol); C_7H_8O; [108-39-4]	Simnick, J. J.; Sebastian, H. M.; Lin, H. M.; Chao, K. C. *Fluid Phase Equilibria* 1979, *3*, 145-154.

VARIABLES:	PREPARED BY:
Temperature, pressure	C. L. Young

EXPERIMENTAL VALUES:

T/K	P/MPa	P/atm	Mole fraction of methane in liquid, x_{CH_4}	in gas, y_{CH_4}
462.25	2.08	20.5	0.0198	0.9579
	3.05	30.1	0.0288	0.9695
	5.11	50.4	0.0489	0.9785
	10.05	99.2	0.0930	0.9836
	15.28	150.8	0.1382	0.9840
	20.17	199.1	0.1768	0.9825
	25.27	249.4	0.2166	0.9806
542.65	2.00	19.7	0.0181	0.7435
	3.03	29.9	0.0301	0.8158
	5.06	49.9	0.0533	0.8764
	10.13	100.0	0.1111	0.9172
	15.15	149.5	0.1684	0.9262
	20.23	199.7	0.2199	0.9261
	25.20	248.7	0.2746	0.9214
623.25	3.07	30.3	0.0224	0.3912
	5.08	50.1	0.0538	0.5699
	10.18	100.5	0.1330	0.7057
	15.16	149.6	0.2136	0.7385
	20.35	200.8	0.3019	0.7333
	22.72	224.2	0.3579	0.7171
	25.33	250.0	0.4888	0.6857
663.35	5.12	50.5	0.0465	0.3265
	10.08	99.5	0.1477	0.4968
	12.91	127.4	0.2135	0.5198
	15.25	150.5	0.3207	0.4809

AUXILIARY INFORMATION

METHOD/APPARATUS/PROCEDURE:	SOURCE AND PURITY OF MATERIALS:
Flow apparatus with both liquid and gaseous components continually passing into a mixing tube and then into a cell in which phases separated under gravity. Liquid sample removed from bottom of cell and vapor sample from top of cell. Composition of samples found by stripping out gas and estimating amount of solvent gravimetrically. Temperature measured with thermo-couple and pressure with Bourdon gauge. Details in ref. (1).	1. Matheson sample, purity better than 99 mole per cent. 2. Aldrich Chemical Co., minimum purity 99 mole per cent. Distilled.

ESTIMATED ERROR:

$\delta T/K = \pm 0.2$; $\delta P/MPa = \pm 0.02$;

δx_{CH_4}, $\delta y_{CH_4} = \pm 2\%$.

REFERENCES:

1. Simnick, J. J.; Lawson, C. C.;
   Lin, H. M.; Chao, K. C.
   Am. Inst. Chem. Engnrs. J.
   1977, *23*, 469.

COMPONENTS:	ORIGINAL MEASUREMENTS:
1. Methane; CH_4; [74-82-8] 2. 2,5,8,11,14 - Pentaoxapenta- decane, (Tetramethylene glycol dimethyl ether) $C_{10}H_{22}O_5$; [143-24-8]	Zubchenko, Yu.P.; Shakhova, S.F. *Tr.N.-i.i Proekt. In-ta Azot Prom-sti i Produktov Organ. Sinteza* <u>1975</u>, (33), 13-15.

VARIABLES:	PREPARED BY:
Pressure	C.L. Young

EXPERIMENTAL VALUES:

T/K	P/atm	P/MPa	α+ vol/vol	Mole fraction of methane in liquid, x_{CH_4}
313.15	24.2	2.45	6.18	0.0502
	37.2	3.77	9.17	0.0727
	54.3	5.50	13.2	0.101
	69.9	7.08	17.1	0.127
	83.3	8.44	20.1	0.147
	84.0	8.51	20.4	0.148
	85.7	8.68	20.7	0.150

+   quoted in original paper, appears to be
    volume of gas at T/K = 273/15 and P =
    1 atmosphere absorbed by unit volume of
    liquid at room temperature.

AUXILIARY INFORMATION

METHOD/APPARATUS/PROCEDURE:	SOURCE AND PURITY OF MATERIALS:
Mixture stirred by ball in rocking autoclave. Sample of liquid analysed by a volumetric method. Details in ref. (1).	1. Purity 97.8 mole per cent. 2. No details given.

ESTIMATED ERROR:
$\delta T/K = \pm 0.1$; $\delta P/atm = \pm 0.1$
$\delta x_{CH_4} = \pm 5\%$.
(estimated by compiler).

REFERENCES:
1. Shakhova, S.F.; Zubchenko,
   Yu.P.; Kaplan, L.K.
   Khim. Prom. <u>1973</u>, *5*, 108.

COMPONENTS:	ORIGINAL MEASUREMENTS:
(1) Methane; CH_4; [74-82-8] (2) Hexadecafluoroheptane or per- fluoroheptane; C_7F_{16}; [335-57-9]	Kobatake, Y.; Hildebrand, J. H. J. Phys. Chem. 1961, 65, 331 - 335.

VARIABLES:	PREPARED BY:
T/K: 291.07 - 303.16 P/kPa: 101.325 (1 atm)	M. E. Derrick H. L. Clever

EXPERIMENTAL VALUES:

Temperature		Mol Fraction	Bunsen Coefficient	Ostwald Coefficient
$t/°C$	T/K	$10^3 x_1$	α/cm^3 (STP) cm^{-3} atm^{-1}	L/cm^3 cm^{-3}
17.92	291.07	8.610	0.8701	0.9272
21.75	294.90	8.414	0.8450	0.9123
25.00	298.15	8.262[1]	0.8253	0.9008
25.68	298.83	8.222	0.8204	0.8975
30.01	303.16	8.026	0.7951	0.8825

[1] Possibly a smoothed value of the authors.

The Bunsen and Ostwald coefficients were calculated by the compiler.

Smoothed Data: For use between 291.07 and 303.16 K.

$$\ln x_1 = -6.5150 + 5.1234/(T/100K)$$

The standard error about the regression line is 3.31×10^{-6}.

T/K	Mol Fraction $10^3 x_1$
293.15	8.503
298.15	8.258
303.15	8.027

AUXILIARY INFORMATION

METHOD/APPARATUS/PROCEDURE:

The apparatus consists of a gas measuring buret, an absorption pipet, and
a reservoir for the solvent. The
buret is thermostated at 25°C, the
pipet at any temperature from 5 to 30
°C. The pipet contains an iron bar
in glass for magnetic stirring. The
pure solvent is degassed by freezing
with liquid nitrogen, evacuating,
then boiling with a heat lamp. The
degassing process is repeated three
times. The solvent is flowed into
the pipet where it is again boiled
for final degassing. Manipulation of
the apparatus is such that the solvent never comes in contact with
stopcock grease. The liquid in the
pipet is sealed off by mercury. Its
volume is the difference between the
capacity of the pipet and the volume
of mercury that confines it. Gas is
admitted into the pipet. Its exact
amount is determined by P-V measurements in the buret before and after
introduction of the gas into the pipet. The stirrer is set in motion.
Equilibrium is attained within 24 hours.

SOURCE AND PURITY OF MATERIALS:

(1) Methane. Matheson Co., Inc.
Research grade. Dried by pas-
sage over P_2O_5 followed by
multiple trap vaporization and
evacuation at liquid N_2
temperature.

(2) Hexadecafluoroheptane. Source
not given. Purified by method of
Glew and Reeves, J. Phys. Chem.
1956, 60, 615.

ESTIMATED ERROR:

$$\delta T/K = 0.02$$
$$\delta x_1/x_1 = 0.003$$

REFERENCES:

COMPONENTS:	ORIGINAL MEASUREMENTS:
(1) Methane; CH_4; [74-82-8] (2) Hexafluorobenzene; C_6F_6; [392-56-3]	Evans, D. F.; Battino, R. *J. Chem. Thermodyn.* 1971, *3*, 753-760.

VARIABLES:	PREPARED BY:
T/K: 283.20 - 297.97 p_1/kPa: 101.325 (1 atm)	H. L. Clever

EXPERIMENTAL VALUES:

$t/°C$	T/K	Mol Fraction $10^3 x_1$	Bunsen Coefficient α/cm^3 (STP) $cm^{-3} atm^{-1}$	Ostwald Coefficient $L/cm^3 cm^{-3}$
10.05	283.20	4.076	0.809	0.839
10.08	283.23	4.071	0.808	0.838
24.67	297.82	3.844	0.747	0.815
24.82	297.97	3.848	0.748	0.816

The Bunsen coefficients were calculated by the compiler.

The solubility values were adjusted to an oxygen partial pressure of 101.325 kPa (1 atm) by Henry's law.

Smoothed Data: For use between 283.15 and 298.15 K.

$$\ln x_1 = -6.6693 + 3.3025/(T/100 \text{ K})$$

The standard error about the regression line is 3.81×10^{-6}.

T/K	Mol Fraction $10^3 x_1$
283.15	4.075
288.15	3.993
293.15	3.916
298.15	3.842

AUXILIARY INFORMATION

METHOD/APPARATUS/PROCEDURE:

The solubility apparatus is based on the design of Morrison and Billett (1) and the version used is described by Battino, Evans, and Danforth (2). The degassing apparatus is that described by Battino, Banzhof, Bogan, and Wilhelm (3).

Degassing. Up to 500 cm^3 of solvent is placed in a flask of such size that the liquid is about 4 cm deep. The liquid is rapidly stirred, and vacuum is intermittently applied through a liquid N_2 trap until the permanent gas residual pressure drops to 5 microns.

Solubility Determination. The degassed solvent is passed in a thin film down a glass helical tube containing solute gas plus the solvent vapor at a total pressure of one atm. The volume of gas absorbed is found by difference between the initial and final volumes in the buret system. The solvent is collected in a tared flask and weighed.

SOURCE AND PURITY OF MATERIALS:

(1) Methane. Either Air Products and Chemicals Inc. or the Matheson Co., Inc. Purest grade available. Minimum purity 99.0 mole per cent (usually > 99.9 mole per cent).

(2) Hexafluorobenzene. Imperiel Smelting Co., Avnomouth, U.K. GC purity 99.7%, density, $\rho_{298.15} = 1.60596$ g cm^{-3}. Purification described *Anal. Chem.* 1968, *40*, 224.

ESTIMATED ERROR: $\delta T/K = 0.03$
$\delta p/mmHg = 0.5$
$\delta x_1/x_1 = 0.005$

REFERENCES:
1. Morrison, T. J.; Billett, F.
   J. Chem. Soc. 1948, 2033.
2. Battino,R.;Evans,F.D.;Danforth,W.F.
   J.Am.Oil Chem.Soc. 1968, *45*, 830.
3. Battino,R.; Banzhof, M.;
   Bogan, M.; Wilhelm, E.
   Anal. Chem. 1971, *43*, 806.

COMPONENTS:	ORIGINAL MEASUREMENTS:
(1) Methane; CH_4; [74-82-8] (2) 1,1,2-Trichloro-1,2,2-trifluoro-ethane; $C_2Cl_3F_3$; [76-13-1]	Hiraoka, H.; Hildebrand, J. H. *J. Phys. Chem.* <u>1964</u>, *68*, 213-214.

VARIABLES:	PREPARED BY:
T/K = 277.15 - 308.15 p_1/kPa = 101.325 (1 atm)	M. E. Derrick H. L. Clever

EXPERIMENTAL VALUES:

Temperature		Mol Fraction	Bunsen Coefficient	Ostwald Coefficient
t/°C	T/K	$10^3 x_1$	α/cm^3(STP)cm^{-3}atm^{-1}	L/cm^{3}cm^{-3}
4.00	277.15	5.651	1.09	1.11
14.90	288.05	5.278	1.01	1.06
25.09	298.24	4.978	0.935	1.02
35.00	308.15	4.872	0.902	1.02

The Bunsen and Ostwald coefficients were calculated by the compiler assuming ideal gas behavior.

Smoothed Data: For use between 277.15 and 308.15 K.

$$\ln x_1 = -6.6987 + 4.2023/(T/100\ K)$$

The standard error about the regression line is 6.66 x 10^{-5}.

T/K	Mol Fraction $10^3 x_1$
278.15	5.584
288.15	5.299
298.15	5.046
308.15	4.820

AUXILIARY INFORMATION

METHOD/APPARATUS/PROCEDURE:	SOURCE AND PURITY OF MATERIALS:
The apparatus consists of a gas measuring buret, an absorption pipet, and a reservoir for the solvent. The buret is thermostated at 25°C, the pipet at any temperature from 5 to 30°C. The pipet contains an iron bar in glass for magnetic stirring. The pure solvent is degassed by freezing with liquid nitrogen, evacuating, then boiling with a heat lamp. The degassing process is repeated three times. The solvent is flowed into the pipet where it is again boiled for final degassing. Manipulation of the apparatus is such that the solvent never comes in contact with stopcock grease. The liquid in the pipet is sealed off by mercury. Its volume is the difference between the capacity of the pipet and the volume of mercury that confines it. Gas is admitted into the pipet. Its exact amount is determined by P-V measurements in the buret before and after	(1) Methane. Phillips Petroleum Co. Gas passed through a cold trap. (2) 1,1,2-Trichloro-1,2,2-trifluoro-ethane. Union Carbide Co. Distilled, purity checked by ultraviolet absorbance.

	ESTIMATED ERROR:
	δT/K = 0.02 $\delta x_1/x_1$ = 0.003

	REFERENCES:
	1. Kobatake, Y.; Hildebrand, J. H. *J. Phys. Chem.* <u>1961</u>, *65*, 331.

introduction of the gas into the pipet. The stirrer is set in motion.
Equilibrium is attained within 24 hours.

COMPONENTS:	EVALUATOR:
(1) Methane; CH_4; [74-82-8]	H. Lawrence Clever Department of Chemistry
(2) Tetrachloromethane; CCl_4; [56-23-5]	Emory University Atlanta, GA 30322 USA 1985, May

CRITICAL EVALUATION:

Horiuti (ref 1) reports solubility values at five temperatures between 253.35 and 333.15 K. Tomonaga *et al.* (ref 2) report solubility values at temperatures of 282.71, 298.14 and 308.15 K. Their mole fraction solubility values were calculated from Henry's constants corrected for non-ideal behavior at the vapor pressure of the solvent. They report three values at 308.15 K which average $(2.786 \pm 0.020) \times 10^{-3}$ mole fraction at atmospheric pressure. Both laboratories have reputations for reliable work. Both data sets are classed as tentative.

The Tominaga *et al.* average value at 308.15 K was used twice and all of the other experimental values from both papers used once in a linear regression to obtain the equation for use over the 253 to 333 K interval of

$$\ln x_1 = -9.56027 + 7.08799/(T/100\ K) + 1.2145 \ln (T/100\ K)$$

with a standard error about the regression line of 2.52×10^{-5}.

All of the Horiuti data were within 0.25 percent of the regression line except the 313.15 K value which was 0.44 % smaller. The Tominaga *et al.* values were 0.29 % larger, 1.57 % smaller and 1.01 % larger at the temperatures of 282.71, 298.14, and 308.15 K, respectively.

The thermodynamic changes for the transfer of one mole of methane from the gas at 0.101325 MPa to the infinitely dilute solution were calculated from the equation constants to be:

T/K	$\Delta H_1^0/kJ\ mol^{-1}$	$\Delta S_1^0/J\ K^{-1}\ mol^{-1}$	$\Delta C_{p1}^0/J\ K^{-1}\ mol^{-1}$
253.15	-3.34	-60.0	10.1
273.15	-3.13	-59.8	10.1
298.15	-2.88	-58.4	10.1

Smoothed values of the mole fraction solubility are in Table 1.

Table 1. Solubility of methane in tetrachloromethane. Tentative values of the mole fraction solubility as a function of temperature at a methane partial pressure of 0.101325 MPa.

T/K	$10^3 x_1$	T/K	$10^3 x_1$
253.15	3.580	298.15	2.862
263.15	3.374		
273.15	3.199	303.15	2.808
283.15	3.049	313.15	2.711
293.15	2.920	323.15	2.626
		333.15	2.551

REFERENCES:

1. Horiuti, J. *Sci. Pap. Inst. Phys. Chem. Res. (Jpn)* <u>1931/32</u>, *17*, 125-256.

2. Tominaga, T.; Battino, R.; Gorowara, B.; Dixon, R. D.; Wilhelm, E.
   J. Chem. Eng. Data <u>1986</u>, *31*,

COMPONENTS:	ORIGINAL MEASUREMENTS:
(1) Methane; CH_4; [74-82-8]	Horiuti, J.
(2) Tetrachloromethane or carbon tetrachloride; CCl_4; [56-23-5]	*Sci. Pap. Inst. Phys. Chem. Res. (Jpn)* 1931/32, *17*, 125 - 256.

VARIABLES:	PREPARED BY:
T/K: 253.35 - 333.15 p_1/kPa: 101.325 (1 atm)	M. E. Derrick H. L. Clever

EXPERIMENTAL VALUES:

T/K	Mol Fraction $10^3 x_1$	Bunsen Coefficient α/cm^3(STP)cm^{-3}atm^{-1}	Ostwald Coefficient L/cm^{3}cm^{-3}
253.35	3.578	0.8743	0.8109
273.15	3.193	0.7621	0.7621
293.15	2.919	0.6775	0.7271
313.15	2.699	0.6133	0.7031
333.15	2.545	0.5638	0.6876

The mole fraction and Bunsen coefficient values were calculated by the compiler with the assumption the gas is ideal and that Henry's law is obeyed.

Smoothed Data: For use between 253.35 and 333.15 K.

$$\ln x_1 = -9.7099 + 7.3146/(T/100K) + 1.2798 \ln (T/100K)$$

The standard error about the regression line is 5.28×10^{-6}.

T/K	Mol Fraction $10^3 x_1$
258.15	3.473
273.15	3.195
288.15	2.977
298.15	2.856
308.15	2.751
318.15	2.660
333.15	2.544

AUXILIARY INFORMATION

METHOD/APPARATUS/PROCEDURE:	SOURCE AND PURITY OF MATERIALS:
The apparatus consists of a gas buret, a solvent reservoir, and an absorption pipet. The volume of the pipet is determined at various meniscus heights by weighing a quantity of water. The meniscus height is read with a cathetometer. The dry gas is introduced into the degassed solvent. The gas and solvent are mixed with a magnetic stirrer until saturation. Care is taken to prevent solvent vapor from mixing with the solute gas in the gas buret. The volume of gas is determined from the gas buret readings, the volume of solvent is determined from the meniscus height in the absorption pipet.	(1) Methane. Aluminum carbide was prepared from aluminum and soot carbon. The aluminum carbide was treated with hot water. The gas evolved was scrubbed to remove impurities, dried and fractionated. Final product had a density, ρ/g dm^{-3} = 0.7168±0.0003 at normal conditions. (2) Tetrachloromethane. Kahlbaum. Dried over P_2O_5 and distilled. Boiling point(760mmHg) 76.74°C.

ESTIMATED ERROR:

$$\delta T/K = 0.05$$
$$\delta x_1/x_1 = 0.01$$

REFERENCES:

COMPONENTS:	ORIGINAL MEASUREMENTS:
(1) Methane; CH_4; [74-82-8] (2) Tetrachloromethane or carbon tetrachloride; CCl_4; [56-23-5]	Tominaga, T.; Battino, R.; Gorowara, B.; Dixon, R. D.; Wilhelm, E. *J. Chem. Eng. Data* <u>1986</u>, *31*,
VARIABLES: $T/K = 282.71 - 308.15$ $p_1/kPa = 101.325$	PREPARED BY: H. L. Clever

EXPERIMENTAL VALUES:

T/K	Mol Fraction $10^3 x_1$	Ostwald Coefficient $L/cm^3 \ cm^{-3}$	Henry's Constant $10^{-6} H/Pa$
282.71	3.064	0.7414	33.07
298.14	2.818	0.7060	35.95
308.15	2.762	0.7060	36.69
	2.786	0.7122	36.37
	2.810	0.7180	36.06

The mole fraction solubility at 101325 Pa was calculated from the author's Henry's constant by the compiler with no corrections.

Henry's constant $H/Pa = (p_1/Pa)/x_1$.

101325 Pa $\equiv$ 1 atm

AUXILIARY INFORMATION

METHOD/APPARATUS/PROCEDURE:	SOURCE AND PURITY OF MATERIALS:
The solubility apparatus is based on the design of Ben-Naim and Baer (ref 1). The degassing apparatus is that described by Battino *et al.* (ref 2). Degassing. Up to 500 cm^3 of solvent is placed in a flask of such size that the liquid si about 4 cm deep. The liquid is rapidly stirred, and a vacuum is intermittently applied through a liquid N_2 trap until the permanent gas residual pressure drops to 5 microns. Solubility Determination. Ben-Naim and Baer's procedure is used. The gas is liquid vapor saturated, dissolution is usually comlete within 10-20 minutes. The mixing chamber volume are about 26, 65, 380, and 1650 cm^3 calibrated to ± 0.01 cm^3. The pressure is maintained constant and the volume changed by a microprocessor controled steping motor operating a piston in a precision bore tube.	(1) Methane. Matheson Co., Inc. 99.97 minimum mole percent. (2) Tetrachloromethane. Fisher. Certified grade, 99 mol percent. Distilled through a 1.2 m packed column, middle 80 % stored protected from light until use.
	ESTIMATED ERROR: $\delta x_1/x_1 = \pm \ 0.008$
	REFERENCES: 1. Ben-Naim, A.; Baer, S. *Trans. Faraday Soc.* <u>1963</u>, *59*, 2935. 2. Battino, R.; Banzhof, M.; Bogan, M.; Wilhelm, E. *Anal. Chem.* <u>1971</u>, *43*, 806.

COMPONENTS:	ORIGINAL MEASUREMENTS:
1. Methane; CH₄; [74-82-8] 2. Dichlorodifluoromethane (Freon 12); CCl₂F₂; [75-71-8]	Yorizane, M.; Yoshimura, S.; Masuoka, H.; Miyano, Y.; Kakimoto, Y. *J. Chem. Eng. Data* <u>1985</u>, *30*, 174-176.
VARIABLES:	PREPARED BY: C. L. Young

EXPERIMENTAL VALUES:

T/K	P/MPa	Mole fraction of methane in vapor, y_{CH_4}	in liquid, x_{CH_4}	T/K	P/MPa	Mole fraction of methane in vapor, y_{CH_4}	in liquid, x_{CH_4}
263.2	1.36	0.091	0.773	298.2	1.99	0.104	0.512
	2.79	0.221	0.866		2.96	0.147	0.630
	3.87	0.305	0.893		4.06	0.222	0.723
	5.11	0.371	0.904		4.51	0.246	0.759
	6.28	0.434	0.907		4.90	0.272	0.791
	7.29	0.498	0.899		5.61	0.327	0.801
	8.41	0.577	0.892		6.44	0.365	0.816
	9.14	0.635	0.875		6.99	0.401	0.833
	9.60	0.655	0.872		7.40	0.431	0.829
	9.93	0.753[a]	critical		7.95	0.465	0.828
273.2	1.53	0.088	0.785		8.51	0.497	0.828
	1.75	0.107	0.816		8.98	0.562	0.825
	2.12	0.136	0.844		9.27	0.547	0.818
	3.00	0.203	0.877		9.79	0.604	0.815
	3.79	0.256	0.891		9.98	0.608	0.811
	4.43	0.338	0.901		10.25	0.647	0.796
	5.72	0.400	0.914		10.67	0.740[a]	critical
	7.49	0.506	0.896				
	8.30	0.541	0.884				
	8.59	0.566	0.887				
	9.33	0.615	0.866		[a] Estimated values.		
	11.16	0.778[a]	critical				

AUXILIARY INFORMATION

METHOD APPARATUS/PROCEDURE:	SOURCE AND PURITY OF MATERIALS:
Apparatus consisted of two similar equilibrium cells, one fixed in position, the other could be moved so that liquid and vapor flowed between cells. Samples from cells analysed using gas chromatography. Pressure measured with a Bourdon gauge and temperature with a standard mercury thermometer. Details of apparatus and procedure in source.	1. Purity 99.9 volume per cent. 2. Purity 99.95 volume per cent.
	ESTIMATED ERROR: $\delta T/K = \pm0.1$; $\delta P/P = \pm0.005$.
	REFERENCES:

COMPONENTS:	ORIGINAL MEASUREMENTS:
1. Methane; CH_4; [74-82-8] 2. Chlorodifluoromethane; $CHClF_2$; [75-45-6]	Nohka, J.; Sarashina, E.; Arai, Y.; Saito, S. *J. Chem. Eng. Japan*, <u>1973</u>, *6*, 10-17

VARIABLES:	PREPARED BY:
Temperature, pressure	C.L. Young

EXPERIMENTAL VALUES:

T/K	$p/10^5$Pa	Mole fraction of methane in liquid, x_{CH_4}	in gas, y_{CH_4}
273.15	16.4	0.0536	0.658
	30.0	0.124	0.794
	41.8	0.187	0.831
	63.4	0.307	0.852
	81.1	0.424	0.853
	92.9	0.505	0.838
	98.8	0.546	0.821
	105.4	0.615	0.778
298.15	20.3	0.0429	0.437
	40.5	0.142	0.659
	57.6	0.232	0.730
	81.1	0.358	0.754
	92.0	0.424	0.735
	95.3	0.448	0.727
	101.3	0.517	0.701
323.15	30.4	0.0488	0.305
	40.5	0.0969	0.436
	48.6	0.132	0.502
	62.8	0.207	0.552
	75.3	0.272	0.569
	84.7	0.326	0.567
	92.9	0.400	0.528
348.15	45.8	0.0595	0.202
	55.7	0.107	0.268
	65.6	0.166	0.305
	70.9	0.204	0.322

AUXILIARY INFORMATION

METHOD/APPARATUS/PROCEDURE:	SOURCE AND PURITY OF MATERIALS:
Static cell fitted with magnetic stirrer. Temperature measured with liquid in glass thermometer and pressure measured with Bourdon gauge. After equilibrium established vapor and liquid samples analysed by gas chromatography. Details in ref. 1 and 2.	1. No details given. 2. Purity better than 99.9 mole %.

ESTIMATED ERROR:

δT/K = ±0.1; $\delta P/10^5$Pa = ±0.1;
δx_{CH_4}, δy_{CH_4} = ±1%
(estimated by compiler).

REFERENCES:
1. Kaminishi, G.; Arai, Y.; Saito,
   S.; Maeda, S.
   J. Chem. Eng. Japan, <u>1968</u>, *1*, 109.

2. Sarashina, E.; Arai, Y.; Saito,
   S.
   J. Chem. Eng. Japan. <u>1971</u>, *4*, 377.

COMPONENTS:	ORIGINAL MEASUREMENTS:
1. Methane; CH_4; [74-82-8] 2. Chlorodifluoromethane (Freon 22); $CHClF_2$; [75-45-6]	Yorizane, M.; Yoshimura, S.; Masuoka, H.; Miyano, Y.; Kakimoto, Y. *J. Chem. Eng. Data* 1985, *30*, 174-176.

VARIABLES:	PREPARED BY:
	C. L. Young

EXPERIMENTAL VALUES:

T/K	P/MPa	Mole fraction of methane in vapor, y_{CH_4}	in liquid, x_{CH_4}	T/K	P/MPa	Mole fraction of methane in vapor, y_{CH_4}	in liquid, x_{CH_4}
263.2	1.12	0.051	0.617	273.2	8.83	0.458	0.838
	2.03	0.096	0.775		9.82	0.537	0.828
	3.21	0.172	0.843		10.27	0.580	0.818
	3.99	0.195	0.859		11.08	0.711^a	critical
	5.02	0.276	0.881	298.2	1.92	0.040	0.401
	8.00	0.434	0.880		2.72	0.072	0.507
	8.86	0.480	0.852		4.24	0.157	0.665
	9.25	0.510	0.850		5.95	0.219	0.716
	9.80	0.540	0.844		7.28	0.296	0.721
	10.21	0.601	0.819		8.20	0.355	0.721
	10.48	0.631	0.760		9.13	0.380	0.701
	10.67	0.712^a	critical		10.01	0.461	0.680
					10.36	0.517	0.644
273.2	3.18	0.147	0.806		10.76	0.592^a	critical
	4.11	0.195	0.830				
	4.98	0.239	0.847				
	6.67		0.853				
	8.32	0.445	0.831				

[a] Estimated values.

AUXILIARY INFORMATION

METHOD APPARATUS/PROCEDURE:	SOURCE AND PURITY OF MATERIALS:
Apparatus consisted of two similar equilibrium cells, one fixed in position, the other could be moved so that liquid and vapor flowed between cells. Samples from cells analysed using gas chromatography. Pressure measured with a Bourdon gauge and temperature with a standard mercury thermometer. Details of apparatus and procedure in source.	1. Purity 99.9 volume per cent. 2. Purity 99.95 volume per cent.
	ESTIMATED ERROR: $\delta T/K = \pm 0.1$; $\delta P/P = \pm 0.005$.
	REFERENCES:

COMPONENTS:	ORIGINAL MEASUREMENTS:
(1) Methane; CH_4; [74-82-8] (2) 1-Chlorohexane; $C_6H_{11}Cl$; [544-10-5]	Guerry, D. Jr. Ph.D. thesis, 1944 Vanderbilt University Nashville, TN Thesis Director: L. J. Bircher

| VARIABLES:
     T/K: 293.15, 298.15
     P/kPa: 101.325 (1 atm) | PREPARED BY:

     H. L. Clever |

EXPERIMENTAL VALUES:

T/K	Mol Fraction x_1 x 10^4	Bunsen Coefficient α	Ostwald Coefficient L
293.15	31.9	0.522	0.560
298.15	31.1	0.506	0.553

The Ostwald coefficients were calculated by the compiler.

AUXILIARY INFORMATION

METHOD/APPARATUS/PROCEDURE:	
A Van Slyke-Neill Manometric Apparatus manufactured by the Eimer and Amend Co. was used. The procedure of Van Slyke (1) for pure liquids was modified (2) so that small solvent samples (2 cm^3) could be used with almost complete recovery of the sample. An improved temperature control system was used.	

SOURCE AND PURITY OF MATERIALS:	ESTIMATED ERROR:
(1) Methane. Prepared by hydrolysis of crystaline methyl Grignard reagent. Passed through conc. H_2SO_4, solid KOH, and Dririte.	$\delta T/K = 0.05$

| (2) 1-Chlorohexane. Eastman Kodak Co. Purified, distilled from P_2O_5 in a N_2 atm. B.p. (746.6 mmHg) t/°C 134.66 (corr.). Refractive index, density, and vapor pressure data are in the thesis. | REFERENCES:
 1. Van Slyke, D. D.
    J. Biol. Chem. 1939, *130*, 545.

 2. Ijams, C. C.
    Ph.D. thesis, 1941
    Vanderbilt University |

COMPONENTS:	EVALUATOR:
1. Methane; CH_4; [74-82-8] 2. Chlorobenzene; C_6H_5Cl; [108-90-7]	Colin L. Young Department of Physical Chemistry, University of Melbourne. Parkville, Victoria, 3052 Australia. February 1986.

EVALUATION:

     This system has been studied by three groups of workers and there is good consistency between the three sets of data. The data of Horiuti (1) is the most extensive covering the temperature range 232 K to 373 K. The more recent data of Lopez et al. (2) is in good agreement over the temperature range studied of 263 K to 303 K. The data of Berlin et al. (3) were determined at elevated pressures at 293.2 K. This latter set of data are not of high precision but, when extrapolated assuming Henry´s law to be obeyed, yield mole fraction solubilities at 1 atmosphere partial pressure which are consistent with values given by the other two groups. Horiuti (1) data are classified as recommended and thought to be accurate to better than two per cent.

References.

1. Horiuti, J.
   Sci. Pap. Inst. Phys. Chem. Res. (Jpn), <u>1931/32</u>, *17*, 125.
2. Lopez, M. C.; Gallardo, M. A.; Urieta, J. S.; Gutierrez Losa, C.;
   Int. Conf. Thermodyn. Solns. Nonelectrolytes, <u>1984</u>, No 127.
3. Berlin, M. A.; Pluzhnikova, M. F.; Stepanova, I. N.; Potapov, V. F.;
Vasil´eva, N. A.; Tsybnlevskii, A. M.;
   Zh. Prikl. Khim., <u>1980</u>, *53*, 1661.

COMPONENTS:	ORIGINAL MEASUREMENTS:
(1) Methane; CH_4; [74-82-8] (2) Chlorobenzene; C_6H_5Cl; [108-90-7]	Horiuti, J. *Sci. Pap. Inst. Phys. Chem. Res.* *(Jpn)* 1931/32, *17*, 125 - 256.

VARIABLES:
    T/K: 232.35 - 372.75
    p_1/kPa: 101.325 (1 atm)
	PREPARED BY:
	
	    M. E. Derrick
	    H. L. Clever

EXPERIMENTAL VALUES:

T/K	Mol Fraction $10^3 x_1$	Bunsen Coefficient α/cm^3(STP)cm^{-3}atm^{-1}	Ostwald Coefficient L/cm^{3}cm^{-3}
232.35	2.864	0.6704	0.5703
252.65	2.477	0.5686	0.5259
273.15	2.211	0.4976	0.4976
293.15	2.029	0.4480	0.4808
303.15	1.949	0.4260	0.4728
333.15	1.817	0.3852	0.4698
353.25	1.748	0.3631	0.4696
372.75	1.710	0.3479	0.4748

The mole fraction and Bunsen coefficient values were calculated by the compiler with the assumption the gas is ideal and that Henry's law is obeyed.

Smoothed Data: For use between 232.35 and 372.75 K.

$$\ln x_1 = -11.8817 + 9.6607/(T/100K) + 2.2178 \ln (T/100K)$$

The standard error about the regression line is 6.08×10^{-6}.

T/K	Mol Fraction $10^3 x_1$	T/K	Mol Fraction $10^3 x_1$
243.15	2.637	318.15	1.876
258.15	2.391	328.15	1.832
273.15	2.206	343.15	1.779
288.15	2.066	358.15	1.738
298.15	1.992	373.15	1.708
308.15	1.929		

AUXILIARY INFORMATION

METHOD/APPARATUS/PROCEDURE:

The apparatus consists of a gas buret, a solvent reservoir, and an absorption pipet. The volume of the pipet is determined at various meniscus heights by weighing a quantity of water. The meniscus height is read with a cathetometer.

The dry gas is introduced into the degassed solvent. The gas and solvent are mixed with a magnetic stirrer until saturation. Care is taken to prevent solvent vapor from mixing with the solute gas in the gas buret. The volume of gas is determined from the gas buret readings, the volume of solvent is determined from the meniscus height in the absorption pipet.

SOURCE AND PURITY OF MATERIALS:

(1) Methane. Aluminum carbide was prepared from aluminum and soot carbon. The aluminum carbide was treated with hot water. The gas evolved was scrubbed to remove impurities, dried and fractionated. Final product had a density, ρ/g dm^{-3} = 0.7168±0.0003 at normal conditions.

(2) Chlorobenzene. Kahlbaum. Dried and distilled. Boiling point (760 mmHg) 131.96°C.

ESTIMATED ERROR:

$$\delta T/K = 0.05$$
$$\delta x_1/x_1 = 0.01$$

REFERENCES:

COMPONENTS:	ORIGINAL MEASUREMENTS:
1. Methane; CH_4; [74-82-8]	Berlin, M. A.; Pluzhnikova, M. F.; Stepanova, I. N.; Potapov, V. F.; Vasil'eva, N. A.; Tsybnlevskii, A. M.
2. Chlorobenzene; C_6H_5Cl; [108-90-7]	*Zh. Prikl. Khim.* 1980, *53*, 1661-3.

VARIABLES:	PREPARED BY:
	C. L. Young

EXPERIMENTAL VALUES:

T/K	P/MPa	α[a]	Mole fraction of methane[b] x_{CH_4}
293.2	1.0	0.00	–
	6.0	30.75	0.1152

[a] volume of methane measured at 293.2 K and 1 atmosphere pressure dissolved by unit volume of liquid.

[b] calculated by compiler assuming molar volume of methane at 293.2 K and 1 atmosphere is 24.04 L.

AUXILIARY INFORMATION

METHOD/APPARATUS/PROCEDURE:	SOURCE AND PURITY OF MATERIALS:
A gas chromatographic method. No details given except ref. (1) which contains little additional information.	1. Purity about 99.6-99.8 mole per cent. 2. Purified, final purity checked by refractive index measurements.

	ESTIMATED ERROR:

	REFERENCES:
	1. Berlin, M. A.; Pluzhnikova, M. F.; Stepanova, I. N.; Tsybnlevskii, A. M. *Zh. Fiz. Khim.* 1977, *51*, 767.

COMPONENTS:	ORIGINAL MEASUREMENTS:
(1) Methane; CH_4; [74-82-8] (2) Chlorobenzene; C_6H_5Cl; [108-90-7]	López, M. C.; Gallardo, M. A.; Urieta, J. S.; Gutièrrez Losa, C. Int. Conf. Thermodyn. Solutions of Nonelectrolytes, <u>1984</u>, Paper No. 127.

VARIABLES:	PREPARED BY:
T/K = 263.15 - 303.15 p_1/kPa = 101.3	H. L. Clever

EXPERIMENTAL VALUES:

T/K	Mol Fraction $10^4 x_1$
263.15	23.2
273.15	22.1
283.15	21.0
293.15	20.1
303.15	19.2

The authors fit their data to the equation

$-\ln x_1 = -1.44 \ln (T/K) + 1.35$

From which they obtained thethermodynamic changes

ΔH_1^0/kJ mol^{-1} = - 3.34   and

ΔS_1^0/J K^{-1} mol^{-1} = - 63.

AUXILIARY INFORMATION

METHOD/APPARATUS/PROCEDURE:	SOURCE AND PURITY OF MATERIALS:
The solubility apparatus was similar to that used by Ben-Naim and Baer (ref 1). It consisted of a gas buret, mercury manometer, and solution vessel. The solvent was degassed in the solution vessel. Measurements were carried out on the vapor saturated gas.	(1) Methane. Sociedad Espanol del Oxigeno. Stated to be 99.95 per cent pure. (2) Chlorobenzene.
	ESTIMATED ERROR: $\delta T/K$ = ± 0.1 $\delta x_1/x_1$ = ± 0.01
	REFERENCES: 1. Ben-Naim, A. Baer, S. *Trans. Faraday Soc.* <u>1963</u>, *59*, 2735.

COMPONENTS:	ORIGINAL MEASUREMENTS:
1. Methane; CH_4; [74-82-8] 2. (1-chloroethyl)-benzene; C_7H_7Cl; [106-43-4]	Berlin, M. A.; Pluzhnikova, M. F.; Stepanova, I. N.; Potapov, V. F.; Vasil'eva, N. A.; Tsybnlevskii, A. M. *Zh. Prikl. Khim.* <u>1980</u>, *53*, 1661-3.

VARIABLES:	PREPARED BY:
	C. L. Young

EXPERIMENTAL VALUES:

T/K	P/MPa	α^a	Mole fraction of methane[b] x_{CH_4}
293.2	1.0	0.00	–
	6.0	7.40	0.039

[a] volume of methane measured at 293.2 K and 1 atmosphere pressure
dissolved by unit volume of liquid.

[b] calculated by compiler assuming molar volume of methane at 293.2 K
and 1 atmosphere is 24.04 L.

AUXILIARY INFORMATION

METHOD/APPARATUS/PROCEDURE:	SOURCE AND PURITY OF MATERIALS:
A gas chromatographic method. No details given except ref. (1) which contains little additional information.	1. Purity about 99.6-99.8 mole per cent. 2. Purified, final purity checked by refractive index measurements.
	ESTIMATED ERROR:

	REFERENCES:
	1. Berlin, M. A.; Pluzhnikova, M. F.;
	   Stepanova, I. N.;
	   Tsybnlevskii, A. M.
	   Zh. Fiz. Khim.
	   <u>1977</u>, *51*, 767.

COMPONENTS:	ORIGINAL MEASUREMENTS:
1. Methane; CH_4; [74-82-8] 2. 1-Bromooctane; $C_8H_{17}Br$; [111-83-1]	Berlin, M. A.; Pluzhnikova, M. F.; Stepanova, N. I.; Potapov, V. F.; Vasil'eva, N. A.; Tsybnlevskii, A. M. *Zh. Prikl. Khim.* 1980, *53*, 1661-3.

VARIABLES:	PREPARED BY:
	C. L. Young

EXPERIMENTAL VALUES:

T/K	P/MPa	α[a]	Mole fraction of methane[b] x_{CH_4}
293.2	1.0	1.9	0.014
	2.0	2.8	0.020
	3.5	6.9	0.047
	6.0	12.0	0.080

[a] volume of methane measured at 293.2 K and 1 atmosphere pressure dissolved by unit volume of liquid.

[b] calculated by compiler assuming molar volume of methane at 293.2 K and 1 atmosphere is 24.04 L.

AUXILIARY INFORMATION

METHOD/APPARATUS/PROCEDURE:	SOURCE AND PURITY OF MATERIALS:
A gas chromatographic method. No details given except ref. (1) which contains little additional information.	1. Purity about 99.6-99.8 mole per cent. 2. Purified, final purity checked by refractive index measurements.
	ESTIMATED ERROR:

	REFERENCES:
	1. Berlin, M. A.; Pluzhnikova, M.F.;
	   Stepanova, I. N.;
	   Tsybnlevskii, A. M.
	   Zh. Fiz. Khim.
	   1977, *51*, 767.

COMPONENTS:	ORIGINAL MEASUREMENTS:
1. Methane; CH_4; [74-82-8]	Berlin, M. A.; Pluzhnikova, M. F.; Stepanova, I. N.; Potapov, V. F.; Vasil'eva, N. A.; Tsybnlevskii, A. M.
2. 1-Choorooctane; $C_8H_{17}Cl$; [111-85-3]	*Zh. Prikl. Khim.* 1980, *53*, 1661-3.

VARIABLES:	PREPARED BY:
	C. L. Young

EXPERIMENTAL VALUES:

T/K	P/MPa	α[a]	Mole fraction of methane[b] x_{CH_4}
293.2	1.0	3.4	0.023
	2.0	8.6	0.057
	3.5	10.6	0.070
	6.0	17.3	0.109

[a] volume of methane measured at 293.2 K and 1 atmosphere pressure dissolved by unit volume of liquid.

[b] calculated by compiler assuming molar volume of methane at 293.2 K and 1 atmosphere is 24.04 L.

AUXILIARY INFORMATION

METHOD/APPARATUS/PROCEDURE:	SOURCE AND PURITY OF MATERIALS:
A gas chromatographic method. No details given except ref. (1) which contains little additional information.	1. Purity about 99.6-99.8 mole per cent. 2. Purified, final purity checked by refractive index measurements.
	ESTIMATED ERROR:
	REFERENCES: 1. Berlin, M. A.; Pluzhnikova, M. F.; Stepanova, I. N.; Tsybnlevskii, A. M. *Zh. Fiz. Khim.* 1977, *51*, 767.

COMPONENTS:	ORIGINAL MEASUREMENTS:
1. Methane; CH₄; [74-82-8] 2. 1-Iodooctane; $C_8H_{17}I$; [629-27-6]	Berlin, M. A.; Pluzhnikova, M. F.; Stepanova, I. N.; Potapov, V. F.; Vasil'eva, N. A.; Tsybnlevskii, A. M. *Zh. Prikl. Khim.* <u>1980</u>, *53*, 1661-3.
VARIABLES:	PREPARED BY: C. L. Young

EXPERIMENTAL VALUES:

T/K	P/MPa	α[a]	Mole fraction of methane[b] x_{CH_4}
293.2	1.0	1.1	0.008
	2.0	1.7	0.013
	3.5	6.9	0.049
	6.0	12.0	0.083

[a] volume of methane measured at 293.2 K and 1 atmosphere pressure dissolved by unit volume of liquid.

[b] calculated by compiler assuming molar volume of methane at 293.2 K and 1 atmosphere is 24.04 L.

AUXILIARY INFORMATION

METHOD/APPARATUS/PROCEDURE:	SOURCE AND PURITY OF MATERIALS:
A gas chromatographic method. No details given except ref. (1) which contains little additional information.	1. Purity about 99.6-99.8 mole per cent. 2. Purified, final purity checked by refractive index measurements.
	ESTIMATED ERROR:
	REFERENCES: 1. Berlin, M. A.; Pluzhnikova, M. F.; Stepanova, I. N.; Tsybnlevskii, A. M. *Zh. Fiz. Khim.* <u>1977</u>, *51*, 767.

COMPONENTS:	ORIGINAL MEASUREMENTS:
1. Methane; CH₄; [74-82-8] 2. 2-Iodooctane; C₈H₁₇I; [557-36-8]	Berlin, M. A.; Pluzhnikova, M. F.; Stepanova, I. N.; Potapov, V. F.; Vasil'eva, N. A.; Tsybnlevskii, A.M. *Zh. Prikl. Khim.* <u>1980</u>, *53*, 1661-3.

VARIABLES:	PREPARED BY:
	C. L. Young

EXPERIMENTAL VALUES:

T/K	P/MPa	α [a]	Mole fraction of methane[b] x_{CH_4}
293.2	1.0	1.5	0.011
	2.0	1.4	0.010
	3.5	5.5	0.040
	6.0	4.2	0.031

[a] volume of methane measured at 293.2 K and 1 atmosphere pressure dissolved by unit volume of liquid.

[b] calculated by compiler assuming molar volume of methane at 293.2 K and 1 atmosphere is 24.04 L.

AUXILIARY INFORMATION

METHOD/APPARATUS/PROCEDURE:	SOURCE AND PURITY OF MATERIALS:
A gas chromatographic method. No details given except ref. (1) which contains little additional information.	1. Purity about 99.6-99.8 mole per cent. 2. Purified, final purity checked by refractive index measurements.
	ESTIMATED ERROR:
	REFERENCES: 1. Berlin, M. A.; Pluzhnikova, M. F.; Stepanova, I. N.; Tsybnlevskii, A. M. *Zh. Fiz. Khim.* <u>1977</u>, *51*, 767.

COMPONENTS:	ORIGINAL MEASUREMENTS:
1. Methane; CH_4; [74-82-8] 2. 1-Chloronaphthalene; $C_{10}H_7Cl$; [90-13-1]	Berlin, M. A.; Pluzhnikova, M. F.; Stepanova, I. N.; Potapov, V. F.; Vasil'eva, N. A.; Tsybnlevskii, A. M. *Zh. Prikl. Khim.* <u>1980</u>, *53*, 1661-3.

VARIABLES:	PREPARED BY:
	C. L. Young

EXPERIMENTAL VALUES:

T/K	P/MPa	α[a]	Mole fraction of methane[b] x_{CH_4}
293.2	1.0	3.95 (?)	0.0219
	6.0	2.28 (?)	0.0128

[a] volume of methane measured at 293.2 K and 1 atmosphere pressure dissolved by unit volume of liquid.

[b] calculated by compiler assuming molar volume of methane at 293.2 K and 1 atmosphere is 24.04 L.

AUXILIARY INFORMATION

METHOD/APPARATUS/PROCEDURE:	SOURCE AND PURITY OF MATERIALS:
A gas chromatographic method. No details given except ref. (1) which contains little additional information.	1. Purity about 99.6-99.8 mole per cent. 2. Purified, final purity checked by refractive index measurements.
	ESTIMATED ERROR:
	REFERENCES: 1. Berlin, M. A.; Pluzhnikova, M. F.; Stepanova, I. N.; Tsybnlevskii, A. M. *Zh. Fiz. Khim.* <u>1977</u>, *51*, 767.

COMPONENTS:	EVALUATOR:

(1) Methane; CH_4; [74-82-8]
 	
(2) Sulfur compounds
    Carbon disulfide; CS_2; [75-15-0]
    Sulfinylbismethane or dimethyl-
    sulfoxide; C_2H_6SO; [67-68-5]
	H. Lawrence Clever
	Chemistry Department
	Emory University
	Atlanta, GA 30322 USA
	
	1985, April

CRITICAL EVALUATION:

    Both Kobatake and Hildebrand (ref 1) and Powell (ref 2) report the solubility of methane in carbon disulfide as a function of temperature. The studies were carried out in the same laboratory. There is a brief comment in the Powell paper saying his apparatus is capapble of better acuracy than the earlier work. Powells mole fraction solubility values range from 4.7 to 1.3 per cent smaller than the Kobatake and Hildebrand values as the temperature increases from 288 to 308 K. We class both data sets as tentative, but prefer the Powell data on the basis of his statement.

    Powell gives only the 298.15 K solubility value in his paper along with the value of 'R(slope)' from the log x_1 vs. log T straight line. The solubility values calculated from his information are consistent with the thermodynamic changes for the transfer of one mole of methane from the gas at 0.101325 MPa to the infinitely dilute solution of

$$\Delta H_1^0 / kJ \text{ mol}^{-1} = -1.04 \quad \text{and} \quad \Delta S_1^0 / J \text{ K}^{-1} \text{ mol}^{-1} = -58.9$$

The smoothed solubility values were calculated form the information in Powell's paper.

Table 1. The solubility of methane in carbon disulfide. Tentative
         values of the mole fraction solubility as a function of
         temperature at a methane partial pressure of 0.101325 MPa.

T/K	$10^3 x_1$	T/K	$10^3 x_1$
273.15	1.322	293.15	1.281
278.15	1.311	298.15	1.272
283.15	1.301	303.15	1.263
288.15	1.291	308.15	1.254

    Dymond (ref 3) and Lenoir et al. (ref 4) report the solubility of methane in sulfinylbismethane (dimethylsulfoxide) at 298.15 K by different methods. Dymond used a volumetric method. Lenoir et al. used a GLC-retention time method with the methane at a relatively low partial pressure.

    The mole fraction solubility values calculated at 298.15 K for a methane partial pressure of 0.101325 MPa are:

    Dymond 3.86×10^{-4}
    Lenoir et al. 4.10×10^{-4}

The values differ by about 6 percent which is satisfactory for such different methods. Both values are classed as tentative, but the Dymond value is preferred when actually working at atmospheric pressure (0.101325 MPa).

REFERENCES:

1. Kobatake, Y.; Hildebrand, J. H. J. Phys. Chem. 1961, 65, 331-5.

2. Powell, R. J. J. Chem. Eng. Data 1972, 17, 302-4.

3. Dymond, J. H. J. Phys. Chem. 1967, 71, 1829-31.

4. Lenoir, J.-Y.; Renault, P.; Renon, H. J. Chem. Eng. Data 1971, 16, 340-2.

COMPONENTS:	ORIGINAL MEASUREMENTS:
(1) Methane; CH_4; [74-82-8] (2) Carbon disulfide; CS_2; [75-15-0]	Powell, R. J. *J. Chem. Eng. Data* <u>1972</u>, *17*, 302 - 304.
VARIABLES: T/K: 273.15 - 303.15 p_1/kPa: 101.325 (1 atm)	PREPARED BY: P. L. Long H. L. Clever

EXPERIMENTAL VALUES:

T/K	Mol Fraction $10^4 x_1$	Bunsen Coefficient α/cm^3 (STP) cm^{-3} atm^{-1}	Ostwald Coefficient L/cm^3 cm^{-3}	$N = R \dfrac{\Delta\log x_1}{\Delta\log T}$
298.15	12.72	0.471	0.514	-0.87

The Bunsen and Ostwald coefficients were calculated by the compiler.

The author states that the solubility measurements were made over the temperature interval of about 273.15 to 303.15 K, but only the solubility value at 298.15 K was given in the paper. The slope, $N = R(\Delta\log x_1/\Delta\log T)$, was given.

Smoothed Data: For use between 273.15 and 303.15 K

The smoothed data were calculated by the compiler from the slope, N, in the form

$$\log x_1 = \log (12.72 \times 10^{-4}) - (0.87/R) \log (T/298.15)$$

with $R = 1.9872$ cal K^{-1} mol^{-1}.

T/K	Mol Fraction $10^3 x_1$
273.15	1.322
278.15	1.311
283.15	1.301
288.15	1.291
293.15	1.281
298.15	1.272
303.15	1.263

AUXILIARY INFORMATION

METHOD/APPARATUS/PROCEDURE:	SOURCE AND PURITY OF MATERIALS:
The apparatus is the Dymond and Hildebrand (1) apparatus which uses an all glass pumping system to spray slugs of degassed solvent into the gas. The amount of gas dissolved is calculated from the initial and final pressures. The solvent is degassed by freezing, pumping, and followed by boiling under reduced pressure.	(1) Methane. Source not given. Stated to be manufacturer's research grade, dried over $CaCl_2$ before use. (2) Carbon disulfide. Source not given. Stated to be manufacturer's spectrochemical grade.

ESTIMATED ERROR:

$$\delta x_1/x_1 = \pm\ 0.002$$
$$\delta N/cal\ K^{-1}\ mol^{-1} = \pm\ 0.1$$

REFERENCES:
1. Dymond, J. H.; Hildebrand, J. H. *Ind. Eng. Chem. Fundam.* <u>1967</u>, *6*, 130.

COMPONENTS:	ORIGINAL MEASUREMENTS:
(1) Methane; CH_4; [74-82-8] (2) Carbon disulfide; CS_2; [75-15-0]	Kobatake, Y.; Hildebrand, J. H. *J. Phys. Chem.* <u>1961</u>, *65*, 331 - 335.

VARIABLES:
       T/K: 288.16 - 307.95
      P/kPa: 101.325 (1 atm)
	PREPARED BY:
	     M. E. Derrick
	     H. L. Clever

EXPERIMENTAL VALUES:

Temperature		Mol Fraction	Bunsen Coefficient	Ostwald Coefficient
$t/°C$	T/K	$10^2 x_1$	α/cm^3 (STP) cm^{-3} atm^{-1}	L/cm^3 cm^{-3}
15.01	288.16	1.351	0.506	0.534
25.00	298.15	1.312	0.486	0.530
34.80	307.95	1.269	0.464	0.523

The Bunsen and Ostwald coefficients were calculated by the compiler.

Smoothed Data: For use between 288.16 and 307.95 K.

$$\ln x_1 = -7.5787 + 2.8034/(T/100K)$$

The standard error about the regression line is 3.59×10^{-6}.

T/K	Mol Fraction $10^3 x_1$
288.15	1.352
298.15	1.309
308.15	1.270

AUXILIARY INFORMATION

METHOD/APPARATUS/PROCEDURE:	SOURCE AND PURITY OF MATERIALS:
The apparatus consists of a gas measuring buret, an absorption pipet, and a reservoir for the solvent. The buret is thermostated at 25°C, the pipet at any temperature from 5 to 30 °C. The pipet contains an iron bar in glass for magnetic stirring. The pure solvent is degassed by freezing with liquid nitrogen, evacuating, then boiling with a heat lamp. The degassing process is repeated three times. The solvent is flowed into the pipet where it is again boiled for final degassing. Manipulation of the apparatus is such that the solvent never comes in contact with stopcock grease. The liquid in the pipet is sealed off by mercury. Its volume is the difference between the capacity of the pipet and the volume of mercury that confines it. Gas is admitted into the pipet. Its exact amount is determined by P-V measurements in the buret before and after	(1) Methane. Matheson Co., Inc. Research grade. Dried by passage over P_2O_5 followed by multiple trap vaporization and evacuation at liquid N_2 temperature. (2) Carbon disulfide. Mallinckrodt Chemical Works. Analytical Reagent grade. Shaken successively with Hg and $HgCl_2$, filtered, distilled, and stored over Hg more than 5 days before use.

	ESTIMATED ERROR:
	      $\delta T/K = 0.02$
	      $\delta x_1/x_1 = 0.003$

	REFERENCES:

introduction of the gas into the pipet. The stirrer is set in motion.
Equilibrium is attained within 24 hours.

COMPONENTS:	ORIGINAL MEASUREMENTS:
(1) Methane; CH_4; [74-82-8] (2) Sulfinylbismethane or dimethyl sulfoxide; C_2H_6OS (CH_3SOCH_3); [67-68-5]	Dymond, J. H. *J. Phys. Chem.* 1967, *71*, 1829-1831.

VARIABLES:
    T/K: 298.15
    p/kPa: 101.325 (1 atm)
	PREPARED BY:
	    M. E. Derrick
	    H. L. Clever

EXPERIMENTAL VALUES:

T/K	Mol Fraction $10^4 x_1$	Bunsen Coefficient α/cm^3(STP)cm^{-3}atm^{-1}	Ostwald Coefficient L/cm^{3}cm^{-3}
298.15	3.86	0.121	0.132

The Bunsen and Ostwald coefficients were calculated by the compiler.

AUXILIARY INFORMATION

METHOD/APPARATUS/PROCEDURE:	SOURCE AND PURITY OF MATERIALS:
The liquid is saturated with the gas at a gas partial pressure of 1 atm. The apparatus is that described by Dymond and Hildebrand (1). The apparatus uses an all-glass pumping system to spray slugs of degassed solvent into the gas. The amount of gas dissolved is calculated from the initial and final gas pressure.	(1) Methane. Phillips Petroleum Co. Dried. (2) Dimethylsulfoxide. Matheson, Coleman and Bell Co. Spectro-quality. Dried and fractionally frozen. m.p. 18.37°C.
	ESTIMATED ERROR:
	REFERENCES: 1. Dymond, J.; Hildebrand, J. H. *Ind. Eng. Chem. Fundam.* 1967, *6*, 130.

COMPONENTS:	ORIGINAL MEASUREMENTS:
1. Methane; CH₄; [74-82-8] 2. Sulfinylbismethane, (Dimethyl-sulfoxide); C₂H₆SO; [67-68-5]	Lenoir, J-Y; Renault, P.; Renon, H. *J. Chem. Eng. Data*, <u>1971</u>, *16*, 340-2.

VARIABLES:	PREPARED BY:
	C. L. Young

EXPERIMENTAL VALUES:

T/K	Henry's constant H_{CH_4}/atm	Mole fraction at 1 atm* x_{CH_4}
298.2	2440	0.000410

* Calculated by compiler assuming a linear function of P_{CH_4} vs x_{CH_4}, i.e., x_{CH_4} (1 atm) = $1/H_{CH_4}$.

AUXILIARY INFORMATION

METHOD/APPARATUS/PROCEDURE:	SOURCE AND PURITY OF MATERIALS:
A conventional gas-liquid chromatographic unit fitted with a thermal conductivity detector was used. The carrier gas was helium. The value of Henry's law constant was calculated from the retnetion time. The value applies to very low partial pressures of gas and there may be a substantial difference from that measured at 1 atm. pressure. There is also considerable uncertainty in the value of Henry's constant since surface adsorption was not allowed for although its possible existence was noted.	(1) L'Air Liquide sample, minimum purity 99.9 moler per cent. (2) Touzart and Matignon or Serlabo sample, purity 99 mole per cent.
	ESTIMATED ERROR: $\delta T/K = \pm 0.1$; δH/atm = $\pm 6\%$ (estimated by compiler).
	REFERENCES:

COMPONENTS:	ORIGINAL MEASUREMENTS:
(1) Methane; CH_4; [74-82-8] (2) Cyclic amines; C_4H_9N, C_5H_5N, and $C_5H_{10}N$	Guerry, D. Jr. Ph.D. thesis, 1944 Vanderbilt University Nashville, TN Thesis Director: L. J. Bircher

VARIABLES:	PREPARED BY:
T/K: 293.15, 298.15 P/kPa: 101.325 (1 atm)	H. L. Clever

EXPERIMENTAL VALUES:

T/K	Mol Fraction x_1 x 10^4	Bunsen Coefficient α	Ostwald Coefficient L
Pyrrolidine; C_4H_9N; [123-75-1]			
293.15	14.4	0.389	0.417
298.15	14.1	0.379	0.414
Pyridine; C_5H_5N; [110-86-1]			
293.15	11.2	0.313	0.336
298.15	11.2	0.310	0.338
Piperidine; $C_5H_{11}N$; [110-89-1]			
293.15	18.8	0.427^1	0.459
298.15	19.0	0.430	0.469

The Ostwald coefficients were calculated by the compiler.

[1] The value in the published abstract of the thesis is 0.472. However, the value 0.427 is consistent with the mole fraction value.

AUXILIARY INFORMATION

METHOD/APPARATUS/PROCEDURE:

A Van Slyke-Neill Manometric Apparatus manufactured by the Eimer and Amend Co. was used.

The procedure of Van Slyke (1) for pure liquids was modified (2) so that small solvent samples (2 cm^3) could be used with almost complete recovery of the sample.

An improved temperature control system was used.

SOURCE AND PURITY OF MATERIALS:

(1) Methane. Prepared by hydrolysis of crystaline methyl Grignard reagent. Passed through conc. H_2SO_4, solid KOH, and Dririte.

(2) Cyclic amines. The pyridine and pyrrolidine were distilled from BaO under a N_2 atmosphere. The piperidine was distilled from KOH under a N_2 atmosphere. Experimental data on refractive index, density and vapor pressure are in the thesis.

SOURCE AND PURITY OF MATERIALS:

Pyrrolidine. Pyrrole was prepared and catalytically reduced to pyrrolidine. B.p. (750 mmHg) t/°C 88.12 - 88.26 (corr.).

Pyridine. Mallincrodt Chemical Co. Purified and distilled. B.p. (743.9 mmHg) t/°C 114.96 - 115.06 (corr.).

Piperidine. Part was a commercial sample (Eastman Kodak Co.), part prepared by reduction of pyridine. B.p. (752.4 mmHg) t/°C 106.00 - 106.17.

ESTIMATED ERROR:

$\delta T/K = 0.05$

REFERENCES:

1. Van Slyke, D. D.
 J. Biol. Chem. 1939, *130*, 545.

2. Ijams, C. C.
 Ph.D. thesis, 1941
 Vanderbilt University

COMPONENTS:	ORIGINAL MEASUREMENTS:
1. Methane; CH_4; [74-82-8] 2. Cyclohexylamine; $C_6H_{13}N$; [108-91-8]	Keevil, T.A.; Taylor, D.R.; Streitwieser, A. *J. Chem. Engng. Data.* <u>1978</u>,*23*, 237-239.
VARIABLES:	PREPARED BY: C.L. Young

EXPERIMENTAL VALUES:

Partial pressure of methane = 1 atm = 101.3 kPa.

T/K	Mole fraction of methane, x_{CH_4}
303.2	0.00192

AUXILIARY INFORMATION

METHOD/APPARATUS/PROCEDURE:	SOURCE AND PURITY OF MATERIALS:
Volumetric apparatus of moderate accuracy. Solvent confined to glass bulb and known amount of gas added. Pressure measured using a mercury manometer together with a null point manometer in which the gas pressure was balanced by dry air. Details in source.	1. No details given. 2. Degassed and dried over lithium cyclohexylamide.
	ESTIMATED ERROR: $\delta T/K = \pm 0.1$; $\delta x_{CH_4} = \pm 1\%$
	REFERENCES:

COMPONENTS:	ORIGINAL MEASUREMENTS:
1. Methane; CH_4; [74-82-8] 2. Benzenamine; (Aniline); C_6H_7N; [62-53-3]	Lenoir, J-Y.; Renault, P.; Renon, H. *J. Chem. Eng. Data,* 1971, *16*, 340-2.

VARIABLES:	PREPARED BY:
	C. L. Young

EXPERIMENTAL VALUES:

T/K	Henry's constant H_{CH_4}/atm	Mole fraction at 1 atm* x_{CH_4}
298.2	1580	0.000633

* Calculated by compiler assuming a linear function of P_{CH_4} vs x_{CH_4}, i.e., x_{CH_4} (1 atm) $= 1/H_{CH_4}$.

AUXILIARY INFORMATION

METHOD/APPARATUS/PROCEDURE:	SOURCE AND PURITY OF MATERIALS:
A conventional gas-liquid chromatographic unit fitted with a thermal conductivity detector was used. The carrier gas was helium. The value of Henry's law constant was calculated from the retention time. The value applies to very low partial pressures of gas and there may be a substantial difference from that measured at 1 atm. pressure. There is also considerable uncertainty in the value of Henry's constant since surface adsorption was not allowed for although its possible existence was noted.	(1) L'Air Liquide sample, minimum purity 99.9 mole per cent. (2) Touzart and Matignon or Serlabo sample, purity 99 mole per cent.
	ESTIMATED ERROR: $\delta T/K = \pm 0.1$; δH/atm $= \pm 6\%$ (estimated by compiler).
	REFERENCES:

COMPONENTS:	ORIGINAL MEASUREMENTS:
1. Methane; CH_4; [74-82-8] 2. Quinoline; C_9H_7N; [91-22-5]	Simnick, J.J.; Sebastian, H.M.; Lin, H-M.; Chao, K-C. *J. Chem. Engng. Data.* <u>1979</u>, *24*, 239-240

VARIABLES:	PREPARED BY:
Temperature, pressure	C.L. Young

EXPERIMENTAL VALUES:

T/K	p/atm	p^+/MPa	Mole fraction of methane in liquid, x_{CH_2}	in gas, y_{CH_2}
462.75	19.91	2.017	0.0197	0.9815
	29.98	3.038	0.0297	0.9865
	49.9	5.06	0.0488	0.9901
	100.2	10.15	0.0945	0.9919
	149.4	15.14	0.1352	0.9919
	200.0	20.27	0.1750	0.9904
	249.7	25.30	0.2112	0.9888
542.85	19.90	2.016	0.0206	0.8876
	29.91	3.031	0.0311	0.9187
	50.0	5.07	0.0529	0.9449
	99.3	10.06	0.1036	0.9612
	152.4	15.44	0.1571	0.9643
	199.4	20.20	0.2025	0.9633
	249.1	25.24	0.2455	0.9622
622.65	19.83	2.009	0.0170	0.5838
	30.11	3.051	0.0304	0.7052
	49.8	5.05	0.0557	0.7999
	99.9	10.12	0.1191	0.8692
	151.3	15.33	0.1831	0.8866
	199.5	20.21	0.2410	0.8886
	249.0	25.23	0.3010	0.8837

+ calculated by compiler.

AUXILIARY INFORMATION

METHOD/APPARATUS/PROCEDURE:	SOURCE AND PURITY OF MATERIALS:
Flow apparatus with both liquid and gas components continually passing into a mixing tube and then into a cell in which the phases separated under gravity. Liquid sample removed from bottom of cell and vapor sample from top of cell. Composition of samples estimated using gas chromatography. Temperature measured with thermocouple and pressure with Bourdon gauge. Details in ref. (1).	1. Matheson sample, purity better than 99 mole per cent. 2. Fisher Scientific Co. sample, distilled over zinc under helium, purity better than 99 mole per cent.
	ESTIMATED ERROR: $\delta T/K = \pm 0.1$-02; $\delta p/MPa = \pm 0.5\%$. δx_{CH_2}, $\delta y_{CH_2} = \pm 2\%$.
	REFERENCES: 1. Simnick, J.J.; Lawson, C.C.; Lin, H-M.; Chao, K-C. *Am. Inst. Chem. Engnrs. J.* <u>1977</u>, *23*, 469.

COMPONENTS:	ORIGINAL MEASUREMENTS:
1. Methane; CH_4; [74-82-8]	Simnick, J.J.; Sebastian, H.M.; Lin, H-M.; Chao, K-C.
2. Quinoline; C_9H_7N; [91-22-5]	*J. Chem. Engng. Data.* <u>1979</u>, *24*, 239-240.

EXPERIMENTAL VALUES:

			Mole fraction of methane	
T/K	p/atm	p^+/MPa	in liquid, x_{CH_2}	in gas, y_{CH_2}
702.85	30.03	3.043	0.0170	0.2716
	50.4	5.11	0.0514	0.4770
	99.9	10.12	0.1379	0.6381
	152.0	15.40	0.2283	0.6802
	175.2	17.75	0.2767	0.6752
	200.2	20.29	0.3311	0.6639
	220.2	22.31	0.4027	0.6391

+ calculated by compiler.

COMPONENTS:	EVALUATOR:
1. Methane; CH_4; [74-82-8] 2. 1-Methyl-2-pyrrolidinone; C_5H_9NO; [872-50-4]	Colin L. Young Department of Physical Chemistry, University of Melbourne. Parkville, Victoria, 3052 Australia. February 1986.

EVALUATION:

There appears to be large discrepancies between the mole fraction solubilities for this mixture as reported by the various workers (1-4). The datum of Wu et al.(4) at 298.15 K is in reasonable agreement with the datum of Lenoir et al.(2) at the same temperature. The high pressure data of Shakhova and Zubchenko (1) appears to be consistent with that of Wu et al.(4) and Lenoir et al.(2) but the necessary assumption that Henry´s law is valid between 0.1 and 2.1 MPa does not allow closer comparison. All three sets of data are classified as tentative. The data of Murrieta-Guevara and Rodriguez (3) disagrees with the previous three workers data being of the order of 80-100 per cent larger. Their data are classified as doubtful.

References

1. Shakhova, S. F.; Zubchenko, Yu. P.
   Khim. Prom., <u>1973</u>, *49*, 595
2. Lenoir, J-Y.; Renault, P.; Renon, H.
   J. Chem. Eng. Data <u>1971</u>, *16*, 340.
3. Murrieta-Guevara, F.; Rodriguez, A. T.
   J. Chem. Eng. Data <u>1984</u>, *29*, 456.
4. Wu, Z.; Zeck, S.; Knapp, H.
   Ber. Bunsenges. Phys. Chem., <u>1985</u>, *89*, 1009.

COMPONENTS:	ORIGINAL MEASUREMENTS:
1. Methane; CH₄; [74-82-8] 2. 1-methyl-2-pyrrolidone; C₅H₉NO: [872-50-4]	Shakhova, S.F.; Zubchenko, Yu.P. *Khim. Prom.* <u>1973</u>, *49*, 595-6

VARIABLES:	PREPARED BY:
Temperature, pressure	C.L. Young

EXPERIMENTAL VALUES:

T/K	P/MPa	Mole fraction of methane in liquid, x_{CH_4}	α^+ vol/vol
298.15	2.087	0.0190	4.5
	2.969	0.0269	6.4
	3.334	0.0293	7.0
	4.316	0.0374	9.0
	5.249	0.0465	11.3
	6.080	0.0515	12.6
	7.255	0.0596	14.7
	7.964	0.0661	16.4
	8.045	0.0664	16.5
	8.339	0.0672	16.7
	8.481	0.0672	16.7
	10.254	0.0783	19.7
343.15	2.158	0.0186	4.4
	3.587	0.0305	7.3
	3.809	0.0317	7.6
	6.870	0.0535	13.1
	6.870	0.0550	13.5
	8.734	0.0668	16.6
	9.372	0.0713	17.8
	9.859	0.0735	18.4
	11.136	0.0820	20.7
	11.571	0.0856	21.7
373.15	2.158	0.0182	4.3
	4.904	0.0386	9.3

AUXILIARY INFORMATION

METHOD/APPARATUS/PROCEDURE:	SOURCE AND PURITY OF MATERIALS:
Rocking autoclave. Mixture stirred by ball in rocking autoclave. Samples of liquid analysed by volumetric method. Details in ref. (1).	1. Purity 97.8 mole per cent. 2. No details given.

ESTIMATED ERROR:

$\delta T/K = \pm 0.1$; $\delta P/MPa = \pm 0.01$;
$\delta x_{CH_4} = \pm 5\%$.
(estimated by compiler)

REFERENCES:

1. Shakhova, S.F.; Zubchenko, Yu.P.; Kaplan, L.K.

   Khim. Prom. <u>1973</u>, *5*, 108.

COMPONENTS:	ORIGINAL MEASUREMENTS:
1. Methane; CH_4; [74-82-8]	Shakhova, S.F.; Zubchenko, Yu.P.
2. 1-methyl-2-pyrrolidone; C_5H_9NO; [872-50-4]	*Khim. Prom.*, <u>1973</u>, *49*, 595-6.

EXPERIMENTAL VALUES:

T/K	P/MPa	Mole fraction of methane in liquid, x_{CH_4}	α^+ vol/vol
373.15	5.350	0.0421	10.2
	7.407	0.0569	14.0
	9.221	0.0668	16.6
	9.393	0.0687	17.1
	10.305	0.0750	18.8
	10.375	0.0750	18.8

+ quoted in original paper, appears to be volume of
  gas at T/K = 273.15 and P = 1 atmosphere absorbed
  by unit volume of liquid at room temperature.

COMPONENTS:	ORIGINAL MEASUREMENTS:
1. Methane; CH_4; [74-82-8] 2. 1-Methyl-2-pyrrolidinone; C_5H_9NO; [872-50-4]	Lenoir, J-Y.; Renault, P.; Renon, H. *J. Chem. Eng. Data* <u>1971</u>, *16*, 340-2

VARIABLES:	PREPARED BY:
	C.L. Young

EXPERIMENTAL VALUES:

T/K	Henry's Constant H_{CH_4}/atm	Mole fraction at 1 atm* x_{CH_4}
298.15	1020	0.000980

* Calculated by compiler assuming a linear function of p_{CH_4} vs x_{CH_4},
  ie. x_{CH_4} (1 atm) = $1/H_{CH_4}$

AUXILIARY INFORMATION

METHOD/APPARATUS/PROCEDURE:	SOURCE AND PURITY OF MATERIALS:
A conventional gas-liquid chromato-graphic unit fitted with a thermal conductivity detector was used. The carrier gas was helium. The value of Henry's law constant was calculated from the retention time. The value applies to very low partial pressures of gas and there may be a substantial difference from that measured at 1 atm. pressure. There is also considerable uncertainty in the value of Henry's constant since surface adsorption was not allowed for although its possible existence was noted.	(1) L'Air Liquide sample, minimum purity 99.9 mole per cent. (2) Touzart and Matignon or Serlabo sample, purity 99 mole per cent.
	ESTIMATED ERROR: $\delta T/K = \pm 0.1$; δH/atm = ± 6% (estimated by compiler).
	REFERENCES:

COMPONENTS:	ORIGINAL MEASUREMENTS:
(1) Methane; CH_4; [74-82-8] (2) 2-Amino-ethanol or monoethanol-amine; C_2H_7NO; [141-43-5] (3) 1-Methyl-2-pyrrolidinone or *N*-methylpyrrolidone; C_5H_9NO; [872-50-4]	Murrieta-Guevara, F.; Rodriguez, A.T. *J. Chem. Eng. Data* <u>1984</u>, *29*, 456 - 60.

VARIABLES:	PREPARED BY:
$T/K = 298.2$ $p_1/MPa = 0.0261 - 0.1972$	H. L. Clever

EXPERIMENTAL VALUES:

Temperature		Pressure		Monoethanol-amine	Methane
$t/^0C$	T/K	p_1/atm	p_1/MPa	wt %	$10^3 x_1$
25.0	298.2	0.269	0.0273	0	0.6
		0.469	0.0475		1.0
		0.687	0.0696		1.3
		0.937	0.0949		1.7
		1.213	0.1229		2.0
		1.578	0.1599		2.4
		1.942	0.1968		2.7
25.0	298.2	0.258	0.0261	5.1	0.7
		0.466	0.0472		1.4
		0.686	0.0695		2.1
		0.937	0.0949		2.6
		1.247	0.1264		3.3
		1.594	0.1611		4.0
		1.938	0.1964		4.6
25.0	298.2	0.274	0.0278	14.3	0.6
		0.486	0.0492		1.3
		0.702	0.0711		1.7
		0.948	0.0961		2.2
		1.158	0.1173		2.4
		1.468	0.1487		3.0
		1.694	0.1716		3.3
		1.946	0.1972		3.4

METHOD/APPARATUS/PROCEDURE:	SOURCE AND PURITY OF MATERIALS:
The apparatus was a liquid-vapor equilibrium system with circulation of the gas phase. The 170 cm³ equilibrium cell was made of glass with a gas inlet tube ending in a fritted glass disk at the bottom of the cell. The solvent was placed in the cell and weighed. The degassing was carried out *in situ* by freezing-evacuating-thawing cycles. A known amount of solute gas was added to the system at thermal equilibrium and a pump started to circulate the vapor phase. Equilibrium was attained in 30 minutes. The equilibrium pressure in the cell was measured with a calibrated stainless steel pressure transducer used in absolute fashion. Purification of the solvents is described in (ref 1).	(1) Methane. Matheson Co., Inc. Stated to be 99.99 mol %. (2) Monoethanolamine. J.T.Baker Co. 99.56 mol %. Fractionated and dried. (3) N-Methylpyrrolidone. Matheson, Coleman and Bell. 98 mol %. Fractionated and dried, GLC purity better than 99.5 mol %.
	ESTIMATED ERROR: $\delta T/K = \pm 0.1$ $\delta p_1/p_1 = \pm 0.001$ $\delta x_1/x_1 = \pm 0.10$ (compiler)
	REFERENCES: 1. Murrieta-Guevara,F.; Rodriguez,A. *J. Chem. Eng. Data* <u>1984</u>, *29*, 204.

COMPONENTS:	ORIGINAL MEASUREMENTS:
(1) Methane; CH_4; [74-82-8] (2) 2,2'-Iminobis-ethanol or diethan-olamine; $C_4H_{11}NO_2$; [111-42-2] (3) 1-Methyl-2-pyrrolidinone or N-methylpyrrolidone; C_5H_9NO; [872-50-4]	Murrieta-Guevara, F.; Rodriguez, T. A. *J. Chem. Eng. Data* <u>1984</u>, *29*, 456 - 60.

VARIABLES:	PREPARED BY:
T/K = 298.2 p_1/MPa = 0.0273 - 0.1981	H. L. Clever

EXPERIMENTAL VALUES:

Temperature		Pressure		Diethanol-amine	Methane
$t/°C$	T/K	p_1/atm	p_1/MPa	wt %	$10^3 x_1$
25.0	298.2	0.269	0.0273	0	0.6
		0.469	0.0475		1.0
		0.687	0.0696		1.3
		0.937	0.0949		1.7
		1.213	0.1229		2.0
		1.578	0.1599		2.4
		1.942	0.1968		2.7
25.0	298.2	0.270	0.0274	14.3	0.6
		0.468	0.0474		1.0
		0.721	0.0732		1.4
		0.937	0.0949		1.7
		1.238	0.1254		2.2
		1.582	0.1603		2.8
		1.955	0.1981		3.2

AUXILIARY INFORMATION

METHOD/APPARATUS/PROCEDURE:	SOURCE AND PURITY OF MATERIALS:
The apparatus was a liquid-vapor equilibrium system with circulation of the gas phase. The 170 cm³ equilibrium cell was made of Pyrex glass with a gas inlet tube ending in a fritted glass disk at the bottom of the cell. The solvent was placed in the cell and weighed. The degassing was carried out *in situ* by freezing-evacuating-thawing cycles. A known amount of solute gas was added to the system already at thermal equilibrium and the vapor phase circulated by a magnetic pump. Equilibrium pressure was attained within 30 minutes. The equilibrium pressure was measured with a calibrated stainless steel pressure transducer used in the absolute fashion. Purification of the solvents is described in (ref 1).	(1) Methane. Matheson Co., Inc. Stated to be 99.99 mol %. (2) Diethanolamine. J. T. Baker Co. 98.5 mol %. (3) N-Methylpyrrolidone. Matheson, Coleman and Bell. 98 mol %. Both solvents fractionated and dried. GLC purity then better than 99.5 mol %. See (ref 1).

ESTIMATED ERROR:

$\delta T/K$ = ± 0.1
$\delta p_1/p_1$ = ± 0.001
$\delta x_1/x_1$ = ± 0.10   (compiler)

REFERENCES:

1. Murrieta-Guevara,F.; Rodriguez,A.T. *J. Chem. Eng. Data* <u>1984</u>, *29*, 204.

COMPONENTS:	ORIGINAL MEASUREMENTS:
1. Methane; CH_4; [74-82-8] 2. 1-Methyl-2-pyrrolidinone; C_5H_9NO ; [872-50-4]	Wu, Z.; Zeck, S.; Knapp, H. *Ber. Bunsenges. Phys. Chem.*, 1985, *89*, 1009-1013.

VARIABLES:	PREPARED BY:
Composition of solvent	C. L. Young.

EXPERIMENTAL VALUES:

T/K	Henry's constant /MPa	Ostwald coefficient, L	Mole fraction[a] of methane $x \times 10^4$
298.15	39.060	0.0349	0.2594

[a] Calculated by compiler for a partial pressure of 1 atmosphere

AUXILIARY INFORMATION

METHOD/APPARATUS/PROCEDURE:	SOURCE AND PURITY OF MATERIALS:
Precision volumetric apparatus described in detail in ref. (1). Pressure measured with mercury manometer.	1. Purity better than 99 volume per cent. 2. Merck sample, dried with molecular sieve 4 X. Final water content less than 0.01 mass per cent, purity 99.9 mole per cent by GC.

ESTIMATED ERROR:
$\partial T/K = \pm 0.01$; $\partial P/Pa = \pm 50$;
$\partial x = \pm 0.005$

REFERENCES:
1. Zeck, S.; Dissertation, TU Berlin, 1985.

COMPONENTS:	ORIGINAL MEASUREMENTS:
(1) Methane; CH_4; [74-82-8] (2) N-Methylformamide; C_2H_5NO; [123-39-7]	de Ligny, C. L.; Denessen, H.J.M. Alfenaar, M. *Recl. Trav. Chim. Pays-Bas* <u>1971</u>, *90*, 1265-1284.

VARIABLES:	PREPARED BY:
T/K: 298.15 p_1/kPa: 101.325 (1 atm)	H. L. Clever

EXPERIMENTAL VALUES:

T/K	Molality $10^3 m_1$/mol kg^{-1}	Mol Fraction $10^4 x_1$	Bunsen Coefficient α/cm^3(STP)cm^{-3}atm^{-1}	Ostwald Coefficient L/cm^{3}cm^{-3}
298.15	8.32 ± 0.06	4.92	0.186	0.204

The mole fraction, Bunsen and Ostwald coefficients values were calculated by the compiler assuming ideal gas behavior.

AUXILIARY INFORMATION

METHOD/APPARATUS/PROCEDURE:	SOURCE AND PURITY OF MATERIALS:
Details of the method are given in an earlier paper (1). The solvent is saturated with gas in a special two cell vessel in which the gas is pre-saturated with solvent vapor. A one cm^3 sample of the gas-saturated liquid is taken and injected into a Becker gas chromatograph equipped with a stripping vessel mounted in front of a 15% carbowax-on-celite 1 m column. The carrier gas is oxygen free helium. A katharometer detector is used.	(1) Methane. Baker Chemical Co. Ultra pure. (2) N-Methylformamide. Source not given. Purified by the method of Verhoek (2). Water content < 0.02% (Fisher titration), no other impurities detected by GLC.

ESTIMATED ERROR:

$$\delta m_1/m_1 = \quad 0.01$$

REFERENCES:
1. de Ligny, C.L.; van der Veen, N.G. *Recl. Trav. Chim. Pays-Bas* <u>1971</u>, *90*, 984.

2. Verhoek, F. H. *J. Am. Chem. Soc.* <u>1936</u>, *58*, 2577.

COMPONENTS:	ORIGINAL MEASUREMENTS:
(1) Methane; CH_4; [74-82-8] (2) N,N-Dimethylformamide; C_3H_7NO; [68-12-2]	Haidegger, E.; Szebenyi, I.; Szekely, A. *Magy. Kem. Foly.* <u>1958</u>, *64*, 365-71.

VARIABLES:	PREPARED BY:
$T/K = 274.15 - 313.15$ $p_1/kPa = 26.66 - 119.99$	H. L. Clever

EXPERIMENTAL VALUES:

Temperature		Pressure	Absorption Coefficient	Bunsen Coefficient
$t/^0C$	T/K	p_1/mmHg	$/cm^3(STP)cm^{-3}$	$\alpha/cm^3(STP)cm^{-3}atm^{-1}$
1	274.15	200	0.10	0.38
		400	0.21	0.40
		600	0.32	0.405
		760	0.42	0.42
		900	0.49	0.41
5	278.15	200	0.08	0.30
		400	0.19	0.36
		600	0.31	0.39
		760	0.39	0.39
		900	0.46	0.39
20	293.15	200	0.07	0.27
		400	0.16	0.30
		600	0.25	0.32
		760	0.31	0.31
		900	0.38	0.32
40	313.15	200	0.04	0.15
		400	0.09	0.17
		600	0.16	0.20
		760	0.21	0.21
		900	0.25	0.21

The compiler calculated the Bunsen coefficients.

AUXILIARY INFORMATION

METHOD/APPARATUS/PROCEDURE:	SOURCE AND PURITY OF MATERIALS:
The apparatus consists of an absorption flask in a thermostated bath and an water-jacketted buret.	(1) Methane. No information. (2) N,N-Dimethylformamide. Distilled, dried. Refractive index $n_D^{25} = 1.4265$, density $\rho_4^{25} = 0.9451$ g cm^{-3}. The water content was 0.2 wt percent.
	ESTIMATED ERROR: $\delta\alpha/\alpha = \pm 0.05$ (compiler) At pressures 600 mmHg and above.
	REFERENCES:

COMPONENTS:	ORIGINAL MEASUREMENTS:
(1) Methane; CH_4; [74-82-8] (2) 1,1,2,2,3,3,4,4,4-Nonafluoro-N, N-bis(nonafluorobutyl)-1-butanamine or perfluorotributyl-amine; $(C_4F_9)_3N$; [311-89-7]	Powell, R. J. J. Chem. Eng. Data <u>1972</u>, 17, 302 - 304.

VARIABLES:	PREPARED BY:
T/K: 288.15 - 318.15 p_1/kPa: 101.325 (1 atm)	P. L. Long H. L. Clever

EXPERIMENTAL VALUES:

T/K	Mol Fraction $10^4 x_1$	Bunsen Coefficient α/cm^3 (STP) cm^{-3} atm^{-1}	Ostwald Coefficient L/cm^3 cm^{-3}	$N = R \dfrac{\Delta \log x_1}{\Delta \log T}$
298.15	68.83	0.435	0.475	-1.88

The Bunsen and Ostwald coefficients were calculated by the compiler.

The author states that the solubility measurements were made over the temperature interval of about 288.15 to 318.15 K, but only the solubility value at 298.15 K was given in the paper. The slope, $N=R(\Delta \log x_1/\Delta \log T)$, was given.

Smoothed Data: For use between 288.15 and 303.15 K

The smoothed data were calculated by the compiler from the slope, N, in the form

$$\log x_1 = \log (68.83 \times 10^{-4}) - (1.88/R) \log (T/298.15)$$

with R = 1.9872 cal K^{-1} mol^{-1}.

T/K	Mol Fraction $10^3 x_1$
288.15	7.109
293.15	6.994
298.15	6.883
303.15	6.776
308.15	6.671
313.15	6.571
318.15	6.473

AUXILIARY INFORMATION

METHOD/APPARATUS/PROCEDURE:	SOURCE AND PURITY OF MATERIALS:
The apparatus is the Dymond and Hildebrand (1) apparatus which uses an all glass pumping system to spray slugs of degassed solvent into the gas. The amount of gas dissolved is calculated from the initial and final pressures. The solvent is degassed by freezing, pumping, and followed by boiling under reduced pressure.	(1) Methane. Source not given. Stated to be manufacturer's research grade, dried over $CaCl_2$ before use. (2) 1,1,2,2,3,3,4,4,4-Nonafluoro-N, N-bis(nonafluorobutyl)-1-butanamine. Minnesota Mining & Manufacturing Co. Distilled, used portion boiling between 447.85-448.64 K which gave a single GLC peak. $\rho_{298.15}$ = 1.880 g cm^{-3}.

ESTIMATED ERROR:
$$\delta x_1/x_1 = \pm 0.002$$
δN/cal K^{-1} mol^{-1} = ± 0.1

REFERENCES:

1. Dymond, J. H.; Hildebrand, J. H. Ind. Eng. Chem. Fundam. <u>1967</u>, 6, 130.

COMPONENTS:	ORIGINAL MEASUREMENTS:
1. Methane; CH_4; [74-82-8] 2. Nitrobenzene; $C_6H_5NO_2$; [98-95-3]	Lenoir, J-Y.; Renault, P.; Renon, H. *J. Chem. Eng. Data*, <u>1971</u>, *16*, 340-2.

VARIABLES:	PREPARED BY:
	C. L. Young

EXPERIMENTAL VALUES:

T/K	Henry's constant H_{CH_4}/atm	Mole fraction at 1 atm* x_{CH_4}
298.2	940	0.00106

* Calculated by compiler assuming a linear function of P_{CH_4} vs x_{CH_4}, i.e., x_{CH_4}(1 atm) = $1/H_{CH_4}$.

AUXILIARY INFORMATION

METHOD/APPARATUS/PROCEDURE:	SOURCE AND PURITY OF MATERIALS:
A conventional gas-liquid chromatographic unit fitted with a thermal conductivity detector was used. The carrier gas was helium. The value of Henry's law constant was calculated from the retention time. The value applies to very low partial pressures of gas and there may be a substantial difference from that measured at 1 atm. pressure. There is also considerable uncertainty in the value of Henry's constant since surface adsorption was not allowed for although its possible existence was noted.	(1) L'Air Liquide sample, minimum purity 99.9 mole per cent. (2) Touzart and Matignon or Serlabo sample, purity 99 mole per cent.
	ESTIMATED ERROR: $\delta T/K = \pm0.1$; δH/atm $= \pm6\%$ (estimated by compiler).
	REFERENCES:

COMPONENTS:	ORIGINAL MEASUREMENTS:
(1) Methane; CH_4; [74-82-8] (2) Ethylamine, nitrate or ethyl ammonium nitrate; $C_2H_7N.HNO_3$; [22113-86-6]	Evans, D. F.; Chen, S.-H.; Schriver, G. W.; Arnett, E. M. J. Am. Chem. Soc. <u>1981</u>, 103, 481-2.
VARIABLES: $T/K = 288.15 - 313.15$	PREPARED BY: H. L. Clever

EXPERIMENTAL VALUES:

T/K^a	H/atm	H/MPa^a	$10^4 x_1{}^a$
298.15	4750 ± 110	480 ± 10	2.11

^{a} Calculated by compiler.

The authors state that measurements were made at 15, 25, and 40 ^{0}C, but only the 25 ^{0}C value is given in the paper.

Henry's constant is defined as H/atm $= (p_1$/atm$)/x_1$.

The authors give the folowing values for the thermodynamic changes for transfer of one mole of methane from the gas at 0.101325 MPa to the infinitely dilute solution:

$$\Delta G_1^0 \text{/kcal mol}^{-1} = 5.024 \pm 0.073,$$

$$\Delta H_1^0 \text{/kcal mol}^{-1} = -0.057 \pm 0.058, \text{ and}$$

$$\Delta S_1^0 \text{/cal K}^{-1}\text{mol}^{-1} = -17.04 \pm 0.15.$$

The Henry's constant given in the paper is apparently the experimental value at 298.15 K. Values of Henry's constant calculated by the compiler from the thermodynamic information above at 288.15, 298.15, and 313.15 K are 4794, 4815, and 4833 atm, respectively. The Henry constant increases 0.8 percent and the mole fraction solubility at 0.101325 MPa decreases 0.8 percent as the temperature changes from 288.15 to 313.15 K.

AUXILIARY INFORMATION

METHOD/APPARATUS/PROCEDURE:	SOURCE AND PURITY OF MATERIALS:
The solubility measurement was carried out in an apparatus described earlier for vapor pressure measurement (ref 1). The apparatus consists of an equilibration vessel, a fused quartz pressure gage, and a vacuum system. The solvent is degassed by a freeze thaw cycle in the equilibration vessel. Known amounts of gas were added to the system and the system was stirred for at least an hour after constant pressure was achieved. The vapor pressure of the solvent was unmeasurably low in the apparatus used.	(1) Methane. Source not given. Stated to be > 99.9 percent. (2) Ethyl ammonium nitrate. No information. In (ref 2) it is stated the substance is prepared from ethylamine and nitric acid, concentrated and dried on a rotary evaporator, and had a melting point about 14 ^{0}C.

ESTIMATED ERROR:

$$\delta T/K = \pm 0.001$$
$$\delta p_1\text{/mmHg} = \pm 0.1$$
$$\delta H/H = \pm 0.03$$

REFERENCES:
1. Arnett, E. M.; Chawla, B. J. Am. Chem. Soc. <u>1979</u>, 101, 7141.
2. Evans,D.F.; Yamauchi,A.;Roman,R.; Casassa,E.Z. J. Coll. Interface Sci. <u>1982</u>, 88, 89.

COMPONENTS:	EVALUATOR:
(1) Methane; CH_4; [74-82-8]	H. Lawrence Clever
	Chemistry Department
(2) Octamethylcyclotetrasiloxane;	Emory University
$C_8H_{24}O_4Si_4$; [556-67-2]	Atlanta, GA 30322 USA
	1985, April

CRITICAL EVALUATION:

Chappelow and Prausnitz (ref 1) report Henry's constants for the system at 25 degree intervals from 300 to 425 K. The Henry's constants were converted to mole fraction solubilities at a methane pressure of 0.101325 MPa assuming a linear relationship between pressure and mole fraction. The solubilities go through a minimum between 375 and 400 K. Wilcock *et al.* (ref 2) report solubilities at three temperatures between 292.15 and 313.15 K. Both laboratories used a volumetric method. The smoothed mole fraction solubilities from the two papers agree within 0.5 percent over the 292 to 325 K range.

The data from both papers are classed as tentative. The data have been treated in two ways. First, the mole fraction solubilities between 292.15 and 350 K were fit by a linear regression to the two constant equation

$$\ln x_1 = -6.04961 + 4.12303/(T/100\ K)$$

with a standard error about the regression line of 5.37×10^{-5}. Second, all of the data from both papers were treated by a linear regression to obtain a four constant equation that gives a minimum solubility near 393 K.

$$\ln x_1 = 52.96522 - 77.0211/(T/100\ K) - 50.6322 \ln (T/100\ K) + 7.8863(T/100)$$

with a standard error about the regression line of 5.73×10^{-5}.

The two equations give mole fraction solubilities over the 292.15 to 348.15 K interval that agree within 0.2 percent. The thermodynamic changes for the transfer of one mole of methane from the gas at a methane pressure of 0.101325 MPa to the infinitely dilute solution are

$$\Delta H_1^0/kJ\ mol^{-1} = -3.43 \quad and \quad \Delta S_1^0/J\ K^{-1}\ mol^{-1} = -50.3$$

from the two constant equation. The four constant equation gives values of ΔH_1^0, ΔS_1^0, and ΔC_{p1}^0 which are functions of temperature of

T/K	$\Delta H_1^0/kJ\ mol^{-1}$	$\Delta S_1^0/J\ K^{-1}\ mol^{-1}$	$\Delta C_{p1}^0/J\ K^{-1}\ mol^{-1}$
293.15	-3.02	-48.9	-36.5
308.15	-3.42	-50.3	-16.9
323.15	-3.53	-50.6	+ 2.8
373.15	-1.75	-45.6	68.4
398.15	+0.37	-40.1	101.2
423.15	3.31	-33.0	133.9

It is unusual to find solubility data that requires more than a three constant equation to represent the temperature dependence, but this is such a case. Smoothed values of the mole fraction solubility from both equations are in Table 1.

Table 1. The solubility of methane in octamethylcyclotetrasiloxane.
         Tentative values of the mole fraction solubility as a function
         of temperature at a methane partial pressure of 0.101325 MPa.

T/K	$10^3 x_1$ 2 const eqn	$10^3 x_1$ 4 const eqn	T/K	$10^3 x_1$ 4 const eqn
293.15	9.63	9.61	353.15	7.61
298.15	9.40	9.40	363.15	7.43
303.15	9.19	9.20	373.15	7.29
308.15	8.99	9.00	383.15	7.21
313.15	8.80	8.81	393.15	7.18
318.15	8.62	8.63	403.15	7.20
323.15	8.45	8.45	413.15	7.27
			423.15	7.40
348.15	7.71	7.72		

REFERENCES:

1. Chappelow, C. C.; Prausnitz, J. M. *Am. Inst. Chem. Eng. J.* **1974**, *20*, 1097 - 1104.
2. Wilcock, R. J.; McHale, J. L.; Battino, R.; Wilhelm, E. *Fluid Phase Equilib.* **1978**, *2*, 225 - 30.

COMPONENTS:	ORIGINAL MEASUREMENTS:
(1) Methane; CH_4; [74-84-8] (2) Octamethylcyclotetrasiloxane; $C_8H_{24}O_4Si_4$; [556-67-2]	Wilcock, R. J.; McHale, J. L.; Battino, B.; Wilhelm, E. *Fluid Phase Equilib.* <u>1978</u>, *2,* 225-230.
VARIABLES: T/K: 292.15 - 313.04 p/kPa: 101.325 (1 atm)	PREPARED BY: H. L. Clever

EXPERIMENTAL VALUES:

T/K	Mol Fraction $10^3 x_1$	Bunsen Coefficient α/cm^3(STP)cm^{-3}atm^{-1}	Ostwald Coefficient L/cm^{3}cm^{-3}
292.15	9.647	0.7032	0.7521
298.01	9.346	0.6764	0.7380
313.04	8.880	0.6310	0.7231

The solubility values were adjusted to a gas partial pressure of 101.325 kPa by Henry's law.

The Bunsen coefficients were calculated by the compiler.

Smoothed Data: For use between 292.15 and 313.04 K.

$$\ln x_1 = -5.8575 + 3.5442/(T/100K)$$

The standard error about the regression line 5.46 x 10^{-5}.

T/K	Mol Fraction $10^3 x_1$
298.15	9.384
308.15	9.029

AUXILIARY INFORMATION

METHOD/APPARATUS/PROCEDURE:	SOURCE AND PURITY OF MATERIALS:
The apparatus is based on the design of Morrison and Billett (1), and the version used is described by Battino, Evans, and Danforth (2). The degassing apparatus and procedure are described by Battino, Banzhof, Bogan, and Wilhelm (3). Degassing. Up to 500 cm^3 of solvent is placed in a flask of such size that the liquid is about 4 cm deep. The liquid is rapidly stirred, and vacuum is applied intermittently through a liquid N$_2$ trap until the permanent gas residual pressure drops to 5 microns. Solubility Determination. The degassed solvent is passed in a thin film down a glass spiral tube containing the solute gas plus the solvent vapor at a total pressure of one atm. The volume of gas absorbed is found by difference between the initial and final volumes in the buret system. The solvent is collected in a tared flask and weighed.	(1) Methane. Matheson Co., Inc. Stated to be 99.97 mole percent minimum purity. (2) Octamethylcyclotetrasiloxane. General Electric Co. Distilled, density at 298.15 K was 0.9500 g cm^{-3}.
	ESTIMATED ERROR: δT/K = 0.03 δP/mmHg = 0.5 $\delta x_1/x_1$ = 0.1
	REFERENCES: 1. Morrison, T. J.; Billett, F. *J. Chem. Soc.* <u>1948</u>, 2033. 2. Battino,R.;Evans,F.D.;Danforth,W.F. *J.Am.Oil Chem.Soc.* <u>1968</u>, *45*, 830. 3. Battino, R.; Banzhof, M.; Bogan, M.; Wilhelm, E. *Anal. Chem.* <u>1971</u>, *43*, 806.

COMPONENTS:	ORIGINAL MEASUREMENTS:
1. Methane; CH₄; [74-82-8] 2. Octamethylcyclotetrasiloxane; $C_8H_{24}O_4Si_4$; [556-67-2]	Chappelow, C.C.; Prausnitz, J.M. *Am. Inst. Chem. Engnrs. J.* <u>1974</u>, *20*, 1097-1104.

VARIABLES:
  Temperature
	PREPARED BY:
	  C.L. Young

EXPERIMENTAL VALUES:

T/K	Henry's Constant[a] /atm	Mole fraction[b] of methane at 1 atm partial pressure, x_{CH_4}
300	107	0.00935
325	119	0.00840
350	131	0.00763
375	138	0.00725
400	138	0.00725
425	135	0.00741

a. Authors stated measurements were made at several pressures and values of solubility used were all within the Henry's Law region.

b. Calculated by compiler assuming linear relationship between mole fraction and pressure.

AUXILIARY INFORMATION

METHOD/APPARATUS/PROCEDURE:	SOURCE AND PURITY OF MATERIALS:
Volumetric apparatus similar to that described by Dymond and Hildebrand (1). Pressure measured with a null detector and precision gauge. Details in ref . (2).	Solvent degassed, no other details given.

ESTIMATED ERROR:

$\delta T/K = \pm 0.1$; $\delta x_{CH_4} = \pm 1\%$

REFERENCES:
1. Dymond, J.; Hildebrand, J.H. *Ind.Chem.Eng.Fundam.*<u>1967</u>,*6*,130.
2. Cukor, P.M.; Prausnitz, J.M. *Ind.Chem.Eng.Fundam.*<u>1971</u>,*10*,638.

COMPONENTS:	ORIGINAL MEASUREMENTS:
1. Methane; CH_4; [74-82-8] 2. Esters of phosphoric acid	Lenoir, J-Y.; Reanult, P.; Renon, H. *J. Chem. Eng. Data*, <u>1971</u>, *16*, 340-2.

VARIABLES:	PREPARED BY:
Temperature	C. L. Young

EXPERIMENTAL VALUES:

T/K	Henry's constant H_{CH_4}/atm	Mole fraction at 1 atm* x_{CH_4}
Phosphoric acid, triethyl ester; $C_6H_{15}O_4P$; [78-40-0]		
325.2	705	0.00142
Phosphoric acid, tripropyl ester; $C_9H_{21}O_4P$; [513-08-6]		
298.2	257	0.00389
323.2	306	0.00327
343.2	325	0.00308
Phosphoric acid, tributyl ester; $C_{12}H_{27}O_4P$; [126-73-8]		
325.2	224	0.00446
Phosphoric acid, tris(2-methylpropyl)ester; $C_{12}H_{27}O_4P$; [126-71-6]		
325.2	190	0.00526

* Calculated by compiler assuming a linear function of P_{CH_4} vs x_{CH_4}, i.e., $x_{CH_4}(1\,atm) = 1/H_{CH_4}$.

AUXILIARY INFORMATION

METHOD/APPARATUS/PROCEDURE:	SOURCE AND PURITY OF MATERIALS:
A conventional gas-liquid chromatographic unit fitted with a thermal conductivity detector was used. The carrier gas was helium. The value of Henry's law constant was calculated from the retention time. The value applies to very low partial pressures of gas and there may be a substantial difference from that measured at 1 atm. pressure. There is also considerable uncertainty in the value of Henry's constant since surface adsorption was not allowed for although its possible existence was noted.	(1) L'Air Liquide sample, minimum purity 99.9 mole per cent. (2) Touzart and Matignon or Serlabo sample, purity 99 mole per cent.
	ESTIMATED ERROR: $\delta T/K = \pm 0.1$; $\delta H/atm = \pm 6\%$ (estimated by compiler).
	REFERENCES:

COMPONENTS:	ORIGINAL MEASUREMENTS:
1. Methane; CH_4; [74-82-8] 2. Phosphoric acid tributyl ester (Tributyl phosphate); $(C_4H_9)_3PO_4$; [126-73-8]	Shakhova, S.F.; Zubchenko, Yu.P. *Khim. Prom.* 1973, *49*, 595-6.

VARIABLES:	PREPARED BY:
Temperature, pressure	C.L. Young

EXPERIMENTAL VALUES:

T/K	$P/10^5$ Pa	Mole fraction of methane in liquid, x_{CH_4}	α^+ vol/vol
298.15	37.49	0.1475	14.9
	47.52	0.1760	18.4
	58.36	0.2194	24.2
	67.38	0.2305	25.8
	74.58	0.2400	27.2
	90.02	0.2854	34.4
	100.72	0.3005	37.0
323.15	38.81	0.1372	13.7
	48.03	0.1600	16.4
	72.14	0.2186	24.1
	93.22	0.2653	31.1
	95.14	0.2678	31.5
	103.15	0.2848	34.3
	105.48	0.2913	35.4
	110.64	0.2901	35.2
343.15	43.37	0.1346	13.4
	53.50	0.1591	16.3
	66.27	0.1915	20.4
	82.38	0.2326	26.1
	96.16	0.2577	29.9
	98.29	0.2603	30.3
	104.97	0.2715	32.1

+ values quoted in original paper, appear to be the volume of gas at T/K = 273.15 and P = 1 atmosphere absorbed by unit volume of liquid at room temperature.

AUXILIARY INFORMATION

METHOD/APPARATUS/PROCEDURE:	SOURCE AND PURITY OF MATERIALS:
Rocking autoclave. Mixture stirred by ball in rocking autoclave. Sample of liquid analysed by volumetric method. Details in source.	1. Purity 97.8 mole per cent. 2. No details given.

ESTIMATED ERROR:

$\delta T/K = \pm 0.1$; $\delta P/10^5$Pa $= \pm 0.1$;

$\delta x_{CH_4} = \pm 5\%$.

(estimated by compiler)

REFERENCES:

1. Shakhova, S.F. Zubchenko, Yu.P. Kaplan, L.K.

    Khim. Prom. 1973, *5*, 102.

COMPONENTS:	ORIGINAL MEASUREMENTS:
1. Methane; CH_4: [74-82-8] 2. Hexamethylphosphoric triamide; $C_8H_{24}N_3OP$; [680-31-9]	Lenoir, J-Y.; Renault, P.; Renon, H. *J. Chem. Eng. Data,* <u>1971</u>, *16*, 340-2.

VARIABLES:	PREPARED BY:
	C. L. Young

EXPERIMENTAL VALUES:

T/K	Henry's constant H_{CH_4}/atm	Mole fraction at 1 atm* x_{CH_4}
298.2	471	0.00212

* Calculated by compiler assuming a linear function of P_{CH_4} vs x_{CH_4}, i.e., x_{CH_4} (1 atm) = $1/H_{CH_4}$.

AUXILIARY INFORMATION

METHOD/APPARATUS/PROCEDURE:	SOURCE AND PURITY OF MATERIALS:
A conventional gas-liquid chromatographic unit fitted with a thermal conductivity detector was used. The carrier gas was helium. The value of Henry's law constant was calculated from the retention time. The value applies to very low partial pressures of gas and there may be a substantial difference from that measured at 1 atm. pressure. There is also considerable uncertainty in the value of Henry's constant since surface adsorption was not allowed for although its possible existence was noted.	(1) L'Air Liquide sample, minimum purity 99.9 mole per cent. (2) Touzart and Matignon or Serlabo sample, purity 99 mole per cent.
	ESTIMATED ERROR: $\delta T/K = \pm 0.1$; δH/atm = $\pm 6\%$ (estimated by compiler).
	REFERENCES:

COMPONENTS:	ORIGINAL MEASUREMENTS:
1. Methane; CH_4; [74-82-8]	Miller, K. W.; Hammond, L.; Porter, E. G.
2. Phospholipids	*Chem. Phys. Lipids* 1977, *20*, 229-241.

VARIABLES:	PREPARED BY:
	C. L. Young

EXPERIMENTAL VALUES:

$$T/K = 298.4 \quad t/°C = 25.2$$

96 mole per cent egg phosphatidylcholine
+ 4 mole per cent egg phosphatidic acid sonicated vesticles

Bunsen coefficient 0.20

68.2 mole per cent egg phosphatidylcholine
+ 2.8 mole per cent egg phosphatidic acid sonicated vesticles
+ 29 mole per cent cholesterol

Bunsen coefficient 0.18

AUXILIARY INFORMATION

METHOD/APPARATUS/PROCEDURE:	SOURCE AND PURITY OF MATERIALS:
Samples of lipids were prepared as a translucent aqueous suspension containing up to 32 mg/ml of phospholipids. Samples saturated with gas at ambient pressure and then analysed by stripping out gas. Gas so obtained was analysed by gas chromatography using helium as a carrier gas and a Poropak Q column. Details in source. Bunsen coefficient calculated from experimental data on lipid solution and of pure water.	1. Matheson Gas Products sample, purity 99 mole per cent. 2. Grade 1 samples from Lipid Products, Nutford, England.
	ESTIMATED ERROR: $\delta T/K = \pm 0.05$; $\delta p/kPa = \pm 0.5\%$; $\delta \alpha/\alpha = \pm 8\%$ (estimated by compiler).
	REFERENCES:

COMPONENTS:	ORIGINAL MEASUREMENTS:
1. Methane; CH_4; [74-82-8] 2. Rabbit brain and blood and saline solution.	Ohta, Y.; Ar, A.; Farhi, L.E. *J. Appl. Physiology*, 1979, *46*, 1169-1170.

VARIABLES:	PREPARED BY:
	C.L. Young

EXPERIMENTAL VALUES:

T/K	Bunsen coefficient, α	No. of animals
	"Saline"	
310.15	0.0256 ± 0.0003	5
	Blood	
310.15	0.0334 ± 0.0002	5
	Brain	
310.15	0.0361 ± 0.0004	5

The partial pressure of methane was not given but was considerable less than one atmosphere.

AUXILIARY INFORMATION

METHOD/APPARATUS/PROCEDURE:	SOURCE AND PURITY OF MATERIALS:
Saline, rabbit blood and brain were saturated by passing humidified gas through three vessels in series. Brain was prepared by manually squeezing out blood from the brain of a freshly killed rabbit. Volume of brain determined by saline displacement. The tissue was homogenised and diluted with an equal volume of 5% low foam detergent. Blood sample was heparinized. Samples of each of the three solutions were analysed by GC using helium carrier gas, a molecular sieve column and a thermal conductivity detector.	See under method.
	ESTIMATED ERROR: $\delta T/K = \pm 0.1$
	REFERENCES:

COMPONENTS:	ORIGINAL MEASUREMENTS:
1. Methane; CH$_4$; [74-82-8] 2. Dog blood and skeletal muscle	Meyer, M.; Tebbe, U.; Püper, J. *Pflügers. Arch.* <u>1980</u>, *384*, 131-4.

VARIABLES:	PREPARED BY:
	C. L. Young

EXPERIMENTAL VALUES:

T/K = 310   P/kPa = 101.3

Solvent	No. of determinations	No. of dogs	Bunsen Coefficient	S[a]
Water [b]	12	–	0.0260	11.47 ± 0.09
Saline [c]	12	–	0.0232	10.20 ± 0.10
Blood	50	10	0.0260	11.44 ± 0.30
Plasma	30	10	0.0227	9.99 ± 0.21
Red cells	–	10	0.0300	13.21 ± 0.47
Muscle	39	13	0.0271	11.95 ± 0.40

[a] Solubility in units of μmol dm^{-3} kPa^{-1}.

[b] Data also reported in ref. (1).

[c] Normal saline containing 0.154 mol/dm^3 (water).

(cont.)

AUXILIARY INFORMATION

METHOD/APPARATUS/PROCEDURE:	SOURCE AND PURITY OF MATERIALS:
Method involved equilibration of solvent with humidified gas at stated temperature and pressure and subsequent estimation of the amount of gas dissolved in a 2.5 cm^3 sample. The gas dissolved was estimated using an equilibration technique for partial extraction of gas. Quantitative analysis of extracted gas was performed by GC using helium as carrier gas. Details in ref. (1).	1. No details given.
	ESTIMATED ERROR: δT/K = ±0.5 (estimated by compiler).
	REFERENCES: 1. Meyer, M. *Pflügers. Arch.* <u>1978</u>, *375*, 161.

COMPONENTS:

1. Methane; CH_4; [74-82-8]

2. Dog blood and skeletal muscle

ORIGINAL MEASUREMENTS:

Meyer, M.; Tebbe, U.;
Püper, J.
Pflügers. Arch.
<u>1980</u>, *384*, 131-4.

EXPERIMENTAL VALUES:

Heparinized blood samples were from mongrel dogs (fasting for 16 hrs).

Plasma obtained by centrifugation of whole blood.   No sign of
hemolysis was observed.

Solubility in red cells was calculated from the values for whole
blood and plasma of the same animal by volume-weighted subtraction.

Muscle was gastrocnemius muscle excised from dogs, which had been
anesthetized for about 6-8 hr and killed by bleeding.   Blood allowed
to drain from major vessel.   Muscle samples homogenized.

Composition of dog blood (mean values ± SD)

Hematocrit %	45 ± 4.5
Hemoglobin (g/100 ml blood)	16.9 ± 1.6
Plasma protein (g/100 ml plasma)	6.2 ± 0.5
Total lipids (mg/100 ml plasma)	519 ± 118
Triglycerides (mg/100 ml plasma)	108 ± 82
Cholesterol (mg/100 ml plasma)	202 ± 68

COMPONENTS:	ORIGINAL MEASUREMENTS:
(1) Methane; CH_4; [74-82-8] (2) Olive oil	Campos-Carles, A.; Kawashiro, T.; Piiper, J. *Pflugers Arch.* <u>1975</u>, *359*, 209-18.

VARIABLES:	PREPARED BY:
$T/K = 310.15$	H. L. Clever

EXPERIMENTAL VALUES:

Temperature		Solubility Coefficient	Mol Fraction
$t/^0C$	T/K	$/\mu mol\ dm^{-3}mmHg^{-1}$	$10^3 x_1$
37	310.15	16.0 ± 0.1	11.8

The compiler calculated the mole fraction solubility at 101.325 kPa partial pressure methane (760 mmHg).

An olive oil molecular weight of 884 and a density of 0.8979 were used. See Battino, R.; Evans, F. D.; Danforth, W. F. *J. Am. Oil Chem. Soc.* <u>1968</u>, *45*, 830.

AUXILIARY INFORMATION

METHOD/APPARATUS/PROCEDURE:	SOURCE AND PURITY OF MATERIALS:
Used a tonometer and extraction apparatus as described by Farhi (ref 1,2), and gas chromatography. The solubility value is the mean of 8 determinations ± standard error.	(1) Methane. Source not given. Sample stated to be 99.9 or better purity. (2) Olive oil.
	ESTIMATED ERROR:

	REFERENCES:
	1. Farhi, L. E.
	    J. Appl. Physiol. <u>1965</u>, *20*, 1098.
	
	2. Farhi, L. E.; Edwards, A. W. T.; Homma, T.
	    J. Appl. Physiol. <u>1963</u>, *18*, 97.

COMPONENTS:	ORIGINAL MEASUREMENTS:
(1) Methane; CH_4; [74-82-8] (2) Rat abdominal muscle	Campos Carles, A.; Kawashiro, T.; Piiper, J. *Pflugers Arch.* <u>1975</u>, *359*, 209-18.

VARIABLES: $T/K = 310.15$	PREPARED BY: H. L. Clever

EXPERIMENTAL VALUES:

Temperature		Solubility Coefficient	Bunsen Coefficient
$t/^0C$	T/K	$/\mu mol\ dm^{-3} mmHg^{-1}$	$\alpha/cm^3 (STP)\ cm^{-3}\ atm^{-1}$
37	310.15	2.25 ± 0.11 2.42 (corrected)	0.0412

Solubility coefficient the mean of 10 measurements ± standard error.

In another paper (ref 1) the authors report diffusion coefficients of CH_4 in rat skeletal muscle at 37 ^0C.

Krogh's diffusion constant

$10^9 K/mmol\ min^{-1}\ cm^{-1}\ mmHg^{-1} = 1.27 ± 0.03$

Diffusion coefficient

$10^6 D/cm^2\ s^{-1} = 8.72$

AUXILIARY INFORMATION

METHOD/APPARATUS/PROCEDURE:	SOURCE AND PURITY OF MATERIALS:
The methane, saturated with water vapor, was led through an equilibration chamber for 2 h at a rate of 8 ml m^{-1}. The muscle sample rested on a screen in the chamber sothat it was exposed to the gas on all sides. After equilibration the muscle sample was transferred to an extraction chamber filled with room air for the same length of time as the gas equilibration. The gas in the chamber was forced into a gas chromatograph by mercury entering the chamber. Correction factors were applied for unextracted gas and gas lost during transfer between chambers.	(1) Methane. Source not given. Stated to be better than 99.9 per cent pure. (2) Rat abdominal muscle. A flat muscle sheet of about 1.6 g, 1.4 mm thickness, and 10 cm^2 area was excised from rats weighing 250 to 430 g.
	ESTIMATED ERROR:
	REFERENCES: 1. Kawshiro, T.; Campos Carles, A. Perry, S. F.; Piiper, J. *Pflugers Arch.* <u>1975</u>, *359*, 219.

COMPONENTS:	ORIGINAL MEASUREMENTS:
1. Methane; CH_4; [74-82-8] 2. Ammonia; NH_3; [7664-41-7]	Kaminishi, G. *Kogyo Kogaku Zaashi*, <u>1965</u>, *68*, 419-23.

VARIABLES:	PREPARED BY:
Temperature, pressure	C.L. Young

EXPERIMENTAL VALUES:

T/K	$P/10^5 Pa$	Mole fraction of methane in liquid, x_{CH_4}
273.15	50.8	0.0128
	101.8	0.0238
	150.9	0.0315
	199.9	0.0366
298.15	50.8	0.0152
	101.8	0.0322
	150.9	0.0468
	199.9	0.0583
323.15	50.8	0.0145
	101.8	0.0386
	150.9	0.0644
	199.9	0.0901

AUXILIARY INFORMATION

METHOD/APPARATUS/PROCEDURE:	SOURCE AND PURITY OF MATERIALS:
Static equilibrium cell with agitator. Pressure measured with Bourdon gauge. Liquid ammonia placed in cell and then methane pressurized into cell. After equilibrium established liquid sample removed and analysed by volumetric and gravimetric techniques. Details in source.	1. Takachiho Co. sample, purity 99.9 mole per cent. 2. Distilled four times, no other details given.
	ESTIMATED ERROR: $\delta T/K = \pm 0.1$; $\delta P/10^5 Pa = \pm 0.1$; $\delta x_{CH_4} = \pm 1\%$. (estimated by compiler).
	REFERENCES:

COMPONENTS:	EVALUATOR:
1. Methane; CH_4; [74-82-8]	Colin L. Young,
	School of Chemistry,
2. Carbon dioxide; CO_2; [124-38-9]	University of Melbourne,
	Parkville, Victoria 3052,
	Australia.
	March 1982

EVALUATION:

This system has been investigated by a number of workers but there is still a need for a definitive study. Kidnay and coworkers (1), (2) have investigated this system in the temperature range 230 K to 270 K and these data are classified as tentative. The data of Davalos *et al.* (1) at 270 K have a small error in the measurements above 50 atmospheres for the vapor phase composition and therefore the more recent measurements (2) are preferred in this range of pressure.

Kobayashi and coworkers (3) investigated the dew points in this system but since no liquid phase compositions were measured, this work is not considered further here. Neumann and Walch (4) and Sterner (5) presented their data in graphical form and they are not considered further here.

There appears to be fairly good agreement between the limited data of Kaminishi and coworkers (6), (7) and those of Kidnay and coworkers (1), (2) although the former workers' data scatter more. A very detailed comparison is not possible because the isotherm temperatures are different. The data of Donnelly and Katz (8) cover a wider temperature range than those of Kidnay and coworkers and their data show more scatter. However there is fair agreement between the two sets of data although as above a very detailed comparison is not possible because the temperature isotherms are different.

The data of Donnelly and Katz (8) and Kaminishi and coworkers (6), (7) are classified as tentative.

References

1. Davalos, J.; Anderson, W. R.; Phelps, R. E.; Kidnay, A. J.
   J. Chem. Engng. Data, <u>1976</u>, *21*, 81.

2. Somait, F. A.; Kidnay, A. J. *J. Chem. Engng. Data*, <u>1978</u>, *23*, 301.

3. Hwang, S.-C.; Lin, H.-M.; Chappelear, P. S.; Kobayashi, R.
   J. Chem. Engng. Data, <u>1976</u>, *21*, 403.

4. Neumann, A.; Walch, W. *Chem. Ing.-Tech.*, <u>1968</u>, *40*, 241.

5. Sterner, C. J. *Adv. Cryogen. Eng.*, <u>1961</u>, *6*, 467.

6. Kaminishi, G.; Arai, Y.; Saito, S.; Maeda, A.
   J. Chem. Eng. Japan, <u>1968</u>, *1*, 109.

7. Arai, Y.; Kaminishi, G.; Saito, S. *J. Chem. Eng. Japan*, <u>1971</u>, *4*, 113.

8. Donnelly, H. G.; Katz, D. L. *Ind. Eng. Chem.*, <u>1954</u>, *46*, 511.

COMPONENTS:	ORIGINAL MEASUREMENTS:
1. Methane; CH_4; [74-82-8] 2. Carbon dioxide; CO_2; [124-38-9]	Donnelly, H. G.; Katz, D. L. *Ind. Eng. Chem.* 1954, *46*, 511-7.

VARIABLES:	PREPARED BY:
Temperature, pressure	C. L. Young

EXPERIMENTAL VALUES:

T/K	P/MPa	Mole fraction of methane in liquid, x_{CH_4}	in gas, y_{CH_4}	T/K	P/MPa	Mole fraction of methane in liquid, x_{CH_4}	in gas, y_{CH_4}
271.5	5.05	0.0685	0.253	241.5	6.27	0.286	0.676
	5.59	0.0865	0.30		6.67	0.273	0.679
	6.00	0.103	0.329		6.84	0.322	0.686
	6.30	0.1225	-		7.57	0.426	0.680
	6.81	0.16	0.367		7.78	-	0.676
	6.84	0.157	0.369		7.90	0.501	0.672
	7.25	0.165	0.387	223.7	1.48	0.0435	0.509
	7.64	0.191	0.39		3.43	0.1465	0.751
259.8	3.19	0.0315	0.1885		4.01	0.172	0.777
	3.47	0.036	0.235		4.04	0.2043	0.772
	3.69	0.051	0.266		5.39	0.312	0.796
	4.03	0.053	0.306		5.86	0.408	-
	5.05	0.1095	0.425		6.01	0.420	0.797
	6.03	0.1665	0.484		6.20	0.468	0.805
	6.16	-	0.52		6.37	0.483	0.783
	6.81	0.224	0.505		6.54	0.521	0.790
	6.85	0.223	0.509	219.3	1.11	-	0.477
	7.08	0.230	0.495		2.25	-	0.717
241.5	2.39	0.0413	0.404		3.56	-	0.789
	3.10	0.0895	0.521		4.54	0.261	0.813
	4.07	0.134	0.605		5.33	0.347	0.818
	4.70	0.166	0.629		5.60	-	0.822
	5.26	0.191	0.652		6.43	0.663	0.833
	5.59	-	0.658			(cont.)	

AUXILIARY INFORMATION

METHOD/APPARATUS/PROCEDURE:	SOURCE AND PURITY OF MATERIALS:
Recirculating vapor flow apparatus. Composition of co-existing phases determined by analysis. Samples expanded to atmospheric pressure and carbon dioxide dissolved in sodium hydroxide solution. Details in source and ref. (1).	1. Phillips Petroleum sample. 2. No details given.

ESTIMATED ERROR:
$\delta T/K = \pm 0.1$; $\delta P/MPa = \pm 0.015$; $\delta x_{CH_4} = \pm 0.01$ (estimated by compiler).

REFERENCES:
1. Aroyan, H. J.; Katz, D. L.
   Ind. Eng. Chem. 1951, *48*, 185.

COMPONENTS: ORIGINAL MEASUREMENTS:

1. Methane; CH_4; [74-82-8]   Donnelly, H. G.; Katz, D. L.

2. Carbon dioxide; CO_2; [124-38-9]   *Ind. Eng. Chem.* <u>1954</u>, *46*, 511-7.

EXPERIMENTAL VALUES:

T/K	P/MPa	Mole fraction of methane in liquid, x_{CH_4}	in gas, y_{CH_4}	T/K	P/MPa	Mole fraction of methane in liquid, x_{CH_4}	in gas, y_{CH_4}
209.3	4.78	–	0.877	209.3	5.35	0.648	0.882
	4.96	–	0.881		5.55	0.792	0.900
	5.12	0.535	0.879	199.8	4.49	0.771	0.926
	5.21	0.607	0.879		4.98	0.916	0.946

COMPONENTS:	ORIGINAL MEASUREMENTS:
1. Methane; CH_4; [74-82-8] 2. Carbon dioxide; CO_2; [124-38-9]	Kaminishi, G.; Arai, Y.; Saito, S.; Maeda, A. *J. Chem. Eng. Japan* 1968, *1*, 109-116.

VARIABLES:	PREPARED BY:
Temperature, pressure	C. L. Young

EXPERIMENTAL VALUES:

T/K	P/MPa	Mole fraction of methane	
		in liquid, x_{CH_4}	in gas, y_{CH_4}
233.15	3.70	–	0.657
	5.27	0.251	0.717
	6.20	0.360	0.727
	6.81	0.450	–
	7.19	0.519	0.703
253.15	3.70	–	0.396
	5.20	0.150	0.515
	6.20	0.213	0.556
	7.19	0.294	0.566
	7.80	0.357	0.563
	8.11	0.400	0.540
273.15	5.27	0.070	0.254
	6.20	–	0.322
	7.70	0.204	0.370
	8.19	0.246	0.367
283.15	6.20	0.069	0.188
	7.19	0.115	0.240
	8.19	0.177	0.250

AUXILIARY INFORMATION

METHOD/APPARATUS/PROCEDURE:	SOURCE AND PURITY OF MATERIALS:
Static cell fitted with magnetic sturrer. Temperature measured with liquid in glass thermometer and pressure measured with Bourdon gauge. After equilibrium established vapor and liquid samples analysed by a volumetric technique. Carbon dioxide was absorbed in potassium hydroxide soln.	1. Takachiho Chemical Industry Co. purity better than 99.9 mole per cent. 2. No details given.
	ESTIMATED ERROR: $\delta T/K = \pm 0.1$; $\delta P/MPa = \pm 0.01$; δx_{CH_4}, $\delta y_{CH_4} = \pm 1\%$.
	REFERENCES:

COMPONENTS:	ORIGINAL MEASUREMENTS:
1. Methane; CH_4; [74-82-8] 2. Carbon dioxide; CO_2; [124-38-9]	Arai, Y.; Kaminishi, G.; Saito, S. *J. Chem. Eng. Japan* 1971, 4, 113-122.
VARIABLES: Temperature, pressure	PREPARED BY: C. L. Young

EXPERIMENTAL VALUES:

T/K	P/MPa	Mole fraction of methane in liquid, x_{CH_4}	in vapor, y_{CH_4}
253.15	2.63	–	0.204
	3.14	–	0.302
	4.38	–	0.444
	5.33	–	0.499
	6.23	0.204	–
	6.87	–	0.551
	7.42	0.302	–
	7.59	–	0.551*
	8.29	0.444	0.499*
273.15	4.20	–	0.129
	5.05	–	0.220
	5.71	–	0.276
	6.51	–	0.321
	6.63	0.129	–
	6.98	–	0.340
	7.43	–	0.349
	8.13	0.220	–
	8.28	–	0.349*
	8.38	–	0.340*
	8.52	0.276	0.321*

                                             (cont.)

* retrograde condensation point.

AUXILIARY INFORMATION

METHOD/APPARATUS/PROCEDURE:	SOURCE AND PURITY OF MATERIALS:
Bubble point-dew point apparatus consisting of glass capillary cell fitted with magnetic stirrer. Pressure measured with a dead weight gauge. Temperature measured with a mercury in glass thermometer. Mixtures of known composition charged into cell. Bubble point determined from plots of volume against pressure. Dew point determined visually. Details in source.	1. Tanachiko Chemical Industry Co. sample, purity 99.64 mole per cent. 2. Showa Tansan Industry Co. sample, purity 99.9 mole per cent.
	ESTIMATED ERROR: $\delta T/K = \pm 0.01$; $\delta P/MPa = \pm 0.01$; δx_{CH_4}, $\delta y_{CH_4} = \pm 2\%$ (estimated by compiler).
	REFERENCES:

COMPONENTS:	ORIGINAL MEASUREMENTS:
1. Methane; CH_4; [74-82-8]	Arai, Y.; Kaminishi, G.; Saito, S.
2. Carbon dioxide; CO_2; [124-38-9]	*J. Chem. Eng. Japan* 1971, *4*, 113-122.

EXPERIMENTAL VALUES:

T/K	P/MPa	Mole fraction of methane in liquid, x_{CH_4}	in vapor, y_{CH_4}
288.15	5.45	–	0.043
	5.95	–	0.091
	6.25	0.043	–
	6.26	–	0.116
	6.71	–	0.146
	7.06	–	0.167
	7.30	0.091	–
	7.40	–	0.182
	7.75	0.116	–
	8.11	0.146	–
	8.12	–	0.182*
	8.15	0.167**	–

*   retrograde condensation point.

** critical opalescence was clearly observed.

COMPONENTS:	ORIGINAL MEASUREMENTS:
1. Methane; CH_4; [74-82-8] 2. Carbon dioxide; CO_2; [124-38-9]	Davalos, J.; Anderson, W.R.; Phelps, R.E.; Kidnay, A.J. J. Chem. Engng. Data. 1976, 21, 81-4
VARIABLES: Temperature, pressure	PREPARED BY: C.L. Young

EXPERIMENTAL VALUES:

T/K	$P/10^5 Pa$	Mole fraction of methane in liquid, x_{CH_4}	in vapor, y_{CH_4}
230.00	15.20	0.027	0.399
	20.26	0.050	0.525
	32.42	0.115	0.683
	40.53	0.170	0.728
	48.64	0.235	0.751
	55.73	0.318	0.764
	61.91	0.397	0.752
	62.82	0.394	0.762
	65.86	0.472	0.757
	68.90	0.534	0.751
	69.35	0.526	0.732
	70.00	0.543	0.730
	70.73	0.561	0.725
	71.49	0.584	0.716
250.00	20.26	0.010	0.104
	23.63	0.023	0.223
	25.01	–	0.254
	30.40	0.053	0.361
	40.53	0.105	0.491
	50.66	0.166	0.575
	60.79	0.237	0.605
	70.93	0.326	0.615
	78.02	0.400	0.605
	79.54	0.405	0.564
	80.94	0.446	0.558

AUXILIARY INFORMATION

METHOD/APPARATUS/PROCEDURE:	SOURCE AND PURITY OF MATERIALS:
Recirculating vapor flow apparatus. Temperature measured with platinum resistance thermometer. Pressure measured with Bourdon gauge. Gas and liquid samples analysed by gas chromatography using a thermal conductivity detector. Details in source and ref. (1).	1. Matheson ultra high purity sample, maximum impurity 0.03 mole per cent. 2. Purity better than 99.9 mole per cent.

ESTIMATED ERROR:
$\delta T/K = \pm 0.01$; $\delta P/10^5 Pa = \pm 0.03$ up to 3.5 MPa, ± 0.05 above 3.5 MPa; δx_{CH_4}, $\delta y_{CH_4} = \pm 1.5\%$

REFERENCES:
1. Miller, R.C.; Kidnay, A.J.; Hiza, M.J.

J. Chem. Thermodyn. 1972, 4, 807

COMPONENTS:	ORIGINAL MEASUREMENTS:
1. Methane; CH_4; [74-82-8]	Davalos, J.; Anderson, W.R.; Phelps, R.E.; Kidnay, A.J.
2. Carbon dioxide; CO_2; [124-38-9]	*J. Chem. Engng. Data.* 1976, *21*, 81-84.

EXPERIMENTAL VALUES:

		Mole fraction of methane	
T/K	$P/10^5$Pa	in liquid, x_{CH_4}	in vapor, y_{CH_4}
270.00	35.55	0.014	0.083
	37.01	0.018	0.108
	40.28	0.032	0.162
	42.14	0.040	0.190
	50.63	0.077	0.282
	58.58	0.113	0.353
	70.21	0.166	0.405
	80.63	0.260	0.411
	85.19	0.319	0.375

COMPONENTS:	ORIGINAL MEASUREMENTS:
1. Methane; CH_4; [74-82-8] 2. Carbon dioxide; CO_2; [124-38-9]	Mraw, S. C.; Hwang, S.-C.; Kobayashi, R. *J. Chem. Engng. Data* <u>1978</u>, *23*, 135-139.
VARIABLES:	PREPARED BY: C. L. Young

EXPERIMENTAL VALUES:

T/K	P/psi	P/MPa	Mole fraction [a] in liquid, x_{CH_4}	in vapor, y_{CH_4}
219.26	84.4	0.582	0.0000	–
	396.0	2.730	0.1035	–
	495.1	3.414	0.1507	–
	601.7	4.149	0.2189	–
	698.0	4.813	0.3028	–
	757.0	5.219	0.3735	–
	807.9	5.570	0.4488	–
	852.2	5.876	0.5265	–
	880.7	6.072	0.5825	–
	909.2	6.269	0.6394	–
	925.1	6.378	0.6758	–
	932.8	6.431	0.6939	–
	935.9	6.453	0.6031	
	84.4	0.582	–	0.0000
	125.6	0.866	–	0.3111
	147.9	1.020	–	0.4049
	186.1	1.283	–	0.5141
	225.2	1.553	–	0.5869
	297.9	2.054	–	0.6720
	399.8	2.757	–	0.7380
	515.7	3.556	–	0.7778
	615.7	4.245	–	0.7979
	717.8	4.949	–	0.8089

(cont.)

AUXILIARY INFORMATION

METHOD/APPARATUS/PROCEDURE:	SOURCE AND PURITY OF MATERIALS:
Recirculating vapor flow apparatus. Temperature measured with platinum resistance thermometer. Pressure measured with Bourdon gauge. Gas added to cell at low temperature and vapor recirculated until equilibrium established. Samples analysed using gas chromatography. Details in source and ref. (1).	1. Matheson sample, purity at least 99.99 mole per cent. 2. Coleman Instrument grade, purity at least 99.99 mole per cent.

ESTIMATED ERROR:

$\delta T/K = \pm 0.01$; $\delta P/MPa = \pm 0.02$;

δx_{CH_4}, $\delta y_{CH_4} = \pm 2\%$.

REFERENCES:

1. Chu, T. C.; Chen, R. J. J.; Chappelear, P. S.; Kobayashi, R. *J. Chem. Engng. Data* <u>1976</u>, *21*, 41.

COMPONENTS:

1. Methane; CH_4; [74-82-8]

2. Carbon dioxide; CO_2; [124-38-9]

ORIGINAL MEASUREMENTS:

Mraw, S. C.; Hwang, S.-C.;
Kobayashi, R.

J. Chem. Engng. Data
1978, *23*, 135-139.

EXPERIMENTAL VALUES:

T/K	P/psi	P/MPa	Mole fraction [a] in liquid, x_{CH_4}	in vapor, y_{CH_4}
219.26	805.8	5.556	–	0.8116
	870.9	6.005	–	0.8075
	919.9	6.342	–	0.7954
	940.8	6.487	–	0.7483
210.15	625.0	4.309	0.3175	–
	675.0	4.654	0.4056	–
	725.0	4.999	0.5205	–
	776.4	5.353	0.6522	–
	809.7	5.583	0.7239	–
	839.4	5.787	0.7848	–
	799.0	5.509	–	0.8588
203.15	710.1	4.896	0.7491	–
	720.0	4.964	0.7683	–
	730.0	5.033	0.7879	–
	740.0	5.102	0.8051	–
	750.0	5.171	0.8229	–
	760.3	5.242	0.8407	–
	770.3	5.311	0.8593	–
	750.0	5.171	–	0.8975
	770.2	5.310	–	0.8036
	775.2	5.345	–	0.8784
193.15	625.0	4.309	0.8917	–
	640.0	4.413	0.9136	–
	655.0	4.516	0.9328	–
	670.0	4.619	0.9509	–
	680.0	4.688	0.9622	–
	685.0	4.723	0.9674	–
	625.0	4.309	–	0.9490
	639.9	4.412	–	0.9543
	650.0	4.482	–	0.9582
	670.0	4.619	–	0.9669
183.15	494.2	3.407	0.9453	–
	500.0	3.447	0.9549	–
	504.9	3.481	0.9630	–
	513.0	3.537	0.9762	–
	520.0	3.585	0.9870	–
	524.9	3.619	0.9942	–
	528	3.64	1.0000	–
173.15	362.3	2.498	0.9630	–
	366.4	2.526	0.9730	–
	370.4	2.554	0.9828	–
	374.0	2.579	0.9917	–
	377.8	2.605	1.0000	–
153.15	172.4	1.189	0.9941	–
	173.3	1.195	0.99895	–
	172.4	1.189	–	0.99901
	173.3	1.195	–	0.99984

[a] Mole fraction of carbon dioxide given in source.

COMPONENTS:	ORIGINAL MEASUREMENTS:
1. Methane; CH_4; [74-82-8] 2. Carbon dioxide, CO_2; [124-38-9]	Somait, F.A.; Kidnay, A.J. *J. Chem. Engng. Data.* <u>1978</u>, *23*,301-5.

VARIABLES:	PREPARED BY:
Temperature	C.L. Young

EXPERIMENTAL VALUES:

T/K	P/bar	Mole fraction of methane in liquid, x_{CH_4}	in gas, y_{CH_4}
270.00	31.99	0.0000	0.0000
	38.35	0.0237	0.1269
	43.92	0.0455	0.2082
	48.79	0.0666	0.2629
	52.69	0.0838	0.2962
	60.80	0.1226	0.3519
	66.27	0.1533	0.3774
	74.96	-	0.3983
	77.01	0.2224	0.4006
	80.86	-	0.3969
	82.48	0.2740	0.3895
	82.83	-	0.3875
	84.32	0.3060	0.3726

AUXILIARY INFORMATION

METHOD:/APPARATUS/PROCEDURE:	SOURCE AND PURITY OF MATERIALS:
Recirculating vapor flow apparatus with diaphragm pump. Temperature measured with platinum resistance thermometer and pressure with Bourdon gauges. Cell stirred with two propeller stirrer. Vapor and liquid samples analysed by gas chromotography using a thermal conductivity detector. Details in source.	No details given
	ESTIMATED ERROR:
	$\delta T/K = \pm 0.02$; δP/bar $= \pm 0.015$ up to 100 bar; ± 0.1 above 100 bar; δx_{CH_4}, δy_{CH_4}, $= \pm 0.002$.
	REFERENCES:

COMPONENTS:	ORIGINAL MEASUREMENTS:
1. Methane; CH_4; [74-82-8] 2. Carbon dioxide; CO_2; [124-38-9]	Al-Sahhaf, T. A.; Kidnay, A. J.; Sloan, E. D. *Ind. Eng. Chem. Fundam.* **1983**, *22*, 372-380.

VARIABLES:	
Temperature, pressure	C. L. Young

EXPERIMENTAL VALUES:

T/K	P/MPa	Mole fraction of methane in liquid, x_{CH_4}	Mole fraction of methane in vapor, y_{CH_4}	T/K	P/MPa	Mole fraction of methane in liquid, x_{CH_4}	Mole fraction of methane in vapor, y_{CH_4}
219.26	0.581	0.0000	0.0000	240.00	7.397	0.4525	0.6624
	0.943		0.3529		7.594	0.4914	0.6473
	1.398	0.0339	0.5457		7.660	0.5048	0.6388
	1.900	0.0579	0.6493		7.772	0.5475	0.6199
	2.432	0.0859	0.7103				
	3.120	0.1308	0.7592	270.00	3.203	0.0000	0.0000
	3.861	0.1923	0.7890		3.673	0.0183	0.1048
					3.810	0.0235	0.1273
240.00	1.287	0.0000	0.0000		4.387	0.0457	0.2086
	2.104	0.0328	0.3480		5.183	0.0793	0.2885
	3.059	0.0761	0.5160		5.943	0.1167	0.3451
	4.089	0.1315	0.6050		6.156	0.1261	0.3543
	5.198	0.2074	0.6550		7.083	0.1785	0.3901
	6.120	0.2873	0.6742		7.969	0.2463	0.3980
	6.657	0.3475	0.6771		8.223	0.2737	0.3916
	7.073	0.4016	0.6728		8.415	0.3060	0.3742
	7.265	0.4315	0.6672		8.511		0.3532

AUXILIARY INFORMATION

METHOD/APPARATUS/PROCEDURE:	SOURCE AND PURITY OF MATERIALS:
Recirculating vapor flow apparatus with diaphragm pump. Temperature measured with platinum resistance thermometer and pressure with Bourdon gauges. Cell stirred with double propeller stirrer. Vapor and liquid samples analysed by gas chromatography using a thermal conductivity detector. Details in ref. (1).	1. Linde ultrahigh purity grade. Purity 99.97 mole per cent. 2. Linde "Coleman" grade. Purity 99.991 mole per cent.
	ESTIMATED ERROR: $\delta T/K = \pm 0.02$; $\delta P/MPa = \pm 0.003$ up to 3 MPa, ± 0.01 up to 10 MPa, ± 0.02 above 10 MPa; δx, $\delta y = \pm 0.002$.

	REFERENCES:
	1. Somait, F.; Kidnay, A. J.
	   J. Chem. Eng. Data
	   1978, *23*, 301.

COMPONENTS:	ORIGINAL MEASUREMENTS:
1. Methane; CH_4; [74-82-8] 2. Carbonyl sulfide; COS; [463-58-1]	Senturk, N. H.; Kalra, H.; Robinson, D. B. *J. Chem. Engng. Data* 1979, *24*, 311-313.

VARIABLES:	PREPARED BY:
Temperature, pressure	C. L. Young

EXPERIMENTAL VALUES:

T/K	P/MPa	Mole fraction of methane in liquid, x_{CH_4}	in vapor, y_{CH_4}
298.15	2.47	0.0446	0.4318
	3.59	–	0.5637
	3.65	0.0884	–
	5.57	0.1850	0.6630
	7.27	0.2653	0.6867
	9.16	0.3660	0.6823
	10.72	0.4866	0.6172
323.15	3.37	0.0411	0.2644
	4.62	0.0915	0.4167
	6.21	0.1613	0.5063
	7.57	0.2240	0.5370
	8.73	0.2878	0.5400
	9.62	0.3438	0.5278
	10.16	0.3889	0.4989
348.15	4.53	0.0276	0.1242
	5.38	0.0610	0.2123
	6.28	0.0962	0.2739
	7.40	0.1502	0.3245
	8.33	0.2026	0.3344
	8.62	0.2208	0.3289
373.15	6.06	0.0102	0.0205
	6.50	0.0313	0.0549
	6.71	0.0421	0.0574

AUXILIARY INFORMATION

METHOD/APPARATUS/PROCEDURE:	SOURCE AND PURITY OF MATERIALS:
Cell fitted with two movable pistons which enabled cell contents to be circulated in external line. Fitted with optical system which allowed measurement of refractive index. Temperature measured with iron-constantan thermocouple and pressure with Bourdon gauge. Components charged into cell, mixed by piston movement. Samples withdrawn and analysed by gas chromatography. Details in source and ref. (1).	1. Matheson, ultra high purity sample, purity 99.97 mole per cent. 2. Matheson special sample purity 99.7 mole per cent. Major impurities were hydrogen sulfide 0.03%, carbon disulfide 0.05%, carbon dioxide 0.21%, traces of nitrogen and carbon monoxide.

ESTIMATED ERROR:

$\delta T/K = \pm 0.05$; $\delta P/MPa = \pm 0.02$;

δx_{CH_4}, $\delta y_{CH_4} = \pm 0.003$ to 0.005.

REFERENCES:

1. Besserer, G. J.; Robinson, D. B.
   Can. J. Chem. Eng.
   1971, *49*, 651.

COMPONENTS:	EVALUATOR:
1. Methane; CH_4; [74-82-8] 2. Hydrogen sulfide; H_2S; [7783-06-4]	Colin L. Young, School of Chemistry, University of Melbourne, Parkville, Victoria 3052, Australia. March 1982

EVALUATION:

   This system has been investigated by three groups of workers.   The
data of Reamer *et al.* (1) covers the temperature range 277.6 K to 444.3 K
and their data are classified as tentative.   The data of Kohn and Kurata
(2) are, in general, in good agreement with those of Reamer *et al.* (1)
(except at 277.6 K and the highest pressure), and are therefore also
classified as tentative.   The data of Robinson and coworkers (3), (4) are
in reasonable agreement with the data of Reamer *et al.* (1) at 277.6 K and
344.3 K but only fair agreement at 310.9 K.   Because of the very limited
nature of the data of Robinson and coworkers, detailed comparison is not
possible.

References

1.   Reamer, H. H.;   Sage, B. H.;   Lacey, W. N.
     Ind. Eng. Chem., 1951, *43*, 976.

2.   Kohn, J. P.;   Kurata, F.
     Am. Inst. Chem. Engnrs. J., 1958, *4*, 211.

3.   Robinson, D. B.;   Lorenzo, A. P.;   Macrygeorgos, C. A.
     Can. J. Chem. Engng., 1959, *37*, 212.

4.   Robinson, D. B.;   Bailey, J. A.
     Can. J. Chem. Engng., 1957, *35*, 151.

COMPONENTS:	ORIGINAL MEASUREMENTS:
1. Methane; CH₄; [74-82-8] 2. Hydrogen sulfide; H₂S; [7783-06-4]	Reamer, H. H.; Sage, B. H.; Lacey, W. N. *Ind. Eng. Chem.* <u>1951</u>, *43*, 976-981.
VARIABLES: Temperature, pressure	PREPARED BY: C. L. Young

EXPERIMENTAL VALUES:

T/K	P/MPa	Mole fraction of methane in liquid, x_{CH_4}	in gas, y_{CH_4}	T/K	P/MPa	Mole fraction of methane in liquid, x_{CH_4}	in gas, y_{CH_4}
277.6	1.38	0.0057	0.1371	277.6	13.44	0.5500	0.5500
	1.72	0.0132	0.2783	310.9	2.76	0.0007	0.0117
	2.07	0.0212	0.3896		3.10	0.0067	0.0963
	2.41	0.0284	0.4604		3.45	0.0128	0.1642
	2.76	0.0354	0.5126		3.79	0.0190	0.2203
	3.10	0.0424	0.5551		4.14	0.0255	0.2688
	3.45	0.0493	0.5879		4.83	0.0385	0.3416
	4.14	0.0636	0.6394		5.52	0.0523	0.3976
	4.83	0.0783	0.6755		6.21	0.0670	0.4396
	5.52	0.0930	0.6989		6.89	0.0828	0.4707
	6.21	0.1083	0.7141		7.58	0.0996	0.4923
	6.89	0.1250	0.7242		8.27	0.1182	0.5079
	7.58	0.1433	0.7299		8.62	0.1282	0.5130
	8.27	0.1635	0.7321		8.96	0.1390	0.5182
	8.62	0.1750	0.7319		9.66	0.1620	0.5240
	8.96	0.1868	0.7306		10.34	0.1885	0.5255
	9.65	0.2137	0.7262		11.03	0.2192	0.5195
	10.34	0.2450	0.7185		11.72	0.2532	0.5058
	11.03	0.2798	0.7075		12.07	0.2725	0.4947
	11.72	0.3240	0.6931		12.41	0.2940	0.4797
	12.07	0.3492	0.6828		12.76	0.3185	0.4580
	12.41	0.3758	0.6686		13.10	0.3578	0.4190
	13.10	0.4401	0.6130		(cont.)		

AUXILIARY INFORMATION

METHOD/APPARATUS/PROCEDURE:	SOURCE AND PURITY OF MATERIALS:
PVT cell charged with mixture of known composition. Pressure measured with pressure balance. Temperature measured using resistance thermometer. Bubble and dew points determined for various compositions from discontinuity in slope of pv isotherm. Co-existing phase compositions determined by graphical means. Details in ref. (1).	1. Crude sample treated for removal of alkanes, CO_2 and water; final purity 99.9 mole per cent. 2. Crude sample purified and twice sublimed.

ESTIMATED ERROR:

$\delta T/K = \pm 0.01$; δP/MPa $= \pm 0.01$;

δx_{CH_4}, $\delta y_{CH_4} = \pm 0.003$.

REFERENCES:

1. Sage, B. H.; Lacey, W. N.
   Trans. Am. Inst. Mining and Met.
   Engnrs. <u>1940</u>, *136*, 136.

COMPONENTS:	ORIGINAL MEASUREMENTS:
1. Methane; CH_4; [74-82-8]	Reamer, H. H.; Sage, B. H.; Lacey, W. N.
2. Hydrogen sulfide; H_2S; [7783-06-4]	*Ind. Eng. Chem.* <u>1951</u>, *43*, 976-981.

EXPERIMENTAL VALUES:

T/K	P/MPa	Mole fraction of methane in liquid, x_{CH_4}	in gas, y_{CH_4}	T/K	P/MPa	Mole fraction of methane in liquid, x_{CH_4}	in gas, y_{CH_4}
310.9	13.15	0.3880	0.3880	344.3	9.66	0.1021	0.2811
344.3	5.52	0.0031	0.0196		10.34	0.1245	0.2775
	5.86	0.0098	0.0592		11.03	0.1547	0.2580
	6.21	0.0167	0.0946		11.38	0.1830	0.2295
	6.89	0.0309	0.1553		11.45	0.2090	0.2090
	7.58	0.0459	0.2021	444.3	6.27	–	0.1000
	8.27	0.0622	0.2367		7.54	–	0.2000
	8.62	0.0720	0.2534		9.58	0.1000	–
	8.96	0.0814	0.2646		11.45	0.2000	–

COMPONENTS:	ORIGINAL MEASUREMENTS:
1. Methane; CH₄; [74-82-8] 2. Hydrogen sulfide; H₂S; [7783-06-4]	Robinson, D. B.; Bailey, J. A. *Can. J. Chem. Engng.* 1957, *35*, 151-158.

VARIABLES:	PREPARED BY:
	C. L. Young

EXPERIMENTAL VALUES:

T/K	T/°F	P/psi	P/MPa	Mole fraction of methane in liquid, x_{CH_4}	in vapor, y_{CH_4}
310.9	100	600	4.14	0.033	0.290
		1200	8.27	0.109[†]	0.510
		1600	11.03	0.260	0.478

[†] extrapolated from ternary data.

AUXILIARY INFORMATION

METHOD/APPARATUS/PROCEDURE:	SOURCE AND PURITY OF MATERIALS:
Static equilibrium cell fitted with glass window. Temperature measured with copper-constantan thermocouples. Pressure measured with Bourdon gauge. Samples of gas and liquid analysed by absorbing hydrogen sulfide in potassium hydroxide solution and measuring the amount of methane volumetrically.	1. Phillips Petroleum Co. research grade sample, purity 99.7 mole per cent. 2. Matheson Co. sample, purity 98.8 mole per cent, 0.4 mole per cent methyl chloride, 0.3 mole per cent methyl mercaptan, 0.2 mole per cent nitrogen and 0.3 mole per cent carbon dioxide.

ESTIMATED ERROR:

$\delta T/K = \pm 0.1$; $\delta P/MPa = \pm 1\%$;

δx_{CH_4}, δy_{CH_4} = ±0.002 (estimated by compiler).

REFERENCES:

COMPONENTS:	ORIGINAL MEASUREMENTS:
1. Methane; CH_4; [74-82-8] 2. Hydrogen sulfide; H_2S; [7783-06-4]	Kohn, J.P.; Kurata, F. *Am. Inst. Chem. Engnrs. J.* <u>1958</u>, *4*, 211-217.
VARIABLES: Temperature, pressure	PREPARED BY: C.L. Young

EXPERIMENTAL VALUES:

T/K	$P/10^5$Pa	Mole fraction of methane in liquid, x_{CH_4}	in vapor, y_{CH_4}
277.6	13.79	0.005	0.140
255.4	13.79	0.009	0.522
233.2	13.79	0.0125	0.754
210.9	13.79	0.018	0.906
188.7	13.79	0.025	0.968
299.8	27.56	0.016	0.251
277.6	27.56	0.035	0.513
255.4	27.56	0.045	0.720
233.2	27.56	0.054	0.862
210.9	27.56	0.062	0.935
188.7	27.56	0.066	0.969
322.0	41.37	0.015	0.143
299.8	41.37	0.038	0.406
277.6	41.37	0.058	0.628
255.4	41.37	0.072	0.792
233.2	41.37	0.083	0.900
210.9	41.37	0.093	0.947
199.8	41.37	0.101	0.955
338.7	55.16	0.010	0.104
322.0	55.16	0.035	0.290
310.9	55.16	0.051	0.405
299.8	55.16	0.068	0.501
288.7	55.16	0.086	0.596
277.6	55.16	0.090	0.690

AUXILIARY INFORMATION

METHOD/APPARATUS/PROCEDURE:	SOURCE AND PURITY OF MATERIALS:
Bubble point-dew point apparatus with borosilicate glass cell. Temperature measured with platinum resistance thermometer. Pressure measured with Bourdon gauge. Analysis of phases carried out using gas density measurement. Some details in source and ref. (1).	1. Pure grade Phillips Petroleum sample minimum purity 99 mole per cent. Purified by passing through alumina and then activated charcoal in dry-ice acetone bath. Final purity about 99.7 mole per cent. 2. Matheson Company sample purity 99 mole per cent. Purified as methane, final purity about 99.9 mole per cent.
	ESTIMATED ERROR: $\delta T/K = \pm 0.06$; $\delta /MPa = \pm 0.015$; $\delta x_{CH_4}, \delta y_{CH_4} < 0.005$
	REFERENCES: 1. Kohn, J.P.; Kurata, F. *Petroleum Process.* <u>1956</u>, *11*, 57.

COMPONENTS:	ORIGINAL MEASUREMENTS:
1. Methane; CH_4; [74-82-8]	Kohn, J.P.; Kurata, F.
2. Hydrogen sulfide; H_2S; [7783-06-4]	*Am. Inst. Chem. Engnrs. J.* 1958, *4*, 211-217.

EXPERIMENTAL VALUES:

T/K	$P/10^5$Pa	Mole fraction of methane in liquid x_{CH_4}	in vapor, y_{CH_4}
366.5	82.74	0.006	0.025
344.3	82.74	0.052	0.238
322.0	82.74	0.095	0.432
299.8	82.74	0.135	0.582
277.6	82.74	0.162	0.727
344.3	110.3	0.155	0.250
322.0	110.3	0.184	0.441
310.9	110.3	0.200	0.520
299.8	110.3	0.213	0.586
277.6	110.3	0.238	0.705

COMPONENTS:	ORIGINAL MEASUREMENTS:
1. Methane; CH_4; [74-82-8] 2. Hydrogen sulfide; H_2S; [7783-06-4]	Robinson, D. B.; Lorenzo, A. P.; Macrygeorgos, C. A. *Can. J. Chem. Engng.* *1959*, *37*, 212-217.
VARIABLES:	PREPARED BY: C. L. Young

EXPERIMENTAL VALUES:

T/K	T/°F	P/MPa	P/psi	Mole fraction of methane in liquid, x_{CH_4}	in vapor, y_{CH_4}
277.6	40	2.76	400	0.023	0.511
		6.89	1000	0.122	0.716
		11.03	1600	0.256	0.708
344.3	160	6.89	1000	0.032	0.156
		11.03	1600	0.159	0.265

AUXILIARY INFORMATION

METHOD/APPARATUS/PROCEDURE:	SOURCE AND PURITY OF MATERIALS:
Static equilibrium cell fitted with glass window. Temperature measured with copper-constantan thermocouples. Pressure measured with Bourdon gauge. Samples of gas and liquid analysed by absorbing hydrogen sulfide in potassium hydroxide solution and measuring the amount of methane volumetrically.	No details given.
	ESTIMATED ERROR: $\delta T/K = \pm 0.1$; $\delta P/MPa = \pm 1\%$; δx_{CH_4}, $\delta y_{CH_4} = \pm 0.002$ (estimated by compiler).
	REFERENCES:

COMPONENTS:	ORIGINAL MEASUREMENTS:
1. Methane; CH_4; [74-82-8] 2. Hydrogen sulfide; H_2S; [7783-06-4] 3. Carbon dioxide; CO_2; [124-38-9]	Robinson, D. B.; Bailey, J. A. *Can. J. Chem. Engng.* 1957, *35*, 151-158.

VARIABLES:	PREPARED BY:
	C. L. Young

EXPERIMENTAL VALUES:

T/K = 310.9   T/°F = 100

Mole fractions

P/MPa	P/psi	in liquid			in vapor		
		x_{H_2S}	x_{CO_2}	x_{CH_4}	y_{H_2S}	y_{CO_2}	y_{CH_4}
4.14	600	0.840	0.160	0.0	0.608	0.392	0.0
		0.810	0.170	0.020	0.660	0.222	0.118
		0.905	0.069	0.026	0.667	0.206	0.127
		0.937	0.044	0.019	0.684	0.115	0.201
		0.891	0.101	0.008	0.644	0.259	0.097
		0.815	0.185	0.0	0.630	0.370	0.0
		0.870	0.122	0.008	0.630	0.317	0.053
		0.967	0.0	0.033	0.710	0.0	0.290
		0.955	0.029	0.061	0.690	0.066	0.244
8.27	1200	0.891	0.0	0.109	0.490	0.0	0.510
		0.843	0.042	0.115	0.470	0.095	0.435
		0.747	0.137	0.116	0.445	0.201	0.354
		0.683	0.213	0.104	0.418	0.278	0.304
		0.615	0.284	0.101	0.411	0.332	0.257
		0.598	0.305	0.097	0.407	0.347	0.246
		0.552	0.366	0.082	0.368	0.422	0.210
		0.416	0.458	0.066	0.348	0.498	0.155
		0.422	0.520	0.058	0.314	0.565	0.121
		0.372	0.563	0.065	0.316	0.579	0.105
12.41	1800	0.740	0.0	0.260	0.522	0.0	0.478
		0.706	0.036	0.258	0.530	0.047	0.423
		0.655	0.050	0.285	0.520	0.079	0.401

AUXILIARY INFORMATION

METHOD/APPARATUS/PROCEDURE:	SOURCE AND PURITY OF MATERIALS:
Static equilibrium cell fitted with glass window. Temperature measured with copper-constantan thermocouples. Pressure measured with Bourdon gauge. Samples of gas and liquid analysed by absorbing hydrogen sulfide and carbon dioxide in potassium hydroxide solution and measuring the amount of methane volumetrically. Hydrogen sulfide in sample estimated by reacting with iodine solution, excess iodine being detected by the blue coloration produced with starch solution.	1. Phillips Petroleum Co. research grade sample, purity 99.7 mole per cent. 2. Matheson Co. sample, purity 98.8 mole per cent, 0.4 mole per cent methyl chloride, 0.3 mole per cent methyl mercaptan, 0.2 mole per cent nitrogen and 0.3 mole per cent carbon dioxide. 3. Purity 99.7 mole per cent.

ESTIMATED ERROR:

$\delta T/K = \pm 0.1$; $\delta P/MPa = \pm 1\%$;

δx, $\delta y = \pm 0.002$ (estimated by compiler).

REFERENCES:

COMPONENTS:	ORIGINAL MEASUREMENTS:
1. Methane; CH_4; [74-82-8] 2. Hydrogen sulfide; H_2S; [7783-06-4] 3. Carbon dioxide; CO_2; [124-38-9]	Robinson, D. B.; Lorenzo, A. P.; Macrygeorgos, C. A. *Can. J. Chem. Engng.* __1959__, *37*, 212-217.

VARIABLES:	PREPARED BY:
	C. L. Young

EXPERIMENTAL VALUES:

T/K (T/°F)	P/MPa (P/psi)	Mole fractions					
		in liquid			in vapor		
		x_{H_2S}	x_{CO_2}	x_{CH_4}	y_{H_2S}	y_{CO_2}	y_{CH_4}
277.6 (40)	2.76 (400)	0.977	0.0	0.023	0.489	0.0	0.511
		0.940	0.027	0.033	0.477	0.131	0.392
		0.889	0.070	0.041	0.454	0.192	0.354
		0.832	0.129	0.039	0.429	0.253	0.318
		0.819	0.148	0.033	0.423	0.352	0.225
		0.702	0.280	0.018	0.404	0.450	0.146
		0.635	0.350	0.015	0.416	0.503	0.081
		0.597	0.403	0.0	0.365	0.635	0.0
	6.89 (1000)	0.878	0.0	0.122	0.284	0.0	0.716
		0.756	0.090	0.154	0.291	0.139	0.570
		0.625	0.221	0.154	0.259	0.266	0.475
		0.412	0.414	0.174	0.229	0.353	0.418
		0.284	0.537	0.179	0.179	0.471	0.350
		0.144	0.680	0.176	0.104	0.567	0.329
		0.0	0.867	0.133	0.0	0.655	0.345
	11.03 (1600)	0.744	0.0	0.256	0.292	0.0	0.708
		0.643	0.063	0.294	0.311	0.087	0.602
		0.488	0.135	0.377	0.340	0.142	0.518
344.3 (160)	6.89 (1000)	0.968	0.0	0.032	0.844	0.0	0.156
		0.936	0.031	0.033	0.836	0.061	0.103
		0.911	0.063	0.026	0.822	0.118	0.060
	(cont.)	0.891	0.109	0.0	0.811	0.189	0.0

AUXILIARY INFORMATION

METHOD/APPARATUS/PROCEDURE:	SOURCE AND PURITY OF MATERIALS:
Static equilibrium cell fitted with glass window. Temperature measured with copper-constantan thermocouples. Pressure measured with Bourdon gauge. Samples of gas and liquid analysed by absorbing hydrogen sulfide and carbon dioxide in potassium hydroxide solution and measuring the amount of methane volumetrically. Hydrogen sulfide in sample estimated by reacting with iodine solution, excess iodine being detected by the blue coloration produced with starch solution.	No details given.
	ESTIMATED ERROR:
	$\delta T/K = \pm 0.1$; $\delta P/MPa = \pm 1\%$; δx, $\delta y = \pm 0.002$ (estimated by compiler).
	REFERENCES:

COMPONENTS:

1. Methane; CH_4; [74-82-8]

2. Hydrogen sulfide; H_2S; [7783-06-4]

3. Carbon dioxide; CO_2; [124-38-9]

ORIGINAL MEASUREMENTS:

Robinson, D. B.; Lorenzo, A. P.; Macrygeorgos, C. A.

Can. J. Chem. Engng.

<u>1959</u>, *37*, 212-217.

EXPERIMENTAL VALUES:

T/K (T/°F)	P/MPa (P/psi)	Mole fractions					
		in liquid			in vapor		
		x_{H_2S}	x_{CO_2}	x_{CH_4}	y_{H_2S}	y_{CO_2}	y_{CH_4}
344.3 (160)	9.17 (1330)	0.903	0.023	0.074	0.724	0.066	0.210
		0.848	0.064	0.088	0.704	0.126	0.170
		0.770	0.123	0.107	0.699	0.162	0.139
		0.718	0.197	0.085	0.676	0.224	0.100
	11.03 (1600)	0.841	0.0	0.159	0.735	0.0	0.265
		0.818	0.018	0.164	0.737	0.027	0.236

COMPONENTS:	ORIGINAL MEASUREMENTS:
1. Methane; CH_4; [74-82-8] 2. Carbon dioxide; CO_2; [124-38-9] 3. Hydrogen sulfide; H_2S; [7783-06-4]	Hensel, W. E. Jr.; Massoth, F. E. *J. Chem. Eng. Data* <u>1964</u>, *9*, 352-356.
VARIABLES:	PREPARED BY: C. L. Young

EXPERIMENTAL VALUES:

Mole fractions

T/K	P/psi	P/MPa	x_{CH_4}	x_{CO_2}	x_{H_2S}	y_{CH_4}	y_{CO_2}	y_{H_2S}
			in liquid			in vapor		
238.8	700	4.82	0.128	0.114	0.758	0.860	0.024	0.116
	500	3.45	0.095	0.134	0.771	0.750	0.095	0.155
	300	2.07	0.048	0.127	0.825	0.652	0.140	0.209
	700	4.82	0.172	0.471	0.357	0.677	0.226	0.097
	500	3.45	0.114	0.511	0.375	0.600	0.289	0.111
	700	4.82	0.194	0.662	0.144	0.651	0.305	0.043
	500	3.45	0.118	0.709	0.175	0.579	0.368	0.053
	300	2.07	0.056	0.776	0.169	0.382	0.543	0.078
222.2	700	4.82	0.126	0.122	0.751	0.893	0.026	0.085
	500	3.45	0.096	0.108	0.796	0.828	0.063	0.110
	300	2.07	0.058	0.130	0.812	0.798	0.085	0.117
	700	4.82	0.218	0.415	0.366	0.856	0.085	0.058
	500	3.45	0.128	0.466	0.406	0.764	0.177	0.059
	300	2.07	0.082	0.470	0.447	0.705	0.219	0.076
	700	4.82	0.269	0.606	0.125	0.777	0.195	0.028
	500	3.45	0.163	0.684	0.153	0.744	0.230	0.026
	300	2.07	0.090	0.732	0.178	0.651	0.309	0.040

AUXILIARY INFORMATION

METHOD/APPARATUS/PROCEDURE:	SOURCE AND PURITY OF MATERIALS:

Static equilibrium cell. Samples analysed using gas chromatography. Pressure measured using Bourdon gauges. Temperatures were measured using an alcohol thermometer which was checked against a copper-constantan thermocouple.	1 and 3. Matheson C.P. grade
	   sample, purities 99.0 and 99.5
	   mole per cent, respectively.
	
	2. Matheson "bone dry" grade,
	   purity 99.95 per cent.

	ESTIMATED ERROR: $\delta T/K = \pm 1$; δx, $\delta y = \pm 2\%$.
	REFERENCES:

COMPONENTS:	ORIGINAL MEASUREMENTS:
1. Methane; CH_4; [74-82-8] 2. Nitrous oxide; N_2O; [10024-97-2]	Zeininger, H. *Chemie-Ing.-Techn.* <u>1972</u>, *44*, 607-12.

VARIABLES:	PREPARED BY:
Temperature, pressure	C. L. Young

EXPERIMENTAL VALUES:

T/K	$P/10^5$Pa	Mole fraction of methane in liquid, x_{CH_4}	in gas, y_{CH_4}	T/K	$P/10^5$Pa	Mole fraction of methane in liquid, x_{CH_4}	in gas, y_{CH_4}
213.15	4.2	0.0018	0.060	213.15	26.4	0.262	0.800
	4.3	0.0024	0.067		31.7	0.320	0.828
	4.6	0.003	0.097		37.7	0.413	0.830
	5.2	0.007	0.186		41.5	0.493	0.795
	5.7	0.009	0.240		42.4	0.577	0.777
	6.1	0.011	0.244	233.15	9.7	0.003	0.064
	6.4	0.014	0.283		10.0	0.006	0.095
	7.1	0.019	0.319		10.9	0.011	0.162
	7.3	0.016	0.368		11.0	0.010	0.147
	8.3	0.030	0.441		11.6	0.019	0.199
	9.1	0.030	0.494		12.4	0.017	0.205
	10.9	0.040	0.534		12.6	0.024	0.262
	11.1	0.049	0.548		13.2	0.028	0.317
	12.8	0.071	0.625		13.4	0.031	0.322
	15.0	0.087	0.653		14.2	0.032	0.285
	15.3	0.078	0.667		16.5	0.050	0.392
	17.4	0.099	0.696		21.6	0.084	0.510
	17.4	0.111	0.708		25.3	0.101	0.554
	21.0	0.154	0.753		32.1	0.151	0.633
	24.0	0.178	0.766		35.7	0.188	0.671

(cont.)

AUXILIARY INFORMATION

METHOD/APPARATUS/PROCEDURE:	SOURCE AND PURITY OF MATERIALS:
Static equilibrium cell stirred with a steel ball. Samples of gas and liquid phases removed and analysed by mass spectrometry. Care was taken to avoid large changes in pressure during sampling by taking small samples. Details in source.	Nitrous oxide was "pure" as determined by gas chromatography.

ESTIMATED ERROR:
$$\delta T/K = \pm0.3; \quad \delta P/10^5 Pa = \pm0.2;$$
$$\delta x_{CH_4} = \pm0.011; \quad \delta y_{CH_4} = \pm0.016.$$

REFERENCES:

COMPONENTS:	ORIGINAL MEASUREMENTS:
1. Methane; CH_4; [74-82-8]	Zeininger, H.
	Chemie-Ing.-Techn.
2. Nitrous oxide; N_2O;	1972, *44*, 607-12.
[10024-97-2]	

EXPERIMENTAL VALUES:

T/K	$P/10^5$Pa	Mole fraction of methane in liquid, x_{CH_4}	in gas, y_{CH_4}	T/K	$P/10^5$Pa	Mole fraction of methane in liquid, x_{CH_4}	in gas, y_{CH_4}
233.15	41.7	0.256	0.673	253.15	33.7	0.098	0.398
	43.8	0.265	0.709		36.6	0.098	0.398
	46.2	0.411	0.635		39.9	0.122	0.436
253.15	18.5	0.003	0.029		43.8	0.143	0.437
	23.3	0.026	0.202		44.8	0.147	0.444
	27.8	0.050	0.308		50.5	0.158	0.430
	30.8	0.072	0.359		52.7	0.240	0.411

COMPONENTS:	ORIGINAL MEASUREMENTS:
1. Methane; CH₄; [74-82-8] 2. Sulfur Dioxide; SO₂; [7446-09-5]	Dean, M. R.; Walls, W. S. *Ind. Eng. Chem.* <u>1947</u>, *39*, 1049-1051.

VARIABLES:	PREPARED BY:
Temperature, pressure	C. L. Young

EXPERIMENTAL VALUES:

T/K	P/MPa	Mole fraction of methane in liquid, x_{CH_4}	in gas, y_{CH_4}
301.48	3.55	0.0348	0.839
301.48	1.74	0.0152	0.708
241.10	3.55	0.0326	0.987
241.10	1.90	0.0176	0.979
241.10	1.72	0.0165	0.972

AUXILIARY INFORMATION

METHOD/APPARATUS/PROCEDURE:

Twin steel static cell. Pressure and volume of cell varied by introducing mercury. Pressure measured with Bourdon gauge. Analysis of samples of both gas and liquid phases carried out by Orsat gas analysis. Details in source.

SOURCE AND PURITY OF MATERIALS:

1. Phillips Petroleum sample, purity about 99.5 mole per cent.
2. Refrigeration grade sample from Virginia Smelting Co. Purity about 99.6 mole per cent.

ESTIMATED ERROR:
$\delta T/K = \pm 0.1$; $\delta P/MPa = \pm 0.01$;
$\delta x_{CH_4} = \pm 0.0003$; $\delta y_{CH_4} = \pm 0.002$
(estimated by compiler).

REFERENCES:

M—Z*

SYSTEM INDEX

Page numbers preceded by E refer to evaluation texts whereas those not preceded by E refer to compilation tables. All compounds are listed as in Chemical Abstracts. For example methyl acetate is listed as acetic acid, methyl ester.

A

Acetic acid	580
Acetic acid, methyl ester	662
Acetone see 2-propanone	
2-Aminoethanol see ethanol, 2-amino-	
Ammonia	736
Ammonium bromide (aqueous)	E59, 90
Ammonium chloride (aqueous)	E59, 89
Amyl alcohol see l-pentanol	
iso-Amyl alcohol see l-butanol, 3-methyl-	
Aniline see Benzenamine	
Anisole see benzene, methoxy-	

B

Benzenamine	580, 709
Benzene	E493, E494, 498-501, E511, 512-521, 580
Benzene (multicomponent)	551, 579
Benzene, butyl-	
Benzene, butyl- (ternary)	540, 542
Benzene, chloro-	E692, 693-695
Benzene, (l-chloroethyl)-	696
Benzene, dimethyl-	580
Benzene, 1,2-dimethyl-	E495, E496, 504
Benzene, 1,2-dimethyl- (multicomponent)	579
Benzene, 1,3-dimethyl-	E495, E496, 505, 506, E533, 534-536
Benzene, 1,4-dimethyl-	E495, E496, 507
Benzene, 1,4-dimethyl- (multicomponent)	579
Benzene, hexafluoro-	683
Benzene, methoxy -	E643, E644, 654
Benzene, methyl-	E494, 502, 503, E522, 523-532
Benzene, methyl- (multicomponent and ternary)	553-556, 579
Benzene, 1,1-methylenebis-	E496, 509, 548, 549
Benzene, nitro-	580, 722
Benzene, 1,3,5-trimethyl-	537-539
Benzene, 1,3,5-trimethyl- (multicomponent)	579
Benzenemethanol	677
Benzyl alcohol see benzenemethanol	
Bicyclo(3.1.1)hept-2-ene, 2,6,6-trimethyl-	
l,l-Bicyclohexyl	E453, 464
l,l-Bicyclohexyl (multicomponent)	487
l,l´-Biphenyl (multicomponent)	579
Bitumen, athabasca	576, 577
Brine, synthetic	144, 145
Bromide, sodium see sodium bromide	
l-Bromooctane see octane, l-bromo-	
l-Butanamine, 1,1,2,2,3,3,4,4,4-nonafluoro-N,N-bis(nonafluorobutyl)-	721
l-Butanamine, N,N,N-tributyl-, bromide (aqueous)	E61, E62, 96-98
Butane	E281, 282-296
Butane (aqueous)	196-199
Butane (ternary)	387-389, 407-409, 420-422, 436-440, 447-449
Butane, 2,2-dimethyl-	E212, E213, 223
Butane, 2-methyl-	317-318
Butane, 2-methyl- (ternary)	411
Butane, 2,2,3-trimethyl- (ternary)	551, 552
l-Butanol	E582-E584, E589, E590, 611-613
l-Butanol, 3-methyl-	580, E591, 619
iso-Butanol see l-propanol, 2-methyl-	

P

Q

REGISTRY NUMBER INDEX

Page numbers preceded by E refer to evaluation text whereas those not preceded by E refer to compiled tables.

50-00-0	172
50-01-1	E64, 103
50-99-7	183
56-23-5	E685, 686, 687
57-09-0	100
57-13-6	177, 178
57-50-1	180, 181
60-29-7	580, E640, E641, 645, 646
62-53-3	580, 709
64-17-5	174, 580, E582-E585, 601-605, 635
64-20-0	E59, E60, 91
67-56-1	173, 580, E582-E585, 596-600, E629, 630-634
67-63-0	E582-E584, E588, E589, 610, 636
67-64-1	580, E655, 656-659
67-66-3	580
67-68-5	182, E702, 705, 706
68-12-2	720
71-23-8	175, E582-E564, E586, E587, 606-609
71-36-3	E582-E584, E589, E590, 611-613
71-41-0	E582-E584, E590, E591, 615-618
71-43-2	E493, E494, 498-501, E511, 512-521, 551, 578-580
71-91-0	E60, 92-94
74-84-0	194, 195, E248, E249, 250-265, 378-386, 483, 484, 553-555
74-85-1	E477, 478-490
74-98-6	194, 195, E266, 267-280, 378-382, 387-406, 553-555, 556
75-15-0	580, E702, 703, 704
75-21-8	648, 649
75-28-5	E297, 298-302, 485-490
75-45-6	689, 690
75-71-8	688
75-83-2	E212, E213, 223
76-13-1	684
78-40-0	727
78-78-4	317-319, 411
78-83-1	E590, 614
79-20-9	662
80-56-5	E491, 492
83-32-9	578, 579
85-01-8	541, 542, 578, 579
86-73-7	578, 579
90-12-0	545-547, 553-555, 578, 579
90-13-1	701
91-17-8	E497, 510
91-20-3	578, 579
91-22-5	710, 711
91-51-3	E453, 464, 487
92-52-4	578, 579
95-47-6	E495, E496, 504, 578, 579
96-14-0	336, 337
98-95-3	580, 722
100-51-6	677
100-66-3	E643, E644, 654
100-97-0	179
101-81-5	E496, 509, 548, 549
102-76-1	667, 679

AUTHOR INDEX

Page numbers preceded by E refer to evaluation texts whereas page numbers not preceded by E refer to compiled tables.